D1694148

Ullmann's Polymers and Plastics

Excellence since 1914
Built from generations of expertise, for generations to come.

ULLMANN'S ENCYCLOPEDIA OF INDUSTRIAL CHEMISTRY

If you want to learn more about novel technologies in biochemistry or nanotechnology or discover unexpected new aspects of seemingly totally familiar processes in industrial chemistry – ULLMANN'S Encyclopedia of Industrial Chemistry is your first choice.

For over 100 years now, this reference provides top-notch information on the most diverse fields of industrial chemistry and chemical engineering.

When it comes to definite works on industrial chemistry, it has always been ULLMANN'S.

Welcome to the ULLMANN'S ACADEMY

Kick-start your career!

Key topics in industrial chemistry explained by the ULLMANN'S Encyclopedia experts – for teaching and learning, or for simply refreshing your knowledge.

What's new?

The Smart Article introduces new and enhanced article tools for chemistry content. It is now available within ULLMANN'S.

For further information on the features and functions available, go to wileyonlinelibrary.com/thesmartarticle

Only interested in a specific topic?

ULLMANN'S Energy
3 Volume Set • ISBN: 978-3-527-33370-7

ULLMANN'S Fine Chemicals
3 Volume Set • ISBN: 978-3-527-33477-3

ULLMANN'S Fibers
2 Volumes • ISBN: 978-3-527-31772-1

ULLMANN'S Modeling and Simulation
ISBN: 978-3-527-31605-2

ULLMANN'S Renewable Resources
ISBN: 978-3-527-33369-1

ULLMANN'S Agrochemicals
2 Volume Set • ISBN: 978-3-527-31604-5

ULLMANN'S Biotechnology and Biochemical Engineering
2 Volume Set • ISBN: 978-3-527-31603-8

ULLMANN'S Reaction Engineering
2 Volume Set • ISBN: 978-3-527-33371-4

ULLMANN'S Industrial Toxicology
2 Volume Set • ISBN: 978-3-527-31247-4

ULLMANN'S Chemical Engineering and Plant Design
2 Volume Set • ISBN: 978-3-527-31111-8

The information you need in the format you want.

DVD
- Released once a year.
- Fully networkable for up to 200 users.
- Time-limited access for up to 14 months (expires March 2016).
- 2015 Edition
  ISBN: 978-3-527-33754-5

Online
- Over 1,150 articles available online.
- Over 3,000 authors from over 30 countries have contributed.
- Offers flexible access 24/7 from your library, home, or on the road.
- Updated 4 times per year.
  ISBN: 978-3-527-30673-2

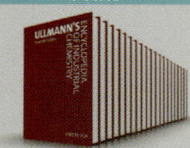

Print
- Available as a comprehensive 40 Volume-Set.
- The 7th Edition published in Aug 2011.
- ISBN: 978-3-527-32943-4

Visit our website to

Find sample chapters

Learn more about the history, the Editor-in-Chief and the Editorial Advisory Board

Discover the ULLMANN'S ACADEMY

and more...

wileyonlinelibrary.com/ref/ullmanns

Chemistry that delivers... Continuous product innovation

Create Innovate Inspire

WILEY-VCH   **WILEY**

Ullmann's Polymers and Plastics

Products and Processes

Volume 3

WILEY-VCH
Verlag GmbH & Co. KGaA

Editor in Chief:

Dr. Barbara Elvers, Hamburg, Germany

All books published by **Wiley-VCH** are carefully produced. Nevertheless, authors, editors, and publisher do not warrant the information contained in these books, including this book, to be free of errors. Readers are advised to keep in mind that statements, data, illustrations, procedural details or other items may inadvertently be inaccurate.

Library of Congress Card No.:
applied for

British Library Cataloguing-in-Publication Data
A catalogue record for this book is available from the British Library.

Bibliographic information published by the Deutsche Nationalbibliothek
The Deutsche Nationalbibliothek lists this publication in the Deutsche Nationalbibliografie; detailed bibliographic data are available on the Internet at <http://dnb.d-nb.de>.

© 2016 Wiley-VCH Verlag GmbH & Co. KGaA, Boschstr. 12, 69469 Weinheim, Germany

All rights reserved (including those of translation into other languages). No part of this book may be reproduced in any form – by photoprinting, microfilm, or any other means – nor transmitted or translated into a machine language without written permission from the publishers. Registered names, trademarks, etc. used in this book, even when not specifically marked as such, are not to be considered unprotected by law.

Print ISBN: 978-3-527-33823-8
ePDF ISBN: 978-3-527-68595-0
ePub ISBN: 978-3-527-68596-7
Mobi ISBN: 978-3-527-68597-4

Cover Design Grafik-Design Schulz, Fußgönheim, Germany
Typesetting Thomson Digital, Noida, India
Printing and Binding Markono Print Media Pte Ltd, Singapore

Printed on acid-free paper

Preface

This handbook features selected articles from the 7th edition of *ULLMANN'S Encyclopedia of Industrial Chemistry*, including newly written articles that have not been published in a printed edition before. True to the tradition of the ULLMANN'S Encyclopedia, polymers and plastics are addressed from an industrial perspective, including production figures, quality standards and patent protection issues where appropriate. Safety and environmental aspects which are a key concern for modern process industries are likewise considered.

More content on related topics can be found in the complete edition of the ULLMANN'S Encyclopedia.

About ULLMANN'S

ULLMANN'S Encyclopedia is the world's largest reference in applied chemistry, industrial chemistry, and chemical engineering. In its current edition, the Encyclopedia contains more than 30,000 pages, 15,000 tables, 25,000 figures, and innumerable literature sources and cross-references, offering a wealth of comprehensive and well-structured information on all facets of industrial chemistry.

1,100 major articles cover the following main areas:

- Agrochemicals
- Analytical Techniques
- Biochemistry and Biotechnology
- Chemical Reactions
- Dyes and Pigments
- Energy
- Environmental Protection and Industrial Safety
- Fat, Oil, Food and Feed, Cosmetics
- Inorganic Chemicals
- Materials
- Metals and Alloys
- Organic Chemicals
- Pharmaceuticals
- Polymers and Plastics
- Processes and Process Engineering
- Renewable Resources
- Special Topics

First published in 1914 by Professor Fritz Ullmann in Berlin, the *Enzyklopädie der Technischen Chemie* (as the German title read) quickly became the standard reference work in industrial chemistry. Generations of chemists have since relied on ULLMANN'S as their prime reference source. Three further German editions followed in 1928–1932, 1951–1970, and in 1972–1984. From 1985 to 1996, the 5th edition of ULLMANN'S Encyclopedia of Industrial Chemistry was the first edition to be published in English rather than German language. So far, two more complete English editions have been published in print; the 6th edition of 40 volumes in 2002, and the 7th edition in 2011, again comprising 40 volumes. In addition, a number of smaller topic-oriented editions have been published.

Since 1997, *ULLMANN'S Encyclopedia of Industrial Chemistry* has also been available in electronic format, first in a CD-ROM edition and, since 2000, in an enhanced online edition. Both electronic editions feature powerful search and navigation functions as well as regular content updates.

Contents

Volume 1 ....
Symbols and Units IX
Conversion Factors XI
Abbreviations XIII
Country Codes XVIII
Periodic Table of Elements XIX

Part 1: Fundamentals 1
Plastics, General Survey, 1. Definition,
    Molecular Structure and Properties 3
Plastics, General Survey, 2. Production of
    Polymers and Plastics 149
Plastics, General Survey, 3. Supermolecular
    Structures 187
Plastics, General Survey, 4. Polymer
    Composites 205
Plastics, General Survey, 5. Plastics and
    Sustainability 223
Plastics, Analysis 231
Polymerization Processes, 1. Fundamentals ... 265
Polymerization Processes, 2. Modeling of
    Processes and Reactors 315
Plastics, Processing, 1. Processing of
    Thermoplastics 367
Plastics, Processing, 2. Processing of
    Thermosets 407
Plastics Processing, 3. Machining, Bonding,
    Surface Treatment 439
Plastics, Properties and Testing 471
Plastics, Additives 527
Plasticizers 581

Volume 2 ....
Part 2: Organic Polymers 601
Fluoropolymers, Organic 603
Polyacrylamides and Poly(Acrylic Acids) ... 659
Polyacrylates 675
Polyamides 697
Polyaspartates and Polysuccinimide 733
Polybutenes 747
Polycarbonates 763
Polyester Resins, Unsaturated 781
Polyesters 791
Polyethylene 817
Polyimides 859
Polymethacrylates 885

Polyoxyalkylenes 899
Polyoxymethylenes 911
Poly(Phenylene Oxides) 927
Polypropylene 937

Volume 3 ....
Polystyrene and Styrene Copolymers 981
Polyureas 1029
Polyurethanes 1051
Poly(Vinyl Chloride) 1111
Polyvinyl Compounds, Others 1141
Poly(Vinyl Esters) 1165
Poly(Vinyl Ethers) 1175
Poly(Vinylidene Chloride) 1181
Polymer Blends 1197
Polymers, Biodegradable 1231
Polymers, Electrically Conducting 1261
Polymers, High-Temperature 1281
Reinforced Plastics 1325
Specialty Plastics 1343
Thermoplastic Elastomers 1365

Volume 4 ....
Part 3: Films, Fibers, Foams 1405
Films 1407
Fibers, 4. Polyamide Fibers 1435
Fibers, 5. Polyester Fibers 1453
Fibers, 6. Polyurethane Fibers 1487
Fibers, 7. Polyolefin Fibers 1495
Fibers, 8. Polyacrylonitrile Fibers 1513
Fibers, 9. Polyvinyl Fibers 1529
Fibers, 10. Polytetrafluoroethylene Fibers 1539
High-Performance Fibers 1541
Foamed Plastics 1563

Part 4: Resins 1595
Alkyd Resins 1597
Amino Resins 1615
Epoxy Resins 1643
Phenolic Resins 1733
Resins, Synthetic 1751

Part 5: Inorganic Polymers 1775
Inorganic Polymers 1777

Author Index 1817
Subject Index 1823

Symbols and Units

Symbols and units agree with SI standards (for conversion factors see page XI). The following list gives the most important symbols used in the encyclopedia. Articles with many specific units and symbols have a similar list as front matter.

Symbol	Unit	Physical Quantity
a_B		activity of substance B
A_r		relative atomic mass (atomic weight)
A	m^2	area
c_B	mol/m^3, mol/L (M)	concentration of substance B
C	C/V	electric capacity
c_p, c_v	$J\,kg^{-1}\,K^{-1}$	specific heat capacity
d	cm, m	diameter
d		relative density (ϱ/ϱ_{water})
D	m^2/s	diffusion coefficient
D	Gy (=J/kg)	absorbed dose
e	C	elementary charge
E	J	energy
E	V/m	electric field strength
E	V	electromotive force
E_A	J	activation energy
f		activity coefficient
F	C/mol	Faraday constant
F	N	force
g	m/s^2	acceleration due to gravity
G	J	Gibbs free energy
h	m	height
$\hbar$	$W \cdot s^2$	Planck constant
H	J	enthalpy
I	A	electric current
I	cd	luminous intensity
k	(variable)	rate constant of a chemical reaction
k	J/K	Boltzmann constant
K	(variable)	equilibrium constant
l	m	length
m	g, kg, t	mass
M_r		relative molecular mass (molecular weight)
n_D^{20}		refractive index (sodium D-line, 20 °C)
n	mol	amount of substance
N_A	mol^{-1}	Avogadro constant ($6.023 \times 10^{23}\,mol^{-1}$)
P	Pa, bar*	pressure
Q	J	quantity of heat
r	m	radius
R	$J\,K^{-1}\,mol^{-1}$	gas constant
R	Ω	electric resistance
S	J/K	entropy
t	s, min, h, d, month, a	time
t	°C	temperature
T	K	absolute temperature
u	m/s	velocity
U	V	electric potential

Symbols and Units (Continued from p. IX)

Symbol	Unit	Physical Quantity
U	J	internal energy
V	m³, L, mL, μL	volume
w		mass fraction
W	J	work
x_B		mole fraction of substance B
Z		proton number, atomic number
α		cubic expansion coefficient
α	Wm^{-2}K^{-1}	heat-transfer coefficient (heat-transfer number)
α		degree of dissociation of electrolyte
$[\alpha]$	10^{-2}deg cm²g^{-1}	specific rotation
η	Pa·s	dynamic viscosity
θ	°C	temperature
$\varkappa$		c_p/c_v
λ	Wm^{-1}K^{-1}	thermal conductivity
λ	nm, m	wavelength
μ		chemical potential
ν	Hz, s^{-1}	frequency
ν	m²/s	kinematic viscosity (η/ϱ)
π	Pa	osmotic pressure
ϱ	g/cm³	density
σ	N/m	surface tension
τ	Pa (N/m²)	shear stress
φ		volume fraction
χ	Pa^{-1} (m²/N)	compressibility

*The official unit of pressure is the pascal (Pa).

Conversion Factors

SI unit	Non-SI unit	From SI to non-SI multiply by
Mass		
kg	pound (avoirdupois)	2.205
kg	ton (long)	9.842×10^{-4}
kg	ton (short)	1.102×10^{-3}
Volume		
m^3	cubic inch	6.102×10^4
m^3	cubic foot	35.315
m^3	gallon (U.S., liquid)	2.642×10^2
m^3	gallon (Imperial)	2.200×10^2
Temperature		
°C	°F	°C × 1.8 + 32
Force		
N	dyne	1.0×10^5
Energy, Work		
J	Btu (int.)	9.480×10^{-4}
J	cal (int.)	2.389×10^{-1}
J	eV	6.242×10^{18}
J	erg	1.0×10^7
J	kW·h	2.778×10^{-7}
J	kp·m	1.020×10^{-1}
Pressure		
MPa	at	10.20
MPa	atm	9.869
MPa	bar	10
kPa	mbar	10
kPa	mm Hg	7.502
kPa	psi	0.145
kPa	torr	7.502

Powers of Ten

E (exa)	10^{18}		d (deci)	10^{-1}
P (peta)	10^{15}		c (centi)	10^{-2}
T (tera)	10^{12}		m (milli)	10^{-3}
G (giga)	10^{9}		μ (micro)	10^{-6}
M (mega)	10^{6}		n (nano)	10^{-9}
k (kilo)	10^{3}		p (pico)	10^{-12}
h (hecto)	10^{2}		f (femto)	10^{-15}
da (deca)	10		a (atto)	10^{-18}

Abbreviations

The following is a list of the abbreviations used in the text. Common terms, the names of publications and institutions, and legal agreements are included along with their full identities. Other abbreviations will be defined wherever they first occur in an article. For further abbreviations, see page IX, Symbols and Units; page XVII, Frequently Cited Companies (Abbreviations), and page XVIII, Country Codes in patent references. The names of periodical publications are abbreviated exactly as done by Chemical Abstracts Service.

abs.	absolute	BGA	Bundesgesundheitsamt (Federal Republic of Germany)
a.c.	alternating current		
ACGIH	American Conference of Governmental Industrial Hygienists	BGB1.	Bundesgesetzblatt (Federal Republic of Germany)
ACS	American Chemical Society	BIOS	British Intelligence Objectives Subcommittee Report (see also FIAT)
ADI	acceptable daily intake		
ADN	accord européen relatif au transport international des marchandises dangereuses par voie de navigation interieure (European agreement concerning the international transportation of dangerous goods by inland waterways)	BOD	biological oxygen demand
		bp	boiling point
		B.P.	British Pharmacopeia
		BS	British Standard
		ca.	circa
		calcd.	calculated
ADNR	ADN par le Rhin (regulation concerning the transportation of dangerous goods on the Rhine and all national waterways of the countries concerned)	CAS	Chemical Abstracts Service
		cat.	catalyst, catalyzed
		CEN	Comité Européen de Normalisation
		cf.	compare
ADP	adenosine 5′-diphosphate	CFR	Code of Federal Regulations (United States)
ADR	accord européen relatif au transport international des marchandises dangereuses par route (European agreement concerning the international transportation of dangerous goods by road)	cfu	colony forming units
		Chap.	chapter
		ChemG	Chemikaliengesetz (Federal Republic of Germany)
AEC	Atomic Energy Commission (United States)	C.I.	Colour Index
		CIOS	Combined Intelligence Objectives Subcommitee Report (see also FIAT)
a.i.	active ingredient		
AIChE	American Institute of Chemical Engineers	CLP	Classification, Labelling and Packaging
		CNS	central nervous system
AIME	American Institute of Mining, Metallurgical, and Petroleum Engineers	Co.	Company
		COD	chemical oxygen demand
ANSI	American National Standards Institute	conc.	concentrated
AMP	adenosine 5′-monophosphate	const.	constant
APhA	American Pharmaceutical Association	Corp.	Corporation
API	American Petroleum Institute	crit.	critical
ASTM	American Society for Testing and Materials	CSA	Chemical Safety Assessment according to REACH
ATP	adenosine 5′-triphosphate	CSR	Chemical Safety Report according to REACH
BAM	Bundesanstalt für Materialprüfung (Federal Republic of Germany)		
		CTFA	The Cosmetic, Toiletry and Fragrance Association (United States)
BAT	Biologischer Arbeitsstofftoleranzwert (biological tolerance value for a working material, established by MAK Commission, see MAK)	DAB	Deutsches Arzneibuch, Deutscher Apotheker-Verlag, Stuttgart
		d.c.	direct current
		decomp.	decompose, decomposition
Beilstein	Beilstein's Handbook of Organic Chemistry, Springer, Berlin – Heidelberg – New York	DFG	Deutsche Forschungsgemeinschaft (German Science Foundation)
BET	Brunauer – Emmett – Teller	dil.	dilute, diluted

DIN	Deutsche Industrienorm (Federal Republic of Germany)		(regulation in the Federal Republic of Germany concerning the transportation of dangerous goods by rail)
DMF	dimethylformamide		
DNA	deoxyribonucleic acid	GGVS	Verordnung in der Bundesrepublik Deutschland über die Beförderung gefährlicher Güter auf der Straße (regulation in the Federal Republic of Germany concerning the transportation of dangerous goods by road)
DOE	Department of Energy (United States)		
DOT	Department of Transportation – Materials Transportation Bureau (United States)		
DTA	differential thermal analysis		
EC	effective concentration	GGVSee	Verordnung in der Bundesrepublik Deutschland über die Beförderung gefährlicher Güter mit Seeschiffen (regulation in the Federal Republic of Germany concerning the transportation of dangerous goods by sea-going vessels)
EC	European Community		
ed.	editor, edition, edited		
e.g.	for example		
emf	electromotive force		
EmS	Emergency Schedule		
EN	European Standard (European Community)	GHS	Globally Harmonised System of Chemicals (internationally agreed-upon system, created by the UN, designed to replace the various classification and labeling standards used in different countries by using consistent criteria for classification and labeling on a global level)
EPA	Environmental Protection Agency (United States)		
EPR	electron paramagnetic resonance		
Eq.	equation		
ESCA	electron spectroscopy for chemical analysis		
esp.	especially	GLC	gas-liquid chromatography
ESR	electron spin resonance	Gmelin	Gmelin's Handbook of Inorganic Chemistry, 8th ed., Springer, Berlin – Heidelberg – New York
Et	ethyl substituent ($-C_2H_5$)		
et al.	and others		
etc.	et cetera	GRAS	generally recognized as safe
EVO	Eisenbahnverkehrsordnung (Federal Republic of Germany)	Hal	halogen substituent ($-F, -Cl, -Br, -I$)
		Houben-Weyl	Methoden der organischen Chemie, 4th ed., Georg Thieme Verlag, Stuttgart
exp (...)	$e^{(...)}$, mathematical exponent		
FAO	Food and Agriculture Organization (United Nations)		
		HPLC	high performance liquid chromatography
FDA	Food and Drug Administration (United States)		
		H statement	hazard statement in GHS
FD&C	Food, Drug and Cosmetic Act (United States)	IAEA	International Atomic Energy Agency
		IARC	International Agency for Research on Cancer, Lyon, France
FHSA	Federal Hazardous Substances Act (United States)		
		IATA-DGR	International Air Transport Association, Dangerous Goods Regulations
FIAT	Field Information Agency, Technical (United States reports on the chemical industry in Germany, 1945)		
		ICAO	International Civil Aviation Organization
Fig.	figure	i.e.	that is
fp	freezing point	i.m.	intramuscular
Friedländer	P. Friedländer, Fortschritte der Teerfarbenfabrikation und verwandter Industriezweige Vol. 1–25, Springer, Berlin 1888–1942	IMDG	International Maritime Dangerous Goods Code
		IMO	Inter-Governmental Maritime Consultive Organization (in the past: IMCO)
FT	Fourier transform		
(g)	gas, gaseous	Inst.	Institute
GC	gas chromatography	i.p.	intraperitoneal
GefStoffV	Gefahrstoffverordnung (regulations in the Federal Republic of Germany concerning hazardous substances)	IR	infrared
		ISO	International Organization for Standardization
GGVE	Verordnung in der Bundesrepublik Deutschland über die Beförderung gefährlicher Güter mit der Eisenbahn	IUPAC	International Union of Pure and Applied Chemistry
		i.v.	intravenous

Abbreviation	Definition
Kirk-Othmer	Encyclopedia of Chemical Technology, 3rd ed., 1991–1998, 5th ed., 2004–2007, John Wiley & Sons, Hoboken
(l)	liquid
Landolt-Börnstein	Zahlenwerte u. Funktionen aus Physik, Chemie, Astronomie, Geophysik u. Technik, Springer, Heidelberg 1950–1980; Zahlenwerte und Funktionen aus Naturwissenschaften und Technik, Neue Serie, Springer, Heidelberg, since 1961
LC_{50}	lethal concentration for 50 % of the test animals
LCLo	lowest published lethal concentration
LD_{50}	lethal dose for 50 % of the test animals
LDLo	lowest published lethal dose
ln	logarithm (base e)
LNG	liquefied natural gas
log	logarithm (base 10)
LPG	liquefied petroleum gas
M	mol/L
M	metal (in chemical formulas)
MAK	Maximale Arbeitsplatzkonzentration (maximum concentration at the workplace in the Federal Republic of Germany); cf. Deutsche Forschungsgemeinschaft (ed.): Maximale Arbeitsplatzkonzentrationen (MAK) und Biologische Arbeitsstofftoleranzwerte (BAT), WILEY-VCH Verlag, Weinheim (published annually)
max.	maximum
MCA	Manufacturing Chemists Association (United States)
Me	methyl substituent (–CH_3)
Methodicum Chimicum	Methodicum Chimicum, Georg Thieme Verlag, Stuttgart
MFAG	Medical First Aid Guide for Use in Accidents Involving Dangerous Goods
MIK	maximale Immissionskonzentration (maximum immission concentration)
min.	minimum
mp	melting point
MS	mass spectrum, mass spectrometry
NAS	National Academy of Sciences (United States)
NASA	National Aeronautics and Space Administration (United States)
NBS	National Bureau of Standards (United States)
NCTC	National Collection of Type Cultures (United States)
NIH	National Institutes of Health (United States)
NIOSH	National Institute for Occupational Safety and Health (United States)
NMR	nuclear magnetic resonance
no.	number
NOEL	no observed effect level
NRC	Nuclear Regulatory Commission (United States)
NRDC	National Research Development Corporation (United States)
NSC	National Service Center (United States)
NSF	National Science Foundation (United States)
NTSB	National Transportation Safety Board (United States)
OECD	Organization for Economic Cooperation and Development
OSHA	Occupational Safety and Health Administration (United States)
p., pp.	page, pages
Patty	G.D. Clayton, F.E. Clayton (eds.): Patty's Industrial Hygiene and Toxicology, 3rd ed., Wiley Interscience, New York
PB report	Publication Board Report (U.S. Department of Commerce, Scientific and Industrial Reports)
PEL	permitted exposure limit
Ph	phenyl substituent (—C_6H_5)
Ph. Eur.	European Pharmacopoeia, Council of Europe, Strasbourg
phr	part per hundred rubber (resin)
PNS	peripheral nervous system
ppm	parts per million
P statement	precautionary statement in GHS
q.v.	which see (quod vide)
REACH	Registration, Evaluation, Authorisation and Restriction of Chemicals (EU regulation addressing the production and use of chemical substances, and their potential impacts on both human health and the environment)
ref.	refer, reference
resp.	respectively
R_f	retention factor (TLC)
R.H.	relative humidity
RID	réglement international concernant le transport des marchandises dangereuses par chemin de fer (international convention concerning the transportation of dangerous goods by rail)
RNA	ribonucleic acid
R phrase (R-Satz)	risk phrase according to ChemG and GefStoffV (Federal Republic of Germany)
rpm	revolutions per minute
RTECS	Registry of Toxic Effects of Chemical Substances, edited by the National Institute of Occupational Safety and Health (United States)
(s)	solid

SAE	Society of Automotive Engineers (United States)		der Technischen Chemie, 4th ed., Verlag Chemie, Weinheim 1972–1984; 3rd ed., Urban und Schwarzenberg, München 1951–1970
SAICM	Strategic Approach on International Chemicals Management (international framework to foster the sound management of chemicals)	USAEC	United States Atomic Energy Commission
s.c.	subcutaneous	USAN	United States Adopted Names
SI	International System of Units	USD	United States Dispensatory
SIMS	secondary ion mass spectrometry	USDA	United States Department of Agriculture
S phrase (S-Satz)	safety phrase according to ChemG and GefStoffV (Federal Republic of Germany)	U.S.P.	United States Pharmacopeia
		UV	ultraviolet
STEL	Short Term Exposure Limit (see TLV)	UVV	Unfallverhütungsvorschriften der Berufsgenossenschaft (workplace safety regulations in the Federal Republic of Germany)
STP	standard temperature and pressure (0°C, 101.325 kPa)		
T_g	glass transition temperature	VbF	Verordnung in der Bundesrepublik Deutschland über die Errichtung und den Betrieb von Anlagen zur Lagerung, Abfüllung und Beförderung brennbarer Flüssigkeiten (regulation in the Federal Republic of Germany concerning the construction and operation of plants for storage, filling, and transportation of flammable liquids; classification according to the flash point of liquids, in accordance with the classification in the United States)
TA Luft	Technische Anleitung zur Reinhaltung der Luft (clean air regulation in Federal Republic of Germany)		
TA Lärm	Technische Anleitung zum Schutz gegen Lärm (low noise regulation in Federal Republic of Germany)		
TDLo	lowest published toxic dose		
THF	tetrahydrofuran		
TLC	thin layer chromatography		
TLV	Threshold Limit Value (TWA and STEL); published annually by the American Conference of Governmental Industrial Hygienists (ACGIH), Cincinnati, Ohio		
		VDE	Verband Deutscher Elektroingenieure (Federal Republic of Germany)
		VDI	Verein Deutscher Ingenieure (Federal Republic of Germany)
TOD	total oxygen demand	vol	volume
TRK	Technische Richtkonzentration (lowest technically feasible level)	vol.	volume (of a series of books)
		vs.	versus
TSCA	Toxic Substances Control Act (United States)	WGK	Wassergefährdungsklasse (water hazard class)
TÜV	Technischer Überwachungsverein (Technical Control Board of the Federal Republic of Germany)	WHO	World Health Organization (United Nations)
TWA	Time Weighted Average	Winnacker-Küchler	Chemische Technologie, 4th ed., Carl Hanser Verlag, München, 1982-1986; Winnacker-Küchler, Chemische Technik: Prozesse und Produkte, Wiley-VCH, Weinheim, 2003–2006
UBA	Umweltbundesamt (Federal Environmental Agency)		
Ullmann	Ullmann's Encyclopedia of Industrial Chemistry, 6th ed., Wiley-VCH, Weinheim 2002; Ullmann's Encyclopedia of Industrial Chemistry, 5th ed., VCH Verlagsgesellschaft, Weinheim 1985–1996; Ullmanns Encyklopädie		
		wt	weight
		$	U.S. dollar, unless otherwise stated

Frequently Cited Companies (Abbreviations)

Air Products	Air Products and Chemicals	IFP	Institut Français du Pétrole
Akzo	Algemene Koninklijke Zout Organon	INCO	International Nickel Company
		3M	Minnesota Mining and Manufacturing Company
Alcoa	Aluminum Company of America	Mitsubishi Chemical	Mitsubishi Chemical Industries
Allied	Allied Corporation		
Amer. Cyanamid	American Cyanamid Company	Monsanto	Monsanto Company
		Nippon Shokubai	Nippon Shokubai Kagaku Kogyo
BASF	BASF Aktiengesellschaft		
Bayer	Bayer AG	PCUK	Pechiney Ugine Kuhlmann
BP	British Petroleum Company	PPG	Pittsburg Plate Glass Industries
Celanese	Celanese Corporation	Searle	G.D. Searle & Company
Daicel	Daicel Chemical Industries	SKF	Smith Kline & French Laboratories
Dainippon	Dainippon Ink and Chemicals Inc.	SNAM	Societá Nazionale Metandotti
Dow Chemical	The Dow Chemical Company	Sohio	Standard Oil of Ohio
		Stauffer	Stauffer Chemical Company
DSM	Dutch Staats Mijnen	Sumitomo	Sumitomo Chemical Company
Du Pont	E.I. du Pont de Nemours & Company	Toray	Toray Industries Inc.
Exxon	Exxon Corporation	UCB	Union Chimique Belge
FMC	Food Machinery & Chemical Corporation	Union Carbide	Union Carbide Corporation
GAF	General Aniline & Film Corporation	UOP	Universal Oil Products Company
W.R. Grace	W.R. Grace & Company	VEBA	Vereinigte Elektrizitäts- und Bergwerks-AG
Hoechst	Hoechst Aktiengesellschaft	Wacker	Wacker Chemie GmbH
IBM	International Business Machines Corporation		
ICI	Imperial Chemical Industries		

Country Codes

The following list contains a selection of standard country codes used in the patent references.

AT	Austria	IL	Israel
AU	Australia	IT	Italy
BE	Belgium	JP	Japan*
BG	Bulgaria	LU	Luxembourg
BR	Brazil	MA	Morocco
CA	Canada	NL	Netherlands*
CH	Switzerland	NO	Norway
CS	Czechoslovakia	NZ	New Zealand
DD	German Democratic Republic	PL	Poland
DE	Federal Republic of Germany (and Germany before 1949)*	PT	Portugal
		SE	Sweden
DK	Denmark	SU	Soviet Union
ES	Spain	US	United States of America
FI	Finland	YU	Yugoslavia
FR	France	ZA	South Africa
GB	United Kingdom	EP	European Patent Office*
GR	Greece	WO	World Intellectual Property Organization
HU	Hungary		
ID	Indonesia		

*For Europe, Federal Republic of Germany, Japan, and the Netherlands, the type of patent is specified: EP (patent), EP-A (application), DE (patent), DE-OS (Offenlegungsschrift), DE-AS (Auslegeschrift), JP (patent), JP-Kokai (Kokai tokkyo koho), NL (patent), and NL-A (application).

Periodic Table of Elements

element symbol, atomic number, and relative atomic mass (atomic weight)

- 1A   "European" group designation and old IUPAC recommendation
- 1   group designation to 1986 IUPAC proposal
- IA   "American" group designation, also used by the Chemical Abstracts Service until the end of 1986
- [a] provisional IUPAC symbol
- * radioactive element; mass of most important isotope given.

1A 1 IA	2A 2 IIA	3A 3 IIIB	4A 4 IVB	5A 5 VB	6A 6 VIB	7A 7 VIIB	8 8 VIII	8 9 VIII	8 10 VIII	1B 11 IB	2B 12 IIB	3B 13 IIIA	4B 14 IVA	5B 15 VA	6B 16 VIA	7B 17 VIA	0 18 VIIIA
1 H 1.0079																	2 He 4.0026
3 Li 6.941	4 Be 9.0122											5 B 10.811	6 C 12.011	7 N 14.007	8 O 15.999	9 F 18.998	10 Ne 20.180
11 Na 22.990	12 Mg 24.305											13 Al 26.982	14 Si 28.086	15 P 30.974	16 S 32.066	17 Cl 35.453	18 Ar 39.948
19 K 39.098	20 Ca 40.078	21 Sc 44.956	22 Ti 47.867	23 V 50.942	24 Cr 51.996	25 Mn 54.938	26 Fe 55.845	27 Co 58.933	28 Ni 58.693	29 Cu 63.546	30 Zn 65.409	31 Ga 69.723	32 Ge 72.61	33 As 74.922	34 Se 78.96	35 Br 79.904	36 Kr 83.80
37 Rb 85.468	38 Sr 87.62	39 Y 88.906	40 Zr 91.224	41 Nb 92.906	42 Mo 95.94	43 Tc* 98.906	44 Ru 101.07	45 Rh 102.91	46 Pd 106.42	47 Ag 107.87	48 Cd 112.41	49 In 114.82	50 Sn 118.71	51 Sb 121.76	52 Te 127.60	53 I 126.90	54 Xe 131.29
55 Cs 132.91	56 Ba 137.33		72 Hf 178.49	73 Ta 180.95	74 W 183.84	75 Re 186.21	76 Os 190.23	77 Ir 192.22	78 Pt 195.08	79 Au 196.97	80 Hg 200.59	81 Tl 204.38	82 Pb 207.2	83 Bi 208.98	84 Po* 208.98	85 At* 209.99	86 Rn* 222.02
87 Fr* 223.02	88 Ra* 226.03		104 Rf* 261.11	105 Db* 262.11	106 Sg 	107 Bh	108 Hs	109 Mt	110 Ds	111 Rg	112 Cn	113 Uut[a]	114 Fl	115 Uup[a]	116 Lv		118 Uuo[a]

57 La 138.91	58 Ce 140.12	59 Pr 140.91	60 Nd 144.24	61 Pm* 146.92	62 Sm 150.36	63 Eu 151.97	64 Gd 157.25	65 Tb 158.93	66 Dy 162.50	67 Ho 164.93	68 Er 167.26	69 Tm 168.93	70 Yb 173.04	71 Lu 174.97
89 Ac* 227.03	90 Th* 232.04	91 Pa* 231.04	92 U* 238.03	93 Np* 237.05	94 Pu* 244.06	95 Am* 243.06	96 Cm* 247.07	97 Bk* 247.07	98 Cf* 251.08	99 Es* 252.08	100 Fm* 257.10	101 Md* 258.10	102 No* 259.10	103 Lr* 260.11

Polystyrene and Styrene Copolymers

JÜRGEN MAUL, Hüls AG, Marl, Germany

BRUCE G. FRUSHOUR, Monsanto Chemical Company, Springfield, MA 01150, United States

JEFFREY R. KONTOFF, Monsanto Chemical Company, Springfield, MA 01150, United States

HERBERT EICHENAUER, Bayer AG, Dormagen, Germany

KARL-HEINZ OTT, Bayer AG, Dormagen, Germany

CHRISTIAN SCHADE, BASF AG, Ludwigshafen, Germany

1.	Polystyrene	981
1.1.	Introduction	981
1.2.	Production	983
1.2.1.	Bulk Polymerization	984
1.2.2.	Suspension Polymerization	986
1.3.	Properties	987
1.3.1.	Chemical Properties	987
1.3.2.	Physical and Processing Properties	987
1.4.	Processing and Uses	990
1.5.	Quality Specifications	990
1.6.	Storage and Transportation	991
1.7.	Recycling	991
1.8.	Environmental Aspects	991
2.	Styrene–Acrylonitrile (SAN) Copolymers	992
2.1.	Production	992
2.2.	Properties	994
2.3.	Processing	995
2.4.	Uses	996
2.5.	Blending of SAN	996
3.	Barrier Resins	996
4.	Other Copolymers	997
4.1.	α-Methylstyrene–Acrylonitrile Copolymers	998
4.2.	Styrene–Methyl Methacrylate Copolymers	998
4.3.	Styrene–Maleic Anhydride Copolymers	998
4.4.	Styrene–Maleimide Copolymers	999
4.5.	Styrene–Acrylate Copolymers	999
5.	Acrylonitrile–Butadiene–Styrene (ABS) Polymers	999
5.1.	Definition and Structure	999
5.1.1.	Historical Aspects	999
5.1.2.	Structural Principles	1000
5.1.3.	Synthesis of the Two-Phase Structure	1001
5.1.4.	Properties of the Resin Matrix	1002
5.2.	Structure–Property Relationships	1003
5.3.	Production of ABS Polymers	1004
5.3.1.	ABS Production by Emulsion Polymerization	1005
5.3.2.	Production of the Matrix Copolymer	1009
5.3.3.	ABS Production by Bulk Polymerization	1011
5.3.4.	ABS Production by Bulk Suspension Polymerization	1013
5.3.5.	Process Combinations	1014
5.3.6.	Additives	1015
5.4.	Quality Assurance and Standardization	1015
5.5.	Properties	1016
5.6.	Special Product Modifications	1018
5.7.	Legal Aspects	1018
5.8.	Storage and Transportation	1019
5.9.	Uses	1019
5.10.	Economic Aspects of Acrylonitrile-Based Styrene Copolymers	1019
5.11.	Recycling	1020
5.12.	ABS-Analogous Systems	1020
5.12.1.	ASA, AES, and ACS Polymers	1020
5.12.2.	MBS and MABS Polymers	1022
5.13.	ABS Blends	1023
	References	1024

1. Polystyrene

1.1. Introduction

Polystyrene (PS) belongs to the group of standard thermoplastics that also includes polyethylene, polypropylene, and poly(vinyl chloride). Because of its special properties, polystyrene can be used in an extremely wide range of applications. The annual consumption worldwide in 2004 was ca. 11.5×10^6 t; thus polystyrene is one of the most quantitatively important chemicals.

Historical Aspects. Styrene [*100-42-4*] (known as styrax) was first isolated in 1831 by BONASTRE from the resin of the amber tree. In 1839 E. SIMON, who also first described the polymer, gave the monomer its name. He observed that styrene was slowly converted into a viscous solution on standing. Around 1925 the development of an industrial production process for polystyrene began; this work achieved success in the plants of the IG Farbenindustrie in Germany in 1930 [1]. In the United States polystyrene was first produced on a commercial scale in 1938 by the Dow Chemical Company.

Polystyrene and Impact-Resistant Polystyrene. Chapter 1 deals with the clear polystyrene molding materials [*9003-53-6*] and rubber modified polystyrene [*9003-55-8*], also known as styrene–butadiene molding material.

Polystyrene molding materials are hard, transparent materials with a high gloss. They are most commonly described as general-purpose polystyrene (GPPS) but terms such as standard polystyrene, normal polystyrene, clear polystyrene, crystal polystyrene, or styrene homopolymer are also in use. In this article the definition polystyrene molding material (PS) is used according to ISO 1622-2.

Below 100 °C PS molding materials solidify to give a glasslike material with adequate mechanical strength, good dielectric properties, and resistance toward a large number of chemicals for many areas of application.

Above its softening point clear polystyrene occurs as a melt which can be readily processed by techniques such as injection molding or extrusion. Small quantities of lubricants can be used internally or externally as processing aids. The addition of antistatic agents, UV stabilizers, glass fibers, or colorants is also common.

The mechanical properties of the relatively brittle PS molding materials can be considerably improved by adding rubbers, generally polybutadiene [*106-99-0*]. Styrene–butadiene molding materials are generally referred to as high-impact polystyrene (HIPS), this term is used here. They are also known as toughened PS or rubber-modified PS; ISO 2897-2 defines them as impact-resistant polystyrene (IPS).

Early production processes for HIPS were based on mixing PS molding materials with a rubber component. Polymerization of a styrene–

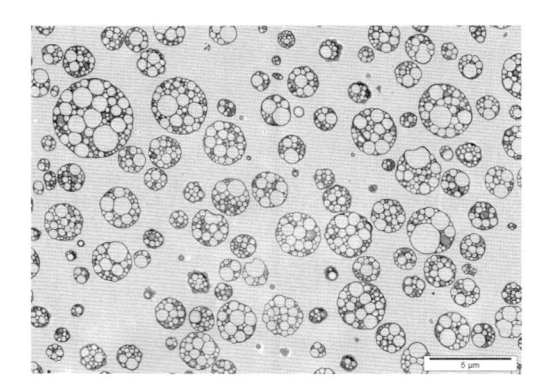

Figure 1. Electron micrograph of HIPS molding material
The dispersed rubber particles (dark as a result of OsO$_4$ treatment) are embedded in the polystyrene matrix (lighter regions). The rubber particles contain lighter polystyrene inclusions
Courtesy of W. HECKMANN.

polybutadiene solution is, however, much more effective. A two-phase system is formed due to the immiscibility of polystyrene and polybutadiene. Polystyrene forms the continuous phase (matrix) and polybutadiene the disperse phase. The globular rubber particles contain small inclusions of polystyrene (Fig. 1).

The additives commonly used with PS molding materials can also be employed in HIPS. Paraffinic white oil is commonly used as lubricant, zinc stearate as mold release agent. Antioxidants are used for rubber stabilization. Flame retardants, antistatic agents, and other additives are added for special applications.

The rubber particles in HIPS generally have a mean diameter of 0.5–9 μm. They thus scatter visible light and the transparency of the PS molding materials is lost. Sequential anionic block copolymerization of styrene and butadiene allows the production of transparent [3, 4]. Phase separation of the different blocks occurs. Typically lamellar or similar impact resistant molding materials, also called styrene block copolymers (SBC) structures are formed instead of discrete particles (Fig. 2). The thickness of the layers is far less than the particle size in HIPS, so that light scattering does not occur at the phase boundaries.

Styrene Copolymers. Styrene can be copolymerized with many other monomers. Styrene–acrylonitrile molding materials, in particular, have achieved great economic importance in transparent and rubber-modified forms (SAN

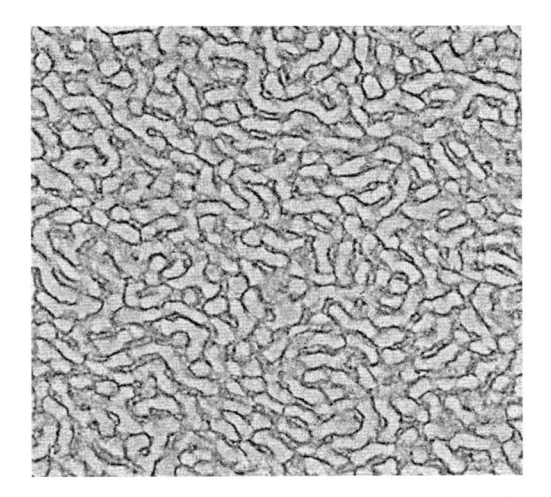

Figure 2. Electron micrograph of a transparent IPS molding material (SBC, contrasting as in Fig. 1). The size of the square is approx. 1.5 μm × 1.5 μm.
Courtesy of W. HECKMANN.

[9003-54-7], Chap. 2; ABS [9003-56-9], Chap. 5). Compared with the pure styrene polymers they have advantages with regard to hardness, strength, and resistance to heat distortion and environmental stress cracking. However, these advantages are offset by a higher price and more difficult processing.

Copolymers of styrene and maleic anhydride [9011-13-6] (Section 4.3) have a softening point that is up to 30 °C higher than that of PS molding materials. These products are used in the form of foams in the automotive industry (e.g, as interior headliners).

Polystyrene Foams. Expandable polystyrene (EPS) is the starting material for PS hard foam materials. It is produced from polystyrene by the addition of ca. 6 % of a low-boiling hydrocarbon (e.g., pentane) as the foaming agent. Extruded polystyrene (XPS) foams are produced from polystyrene and a blowing agent, e.g., carbon dioxide. Foamed polystyrene is discussed in detail elsewhere, →Foamed PlasticsFoamed Plastics.

1.2. Production

Styrene. The properties and production of the raw material styrene are described elsewhere (→ →StyreneStyrene). The styrene used for polymerization should have a purity greater than 99.6 % because the contaminants arising from the production process, mainly ethylbenzene, cumene, and xylenes, affect the molecular mass of the polystyrene. Commonly, purities of 99.9 % or more are used.

For storage and transport styrene is stabilized with inhibitors such as 4-*tert*-butylcatechol to prevent polymerization at low temperatures. These inhibitory radical scavengers must be removed by distillation or adsorption on alumina.

Polymerization. Styrene can function as an electron donor or an electron acceptor. It can therefore be polymerized by radical, cationic, or anionic mechanisms, or by coordination propagation steps.

The industrial polymerization of styrene to PS and HIPS molding materials is carried out exclusively by a free radical mechanism. The chain reaction does not necessarily have to be started by the addition of radical-forming, readily decomposing initiators because styrene itself can form polymerization-initiating radicals [5]. The propagation mechanism for the chain growth proceeds by addition of further monomer to the radical chain end. Growth of the radical chain is mainly terminated by recombination. With increasing temperature, disproportion and back-biting (see → a21_487-sec4-0002 Polyolefins 1.3.1.1. IntroductionPolyolefins, Section 1.3.1.1.) followed by β-scission [6] play an increasing role. The growth of a chain proceeds within seconds; the gross rate is largely determined by the radical initiation reaction. Average molecular masses between 100 000 and 400 000 are obtained within a short time. Molecular masses are mainly controlled by reaction temperature, chain-transfer agents — including solvents — and by initiators. The kinetics of the process are well understood [7] and may be modeled by commercially available software.

Cationic polymerization of styrene (e.g., by strong Lewis acids such as $AlCl_3$, $HClO_3$, or BF_3) is unimportant industrially because control of the molecular mass is difficult. Very small quantities of very low molecular mass polystyrene are synthesized via such routes.

Transparent HIPS (SBC) is, however, produced industrially by anionic block copolymerization of butadiene and styrene. Organolithium compounds (e.g., *n*-butyllithium) are used as initiators. The monomers and solvent used in anionic polymerization have extremely high purity requirements. The polymerization gross rate is higher than that of the radical chain reaction due to the simultaneous growth of all initiated chains. Polymerization is carried out at low temperature in solution to enable the reaction to be controlled.

The coordination polymerization of styrene with conventional Ziegler catalysts produces a sterically ordered, isotactic polymer. These highly ordered chains can form polymer crystals. The melting temperature of the crystalline zones of isotactic polystyrene is ca. 230 °C. The rate of crystallization is, however, so low that this property cannot be exploited industrially [8].

Syndiotactic polystyrene crystallizes very rapidly and is produced by using special titanium–aluminoxan catalysts [9, 10]. The melting temperature of the syndiotactic sequences is ca. 270 °C. The material has found limited use in heat-resistant applications, often glass-fiber reinforced or blended with polyamide.

HIPS Synthesis. Rubbers used for HIPS synthesis are medium-*cis*-polybutadienes with a glass transition temperature around –85 °C. They are obtained by butyllithium-initiated anionic polymerization of butadiene. Usually the polybutadienes are long-chain branched to prevent cold flow during storage. Other rubbers (e.g., high-*cis* grades, styrene–butadiene block copolymers, EPDM rubbers) may be also employed, but have less significance.

The synthesis of HIPS requires careful control of the rubber particle size and morphology. Initially, the solution consists of polybutadiene rubber dissolved in styrene and further additives such as diluents, chain-transfer agents, or initiators. With the onset of polymerization, the polystyrene forms a separate phase. Styrene and optional diluents are the common solvents for the elastomer and the polystyrene phase (oil-in-oil emulsion). The polystyrene phase grows at the expense of the rubber phase with further conversion. The rubber phase initially is the continuous phase and remains so if unstirred. In a stirred reactor, phase inversion occurs, largely depending on the phase volume ratios and viscosities of the two phases. The reaction conditions during phase inversion determine the rubber morphology and particle size to a large extent.

Simple blending of polystyrene and polybutadiene leads to a mechanically inferior product. The mechanical strength of HIPS is supported by polystyrene molecules, which are grafted onto rubber molecules. Grafting occurs especially at low styrene conversions, when the volume of the polystyrene phase is still small. Quantity and nature of these copolymers have big impact on the particle size and morphology of the HIPS rubber particles [11, 15].

1.2.1. Bulk Polymerization

Currently, PS and HIPS molding materials are mainly produced by continuous polymerization processes. The polystyrene product has consistent quality, high purity, and a low residual monomer content (generally less than 500 ppm monomeric styrene). The most difficult task from an industrial standpoint is the control of the highly viscous melt and the removal of the heat of polymerization (71 kJ/mol).

Solvents such as toluene or ethylbenzene are frequently present in small proportions (typically 5–15 %) to provide better control of the polymerization rate, the viscosity of the polymer melt, and the cross-linking of the rubber phase. This technique is known as *solution polymerization*. Due to the limited quantities of the diluent, the processes are often also referred to as *continuous bulk polymerization*. The solvent and the unreacted monomer must be removed by degassing. They are generally recovered, optionally purified and transferred back to the feed. In a running continuous operation, solvents are in essence introduced via styrene impurities (mostly ethylbenzene).

The plant setup generally comprises a polymerization section, a degassing and solvent recovery section, and a pelletizing section.

The most common reactor for *PS molding materials* is an ebullient single stirred-tank reactor. A second stirred-tank reactor or a tubular plug-flow reactor may be used to increase conversion. PS molding materials may also be produced in more complex reactor configurations like HIPS units.

HIPS plants generally consist of a series of three to five reactors, split into a prepolymerization and postpolymerization section. In the prepolymerization section phase inversion takes place, and rubber morphology and particle size are largely adjusted. Various reactor designs have been proposed for this part of the plant. Common are one or two stirred-tank reactors, tower reactors, loop reactors with static mixers, or combinations of these. The reaction is typically carried out at 100–150 °C to 15–30 % conversion. The postpolymerization part commonly consists of several plug-flow type reactors to achieve high styrene conversions, but continuous stirred-tank reactors may be used as well. Common temperatures are 140–190 °C in this part of the plant. Figure 3 shows several common reactor configurations.

In the *continuous stirred-tank reactor* part of the heat is removed by means of reflux cooling. The heat of polymerization causes part of the styrene to evaporate. The gaseous monomer is condensed in condensers with large surface areas and recycled back to the reactor. The prerequisite for this type of heat removal is effective mixing of the reactor contents. Anchor stirrers, impellers, or helical ribbon agitators are commonly used. The reaction temperature is the most important parameter affecting the molecular mass of the polystyrene and thus the properties of the product. It is controlled by varying the

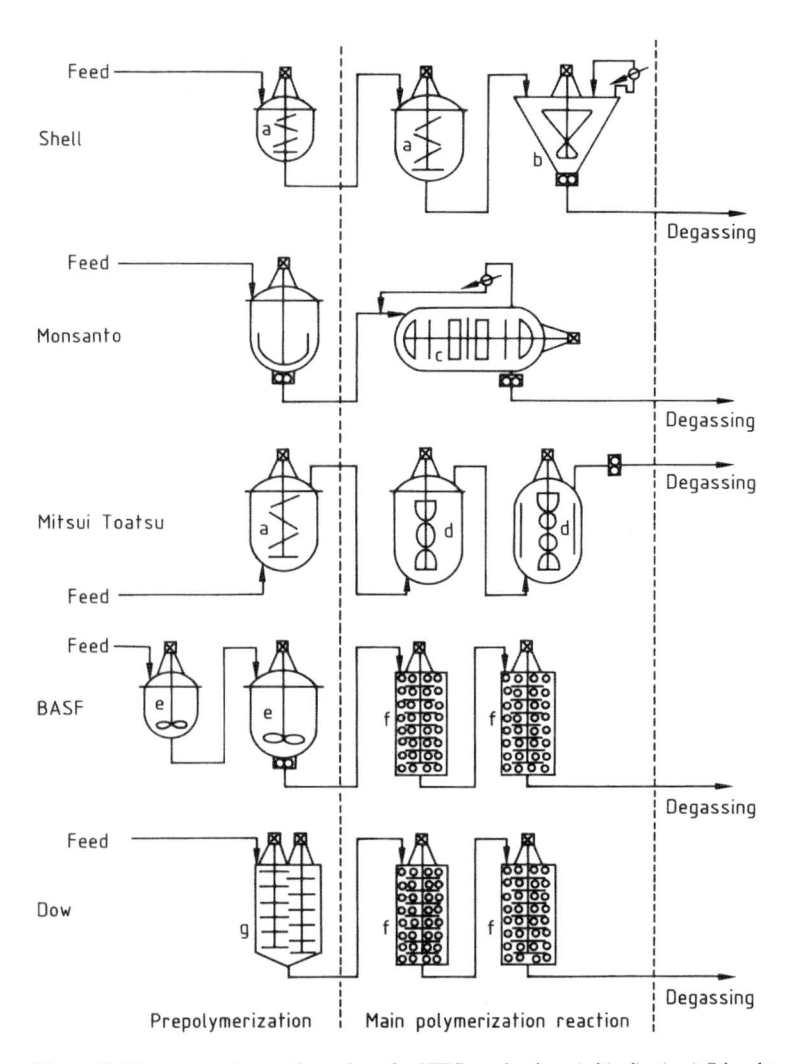

Figure 3. Various reactor configurations for HIPS production a), b), d), e), g) Stirred-tank reactors; c) Horizontal plug-flow reactor; f) Tower reactors
Reproduced from [14] with permission of Hanser Publishers.

pressure and thus the boiling point of styrene in the stirred tank. The temperatures in the reactors are between 100 and 170 °C, corresponding to pressures of ca. 0.5–2 bar.

The reactor feed can also be used to remove heat of polymerization. Reactors converting more than 40 % of the styrene feed usually need additional cooling. The lower solid content of the HIPS prepolymerization reactors, on the other hand, necessitates preheating of the reactor feed.

Tower reactors are equipped with special baffles in which a cooling agent circulates to ensure the removal of the heat of polymerization. Different temperature zones can be set up in a reactor by adjusting the temperature of the cooling agent. The tower reactors are also equipped with agitator blades.

Tubular reactors with static mixing elements and cooling are also in use. Adiabatic or isothermal tubular reactors are sometimes used for high conversions. Also known are axially segmented and agitated tubular reactors operating under reflux.

The polystyrene melt leaving the reactor section has a concentration of 60–90 %. Unreacted styrene monomer and solvent are removed by flash devolatilization. The polymer is fed into an evacuated, heated vessel where the volatile components evaporate, the melt is then pumped out. The styrene–solvent mixture formed is recycled to the first reactor. A small fraction must be discarded, however, to avoid increasing the concentration of styrene contaminants.

A second flash tank or another degassing unit (e.g., falling strand degassers, steam strippers, thin-film evaporators, degassing extruders) may be used to further decrease content of volatile compounds.

The polymer melts can be mixed with additives (e.g., dyes, stabilizers, or lubricants) by using static or mechanical mixing equipment.

Degassing of the polystyrene melt is followed by pelletization. In extrusion pelletization the polymer is forced through a perforated disk in continuous strands. The necessary pressure is applied by means of the discharge pump in the degassing unit or by an extruder. The strands are cooled in a water bath and cut into cylindrical pellets.

Underwater and water-ring pelletizing systems are also used and yield lentil-shaped pellets.

The pellets can be coated with an externally applied lubricant (e.g., zinc stearate or N,N'-ethylene-bis(stearamide)) to improve the sliding properties.

1.2.2. Suspension Polymerization

Suspension processes are chiefly used for the production of expandable polystyrene. PS molding materials may be produced by this process when the blowing agent is omitted. The combined bulk–suspension process is sometimes still used for HIPS production.

In suspension polymerization styrene is dispersed in water in the form of small droplets by stirring. The addition of suspension agents such as poly(vinyl pyrrolidone), poly (vinyl alcohol), alkaline-earth phosphates, or celluloses is necessary to stabilize the organic phase. Polymerization must be started with radical-forming initiators because the process temperatures are considerably lower than those of continuous bulk polymerization (80–110 °C).

The main advantage of suspension polymerization is that the heat of polymerization can easily be removed via the aqueous phase. A high degree of conversion can also be achieved.

Generally, however, the polystyrene beads must be repelletized after separation of the water and drying to satisfy market requirements with regard to processing of the molding materials (e.g., injection molding equipment). Additives and dyes can be incorporated during regranulation.

In the case of HIPS, suspension polymerization must be preceded by bulk polymerization to ensure that rubber particles are formed. The reaction can only be carried out in suspension after the formation of the disperse polybutadiene phase. The product has comparable material properties to HIPS produced by the bulk process.

The batch suspension process for styrene polymerization has been gradually replaced by the more economically favorable continuous bulk processes. Very few of the older plants based on batch suspension technology are still run commercially.

1.3. Properties

1.3.1. Chemical Properties

The chemical properties of atactic PS and HIPS are similar.

Under certain conditions oxidizing acids lead to oxidative degradation. However, the molding materials are resistant to dilute acid, alkali, and salt solutions. Exposure to UV radiation in the presence of oxygen causes yellowing and embrittlement, particularly in HIPS modified with polybutadiene. Reactions at the remaining double bonds bring about a gradual loss in elastic properties of the soft components. This damage can be suppressed but not completely inhibited by the addition of light stabilizers. Polystyrene is therefore not generally used in areas of application where weather resistance is required.

The replacement of polybutadiene rubber by EPDM rubber and the addition of special antioxidants lengthen the service lifetime of materials exposed to weathering by a factor of 15–20.

Oils and fats (particularly important in the food packaging sector) produce stress cracking. This can, however, be effectively counteracted by a range of measures [13], for example, by moderately increasing the average particle size to more than 5 μm.

The self-ignition temperature of PS and HIPS is 490 °C. Polystyrene articles are generally classified as being moderately flammable (DIN 4102) and as slow-burning (Underwriters Laboratories UL 94 HB). The flammability of special HIPS molding materials can be reduced considerably with additives (e.g., organobromine compounds) to give products that are classified as having a low flammability. The main area of application for these products is in the electrical sector. They are also increasingly being used for housings for technical appliances (e.g., photocopiers, televisions).

1.3.2. Physical and Processing Properties

Polystyrene, being a linear hydrocarbon chain with phenyl side groups linked, on average, to every second carbon atom, dissolves in aromatic hydrocarbons (toluene, ethylbenzene), halogenated hydrocarbons, aliphatic ethers (tetrahydrofuran), esters (ethyl acetate), and ketones (butanone), but not in aliphatic hydrocarbons (hexane) and alcohols (methanol). A summary of the solubility behavior of polystyrene can be found in [12].

The most important physical data for PS and HIPS are given in Table 1.

Special PS molding materials are available with specific property profiles for different processing technologies such as injection molding, extrusion, or thermoforming.

The mechanical and rheological behavior of polystyrene is predominantly determined by its average molecular mass (normally 150 000–400 000). The strength improves with increasing chain length but the melt viscosity increases making processing more difficult. Lubricants such as butyl stearate or paraffinic oil can be added to lower the melt viscosity of PS and HIPS. However these agents also lower the softening point.

PS grades with very good or good flow properties are mainly used for processing in injection molding machines. High molecular mass grades with higher strength and resistance to heat distortion are preferred for extrusion of films and sheets.

In extrusion and injection molding, the polystyrene melt is forced through narrow nozzles or slits. The polymer molecules orientate themselves parallel to the flow direction due to the high shear rate. The melt often solidifies in the mold so quickly that this initial orientation cannot be reversed by molecular diffusion. Polystyrene end products may therefore display anisotropy of their mechanical properties parallel and perpendicular to the processing direction. Tensile properties are much increased in parallel to the processing direction, while they are generally weakened perpendicularly to it. Biaxially oriented polystyrene has increased mechanical strength over the entire film. It may dissipate mechanical energy by a so-called "shear yielding" rather than the more common and less efficient "crazing" mechanism.

The impact resistance of HIPS is higher than that of PS due to the rubber component, but transparency is lost [exception: transparent styrene–butadiene multiblock copolymers (SBC) produced by anionic polymerization (Section 1.1)]. Processing properties are not altered significantly by the polybutadiene phase. However,

Table 1. Physical properties of PS and HIPS resins

Property	Test method	Typical values	
		PS resins	HIPS resins
Density, g/cm^3	DIN 53 479 ASTM D 792	1.05	1.03–1.05
Melt flow index (200 °C/5 kg), g/10 min	DIN 53 735 ASTM D 1238/G	1–30	2–20
Tensile modulus, MPa	DIN 53 457 ASTM D 638	3000–3400	1600–2800
Yield tensile strength, MPa	DIN 53 455 ASTM D 638	35–60	20–45
Tensile elongation at break, %	DIN 53 455 ASTM D 638	1.5–3	25–70
Ball indentation hardness, MPa	DIN 53 456 ISO 2039	150–160	75–140
Flexural strength, MPa	DIN 53 452 ASTM D 790/1	60–120	35–80
Charpy impact strength, kJ/m^2	ISO 179	10–25	35–no break
Notched Charpy impact strength, kJ/m^2	DIN 53 453	1–2	5–17
Izod impact strength, kJ/m^2	ISO 180/1 A	10–25	35–no break
Notched Izod impact strength, J/m	ASTM D 256/A	15–30	50–150
Vicat softening temperature, °C			
VST/B (50°/h, 5 kg)	DIN 53 460	78–102	75–97
VST/A (120°/h, 1 kg)	ASTM D 1525/B	85–110	80–100
Thermal conductivity, W K^{-1} m^{-1}	DIN 52 612	0.16	0.16
Dielectric constant (1 MHz)	DIN 53 483 ASTM D 150	2.5	2.5–2.7
Dissipation factor (1 MHz)	DIN 53 483 ASTM D 150	$(1–4) \times 10^{-4}$	$(2–8) \times 10^{-4}$
Thermal coefficient of linear expansion, K^{-1}	DIN 53 752	$(6–8) \times 10^{-5}$	$(8–10) \times 10^{-5}$
Refractive index	DIN 53 491	1.59	

the softening point, the modulus of elasticity, and the hardness are lower than those of PS molding materials.

In HIPS, the rubber phase (generally polybutadiene) is distributed in the form of discrete particles in the polystyrene matrix. The most important parameters affecting the properties of the end products are the rubber content, morphology, average particle size, particle-size distribution, and the degree of cross-linking of the polybutadiene phase (Fig. 4).

Rubber Content. The HIPS grades from different producers may be classified according to their polybutadiene (PB) content: semi-impact-resistant (ca. 3–6% PB) and impact-resistant (6–10% PB). The real criterion is not the rubber content, but the gel content that is obtained with a given rubber content (i.e., the mass fraction of the disperse phase including the polystyrene enclosed within the particles and grafted onto the polybutadiene chains) [14]. The gel content is roughly three to four times the quantity of rubber added.

Rubber Morphology. A wide range of different particle morphologies can be obtained depending on the chemical composition of the rubber and the polymerization conditions. Standard HIPS grades containing polybutadiene usually consist of cellular particles (Fig. 1). If part of the polybutadiene is replaced by styrene–butadiene block copolymers, coils, mazes, shell, or capsule particles are formed [15].

Particle Size and Particle-Size Distribution. The particle-size spectrum of the rubber phase is the factor determining the impact resistance of the molding materials. Commercial grades have average particle sizes in the range 0.5–9 µm. The highest mechanical efficiency is in the range 1.5–3 µm. If the particles are significantly larger, less crazes and less cavitation can be supported by the same rubber quantity [16, 17]. Grades with particles sizes > 5 µm have increased environmental stress crack resistance. Rubber particles smaller than 1 µm are less efficient in craze initiation. Such HIPS

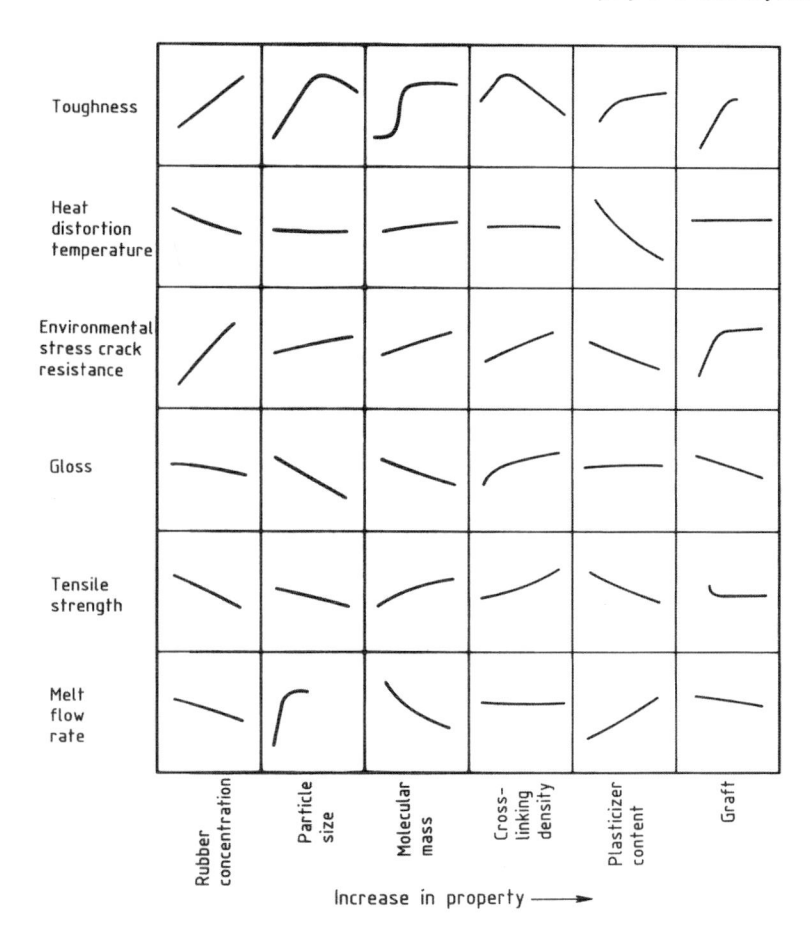

Figure 4. Summary of property relationships of commercial mass-produced HIPS
Reproduced from [12] with permission of The Dow Chemical Company.

grades have reduced impact strength and are used for high gloss applications.

The average particle size depends on the polymerization conditions during phase inversion. The most important parameters listed below can be modified as required:

1. The ratio between the solution viscosities of the disperse polybutadiene and homogeneous polystyrene phases
2. The shear force applied to the polymerizing material via the agitator
3. The surface tension between the two phases [18]—largely affected by the degree of grafting

Degree of Cross-Linking. Another important parameter is the cross-linking density of the polybutadiene phase. The rubber chains must be cross-linked with one another to a certain degree so that they can resist the shear forces used in subsequent processing, thus guaranteeing a stable particle form. However, the cross-linking density must not be too high because this would lead to the loss of the desired rubber-elastic properties. Cross-linking is a thermally initiated radical reaction and occurs mainly at high degrees of conversion and in the degassing section of the polymerization plant. The parameters in this process step (residence time, temperature) must therefore be adjusted to give the desired degree of cross-linking in the end product.

1.4. Processing and Uses

PS and HIPS molding materials are processed using molding processes generally used for thermoplastics (e.g., injection molding, extrusion, or thermoforming). About 50 % of the total quantity is processed with screw injection molding machines at 200–280 °C.

Films and sheets are produced by extrusion at ca. 200 °C with broad-slit dies. Thermoforming is used to produce articles such as refrigerator linings from sheets and yogurt cups and disposable tableware from films.

PS and HIPS molding materials are mutually compatible and can be mixed with one another in all proportions. The properties of the mixtures depend on the proportions of the components.

The main areas of application of PS and HIPS molding materials are in packaging, technical items, household and consumer goods, and refrigeration equipment. Roughly two thirds of PS and HIPS are consumed for packaging and for technical items. Packaging accounts for more than 50 % of the consumption in Western Europe and the United States, where technical items represent 20–25 % of the market. In Asia, 20–25 % of the material goes to the packaging sector, while more than 50 % are used for technical items. Asia accounts for more than 40 % of the total polystyrene consumption, the NAFTA region and the European Union for about 25 % each (2005). Recent annual growth rates have been ca. 2 %.

Technical items include housings for television, computers, printers, air conditioners and the like, electrical appliances, computer accessories, and sanitary ware.

Household and consumer goods include toys, containers, furniture, disposable tableware, and cups for drink vending machines. Polystyrene is used for the packaging of compact discs, audiocassettes, foods and pharmaceuticals. Polystyrene is especially suitable for the packaging of dairy products. Biaxially oriented polystyrene and PS blown film are used for display packaging, wrap film, and envelope windows.

PS and HIPS molding materials do not affect odor or taste and are licensed worldwide for contact with foods. Legal requirements in individual countries specify permitted concentrations of materials used in production and processing. All standard commercial PS and HIPS molding materials comply with these guidelines.

Trade Names and Producers. PS and HIPS molding materials are produced by more than 100 raw materials manufacturers throughout the world who offer an extensive range of products classified according to strength, processing behavior, and dimensional stability when subjected to heat. Total installed capacity is about 15×10^6 t. Examples of trade names are given in alphabetical order by producer below:

Polystyrol	BASF Aktiengesellschaft, Germany
Starex	Cheil Industries, Korea
Polyrex	Chi Mei, Taiwan
-	Chevron–Phillips Chemical Comp.
Styron	Dow Chemical, USA
Edistir	Enichem Polimeri, Italy
-	Estizulia, Venezuela
-	Innova, Brazil
-	Ineos, Great Britain
-	Kumho, Korea
LG	LG Chemicals, Korea
-	Japan Polystyrene
-	Nizhnekamskneftekhim, Russia
-	Nova Chemicals, Canada
-	Resirene, Mexico
-	PS Japan
SC	Supreme Petrochem. Ltd., India
PS	Sabic, Saudi Arabia
-	Secco, China
-	Shantou, China
Lacqrene	Total, France
-	Videolar, Brazil
-	Wuxi Wei Da, China

The producers (and trade names) of transparent HIPS molding materials are Asahi Chemicals, Japan (Asaflex), BASF, Germany (Styrolux); Chevron–Phillips Chemicals, USA (K-resin); Chi Mei, Taiwan (Kibiton), Denka, Japan (Clearen), and Kraton Polymers, USA (Kraton). Total installed capacity is about 0.5×10^6 t (2006).

Syndiotactic polystyrene is produced by Idemitsu Petrochemical, Japan, under the trade name Xarec. Installed capacity is 5000 t/a (2001).

1.5. Quality Specifications

Most PS and HIPS molding materials are processed automatically which means that product quality must be as uniform as possible. Quality assurance systems (ISO 9000–9004) are common in the production of molding materials to guarantee maximum uniformity. The ISO standard specifies the quality of all raw materials, the production conditions, and the properties of the end products.

The rheological, mechanical, thermal, and electrical properties of the molding materials are defined in ISO 1622-2 (PS) and 2897-2 (HIPS). The Vicat softening temperature (VST B/50, ISO 306), the melt flow index (MFI 200/5, ISO 1133), and the impact resistance (Charpy, ISO 179) are among the most important properties. National standards such as ASTM or DIN (Table 1) largely correspond to the ISO standards.

1.6. Storage and Transportation

The storage and transportation of PS and HIPS molding materials do not pose any special problems. They are subject to the general fire protection requirements applicable to solid combustible materials such as wood. Roofed sheds or silos are suitable for storage. Polystyrene is not moisture-sensitive, but must be dry for processing. PS and HIPS molding materials are transported in paper or polyethylene bags, in large cardboard containers (capacity ca. 1 t), in silo vehicles or containers. In the NAFTA region, rail cars (hopper cars, capacity 80 t) are also used.

1.7. Recycling

Waste PS and HIPS can be reutilized in various ways. In *material recycling* the molding materials are broken into smaller pieces, melted, and reprocessed to give finished products. *Chemical recycling* involves the breaking up of the polymer chain into lower molecular mass components which can be used as raw materials for chemical processes. In *thermal recycling* the plastic is incinerated and the liberated energy is used, for example, as process heat or for general heating purposes. See also → → Plastics, RecyclingPlastics, Recycling.

Material Recycling. Recycling of industrial waste, such as trims and scrap, is not problematic. Polystyrene end products can also be recycled after use provided that the material can be cleaned and sorted into grades. In no case, however, should food packaging be produced from this recycled material. Some countries stipulate legal quotas for material recycling of plastics.

Chemical Recycling. Two processes are in the development stage: pyrolysis and cleavage by hydrogenation cracking. Both processes can be applied to pure plastics fractions (e.g., polystyrene) and mixtures of plastics.

Pyrolysis involves the cleavage of long-chain molecules at high temperatures (500–900 °C) with the exclusion of oxygen [22]. In hydrogenating cleavage the polymer chains are broken up under higher pressure at ca. 400 °C, and in the presence of hydrogen, into volatile saturated compounds. Both processes yield a nonspecific spectrum of products (60–80 % of styrene and ethylbenzene, respectively), which can be used as feedstock for other petrochemical processes.

Thermal Recycling. In the presence of sufficient oxygen, polystyrene burns almost completely to give carbon dioxide, water, and a small amount of nontoxic ash residue. The enthalpy of combustion (41 MJ/kg) is close to the energy content of fuel oil (43 MJ/kg). In modern waste incineration plants, polystyrene comprises part of the plastics fraction of household waste, adherence to the necessary combustion conditions is guaranteed.

1.8. Environmental Aspects

PS and HIPS molding materials are not water soluble. No gaseous or liquid components are liberated during transportation and storage. Liquid and gaseous styrene must, however, be handled safely during production processes. Polymerization is therefore carried out in closed systems. The maximum permitted workplace concentration of styrene is between 10 and 50 ppm (42–210 mg/m^3) in all important producer countries.

In bulk polymerization processes, the product only comes into contact with water during the cooling of the extruded strands. Contamination of the water does not occur. In the suspension process, polymerization takes place in an aqueous medium and the wastewater must be treated by using standard sewage technology.

In the processing of PS and HIPS, small quantities of styrene are liberated above ca. 280 °C as a result of the residual monomer content and of thermal degradation. The styrene can be removed by special ventilation devices; however, normal room ventilation is usually adequate to ensure compliance with workplace concentration limits.

2. Styrene–Acrylonitrile (SAN) Copolymers

Many applications requiring stiffness and transparency are served by polystyrene, a moderately priced commodity thermoplastic (Chap. 1). A mature market has also continued to thrive for resins offering a better balance of optical clarity, rigidity, good chemical and heat resistance, high strength, and flexible processing. One class of thermoplastics having these properties is the styrenic copolymers. The following four copolymers continue to have growing commercial importance:

1. Styrene–acrylonitrile (Chap. 2)
2. α-Methylstyrene–acrylonitrile (Section 4.1)
3. Styrene–methyl methacrylate (Section 4.2)
4. Styrene–maleic anhydride (Section 4.3)

These copolymers are produced commercially by free-radical polymerization [23]. Their specific properties are determined by their molecular mass and composition. The initial concentration of the two monomers (M_1 and M_2) prior to initiation of polymerization must be carefully chosen to give the required copolymer composition for the finished product. Techniques are available for calculating the correct monomer concentration that utilize parameters known as reactivity ratios (r) that reflect the relative reactivity of each monomer with the free-radical end of the growing chain [24]. The dependence of the copolymer composition on the monomer concentration for each of the above styrene copolymers is given in Figure 5.

The styrene copolymers having the greatest commercial importance are the styrene–acrylonitrile (SAN) copolymers [9003-54-7].

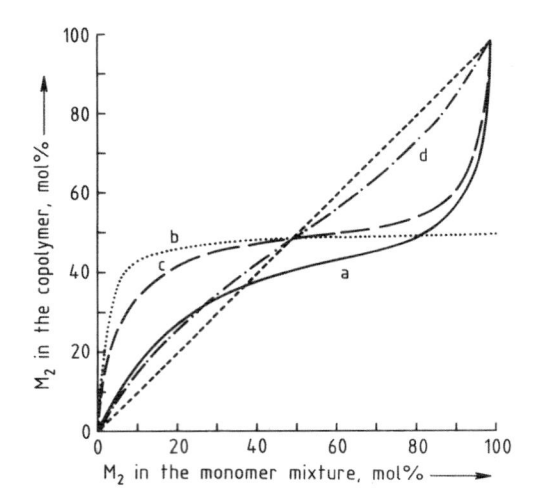

Figure 5. Composition diagram for common styrene copolymers a) Styrene (M_1) – acrylonitrile (M_2) ($r_1 = 0.40$, $r_2 = 0.04$); b) Styrene (M_1) – maleic anhydride (M_2) ($r_1 = 0.04$, $r_2 = 0$); c) α-Methylstyrene (M_1) – acrylonitrile (M_2) ($r_1 = 0.10$, $r_2 = 0.06$); d) Styrene (M_1) – methyl methacrylate (M_2) ($r_1 = 0.59$, $r_2 = 0.54$)

Incorporation of acrylonitrile improves heat resistance, toughness, chemical resistance, and barrier properties when compared against polystyrene. The SAN polymers can be used by themselves in applications requiring transparency, or they can be toughened by incorporation of rubber particles. Use of butadiene rubber gives acrylonitrile–butadiene–styrene (ABS) polymers (see Chap. 5).

The acrylonitrile (AN) levels in commercial SAN polymers fall into two ranges. The lower range (20–35 wt %) covers most injection-molded articles, and the higher range (60–85 wt %) represents a small but growing market for barrier plastics where low gas permeability is needed for packaging of food and beverages (see Chap. 3).

2.1. Production

The relative reactivities of styrene and acrylonitrile differ. Therefore, the polymer composition also differs from the monomer composition, except at the azeotrope (ca. 24 wt % acrylonitrile). This, along with the fact that acrylonitrile is soluble in water, requires special considerations for the various production methods.

Using reactivity ratios [25] and Equation (1) [26] for two monomers M_1 and M_2:

$$C = \frac{P - 1 + \sqrt{(1-P)^2 + 4Pr_1r_2}}{2r_1} \quad (1)$$

C = molar ratio M_1/M_2 in the monomer mixture
P = molar ratio M_1/M_2 in the copolymer
r_1 = reactivity ratio for monomer 1
r_2 = reactivity ratio for monomer 2

the monomer–polymer composition curve can be determined for SAN and other common copolymer pairs (Fig. 5). Reactivity ratios (styrene: $r_1 = 0.4$ and acrylonitrile, $r_2 = 0.04$; Fig. 5.) imply that a mixture of about 75 wt % styrene and 25 wt % acrylonitrile shows no compositional drift ("azeotropic composition").

Reactivity ratios are a function of temperature [27] and the composition curves should be adjusted if the temperature range during manufacture is wide. This is not as critical when the azeotrope composition is approached.

Maintaining tight composition control of SAN is important to obtain acceptable optical and mechanical properties. Differences of a few weight percent in the SAN composition can result in incompatibility [28]. For example, addition of 2 wt % SAN with 32 % acrylonitrile to SAN with 24 % acrylonitrile leads to hazy material. As conversion increases, severe composition drift occurs (Fig. 6) unless corrective actions are taken to replenish the more reactive monomer [29].

Commercially, SAN is produced worldwide by emulsion, suspension, and continuous bulk technologies [30]. These methods are described in detail elsewhere, →Polymerization Processes-Polymerization Processes. The majority (85 %) of SAN is used captively as the matrix component for ABS. Most SANs used for the manufacture of ABS contain 24–32 wt % acrylonitrile. The properties of ABS can be influenced by the composition and molecular mass distribution of the SAN, which are manipulated by the polymerization method employed to make the SAN. This is discussed in detail in Section 5.3.2.

SAN materials sold as finished products are almost exclusively made by continuous bulk methods which results in superior optical

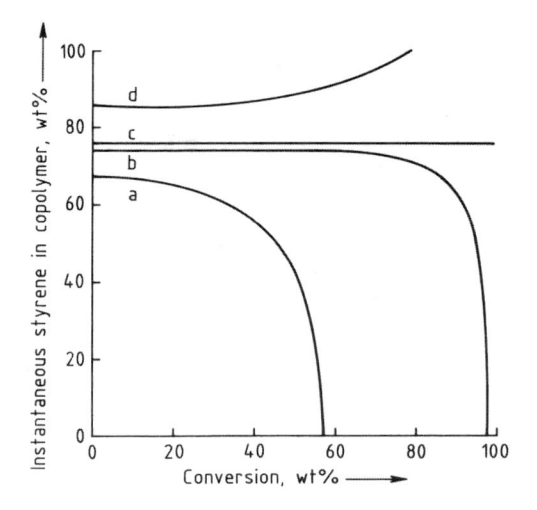

Figure 6. Approximate compositions of SAN copolymers formed at different conversions, starting with various monomer mixtures [29]
Styrene : acrylonitrile weight ratio a) 35/65; b) 70/30; c) 76/24; d) 90/10

properties. Also, continuous bulk technology is the most cost efficient and generates the least waste of the three methods.

Figure 7 shows an example of the manufacture of SAN via continuous bulk polymerization. The process consists of continuous feeds to one or more reactors (a), followed by devolatilization units (f, g) and pelletizer (j). Unreacted monomers and solvent are recycled to maintain conversion and composition at desired levels. The technology is very similar to that employed for PS molding materials (see Section 1.2.1).

The reactor(s) can be plug flow or partially filled types, cooled by jacket and/or reflux [31, 32]. Adequate agitation is critical for proper temperature and composition control. Polymerization temperatures range from 80–170 °C, depending on the process configuration and whether chemical or thermal initiation is used. Conversion ranges from 30 to 70 %.

Devolatilization is usually accomplished via wiped-film or falling-strand type units, but can also be done with a devolatilizing extruder. To achieve the low levels of residual monomer required to meet food contact applications, devolatilization is usually done in a two-stage process with different vacuum levels.

2.2. Properties

SAN copolymers with acrylonitrile contents between 20 and 35 wt % are transparent plastics having better mechanical and chemical resistance properties than polystyrene, thus enabling SAN to command a premium price. Properties for commercial grades of polystyrene and SAN are compared in Table 2 [33]. Next to a higher rigidity and impact resistance, the SANs are stronger and more scratch resistant and have a better heat resistance.

SAN can be reinforced with glass fiber to give higher stiffness, breaking strength, and notched impact strength [34, 35]. The effect of adding 20 % glass fiber to SAN is shown in Table 3 [36].

The conventional method for making glass-fiber-reinforced SAN is to mix the polymer pellets and glass fiber in an extruder, followed by injection molding of the glass-filled pellets. This approach suffers from the inadvertent reduction in glass-fiber length (12 mm) to ca. 0.4–0.5 mm.

An alternative approach is to make a glass concentrate (collimate) by polymerizing the SAN in the presence of the glass fibers, and then blending the collimate with SAN pellets. Commercial collimates are available containing 80 % glass [37]. The glass fiber length in the final product is about 1 mm, and superior reinforcing properties are claimed with the collimate approach. The SAN collimate can be added to other thermoplastic resins.

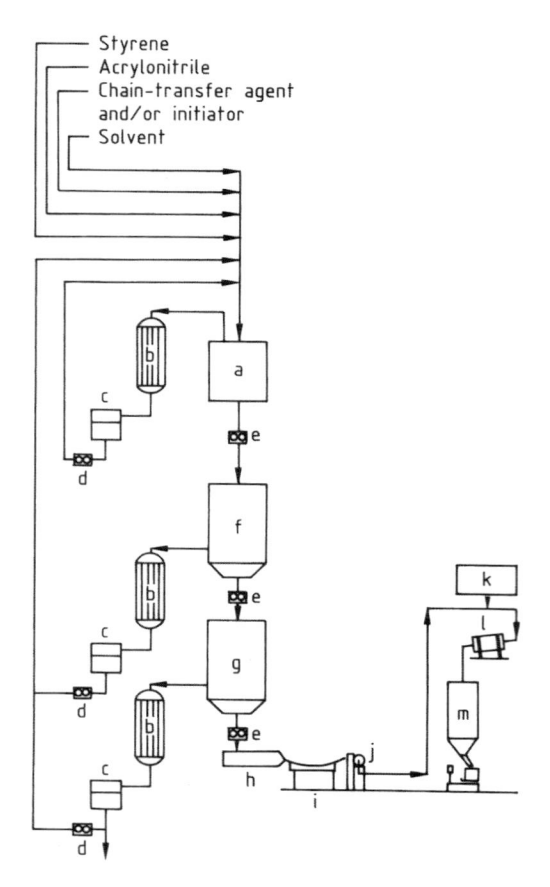

Figure 7. Process for the continuous bulk and solution polymerization of SAN [31] a) Reactor; b) Condenser; c) Condensate receiver; d) Recycle pump; e) Melt pump; f) First-stage devolatilizer; g) Second-stage devolatilizer; h) Discharge screw pump; i) Water bath; j) Pelletizer; k) External lube; l) Screen; m) Silo

Table 2. Comparison of the physical properties of standard polystyrene and SAN copolymers

Property	DIN test method	Standard polystyrene	SAN copolymers (20–35 wt % acrylonitrile)
Density, g/cm^3	53 479	1.05	1.08
Tensile strength, MPa	54 355	50–65	75–85
Tensile elongation to fail, %	53 455	2–4	5
Flexural strength, MPa	53 452	100–110	110–140
Flexural modulus, MPa	53 457	3200–3400	3600–3800
Impact resistance, kJ/m^2	53 453	15–20	20–25
Notched impact test, kJ/m^2	53 453	2–3	3–4
Heat resistance, °C	53 461, version B	80–100	90–105
Thermal conductivity, kJ m^{-1} h^{-1} K^{-1}		0.71	1.17
Linear thermal expansion coefficient, K^{-1}		$(6-8) \times 10^{-5}$	$(6-8) \times 10^{-5}$
Dielectric constant	53 483	2.5–2.7	2.9–3.1
Dielectric loss factor	53 483	$(1.8-2.1) \times 10^{-4}$	$(6-9) \times 10^{-3}$
Specific resistance, Ω · cm	53 482	$> 10^{16}$	$> 10^{16}$
Dielectric strength, kV/mm	53 481	> 100	> 50
Water absorption, 24 h, %	53 472	0.1	0.2–0.6
Processing temperature, °C		150–260	175–300

Table 3. Comparison of properties of filled and unfilled SAN

Property	Unfilled	Filled with 20 % glass fiber
Tensile strength, MPa	69–76	74–83
Elongation at break, %	2–3	1.4–1.9
Tensile modulus, GPa	3.3–3.9	8.3–11.7
Notched Izod impact strength, J/m	21.4–32	133–160
Heat deflection temperature, °C at 1.82 MPa load	101–104	99–110
Density, g/cm^3	1.07–1.08	1.22–1.40

The heat resistance (softening temperature) of typical commercial grades of SAN is ca. 5–6 °C higher than that of polystyrene (Table 2), thus allowing SAN to resist softening when exposed to boiling water. The softening of styrene polymers (and copolymers) is a consequence of the glass to rubber transition that occurs when glassy polymers are heated [38].

Addition of acrylonitrile to the styrene chain continuously increases the glass transition temperature (T_g) until a level of approximately 35 wt % acrylonitrile is reached. The T_g then falls with further addition of acrylonitrile [39]. The location and width of the glass transition for any composition is affected by the polydispersity of the molecular mass.

Commercial SANs contain oligomers left from polymerization which lower the T_g via plasticization. Removal of the oligomers in a commercial SAN (24 % acrylonitrile) increases the glass transition from 108.5 to 114.5 °C [40]. Therefore, any commercial SAN that is carefully treated to remove oligomers will benefit by having a higher heat resistance. The oligomers can be reduced by polymerizing at low temperatures (<120 °C) or by stripping them out during devolatization and subsequent purging from the recycle stream. If a large increase in heat resistance over polystyrene is needed, it is preferable to use the α-methylstyrene–acrylonitrile copolymers (see Section 4.1).

SAN has a slight yellow color in comparison to clear plastics such as polystyrene, acrylics, and polycarbonate. This color arises from cyclic chromophores associated with the acrylonitrile segments, and it becomes stronger as the acrylontrile content increases. Some new commercial, "water-white" grades have been introduced for applications requiring minimum yellowness [27, 28]. The reduction in color is brought about by lowering the level of acrylonitrile and fine-tuning the polymerization process to reduce haze and eliminate contaminants.

One negative consequence of the addition of the acrylonitrile to the styrene chain is an increase in melt viscosity. The nitrile groups in acrylonitrile are strong dipoles, and, as the acrylonitrile level increases, the dipolar attraction between the acrylonitrile groups increases the melt viscosity. This can be compensated for by increasing the processing temperature (Table 2). However, raising the temperature also increases the yellow color, so a compromise must be reached between processability, properties, and color.

The polar nature of the acrylonitrile also results in slight deterioration of electrical properties in comparison to polystyrene (Table 2). SAN still has adequate properties as an insulation material.

Excellent chemical resistance is a very important feature of SAN resins. They are resistant to aliphatic hydrocarbons, nonoxidizing acids, alkali, vegetable oils, foods, some alcohols, and detergents. Many of these materials readily attack polystyrene, but the polarity of the acrylonitrile groups in SAN is responsible for better resistance. SAN resins are attacked by some aromatic hydrocarbons, ketones, esters, and chlorinated hydrocarbons [41]. The moisture absorption of SAN is also slightly higher (Table 2), again because of the polarity of acrylonitrile.

The UV resistance of SAN is not exceptionally good, due to the styrene unit. However, the addition of UV stabilizers improves the UV resistance to the point that some SAN grades are used for glazing and other outdoor applications [41].

2.3. Processing

SAN resins can be processed by all methods common to thermoplastics, with injection molding being the most prevalent [42–44]. The resins are slightly hygroscopic and should therefore be dried prior to processing. Since SAN resins are immiscible with most other polymers it is important that the equipment be thoroughly purged prior to processing.

Injection molding barrel operating temperatures are typically ca. 220–280 °C, which is about 10–20 °C higher than is used for polystyrene. Yellowing and some mechanical property deterioration occurs if the temperature is too high or if too much regrind is used [45].

The mold should be filled as rapidly as possible. This can be accomplished by using large nozzles and gates, high pressure, fast ram, and adequate mold vents. Warm molds are necessary to achieve maximum surface gloss and improve impact strength—the warm mold lowers the cooling rate and allows some of the flow orientation to relax before the part reaches the glassy state. Oven annealing of the part also reduces this orientation and may be required for special applications involving high mechanical stresses and pressures.

Finishing operations applicable to SAN include hot stamping, painting, screen printing, sonic welding, machining, vacuum metallizing, solvent welding, and most other operations used for ABS, acrylics, and polystyrene [46]. The good chemical resistance of SAN facilitates finishing operations requiring chemical exposure. Coloring and provision of antistatic agents, stabilizers, and flame retardants, as well as the reuse of scrap, are handled as with polystyrene (see Chap. 1).

2.4. Uses

The largest end use for SAN is as the rigid component for ABS (Chap. 5). All major ABS producers manufacture SAN for this captive use. SAN is used in the merchant market for a range of applications such as cookware, transparent parts in electronics and electrical appliances, instrument panels, sanitary and medical goods or cosmetic packaging. Chemical resistance and transparency are primary reasons for use [41, 42, 46].

The largest producers of SAN and their trade names are summarized in Table 4.

Table 4. Major producers and trade names of SAN

Producer	Trade name
BASF (Germany)	Luran
Daicel (Japan)	Cevian
Dow Chemical (USA)	Tyril
Techno Polymer Co. (Japan)	Sanrex
Polimeri (Italy)	Kostil

2.5. Blending of SAN

A new resin with unique properties may be obtained by blending (alloying) two or more polymers. Many studies have focused on SAN-based polymers as a major component of the alloy.

Commercial efforts have largely involved blending of ABS with major commercial polymers such as poly(ethylene terephthalate), polycarbonate, poly(vinyl chloride), and polyamide nylon to make resins with improved impact strength, heat resistance, and chemical resistance when compared to the individual polymers (see Section 5.12) [47]. These blends are not transparent because the rubber incorporated into the ABS for impact strength scatters light. The SAN does not become miscible with the other polymer. Rather, distinct domains of ABS and the other polymer phases remain, and the key to developing good mechanical properties is to reduce the domain size and promote good interphase adhesion [48].

Blending of SAN with other polymers to yield a transparent resin would require that the component polymers be thermodynamically miscible, alternatively the domains would have to stay below the critical size for light scattering.

Although polymers are generally not mutually miscible [47], SAN polymers show limited miscibility with poly(methyl methacrylate) [49], styrene–maleimide copolymers [50], styrene–maleic anhydride copolymers [51, 52], polycaprolactone [53], and poly(vinyl chloride) [54].

3. Barrier Resins

Acrylonitrile polymers and copolymers afford efficient barriers to gases, such as oxygen and carbon dioxide. Polyacrylonitrile [*25014-41-9*] (PAN) has the lowest permeability of any polymer, but its inherent rigidity and poor processibility (it degrades upon heating prior to melting) has spurred development of high acrylonitrile copolymers containing 60–85 wt % acrylonitrile.

Rubber-modified acrylonitrile–methyl acrylate copolymers are offered under the trade name Barex by Innovene. The acrylonitrile level in the Barex resin is 74 %. The copolymer is

Table 5. Permeation rates to oxygen and carbon dioxide at room temperature for polymers used in food packaging

Polymer	Permeation rate		Ratio CO_2/O_2
	Oxygen [a]	Carbon dioxide [a]	
Poly(vinylidene chloride)	0.0393	0.118	3.0
Barex 210	0.43	1.22	2.8
AN–styrene copolymer, 70–30	0.39	1.18	3.0
Poly(ethylene terephthalate)	2–4 [b]	4.7–7.8 [b]	
Poly(vinyl chloride) unplasticized	3–6 [c]	7.8–16 [c]	
Cellophane (dry, uncoated)	0.05	0.118	2.3
High-density polyethylene	43	118	2.7
Polypropylene	59	177	3.0
Low-density polyethylene	189	591	3.1

[a] $\frac{cm^3 \cdot cm}{m^2 \cdot d \cdot atm} \times 10^5$ at 23 °C, 100 % relative humidity.
[b] Depends on crystallinity and orientation.
[c] Depends on compound formulation.

toughened by addition of styrene–butadiene rubber at a level of 10 %. The resins may be processed by all major thermoplastic processing methods.

Permeation of oxygen into food and beverages adversely affects their taste [55], and storage of carbonated beverages requires low permeation of carbon dioxide out of the bottle. The high acrylonitrile resins have much lower permeation rates than other thermoplastic packaging resins (Table 5). These high-acrylonitrile resins are also quite good barriers to moisture, which is another requirement for food packaging.

The effect of acrylonitrile level on permeation rates for SAN copolymers is shown in Figure 8. The highest practical limit for acrylonitrile content is approximately 80 %, at which point the copolymer begins to take on the pseudocrystalline structure of polyacrylonitrile and becomes difficult to process [56, 57].

Barrier resins are produced by suspension or emulsion polymerization [58, 59]. Maintaining constant composition at high acrylonitrile levels requires special techniques. A certain amount of the more reactive comonomer (e.g., styrene) is held outside the reactor and fed in a controlled manner during the polymerization cycle.

As animal tests have shown acrylonitrile to be a potent carcinogen, levels of acrylonitrile in the containers must be kept exceedingly low [60].

4. Other Copolymers

Producers and trade names of other important styrene copolymers are listed in Table 6.

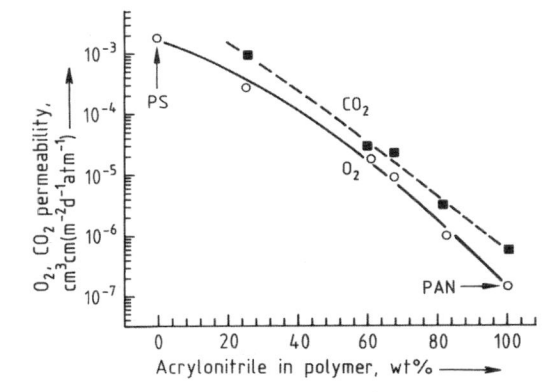

Figure 8. Permeability of oxygen and carbon dioxide through styrene – acrylonitrile copolymers
Values for the two homopolymers, polystyrene (PS) and polyacrylonitrile (PAN), are indicated.

Table 6. Examples of producers and trade names of common thermoplastic styrene copolymers

Producer (country)	Comonomers	Trade name
BASF (Germany)	α-methylstyrene – acrylonitrile	Luran KR 2556
Nova Chemicals (Canada)	styrene – methyl methacrylate	NAS, Zylar
Arco Chemical (USA)	styrene – maleic anhydride	Dylark
Denki Kagaku Kogyo (Japan)	styrene – maleimide	Malecca
Techno Polymer Co. (Japan)	styrene – maleimide	Superex

4.1. α-Methylstyrene–Acrylonitrile Copolymers

The copolymer of α-methylstyrene (AMS) and acrylonitrile [25747-74-4] has a higher heat resistance and softer flow than SAN of the same acrylonitrile content. The heat deflection temperatures of the AMS–AN and SAN copolymers are 104 and 96 °C, respectively [61]. Physical properties and chemical resistance are similar to those of SAN. As with SAN, most AMS–AN copolymers are used as diluents to make high-temperature ABS (Sections 5.1.4 and 5.3.2) and acrylonitrile–styrene–acrylate (ASA) materials. AMS–AN copolymers are extensively compatible with SAN copolymers if they both have an acrylonitrile content of ca. 30 wt % [159]. The AMS copolymer is miscible with poly(vinyl chloride) [62], and could therefore be used to make both toughened and clear blends.

The manufacturing methods used for SAN can also be used for the AMS–AN copolymer. However, AMS reacts more slowly than styrene, greatly increasing the process residence times for a given acrylonitrile content. It is also more difficult to obtain high molecular mass AMS copolymers due to the higher termination rate constant compared with styrene and the strong chain-transfer effect of AMS oligomers and its monomer impurities. AMS-based materials are also yellower than their styrene counterparts.

By utilizing the difference, versus styrene, in reactivity ratio with acrylonitrile, a method for greatly increasing the production rate of AMS–AN copolymer has been developed [63]. Introducing low levels of styrene into the process to make an α-methylstyrene–acrylonitrile–styrene terpolymer is another way to increase reaction rates, with minor sacrifices in heat resistance.

4.2. Styrene–Methyl Methacrylate Copolymers

Methyl methacrylate readily polymerizes with styrene to form a styrene–methyl methacrylate copolymer [25034-86-0], SMMA, with similar physical properties to SAN, but with superior optical properties (low haze and glass-like appearance). Off-azeotrope copolymers are much easier to make, due to the similar reactivities of the monomers (Fig. 5). Commercial materials contain 20–60 % methyl methacrylate and are produced by methods similar to SAN [64, 65]. Methyl methacrylate is more commonly used to make clear ABS-type polymers (see Section 5.12.2).

4.3. Styrene–Maleic Anhydride Copolymers

Styrene and maleic anhydride react readily (Fig. 5) to give a copolymer [9011-13-6], SMA, with alternating monomer units. Reaction rates are significantly higher than for styrene alone or SAN, so polymerization conditions must be carefully chosen to control heat removal. Polystyrene technology is commonly used to produce thermoplastic SMA resins. The continuous stirred tank reactor technology permits synthesis of SMA copolymers with well defined S : MA ratio. The copolymerization parameters otherwise lead to the preferential formation of alternating copolymers.

The primary benefit of maleic anhydride is the greatly increased heat resistance. Also, the cost of this system is low, as both monomers are commodity materials. The products offer high rigidity, are commonly rubber modified, and are easily reinforced with fillers such as glass.

SMA resins generally fall into two categories. Low maleic anhydride (< 25 wt %), high molecular mass thermoplastic materials are produced for molding or extrusion applications, low molecular mass materials with a high maleic anhydride content (25–50 wt %) are produced for alkali-soluble coatings and adhesives.

Thermoplastic SMA grades span a heat resistance range of 110–135 °C (Vicat softening point). Grades developed specifically for expandable foam applications and blends of SMA and polycarbonate for power tool housings are also available [66, 67].

For structurally demanding parts, rubber-modified SMA is reinforced with glass fiber (typically up to 20 %) by melt compounding. These materials are often used for interior automobile applications [67].

SMA resins are unstable at high temperature, liberating carbon dioxide. They must be adequately stabilized and processed at temperatures below 260 °C, otherwise splaying of parts results. The resin is stabilized by incorporating up to 1 % of a hindered phenol antioxidant with a thioester synergist.

4.4. Styrene–Maleimide Copolymers

For applications where heat resistance above the levels available from SMA resins is desirable, copolymers of styrene and maleimides can be used. These polymers are produced by copolymerizing styrene with a maleimide comonomer (e.g., N-phenylmaleimide) or by imidizing SMA with ammonia or an amine.

Polymerization of styrene and maleimides can be accomplished by any of the methods used for SAN. Imidization of SMA is achieved by injecting ammonia or an amine into a vented extruder, or by direct addition to the polymerization [66, 68].

SMI resins are used in a similar manner as SMA. They are rubber modified, glass reinforced, alloyed, or polymerized with termonomers such as acrylonitrile or acrylates. Unlike SMA, they are thermally stable at high temperatures and do not splay.

4.5. Styrene–Acrylate Copolymers

A recyclable, water-soluble copolymer of styrene and acrylate is available from Belland AG [69]. This copolymer contains carboxyl side groups which give it solubility in basic aqueous media. It is recoverable by precipitation via acidification. End uses include typical coatings applications for packaging laminates, book binding glues, and hygiene products.

5. Acrylonitrile–Butadiene–Styrene (ABS) Polymers

5.1. Definition and Structure

ABS [9003-56-9] and HIPS (Chap. 1) represent the industrially most important thermoplastic two-phase systems with an amorphous structure. The ABS polymers are based on three monomers: acrylonitrile (A), butadiene (B), and styrene (S). The polymer components have different chemical compositions and coexist as two separate phases whose

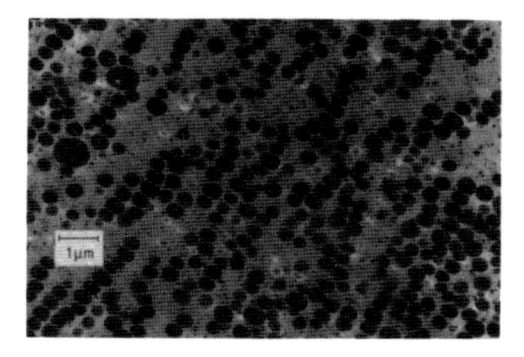

Figure 9. ABS graft polymer showing rubber particles embedded in a styrene – acrylonitrile (SAN) copolymer matrix

compatibility is controlled by their structure and chemical microstructure.

In all classical ABS molding compounds the continuous phase (matrix) consists of copolymers of styrene (or an alkylstyrene) and acrylonitrile. An elastomer based on butadiene forms the disperse phase that is distributed in the continuous phase and has a characteristic morphology (Fig. 9).

ABS is mainly produced via two major industrial processes: Emulsion-ABS by compounding of emulsion-made rubbers into the SAN matrix; and solution-ABS via dissolving a rubber (commonly: butadiene homopolymer) in styrene–acrylonitrile and subsequent bulk polymerization, similar to the HIPS process (see Section 2.1). However, process combinations also exist.

Attempts to synthesize styrene-co-acrylonitrile-block-butadiene copolymers via living free radical polymerization have not resulted in commercial products so far [70].

5.1.1. Historical Aspects

The history of ABS polymers began in the mid-1940s. In attempts to produce bulletproof plastic sheets during the last years of World War II, polymer systems were developed from special butadiene–acrylonitrile copolymers and styrene–acrylonitrile copolymers with high molecular masses. These materials had a high impact resistance on account of their low thermoplastic flow, but could only be processed with extruders. Semifinished products (i.e., predominantly sheets, profiles, and pipes) were the first molded parts to be made from ABS polymers and had dull or matt surfaces.

A drastic increase in the flow of the material allowed processing by injection molding and thus opened up the way for the production of engineering plastics. Improved processability was obtained by the use of graft polymerization. This technique allowed the production of molded parts with glossy or high-gloss surfaces.

Systematic research into ABS polymers has led to the development of other multiphase plastics with a chemically different make-up. These contain other monomer units in addition to or as a replacement for acrylonitrile, butadiene, or styrene.

Recent developments include modifications of the A, B, and S monomer units as well as complex blend systems, in which the ABS polymers are components of polymer mixtures with an extremely wide range of properties. See also, → Polymer BlendsPolymer Blends.

5.1.2. Structural Principles

A knowledge of the structural principles of ABS polymers and the parameters on which they depend is required in order to understand production processes, their relationship to product properties, and the basis for the development of other ABS polymers.

The most important physical properties for the user are good thermoplastic flow, high rigidity, satisfactory heat resistance, high surface gloss (emulsion-ABS), and good (notched) impact strength down to temperatures as low as ca. − 40 °C.

The simultaneous occurrence of high toughness and high rigidity results from proper bonding between the two polymer phases [71–78].

The elastic properties disappear on cooling to just above the glass transition temperature of the elastomer phase. The two-phase system then behaves as a brittle quasi-single-phase system.

The effect of the elastomer phase dispersed in the resin matrix is seen in plots of the modulus of elasticity as a function of temperature (Fig. 10) and of the stress–strain behavior at constant temperature (Fig. 11) [75, 76].

In Figure 11 the ABS polymer (b) absorbs significantly more energy during deformation up to the breaking point than the SAN resin (a). This is responsible for the high toughness of

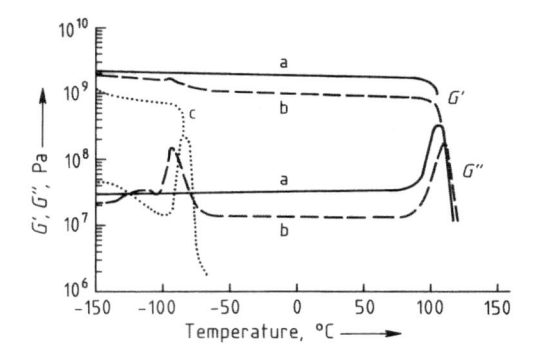

Figure 10. Temperature dependence of the complex shear modulus of ABS graft polymer and its polymer components a) SAN copolymer (matrix); b) ABS graft polymer; c) Polybutadiene (elastomeric graft base)

ABS polymers. The increased deformability in comparison with the pure SAN copolymer and the increased strain in comparison with the pure elastomer component (c) are due to yield processes, in which the applied energy causes the formation of wedge-shaped microcavities (crazes, Fig. 12 A) or initiates shear yielding of the matrix copolymer chains (Fig. 12 B). Both deformation processes are initiated by mechanical loading of the elastomer particles dispersed in the matrix.

The existence of more than one phase is therefore a prerequisite for high toughness. The deformation mechanisms and the effect of different elastomer particles and matrix parameters can be investigated by electron microscopy techniques [79–84].

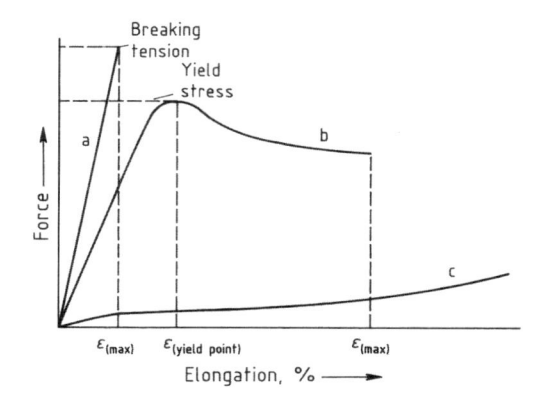

Figure 11. Stress – strain diagram for ABS graft polymer and its polymer components a) SAN copolymer (matrix); b) ABS graft polymer; c) Polybutadiene (elastomeric graft base)

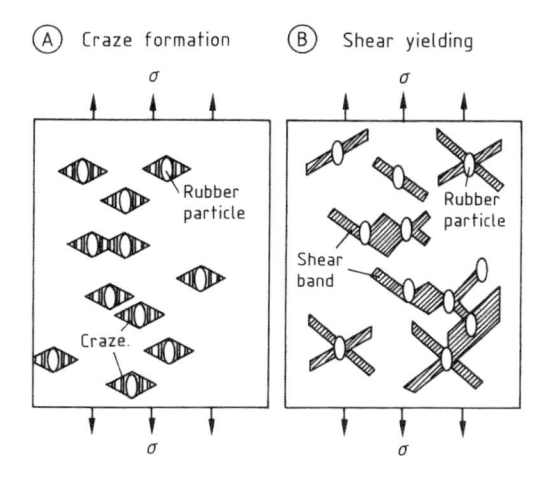

Figure 12. Deformation mechanisms (schematic) in ABS polymers

The glass transition temperature (T_g) depends on the stress frequency (it increases by ca. 30–40 °C after impact). The rubber used as the elastomer phase should therefore have as low a T_g as possible. Apart from special rubbers (e.g., silicone rubbers) polybutadiene is the most suitable. Polybutadiene (T_g − 85 °C) guarantees good low-temperature properties. Copolymers of butadiene with styrene (SBR) or with acrylonitrile (NBR) have much higher T_g values and are mainly used in special types of ABS.

Tough ABS resins require sufficient coupling between the elastomer particles and the resin matrix to ensure that forces can be transferred across the phase interfaces (phase coupling). Optimal phase coupling between the elastomer particles and the matrix is achieved by graft polymerization (Fig. 13). The graft shell

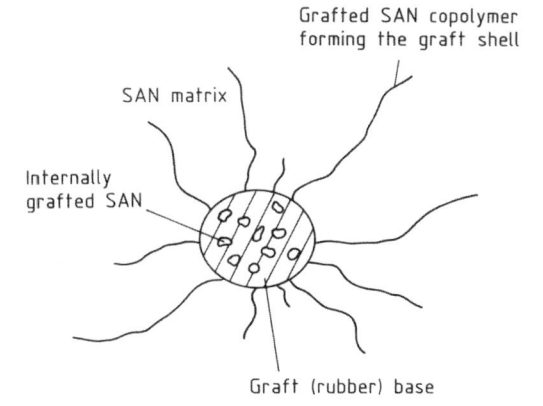

Figure 13. Schematic structure of ABS graft polymers

surrounding the elastomer particles is synthesized (1) from the same monomers that are used to synthesize the matrix resin or (2) from monomers that produce a copolymer which is miscible (compatible) with the matrix.

Various polymerization processes (e.g., emulsion, solution, bulk suspension, and bulk processes) can be used to manufacture ABS molding compounds. These processes are described in detail elsewhere, → → Polymerization ProcessesPolymerization Processes. The producer can also decide whether to use a single-step process involving reaction of polybutadiene and the styrene and acrylonitrile monomers (solution-ABS), or whether to produce the grafted polybutadiene and SAN separately (emulsion-ABS). The second route allows wider variation of individual parameters such as morphology of the elastomer phase (average particle size [85], and degree of cross-linking [75]). The styrene and acrylonitrile can also be copolymerized in different weight ratios [86], and the average molecular mass of the SAN copolymer modified as desired.

Styrene can be replaced partially or completely by α-methylstyrene (AMS). For special requirements (e.g., transparency or high heat resistance), terpolymers are also produced. Monomers used for this purpose, in addition to styrene and acrylonitrile, include methyl methacrylate, maleic anhydride, and N-phenylmaleimide.

5.1.3. Synthesis of the Two-Phase Structure

ABS polymers can be divided into ABS blend types and graft types.

In *ABS blend types*, a rubber copolymer is blended with a SAN copolymer. The rubber copolymer contains a matrix monomer (e.g., acrylonitrile in NBR rubber) or a monomer whose solubility parameters resemble those of the matrix [87–89]. These ABS systems have very little importance today.

Graft ABS polymers contain a polybutadiene graft copolymer as the elastomer component. The graft shell is constructed from the same monomer units as the resin matrix to give optimal phase interaction between the cross-linked rubber particles and the thermoplastic matrix [90, 91].

Grafting Reaction. The radical-initiated grafting of vinyl monomers (e.g., styrene, acrylonitrile, or methyl methacrylate) only produces satisfactory graft yields with a few rubbers. Industrially, butadiene polymers are used almost exclusively as the rubber (graft) base, polybutadiene being the most common because of its low glass transition temperature.

Grafting of vinyl monomers (e.g., styrene) on polybutadiene can take place by two mechanisms [74]:

1. Addition to the double bond

$$-CH_2-CH=CH-CH_2- + R^{\cdot} \longrightarrow$$

$$-CH_2-\overset{R}{\underset{|}{CH}}-\overset{\cdot}{CH}-CH_2- \quad (1)$$

2. Abstraction of hydrogen.

$$-CH_2-CH=CH-CH_2 + R^{\cdot} \longrightarrow \quad (2)$$
$$RH + -\overset{\cdot}{CH}-CH=CH-CH_2- \rightleftharpoons -CH=CH-\overset{\cdot}{CH}-CH_2-$$

where R· is a polymer or initiator radical.

The polybutadiene radicals (R'·) formed according to Equation (1) or (2) react further with monomeric styrene to give the graft copolymer (3):

$$R'' + H_2C=\underset{\underset{C_6H_5}{|}}{CH} \longrightarrow R'-CH_2-\underset{\underset{C_6H_5}{|}}{\overset{\cdot}{CH}} \quad (3)$$

The allyl radical (Eq. 2) is not very reactive because of its high resonance stabilization. Its resonance energy is comparable with that of a styrene radical [92, 93]. Common polybutadiene rubbers are composed of 1,4-*cis*, 1,4-*trans* and 1,2-vinyl units. Grafting occurs mostly via abstraction of the allylic hydrogen atom of the 1,2-vinyl group.

Recently a copolymerization mechanism has also been proposed for the grafting reaction [94, 95].

The kinetics of the radical grafting reaction largely correspond to the normal kinetics of radical polymerization. At low rubber concentrations deviations occur because of cross-linking reactions. Grafting cannot be carried out with 2,2'-azobisisobutyronitrile as an initiator. The resonance stabilization of the $(CH_3)_2(CN)C\cdot$ radical prevents H-abstraction from the rubber backbone [96]. In the thermally initiated grafting of styrene onto polybutadiene the graft yield increases with increasing temperature and with the 1,2-vinyl group content of the graft base [97].

The degree of grafting can be varied within wide limits by factors such as the choice of polymerization process, the graft base : monomer ratio, the degree of conversion, the activating and regulating systems, and the use of redox initiators based on organic peroxides [98].

Cross-Linking of the Rubber Phase. The high hardness and toughness of ABS polymers requires controlled cross-linking of the rubber phase. If cross-linking is insufficient, the rubber particles dispersed in the SAN matrix (particle size ca. 100–1000 nm) are destroyed by the action of shear forces during processing. Rubber which is not cross-linked has no elasticity. The rubber may be cross-linked during the production of the graft base (emulsion polymerization), during the grafting reaction, or in the subsequent workup (bulk polymerization). The swelling index in a good solvent is usually used as a measure of the degree of cross-linking [99]. Quantitative relationships between the degree of swelling and the cross-linking density have also been established [100].

5.1.4. Properties of the Resin Matrix

The resin matrix of ABS polymers is mainly responsible for properties such as processability and heat resistance of the molding compounds produced from it. Processability depends predominantly on molecular mass and molecular mass distribution, whereas heat resistance is mainly determined by the chemical composition (i.e., the monomer units and their arrangement in the polymer chain) and the resulting glass transition temperature.

Table 7 summarizes the important properties of ABS matrix components.

Styrene–Acrylonitrile (SAN) Copolymers. See also Chapter 2. Styrene–acrylonitrile copolymer resins with an acrylonitrile content of ca. 20–35 wt % are usually used for ABS polymers. The weight-average molecular masses are in the range ca. 50 000 to 180 000; the molecular nonuniformities $U = (M_w/M_n) - 1$ are ≥ 2 (for emulsion polymers) and between 1 and 2 (for bulk and suspension polymers); the glass transition temperature is ca. 115 °C. SAN resins can be

Table 7. Properties of ABS matrix resins

Matrix component	Chemical construction	T_g, °C	Concentrations in the ABS blend, wt %	Heat resistance (Vicat B 120), °C	Other properties
SAN	random copolymer S : AN 80–65 : 35–20	115	95–50	104	decrease in thermostability at high acrylonitrile contents
AMS–AN–S copolymer	random copolymer AMS : S : AN 45 : 35 : 20		95–50	108 110	
AMS–AN copolymer	random copolymer AMS : AN ca. 70 : 30	128	95–50	117	depolymerization begins at 280 °C
AMS–AN sequence polymer	AMS : AN ca. 70 : 30 high proportion of AMS pairs	140	95–50	~ 130	toughness lower than with random copolymers
Styrene–AN–NPMI terpolymer	random terpolymer S : AN : NPMI ca. 67 : 28 : 5	140	95–80	~ 130	
Styrene–AN–MA terpolymer					high reactivity of the anhydride ring, heat resistance limited

AN = acrylonitrile; AMS = α-methylstyrene; MA = maleic anhydride; NPMI = N-phenylmaleimide; S = styrene

synthesized by separate copolymerization or during graft polymerization. All ABS polymers also contain ungrafted SAN resin formed during graft polymerization.

α-Methylstyrene–Acrylonitrile (AMS–AN) Copolymers. See also Section 4.1. Copolymers of α-methylstyrene and acrylonitrile are closely related to SAN polymers. Their high glass transition temperatures and their chemical behavior, which is similar to that of SAN resins, yield ABS systems with improved heat resistance. Their thermostability decreases with increasing α-methylstyrene content (back cleavage of triads of α-methylstyrene units).

Resin production is carried out by emulsion [101], solution, or bulk polymerization [102]. Products usually have acrylonitrile contents of ca. 20–35 wt % with weight-average molecular masses of ca. 50 000–150 000.

Special copolymers with a high proportion of α-methylstyrene-containing sequences can be produced by emulsion polymerization and yield ABS polymers with high heat resistance (Vicat B temperature) of up to ca. 130 °C [103].

Other Vinyl Polymers. Copolymers of styrene and methyl methacrylate are used in transparent ABS. 4-Methylstyrene–acrylonitrile copolymers [104], styrene–maleic anhydride copolymers (Section 4.3), polyglutarimides [105], and N-phenylmaleimide–styrene–acrylonitrile terpolymers [106] are used for the production of systems that are analogous to ABS but with high heat resistances.

5.2. Structure–Property Relationships

The individual components described in Sections 5.1.3 and 5.1.4 can be used to draw up structure–property relationships for ABS polymers which, in turn, can be employed to predict properties from a small amount of basic data. The physical behavior of ABS polymers is partially based on the properties of the matrix and the rubber component.

The room-temperature toughness of ABS polymers generally passes through a maximum as the proportion of the rubber phase increases (Fig. 14). In parallel to this the modulus of elasticity, heat resistance, and flow properties decrease [107]. The low-temperature toughness behaves analogously to the room-temperature toughness; its absolute value, however, is determined by the glass transition temperature of the rubber component. Polymer properties are related to the acrylonitrile content of the resin matrix in the range relevant to industry (i.e., ca. 20–35 wt % acrylonitrile). Increasing the acrylonitrile content slightly improves chemical resistance and toughness (maximum at 34 %) but flow properties deteriorate [108].

The most important structural parameters for a given matrix resin and rubber and a constant proportion of the rubber phase are the molecular

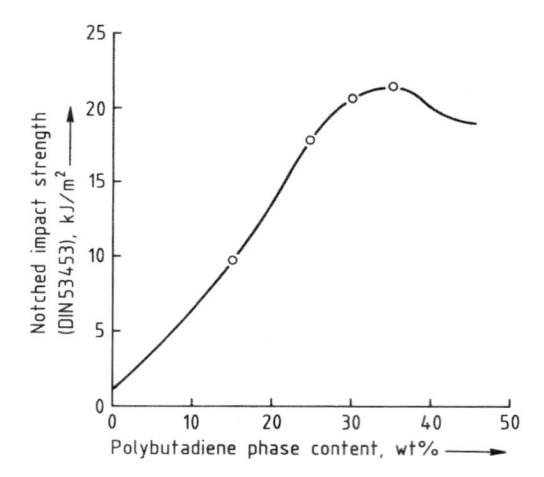

Figure 14. Variation of the toughness of ABS with content of the elastomer (polybutadiene) phase

mass and molecular mass distribution of the matrix resin and the structure of the grafted rubber, especially the particle size, particle size distribution, and degree of grafting and cross-linking.

An increase in the molecular mass of the matrix resin results in an increase in toughness and chemical resistance (environmental stress cracking) but simultaneous deterioration in the thermoplastic flow properties [109]. Broadening of the molecular mass distribution results in easier flow.

The use of cross-linked matrix resins in ABS polymers leads to embrittlement. Even a low degree of cross-linking is detectable because it lowers the surface gloss of molded products.

The properties of ABS polymers are most strongly influenced by the characteristics of the grafted rubber components. Toughness increases with particle diameter.

Grafted rubbers with average particle diameters below ca. 150 nm and a narrow particle size distribution lead to ABS systems which are brittle when subjected to high-velocity impact. In the particle size range of ca. 200–300 nm, the toughening effect of the grafted rubber is very strongly dependent on the degree of cross-linking and the cross-linking structure of the rubber base. A good toughening effect can also be achieved with relatively small particles if the rubber is weakly cross-linked. At particle sizes exceeding 300 nm, grafted rubbers with high toughness are generally obtained. However, with very large, compact particles (emulsion polymers with average particle sizes $\geq$ ca. 450 nm) toughness may decrease due to insufficient coupling to the matrix.

ABS molding compounds produced by bulk polymerization (particle size ca. 1–10 μm) generally have enhanced toughness because of their high proportion of grafting and matrix inclusion within the rubber phase (rubber phase volume).

Thermoplastic flow properties and the surface gloss of ABS polymers are also strongly dependent on the grafted rubber. Increasing particle size for a given rubber content (i.e., smaller number of particles) improves processibility [110], but leads to loss of surface gloss as a result of light scattering [111].

The degree of grafting, the graft density, and the length of the grafted chains of the grafted rubber also affect the properties of the ABS polymers [112]. An increasing degree of grafting improves coupling to the resin matrix resulting in a more uniform distribution within the matrix and a higher toughness. Grafted rubbers with a low degree of grafting tend to agglomerate so that the properties of the molded articles depend increasingly on processing conditions.

At a fixed degree of grafting, the length of the grafted chain and thus the graft density can be varied. Low graft density corresponds to a small number of very long grafted chains and a high graft density to very many short chains. A fine balance during the grafting reaction allows adjustment of the special requirements of ABS molding compounds (e.g., high toughness, improved flow properties).

5.3. Production of ABS Polymers

ABS can be produced by emulsion polymerization, bulk polymerization, or combined processes (Fig. 15).

There are two main types of *emulsion polymerization* which yield ABS graft rubbers with a rubber content of 5–85 wt %.

1. Separate production of the ABS graft rubber and SAN matrix with a subsequent mixing and compounding step
2. Total emulsion process, i.e., production of the molding compound containing both the grafted rubber and the SAN matrix by

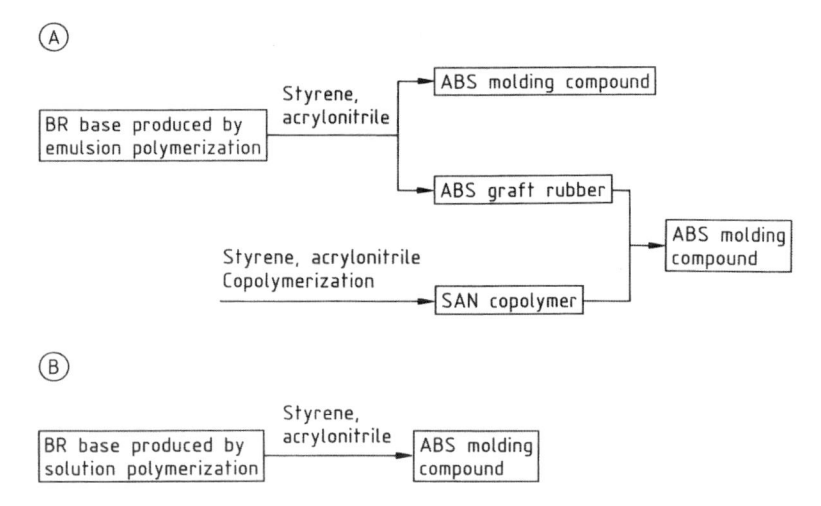

Figure 15. Overview of ABS production processes A) Emulsion polymerization; B) Bulk polymerization

emulsion polymerization (the synthesis of a polybutadiene base precedes the ABS grafted polymer production)

Bulk polymerization starts from a separately produced rubber base (BR or SB solution rubbers). The compound is produced in a bulk process run continuously over several stages. Rubber content is limited to ≤ 20 wt % for viscosity reasons.

The *combined processes* include, e.g., the addition of emulsion graft rubber to bulk ABS molding compounds

The separate production of the grafted rubber and the SAN matrix has the advantage that the properties of each component can be controlled independently and different production processes can be used for the grafted rubber phase and the SAN matrix. SAN matrix copolymers are produced almost exclusively by thermally or radically initiated bulk polymerization [113, 114]. The two-stage bulk suspension process is also still used.

Other styrene copolymers besides SAN can also be combined with the ABS graft rubber (see Section 5.1.4). α-Methylstyrene–acrylonitrile (AMS–AN) copolymers or styrene–acrylonitrile–N-phenylmaleimide terpolymers, for example, give products with a high heat resistance because of their high glass transition temperatures [115, 116]. Some copolymers are only accessible via the separate bulk polymerization route [117]. Bulk or emulsion polymerization is used for AMS–AN copolymers [118, 119]. The bulk process has the advantage of an inherent degassing stage, so that products with very low residual monomer contents and thus high heat resistance can be obtained. However, yields and molecular masses are limited. The advantages of the bulk process lie in the fact that completely continuous processing is possible and that the grafted rubber particles are formed in situ, i.e., at the same time as SAN polymerization. The proportion of internal grafting is thus higher [120, 121] and products have a more favorable notched impact strength–hardness relationship than those produced by emulsion polymerization. The bulk ABS process includes a degassing step which means that unreacted monomers can be directly recycled.

5.3.1. ABS Production by Emulsion Polymerization

The emulsion process is the standard ABS production process but the principle of this method has also been applied to other rubber-modified, two-phase styrene copolymer systems [122–124].

The production of grafted rubbers with high rubber contents and their mixing with a separately produced SAN matrix is used worldwide. The total emulsion process is of minor importance and is not described here.

The production of ABS grafted rubber consists of three stages:

1. Production of the graft base, i.e., a polybutadiene or butadiene copolymer latex with defined particle size, particle size distribution, and gel content
2. Grafting the styrene–acrylonitrile monomer mixture onto the graft base
3. Workup of the grafted rubber latex

The particle size, particle size distribution, and cross-linking density of the graft base; the degree of grafting; the proportion and morphology of the rubber-phase volume; and the molecular mass of the grafted SAN chains largely determine the properties of the resulting grafted rubber system. Bimodal ABS systems have optimal surface properties and toughness synergies. These systems are produced from a graft base with a bimodal particle size distribution, or from two grafted rubbers of differing particle size that are mixed at the latex stage and then worked up together. The particle size of ABS emulsion grafted rubbers lies in the range 50–600 nm, values of 100–400 nm being preferred. Figure 16 shows the production of uni- and bimodal ABS graft rubber systems.

Production of the Graft Base. The production of graft bases mostly takes place batchwise. The butadiene is polymerized at 5–75 °C and ≤12 bar. In the case of butadiene copolymers, the reactivities of the individual monomers must be taken into account. If compositions of the copolymers are chosen which lie outside the "azeotrope region" of the monomer mixture, then high chemical nonuniformity is avoided by using a continuous process and carrying out the polymerization until a given monomer conversion is reached. Unreacted butadiene is fed back into the first reactor of the reactor cascade.

Monomer–water weight ratios of 1 : 0.6 to 1 : 2 are used. Anionic emulsifiers are added in quantities of 1–5 parts per 100 parts of monomer to disperse the monomers and stabilize the polybutadiene latex particles. The alkali salts of disproportionated abietic acid or fatty acids, C_{12}–C_{18} *n*-alkyl- and *n*-alkylarylsulfonates are commonly used as emulsifiers. Emulsifiers are dosed such that the resulting latices have a surface tension of < 0.065 N/m and the particles are extensively coated with emulsifier. Emulsifier systems are preferred which are also used for the subsequent grafting reaction and which readily allow coagulation of the dispersed system on addition of acid or electrolytes to facilitate product isolation.

Polymerization is initiated with alkali peroxydisulfates or combinations of reducing agents with peroxydisulfates or organic (hydro) peroxides. The use of (hydro)peroxide systems allows a sufficiently high reaction rate at low temperature and modification of the polybutadiene microstructure [125].

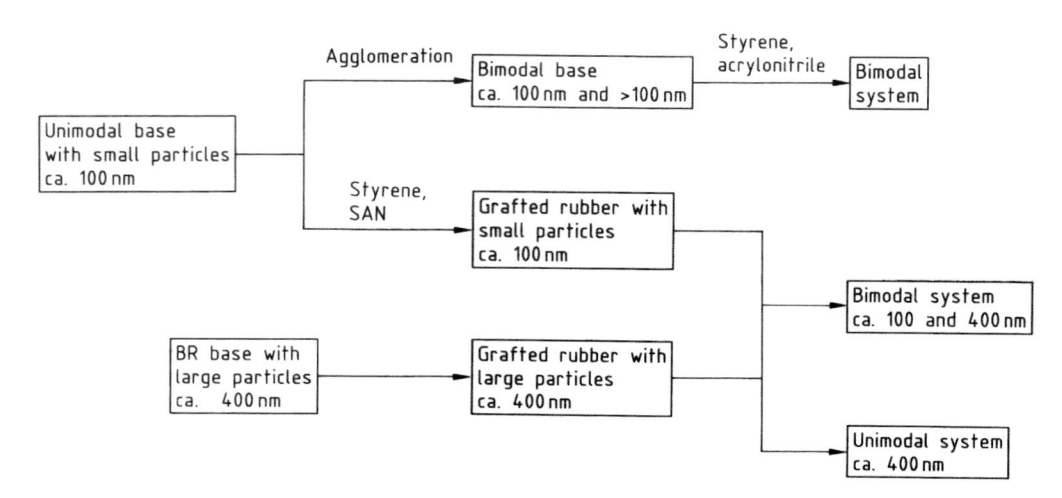

Figure 16. Composition of unimodal and bimodal ABS graft rubbers

Table 8. Production of bimodal ABS graft bases by agglomeration processes

Starting latex (particle size)	Emulsifier	Agglomeration process	Mean final particle size, nm	Reference
BR latex (100 nm)	alkylsulfonate	chemical agglomeration by addition of latices with hydrophilic groups	200 – 500	[127]
Butadiene – acrylonitrile copolymer latices (100 nm)	Na stearate	chemical agglomeration by addition of carboxylic acid anhydrides	200 – 400	[128]
BR latex (100 nm)	Na oleate, cannot be substituted	pressurized agglomeration using Manton – Gaulin homogenizer	200 – 800	[129]
Butadiene – styrene copolymer latex (50 nm)	alkali salts of abietic acid and fatty acids	chemical agglomeration	200	[130]
Butadiene – styrene copolymer latex, polybutadiene	alkali salts of fatty acids and abietic acid	agglomeration by freezing	200 – 400	[131]

The particle size can be controlled by varying the monomer–water ratio, by stepwise addition of emulsifier, and by adjusting the reaction temperature. A much more elegant method involves the use of seed latex technology, mostly in a semibatch process [126].

Polybutadiene bases with a bimodal particle-size distribution can also be produced by chemical or physical agglomeration of latex particles (Table 8). The advantage of this process lies in the fact that latices with a small particle size can be produced much faster than latices with a large particle size.

The molecular mass and cross-linking density of the polybutadiene bases are adjusted by means of modifiers (e.g., n-alkylthiols). Increasing quantities of modifier decrease the average polymer chain length.

Polymerization is usually terminated above 80 % conversion to ensure sufficient cross-linking of the rubber (gel contents > 60 %, preferably > 70 %). No shortstopping agents are used (apart from pure reducing agents) to prevent disruption of the subsequent grafting reaction.

The latex product is devolatilized to remove residual monomers by using vacuum and/or steam in a batch reaction, or continuously in a bubble column.

Grafted Rubber Production. A styrene–acrylonitrile monomer mixture is polymerized in the presence of the graft base. An increase in the ratio of monomers to polybutadiene is accompanied by increases in the net reaction rate, in the molecular masses of the grafted chains and of the "free" SAN copolymer, and in the degree of grafting. Since the reaction is exothermic ($\Delta H = -\,840$ kJ/kg), a monomer-feed process or a completely continuous process is used [132, 133]. If the copolymer compositions lie outside the azeotrope region of styrene and acrylonitrile, an increase in chemical nonuniformity is avoided by adding the more rapidly reacting monomer later (staggered dosing). The kinetics of the reaction are described by the Smith–Ewart theory, [134, 135]. Details of the grafting mechanism are given in [136–139].

The particle size, particle size distribution, and cross-linking density of the graft bases are directly related to the degree of grafting, graft density, rubber phase volume, and morphology of the grafted rubber [140]. The proportion of internal grafting generally increases with increasing particle size and decreasing cross-linking density of the rubber base. At constant particle size and rubber cross-linking density, the degree of grafting and the graft density become the factors determining the properties of the ABS product [77, 141]. The choice of initiator can influence the ratio of SAN copolymerization to graft reaction and the ratio of external to internal grafting. In comparison with monomer-soluble initiators, water-soluble peroxydisulfates lower the proportion of internal grafting and increase the proportion of free SAN copolymer [77].

The degree of grafting can also be affected by stepwise grafting whereby the formation of the graft shell determines the diffusion rate of the monomers [142].

If the weight ratio of monomers to graft base exceeds unity, molecular mass modifiers are usually added to the monomer mixture. They affect the graft reaction and the styrene–acrylonitrile copolymerization, the molecular masses of the free and bound SAN chains, and the degree of grafting decrease.

The emulsifier systems used in the grafting reaction are usually the same as those used in the production of the graft base. High latex stability during the polymerization and ready coagulation during the product workup are also desired.

Stabilization and Workup of the Grafted Rubber. After completion of polymerization, excess (hydro)peroxide initiator is reduced with sodium hydroxymethanesulfinite. Stabilizers are added, generally in the form of a synergistically acting system, to protect the grafted rubber from oxidation during subsequent powder drying and compounding. The stabilizers usually comprise one or more phenolic antioxidants and a heat stabilizer. Long-chain alkyl thiodipropionates, long-chain alkyl phosphites, and low molecular SAN polymers with S-alkyl terminal groups are particularly effective [143–145]. The stabilizer systems should have a low volatility and must fulfil all ecological and toxicological legal requirements. To obtain a homogeneous distribution, the stabilizer system is added to the grafted rubber latex as an aqueous dispersion (0.2–1.5 parts by weight/100 parts of polymer) and coprecipitated with the grafted rubber after mixing.

To isolate the ABS graft rubber from the polymer latex, the emulsifier must be deactivated. Emulsifiers based on alkali carboxylates are coagulated and precipitated by adding dilute acid (acetic or mineral acid) to adjust the pH to below 3. Emulsifiers based on alkyl(aryl) sulfonates are coagulated by adding dilute aqueous solutions of inorganic electrolytes (e.g., $MgSO_4 \cdot 7\,H_2O$ or $CaCl_2$).

Strict control of the precipitation conditions is necessary to obtain a powder with a uniform particle structure and size distribution, from which all water-soluble polymerization and coagulation additives can be easily washed out. Coarse powder particles contain inclusions of precipitants which adversely effect drying and processing. Finely divided powders lead to filtration and wastewater problems. Powder structure, particle size, and particle size distribution can be affected by:

1. Consecutive or parallel addition of two or more grafted rubber latices and the precipitating agent [146]
2. The ratio of latex to precipitant solution
3. The temperature at which the precipitating agent and the grafted rubber latex are charged to the coagulation reactor or coagulation cascade
4. The type of stirring used to mix the process streams, the temperature control in the coagulation reactor cascade, and the average residence time

After coagulation the material is filtered with band, plane, or rotary filters and washed. The filter cake is then fed into a reslurry tank and mixed with fresh water. The slurry is processed in filter presses, extruders, or sieve centrifuges and the wastewater recycled to the process. The powder products have water contents of ca. 15–40 wt %. They are dried to moisture contents of < 1.0 wt % with two-stage paddle dryers, flow dryers, or vacuum tray dryers.

Alternatively, further processing of the moist grafted rubber can be performed in screw extrusion machines. In this case, dewatering leads to grafted rubber granulate [147] or, after mixing with a thermoplastic resin melt, to solid ABS products [148].

During the workup, measures are taken to lower the residual monomer content to comply with legal limits. Examples include latex degassing with a perforated plate column and countercurrent steam, the use of a precipitation–stripping process, and powder workup with a helical sieve followed by compacting and degassing in a twin-screw extruder [149, 150].

Production Process (Fig. 17). The first stage involves production of a polybutadiene latex with a defined particle size. An aqueous soap solution is formed in the polymerization reactor (a) at ca. 65 °C and the molecular mass modifier is added. After flushing with nitrogen, butadiene is pumped in and the polymerization is initiated with potassium peroxydisulfate solution. The reaction rate is determined by the heat removal capacity of the reactor. The maximum conversion rate is ca. 10 %/h. At conversions of ca. 80–90 % the reaction rate decreases; the monomer is then extensively consumed and cross-linking reactions begin. Polymerization is terminated

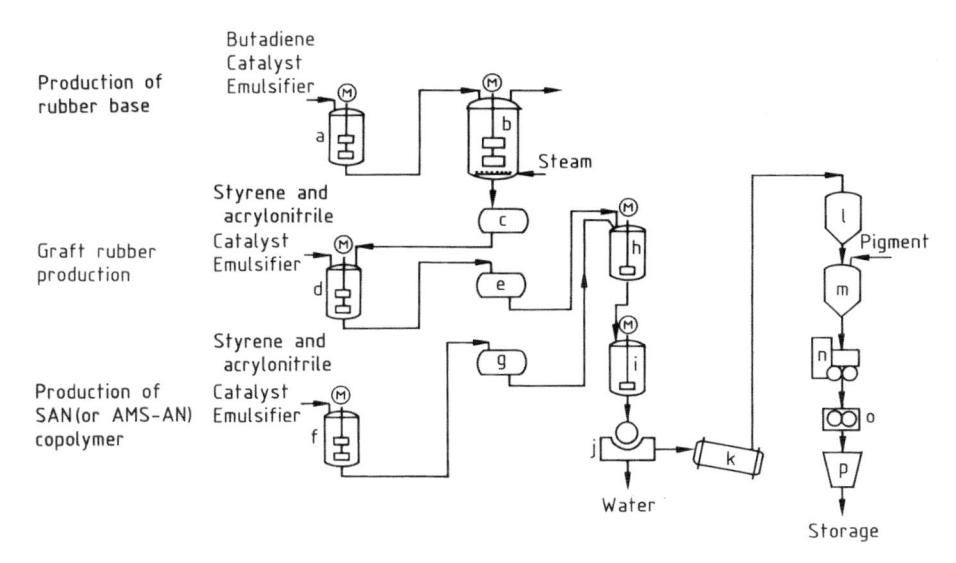

Figure 17. Production of ABS graft rubber and ABS by emulsion polymerization a) Polybutadiene reactor; b) Degassing reactor; c) Polybutadiene latex tank; d) Grafting reactor; e) Graft latex tank; f) SAN reactor; g) SAN latex tank; h) Latex mixing and stabilization tank; i) Coagulation reactor; j) Filter; k) Dryer; l) Powder silo; m) Premixer; n) Internal kneader; o) Rollers; p) Granulator

by lowering the temperature to below 50 °C. Unconverted butadiene is vented and stirred out under vacuum.

In the second stage (the ABS graft rubber production process), the polybutadiene latex, an aqueous potassium persulfate solution, and part of the emulsifier solution are charged to the grafting reactor (d). The temperature is increased to ca. 60 °C and a styrene–acrylonitrile monomer mixture and the remainder of the emulsifier solution are charged to the reactor at a constant rate over 5 h. The reaction temperature is then increased to ca. 65 °C to complete polymerization (conversion $\geq 96\%$, cycle time ca. 8 h, average reactor size is ca. 40 m³). After running through the stabilization tank (h), the latex is coagulated. Washing and drying yield a powder which is subjected to a compounding step. After the addition of lubricants, dyes, and other additives, the polymer is pelletized (p).

5.3.2. Production of the Matrix Copolymer

The properties of the matrix copolymers are summarized in Section 5.1.4.

SAN Copolymers. The main ABS matrix components are styrene–acrylonitrile (SAN) copolymers with an acrylonitrile content of 20–35 wt% and a molecular mass in the range 50 000–180 000.

The production of SAN copolymers can take place by bulk, bulk suspension, or emulsion polymerization. In bulk polymerization the "azeotrope composition" is about 75 wt% styrene and 25 wt% acrylonitrile. In emulsion polymerization the quasi-azeotrope composition is 71.5 wt% styrene and 28.5 wt% acrylonitrile due to monomer partitioning [151].

Chemically and physically uniform SAN copolymers are obtained by *continuous bulk polymerization* (see also Section 2.1) [114, 152]. This also applies to compositions that lie outside the azeotrope region. Compared with suspension and emulsion polymerization, bulk polymerization has environmental advantages: production of wastewater is avoided and virtually all volatile components are removed from the product during workup. A disadvantage is, however, the fact that it is not possible to produce very high molecular masses due to high viscosities. Most of the matrix material is produced by this process.

Uniform SAN copolymers can also be obtained by *bulk suspension polymerization*, within the styrene–acrylonitrile azeotrope range [151]. A SAN prepolymer is first produced which is dispersed in water after addition of water-soluble dispersants and polymerized to give small beads [153]. Unreacted monomers are removed from the beads by means of steam or a degassing extruder before or after the separation of the polymer–water mixture.

In *emulsion polymerization* (Fig. 17, f) the monomers are dispersed in water using an emulsifier. After addition of an activator dispersion, polymerization occurs to give latex particles (diameter ca. 50–200 nm). SAN production by emulsion polymerization is analogous to the production of ABS graft rubbers (See Grafted Ruuber Production, Stablization and workup of the Grafted Rubber) except that polybutadiene latex is not present.

Semibatch or continuous emulsion polymerization can be used to obtain product compositions with extensive chemical uniformity outside the azeotrope region. The semibatch process leads to molecular nonuniformities of ca. 1.5–2. In the continuous process the molecular nonuniformities are in the range 1.5–5 depending on the number of reactors and the average residence time. The emulsion-polymerized SAN copolymers show better flow properties and better resistance to stress cracking in the ABS blend compared with bulk-polymerized SAN. The toughness properties of bulk-polymerized SAN copolymers are not, however, achieved. The emulsion process always offers advantages if high molecular masses are desired [154]. The SAN emulsion copolymers are not usually isolated as such, but are mixed with the ABS grafted rubber latex in the latex step (Fig. 17, h). The mixture is then subjected to workup and residual monomer degassing to give an ABS powder blend.

α-Methylstyrene–Acrylonitrile (AMS–AN) Copolymers. ABS grades with a high heat resistance (HT-ABS) are produced by partial or complete exchange of the SAN matrix with compatible (AMS–AN) copolymers or α-methylstyrene–styrene–acrylonitrile terpolymers [155]. AMS–AN copolymers have a higher glass transition temperature and a higher heat resistance in the ABS blend. This system has an azeotropic mixture (69 wt % α-methylstyrene, 31 wt % acrylonitrile) [156].

Like SAN copolymers, AMS–AN copolymers can be produced by bulk and emulsion polymerization. The AMS–AN system has a much lower reaction rate, however, because of the small AMS growth-rate constant. The reaction rate cannot be increased by raising the temperature because depolymerization occurs [157, 158]. Two alternative approaches are used industrially:

1. A drastic increase in the acrylonitrile concentration of the monomer mixture in the starting phase of the reaction
2. Addition of styrene as a third monomer to give terpolymers with glass transition temperatures between those of SAN and AMS–AN copolymers [159]

The equipment requirements for the production of AMS–AN copolymers are similar to those for SAN bulk polymerization. However, higher average residence times are needed and thermal initiation is not possible. Multistep degassing is used to remove residual monomers and to achieve maximum heat resistance.

The most common route for the production of AMS–AN copolymers is semibatch emulsion polymerization. A high monomer purity ($\geq$ 99.5 %) is necessary. Long incubation times can be overcome by using seed latex technology [101]. Process conditions are similar to those of ABS graft rubber production (Grafted Rubber Production, Stablization and workup of the Grafted Rubber). To achieve monomer conversions of > 95 %, a pH of $\leq$ 6 must be maintained and an appropriate emulsifier chosen. More details on chemical composition are given in [160].

The emulsion copolymer is usually subjected to workup after latex mixing with ABS graft rubbers. A degassing step must be included between polymer workup and product compounding.

The glass transition temperature of the binary AMS–AN copolymer (ca. 128 °C) can be raised to 140–150 °C if the α-methylstyrene concentration at the beginning of emulsion copolymerization is raised above the azeotrope ratio and the reaction is terminated at conversions of $\leq$ 90 %. This increase is due to formation of sequence polymers containing AMS–AMS–AN blocks [161, 162]. The reaction rate and latex stability can be improved by using low concentrations of other monomers [163]. In the ABS blend these AMS–AN sequence polymers produce heat resistance (Vicat B 120) of 130 °C. Toughness diminishes slightly, however, because of incompatibility between the grafted rubber and the matrix.

ABS molding compounds with an AMS–AN matrix have less favorable flow properties than those with a SAN matrix. Although styrene can be replaced by α-methylstyrene in the ABS graft rubber system, this has not been used industrially for processing reasons [164].

Styrene–Acrylonitrile–NPMI Terpolymers. Incorporation of imide comonomers into the SAN matrix yields heat-resistant ABS polymers [165, 166]. *N*-Phenylmaleimide (NPMI) has been used to produce styrene–NPMI–AN terpolymers [167]. Compare Section 5.3.3.

Other Styrene Copolymers. Attempts have been made to replace the SAN matrix by co- or terpolymers other than those described above. Although styrene can be copolymerized with almost all important vinyl monomers [168], none of the efforts made so far has been used industrially. Attempts to shift the heat distortion temperature of ABS polymers into the Vicat B region 120–130 °C by adding styrene–maleic anhydride co- and terpolymers failed because the anhydride ring opened under processing conditions [169, 170].

5.3.3. ABS Production by Bulk Polymerization

ABS production by bulk polymerization (Fig. 15 B) is based on the polymerization of styrene–acrylonitrile mixtures in the presence of a rubber substrate dissolved in this monomer phase. The process can be divided into three steps:

1. The rubber substrate is dissolved in the monomer mixture
2. The rubber–monomer mixture is prepolymerized with continuous mixing (conversion 15–30 wt %)
3. The polymer–monomer mixture is polymerized further by bulk suspension polymerization or by continuous bulk polymerization in a high-viscosity reactor cascade

Unlike the emulsion process (Section 5.3.1), no preformed rubber particles are present at the beginning of the grafting reaction. In emulsion polymerization the particle size distribution is determined by the particle size of the grafting base, in bulk polymerization it is mainly determined by the choice of the rubber base and the reaction conditions during prepolymerization and phase inversion. BR rubbers with a glass transition temperature of ≤ -80 °C are usually used as the grafting base. BR rubbers containing ca. 40 % of units with a 1,4-*cis* configuration are preferred because they do not crystallize and the resulting ABS blend has a good low-temperature toughness. The molecular mass and degree of branching are very important for adjusting the particle size and the rubber phase volume in the prepolymerization stage. Long-chain branched BR rubbers with an average molecular mass of 180 000–260 000 are commonly used [171]. High gloss grades are made from star-branched BR rubbers with low solution viscosity. The rubber content is limited to ca. 20 wt % in ABS bulk processes for viscosity reasons.

The rubber is chopped to pieces and charged to a dissolving vessel. Dissolution in the monomer mixture and optionally a solvent (ethylbenzene, toluene and the like; usually 5–15 %) may take up to 10 h. Acrylonitrile acts as a precipitation agent for the butadiene rubber; with increasing acrylonitrile content of the solution less rubber can be dissolved and additional solvents are necessary. The rubber solution is transferred to a feed tank and continuously fed to the reactor cascade.

The prepolymerization stage is performed with continuous stirring. Phase separation and phase inversion begin during prepolymerization and largely determine the size, structure, and particle size distribution of the rubber phase in the resin matrix and thus the properties of the products. In the radical-initiated prepolymerization step, an oil-in-oil emulsion is formed at conversions of ca. 2 % [172–174]. Monomer-charged SAN droplets are emulsified in the rubber solution by graft SAN polymer chains which are simultaneously formed on the rubber. With increasing SAN formation the phase-volume ratio changes [175]. When the phase-volume ratio reaches unity, phase inversion begins and the rubber phase is then dispersed in the SAN phase [176], i.e., the SAN–styrene–acrylonitrile monomer phase becomes the continuous phase [177]. SAN is enclosed in the newly formed polybutadiene particles in the form of subinclusions. The phase inversion process, which is initially reversible, is terminated at conversions of ca. 12–15 %, depending on the rubber content of the starting solution.

The rubber particle size is determined by three main parameters:

1. The shear forces applied during stirring [178]
2. The viscosity ratio of the two phases [179]
3. The interfacial surface tension between the two phases [180]

Phase inversion can be detected by viscosity measurements. After phase inversion polymerization is continued to a conversion of ca. 25–30 %. The viscosity of the polymerizing mass is then already so high that the phase inversion becomes irreversible.

Another important factor affecting ABS production by bulk polymerization is the degree of grafting [181]. Grafting occurs as a result of hydrogen abstraction in the allyl position, grafting by copolymerization has also been discussed (Section 5.1.3) [93]. A first attempt to calculate the maximum possible degree of grafting as a function of conversion is described in [182]. The degree of grafting depends on the nature of the initiators and modifiers, as well as the different solubilities of styrene and acrylonitrile in the rubber and matrix phases [183]. This difference in solubility lowers the degree of grafting and leads to differing acrylonitrile contents in the grafted rubber and the matrix resin.

The molecular mass and molecular mass distribution are principally determined by process parameters (temperature, initiator concentration, solvent, and chain-transfer agents). Thiols are especially effective as molecular mass modifiers in the initial phase of polymerization, chain-transfer agents (e.g., terpinols) are only effective at higher conversions because of their smaller transfer constants.

Another important parameter is the establishment of defined rubber cross-linking densities. The rubber must be highly cross-linked to avoid crack formation within the particles on continuing "hardening out" of the system. If the rubber phase volume is known, the cross-linking density achieved in ABS polymers can be described quantitatively by a modified Flory–Rehner equation [100]. Since the cross-linking density is thermally controlled, this parameter is mainly determined in the degassing stage of the continuous bulk ABS process by variation of the temperature and the residence time [99, 184].

In the continuous bulk solution process, polymerization is carried out in a cascade of two or more reactors [185, 186]. Solvents or mineral oils are often added to lower the viscosity and improve heat transfer.

Prepolymerization is performed in tower or stirred-tank reactors. Prepolymerization is followed by polymerization in a high-viscosity reactor or a tower reactor cascade [187]. Producers use specially constructed reactors for bulk solution polymerization (Fig. 18) mostly with laminar flow conditions [190]. The average residence times are ca. 10–15 h, and the reaction

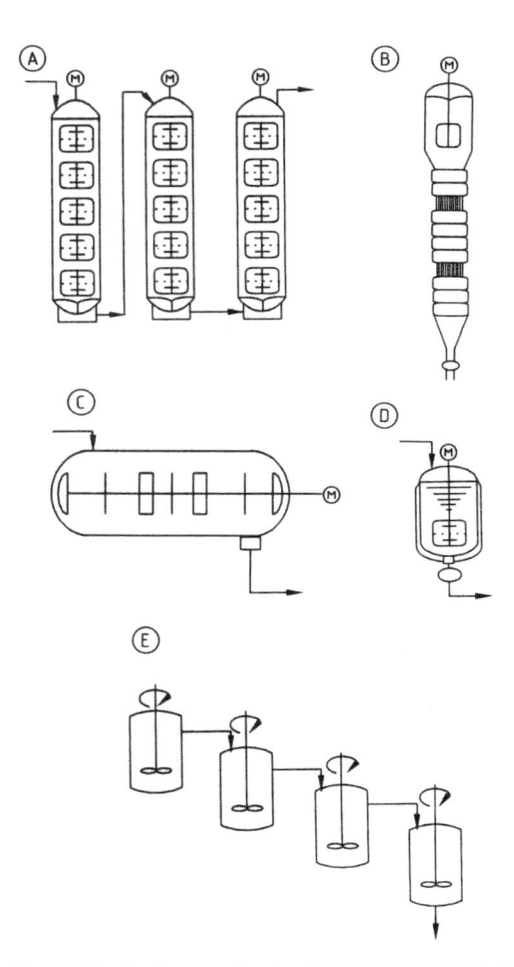

Figure 18. Continuous stirred-tank reactor types [189] A) Tower cascade; B) Stirred tank – tower combination; C) Ring disk; D) Continuous stirred-tank reactor; E) Stirred-tank cascade

temperatures 170–270 °C [191]. Final conversions of 80–85 % are established. Various processes are used to separate the polymer and residual monomer at the end of the bulk polymerization stage: strand degassers, degassing extruders, thin-film evaporators, and spray evaporators. Common are two-stage processes involving a flash tank and a second facility.

The SAN matrix of ABS polymers usually has molecular masses in the range 50 000–170 000, the molecular nonuniformity M_w/M_n is ca. 1.5–3.0. To achieve maximum toughening, an average rubber particle size of ca. 2.5 μm is aimed at. The particles generally have a cellular structure but other structures (e.g., capsules or lamellae) can be obtained by using polybutadiene or butadiene–styrene block copolymer bases or special phase and reaction conditions in the prepolymerization stage [192–194].

Bimodal or multimodal bulk ABS systems are being increasingly described [141, 195, 196]. There are two ways of producing them:

1. Two separately produced prepolymers with different particle sizes and graft structures are mixed with one another before charging to the bulk polymerization stage
2. Two polymers produced in separate prepolymerization and bulk polymerization stages are charged together to a workup zone

In 1990s, intensive research has been carried out on the production of bulk ABS products with a particle size of the rubber phase comparable to that of the emulsion rubber particles ($\leq$ 0.6 μm) [197] with the aim of producing lustrous moldings.

Common to all bulk ABS processes is the fact that the polymers are formed directly as pellets and are also sold in this form for certain uses. More demanding areas of use with defined color requirements need a subsequent compounding stage.

The bulk polymerization process is considered to be more economical than the emulsion and bulk suspension processes because of lower investment costs.

5.3.4. ABS Production by Bulk Suspension Polymerization

After prepolymerization, subsequent polymerization in the suspension process is carried out batchwise in cycles in two or more reactors. Prepolymerization is started as a batch process. After the end of the phase inversion (conversion 25–30 %), the prepolymer is discharged from this reactor and is subjected to further polymerization in a second reactor or reactor cascade. Depending on the initiator system, the polymerization temperature is 80–120 °C [188] and the average residence time 2–4 h.

In the suspension stage the prepolymer is dispersed in the form of droplets (diameter ca. 0.1–1 mm) in an aqueous medium by means of intensive stirring. The water–prepolymer ratio is in the range 3 : 1 to 1 : 1. The suspension is stabilized by adding water-soluble organic polymers, e.g., poly(vinyl alcohol), or colloidal, water-insoluble compounds, e.g., calcium phosphate. Each droplet may be considered as a well-cooled bulk reactor. The bead size depends on the viscosity of the prepolymer syrup, the stirring conditions, the density difference between the organic and the aqueous phase, and the surface tension between the prepolymer syrup and the aqueous medium. The temperature is raised stepwise to 80–170 °C. "Oil-soluble" initiators are used. Since the necessary rubber cross-linking is only achieved at conversions of $\geq$80 %, it is important to adopt a definite temperature–time program. The average residence time of the suspension polymerization stage is ca. 10 h. The polymer dispersion is freed from unreacted monomer by the introduction of steam and transferred to a buffer tank. The polymer is separated from the aqueous phase in centrifuges; the polymer beads have a moisture content of < 10 % and are dried with air in a rotating drier.

Production Process (Fig. 19). The BR or SBR rubber bales are cut into pellets (b) and charged to the dissolving vessel (c). The monomer–modifier mixture (a) is added.

The rubber–monomer mixture is transferred to the prepolymerization reactor (e), which is equipped for mixing highly viscous systems. After addition of the initiator, bulk solution polymerization is started by raising the temperature to ca. 90 °C. After ca. 8 h a conversion of ca. 20–30 % is reached. (The start of phase inversion is recognized by a sharp increase in the solution viscosity). After lowering the temperature to ca. 60 °C the peroxide initiators are added for the subsequent suspension stage. A ca. 20 % poly(vinyl alcohol) solution is often used as suspension agent (f).

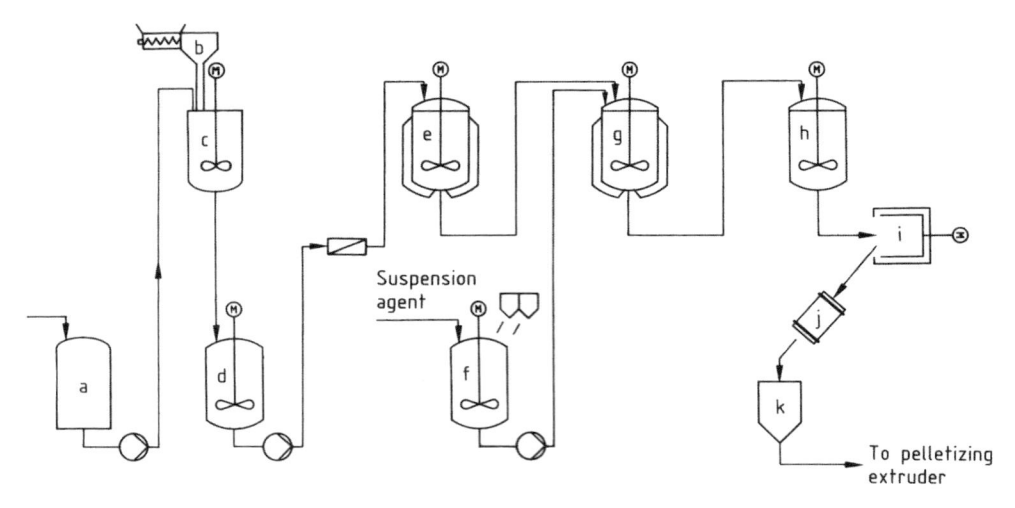

Figure 19. Production of ABS by bulk suspension polymerization [186] a) Monomer feed; b) Rubber chopper; c) Rubber dissolver; d) Solution storage; e) Prepolymerizer; f) Suspension makeup; g) Suspension polymerizer; h) Suspension buffer tank; i) Centrifuge; j) Dryer; k) Bead storage

The prepolymer syrup is heated to 120 °C and then pumped into the suspension polymerizer (g). The ratios of aqueous phase to prepolymer syrup are chosen to give a solids concentration of ca. 40 %. After 3–4 h the temperature is raised to ca. 140 °C and maintained at that level for 3–4 h. At strongly decreasing monomer concentrations the required rubber cross-linking is thus obtained.

The resulting polymer suspension is led via a suspension buffer tank (h) into a sieve centrifuge (i) and separated. The aqueous phase is filtered and transferred to a water treatment plant. The polymer beads are dried in the presence of atmospheric oxygen (k) and then compounded.

5.3.5. Process Combinations

Producers have gone over to combining individual processes or process steps to exploit their advantages. Two possibilities are available:

1. The individual components are produced separately and worked up or compounded together
2. Different polymerization processes follow one another and are used to produce one product

Examples of the first alternative are:

1. An ABS polymer produced by emulsion polymerization is mixed (e.g., in a twin-screw extruder, with a SAN copolymer produced by bulk polymerization [198]).
2. ABS graft rubber is added to the workup stage of a continuous bulk SAN process [199].
3. An ABS polymer produced by bulk suspension polymerization is mixed in the compounding stage with ABS graft rubbers obtained by emulsion polymerization [200, 201]. This process can be used to obtain bimodal bigraft ABS systems [77, 202].

Examples of the second alternative include a latex suspension process (which is also used for high-impact polystyrene [203]) and a bulk dispersion process. Latex suspension polymerization can be performed batchwise or continuously, whereas the bulk dispersion process must be run continuously. Neither variant has achieved industrial importance so far. The bulk dispersion process combines the advantages of emulsion polymerization of a rubber base that has a defined particle size and particle size distribution with a bulk grafting step and a bulk SAN polymerization running parallel to it [204].

Like the bulk dispersion process, the latex suspension process starts from a rubber latex produced by emulsion polymerization. The latex is added to a styrene–acrylonitrile monomer mixture and the swollen mixture then converted into a monomer–polymer–water suspension by addition of dispersing and precipitating agents. Conventional bulk suspension polymerization then follows [205].

The main advantages of these two methods over the emulsion process are: there is no precipitation stage, wastewater contamination is reduced, larger particle sizes can be achieved more easily, and less polymerization auxiliaries are required. Advantages over the conventional bulk process are that the rubber chopping step and the production of a rubber solution are omitted. Because of the defined structure of the rubber particles, higher rubber contents can be used without incurring viscosity and heat removal problems. The critical phase inversion stage is circumvented and the process is reduced to a bulk process.

5.3.6. Additives [137]

See also, → Plastics, AdditivesPlastics, Additives.

Antioxidants. The rubber phase of ABS polymers is degraded by thermooxidation and light in the presence of atmospheric oxygen. Degradation is thought to proceed via a free-radical mechanism [206], resulting in yellowing, deterioration of mechanical properties, and embrittlement. Stabilizers or stabilizer systems are added at concentrations of 0.1–1 wt % to prevent degradation [207, 208] by scavenging radical intermediates or destroying hydroperoxides. They can be divided into sterically hindered phenols, organic compounds with divalent sulfur (thioethers) [143, 209], and organic compounds of trivalent phosphorus (phosphites) [207–210].

Light Stabilizers. ABS polymers undergo degradation reactions when exposed to UV light in the wavelength region 250–320 nm. They can be protected by UV absorbers, mainly derivatives of 2-hydroxyphenylbenzotriazole, 2-hydroxybenzophenone, and cinnamate esters. Hindered amine light stabilizers (HALS) [211] act by scavenging radicals produced during photooxidation [212].

Optimal protection can be attained by combining UV stabilizers and HALS compounds, generally at levels of ca. 0.2–1.5 %.

Lubricants. Long-chain alkyl compounds are often added to ABS polymers as lubricants to improve thermoplastic flow properties and facilitate removal of parts made by injection molding. Fatty acid esters, fatty acid amides, metal stearates (e.g., calcium stearate), or combinations of these substances are used at levels of ca. 0.5–2 %.

Antistatic Agents. Like all plastics of low polarity, ABS polymers tend to build up electrostatic charge on their surfaces which causes adverse effects such as static discharge or dust attraction. Polar antistatic agents are added at levels of 0.5–3 % to prevent this. They can be divided into cationic, anionic, or nonionic compounds [213, 214].

5.4. Quality Assurance and Standardization

Assurance of uniform quality is important both for the raw materials and the final products. A quality assurance scheme is shown in Figure 20. General guidelines on quality assurance, quality assurance programs, and quality audits are laid down in the EC standard EN 29 001 [215]. The test procedures include:

1. Chemical and physical analysis and methods of determination [216, 217]
2. Product-specific approval tests

Analytical procedures are based on appropriate DIN, ISO, and internal company or client-specific standards. They relate to monitoring of the production process and the final products.

Approval tests are based on the properties of the end product. They include mechanical properties (e.g., notched impact strength, hardness), thermal properties (e.g., heat distortion temperature), processing behavior (e.g., flow properties, dispersing), and fire behavior. They are carried out according to relevant national or international standards.

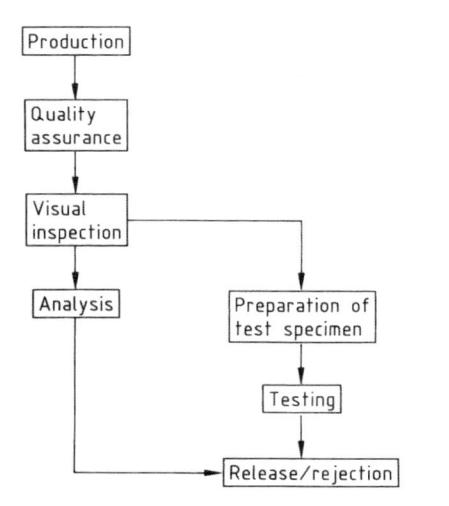

Figure 20. Quality assurance scheme
Visual inspection: Testing the pellets for impurities, pellet shape, and foreign bodies
Analysis: Chemical and physical analysis
Preparation of test specimen: Preparation under standardized conditions and processing parameters
Testing: Product-specific properties
Release/rejection: Authorized by quality assurance department according to prescribed specifications and guidelines

ABS plastics with an elastomer phase based on butadiene and that do not contain fibrous fillers are standardized according to ASTM D 1788–81, DIN 16 772 part 1, and ISO/DIS 2580–1. The large number of ABS types are classified in a grid system (cells), based on three or four properties, each with several ranges of values.

ASTM D 1788–81	
Izod impact strength at 23 °C (D 256)	6
Deflection temperature under load 264 psi (D 648)	6
Tensile stress at yield point (D 638)	6
Density (D 792)	1.0–1.2
Acrylonitrile content, %	min. 13 %
DIN 16 772 Parts 1 and 2 (Sept. 87, Dec. 88)	
Acrylonitrile content of the continuous phase	2
Vicat softening temperature VST B/50	4
Melt flow index, 220/10	4
Notched impact strength at 23 °C	5
ISO 2580–1 (1990)	
Acrylonitrile content of the continuous phase	2
Vicat softening temperature	4
Melt flow index, MFI 220/10	4
Notched impact strength at 23 °C	5
Modulus of elasticity from the bending test (ISO 178)	4

5.5. Properties

ABS polymers have a high toughness (even in the cold), satisfactory rigidity, and good resistance to heat, chemicals, and environmental stress cracking. Molded articles with high dimensional stability and good surface quality can be produced by simple processing techniques. Although each of these properties can be bettered by other thermoplastics, no other system displays such a good combination of technically important properties.

A deeper knowledge of structure–property relationships [77, 218, 219] allows producers to supply a range of commercial products with user-oriented property profiles. Table 9 gives an overview of some commercial grades [220].

The properties of ABS polymers are determined by molecular and morphological parameters. The matrix composition and molecular mass, the type of rubber, the volume ratio of the rubber to the continuous phase, the rubber particle size, the grafted rubber structure, and the additive content are also important.

Standard ABS systems are opaque because their two phases have different refractive indices. Opacity depends on the particle size of the grafted rubber phase and the difference between the refractive indices of the two phases (Mie's theory) [221]. If the particle diameter is small enough compared with the wavelength of visible light, the two phases can be considered as optically homogeneous, and they appear translucent. Completely transparent systems are obtained if the refractive indices are equal. This is achieved, for example, by using a MABS resin matrix [222] or by completely substituting methyl methacrylate for acrylonitrile [223] (see also Section 5.12.2). In bimodal graft rubber systems, depending on ratio of the graft base to the grafted polymer, opacity may be observed despite equal refractive indices. The same applies if grafted rubber particles accumulate to form clusters during processing or tempering because of an incompletely closed graft shell.

In all ABS polymers the continuous phase (SAN and/or AMS–AN copolymer) is responsible for most of the chemical properties. Since only C–C bonds are present in the polymer chains, hydrolysis reactions are unimportant. ABS polymers are therefore generally resistant

Table 9. Physical properties of various ABS types

Modulus of elasticity, N/mm² (ISO 178)	Yield stress, N/mm² (ISO R 527)	Izod notched impact strength, kJ/m² (ISO 180, 80 × 10 × 4 mm)		Ball indentation hardness, N/mm² (ISO 2039)	Vicat softening temperature, °C (ISO 306)	Melt flow rate cm³/10 min (ISO 1133)
		23 °C	− 30 °C			
Injection molding and standard grades						
2500	45	10	4	120	95	45
2400	45	15	7	115	96	31
2200	40	22	8	100	98	17
2100	40	29	12	90	97	5
1700	34	38	25	70	96	3
2400	44	16	5	115	93	37
Heat-resistant grades						
2500	47	18	8	105	103	8
2500	53	15	6	110	108	3
2700	53	15	5	115	113	2
Galvanizable grades						
2100	40	24	10	90	95	25
2800	55	12	5	115	111	4
Flameproofed grades						
2400	45	8	5	110	87	45
2300	43	17	6	105	88	19
Extrusion grades						
2400	44	19	10	105	98	7
1900	38	31	19	80	96	4
Special grades						
2300	49	30	10	105	105	5
2900	50	5	4	130	98	50
1900	39	35	22	78	98	12
2000	38	12	6	90	94	10

to aqueous solutions of salts, acids, or bases. They can absorb up to 1.5 wt % water on storage in aqueous media due to residual emulsifier and the polarity of the nitrile side groups. Under normal conditions, dimensional or property changes are negligible.

Paraffinic hydrocarbons do not dissolve ABS polymers. A certain weight gain can be detected on storage in such media, and depends on the nature and quantity of the rubber phase. ABS polymers show equally good resistance to animal and vegetable fats and to a range of cosmetic creams.

Other organic liquids, such as halogenated hydrocarbons, aromatics, esters, and ketones dissolve the SAN phase. Oxidizing agents, especially inorganic acids, break up the chains and thus degrade the polymer. The resistance tables published by ABS producers only have limited meaning for practical applications. In order to answer specific questions, the running of a creep test under constant deformation, or, better, under constant tension is recommended. States of stress exist in each molded article as a result of its production. Additional stress is often produced by binding elements (e.g., rivets, screws, molded-in or enclosed metal parts) or on use of the article. Environmental stress cracking is low compared with polystyrene, but cannot always be excluded. Sensitivity towards stress cracking can be further lowered by increasing the acrylonitrile content and the molecular mass of the SAN phase, as well as by relaxation processes that are favored by the disperse rubber phase.

The double bonds in the elastic rubber phase (BR rubber) are responsible for the relatively high sensitivity of ABS polymers towards the long-term effects of heat, light, and weathering which result in yellowing and graying of the surface and a decrease in toughness [224]. Although satisfactory protection can be provided by using stabilizer combinations (phenolic antioxidant, a thiodipropionate ester, and/or a phosphite ester) during production and processing at 220–260 °C, long-term stabilization, particularly during simultaneous exposure to UV light and weathering, is difficult. Use of established stabilizer systems allows, however, repeated processing (e.g., recycling) without a serious decrease in toughness. For all indoor applications, ABS is aging-resistant. For outdoor use, adequate resistance to weathering is necessary. Three routes are currently used for this:

1. "Pigmenting" with soot (dark grey shades and black) to absorb UV radiation.
2. Coating semifinished or finished articles with barrier layers having a low oxygen permeability, with less sensitive materials, or with UV-protective paints.
3. Using grafted rubber systems with few or no double bonds. Although these systems do not contain the butadiene monomer unit they are nevertheless members of the ABS family because they have the same basic structure and morphology. They are SAN copolymers that are rendered impact-resistant by a dispersed elastomer phase.

At processing temperatures above 280 °C, ABS polymers lose their toughness because of damage of the rubber phase. Depolymerization begins above 300 °C. In the presence of atmospheric oxygen, decomposition can begin at ca. 280 °C.

5.6. Special Product Modifications

Heat-Resistant ABS Polymers. To increase heat resistance the SAN matrix is generally replaced totally or partially by an AMS–AN copolymer matrix. This is associated with small reductions in toughness, flow properties, and surface quality [225, 226]. To counteract these disadvantages, several methods are available to the producer:

1. Use of an AMS–AN sequence polymer matrix with an increased proportion of AMS–AMS–AN structural units (Vicat B values of ca. 130 °C can be reached) [161, 162]
2. Use of a matrix consisting of a styrene–acrylonitrile–N-phenylmaleimide terpolymer (Vicat B values of 130 °C)
3. Use of blends with polymers having higher heat resistance, e.g., polycarbonates or poly (butylene terephthalate) [227–229]

Commercial high-temperature ABS (HT-ABS) grades may either have good toughness with only a limited increase in the Vicat values or a limited toughness and Vicat values in the range 115–130 °C.

Glass-Fiber-Reinforced ABS Polymers. Reinforcement of ABS with 15–20 wt % glass fiber increases the modulus of elasticity by a factor of 2–2.5 [230]. The reinforced material has a lower thermal expansion coefficient (and thus high dimensional stability) as well as high rigidity, hardness, and heat resistance. Finished articles with smooth surfaces can then be obtained by appropriate coordination of the flow properties, glass fiber length, and glass fiber sizing material.

Flameproofed ABS Polymers. Standard ABS plastics are classified as flammable according to ASTM D 635 (rate of combustion 5–7 cm/min) and as class HB according to the UL 94 test. They also pass the Motor Vehicle Safety Standards (MVSS) test 302 and the so-called Kleinbrenner-Test (limited burning test) DIN 4102 B 2 (see also → Flame RetardantsFlame Retardants, Chap. 2.). For certain applications addition of flame retardants is necessary (mostly halogen compounds in combination with antimony trioxide) to obtain UL 94 V-0 grades.

The brominated diphenyl ethers used in the past as flame retardants have been replaced by other compounds (e.g., brominated oligostyrenes and polycarbonates).

5.7. Legal Aspects

ABS polymers are subject to a large number of legal regulations during production, processing, use, and disposal. A few examples are given here and they relate to the situation in Germany. Similar regulations apply in other countries.

The monomer building blocks A, B, S, and AMS are classified as hazardous substances [231]. Contact with these monomers requires observance of maximum workplace concentrations. See also, → AcrylonitrileAcrylonitrile, Chap. 9.; → ButadieneButadiene, Section

8.3.; → NitrilesNitriles, Section 2.5.; → StyreneStyrene.

Emissions of ABS production plants are controlled by clean air regulations (TA-Luft) and the law governing wastewater discharge (Abwasserabgabengesetz). Waste materials are subject to the waste disposal law (Abfallbeseitigungsgesetz).

In the processing of ABS pellets, traces of volatile components are formed, which may include residual monomers. As in ABS production, the maximum workplace concentrations must be observed.

Finished articles based on ABS polymers can be used in industry without limitation. EC guidelines govern toys and contact with foods [232]. Similar guidelines apply in other countries. These stipulate the monomers permitted for plastics production, polymerization auxiliaries, and additives. They also prescribe limiting values for the use of individual components and for their maximum transfer from the plastic into foods and food substitutes. The maximum proportion of heavy metals in pigments is an important criterion for toys [233].

In the area of disposal, restrictive legal measures are being prepared that prescribe recycling of used components (e.g., in the automotive and electrical sectors).

5.8. Storage and Transportation

ABS pellets can be stored and transported in polyethylene sacks stacked on pellets, in octatainers with polyethylene liners, in big bags, and also in silo containers. Labeling according to GefStoffV and EC guidelines is not necessary. Safety data sheets must be enclosed with the goods for road, rail, or marine transport. The surfaces of the pellets absorb water, which must be removed by predrying (e.g., 2 h at 70 °C) before processing.

ABS powders which are used as additives (e.g., for PVC) must be stored and transported in explosion-proof systems because of their ready oxidizability and large specific surface area. Measures should also be taken against buildup of electrostatic charge. The powders are stored and transported in polyethylene sacks or silo containers. If there is a danger of dust formation, dust masks must be worn.

5.9. Uses

ABS polymers are widely used as engineering materials [234–236]. The main consumers are the automotive industry, the domestic appliances industry, the data technology and telecommunications area, and producers of refrigeration equipment, toys, sports articles, and semifinished articles. ABS polymers have a very favorable price–performance ratio. In the automotive industry ABS often competes with modified polypropylenes, but in other applications it also replaces industrial thermoplastics and thermoplastic blends.

Trade names and producers of ABS polymers follow:

BASF	Terluran
Chi Mei	Polylac
Cheil	Starex
Dow Chemical	Magnum
Formosa Plastics	-
GE Plastics	Cycolac
Lanxess	Novodur
LG Chemicals	-
Techno Polymer	-
Toray Industries	Toyolac

5.10. Economic Aspects of Acrylonitrile-Based Styrene Copolymers

The different polymer production and sales options only permit the joint evaluation of data for all acrylonitrile-based styrene copolymers (ABS, SAN, ASA, AES, and ACS). The major part (> 75 %) is assignable to ABS.

Consumption of acrylonitrile-based styrene copolymers has grown considerably during the last years (Table 10) reflecting their wide use as engineering materials and blend components for other industrial thermoplastics. Nominal

Table 10. Product development of acrylonitrile based styrene copolymers (ABS, SAN, ASA, etc.) (10^3 t)

Region	1995	2000	2005
EU	650	800	800
NAFTA	700	800	800
Asia	2300	3400	4700
Others	200	250	300
Total	3850	5250	6600

Table 11. Worldwide production capacities for acrylonitrile-based styrene copolymers in 2006 (10^6 t)

Company	Region	Capacity
Chi Mei	Asia	1.2
BASF	Asia, EU, NAFTA	0.9
LG Chem	Asia	0.75
Lanxess	Asia, EU, NAFTA	0.7
GE Plastics	Asia, EU, NAFTA	0.6
Formosa Chemicals	Asia	0.45
Cheil Industries	Asia	0.4
Dow Chemical	Asia, EU, NAFTA	0.4
Techno Polymer	Asia	0.3
Grand Pacific	Asia	0.25
Kumho	Asia	0.2
Jilin Chemical	Asia	0.2
others		1.2

production capacities and producers are listed in Table 11. Consumption is estimated to grow by 5–6 % over the next years.

5.11. Recycling

ABS articles can be recycled in two ways:

1. The used articles are returned and used directly in the production process (direct recycling)
2. They are degraded chemically into their individual components (i.e., acrylonitrile, butadiene, styrene) or other organic compounds which are then recycled (chemical recycling) [237]

The interest of ABS producers and consumers is concentrated on the first method. Investigations carried out by raw materials producers and the automotive industry show that sorted ABS polymers are well suited to direct recycling (i.e., defined quantities of recycled material are added to the primary material) [238, 239]. Tests have shown that incorporation of 20 % and 50 % recycled ABS derived from articles sorted into pure grades has very little effect on (notched) impact strength, heat resistance (Vicat B 120), and the melt flow index [240]. The basic operations involved in ABS recycling are summarized in Figure 21.

5.12. ABS-Analogous Systems

ABS-analogous systems are elastomer-modified, two-phase thermoplastic systems with

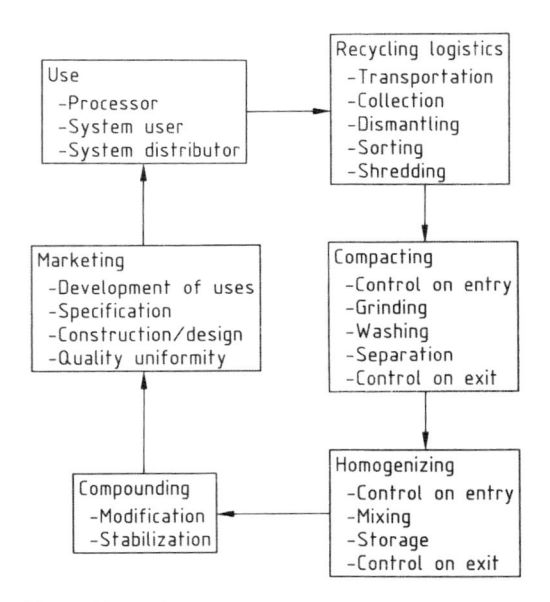

Figure 21. ABS recycling operations
Impact strength of ASA, AES, and ABS

the same basic structure as ABS (i.e., a SAN-grafted rubber phase embedded in a SAN matrix). Production methods and principles are the same as those used as for ABS.

5.12.1. ASA, AES, and ACS Polymers

ASA, AES, and ACS polymers consist of acrylonitrile and styrene grafted onto acrylate, EPDM, or chlorinated polyethylene rubbers, respectively. In these polymers, aging resistance is increased by substituting the polybutadiene rubber of ABS by a rubber base which is only slightly or not at all susceptible to oxidation. Other features of ABS (e.g., toughness–hardness and toughness–heat resistance behavior or the coupling behavior of the grafted rubber) should be retained. These conditions can, however, only be partly satisfied because appropriate "saturated" elastomer candidates behave less favorably than butadiene polymers; they have a higher glass transition temperature and are less prone to grafting reactions. The SAN matrix itself does not have unlimited aging and weather resistance; SAN copolymers undergo oxidative degradation in the presence of air and this is independent of the degradation behavior of the

Table 12. Producers and trade names of two-phase ASA, AES, and ACS plastics

Abbreviation	Matrix	Elastomer base	T_g of the elastomer base, °C	Producer	Trade name
ASA	SAN	acrylate rubber	ca. −48	BASF	Luran S
				Lanxess	Centrex
				GE Plastics	Geloy
				Hitachi Chemicals	Vitax
				UMG ABS Ltd.	Dialac
AES	SAN	EPDM rubber	−50 to −60	Dow Chemical	Rovel
				Polimeri	Koblend
				Techno Polymer	AES
				Nippon A&L	Unibrite
ACS	SAN	chlorinated polyethylene	−20 to −30	Asahi Kasei	Stylac ACS

grafted rubber system [241]. Addition of light stabilizers can delay, but not prevent degradation. To obtain rubber-modified, two-phase plastics that are completely aging- and weather-resistant, another matrix system such as poly(methyl methacrylate) must be used.

Acrylate rubbers (ACM), ethylene–propene rubbers (EPM or EPDM) (→Rubber, 3. SyntheticRubber, 3. Synthetic), chlorinated polyethylene, and other oxidation-resistant rubbers (e.g., silicone rubbers) have been suggested for the production of elastomer-modified, two-phase plastic systems [242, 243]. However, only the first three of these elastomers have been used in aging-resistant ABS polymers. Only ASA [244], AES [245], and ACS polymers are of industrial importance.

ASA production starts from a poly(butyl acrylate) rubber produced by emulsion polymerization and with a uni- or bimodal particle size distribution. To achieve a sufficiently high degree of grafting in the subsequent grafting stage, one of the following grafting bases is used:

1. A terpolymer from *n*-butyl acrylate, vinyl methyl ether, and butadiene [246]
2. An *n*-butyl acrylate–butadiene copolymer [247]
3. A copolymer consisting of butyl acrylate and a monomer acting as a cross-linking or grafting center [248]

Production and workup of ASA graft rubber obtained by emulsion polymerization require the same process parameters and reaction conditions as those described for ABS in Section 5.3.1.

Seed latex technology offers a further possibility, starting from a finely divided polybutadiene seed latex [249].

ASA products are obtained by compounding the grafted polymers and separately produced SAN copolymers with additives and colorants.

AES production usually starts with an EPDM rubber system with 7–12 double bonds for every 1000 carbon atoms [250]. At a rubber content of 10–50 % the AES and grafted rubber are produced by continuous bulk solution polymerization as described for ABS in Section 5.3.3 [251]. The rubber is dissolved in a monomer–solvent mixture and then charged to the first reactor of a bulk solution polymerization plant. The final product can be isolated analogously to ABS but a precipitation stripping process can also be used for grafted rubber systems with high EPDM contents.

ACS production is analogous to AES production, but is declining in importance.

Table 12 lists producers and trade names of ASA, AES, and ACS polymers.

ASA and AES polymers exhibit better aging behavior than ABS polymers (Fig. 22). The

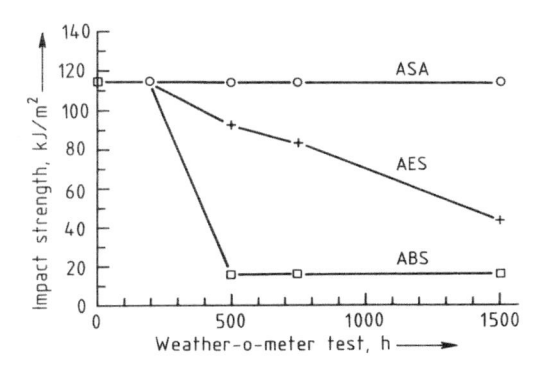

Figure 22. Graft rubber systems after exposure to light (weatherometer test)

principal disadvantages of AES and ASA polymers compared with ABS polymers are their lower low-temperature toughness and lower heat resistance. Deeply colored modifications are difficult to produce (increased opacity due to the different refractive indices of the rubber base and the matrix).

5.12.2. MBS and MABS Polymers

MBS polymers contain a methyl methacrylate (M)–styrene (S) copolymer matrix. The grafted rubber system embedded in the matrix consists of a methyl methacrylate–styrene copolymer shell grafted onto a polybutadiene or butadiene–styrene copolymer core. MABS polymers denote systems in which acrylonitrile is incorporated into the polymer matrix or the grafted rubber shell as an additional monomer. Table 13 lists producers and trade names of MBS and MABS polymers.

The composition of MBS and MABS polymers is usually chosen to ensure good transparency: the refractive indices of the matrix, the grafted rubber shell, and the rubber core should be the same; the grafted rubber particle diameter should be less than half the wavelength of visible light (i.e., <200 nm).

MBS and MABS polymers are seldom used in the pure form; they are generally employed to modify other thermoplastics. They improve processing behavior, notched impact strength, and heat resistance of rigid poly(vinyl chloride) (PVC). The modifiers must be in powder form because rigid PVC is mostly processed as a powder. MBS and MABS polymers are therefore produced by emulsion polymerization or latex suspension polymerization.

MABS polymers used to modify PVC have the following requirements [252]:

1. The particle diameter of the grafted rubber must be <200 nm to ensure high transparency [253]
2. To obtain transparent, colorless polymer blends, the composition of the modifier must be chosen such that its refractive index is the same as that of PVC ($n_D^{20} = 1.54$) [254, 255]
3. The modifier should have high activity at low dosage to minimize loss in modulus and heat resistance

A butadiene–styrene copolymer (weight ratio 75:25) is usually used as the rubber base to achieve a refractive index n_D^{20} of 1.54. Since methyl methacrylate is hydrolyzed relatively rapidly at alkaline pH [256], the grafting reaction is carried out at neutral or weakly acidic pH. Long-chain alkyl or arylsulfonates are used as emulsifiers and yield stable dispersions even below pH 7. Inorganic peroxides (e.g., potassium peroxydisulfate) and redox systems (e.g., cumene hydroperoxide and sodium hydroxymethane-sulfinite) are generally used as initiators [257]. The degree of grafting is controlled by adjusting the ratio of the grafted monomers to the rubber base and by the addition of molecular mass modifiers. The particle size of the grafted rubber is determined by that of the rubber base [120]. This, in turn, is controlled by the water–monomer ratio used in the polymerization, by the emulsifier concentration, or, in seed latex polymerization, by the particle size of the seed latex.

It is advantageous to carry out the grafting reaction in several stages [257, 258], or to use blends of grafted polymers which have a synergistic effect on the toughness of the PVC blend.

The grafted rubber and resin matrix can be synthesized separately and mixed in a workup [259] or compounding stage. Further workup

Table 13. Examples of producers and trade names of ABS, MBS, MABS, and MACR modifiers

Producer	Country	Product	Modifier for[*]	Trade name
BASF	Germany	ABS, MBS, MACR	PVC	Vinuran
Lanxess	Germany	ABS, MBS, MABS, MACR	PVC	Baymod
GE Plastics	United States	ABS, MABS	PVC, PC/PBT	Blendex
Rohm & Haas	United States	MBS, MABS	PVC, PC/PBT	Acryloid/Paraloid
Kaneka	Japan	MBS, MABS	PVC	Kane ACE
Mitsubishi Rayon	Japan	MBS, MABS	PVC, PBT	Metablen
Nippon A & L	Japan	ABS, MBS, MABS	PVC	Kralastic

[*] PBT = poly(butylene terephthalate); PC = polycarbonate; PETP = poly(ethylene terephthalate); PVC = poly(vinyl chloride).

and powder isolation are similar to those of ABS emulsion polymers. Like ABS polymers, MBS and MABS polymers must be protected from attack by atmospheric oxygen by the addition of stabilizers. MBS polymers are being increasingly used for modifying aromatic polyesters and polyester–polycarbonate blends [260, 261]. Since the processing temperature exceeds 280 °C, MACR polymer modifiers are used in which the SBR base is replaced by an acrylate rubber base (see also Table 13). Production requirements for MACR polymers are the same as for ASA and MBS graft rubber systems [262].

5.13. ABS Blends

Polymeric materials with novel properties can be obtained by the combination of known polymers in blends (alloys). Table 14 summarizes ABS-containing polymer blends. See also → Polymer BlendsPolymer Blends, Section 4.3.

Mixtures of ABS with polycarbonate (PC), poly(butylene terephthalate) (PBT), PBT–polycarbonate (PC), PVC, polyamide and thermoplastic polyurethane (TPU) are industrially important, both with regard to quantity and properties [263–265]. Other blends not mentioned in Table 14, (e.g., ABS–PETP and ABS–polysulfone blends) have not yet become important [266]. Blending of polycarbonate with ABS improves heat resistance, processing behavior, low-temperature toughness, and leads to synergistic effects [267, 268].

The combination of ABS–polycarbonate blends with triphenyl phosphate as a flame retardant and polytetrafluoroethylene as an anti-dripping agent yields molding compounds with a low flammability (UL 94 V-0) that are highly suited as housing materials.

ABS–polycarbonate and ASA–polycarbonate blends are predicted to have growth rates of 7–8 % in the next years.

The use of plastics for external automotive parts makes certain demands concerning heat resistance, low-temperature toughness (no spall fracturing), resistance to gasoline, and rigidity. Rubber-modified polymer blends have been developed as a solution to these problems. ABS, PBT, and ABS–PBT–PC blends are widely used. The addition of ABS graft rubbers to PBT improves impact properties [269].

Table 14. Overview of ABS blends

Grafted rubber	Matrix component I	Matrix component II	Phase ratio grafted rubber to matrix I/matrix II	Advantages	Uses
ABS	SAN [a] AMS–AN [c]	BPA–PC [b]	0.05–0.8/0.95–0.2	higher heat resistance, higher toughness than ABS, better processibility and environmental stress cracking behavior than PC	interior car fittings, housings for electronic instruments
ABS	SAN	PBT or PBT–PC [d]	0.25–0.45/0.75–0.55	higher toughness, higher low-temperature toughness than PBT or PC–PBT	car bumpers, housings
ABS		PVC [e]	0.05–0.5/0.95–0.55	higher toughness, higher heat resistance, better processibility than PVC, better flame resistance than ABS	containers, housings
ABS	AMS–AN	PVC			
ABS	SAN	PVC			
ABS	SAN	PUR [f]	0.05–0.25	higher low-temperature toughness, higher rigidity than PUR	ski boots, other injection-molded articles
ABS (modif.) [g]	SAN	PA 6, PA 66	0.05–0.4 0.95–0.6	higher toughness, better dimensional stability than PA, or PA 66	car parts

[a] SAN: Styrene–acrylonitrile copolymer, weight ratio S : AN 0.9–0.6/0.1–0.4;
[b] BPA–PC: Polycarbonate based on bisphenol A (BPA);
[c] AMS–AN: α-Methylstyrene–acrylonitrile copolymer in weight ratio AMS : AN 0.90–0.65/0.10–0.35;
[d] PBT–PC: Blend of poly(butylene terephthalate) and BPA polycarbonate in weight ratio 0.9–0.3/0.1–0.7;
[e] PVC: Rigid PVC, i.e., without addition of plasticizer;
[f] TPU: Thermoplastic polyurethane elastomer, preferably based on methylene diphenylene diisocyanate;
[g] ABS modif.: Incorporation of functional groups into the graft shell to achieve coupling to the amide group.

The addition of 5–20 wt% of ABS graft rubbers to PVC increases its toughness and thus its possible uses. Opaque and transparent blends can be made depending on the size of the grafted rubber particles. The ABS modifier also improves processing behavior, particularly in the production of calendered films [270].

Addition of heat-resistant ABS polymers to rigid PVC (exchange of the SAN matrix for an AMS–AN or styrene–acrylonitrile–NPMI copolymer) increases the Vicat softening temperature and improves toughness. The addition of PVC to ABS polymers improves their flammability behavior and, in combination with antimony trioxide, gives UL 94 V-0 materials. These blends are used on a large scale in the United States, Japan, and the Pacific Rim.

Crashpad sheets are used for covering car dashboards. They contain an ABS–PVC blend as the main component and their flexibility can be altered by extending the blend system with NBR (butadiene–acrylonitrile copolymer) or other polymer plasticizers.

The addition of ABS graft rubbers to thermoplastic polyurethanes in an in-situ reaction [271] can improve their rigidity and low-temperature toughness. Blends of this type (trade name Cycoloy, GE Plastics or Desmopan, Bayer) are widely used in ski boots. Addition of soft TPU types to ABS polymers allows wide variation of low-temperature toughness and processing behavior [272].

Polyamide 6 and polyamide 66 do not exhibit sufficient compatibility with standard ABS grades. This can be overcome by incorporating functional groups (preferably –COOH groups) into the shell of the ABS graft rubber system or using compatibilizing agents [e.g., imidized poly(methyl methacrylate) [273], or maleic anhydride copolymers [274].

ABS polymers and all other thermoplastics mentioned in Table 14 are produced as powders or pellets. Compounding can be carried out in two ways, usually in internal kneaders or twin-screw extruders [275]:

1. Powder mixing with subsequent melt compounding
2. Direct melt compounding in which the individual components are charged simultaneously or consecutively into the compounding equipment

References

General References

1. H. Gausepohl, R. Gellert: Polystyrol,*Kunststoff Handbuch*, vol. 4, Hanser Verlag, München 1996.
2. J. Scheirs, D. Priddy: Modern Styrenic Polymers: Polystyrenes and Styrenic Copolymers, John Wiley & Sons, Chichester 2003.

Specific References

3. Phillips Petroleum, US 3 639 517, 1969 (A. G. Kitchen, F. J. Szalla).
4. BASF, US 4 167 545, 1975 (G. Fahrbach, K. Gerberding, E. Seiler, D. Stein).
5. A. W. Hui, A. E. Hamielec, *J. Appl. Polym. Sci.* **16** (1972) 749–769. K. Kirchner, K. Riederle, *Angew. Makromol. Chem.* **111** (1983) 1–16. K. S. Khuong, W. H. Jones, W. A. Pryor, K. N. Houk, *J. Am. Chem. Soc.* **127** (2005) 1265–1277.
6. J. D. Campbell, F. Teymour, M. Morbidelli, *Macromolecules* **36** (2003) 5491–5501.
7. J. Gao, K. D. Hungenberg, A. Penlidis, *Macromol. Symp.* **206** (2004) 509–522.
8. N. Overbergh et al., *J. Polym. Sci. Polym. Phys. Ed.* **14** (1976) 1177–1186.
9. N. Ishihara et al., *Macromolecules* **19** (1986) 2465–2466.
10. A. Zambelli et al., *Makromol. Chem. Rapid. Commun.* **8** (1987) 277–279.
11. M. Fischer, G. P. Hellmann, *Macromolecules* **29** (1996) 2498–2509.
12. E. R. Moore et al., *Encycl. Polym. Sci. Eng.* **16** (1989) 68.
13. R. A. Bubeck et al., *Polym. Eng. Sci.* **21** (1981) 624–633.
14. A. Echte in C. K. Riew (ed.): *Rubber Toughened Plastics*, ACS, Washington, D.C. 1989, p. 23.
15. A. Echte, *Angew. Makromol. Chem.* **58/59** (1977) 175–198.
16. G. H. Michler, *Kunststoffe* **81** (1991) 449–454.
17. G. H. Michler, *Plaste und Kautschuk* **26** (1979) 680–684.
18. A. Echte in [14] pp. 24–29.
19. U. Reichert, *Kunststoffe* **80** (1990) 1092–1096.
20. J. Le Blanc et al., *Digest of Polym. Developments, Styrenics and Acrylics Series III* **58/59** (1991) May/June, 37.
21. National Polystyrene Recycling Company, company brochure, Washington, D.C. 1989.
22. W. Kaminsky, *Makromol. Chem., Macromol. Symp.* **48/49** (1991) 381–393.
23. J. M. G. Cowie: General Concepts in Addition Polymerization, in *Comprehensive Polymer Science*, vol. 3, part 1, chap. 2, Pergamon Press, Oxford 1989, pp. 1–16.
24. D. M. Tirrell: Copolymer Composition, in *Comprehensive Polymer Science*, vol. 3, part 1, chap. 15, Pergamon Press, Oxford 1989, pp. 195–206.
25. R. Greenly in J. Brandrup, E. Immergut (eds.): *Polymer Handbook*, 3rd ed., John Wiley and Sons, New York 1989, pp. II-216, 221, 222, 228.
26. American Cyanamid: *The Chemistry of Acrylonitrile*, The Beacon Press, New York 1957, p. 33.
27. K. F. O'Driscoll: "Temperature Dependence of Copolymerization Reactivity Ratios," *J. Macromol. Sci. Chem.* **A 3** (1969) no. 2, 307–309.
28. G. E. Moleau, *J. Polym. Sci., Part B* **3** (1965) 1007–1015.

29. C. H. Basdekis: *ABS Plastics*, Reinhold Publishing Corp., New York 1964, p. 47.
30. J. LeBlanc, O. Deex, F. Duston, Y. Phillips inM. Hartung (ed.): *Digest of Polymer Developments—Styrenics and Acrylics, Series 3*, **53** (1989)no. 12, 135–145.
31. Monsanto, *US 3 747 899*, 1971 (L. F. Carter, G. A. Latinen); *DE-OS 2 240 294*, 1972 (G. A. Latinen).
32. Dow Chemical, *US 2 714 101*, 1950 (J. L. Amos, J. C. Frank, K. E. Stober);*US 2 727 884*, 1953 (K. E. Coulter, J. L. McCurdy, D. L. McDonald).
33. *Ullmann*4th ed., **19**, 274.
34. Plastics, ed. 9, Thermoplastics and Thermosets Desk-Top Data Bank, International Plastics Selector, San Diego 1990, pp. 870–877.
35. H. Jenne, *Kunststoffe* **62** (1972) 616; **66** (1976) 581.
36. P. L. Ku: "Polystyrene and Styrene Copolymers; Their Manufacture and Application," *Adv. Polym. Sci.*, **8** (1988) no. 2, 177–196.
37. Monsanto Kasei, Collimate M-800, Technical Data Sheet, Yokkaichi, Japan, 1991.
38. I. M. Ward: *Mechanical Properties of Solid Polymers*, 2nd ed., John Wiley & Sons, New York 1983, p. 168.
39. N. W. Johnston, *Polym. Prepr. (Am. Chem. Soc. Div. Polym. Chem.)* **14** (1973)no. 1, 46.
40. M. J. Guest, J. H. Daly, *European Polymer J.* **25** (1989) 985.
41. T. E. Evans: Styrene–Acrylonitrile, in *Modern Plastics Encyclopedia*, McGraw-Hill, New York 1982–1983,pp. 116–117.
42. F. M. Peng: Acrylontrile Polymers, in *Encyclopedia of Polymer Science and Technology*, vol. 1, 2nd ed.John Wiley & Sons, New York 1985, pp. 455–464.
43. *Plastverarbeiter* **27** (1976) 355–367.
44. F. Forster: Polystyrol, inR. Vieweg,G. Daumiller (eds.): *Kunststoff-Handbuch*, vol. 5, Hanser Verlag, München 1969, p. 230.
45. G. Obieglo, *Kunstst. Ger. Plast.* **69** (1979)no. 7, 8.
46. D. Caldwell, *Modern Plastics Encyclopedia*, vol. **67**,no. 11, McGraw-Hill, New York 1990, pp. 101–102.
47. D. R. Paul, J. W. Barlow, H. Keskkula: Polymer Blends, in *Encyclopedia of Polymer Science and Technology*, 2nd ed., vol. 12, John Wiley and Sons, New York 1978, pp. 399–461.
48. D. R. Paul: Interfacial Agents for Polymer Blends, inD. R. Paul, S. Newman (eds.): *Polymer Blends*, vol. 2, Academic Press, New York 1978, chap. 12.
49. R. Bernstein et al., *Macromolecules* **10** (1977) 681.
50. K. Ogura et al., *Annu. Tech. Conf. Soc. Plast. Eng.* (1987), 1365.
51. W. J. Hall et al., *ACS Symp. Ser.* **229** (1983) 49.
52. J. H. Kim et al., *J. Polym. Sci. Polym. Phys. Ed.* **27** (1989) 223–244.
53. P. B. Rim, J. Runt, *J. Appl. Polym. Sci.* **30** (1985) 1545.
54. J. H. Kim et al., *J. Polym. Sci. Polym. Phys. Ed.* **27** (1989) 2211–2227.
55. I. Louis Gomez: *High Nitrile Polymers for Beverage Container Applications*, Technomic, Lancaster 1990, pp. 17–18.
56. F. M. Peng: Acrylonitrile Polymers, in *Encyclopedia of Polymer Science and Technology*, 2nd ed., vol. 1, John Wiley & Sons, New York 1985, pp. 455–464.
57. B. G. Frushour, R. S. Knorr: Acrylic Fibers. inM. Lewin,E. M. Pearce (eds.): *Handbook of Fiber Science and Technology*, vol. IV, Marcel Dekker, New York 1985, pp. 228–235.
58. Monsanto*US 3 540 577*, 1967 (Q. A. Trementozzi); *DE-OS 2 050 535*, 1971 (C. Lee, S. P. Nemphos).
59. Standard Oil, *DE-OS 1 929 860*, 1970 (J. T. Duke, D. C. Prem); *DE-OS 2 007 519*, 1970 (J. T. Duke).
60. W. A. Combellick: Barrier Polymers, in *Encyclopedia of Polymer Science and Technology*, 2nd ed., vol. 2, John Wiley & Sons, New York 1985, pp. 176–192.
61. Plastics Technology,*Manufacturing Handbook and Buyers' Guide*,1989/1990, Bill Communications, New York 1990, p. 675.
62. J. H. Kim et al., *Polym. Eng. Sci.* **29** (1981) 581.
63. Monsanto, *US 4 874 829*, 1989 (C. E. Schwier, W. C. Wu).
64. Dow Chemical, *US 2 769 804*, 1951 (A. W. Hanson); *US 3 080 348*, 1963 (A. F. Roche).
65. Richardson Company, *US 4 001 484*, 1974 (I. H. Song).
66. J. LeBlanc, O. Deex, F. Duston, inS. Englehart (ed.): *Digest of Polymer Developments—Styrenics and Acrylics, Series 3*, **54/55** (1990),May/June 128–148.
67. A. Wambach: Styrene Maleic Anhydride,, *Modern Plastics Encyclopedia*, McGraw-Hill, New York 1991, pp. 104–106.
68. N. Platzer et al. inM. Hartung (ed.): *Digest of Polymer Developments—Styrenics and Acrylics, Series 3*, **45** (1987/1988) 119–141.
69. *Europa-Chemie* **47** (1991) 2.
70. Dow, *WO 30944,36944*, 1997.
71. C. B. Bucknall, R. R. Smith, *Polymer* **6** (1965) 437.
72. K. Dinges, *Kunststoffe* **56** (1966) 548.
73. J. Frazer, *Chem. Ind. (London)* **33** (1966) 1397.
74. K. Dinges, H. Schuster, *Makromol. Chem.* **101** (1967) 202.
75. L. Bohn, *Angew. Makromol. Chem.* **20** (1971) 129.
76. L. Morbitzer, K. H. Ott, H. Schuster, D. Kranz, *Angew. Makromol. Chem.* **27** (1972) 57.
77. L. Morbitzer, D. Kranz, G. Humme, K. H. Ott, *J. Appl. Polym. Sci.* **20** (1976) 2691.
78. M. Yokouchi, S. Seto, Y. Kobayashi, *J. Appl. Polym. Sci.* **28** (1983) 2209.
79. J. A. Sauer, C. C. Chen, *Polym. Eng. Sci.* **24** (1984) 786.
80. F. Lednicky, Z. Pelzbauer, *Polym. Test.* **7** (1987) 91.
81. M.-J. Zhang, F.-X. Zhi, X.-R. Su, *Polym. Eng. Sci.* **29** (1989) 1142.
82. C. C. Chen, J. A. Sauer, *J. Appl. Polym. Sci.* **40** (1990) 503.
83. G. Michler, *Plaste Kautsch.* **35** (1988) 347.
84. G. H. Michler, *Kunststoffe* **81** (1991) 449.
85. C. F. Pearsons, E. L. Suck, *Adv. Chem. Sci.* **99** (1971) 340.
86. H. Kim, H. Keskkula, D. R. Paul, *Polymer* **32** (1991) 1447.
87. P. J. Flory: *Principles of Polymer Chemistry*, Cornell University Press, New York 1953.
88. S. Krause, *J. Macromol. Sci. Rev. Macromol. Chem.* **7** (1972) 251.
89. M. T. Shaw, *J. Appl. Polym. Sci.* **18** (1974) 449.
90. B. J. Schmitt, *Angew. Chem.* **91** (1979) 286.
91. K. Dinges, *Kunststoffe* **32** (1979) 748.
92. A. Brydon, G. M. Burnett, G. G. Cameron, *J. Polym. Sci. Polym. Chem. Ed.* **11** (1973) 3255; **12** (1974) 1011.
93. V. K. Gupta, G. S. Bhargava, K. K. Bhattacharyya, *J. Macromol. Sci. Chem.* **A 16** (1981) 1107.
94. R. A. Hayes, S. Futamura, *J. Polym. Sci. Polym. Chem. Ed.* **19** (1981) 985.
95. R. A. Hayes, *J. Polym. Sci. Polym. Chem. Ed.* **19** (1981) 993.
96. H. A. J. Battaerd, G. W. Tregear: *Graft Copolymers*, Wiley Interscience, New York 1970.
97. J. P. Fischer, *Angew. Makromol. Chem.* **33** (1973) 35.
98. E. S. Daniels, V. L. Dimonie, M. S. El-Aasser, J. W. Vanderhoff, *J. Appl. Polym. Sci.* **41** (1990) 2463.
99. K. Mc Creedy, H. Keskkula, *Polymer* **20** (1979) 1155.
100. H. J. Karam, L. Tien, *J. Appl. Polym. Sci.* **30** (1985) 1969.
101. Bayer, *EP 331 999*, 1989.
102. Bayer, *EP 255 889*, 1988.

103. N. W. Johnston, *Macromolecules* **6** (1973) 453.
104. Mobil Oil, *EP 003 639*, 1979.
105. Rohm & Haas, *US 4 217 424*, 1979.
106. Denki Kagaku Kogyo, *US 4 404 322*, 1983.
107. M. Bresson, P. Hygounenc, *Rev. Gen. Caoutch. Plast.* **642** (1984) 109.
108. H. Kim, H. Keskkula, D. R. Paul, *Polymer* **31** (1990) 869.
109. J. W. F. Bley, S. A. H. Mohammed, *Polym. Plast. Technol. Eng.* **20** (1983) 161.
110. L. Li, T. Masuda, M. Takahashi, *J. Rheol. (N.Y.)* **34** (1990) 103.
111. F. Lednicky, Z. Pelzbauer, *Angew. Makromol. Chem.* **141** (1986) 151.
112. Y. Aoki, *Macromolecules* **20** (1987) 2208.
113. Bayer, *EP 255 889*, 1988.
114. Dow Chemical, *DE 26 19 969*, 1976.
115. Uniroyal, *US 3 111 501*, 1958.
116. ICI, *US 3 652 726*, 1969.
117. Monsanto, *US 4 374 951*, 1983.
118. Japan Synthetic Rubber *EP 164 874*, 1985.
119. BASF, *DE-AS 2 809 180*, 1980.
120. G. Kämpf, H. Schuster, *Angew. Makromol. Chem.* **27** (1972) 81.
121. E. R. Wagner, L. M. Robeson, *Rubber Chem. Technol.* **43** (1970) 1129.
122. Bayer, *EP 34 748*, 1981.
123. H. Ebneth, *Kunststoffe* **53** (1969) 70.
124. T. O. Purcell, *Encycl. Polym. Sci. Technol.*, **suppl. 1**, 1976, 319.
125. H. F. Mark et al., *Encyclopedia of Polymer Science and Technology*, vol. **2**, Wiley Interscience, New York 1965, p. 696.
126. Bayer, *EP 394 779*, 1990.
127. H. G. Keppler, H. Wesslau, H. Stabenow, *Angew. Makromol. Chem.* **2** (1968) 1.
128. Monsanto, *US 3 558 541*, 1971.
129. ISR, *US 35 73 242*, 1971; *US 3 573 246*, 1971.
130. Chemische Werke Hüls, *DE 1 208 879*, 1966.
131. B. F. Goodrich, *DE-OS 1 494 093*, 1969.
132. US Rubber, *US 2 820 777*, 1955. Röhm, *DE 33 19 340*, 1984.
133. The International Synthetic Rubber, *DE-OS 22 33 287*, 1973.
134. W. V. Smith, R. H. Ewart, *J. Chem. Phys.* **16** (1948) 592.
135. J. L. Gardon, *J. Polym. Sci. Polym. Chem. Ed.* **6** (1968) 623, 665, 687, 2853, 2859.
136. M. R. Granico, D. J. Williams, *J. Polym. Sci. Polym. Chem. Ed.* **8** (1970) 2617.
137. D. J. Williams, *J. Polym. Sci. Polym. Chem. Ed.* **12** (1974) 2133.
138. P. Keusch, R. A. Graff, D. J. Williams, *Macromolecules* **7, 304** (1974).
139. N. Friis, A. E. Hamielec, *J. Polym. Sci. Polym. Chem. Ed.* **11** (1973) 3321.
140. Bayer, *EP 717 077*, 1996.
141. F. Haaf et al., *J. Sci. Ind. Res.* **40** (1981) 659.
142. Japan Synthetic Rubber, *JA 3113/67*, 1967.
143. Bayer, *EP 192 151*, 1986; *EP 225 510*, 1987.
144. R. Gächter, H. Müller: *Taschenbuch der Kunststoff-Additive*, Hanser Verlag, München 1983.
145. J. Stepek, H. Daoust: *Additives for Plastics*, Springer Verlag, New York 1983, p. 167.
146. Bayer, *US 4 522 959*, 1985; *US 4 522 964*, 1985.
147. BASF, *EP 665 095*, 1995.
148. BASF, *EP 734 826*, 1996.
149. A. K. Ghosh, J. T. Lindt: *International Polymer Processing*, vol. **3**, Hanser Verlag, München 1990, p. 195.
150. A. K. Ghosh, J. T. Lindt, *J. Appl. Polym. Sci.* **39** (1990) 1553.
151. C. C. Lin et al., *J. Appl. Polym. Sci.* **26** (1981) 1327.
152. Monsanto Chemical, *US 3 813 369*, 1974. Dow Chemical, *EP 897 165*, 1983.
153. Monsanto, *US 3 708 658*, 1973.
154. Ugine Kuhlmann, *DOS 19 64 915*, 1970. Bayer, *EP 167 772*, 1986.
155. F. E. Karasz, W. J. Macknight, in D. R. Paul, L. H. Sperling (eds.): *Multicomponent Polymer Materials 211 Advances Chemistry Series*, American Chemical Society, Washington, D.C. 1986.
156. A. Rudin, M. C. Samanta, *J. Appl. Polym. Sci.* **24** (1979) 1665.
157. P. Wittmer, *Makromol. Chem.* **103** (1967) 188.
158. P. Wittmer, *Makromol. Chem.* **177** (1976) 991.
159. Borg Warner, *US 3 010 936*, 1961.
160. H. Lange, H. Baumann, *Angew. Makromol. Chem.* **43** (1975) 167.
161. N. W. Johnston, *Appl. Polym. Symp.* **25** (1974) 19.
162. N. W. Johnston, *Polym. Prepr. (Am. Chem. Soc. Div. Polym. Chem.)* **14** (1973) 46.
163. Bayer, *EP 330 039*, 1989.
164. Borg-Warner, *US 2 908 661*, 1959.
165. Nitto Chemical, *EP 177 031*, 1985. Nippon Shokubai K.K., *EP 213 933*, 1986.
166. Mitsui Toatsu Chemicals, *EP 216 524*, 1986.
167. Monsanto, *US 4 567 233*, 1986.
168. C. Walling, *J. Am. Chem. Soc.* **71** (1949) 1930.
169. Monsanto, *US 4 305 869*, 1981; *US 4 197 376*, 1980; *US 4 223 096*, 1980.
170. W. J. Hall, R. L. Kruse, R. A. Mendelson, Q. A. Trementozzi, *Polym. Coat. Appl. Polym. Sci. Prep. Div. Am. Chem. Soc.* **47** (1982) 298.
171. Dow Chemical, *EP 103 657*, 1984.
172. J. L. White, R. D. Patel, *J. Appl. Polym. Sci.* **19** (1975) 1775.
173. G. E. Molau, *J. Polym. Sci. Part A* **3** (1965) 4235.
174. S. Zhiqiang, Y. Huigen, Z. Yuan, *J. Polym. Sci.* **32** (1986) 3349.
175. G. F. Freeguard, M. Karmarkar, *J. Appl. Polym. Sci.* **15** (1971) 1657.
176. G. E. Molau, H. J. Keskkula, *J. Polym. Sci. Part A* **4** (1966) 1595.
177. B. W. Bender, *J. Appl. Polym. Sci.* **9** (1965) 2887.
178. J. L. Amos, *Polym. Eng. Sci.* **14** (1974) 1.
179. F. D. Rumscheidt, S. G. Mason, *J. Colloid Sci.* **16** (1961) 238.
180. R. W. Flumerfelt, *Ind. Eng. Chem. Fundam.* **11** (1972) 312.
181. M. Baer, *J. Appl. Polym. Sci.* **16** (1972) 1109.
182. S. L. Rosen, *J. Appl. Polym. Sci.* **17** (1973) 1805.
183. J. L. Locatelli, G. Riess, *Makromol. Chem.* **175** (1974) 3523; *J. Polym. Sci. Polym. Chem. Ed.* **11** (1973) 3309.
184. A. E. Platt in: *Encyclopedia of Polymer Science and Technology*, vol. 13, Wiley Interscience, New York 1970, p. 156.
185. N. Platzer: "Design of Continuous and Batch Polymerization Processes," *Ind. Eng. Chem.* **62** (1970) 6.
186. H. G. Pohlemann, A. Echte: "Polymer Science Overview," in G. A. Stahl (ed.): *ACS Symp. Ser.* **175** (1981) 265.
187. Dow Chemical, *US 2 727 884*, 1955; *US 2 694 692*, 1954.
188. H. J. Karam, J. C. Bellinger, *Ind. Eng. Chem. Fundam.* **7** (1968) 576.
189. W. H. Ray, *ACS Symp. Ser.* **226** (1983) 101.
190. Toray Ind. K.R., *US 3 981 944*, 1976.
191. Mitsui Toatsu Chemicals, Toyo Engineering Corporation: *Continuous Buck Polymerization Process for ABS Products*, company brochure, Tokyo 1985.
192. A. M. Donald, E. J. Kramer, *J. Appl. Polym. Sci.* **27** (1982) 3729.
193. C. B. Bucknall, P. Davies, I. K. Partidge, *J. Mater. Sci.* **22** (1987) 1341.

194 D. G. Gilbert, A. M. Donald, *J. Mater. Sci.* **21** (1986) 1819.
195 BASF, *US 44 93 922*, 1985.
196 BASF, *EP 507 117*, 1992; Dow Chemical, *EP 412 801*, 1992.
197 BASF, *EP 477 671*, 1992; *EP 477 764*, 1992; *EP 505 798*, 1992. General Electric, *EP 657 479*, 1995. Mitsui Toatsu Chemicals, *US 5 506 304*, 1996.
198 BASF, *DE 20 37 984*, 1972.
199 Monsanto, *US 3 928 495*, 1975.
200 Societa Edison, *FR 14 56 141*, 1966.
201 Monsanto, *US 3 509 237*, 1971.
202 Bayer, *BE 828 318*, 1975; *BE 828 319*, 1975.
203 S. H. Roth, *J. Appl. Polym. Sci.* **18** (1974) 3305.
204 Toray Ind. K.K., *DE-AS 2 044 427*, 1970.
205 D. E. Bell Richardson, *US 3 370 105*, 1968.
206 Y. A. Shlyapnikov, *Russ. Chem. Rev. (Engl. Transl.)* **50** (1981) 581.
207 G. Scott, *J. Appl. Polym. Sci. Appl. Polym. Symp.* **35** (1979) 123.
208 C. R. H. I. de Jonge, *Pure Appl. Chem.* **55** (1983) 1637.
209 Pennwalt, *US 3 652 680*, 1972.
210 K. Schwarzenbach in [144] p. 1.
211 F. Gugumus in [144] p. 101.
212 D. M. Wiles, J. P. P. Jensen, D. J. Carlsson, *Pure Appl. Chem.* **55** (1983) 1651.
213 H. W. Finck in [144] p. 581.
214 BASF, *DE 31 12 428*, 1982; Bayer, *EP 278 347*, 1988; Bayer, *EP 278 349*, 1988.
215 European Standardization Committee, *EN 29 001*, 1987.
216 G. Lawson, J. A. Sidwell, *Plast. Rubber Int.* **8** (1983) 15.
217 J. R. Fried, *Plast. Eng.* **38** (1982) no. 8, 27.
218 C. B. Bucknall: *Toughened Plastics*, Applied Science Publishers, London 1977.
219 D. R. Paul, S. Newman: Polymer Blends, in: *Encyclopedia of Polymer Science and Engineering*, 2nd ed., vol. 12, Wiley, New York 1988, p. 399.
220 Bayer AG: *Novodur*, company brochure no. KU 41 662, ed. 12/95.
221 B. F. Conaghan, S. L. Rosen, *Polym. Eng. Sci.* **12** (1972) 134.
222 H. Ebneth, *Plastverarbeiter* **4** (1968) 1.
223 H. Jenne, *Kunststoffe* **66** (1976) 581.
224 E. Zahn, *Appl. Polym. Symp.* **11** (1969) 209. P. Pagan, *Polym. Paint Colour J.* **117** (1987) 704.
225 K. Stange, *Chem. Ind. (Düsseldorf)* **20** (1968) 804.
226 H. Jenne, *Kunststoffe* **62** (1972) 616; **66** (1976) 581.
227 W. Nouvertne, H. Peters, H. Beicher, *Plastverarbeiter* **33** (1982) 1070.
228 W. Witt, *Kunststoffe* **74** (1984) 592.
229 P. R. Müller, *Kunststoffe* **74** (1984) 569.
230 A. S. Wood, *Mod. Plast. Int.* **20** (1990)June, 58.
231 Verordnung über gefährliche Stoffe, Aug. 26, 1986, BGBL I, p. 1470, and from April 23, 1990, BGBL V, p. 790.
232 EG Kunststoffrichtlinie 90/128, EWG no. 349/26 from Dec. 13, 1990.
233 Positiv-Liste für Additive und Polymerisationshilfsstoffe, Synoptic Document REV 5 (III/3141/89-EN), Aug. 01, 1991.
234 K. Schneider, E. Frohberg, *Kunststoffe* **80** (1990) 1099.
235 R. Theyson, *Kunststoffe* **79** (1989) 913.
236 E. Frohberg, *Kunststoffe* **86** (1996) 1498.
237 S. Menges, R. Fischer, *Kunststoffe* **81** (1991) 6.
238 R. D. Leaversuch, *Mod. Plast. Int.* **21** (1991)Feb., 12.
239 M. de Braaf, Dow Chemical, company brochure 1991.
240 Bayer AG: *Kunststoff-Recycling*,company brochure no. KU 40 237, 11/91.
241 G. M. Ruhnke, L. F. Biritz, *Kunststoffe* **62** (1972) 250.
242 Dow Corning, *DE 2 321 904*, 1973; *US 3 898 300*, 1975.
243 Bayer, *EP 246 537*, 1987; *EP 258 746*, 1988.
244 E. Zahn, H. W. Otto, *Kunststoffe* **57** (1967) 921.
245 Japan Synthetic Rubber, *DE 30 01 766*, 1980; *EP 96 527*, 1983; *US 4 814 388*, 1989. Montedipe, *EP 286 071*, 1988.
246 BASF, *DE 1 174 069*, 1964; *DE 1 180 126*, 1964; *DE 1 182 811*, 1964; *1 238 207*, 1976.
247 Stauffer Chem., *US 3 994 631*, 1976.
248 Rohm and Haas, *US 4 096 202*, 1978.
249 Bayer, *EP 34 748*, 1981.
250 C. L. Meredith, *Rubber Chem. Technol.* **44** (1971) 1130.
251 Japan Synthetic Rubber, *DE 30 36 921*, 1981.
252 Kanegafuchi Chem., *FR 1 356 523*, 1964.
253 Kanegafuchi Chem., *JA 10 735 67*, 1967; *JA 190 535 64*, 1964.
254 American Cyanamid, *GB 994 924*, 1963.
255 Foster Grant, *GB 1 009 360*, 1962.
256 W. C. Mast, C. H. Fischer, *Ind. Eng. Chem.* **41** (1949) 790.
257 Kanegafuchi Chem., *DE-OS 1 570 855*, 1965.
258 Japan Geon K.K., *GB 1 469 868*, 1977.
259 Bayer, *DE 38 13 363*, 1989; *DE 33 12 541*, 1984.
260 Teijin, *US 38 64 428*, 1975.
261 Montedison, *DE 23 48 377*, 1975.
262 J. T. Lutz, *Polym. Plast. Technol. Eng.* **21** (1983) 99.
263 W. Witt, *Kunststoffe* **77** (1987) 1009.
264 D. R. Paul, S. Newman: *Polymer Blends*, vols. I and II, Academic Press, New York 1978.
265 K. Kheit, D. V. Hove, *Polymer Blends and Alloys Symposium*, Luzern,June 1988.
266 Uniroyal, *US 3 555 119*, 1971; *US 3 636 140*, 1972.
267 L. Morbitzer, H. J. Kress, C. Lindner, K. H. Ott, *Angew. Makromol. Chem.* **132** (1985) 19.
268 H. Peters, L. Morbitzer: *Plasticon 81 Polymer Blends*, Plastic and Rubber Institute, University of Warwick,Preprint 29.
269 Bayer, *US 4 292 233*, 1981.
270 Bayer, *EP 101 900*, 1984; *EP 101 904*, 1984; *EP 101 899*, 1984.
271 Bayer, *DE 28 54 407*, 1980.
272 W. J. Farrissey et al. in: *Advances in Polymer Blends and Alloys Technology*, vol. 1, Technomic Publishing Company, Lancaster 1988.
273 Rohm and Haas, *US 4 436 871*, 1984.
274 M. Weber, W. Heckmann, A. Goeldel, *Macromol. Symp.* **233** (2006) 1–10.
275 A. P. Plochocki, S. S. Dagli, H. H. Mack, *Kunststoffe* **78** (1988) 254.

Further Reading

M. Biron: *Thermoplastics and Thermoplastic Composites*, Elsevier, Amsterdam 2006.

S.-S. Chen: *Styrene*, Kirk Othmer Encyclopedia of Chemical Technology, 5th edition, vol. 23, p. 325–357, John Wiley & Sons, Hoboken, NJ, 2007, online: DOI: *10.1002/0471238961. 192025180308051.a01.pub2* (September 2006).

J. G. Drobny: *Handbook of Thermoplastic Elastomers*, William Andrew Pub, Norwich, NY 2007.

J. K. Fink: *Handbook of Engineering and Specialty Thermoplastics*, Wiley, Hoboken 2010.

J. Scheirs,D. B. Priddy (eds.): *Modern Styrenic Polymers*, Wiley, Chichester 2003.

J. Schellenberg (ed.): *Syndiotactic Polystyrene*, Wiley, Hoboken, NJ 2010.

L. H. Sperling: *Introduction to Physical Polymer Science*, 4th ed., Wiley, Hoboken, NJ 2006.

Polyureas

CONSTANTIN I. CHIRIAC, Institute of Macromolecular Chemistry "Petru Poni", Jassy, Romania

FULGA TANASÄ;, Institute of Macromolecular Chemistry "Petru Poni", Jassy, Romania

1.	Introduction	1029		7.1.	Physical Properties	1040
2.	Synthesis	1029		7.2.	Chemical Properties	1041
3.	Polyureas as Starting Materials for			8.	Uses	1042
	Other Polymers	1034		8.1.	Foams	1042
4.	Homopolyureas	1035		8.2.	Moldings	1044
5.	Copolyureas	1035		8.3.	Fibers	1046
5.1.	Simple Copolyureas	1036		8.4.	Films and Membranes	1046
5.2.	Copolyureas Containing Other Functional Groups	1037		8.5.	Coatings	1047
				8.6.	Other Applications	1047
6.	Structure	1040		9.	Safety and Environmental Aspects	1047
7.	Properties	1040			References	1048

1. Introduction

Polyureas are defined as polymers which contain ureylene groups –NHCONH– in the polymer chain. Urea, regarded as the diamide of carbonic acid, was first synthesized by WÖHLER in 1892, but the chemistry and technology of polyureas have a more recent origin.

Linear polyureas are thermoplastic polycondensation products with aromatic (R = arylene) or aliphatic (R = alkylene) structures:

$$\left(\begin{array}{c} O \\ \| \\ N-C-N-R \\ | \quad \quad | \\ H \quad \quad H \end{array} \right)_n$$

Polyureas or copolyureas with aliphatic structures exhibit a difference of 50 – 100 °C between the softening point and the onset of decomposition. They can be used as moldings. Polyureas and copolyureas with aromatic structures have softening points near their decomposition temperatures and are soluble in organic solvents, particularly in aprotic dipolar solvents such as DMF, N-methylpyrrolidinone (NMP), and DMSO. They are used for the preparation of lacquers, varnishes, and coatings [1]. The relationship between the softening points and structure is shown in Table 1.

Polyureas were first produced on a commercial scale at I.G. Farben by employing the reaction between diisocyanates and diamines. Mitsui Toatsu Chemical produces synthetic fibers from nonamethylenediamine and urea. Polyureas and copolyureas, particularly poly(urethaneurea)s, have many practical applications as foams, elastomers, fibers, etc.

Worldwide consumption was below 4.8×10^6 t in 1984 (in Europe, 2×10^6 t) and 5.9×10^6 t in 1995 [4]. More than 75 % of consumption was for foams [5]. Growth is estimated at 3 – 5 % per annum in the period 1995 to 2000. The number of products and producers continues to increase, and new processing technologies for thermosets, such as reaction injection molding (RIM), liquid injection molding (LIM), and others, are being developed.

2. Synthesis

The synthesis of polyureas is based on reactions between polyamines, especially diamines, with

Table 1. Alternating copolyureas derived from diamines and diisocyanates

Diamine	Diisocyanate	Intrinsic viscosity, dL/g	Softening point, °C	Ref.
Piperazine	1,4-cyclohexylene diisocyanate	0.60	324	[2]
2,5-Dimethylpiperazine	1,4-cyclohexylene diisocyanate	0.48	390	[2]
2,5-Dimethylpiperazine	1,2-bis(4-isocyanatophenyl)ethene	0.33	362 (decomp.)	[2]
2,5-Dimethylpiperazine	bis(4-isocyanatophenyl)methane	1.30	350 (decomp.)	[2]
1,4-Phenylenediamine	bis(4-isocyanatophenyl)methane	0.36	400 (decomp.)	[2]
Hexamethylenediamine	4-methyl-1,3-phenylene diisocyanate	0.47	225–230	[3]

aliphatic, aromatic, or heterocyclic structures and carbon dioxide, carbonyl sulfide, carbonic esters, phosgene, urea, urethane, and isocyanates.

From Diisocyanates and Diamines. Polyureas are readily prepared by a polyaddition reaction between diisocyanates and diamines.

$$H_2N-R-NH_2 + OCN-R'-NCO \longrightarrow$$

$$\left(\begin{array}{c} O \\ \parallel \\ N-R-N-C-N-R'-N-C \\ | \quad | \quad | \quad | \\ H \quad H \quad H \quad H \end{array} \right)_n$$

Isocyanates are reactive products, and the reactivity toward compounds containing active hydrogen is as follows: $RNH_2 > ArNH_2 >$ primary $OH >$ water $>$ secondary $OH >$ tertiary $OH >$ phenolic $OH > COOH > RNHCONHR' > RCONHR' > RNHCOOR'$ [6]. An isocyanate group bonded to an aromatic ring is more reactive than one bonded to an aliphatic chain. Many aliphatic amines are extremely reactive with isocyanates, even at low temperatures. Relative reactivities of amines with phenyl isocyanate in diethyl ether at 0°C [7] follow:

NH_3	1.00
Ethylamine	9.72
n-Propylamine	8.22
n-Butylamine	9.17
n-Amylamine	9.17
Aniline	0.53

Secondary aliphatic and primary aromatic amines react similarly. Secondary aromatic amines are less reactive [7].

For the amine – isocyanates reaction, the following mechanism has been proposed [8]:

$$Ar-N=C=O + Ar'-NH_2 \underset{k_2}{\overset{k_1}{\rightleftharpoons}} \left[\begin{array}{c} Ar'-NH_2 \\ | \\ Ar-N=C=O \end{array} \right] \longrightarrow$$

$$\underset{O}{Ar-N=C}\overset{Ar'\overset{+}{\underset{}{N}H_2}}{} + HM \overset{k_3}{\longrightarrow} \underset{O\cdots H}{Ar-N=C}\overset{Ar'\overset{+}{\underset{}{N}}\overset{H}{\underset{H}{}}}{\underset{}{}}|M \longrightarrow$$

$$\underset{O\cdots H}{Ar-N=C}\overset{Ar'\overset{+}{\underset{}{N}}\overset{H}{\underset{}{}}}{\underset{}{}}\overset{-}{M} \longrightarrow \underset{O-H}{Ar-N=C}\overset{Ar'}{\underset{}{NH}} + HM$$

$$\longrightarrow Ar\underset{H}{\underset{|}{N}}\overset{O}{\overset{\parallel}{C}}\underset{H}{\underset{|}{N}}Ar'$$

In this mechanism, HM is a proton acceptor/donor, for example, an amine, urea, or carboxylic acid, which acts as catalyst [9]. The reactions between isocyanates and amines are usually autocatalytic, but they can be catalyzed by many other catalysts, amines, or acids [10], [11].

Weak aromatic carboxylic acids show a high catalytic activity [12]. In benzoic acid derivatives, electron-acceptor substituents such as NO_2, Cl, or OH reduce catalytic activity, whereas electron-donor substituents such as alkyl groups increase it; 2-methylbenzoic acid is the most active catalyst.

Lewis bases such as 1,4-diazabicyclo-[2.2.2.]-octane (DABCO), triethylamine, dimethylbenzylamine, and pyridine also exhibit catalytic activity. 4-(Dimethylamino)pyridine is an exceptional catalyst, not only in reactions with amines (strongly nucleophilic reagents), but even with alcohols (weakly nucleophilic reagents) [13]. For reactions with alcohols, the following mechanism has been proposed [14]:

Insertion catalysts such as dibutyltin dilaurate, tin(II) octanoate, and cobalt naphthenate have also been used in this field [15].

The effect of substituents on the reaction between aromatic isocyanates and aromatic amines has been studied [10]. The reaction rates were determined by refluxing 0.005 mol substituted phenyl isocyanate and 0.005 mol substituted aniline for 1 h in a mixture of toluene and benzene. The urea yields are shown in Table 2.

The effect of substituents on the reactivity of the isocyanate decreases in the order: 2,4-$(NO_2)_2$ > 3,5-$(NO_2)_2$ > 4-NO_2 > 3-NO_2 > 2-NO_2 ≫ H, 4-CH_3, 3-OCH_3, 4-OCH_3. Electron-acceptor substituents in aromatic isocyanates increase the reaction rate. The same substituents in aromatic amines reduce the reaction rate. Steric hindrance is also influential. *Ortho* substitution in aromatic isocyanates or amines reduces the reaction rate.

The reaction between diamines and diisocyanates is usually carried out by interfacial polyaddition [16] or in solution [17]. In the bulk, infusible, insoluble cross-linked polymers can be obtained instead of the desired linear polyureas. Cross-linking occurs by the interaction of the highly reactive free isocyanate groups with active hydrogen atoms from the urea groups [18]. The most important application of this reaction is the synthesis of poly(urethaneurea)s, also called copolyureas.

Another interesting application for polyureas is in nonlinear optics (NLO); the NLO response of polyureas is extremely rapid, because effects occur primarily through electronic polarization [19]. It is now well-known that molecules containing electron-donor and electron-acceptor groups separated by a large conjugated π framework possess large values of second-order molecular hyperpolarizability [19]. For the synthesis of polyureas with these properties, 4-(dialkylamino)-4'-(alkylsulfonyl)azobenzene NLO chromophores have been used, which contain methylsulfonyl groups as electron-acceptor groups, 4-dialkylamino groups as electron-donor groups, and between them azo groups [20–22].

Another interesting application is the synthesis of hyperbranched polyureas (Fig. 1) [23], which have unusual properties [24].

In these syntheses, the most important monomer is 3,5-diaminobenzoyl azide (), which on heating in NMP at 110 °C under argon for 16 h gives unstable 3,5-diaminophenylisocyanate (**5**) and finaly hyperbranched polyureas. Curiously, these hyperbranched polyureas are soluble in aprotic dipolar solvents such as DMSO, NMP, and DMF.

From Diisocyanates and Water. Polyureas can be prepared from diisocyanates and water:

Table 2. Urea yields from aromatic isocyanates and aromatic amines

Isocyanate substituent	Amine substituent								
	2,4-$(NO_2)_2$	2-NO_2	3,5-$(NO_2)_2$	4-NO_2	3-NO_2	3-OCH_3	H	4-CH_3	4-OCH_3
4-OCH_3	–	–	–	–	15	91	90	95	98
4-CH_3	–	–	–	3	14	90	95	98	–
H	0	0	0	0	26	92	92	100	–
3-OCH_3	–	–	–	0	46	95	–	–	–
3-NO_2	0	0.5	4	13	52	96	100	–	–
4-NO_2	0	4	8	14	61	97	100	–	–
3,5-$(NO_2)_2$	0	8	16	54	76	95	100	–	–
2-NO_2	0	0	1	8	60	98	100	–	–
2,4-$(NO_2)_2$	0	45	51	76	84	87	100	–	–

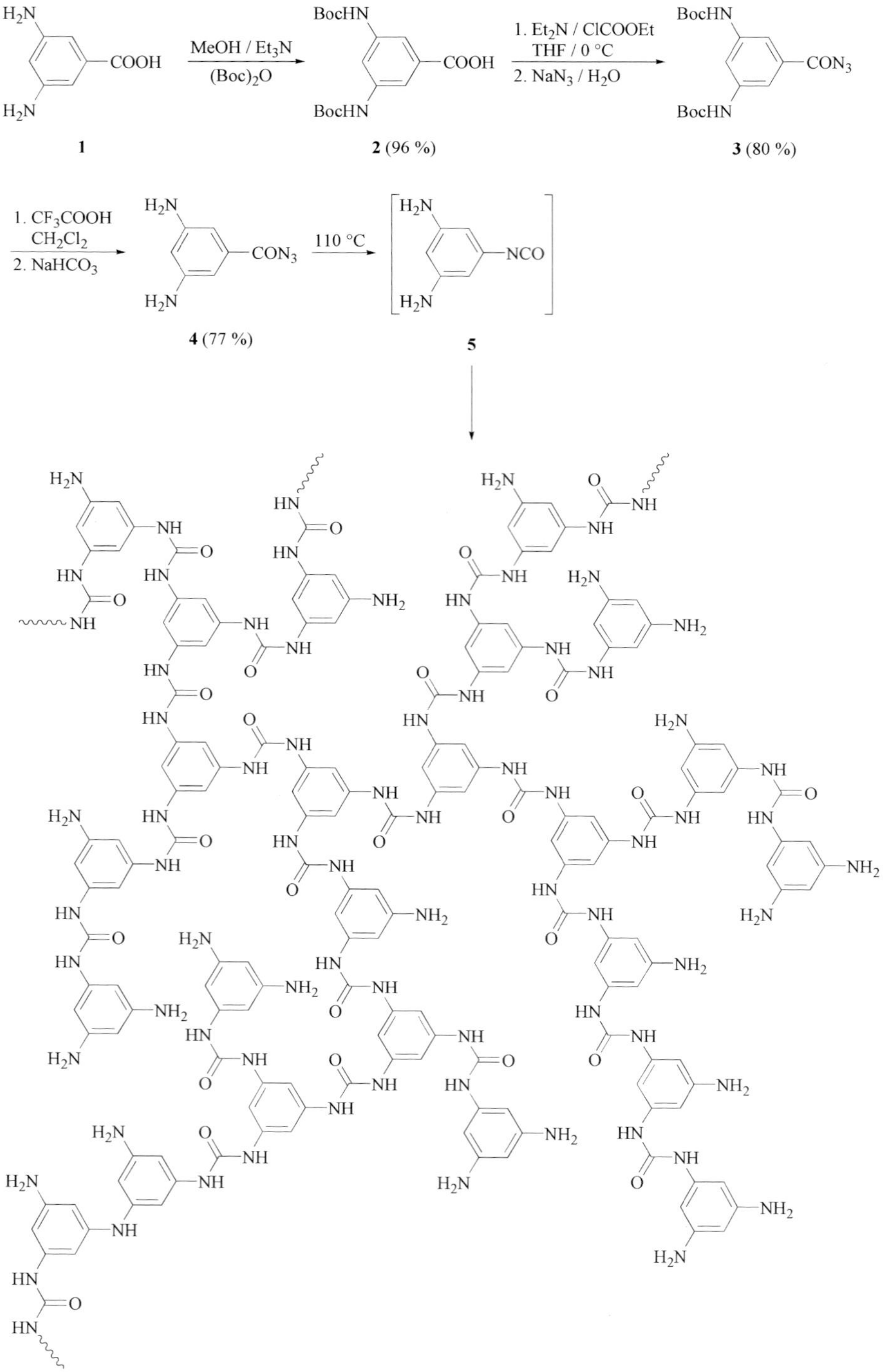

Figure 1. Synthesis of hyperbranched polyureas

$$OCN-R-NCO + 2\,H_2O \longrightarrow HOOC-NH-R-NH-COOH$$

$$\xrightarrow{-2\,CO_2} H_2N-R-NH_2$$

$$H_2N-R-NH_2 + OCN-R-NCO$$

$$\longrightarrow \left(\begin{array}{c} H\ \ \ \ \ O\ \ \ \ \ H\ \ \ \ \ H\ \ \ \ \ O\ \ \ \ \ H \\ -N-R-N-C-N-R-N-C- \end{array} \right)_n$$

The diamine resulting from reaction of water with the diisocyanate reacts with unconverted diisocyanate more rapidly than water, and polyureas are obtained in high yields. The reactivity of water with diisocyanates is similar to that of secondary alcohols [10]. This apparently simple reaction is, in fact, quite complicated:

R—NCO + H₂O
↓
(a) + RNCO → R-NH-CO-O-CO-NH-R → -CO₂ (a) → R-NH-CO-NH-R
↓
R-NH-COOH
(b) ↓ -CO₂   + RNCO (b)
R—NH₂
↓ + RNHCOOH
R-NH-COO⁻ RNH₃⁺

Route (a) is favored when the carbamic acid is stable and reactive toward the isocyanate [25]. Route (b) is favored when the carbamic acid is unstable and decomposes to an amine, which reacts quickly with the isocyanate. Route (c) is favored when the reaction between the carbamic acid and amine is faster than the reaction between the amine and isocyanate. The effect of substituents on the reaction of aromatic isocyanates with water is similar to that observed in the reaction of aromatic isocyanates and amines.

This simple reaction is employed in the manufacture of foams, in which the resulting carbon dioxide acts as a foaming agent. Different amines or amine complexes of metal salts are suitable catalysts for this reaction [26]. With tertiary amines, the following mechanism is proposed [27]:

$$R-N=C=O + \ |N\!-\! \longrightarrow R-N=C\!\begin{smallmatrix}O\\N-\end{smallmatrix} \xrightarrow{H_2O}$$

$$R-N=C\!\begin{smallmatrix}O-H\\N-H\end{smallmatrix} \longrightarrow \left[R-NH-COOH \right] + \ |N\!-\!$$

$$\downarrow -CO_2$$

$$R-NH_2$$

From Diamines and Carbon Dioxide or Its Derivatives. The polycondensation of diamines and carbon dioxide or its derivatives gives polyureas in high yields (X = halogen, OR, NH₂, etc):

$$R-NH_2 + \ \underset{X}{\overset{O}{\underset{\|}{C}}}-X \longrightarrow R-\overset{+}{N}H_2-CO-X \longrightarrow R-NH-CO-X + HX$$

$$R-NH_2 + \ \underset{X}{\overset{O}{\underset{\|}{C}}}-NH-R \longrightarrow R-\overset{+}{N}H_2-C(OX)-NH-R$$

$$\longrightarrow R-NH-CO-NH-R + HX$$

Polyureas are obtained by the polycondensation reaction of *carbon dioxide* and diamines in the presence of diphenyl phosphite in pyridine at 40 °C and 2 MPa. This reaction involves a carbamoyloxy-*N*-phosphonium salt of pyridine as intermediate [28]:

$$CO_2 + R-NH_2 + HO-P(OC_6H_5)_2 \xrightarrow{Py}$$

[pyridinium-P(OC₆H₅)(OH)-O-CO-NH-R] + R—NH₂ ⟶

$$R-NH-CO-NH-R + C_6H_5OH + (HO)_2POC_6H_5$$

Ethylene chlorophosphites with aliphatic structures are also suitable reagents for this reaction under mild conditions, the mechanism of which involves a four-center transition state [29]:

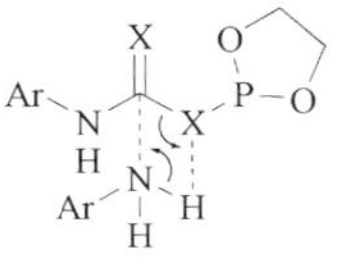

$$\begin{array}{c} Ar-N-C(X)(X-P-O) \\ H \quad N \\ Ar \quad H \end{array} \quad X = O, S$$

Without catalysts, the reaction of carbon dioxide with diamines requires high temperatures and high pressures [30].

Polycondensation of *phosgene* (carbonic acid dichloride) with diamines gives polyureas [31]. Similarly, *bis(carbamyl chloride)s* prepared from diamines and phosgene, can react with diamines of similar or different structures to afford polyureas (a) or alternating copolyureas (b), respectively:

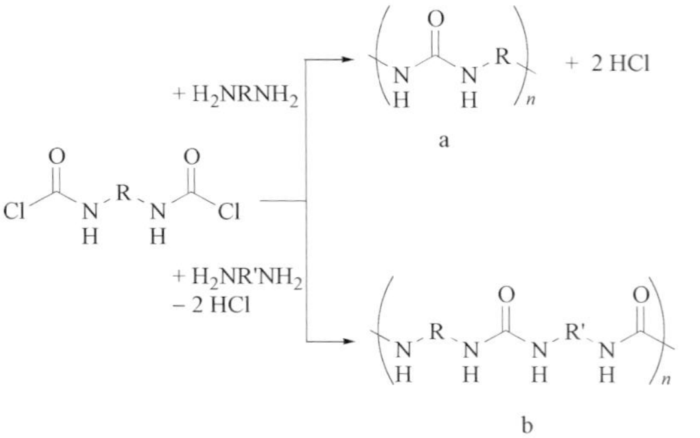

[Reaction scheme showing bis(carbamyl chloride) Cl-CO-NH-R-NH-CO-Cl reacting with H₂NRNH₂ to give polyurea (a) + 2 HCl, or with H₂NR'NH₂ − 2 HCl to give alternating copolyurea (b)]

From aliphatic or aromatic diamines and *urea* at high temperatures, polyureas result with the evolution of ammonia. This reaction can be carried out in the melt [32] or in solution [33].

Diphenyl carbonate does not react with aromatic amines in boiling toluene, but the addition of MgCl₂ as catalyst in pyridine makes this reaction possible at 120 °C [34]. The same reaction is also possible in phenol or cresol, but at 210 – 230 °C [35]. Activated *carbonic esters*, such as bis(p-nitrophenyl)carbonate or bis(2,4-dinitrophenyl)carbonate, can react with diamines at low temperatures to afford polyureas with high viscosities [36]. The rate of reaction is higher for aliphatic diamines than for aromatic diamines [37].

By Other Methods. Diamines can react with N,N'-carbonyldiimidazole in THF under reflux to give polyureas [38]. Polyureas with moderate molecular masses were prepared by the condensation reaction of lithium carbonate with aromatic diamines in presence of triphenyl phosphite and hexachloroethane in pyridine under reflux [39]. By using the same reaction, polyureas with high thermostability result from 2,8-diaminobenzothiophene or 2,8-diaminophenoxanthine and lithium carbonate [40].

A simple and ingenious method involves a controlled decomposition reaction of sebacoyl diazide in water to form poly(n-octylurea) [41].

In an interesting reaction, palladium diacetate catalyzes oxidative polymerization reactions of aliphatic diamines $H_2N(CH_2)_nNH_2$ ($n = 4, 6, 8, 10, 12$) with carbon monoxide in presence of potassium carbonate, triphenyl phosphite, and iodine to afford polyureas [42].

Poly(N,N'-diacylurea)s have been obtained by polyaddition reactions of diamides to bis(N-acylisocyanate)s [43].

3. Polyureas as Starting Materials for Other Polymers

Polytriketoimidazolidines. Polymers with this structure in the macromolecular chains were prepared by reactions between various polyureas and oxalyl chloride in the presence of pyridine [44]; they decompose 70 – 100 °C higher than the corresponding polyureas.

Polyhydantoins. Aromatic polyhydantoins of high molecular mass have been prepared by a polyaddition/cyclocondensation reaction of diethyl *m*-phenylenebisiminoacetate and aromatic diisocyanates in cresol under reflux [45]:

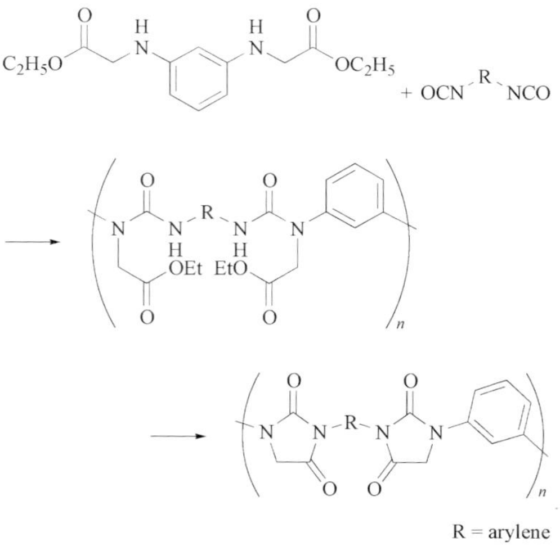

[Reaction scheme: C₂H₅O-CO-CH₂-NH-C₆H₄-NH-CH₂-CO-OC₂H₅ + OCN-R-NCO → intermediate with OEt groups → polyhydantoin]

R = arylene

The polyhydantoin obtained from methylenebis(4-phenylisocyanate) is stable in air up to 400 °C. Aspartate polyhydantoin prepolymers can react with blocked polyisocyanates to give products that are useful for coatings, adhesives, and elastomers [46].

Polyquinazolinediones. The reaction of 3,3′-benzidinedicarboxylic acid and diisocyanates with triethylamine as catalyst and NMP as solvent results in polyureas, which are cyclized by heating at 220 °C to polyquinazolinediones [47]:

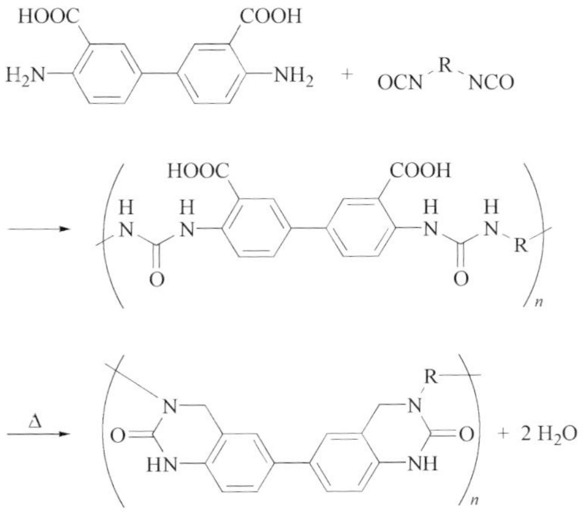

R = arylene

Polyhydrouracils can be obtained by the cyclization of 2-carbomethoxyethyl-substituted polyureas with polyphosphoric acid (PPA) [48]:

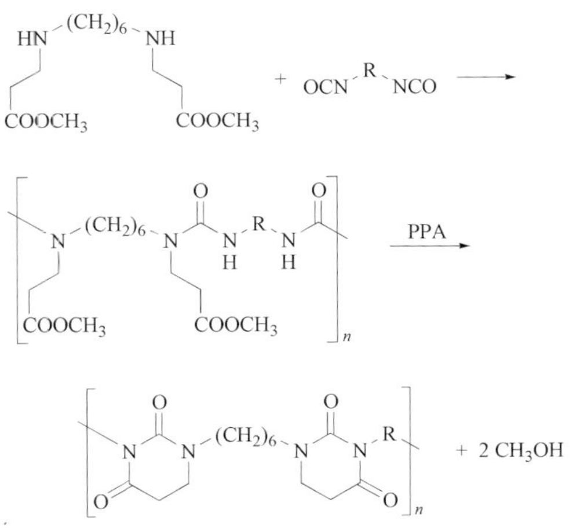

R = alkylene or arylene

Polyhydrouracils decompose at 400 °C; analogous unsubstituted polyureas decompose at 300 °C.

Other Polymers. Poly(imideurea) elastomers with high elasticity and improved thermostability can be obtained by reaction of polyureas (R^1NHCONHR^2NHCONH)$_n$ with aromatic tetracarboxylic dianhydrides (R^1 = aromatic or aliphatic diisocyanate residue; R^2 = oligomer residue containing amino group-terminated elastomer segment; $n = 2 - 500$) [49].

Heat-resistant polycarbodiimide copolymers for electric cable insulation can be prepared from urea copolymers and 3-methyl-1-phospha-3-cyclopentane-1-oxide as dehydrating agent [50]. The resulting copolymers have high thermostability (at 331 °C, losses are 5 %). Soft Cu conductors were coated with these polycarbodiimides with excellent results.

The end groups of aliphatic polyamides can react with polyureas at high temperatures (200 °C) to give copolymers that exhibit improved fiber dyeability [51].

4. Homopolyureas

Homopolyureas, [NHCOHN(R)]$_n$ have polar structures and high softening points. Homopolyureas with aromatic or heterocyclic structures have softening points higher than those with aliphatic backbones. The introduction of a symmetrical 1,4-phenylene ring into the polymer chain raises the softening point. *N*-Alkylation lowers softening points. Aliphatic homopolyureas have softening points of about 200 °C and they can be injection-molded, extruded, and processed like other thermoplastics.

5. Copolyureas

Copolyureas have the general formula [NH(R)NHCONH(R^1)NHCO]$_n$, where R and R^1 can be aliphatic, aromatic, or heterocyclic and contain phosphorus, sulfur, silicon, or halogens (simple copolyureas). Copolyureas may contain other functional groups, such as urethane, amide, imide, carbonate, etc.

5.1. Simple Copolyureas

Copolyureas Containing Silicon. Thermostable polysilylureas with good flexibility have been obtained from diamines $H_2N-R^1-NH_2$ and $O[SiR^2(CH_2)_nHCO]_2$ (where $n = 1 - 3$, R^1 = alkyl or aryl, and $R^2 = CH_3$ or C_6H_5) [52]:

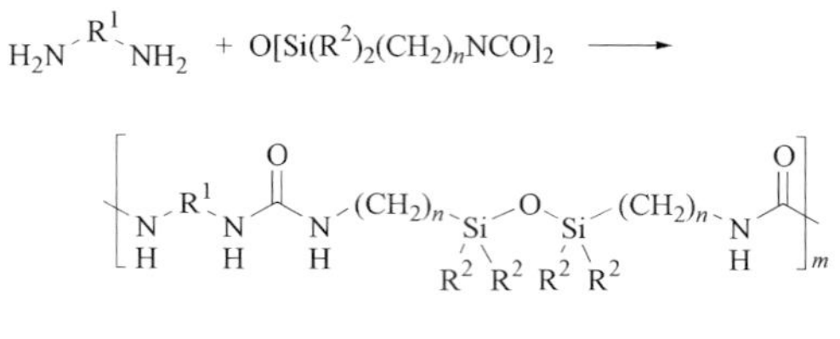

Also, from 1,3-bis(chloromethyldimethyl) siloxane and KOCN the corresponding Si-containing diisocyanate results, which reacts with diamines [53].

Copolyureas Containing Phosphorus. Polyphosphorylureas are prepared from aryl or alkyl phosphoric diisocyanates and diamines by interfacial polyaddition [54]:

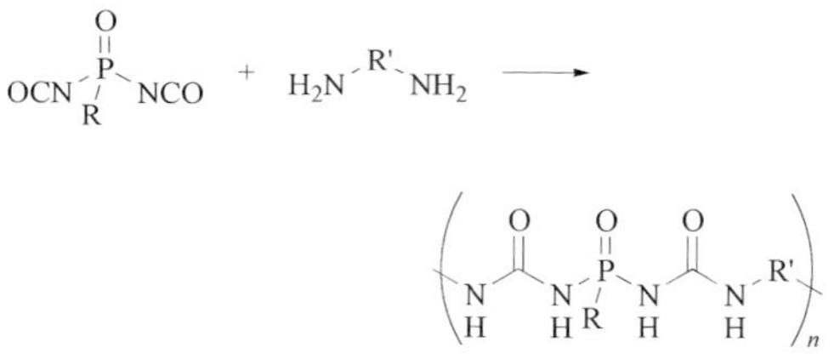

Polyphosphorylureas in which R is C_6H_5 and R' is m-C_6H_4 have a 10 % weight loss at 400 °C and a 40 % weight loss at 600 °C. Flame-retardant copolyureas useful for foams or films are prepared from 1-[(dialkoxyphosphinyl)methyl]-2,4-diaminobenzenes and diisocyanates in NMP at low temperature [55]. By similar methods, phosphorus-containing copolyureas have been prepared from phenyl bis(piperazidophosphate) and diisocyanates [56].

Copolyureas Containing Halogens. Heat-resistant, flame-retardant copolyureas are prepared from fluorinated primary diamines and diisocyanates [57]:

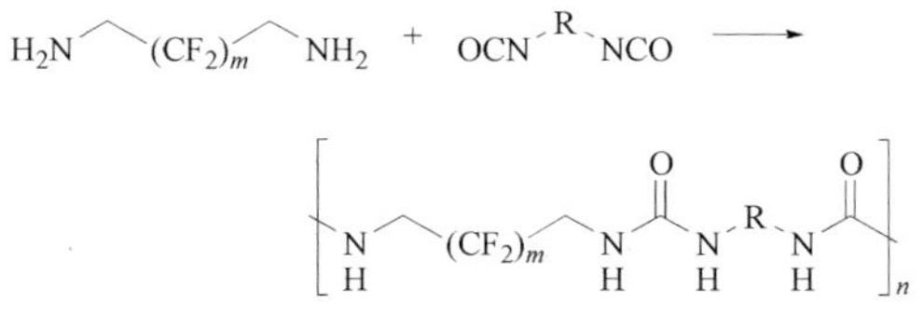

R = alkylene or arylene; $m = 2 - 4$

These copolyureas are useful for oil-resistant coatings and as curing agents for epoxides. Other halogen-containing diamines, such as 1,1,1-trichloro-2,2-bis(4'-chloro-3'-aminophenyl)ethane [58] or 1-(tetrafluoroethoxy)-2,4-phenylenediamine [59] can be used in reaction with diisocyanates, to afford flame-retardant copolyureas.

Copolyureas with Acid or Ester Groups. Polyurea electrolytes are prepared from sodium 3,3'-benzidinedicarboxylate and diisocyanates in the presence of sodium laurate as catalyst [60]:

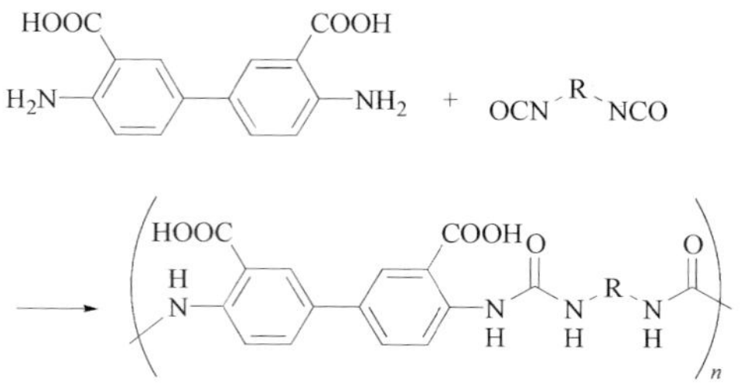

R = alkylene or arylene

From L-lysine diisocyanate methyl ester and optically inactive diamines, optically active copolyureas have been prepared [61]:

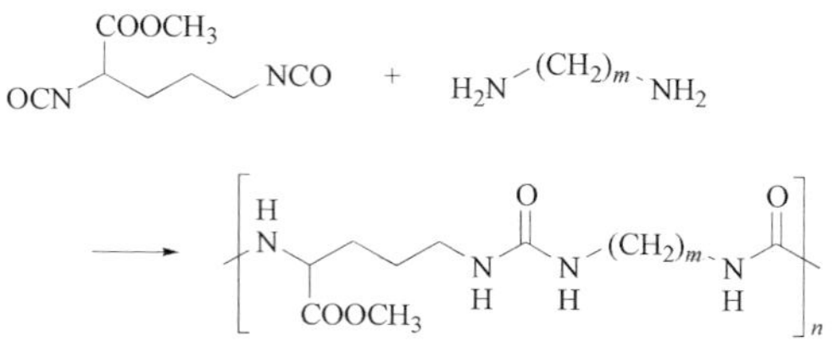

$m = 2 - 6$

Optical studies on these copolyureas revealed that they contain macromolecules with different conformations, ranging from disordered to helical [62]. Treating the polyureas with trifluoroacetic acid causes rupture of hydrogen bonds and reduces the ordering. These polymers can be degraded by chymotripsin or subtilisin [63]. An interesting method for the preparation of these copolyureas involves the polycondensation of L-lysine with 4,4′-dinitrophenylcarbonate [64].

Other Simple Copolyureas. Polycyanoureas with high molecular masses are prepared by a polyaddition of 3,3′-dicyano-4,4′-oxydianiline and diisocyanates in NMP with triethylamine as catalyst [65].

Thermostable copolyureas with oxadiazole rings in the main chain are prepared from bis[5-(3-aminophenyl)-1,3,4-oxadiazol-2-yl]alkanes and diisocyanates in DMF at 60 °C [66].

Polyverdazyl – polyureas are prepared from 1,4-bis[(1,5-diphenyl)-3-(4-aminophenyl)verdazyl-6-yl]butane and diisocyanates [67].

Triazine [68], carborane [69], and ferrocene [70] copolyureas also have high thermostability.

Copolyureas containing crown ethers (cryptates) are used as phase-transfer catalysts for alkylation, esterification, and oxidation. The copolyurea prepared from *trans*-Ar,Ar′-diaminodibenzo-18-crown-6 and TDI has a potassium ion absorption of 45 mg KOH per 100 g polymer, compared to 0.02 mg for the diamine monomer [71].

5.2. Copolyureas Containing Other Functional Groups

Poly(urethaneurea)s are the most important class of copolyureas. They can be prepared from a macrodiol (polyester diols, polyether diols, etc.) and an excess of a diisocyanate; this results in prepolymers with NCO end groups, which react with diamines (chain extenders). At the end of the reaction, a chain terminator (ROH, RHNH$_2$) is added [72], [73]:

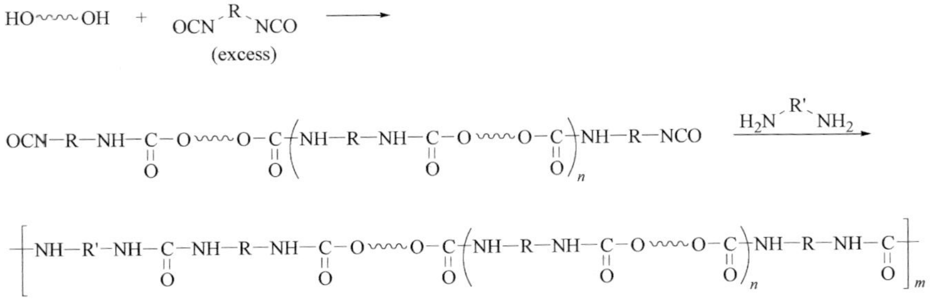

In the scheme presented above, the NCO – NH$_2$ polyaddition reaction is much more rapid than the NCO – OH polyaddition reaction and requires no catalysts. This rapid curing results in short production cycles.

Prepolymers with NCO end groups can also react with water, whereby the resulting carbon dioxide byproduct is used as a foaming agent in the production of poly(urethaneurea) foams. Poly(urethaneurea) copolymers have many practical applications as foams, cellular elastomers, fibers, aqueous dispersions, etc.

Optically active poly(urethaneurea)s were synthesized by solution polyaddition of (1*S*,2*S*)-(+)-2-amino-3-methoxy-1-phenyl-1-propanol with diisocyanates [74].

Poly(urethaneurea)s obtained from D-glucosamine and diisocyanates also have optical activity [75]. These polymers absorb 10 – 24 % water, whereas the poly(urethaneurea)s prepared from HO(CH$_2$)$_5$NH$_2$ and diisocyanates are almost non-hygroscopic; acetylation reduces the hygroscopicity.

Light-resistant poly(urethaneurea)s are prepared from 1,3-bis(isocyanatomethyl)adamantane and polyester diols with diamines as chain extenders [76].

Poly(urethaneurea)s with ionic links in the main chain were synthesized from MDI or 2,4-TDI and 4-hydroxybutyl phthalate as salts with divalent metal ions such as Ca^{2+}, Mn^{2+}, and Pb^{2+} [77].

Poly(AmideUrea)s. Aromatic poly(amideurea)s are prepared from 4-isocyanatobenzoyl chloride and diamines [78] and from 4,4′-diaminocarbanilide and R(COCl)$_2$ [79]:

HO~~~OH + OCN−R−NCO (excess) ⟶

OCN−R−NH−C(=O)−O~~~O−C(=O)−[NH−R−NH−C(=O)−O~~~O−C(=O)]$_n$−NH−R−NCO $\xrightarrow{H_2N-R'-NH_2}$

[−NH−R′−NH−C(=O)−NH−R−NH−C(=O)−O~~~O−C(=O)−(NH−R−NH−C(=O)−O~~~O−C(=O))$_n$−NH−R−NH−C(=O)−]$_m$

R, R′ = alkylene, arylene, heterocycle, etc.

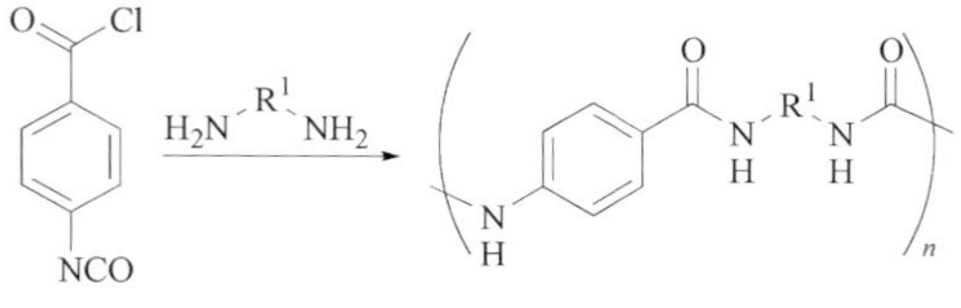

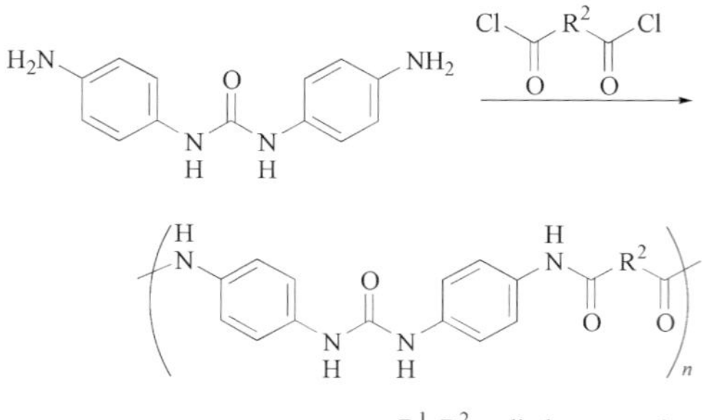

R¹, R² = alkylene or arylene

These poly(amideurea)s have excellent thermal stability and good solubility in aprotic dipolar solvents.

Thermostable poly(amideurea) acids can be prepared from 4,4′-diaminodiphenyl ether, pyromellitic dianhydride, and diisocyanates, which can be used as intermediates in the syntheses of poly(imideurea)s [80]:

Poly(imideurea)s with improved thermal stability are obtained from pyromellitic dianhydride and N,N'-dimethyl-N,N'-bis(4-aminophenyl) urea [81], or from the same dianhydride and $H_2N(NHCONHR)_mNH_2$, where R is $-(CH_2)_{12}-$; $-C_6H_4OC_6H_4-$, $-C_6H_4CH_2C_6H_4-$, or $-C_6H_4SO_2C_6H_4-$ [82]:

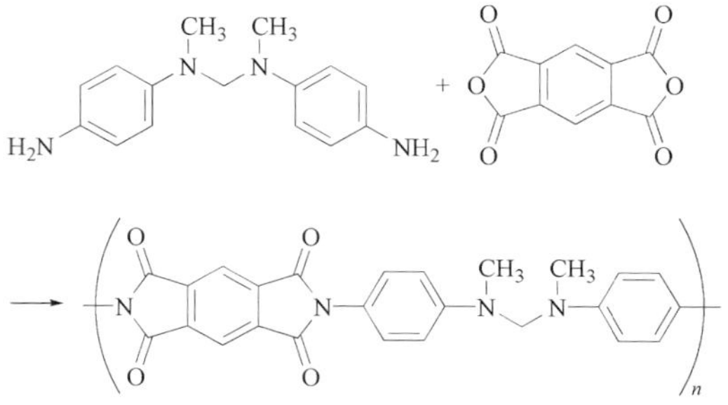

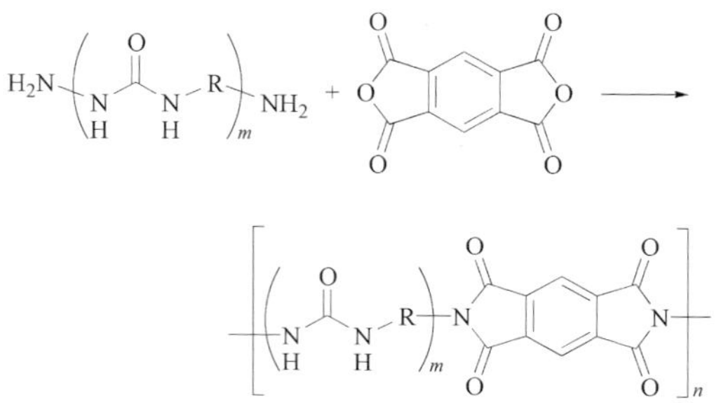

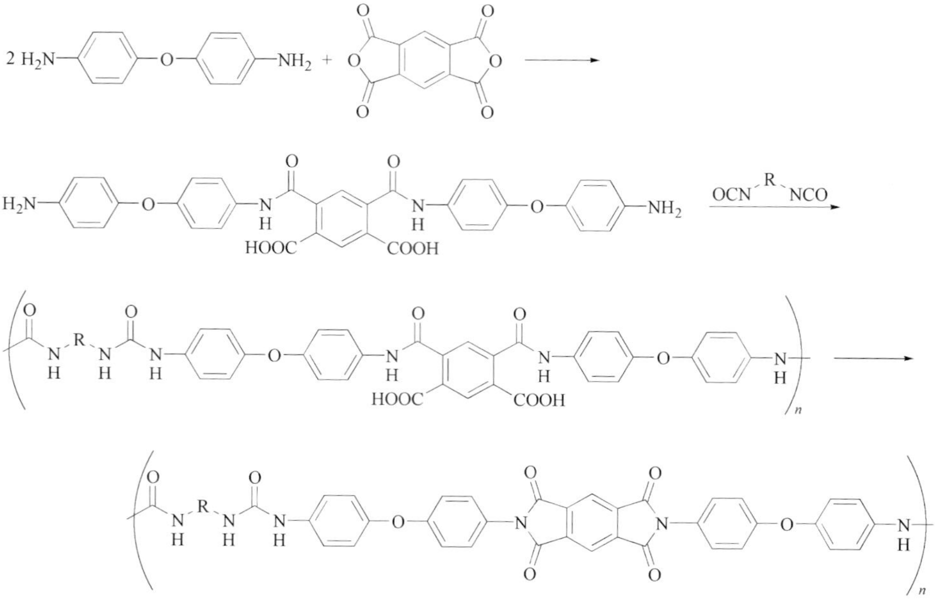

R = arylene

Sequential condensation can result in poly(imideurea)s with thermal stability up to 380 °C:

These poly(imideurea)s can be cyclized to polyiminoquinazolinediones.

Poly(carbonateurea)s can be prepared from bis-(4-aminophenyl) carbonate and diisocyanates (a) [83], or from *p*-isocyanatobenzoyl chloride and hydroquinone, which give a monomer with NCO end groups (b) that reacts with diamines [84].

R = arylene

These poly(carbonateurea)s have properties similar to these of aromatic polyureas, but exhibit higher solubility.

6. Structure

The structure of polyureas and copolyureas controls their chemical and mechanical properties.

Poly(urethaneurea)s contain hard and soft segments. Usually, the hard segments have an aromatic structure and the soft segments an aliphatic structure. During solidification from solution or melt, the hard and soft segments undergo phase separation, which results in elastomeric properties. The properties of poly(urethaneurea)s depend largely on the extent of phase separation, which is due to thermodynamic incompatibility [85]. Compositional variables, such as symmetry in the diisocyanate structure [86], chain extenders, aromatic or aliphatic diamines [87], molecular mass distribution of the soft and hard segments [88], crosslinking [89], and hydrogen bonding [90], have a strong influence on the extent of phase separation.

In each urethane or urea linkage, the NH and CO groups act as proton donor and proton acceptor, respectively. Since the urea group is more polar than the urethane group, the separation of hard and soft segments is more marked, the melting range of the hard segment domains is higher, and the intermolecular (interchain) interactions are stronger. The urea carbonyl groups are considered to be completely hydrogen-bonded [91]; this feature is probably responsible for the increased phase separation in poly(urethaneurea)s.

The presence of hard segments is responsible for the high tensile strength and modulus exhibited by the materials derived from these copolyureas. The soft segments determine the thermoplastic behavior.

Phase separation has been proved by electron microscopy [92], small angle X-ray scattering [93], differential scanning calorimetry [94], and other methods [95]. The same phenomenon was studied by advanced solid-state NMR spectroscopy; the sizes of various domains and the interfaces between them were quantified by spin measurements. Also, the impact of annealing, method of polymerization, and hard segment content on the properties of poly(urethaneurea)s were studied. Major tranformations have been observed after annealing at 190 °C, when the system behavior changes from "soft in hard" to "hard in soft". The highest microphase separation is observed in solution-polymerized samples. These NMR experiments reveal the nonequilibrium nature of RIM systems for poly(urethaneurea)s [96]. For the same system of RIM poly(urethaneurea)s, FT-IR spectrophotometric data suggested that an increase in degree of phase separation and a decrease in the size of hard-segment domains take place with increasing catalyst concentration [97].

Problems of structure arise in the field of polyureas and poly(urethaneurea)s expanded by water for foams. The reaction of isocyanate groups with water to give urea groups can be monitored by IR spectroscopy at 1646 cm^{-1} for carbonyl groups and 3300 – 3320 cm^{-1} for NH groups [98]. The absorption of the carbonyl group from polyureas is not restricted to the region around 1640 cm^{-1} and varies within wide limits according to the degree of association [99]. A band at 1695 cm^{-1} has been attributed to an unassociated urea group in polyureas [100]. The NH absorption at 3320 cm^{-1} indicates that most of the NH groups are hydrogen-bonded, because the free-NH absorption at 3445 cm^{-1} is negligible [91]. This polar structure can be represented by the following structures:

$$\left(\begin{array}{c} O \\ \parallel \\ N-R-N-C-N-R'-N \\ H \quad H \quad H \quad H \end{array} \right)_n \rightleftharpoons \left(\begin{array}{c} O^- \quad\quad O \\ | \quad\quad\quad \parallel \\ N-R-N=C-N-R'-N \\ H \quad H \quad\quad H \quad H \end{array} \right)$$

R, R' = arylene, alkylene, etc.

The high melting points, thermostability, and crystallinity of polyureas may be due to the partial polar structure.

The capacity for hydrogen bonding between urea groups and solvents is as follows: alcohols ≫ THF > CHCl$_3$ > toluene [101].

7. Properties

7.1. Physical Properties

Melting Temperature. The polar structure of polyureas results in high melting points that are higher than those of the corresponding

Table 3. Melting points of homopolyureas

Main chain	mp, °C	Prepared from	Ref.
Aliphatic	180 – 230	diamines and urea	[104]
Aromatic	260 – 320	diamines and carbon dioxide	[105]
Aliphatic/aromatic	200 – 240	aromatic/aliphatic dihalide and alkali metal cyanate	[106]
Naphthalenediyl	350 – 380	diamines and urea	[107]
Heterocyclic	330 – 350	diamines and diisocyanates	[108]

polyimides. Aliphatic polyureas melt at lower temperatures than aromatic polyureas. The melting temperature of aliphatic polyureas decreases with increasing number of methylene groups. Thus, nonamethylenediamine gives polyureas with melting temperatures of about 200 – 210 °C that can be injection-molded and extruded [102]. Alkylation or acylation of the nitrogen atoms lowers the melting points [103]. The copolyureas also melt at lower temperatures. For homopolyureas, melting point variations can be seen in Table 3.

The glass transition temperature T_g also depends on structure. Aliphatic polyureas usually have a T_g below 100 °C, whereas the T_g of aromatic polyureas is above 100 °C.

Solubility. Polyureas having polar structures, are difficult to dissolve in organic solvents. Aliphatic polyureas are soluble in phenol, cresols, acetic acid, and similar solvents. Aromatic polyureas are usually soluble only in aprotic dipolar solvents such as DMF, DMA, and NMP [109]. The symmetry or asymmetry of the aromatic monomers influences the solubility of the polyureas. Copolyureas, particularly aliphatic ones, are usually more soluble than polyureas. The solubility of poly(urethaneurea)s can be improved by reaction with trifluoroacetic anhydride to give polymers with =N–COCF$_3$ groups.

Water Absorption and Permeability. The water absorption of polyureas is low and depends on structure. Water absorption is increased by incorporation of heteroatoms, such as oxygen, sulfur, –COOH, or –SO$_3$H, whereas it decreases with increasing number of carbon atoms. Poly(urethaneurea)s derived from D-glucosamine and diisocyanates have a water absorption of 10 – 24 % [110]. The water permeability of polyurea films is higher than that of polyamide films.

7.2. Chemical Properties

Thermal Stability. Polyureas have a thermal stability higher than that of polyurethanes but lower than that of polyamides [111]. The thermal stability of the urea group is a complex issue. Thus, polyhexamethyleneurea ($mp = 300$ °C) can be heated to 260 °C without decomposition, whereas polydecamethyleneurea ($mp = 230$ °C) decomposes rapidly at 260 °C in the molten state. The mechanism of this thermal decomposition has been studied by mass spectrometry and involves a hydrogen transfer to give isocyanates and amines [112]:

N,N'-Disubstituted polyureas decompose at higher temperatures than unsubstituted polyureas because of the higher energy required for the C–H hydrogen transfer, which involves two competing decompositions [112]:

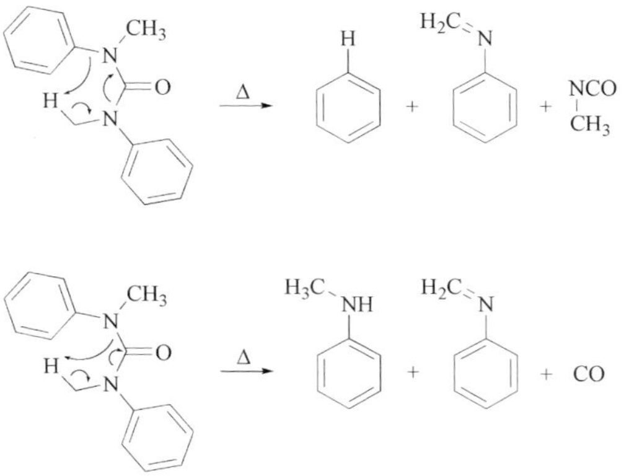

This six-centered transition state is indicated by a peak in the MS (mass spectrometry) spectrum that corresponds to methyl isocyanate [112]. The thermal decomposition of polymers containing a carbonyl group flanked by 1,4-piperazinediyl units involves hydrogen transfer from a methylene group to nitrogen [113]:

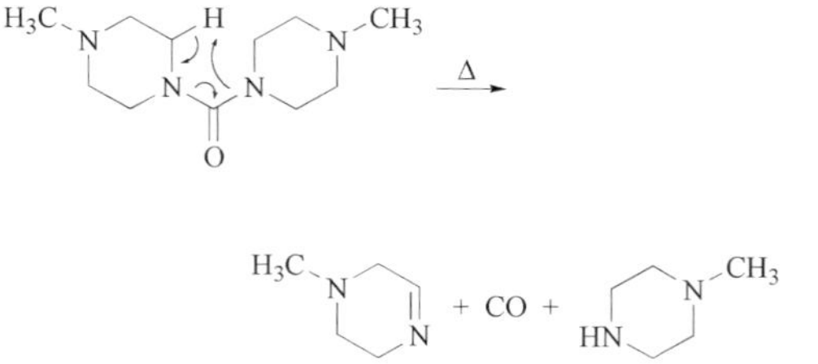

A method has been reported for the synthesis of isocyanates by thermal decomposition of polyureas that involves this interesting reaction [114].

Thermal stability of polyureas can be improved by copolymerization to give poly(phosphorylurea)s, poly(silylurea)s, poly(amideurea)s, poly(imideurea)s, etc. Thermal stabilities of N-substituted polyureas are higher than those of the corresponding unsubstituted polymers.

Hydrolytic Stability. Polyureas have excellent stability to hydrolysis under acidic and alkaline conditions. For example, polynonamethyleneurea fibers are more resistant to acid than nylon 6 or nylon 6,6; resistance to alkalis is higher than that of poly(ethylene terephthalate). In 40 % sulfuric acid or 40 % caustic soda solution at room temperature, these fibers lose less than 10 % of their strength after 10 h [115].

Improved resistance toward acids and bases has been reported for N-substituted polyureas prepared from N,N'-dialkylhexamethylenediamines and phosgene [116].

Biodegradation. Copolyureas containing amino acids or peptides in their chains are degraded by living organisms. Copolyureas containing dipeptides are degraded twice as fast as than those containing amino acids [117]. Similar polyureas are also degraded by chymotrypsin or subtilisin [118].

8. Uses

The urea linkages in poly(urethaneurea)s or polyureas extended with diamines cause an increase in phase separation compared to polyurethanes extended with diols. This phase separation improves various mechanical properties, such as elongation at break, hysteresis, and stress elongation [119].

8.1. Foams

Polyurethane and poly(urethaneurea) foams are generally divided into flexible, semiflexible, and rigid products, which can be produced by one-shot or by prepolymer methods.

In the one-shot method, the components are mixed directly, generally with simultaneous addition of auxiliaries such as catalysts, reinforcing agents, fillers, foam stabilizers, and flame retardants. These systems are exothermic.

In the prepolymer method, NCO prepolymers can be obtained from di- or polyhydroxy compounds, usually hydroxy-terminated polyethers or polyesters, which react with di- or polyisocyanates in excess. These NCO prepolymers are important compounds, because they can be cured with active hydrogen atoms from diamines, diols, water, etc. Many polyurethane and poly(urethaneurea) elastomers are produced via NCO prepolymer intermediates.

The curing of NCO prepolymers with diols is used in the production of polyurethane elastomers.

The same reaction, but with diamines, gives poly (urethaneurea) elastomers. With aliphatic amines (strong nucleophiles), this reaction is too fast to allow satisfactory mixing and a sufficiently long pot life. For this reason, less reactive aromatic diamines are used as chain extenders, in particular sterically hindered diamines or diamines whose nucleophilicity is weakened by electron-withdrawing groups.

Chain extension with diamines has advantages. For example, this curing process occurs rapidly and without additional heating; short production cycles are therefore possible. Usually, the NCO/NH$_2$ reaction ratio is about 1:1. Symmetrical diamines with deactivating substituents are used for elastomer synthesis, and alkyl-substituted toluenediamines, such as diethyltoluenediamine, are usually employed for reactions with NCO prepolymers in the RIM process.

If this reaction is carried out in polar organic solvents, practically any diamine can be used, even especially reactive aliphatic ones such as ethylenediamine.

The NCO/H$_2$O reaction (chemical blowing process) is a very important process for the production of foams. This chemical blowing process can be supplemented by physical blowing with low-boiling liquids as blowing agents, such as trichlorofluoromethane, which can be used to reduce the foam density further.

Organometallic compounds such as organotin compounds, in combination with tertiary amines, have a synergistic effect in the catalysis of the isocyanate – water reaction (Table 4) [120].

Linear or slightly branched polyols with a relatively high molecular mass (2000 – 8000) yield flexible foams. Combinations of these polyols with low molecular mass glycols or amines as chain extenders afford semiflexible foams. Highly branched polyols with a relatively low molecular mass (< 1000) yield rigid foams.

Polyurea foam materials are also prepared from diisocyanates with different structures by reaction with water. For flexible foams, linear or slightly branched prepolymers are used; more highly branched prepolymers produce rigid polyurea foams.

Flexible foams are prepared from diisocyanates, polyfunctional polyols, blowing agents, catalysts, and surfactants. Although various diisocyanates have been used for flexible foams, toluene-2,4-diisocyanate (TDI) is preferred. Modifiers, fillers, and plasticizers are used to confer special properties. The flexibility of these foams depends on the molecular mass of the polyesters and polyethers used. The ether group imparts flexibility, which, in turn, promotes softness, low melting point, low glass transition temperature, and elasticity. Surfactants promote intimate mixing of the components, regulate and control cell size, and impart stability. Many types are used, such as organic nonionic or ionic surfactants and silicones.

The properties of flexible foams also depend on the processing conditions (mixer speed, air-injection rate, etc.). Polyester-based flexible foams have higher tensile strength and elongation-at-break than polyether foams.

Flexible foams are characterized by reversible deformability and open-cell structure that is associated with air permeability. They have many applications, for example, mattress manufacture, automobiles, and aircraft construction.

Semiflexible foams are used almost exclusively as moldings and are 90 % open-celled. They have lower elasticity and good damping properties. In consequence, they are used as protective cushioning for control panels and consoles and for door and pillar paneling.

Rigid foams are prepared similarly to flexible foams but with different starting materials. In rigid foams, groups that restrict chain rotation, particularly aromatic rings, confer stiffness, which increases the melting point and hardness, and reduces elasticity.

Polyether resins used for rigid foams have lower molecular mass (300 – 1000) than those for flexible foams (3000 – 5000). Polyethers

Table 4. Effects of catalysts on relative reactivities of phenyl isocyanate with active hydrogen compounds in dioxane at 70 °C

Catalyst,	Relative reactivity with		
10 mol %	*n*-Butanol	Water	Diphenylurea
None	1.0	1.1	2.2
N-Methylmorpholine	40	25	10
Triethylamine	80	47	4
Tetramethyl-1,3-butanediamine	260	100	12
Triethylenediamine	1.20	380	90
Tributyltin acetate	80	14	8
Dibutyltin diacetate	600	100	12

with aliphatic/aromatic structure, prepared from aromatic diamines and propylene oxide, contribute to the stiffness of rigid poly(urethaneurea) foams.

The production of rigid poly(urethaneurea) foams from hydroxy-terminated resins and isocyanates requires auxiliaries such as blowing agents, catalysts, cell-size regulators, flame retardants, surfactants, and cross-linking agents. As with flexible foams, the reaction between diisocyanates and water generates CO_2, which serves as blowing agent. A halogenated hydrocarbon, such as trichlorofluoromethane, may also be employed, because this inert liquid with bp 23.8 °C is quickly volatilized by heating. The gas bubbles expand the foam and remain in the closed cells of the final product. In addition, CCl_3F has a low thermal conductivity and is nonflammable.

The density of rigid foams depends on the amount of gas trapped in the cells. A commercial rigid foam that contains 3 % solid polymer and 97 % gas has a density of about 16 kg/m^3. These foams are manufactured in such a way as to minimize convection currents [121]. Rigid foams are predominantly of the closed-cell type. For special uses, open-cell types are also produced.

Mechanical properties depend on the composition used in the formulation and the plastic nature of the foam at high temperature. Rigid poly(urethaneurea) foams provide excellent thermal insulation, especially when blown with fluorocarbons; high strength combined with light weight; heat resistance; and adhesion to wood, glass, ceramics, and fibers. The rigid foams can be used between − 200 and + 130 °C, and for short periods they can tolerate + 250 °C. As cross-linked plastics, the rigid foams are largely stable to chemicals, such as dilute acids and alkalis. The chief uses of rigid foams are in the heat and cold insulation field. They are used for insulation of refrigerators, including industrial refrigerators, freezers, and hot water accumulators, and as insulation for industrial tanks, containers, etc.

Polyurea foams are prepared by the reaction of diisocyanates and water; the mechanism is similar to that involved in the preparation of poly(urethaneurea) foams. In this reaction, tertiary amines are used as catalysts. Cu(II) and Zn(II) salts and tertiary amines exhibit a synergistic effect in the catalysis of the MDI − water reaction, which gives polyurea foams with satisfactory properties [122]. In these formulations, when N,N'-dimethylaminoethanol is used as catalyst, better results are obtained with copper nitrate as cocatalyst than with zinc nitrate. Flexible polyurea foams are used for cushions and in the automotive industry, where they provide light weight, high strength, and ease of fabrication. Rigid polyurea foams are used for refrigerators and in constructions.

8.2. Moldings

The production of polyurethanes, poly(urethaneurea)s, and polyureas by reaction injection molding (RIM) is now a well-established process. It offers several advantages for high-volume production of large parts: low pressure (< 343 kPa), low temperatures (54 − 60 °C) and use of reactive liquid intermediates [123]. Since the urethane polymerization process is highly exothermic, minimal heat input is required to maintain tooling temperature during production. With polyols, isocyanates, and extenders, different polymers can be formulated with a wide range of properties, from a flexible elastomer to a rigid plastic. The final product can be solid, microcellular, or foam-modified with a variety of fillers. Glass fiber reinforced RIM (RRIM) provides dimensional stability and high impact strength [124].

Poly(urethaneurea) and polyurea elastomers represent new classes of commercial RIM elastomers. They are obtained from diisocyanates and macrodiols or macrodiamines with hindered aromatic diamines as extenders. Diethyltoluenediamine (DETDA) is used in large quantities as an extender for these elastomers which have high thermostability [125]. Because of the high rate of urea formation, aliphatic diamine extenders have found limited application in RIM.

Poly(urethaneurea) and polyurea elastomers have many practical applications; the most important is in the automotive industry. To reduce the weight of automobiles, high-modulus RIM and RRIM materials suitable for external body panels have been developed from these elastomers. In addition, these elastomers are corrosion- and impact-resistant.

Because of their mechanical and thermal properties, these rigid foams are especially suitable for structural applications and can compete with wood, glass fiber reinforced plastics, and other materials.

High-modulus poly(urethaneurea) RIM elastomers are prepared from a prepolymer of MDI and diamine, using a system termed Bayflex 110 [126]. Some properties of Bayflex 110 follow:

Density	1042 kg/m^3
Flexural modulus	
at -30 °C	829 MPa
at 20 °C	359 MPa
at 65 °C	241 MPa
Tear strength	105 kN/m
Tensile strength	26.2 MPa
Elongation	275 %

These RIM elastomers, manufactured by Mobay Chemicals, have superior physical properties and can be released from the mold after 15 s with excellent green strength.

Polyurea RIM elastomers with high modulus can be prepared from a polyamine (Jeffamine D-2000), polyisocyanate, and an amine-terminated chain extender in an Accuratio V-R100 RIM machine, and released after 1 min. The resulting elastomer is cured at 121 °C for 1 h; properties are given below:

Tensile strength	23 – 28 MPa
Elongation	280 – 360 %
Tear strength	65 – 68 kN/m
Flexural modulus	
at 25 °C	148.2 – 148.9 MPa
at -29 °C	526 – 599 MPa
Heat sag	6.8 – 11.5 mm

The molding behavior of the RIM polyurethane depends on the catalyst [127]. The catalyst used for the isocyanate – water reaction can affect the physical properties of the final product. Because of yellowing under the influence of UV light, the moldings must be coated with a paint of high opacity, unless they are colored black or dark brown in the solid. In the photodegradation of MDI-based poly(urethaneurea) elastomers, photo-Fries products have been observed by UV and IR spectroscopy [128].

Polyols in polyurethane and poly(urethaneurea) elastomers increase the hygroscopicity and affect the physical properties. These elastomers are less stable at elevated temperature than polyurea elastomers. Polyisocyanates react slowly with polyols, which prevents automation of the process.

Advances in RIM technology include the development of new polyurea systems for the production of automobile body panels. These catalyst-free systems are based on amine-terminated polyesters. In automobile body panels, polyurea RIM materials offer greater impact resistance at equivalent glass-flake reinforcement; improvement in thermal stability for less distortion during paint cure cycles; and lower water absorption for better environmental resistance. These polyureas are usually prepared by the reaction of an amine – MDI-based isocyanate with an amine-terminated polyether (instead of the standard hydroxy-terminated polyether) and an aromatic diamine as chain extender.

Polyurea RIM elastomers present the following advantages:

- The autocatalytic reaction of isocyanate with amine is faster than the isocyanate – water reaction and results in high modulus polymers for automotive applications
- Catalysts are not necessary, and the process can be automated
- Decomposition and yellowing of the final product do not occur
- The higher thermostability extends their practical utility

Table 5 shows the development of RIM technology for polyurethane, poly(urethaneurea), and polyurea [129]. Texaco produces a polyether with amino end groups and a molecular mass of ca. 4000 (Jeffamine), which can be used to replace the polyols in polyurethanes and poly(urethaneurea)s. Mobay Chemicals investigated and established the optimum conditions for poly(urethaneurea) systems (Bayflex 110) and polyurea systems (Bayflex 150). Dow Chemicals investigated Spectrim HT systems; ICI developed the polyurethane Rubicon; and Admiral Equipment investigated machines with a high degree of automation adapted for a very fast RIM process.

The PHD (Polyharnstoff Dispersion) polyols constitute an important class of poly

Table 5. Development of polyurethane, poly(urethaneurea) and polyurea RIM elastomers

Parameter	Polyurethanes (1974–1975)	Poly(urethaneurea)s (1980–1984)	Polyureas (1985)	
			at 20 °C	at 70–75 °C
Productivity, kg/s	2.3–3.2	3.2–4.5	5.4–6.8	10.9–13.6
Cream time, s	4–8	2.5–3	2–2.5	1–1.5
Total production cycle*, s	180–420	180	60–70	60

*Rise time.

(urethaneurea)s with applications such as RIM elastomers, which were first prepared at Bayer by a polyaddition reaction in situ from diisocyanates and diamines in a polyol medium [130].

8.3. Fibers

Polyureas can be used for fibers because they have good physical and mechanical properties, melting points over 200 °C, low specific gravity, high stability toward acids and alkalis, and excellent dyeability. For example, polynonamethyleneurea has a melting point of 237 °C and a softening point of about 205–210 °C, which are suitable properties for fibers.

Polyureas with fiber-forming properties are also obtained by the reaction of $H_2N(CH_2)_nNH_2$ and $O[(CH_2)_4NCO]_2$. For $n = 4$, the resulting polyurea melts at 223–225 °C and is suitable for fibers [131].

A copolymer polyester – polycarbonate – polyurea is suitable for spandex fibers with good resistance to environmental degradation [132].

In an interesting continous solventless process, elastomeric filaments and fibers with a poly(urethaneurea) structure have been prepared by spinning in the gas phase from the corresponding monomers [133].

The addition of a polyurea to polyoxymethylene gives heat-stable transparent fibers with good dye affinity [134].

A polyester – cotton blend with improved luster and washfastness was prepared by impregnating the fabric with a diamine and then with a diisocyanate; calendering the fabric formed the polyurea in situ [135].

8.4. Films and Membranes

An interesting application of polyureas is for new nonlinear optical materials prepared as thin films from aromatic polyureas. For example, such thin films were prepared by simultaneous deposition of 4,4′-diaminodiphenylmethane and MDI. Investigations suggest that the poling process of polymerization of oligomers takes place at about 116 °C under a high electric field to give polar urea bonds with a large residual polarization. This polarization determines the pyroelectic and piezoelectric activities after poling treatment [136].

Similar results have been obtained for polyureas prepared from α,α'-dianilino-p-xylene and MDI, whose large second-order optical nonlinearity coupled with their unique optical transparency in the UV region make them potentially useful materials for nonlinear optics [137].

Microporous films for artificial leather can be obtained from poly(urethaneurea)s, provided these materials have adequate mechanical properties and good permeability to water [138]. Large quantities of these polymers are used for artificial leather.

Poly(urethaneurea) membranes with high H_2S/CH_4 selectivity have been obtained [139].

Ultrathin polyurea membranes are employed for water desalination by reverse osmosis. The urea group in polyureas is hydrophilic and resistant to chemicals and organic solvents; thus, these polymers are suitable for ultrafiltration applications [140].

Polyurea films are useful as flexible materials impermeable to air and other gases and for construction of inflatable boats, balloons, and similar applications [141].

Polyisocyanates with the general formula $R(CH_2)_y(NCO)_x$, in which R = alkyl, $y = 0$ or 1, and $x = 2 - 4$, react with polyamines; the resulting polyurea films have good adhesion to steel and aluminum and high resistance to chemicals and solvents [142].

Semipermeable membranes with high salt rejection have been prepared from polyureas [143].

The polyurea films produced by an interfacial polymerization reaction between water and a triisocyanate facilitate the formation of stable emulsions [144].

8.5. Coatings

Coating materials can be made from polyureas or poly(urethaneurea)s in solution or emulsion. A new spraying technology has many practical applications; polymers thus applied show excellent adhesion to a variety of substrates, including sand-blasted or primed steel, aluminum, and concrete. The products include two-component sprayed polyurea elastomers, prepared from amine-terminated polyether resins and polyisocyanates. The polyurea spray elastomers systems require no catalyst and have extremely high reactivity and cure [145]. In other applications, aqueous polyurea dispersions can be used for coatings with improved hardness and solvent resistance [146].

8.6. Biomedical Applications

Several polyurea and poly(urethaneurea) materials exhibit blood compatibility and have found applications in medical equipment and artificial organs [147]. The polymer membranes in pneumatic blood pumps [148] should have high tensile strength, low permanent deformation, low hydrolyzability, and high purity.

The hydrolytic stability of poly(urethaneurea)s in extracellular solutions was studied. After 50 d, failure was observed. After 120 d, resistance decreased by 15 – 20 % due to hydrolytic degradation of the polymers. The rate of hydrolysis was higher in extracellular solutions than in water [149]. Poly(esterurethaneurea)s are not suitable for medical applications because of their hydrolytic instability caused by the ester group.

Cross-linkable polyureas have been prepared for contact lenses [150].

8.7. Other Applications

Polyureas are used as thickeners for lubricating oils; the resulting lubricating greases have long lifetime at high temperatures, high oxidation resistance, good corrosion protection, high structural stability, and good low-temperature properties [151].

Polyureas and poly(urethaneurea)s can be used for the microcapsulation of herbicides, inks, colorants, medicines, and other products [153], [153].

Polyureas with carboxylic or sulfonic groups are good ion exchangers [154].

Polyureas are used for the removal of oils and bitumen from wastewater by filtration [155].

Petroleum production from oil wells is increased by injection into the borehole of a polyamine followed by a polyisocyanate. These react in situ and form polyureas that contribute to the consolidation of the oil sand and prevent clay swelling [156].

Trimellitic anhydride, polyisocyanates, and polyamines give electric insulator compositions for Cu wire with excellent flexibility and resistance to heat, abrasion, solvents, and thermal shock [157].

9. Safety and Environmental Aspects

Due to the wide range of applications of polyureas, copolyureas, and especially poly(urethaneurea)s, millions of people come into contact with them.

Many years' experience in handling the raw and auxiliary materials for these polymers has shown that these products can be handled safely. Air extraction systems must be present at all workplaces where TDI is processed at room temperature or MDI at temperatures above 40 °C. Aminopolyethers are moderately irritating to the eyes and skin; hence, contact with them should be avoided. The same applies to tertiary amines used as catalysts.

Since the carbon dioxide formed in the reaction of water with isocyanates is a natural constituent of air, it presents no problem. The other blowing agents are discharged into the outside air.

Wastes for which no utilization is possible can be disposed of in incineration plants. Granulated RIM wastes can be molded by the flow molding process at high pressure and high temperature to give new moldings. Fine-ground wastes can be used as fillers in production.

The glycolysis of reinforced reaction injection moldings (RRIM) of these polymers with diethylene glycol or dipropylene glycol at high temperatures can be used for the recovery of polyols and aromatic polyamines as suitable products for new RIM [158].

Organic isocyanates can be prepared by thermal decomposition of polyureas at 220 °C [114]. Another route is to utilize the energy by incineration of these polymers, but the calorific yield is small in comparison with other polymers, for example, polyethylene.

References

Specific References

1. Institut de Petrole, des Carburants et Lubrifiants, DE-O 1 946 942, 1968 (J. C. Silion, G. B. Gaudemaris); *Chem. Abstr.* **72** (1970) 123035j.
2. DuPont, US 2 888 438, 1959; 2 975 157, 1961 (M. Katz).
3. S. V. Joshi, A. Rao, *J. Appl. Polym. Sci.* **28** (1983) 1457.
4. *Mater. Plast. Elastomeri*,(1984)April, 217.
5. *Eur. Chem. News* (1985)Oct., 21.
6. J. T. Davis, F. Ebersole, *J. Am. Chem. Soc.* **56** (1934) 885.
7. R. Scholl, K. Holdermann, *Ann.* **345** (1907) 376.
8. R. Craven, *Polym. Prepr. (Am. Chem. Soc., Div. Polym. Chem.)* **16** (1956)no. 2, 3.
9. J. Baker, D. N. Bailey, *J. Chem. Soc.* (1957) 4649.
10. C. A. Naegeli, L. Conrad, *Helv. Chim. Acta* **21** (1938) 1127.
11. BASF, EP-A 896 010, 1999 (D. T. Ulrich).
12. Y. Watabe, M. Ishi, Y. Iseda, *J. Appl. Polym. Sci.* **25** (1980) 2747.
13. E. V. Scriven, *J. Chem. Soc. Rev.* **12** (1983) 136.
14. R. B. Moodie, *J. Chem. Soc. Perkin Trans.* **2** (1981) 664.
15. S. I. Axelrood, C. W. Hamilton, *Ind. Eng. Chem.* **53** (1961) 889.
16. F. Millich, C. Carraher: *Interfacial Synthesis*, vol. **2**, Marcel Dekker, New York 1977.
17. G. S. Kumar, *Macromolecules* **17** (1984) 2463.
18. O. Bayer, *Angew. Chem.* **59** (1947) 263.
19. D. J. Williams, *Angew. Chem., Int. Ed.* **23** (1984) 690.
20. D. P. Loker, *Polym. Prepr. (Am. Chem. Soc., Div. Polym. Chem.)* **39** (1998) 579.
21. A. Ulman, D. J. Williams, C. S. Willand, *J. Am. Chem. Soc.* **112** (1990) 7083.
22. N. Nishi, S. Tokura, *Makromol. Chem.* **192** (1991) 1881.
23. A. Kumar, E. W. Meijer, *Polym. Prepr. (Am. Chem. Soc., Div. Polym. Chem.)* **39** (1998)no. 2, 619.
24. C. J. Hawker, P. Farrington, *J. Am. Chem. Soc.* **117** (1995) 6123.
25. J. A. Saunders, K. C. Frisch: *Polyurethane, Chemistry and Technology, Part 1: Chemistry*, Wiley Interscience, New York 1962.
26. P. Merkaert, R. Jerome, *J. Appl. Polym. Sci.* **27** (1982) 4221.
27. J. M. Borsus, Ph. Tayssie, *J. Appl. Polym. Sci.* **27** (1982) 4029.
28. N. Yamazaki, T. Iguchi, F. Higashi, *J. Polym. Sci., Chem. Ed.* **13** (1975) 785.
29. C. I. Chiriac, *Polym. Bull.* **15** (1986) 65.
30. T. Lieser, H. Gehlen, *Liebigs Ann. Chem.* **556** (1944) 127.
31. DuPont, US 2 816 879, 1957 (E. L. Wittbecker).
32. H. Iiyama, M. Asakura, K. Kimoto, *Kogyo Kagaku Zasshi* **68** (1966) 236.
33. G. S. Kolesnikov, *Vysokomol. Soedin., Ser. A* **11** (1969) 2682.
34. N. Yamazaki, F. Higashi, *J. Polym. Sci., Chem. Ed.* **17** (1979) 835.
35. DuPont, US 2 190 770, 1940 (W. H. Carothers).
36. SU 905 228, 1980 (R. D. Katsarava); *Chem. Abstr.* **97** (1982) 24396n.
37. A. M. Schmuki, W. U. Suter, *Macromolecules* **18** (1985) 823.
38. W. R. Grace, FR 1 299 698, 1962.
39. N. Ogata, M. Watanabe, *J. Polym. Sci., Polym. Lett. Ed.* **24** (1986) 66.
40. M. Srinivasan, *Proc. Matl. Symp. Therm. Anal.* **9** (1993) 167; *Chem. Abstr.* **122** (1995) 315230s.
41. L. Takarsewski, J. Ossowski, W. Pietranek, *Polimery (Warsaw)* **39** (1994) 738; *Chem. Abstr.* **123** (1995) 56774f.
42. S. Ayusman, *Polym. Prepr. (Am. Chem. Soc., Div. Polym. Chem.)* **33** (1992)no. 2, 180; *Chem. Abstr.* **120** (1994) 31397k.
43. S. Kanamaru, T. Takata, *Macromolecules* **27** (1994) 7492.
44. G. Caraculacu, E. Scortanu, A. Caraculacu, *Eur. Polym. J.* **19** (1983) 143; **34** (1998) 1265.
45. Y. Imai, *J. Polym. Sci., Part A-1* **5** (1967) 2289.
46. Bayer, EP-A 743 332, 1996 (J. E. Haakan); *Chem. Abstr.* **126** (1997) 47677m.
47. Toyo Rayon, JP 69 19 556, 1969 (N. Yoda, M. Kurihara).
48. E. Dyer, *J. Polym. Sci., Part A-1* **7** (1969) 833.
49. Asai Shokusan Y. K., JP 11 106 507 20, 1999 (H. Okamoto); *Chem. Abstr.* **130** (1999) 313020q.
50. Mitsui Toatsu Chemicals, JP 08 259 603, 1996 (M. Murata); *Chem. Abstr.* **126** (1997) 48436u.
51. DuPont, US 16 475, 1997 (M. Teasley); *Chem. Abstr.* **127** (1997) 34716u.
52. SU 590 962, 1973 (N. P. Smetankina, N. N. Laskovenko); *Chem. Abstr.* **95** (1981) 8239s.
53. N. Laskovenko, *Vysokomol. Soedin.* **66** (1993) 203; *Chem. Abstr.* **119** (1993) 181375v.
54. C. E. Carraher, C. W. Krueger, *Makromol. Chem.* **133** (1970) 219.
55. J. A. Mikroyannidis, *J. Polym. Sci., Polym. Chem. Ed.* **22** (1984) 3423.
56. Texaco Belgium, GB 1 514 511, 1978 (C. J. Hermans).
57. Aerojet General Corporation, US 5 637 772, 1997 (A. Malik, P. Carlson); *Chem. Abstr.* **127** (1997) 95718t.
58. T. Lesiak, T. Nowakovski, *J. Prakt. Chem.* **321** (1979) 921.
59. G. S. Goldin, *Vysokomol. Soedin., Ser. B* **12** (1970) 307.
60. Toyo Rayon, JP 69 19 555, 1968 (N. Yoda, M. Kurihara).
61. T. Yasugawa, H. Yamaguchi, *J. Polym. Sci., Polym. Chem. Ed.* **17** (1979) 3387.
62. V. V. Korshak et al., *Vysokomol. Soedin., Ser. A* **22** (1980) 41.
63. S. J. Huang, V. K. Leong, *Polym. Prepr. (Am. Chem. Soc., Div. Polym. Chem.)* **20** (1979) 552.
64. R. D. Katsarava, *Makromol. Chem.* **194** (1993) 3209.
65. N. N. Barashkov, E. N. Teleshov, *Vysokomol. Soedin., Ser. A.* **20** (1978) 2749.
66. A. P. Grecov, K. H. Kornev, *Khim. Geterotsikl. Soedin.* **5** (1967) 806; *Chem. Abstr.* **68** (1968) 22287e.
67. F. A. Neugebauer, H. Trischmann, *J. Polym. Sci., Part B* **6** (1968) 255.
68. T. Yuki, S. Sakurai, T. Noguchi, *Bull. Chem. Soc. Jpn.* **43** (1970) 2130.
69. SU 569 584, 1977 (V. V. Korshak, N. I. Bekasova); *Chem. Abstr.* **87** (1977) 185488e.
70. K. Gonsalves, D. Marvin, *J. Am. Chem. Soc.* **106** (1984) 3862.

71. Nippon Soda Co., Ltd.,JP Kokai 77 109 593, 1977 (H. Okamura).
72. DE 120 107, 1976 (K. Walter, H. Mey); *Chem. Abstr.* **86** (1977) 73692n.
73. R. S. Ward, *J. Appl. Polym. Sci.* **27** (1982) 2167.
74. C. Jun, *J. Polym. Sci., Part A: Polym. Chem.* **31** (1993) 1719; *Chem. Abstr.* **119** (1993) 117957j.
75. Y. Iwakura, *Makromol. Chem.* **178** (1977) 2939; **180** (1979) 2331.
76. SU 647 313, 1979 (A. P. Khardin, N. G. Gureev); *Chem. Abstr.* **90** (1979) 205549r.
77. R. A. Prasath, *Macromolecules* **31** (1998) 110.
78. K. Hayashi, S. Hang, *Makromol. Chem.* **86** (1965) 64.
79. Firestone Tire and Rubber Co., US 4 035 344, 1977 (W. J. Spiewak).
80. V. P. Telesov, E. N. Provednikov, *Vysokomol. Soedin., Ser. A* **20** (1978) 1586.
81. Y. Imai, K. Masaaki, *Makromol. Chem. Phys.* **199** (1998) 457.
82. Société Rhodiaceta, DE-O 1 949 281, 1970 (K. Grundschober, C. Carvini); *Chem. Abstr.* **73** (1970) 15534h.
83. O. V. Smirnov, *Vysokomol. Soedin., Ser. A* **11** (1961) 2211; *Chem. Abstr.* **72** (1970) 32287b.
84. Mitsui Toatsu Chemicals, Inc., JP Kokai Tokyo Koho JP 82 40 452, 1982.
85. C. S. Paik Sung et al., *Macromolecules* **12** (1979) 538.
86. C. S. Paik Sung: *Polymer Alloys*, Plenum Press, New York 1977, p. 261.
87. C. S. Wu et al., *Polym. Prepr. (Am. Chem. Soc., Div. Polym. Chem.)* **19** (1978)no. 2, 679.
88. R. W. Seymour, S. C. Cooper, *Polymer* **14** (1973) 255.
89. V. V. Shilov, *J. Appl. Polym. Sci.* **29** (1984) 1912.
90. J. L. Blackwell, K. H. Gardner, *Polymer* **20** (1979) 13.
91. M. Shibayama et al., *Polym. J.* **18** (1986) 719.
92. C. Dequartre, J. P. Pascault, *Angew. Makromol. Chem.* **72** (1978) 11.
93. S. L. Samuels, G. L. Wilkes, *J. Polym. Sci., Part C* **43** (1973) 149.
94. Y. Camberlin, *J. Polym. Sci., Polym. Chem. Ed.* **21** (1983) 415.
95. G. Wilkes: *Adv. Chem.*, 1979, Ser. **176** (Multiphase Polym.) 53; *Chem. Abstr.* **90** (1979) 204991s.
96. S. Lehmann, A. Donald, *J. Polym. Sci., Part B* **36** (1998) 693.
97. L. Ming, S. Kahn, *J. Polym. Sci., Part B* **35** (1997) 865.
98. F. E. Bailey, *Proc. IUPAC, Macromol. Symp.* **28** (1982) 100.
99. J. Hocker, *J. Appl. Polym. Sci.* **25** (1980) 2879.
100. K. Saito, *J. Macromol. Sci. Phys.* **10** (1974) 591.
101. L. Ming, *Polym. Mater. Sci.* **77** (1992) 562; *Chem. Abstr.* **127** (1997) 234890h.
102. Schering, NL-A 6 614 989, 1967; *Chem. Abstr.* **67** (1967) 65023n.
103. S. Caruso et al., *J. Polym. Sci., Polym. Chem. Ed.* **20** (1982) 1690.
104. P. Bornev, R. Paseday, *Makromol. Chem.* **101** (1967) 1.
105. N. Yamazaki, S. Nakahama, *Polym. Prepr. (Am. Chem. Soc., Div. Polym. Chem.)* **20** (1979)no. 1, 146.
106. Marathon Oil, FR 1 545 708, 1968 (A. P. Aragabright, J. V. Sinkey).
107. G. S. Kolesnikov, *Vysokomol. Soedin., Ser. B* **9** (1967) 242.
108. V. P. Telesov, *Vysokomol. Soedin., Ser. A* **20** (1978) 1586.
109. F. Higashi, *Polym. Prepr. Jpn.* **30** (1981) 143.
110. K. Kurita, *Kenkya Hokoku-Asahi* **37** (1980) 59; *Chem. Abstr.* **94** (1981) 209259b.
111. M. Y. Ah, *Han'guk Somyu Konghakhoechi* **23** (1986) 130; *Chem. Abstr.* **105** (1986) 153623s.
112. G. Montando, D. Vitalini, *J. Polym. Sci., Polym. Chem. Ed.* **21** (1983) 3321.
113. S. Foti et al., *Anal. Chem.* **54** (1982) 671.
114. PCT Int. Appl. WO 98 54 129, 1998 (J. Bosman); *Chem. Abstr.* **130** (1999) 14324y.
115. R. W. Moncrieff: *Man-Made Fibers*, John Wiley & Sons Inc., New York 1963, p. 356.
116. Union Carbide, US 3 130 170, 1964 (R. J. Cotter).
117. T. E. Lipatova, *Vysokomol. Soedin., Ser. B* **26** (1984) 149.
118. S. J. Huang, K. W. Leong, *Polym. Prepr. (Am. Chem. Soc., Div. Polym. Chem.)* **20** (1979)no. 2, 552.
119. T. V. Smith, N. H. Sung, *Macromolecules* **13** (1980) 121.
120. K. A. Pigott: *Encyclopedia of Polymer Science and Technology*, vol. **11**, John Wiley & Sons, Inc., New York 1969, p. 524.
121. R. H. Harding, B. F. James, *Modern Plast.* **39** (1962) 133.
122. J. M. Borsus, R. Jerome, *J. Appl. Polym. Sci.* **26** (1981) 3027.
123. G. D. Lewis, *J. Elastomers Plast.* **14** (1982) 3.
124. S. J. Monte, G. Sugerman, *J. Elastomers Plast.* **14** (1982) 34.
125. R. J. G. Dominguez, *Polym. Prepr. (Am. Chem. Soc., Div. Polym. Chem.)* **25** (1984)no. 2, 293.
126. S. H. Mitzger, K. Seel, *J. Cell. Plast.* **17** (1981) 268.
127. R. E. Camargo et al., *Polym. Prepr. (Am. Chem. Soc., Div. Polym. Chem.)* **25** (1984)no. 2, 294.
128. C. E. Hoyle, H. Shah.*Report* (1993), Order No. Ad-A267308; *Chem. Abstr.* **123** (1995) 289207p.
129. *Polyplasti e Plastici Rinforzati* **343** (1986) May, 47.
130. Bayer, DE 39 001, 1963 (E. Müller).
131. Asahi Chemical Industry, JP 67 5437, 1963 (K. Satome, K. Sato).
132. DuPont, WO 98 25 986, 1998 (H. Shingo, T. Matsuda); *Chem. Abstr.* **129** (1998) 82725j.
133. US 5 616 675, 1997 (K. Wilkinson); *Chem. Abstr.* **126** (1997) 306357u.
134. Kurashiki Rayon, JP 69 00 914, 1969 (K. Tanabe, K. Matsubayashi).
135. Nitto Boseki, JP Kokai Tokyo Koho 80 107 578, 1980.
136. S. Wang, *Jpn. Appl. Phys., Part 1* **32** (1993) 2768; *Chem. Abstr.* **119** (1993) 73567a.
137. T. Watanabe, *Synth. Met.* **57** (1993) 3895; *Chem. Abstr.* **120** (1994) 55555m.
138. Farbenfabriken Bayer, GB 1 167 670, 1969, (D. Dieterich, A. Reischil).
139. G. Chaterjee, S. A. Stern, *J. Membr. Sci.* **135** (1997) 99; *Chem. Abstr.* **127** (1997) 294210w.
140. P. Zschocke, H. Stratmann, *Angew. Makromol. Chem.* **73** (1978) 1.
141. Vereinigte Seidenwebereien, DE-O 2 635 114, 1978; *Chem. Abstr.* **88** (1978) 137836p.
142. General Mills, NL-A 6 607 543, 1966; *Chem. Abstr.* **66** (1967) 96307s.
143. Nitto Denko, JP Kokai Tokyo Koho 04 338 225, 1992 (J. Kitamura, A. Ikeda); *Chem. Abstr.* **119** (1993) 50909z.
144. T. Mikami, *J. Dispersion Sci. Technol.* **14** (1993) 71; *Chem. Abstr.* **118** (1993) 82038i.
145. D. Primeaux, *Int. Symp. Exhib.* (1998) 224; *Chem. Abstr.* **129** (1998) 137175n.
146. Bayer Corp., US 5 569 706, 1996 (P. B. Jacobs); *Chem. Abstr.* **125** (1996) 331744e.
147. D. J. Lyman, *J. Polym. Sci., Polym. Lett. Ed.* **18** (1980) 411.
148. H. D. Stenzenberger, D. O. Hummel, *Angew. Makromol. Chem.* **82** (1979) 103.
149. T. K. Khlystalova, M. B. Pestova, *Mech. Kompoz. Mater. Zinatne* **6** (1985) 1096; *Chem. Abstr.* **104** (1986) 136041n.

150. Novartis, WO 99 14 253, 1999 (F. Stockinger); *Chem. Abstr.* **130** (1999) 252850m.
151. H. Zajezierska, S. Patan, *Nafta Katowice Pol.* **37** (1981) 280; *Chem. Abstr.* **96** (1982) 125818m.
152. Monsanto, EP-A 165 227, 1985 (B. G. Beestman).
153. Cassella, DE-O 3 333 654, 1983 (U. Greiner, K. Albert); *Chem. Abstr.* **102** (1985) 221898a.
154. NHK Spring, DE-O 2 807 861, 1977 (T. Iwashita); *Chem. Abstr.* **89** (1978) 216338b.
155. Davy Bamag, DE 2 748 652, 1979 (H. Kreuzer).
156. Armour Industrial Chemical, US 3 533 470, 1970 (B. E. Marsh, B. Mosier).
157. Furukawa Electric, JP Kokai Tokyo Koho 78 99 299, 1977 (K. Yamazaki).
158. Bayer, DE-O 4 217 024, 1992; *Chem. Abstr.* **121** (1994) 11220q.

Further Reading

K. Ashida: *Polyurethane and Related Foams*, CRC Press/Taylor & Francis, Boca Raton, FL 2007.

P. DuboisO. CoulembierJ. M. Raquez (eds.): *Handbook of Ring-Opening Polymerization*, Wiley-VCH, Weinheim 2009.

T. MeyerJ. T. F. Keurentjes (eds.): *Handbook of Polymer Reaction Engineering*, Wiley-VCH, Weinheim 2005.

A. ShuklaG. RavichandranY. Rajapakse (eds.): *Dynamic Failure of Materials and Structures*, Springer, New York, NY 2010.

A. A. Tracton (ed.): *Coatings Materials and Surface Coatings*, CRC Press, Boca Raton, FL 2007.

D. G. Weldon: *Failure Analysis of Paints and Coatings*, rev. ed., Wiley, Chichester 2009.

Z. W. Wicks, F. N. Jones, S. P. Pappas, D. A. Wicks: *Organic Coatings*, 3rd ed., Wiley-Interscience, Hoboken, NJ 2007.

Polyurethanes

NORBERT ADAM, Bayer MaterialScience AG, Leverkusen, Federal Republic of Germany

GEZA AVAR, Bayer MaterialScience AG, Leverkusen, Federal Republic of Germany

HERBERT BLANKENHEIM, Bayer MaterialScience AG, Leverkusen, Federal Republic of Germany

WOLFGANG FRIEDERICHS, Bayer MaterialScience AG, Dormagen, Federal Republic of Germany

MANFRED GIERSIG, Bayer MaterialScience AG, Leverkusen, Federal Republic of Germany

ECKEHARD WEIGAND, Bayer MaterialScience AG, Leverkusen, Federal Republic of Germany

MICHAEL HALFMANN, Bergische Universität Wuppertal, Wuppertal, Federal Republic of Germany

FRIEDRICH-WILHELM WITTBECKER, Bergische Universität Wuppertal, Wuppertal, Federal Republic of Germany

DONALD-RICHARD LARIMER, Bayer MaterialScience AG, Leverkusen, Federal Republic of Germany

UDO MAIER, Bayer MaterialScience AG, Leverkusen, Federal Republic of Germany

SVEN MEYER-AHRENS, Bayer MaterialScience AG, Leverkusen, Federal Republic of Germany

KARL-LUDWIG NOBLE, Bayer MaterialScience AG, Leverkusen, Federal Republic of Germany

HANS-GEORG WUSSOW, Bayer MaterialScience AG, Leverkusen, Federal Republic of Germany

1.	Introduction	1052
2.	Basic Reactions	1053
3.	Starting Materials	1055
3.1.	Polyisocyanates	1055
3.1.1.	Aromatic Polyisocyanates	1056
3.1.2.	Aliphatic Polyisocyanates	1057
3.1.3.	Blocked Isocyanates	1058
3.2.	Polyols	1058
3.2.1.	Polyether Polyols	1058
3.2.2.	Polyester Polyols	1059
3.2.3.	Polycarbonate Polyols	1060
3.2.4.	Other Polyols	1060
3.3.	Diamines and Amino-Terminated Polyethers	1060
3.4.	Special Building Blocks	1061
3.5.	Catalysts and Additives	1061
4.	Structure and Morphology	1063
4.1.	Polyurethanes Without Segmented Structure	1063
4.2.	Polyurethanes with Segmented Structure	1063
4.2.1.	Hard and Soft Segments	1063
4.2.2.	Segregation and Morphology	1064
4.3.	Cross-linking of Polyurethane	1065
4.4.	Polyisocyanurates	1065
5.	Production of Polyurethanes	1065
5.1.	Stoichiometry	1065
5.2.	Reaction without Solvents	1067
5.2.1.	One-Shot Process	1067
5.2.2.	Prepolymer Processes	1067
5.3.	Reaction in Solvents	1068
5.3.1.	One-Component Systems	1068
5.3.2.	Two-Component Systems	1068
5.4.	Reactive One-Pack Systems	1069
5.5.	Other Processes	1069
6.	Processing of Polyurethanes	1070
6.1.	Supply, Storage, and Preparation of Raw Materials	1070
6.2.	Metering and Mixing Technology	1070
6.3.	Processing Plants	1072
7.	Foams	1073
7.1.	Flexible Foams	1074

Ullmann's Polymers and Plastics: Products and Processes
© 2016 Wiley-VCH Verlag GmbH & Co. KGaA, Weinheim
ISBN: 978-3-527-33823-8 / DOI: 10.1002/14356007.a21_665.pub2

7.1.1.	Flexible Slabstock Foam	1076	8.	**Noncellular Polyurethanes**	**1093**	
7.1.1.1.	Raw Materials	1076	8.1.	**Cast Elastomers.**	**1093**	
7.1.1.2.	Production	1076	8.1.1.	Applications	1093	
7.1.1.3.	Properties	1078	8.1.2.	Production	1094	
7.1.1.4.	Trimming and Processing	1080	8.1.3.	Properties	1095	
7.1.1.5.	Applications	1080	8.2.	**Thermoplastic Polyurethane Elastomers (TPU)**	**1095**	
7.1.2.	Molded Flexible Foam	1081				
7.1.2.1.	Production	1082	9.	**Polyurethane Coatings**	**1096**	
7.1.2.2.	Molding Process	1082	10.	**Polyurethane Adhesives**	**1098**	
7.1.2.3.	Properties	1083	11.	**Polyurethane Fibers**	**1099**	
7.1.2.4.	Applications	1084	12.	**Polyurethanes and Isocyanates as Binders**	**1099**	
7.2.	**Semirigid Foams**	**1084**				
7.2.1.	Applications	1084	13.	**Special Products**	**1100**	
7.2.2.	Production	1084	14.	**Safety and Ecology**	**1101**	
7.2.3.	Properties	1085	14.1.	Safety Precautions when Handling the Raw Materials	1101	
7.3.	**Rigid Foams**	**1085**				
7.3.1.	Raw Materials	1085	14.2.	Emissions, Accidental Release, and Waste Disposal	1102	
7.3.2.	Processing	1086				
7.3.3.	Properties	1088	14.3.	Recycling/Recovery of Polyurethanes	1103	
7.3.4.	Special Types	1090				
7.4.	**Integral Skin Foams and RIM Materials**	**1090**	14.4.	Fire Performance of Polyurethanes	1104	
7.4.1.	Applications	1091	15.	**Economic Aspects**	**1105**	
7.4.2.	Production	1091		**References**	**1106**	
7.4.3.	Properties	1092				

1. Introduction

Polyurethane (PUR) is the collective name for an extensive group of polymers with very different compositions and correspondingly varied property profiles. All PURs are built on a common principle: they are produced by the polyaddition process [1, 2] of (poly)isocyanates typically with (poly)alcohols. The characteristic chain link is the urethane group. In most cases two or more closely positioned urethane groups link poly(alkylene ether) and/or polyester sequences with molecular masses between ca. 200 and 6000 g/mol. Thus, the urethane group usually is only present to a minor extent.

PURs also include polymers produced from polyisocyanates that have urea, isocyanurate, or carbodiimide groups as their characteristic structural unit. As a result of the increasing importance of di- and polyamines and water as reactants, most of the industrially produced PURs contain urea groups (including the isocyanurates as cyclic ureas) as their main property-determining structural units.

A PUR is usually understood to be the fully reacted finished product (e.g., foams and coatings). These products are free from monomeric isocyanates, and unconsumed isocyanate groups, which may be present in small amounts, are bound to the PUR matrix. In adhesives and paints, the isocyanate-containing precursors are often called "PUR adhesive," "PUR hardener," or "PUR paint" to indicate the field of application of the end product produced from them.

The diisocyanate polyaddition process for synthesizing PURs is characterized by the following statements:

1. A large number of different reactants (monomers and oligomers) are available.
2. It is an exothermic reaction that can even take place at room temperature. No byproducts need be separated.
3. The reaction rate can be controlled over a wide range by catalysts, and in some cases a specific reaction can be accelerated by careful choice of the catalyst (e.g., tin compounds catalyze the urethane reaction, tertiary amines catalyze the isocyanate-water reaction; see Section 7.1.1).
4. Intermediates (prepolymers) with reactive end groups can be produced.

5. The stoichiometry influences the molecular mass (see Section 5.2.2).
6. Production and composition can be adapted to various process techniques and extremely diverse product requirements.
7. A large number of end products can be made.

Besides their unique properties, one of the main advantages of PURs is that all types of articles (foams, elastomers, thermosets) can be produced starting from liquid reactive components. This has led to their great industrial importance (see Section 5.2.1 and Chap. 15).

PUR products include highly elastic foams (mattresses, cushions, car seats), rigid foams (insulation materials); rigid and flexible moldings with compact skins (window frames, housings, skis, damping units such as car shock absorbers, front- and rear-end body parts, steering wheels, and shoe soles); engineering moldings with a high hardness and elasticity, ski boots, films, hoses, blow-molded parts, noise shields for trucks, seals for stoneware pipes, roller coatings, sealants, grouting compounds, surfacings for sport and play areas, off-road tires, windsurfing equipment, hydrocyclones, fenders, printing rollers, cable sheathing, catheters, high-quality paints, corrosion protection for steel-reinforced concrete, adhesives, textile coatings, high-gloss paper coatings, leather finishes, poromerics, glass fiber sizes, and wool finishing agents. The number and range of applications are constantly growing. The production, properties, technology, and use of PURs are reviewed in [3–13].

2. Basic Reactions [3, 4, 8]

The addition reaction of an alcohol to an isocyanate to give an urethane (carbamate) has been known since 1849 (WURTZ):

$$R-N=C=O + H-O-R' \longrightarrow R-NH-\overset{O}{\underset{\|}{C}}-O-R'$$

The extension of this reaction to di- and polyfunctional isocyanates and hydroxy compounds by OTTO BAYER in 1937 led to the principle of the polyaddition reaction resulting in the formation of linear, branched, or cross-linked polymers:

$$O=C=N-R-N=C=O + H-O-R'-O-H + O=C=N-R-N=C=O + H-O-R'-O-H$$

$$\downarrow$$

$$-O-\overset{O}{\underset{\|}{C}}-NH-R-NH-\overset{O}{\underset{\|}{C}}-O-R'-O-\overset{O}{\underset{\|}{C}}-NH-R-NH-\overset{O}{\underset{\|}{C}}-O-R'-O-$$

In addition to alcohols, the most important group of NCO-reactive compounds are the amines, which lead to the formation of ureas:

$$R-N=C=O + R'-NH_2 \longrightarrow R-NH-\overset{O}{\underset{\|}{C}}-NH-R'$$
Disubstituted urea

$$R-N=C=O + R'-NH-R' \longrightarrow R-NH-\overset{O}{\underset{\|}{C}}-N\overset{R'}{\underset{R'}{\diagdown}}$$
Trisubstituted urea

Ureas are also formed by the reaction of water with isocyanates, in which the carbamic acid formed in the first step of the reaction spontaneously decomposes to an amine with elimination of carbon dioxide. This amine then reacts with excess isocyanate to yield symmetrically substituted ureas:

$$R-N=C=O + H_2O \longrightarrow R-NH-COOH$$
$$\longrightarrow R-NH_2 + CO_2$$

$$R-N=C=O + R-NH_2 \longrightarrow R-NH-\overset{O}{\underset{\|}{C}}-NH-R$$
Disubstituted urea

This reaction is the basic reaction leading to PUR foams, especially flexible foams.

Another process which leads to the formation of carbon dioxide is the reaction of isocyanates with organic acids. The reaction product formed in this way is an amide:

$$R-N=C=O + R'-COOH \longrightarrow R-NH-\underset{\underset{O}{\|}}{C}-R' + CO_2$$
Amide

As the urethanes and ureas formed in the reaction mentioned above still contain reactive hydrogen atoms they can react with excess isocyanate to form allophanates and biurets, respectively.

$R-N=C=O + R-NH-COOR' \longrightarrow$ Allophanate

$R-N=C=O + R-NH-\underset{\underset{O}{\|}}{C}-NH-R' \longrightarrow$ Biuret

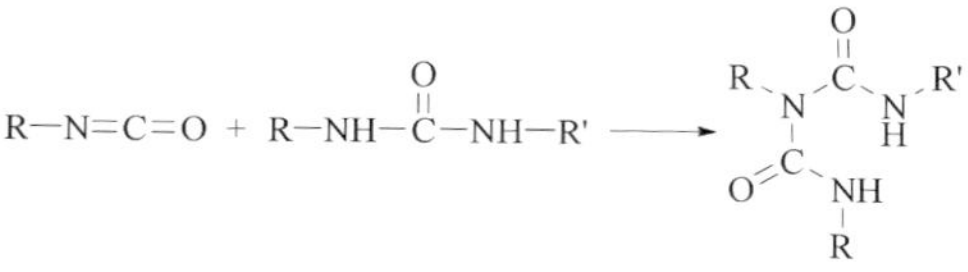

These cross-linking reactions are important in the production of PURs and play a key role in the development of polymeric materials with tailor-made properties.

Apart from these polyaddition reactions, the following reactions involving two or more isocyanate groups are also important in PUR chemistry (catalyst in parentheses):

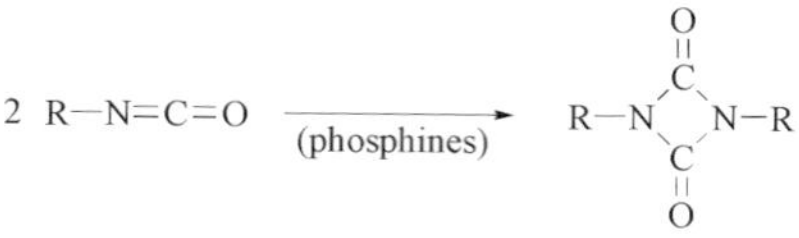

$2\ R-N=C=O \xrightarrow{\text{(phosphines)}}$ Uretdione

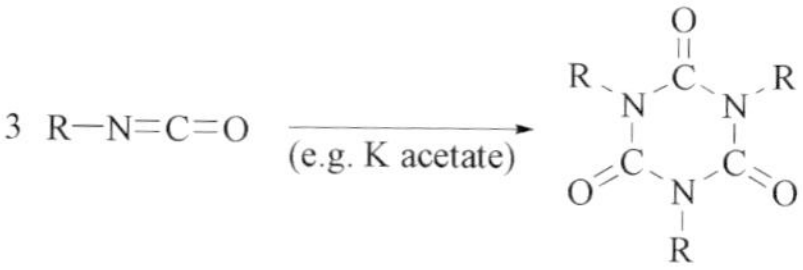

$3\ R-N=C=O \xrightarrow{\text{(e.g. K acetate)}}$ Isocyanurate

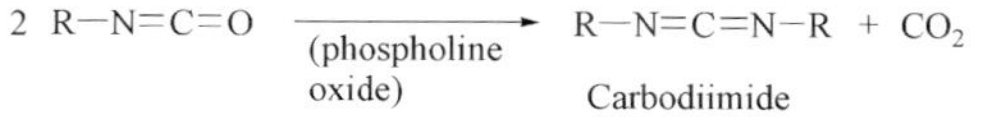

$2\ R-N=C=O \xrightarrow{\text{(phospholine oxide)}} R-N=C=N-R + CO_2$
Carbodiimide

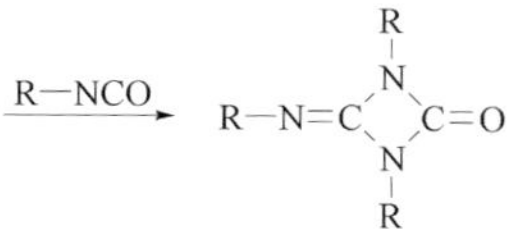

$R-N=C=O \xrightarrow{R-NCO}$ Uretonimine

Due to the high reactivity of the isocyanate group some further reactions are well known but only of minor importance in the production of PURs. Examples are the reactions of amides and epoxides with isocyanates:

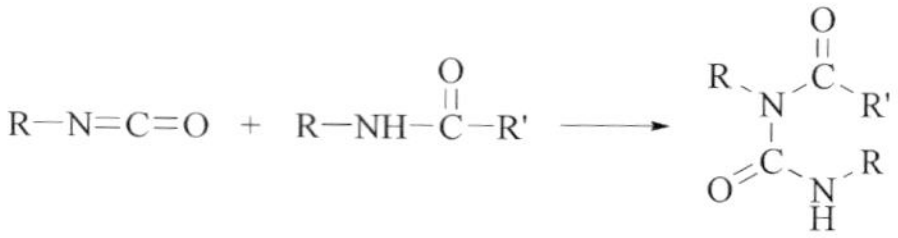

$R-N=C=O + R-NH-\underset{\underset{O}{\|}}{C}-R' \longrightarrow$ Acylurea

$R-N=C=O + \text{(epoxide)} \longrightarrow$ Oxazolidone

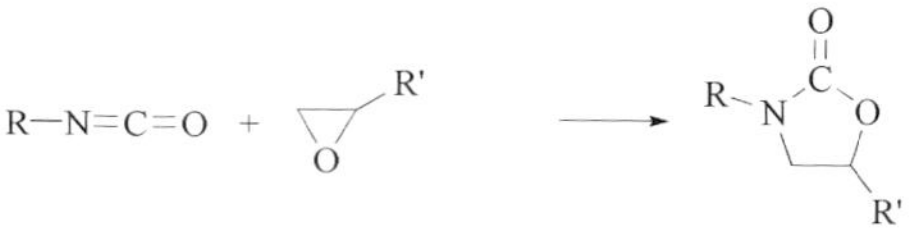

The tremendous number of different possible reactions not only play a role in the production of the final PUR polymers but also in the production of modified polyisocyanates (see Section 3.1) used as raw materials for PURs.

Usually the reactions described above proceed in parallel. For example, the production of foams is based on the combination of the NCO/ROH reaction leading to urethanes and the NCO/H_2O reaction leading to ureas and causing foaming of the reaction mixture by the evolution of carbon dioxide.

Other methods for preparing PURs which are usually not used in industrial applications are the polycondensation of bis-chloroformate esters and diamines, and the reaction of polyamines with carbonylbiscaprolactam (CBC) or polycyclocarbonates.

Thermally stable N-alkyl PURs are formed by the reaction of isocyanates with secondary diamines (e.g., piperazine). Since these products are unable to form hydrogen bonds, their properties are different from those of conventional PURs.

3. Starting Materials [3]

The range of PUR starting materials developed since the early 1950s is based on light petroleum fractions (ethylene, propene, naphtha). To a minor part renewable raw materials (natural products such as sugar, castor oil, etc.) are used [14–16]. The suitability of natural fatty acid esters [17, 18], lignin [19, 20], polysaccharides [19, 21, 22], and chitin as raw materials has been examined for special applications.

3.1. Polyisocyanates

See also → Isocyanates, Organic

More than 90 % of PURs are produced from aromatic polyisocyanates. Compared to aliphatic and cycloaliphatic polyisocyanates, the aromatic polyisocyanates show higher reactivities towards hydroxyl compounds and give PURs with better mechanical properties.

The aliphatic and cycloaliphatic polyisocyanates are used to obtain PURs that do not become discolored on exposure to light or heat. This is particularly important for coating materials (paints, varnishes, textile coatings, and glass-fiber sizes). The stability of aliphatic or cycloaliphatic polyisocyanates PURs can be improved by incorporating urea, biuret, or hydrazide groups into the polymer [23–25].

3.1.1. Aromatic Polyisocyanates

Aromatic polyisocyanates are primarily used for a wide variety of PUR foamed plastics, elastomers, thermosets, and adhesives.

2,4-Toluenediisocyanate [584-84-9] (TDI), 2,4-diisocyanatotoluene (**1a**), bp 121 °C (1.33 kPa), mp 21.8 °C.

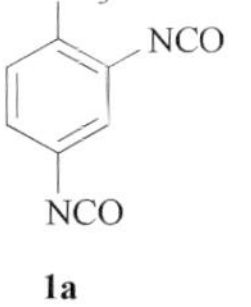

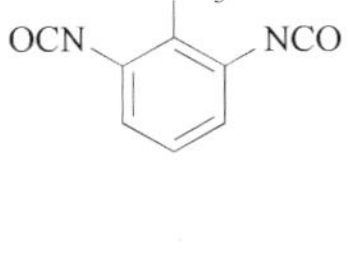

Usually the pure 2,4 isomer or mixtures of 2,4-TDI (**1a**) with 2,6-TDI (**2b**) are used for industrial applications:

TDI 80 (80 % 2,4-TDI, 20 % 2,6-TDI), bp 121 °C (1.33 kPa) mp 13.5 °C, vapor pressure (25 °C) 3 Pa, equilibrium vapor concentration 30 ppm, MAK value 0.005 ppm (according to TRGS 900).

TDI 65 (65 % 2,4-, 35 % 2,6-), bp 121 °C (1.33 kPa), mp 5 °C.

1,3-Bis(3-isocyanato-4-methylphenyl)-2,4-dioxo-1,3-diazetidine [26747-90-0] (**2**) (uretdione of 2,4-TDI, TDI-dimer; Desmodur TT, Bayer MaterialScience), mp 153 °C.

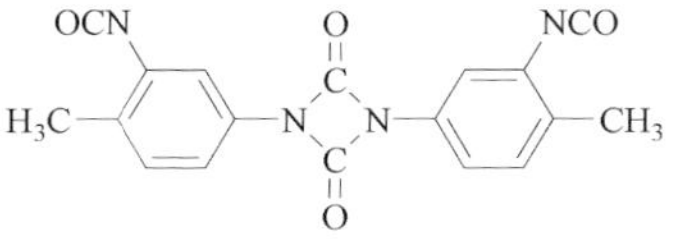

Uretdiones react as blocked isocyanates: the four-membered ring opens at elevated temperature to release the NCO groups.

"TDI-urea" [5206-52-0] (**3**) (TDIH, Bayer MaterialScience) mp 180–184 °C [26, 27]

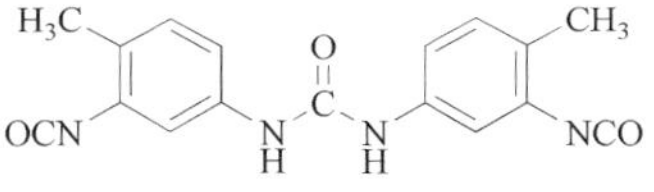

1,3,5-Tris(3-isocyanato-4-methylphenyl)-2,4,6-trioxohexahydro-1,3,5-triazine [20649-21-6] (**4**) (TDI-trimer), ca. 51 % in butyl acetate (Desmodur IL, Bayer MaterialScience): NCO content ca. 8 %, viscosity (20 °C) 2000 ± 500 mPa · s.

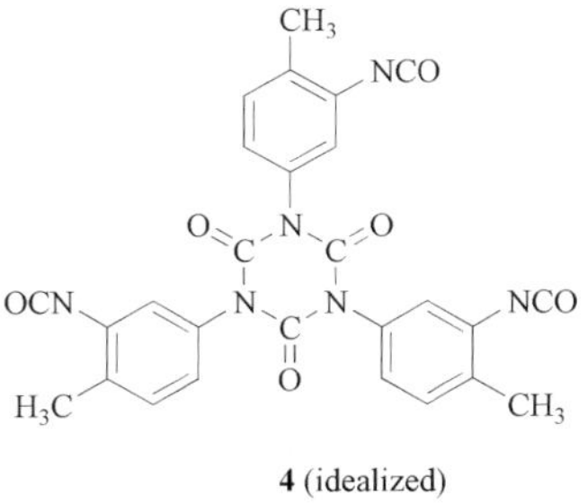

Adduct of TDI and trimethylolpropane [21092-47-7] (**5**), 75 % solution in ethyl acetate (Desmodur L, Bayer MaterialScience): NCO content 13 ± 0.5 %, viscosity (20 °C) 2000 ± 500 mPa · s.

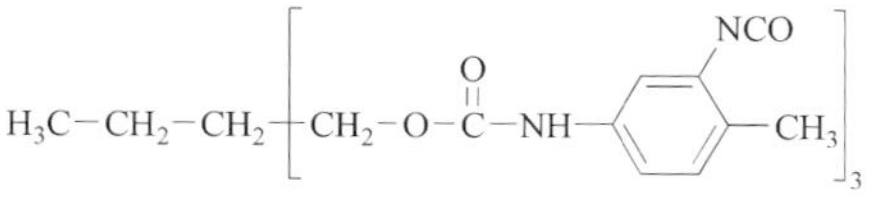

$$H_3C-CH_2-CH_2-[CH_2-O-\overset{O}{\underset{\|}{C}}-NH-C_6H_3(NCO)(CH_3)]_3$$

5

Monomeric MDI, 4,4'-Methylene diphenyl diisocyanate [101-68-8], 4,4'-methylene-bis(phenyl isocyanate), bis(4-isocyanatophenyl)methane, 4,4'-diisocyanatodiphenylmethane (**6a**), Monomer-MDI, bp 208 °C (1 kPa), mp 39.5 °C, vapor pressure (25 °C) <1 mPa, equilibrium vapor concentration 0.009 ppm, MAK value 0.005 ppm (according to TRGS 900)

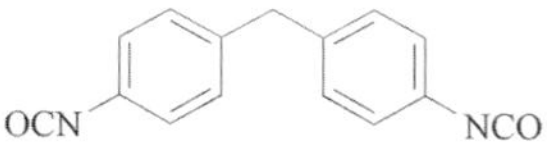

6a OCN—C$_6$H$_4$—CH$_2$—C$_6$H$_4$—NCO

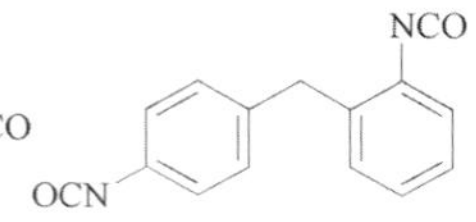

6b OCN—C$_6$H$_4$—CH$_2$—C$_6$H$_4$(NCO) (2,4'-isomer)

60 % 2,4'-MDI (**6a**), 40 % 4,4'-MDI (**6b**) (Desmodur 1806, Bayer MaterialScience), mp 16–17 °C.

Polymeric MDIs (PMDIs) [9016-87-9] (**7**) typically contain 30–70 % dinuclear components (n = 0), 14–40 % trinuclear components (n = 1), and 15–30 % higher nuclear components (n > 1), viscosity 50–20000 mPa · s, NCO content 25–33 %.

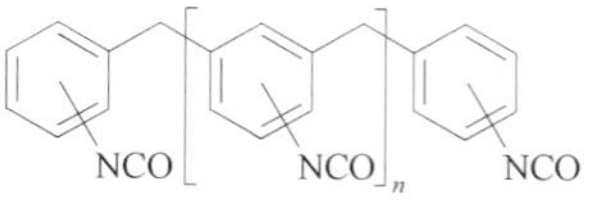

7

Urethanized 4,4'-MDI [60440-22-4] (Desmodur PF, Bayer MaterialScience) mp 10–15 °C.

Carbodiimidized 4,4'-MDI [37353-55-2], mp 10–15 °C (Desmodur CD, Bayer Material Science; for carbodiimidization, see Chap. 2).

1,5-Naphthalene diisocyanate (NDI) [3173-72-6], 1,5-diisocyanatonaphthalene (**8**), mp 127 °C (Desmodur 15, Bayer MaterialScience).

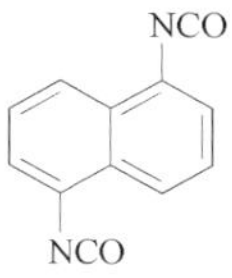

8

Tris(4-isocyanatophenyl)methane [2422-91-5] (**9**), 20 % solution in ethyl acetate, NCO content 7 ± 0.1 %, viscosity (20 °C) ca. 1 mPa · s.

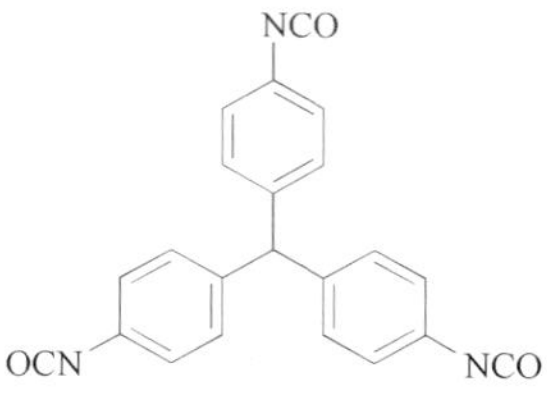

9

Aromatic isocyanates with the NCO group in the benzylic position behave similarly to the (cyclo-)aliphatic isocyanates with regard to reactivity and light stability:

1,3-Bis(1-isocyanato-1-methylethyl)benzene [2778-42-9] (**10**), m-tetramethylxylylene diisocyanate (TMXDI, American Cyanamid) [28, 29], mp –10 °C, bp 150 °C (49.5 kPa).

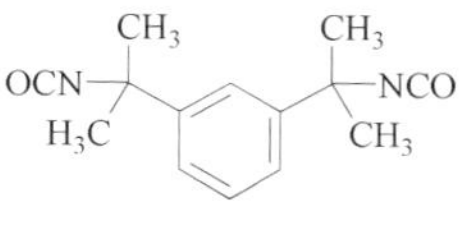

10

3.1.2. Aliphatic Polyisocyanates

Aliphatic and cycloaliphatic polyisocyanates are used to obtain light-stable PURs. This is particularly important for coating materials (paints, varnishes, textile coatings, and glass-fiber sizes).

Hexamethylene diisocyanate (HDI) [822-06-0], 1,6-diisocyanatohexane (**11**) (Desmodur H, Bayer MaterialScience), bp 127 °C (1.33 kPa).

OCN―(CH₂)₆―NCO

11

Biuret based on HDI, *N*-isocyanatohexylaminocarbonyl-*N*, *N'*-bis-(isocyanatohexyl) urea [28182-81-2] (**12**), biuret triisocyanate (technical mixture), ca. 75 % solution in methoxypropyl acetate/xylene (1/1) (Desmodur N75, Bayer MaterialScience), NCO content 16–17 %, viscosity (20 °C) 250 ± 100 mPa · s.

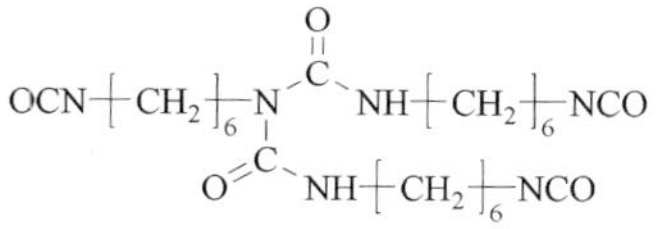

12 (idealized)

Trimer based on HDI, 2,4,6-trioxo-1,3,5-tris (6-isocyanatohexyl)hexahydro-1,3,5-triazine [28182-81-2] (**13**), trimeric hexane diisocyanate, ca. 90 % solution in butyl acetate (Desmodur N3390, Bayer MaterialScience), viscosity (25 °C) 2500 ± 500 mPa · s, NCO content 21.5 %, monomer content < 0.5 %.

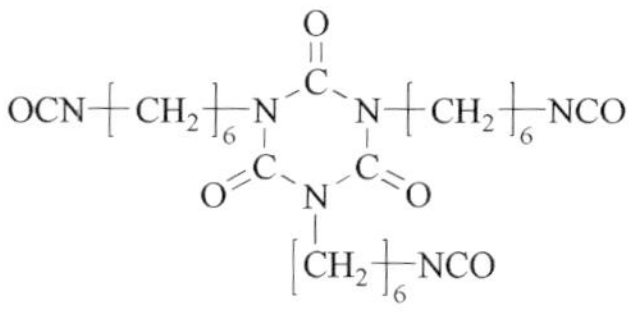

13 (idealized)

Unsymmetrical trimer based on HDI, 2,4-dioxo-3,5-bis(6-isocyanatohexyl)-6-(6-isocyanatohexyl)hexahydro-1-oxa-3,5-diazine [28182-81-2] (**14**), unsymmetrical trimeric hexane diisocyanate [30] (Desmodur XP2410, Bayer MaterialScience), viscosity (23 °C) ca. 700 mPa · s, NCO content ca. 24 %, monomer content <0.3 %.

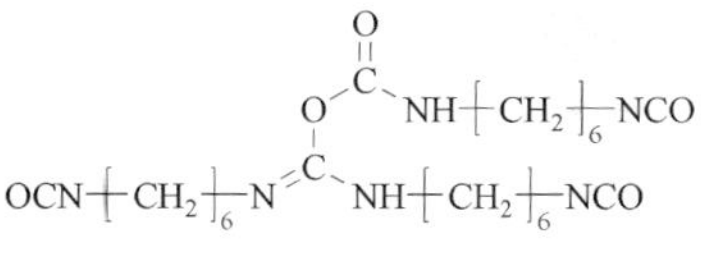

14 (idealized)

Isophorone diisocyanate (IPDI) [4098-71-9], 3,5,5,-trimethyl-1-isocyanato-3-isocyanatomethylcyclohexane (**15**) (IPDI, VEBA), bp 158 °C (1.33 kPa), 70 % cis, 30 % trans [31, 32]. For reactivity, see [33].

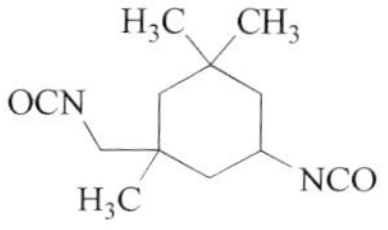

15

Uretdione of IPDI (Crelan TP.LS 2147) [34], mp 105–115 °C, ca. 15.4 % NCO.

Trimer based on IPDI, 2,4,6-trioxo-1,3,5-tris (5-isocyanato-1,3,3-trimethylcyclohexylmethyl) hexahydro-1,3,5-triazine [53880-05-0], ca. 70 % solution in xylene/methoxypropyl acetate (1/1) (Desmodur Z4470, Bayer MaterialScience), softening range 90–110 °C, NCO content 17 %, monomer content < 0.5 %.

1,1-Methylenebis(4-isocyanatocyclohexane), H12-MDI [5124-30-1], 4,4'-diisocyanatodicyclohexylmethane (**16**), trans,trans-4,4'-diisocyanatodicyclohexylmethane, mp 83 °C [35]. Technical isomer mixture (24 % cis, cis, 43 % cis, trans, 20 % trans, trans, 13 % 2,4 and 2,2' isomers): mp 15 °C (Desmodur W, Bayer MaterialScience)

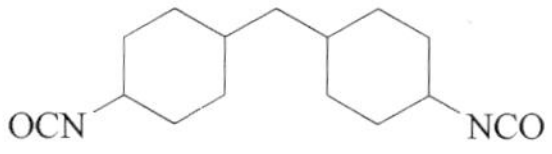

16

3.1.3. Blocked Isocyanates [36–39]

Polyisocyanates or prepolymers containing NCO groups are not stable on storage in the presence of atmospheric humidity or compounds that contain reactive hydrogen atoms.

Reaction of the NCO groups with certain organic compounds that have acidic hydrogen atoms such as phenol, cresol, nonylphenol, caprolactam, and methyl ethyl ketoxime leads to "blocked isocyanates," which are stable (storable) at room temperature and labile at elevated temperatures. These adducts are inert to polyols at room temperature and can be mixed with them in one-pack systems. At higher temperatures, usually at 140–180 °C the blocking agent is released, and the free isocyanate reacts as usual with the NCO-reactive components. Due to their thermal lability, uretdiones and uretonimines also behave like blocked isocyanates.

For the physical blocking of solid polyisocyanates by a diffusion barrier layer, see Section 5.4.

3.2. Polyols

Polyols are the largest group of PUR starting materials. The wide range of properties of the resulting PURs is largely determined by the chemical composition and molecular mass of the polyols. The largest group of polyols used for PURs is the group of polyethers. The volume of polyesters is much smaller, but they nevertheless have become more important, especially for flame-retarded rigid foams (see Section 7.3.4).

3.2.1. Polyether Polyols

See also → Polyoxyalkylenes

Polyethers (polyether polyols) with terminal hydroxyl groups are produced by addition of cyclic ethers, especially propylene oxide (PO, $R'=CH_3$) and ethylene oxide (EO, $R'=H$), to polyfunctional "starter" molecules.

$$R-OH + n \underset{O}{\overset{R'}{\triangle}} \longrightarrow R-O+CH_2-\underset{R'}{\overset{|}{CH}}-O+_n H$$

Usually, the EO and PO units can be structured as homo blocks, produced by batchwise addition of EO and PO to the starter molecule, or as mixed blocks by feeding an EO/PO mixture to the starter. The combination PO block-mixed block-EO block is common.

The structure determines the reactivity, the hydrophobic-hydrophilic properties, the surface activity, and consequently the foaming behavior of the polyethers. Polyols with terminal EO units contain primary hydroxyl groups and consequently have a higher reactivity than polyols with the secondary hydroxyl groups of terminating PO units.

The following starters are typically used industrially: ethylene glycol (EG), 1,2-propanediol (PG), bisphenol A (BPA), trimethylolpropane (TMP), glycerol (GLY), pentaerythritol, sorbitol, sucrose, water, ethylenediamine (EDA), and diaminotoluene (TDA). Monofunctional polyethers are usually started with 1-butanol.

Depending on their structure, two groups of polyethers can be distinguished: long-chain polyols, mainly used for flexible foams, and short-chain polyols, mainly used for rigid foams.

The long-chain polyols typically have hydroxyl numbers below 100 mg KOH/g and functionalities of 2 to 3, which corresponds to molecular weights of 2000 g/mol or more. The properties of some long-chain polyols are given in Table 1.

The short chain polyols usually have OH numbers of 200 mg KOH/g or more, functionalities of up to 6, and molecular weights of 1000 g/mol or less (Table 2).

Polyethers produced from trifunctional starters and bifunctional cyclic ethers (e.g., epoxy resins) are highly branched and, to some extent, intramolecularly cross-linked [40]. Special polyethers contain, for example, quaternary ammonium or sulfonate groups [41, 42].

An important special polyether is polytetrahydrofuran [43], which is described in detail elsewhere (→ Polyoxyalkylenes). It is used mainly for the production of high-grade elastomers for the engineering field and is liquid above 40 °C [44, 45].

A newer process, known as the IMPACT process, uses double metal cyanides (DMC) as catalysts for the production of especially long chain polyether polyols [46–48] that show improved properties compared to polyols

Table 1. Properties of long-chain polyether polyols

Starter	Alkylene oxide	Functionality	M_r, g/mol	OH no., mg KOH/g	Viscosity (25°C), mPas
PG	PO/EO	2	4000	28	870
PG	PO	2	2000	56	310
TMP/PG	PO	2.8	3750	42	630
GLY	PO/EO	3	4800	35	860

Table 2. Properties of short-chain polyether polyols

Starter	Alkylene oxide	Functionality	M_r	OH no., mg KOH/g	Viscosity (25°C), mPa s
TMP	PO	3	435	385	600
TMP	PO	3	305	550	1800
EDA	PO	4	350	620	19200
TDA	PO/EO	4	475	470	8000
Sucrose/PG	PO	3,1	385	450	15000

produced by the classical KOH process. During the addition of propylene oxide to the starter molecule, base-catalyzed isomerization of propylene oxide to allylic alcohol, which itself can act as starter molecule, leads to a mono-ol containing a double bond. These side products lower the average functionality of the polyols and reduce, for example, their stability against oxidation. The IMPACT process leads to polyols with extremely low mono-ol contents and therefore a low level of unsaturation and results in polyols with functionalities very close to the theoretical number. The IMPACT polyols have become more and more important, especially in the production of flexible foams.

Products based on polyethers have good low-temperature behavior and usually high hydrolytic stability. The viscosity of the liquid polyethers is relatively low (40–15000 mPa · s). A disadvantage of polyether polyols is that they undergo thermooxidative degradation on exposure to heat and atmospheric oxygen; UV radiation is mainly responsible for light-induced degradation changes. Both antioxidants and UV stabilizers counteract this degradation. In some special applications polyether polyols containing hydrazide or urea groups, both of which have a stabilizing effect, are used to avoid thermooxidative degradation.

Polyether dispersions (two-phase systems with a solid polymer as the disperse phase) are a further group of polyethers. Mainly styrene-acrylonitrile polymers (SAN polyols), polyurea, or poly(hydrazodicarbonamide) polymers (PHD polyols) are used as fillers. The dispersions are milky white, of higher viscosity than the corresponding base polyols, and completely stable to sedimentation. The average particle diameter is ca. 1 µm.

3.2.2. Polyester Polyols

Compared with polyethers, polyesters with terminal hydroxyl groups show outstanding resistance to light and thermal aging. PURs produced from them are therefore primarily used for paints, coating materials, and, due to their thermal stability, for flame-retarded rigid foams. Special types are suitable for high-grade PUR elastomers [49, 50].

Polyesters are produced by polycondensation of di- and trifunctional polyols with dicarboxylic acids or their anhydrides. Polyols commonly used are ethylene glycol, 1,2-propanediol, 1,4-butanediol, 1,6-hexanediol, neopentyl glycol, diethylene glycol, glycerol, and trimethylolpropane. Common dicarboxylic acids or anhydrides are succinic acid, glutaric acid, adipic acid, phthalic anhydride, isophthalic acid, and terephthalic acid.

The hydrolytic stability of a polyester urethane is determined by the constitution of the polyester. Polyesters synthesized from glycols and dicarboxylic acids that each contain at least five carbon atoms have good hydrolytic stability. This is because the increase in hydrophobicity of the polyester segments repels moisture [49, 50]. The best results have been obtained with polyesters derived from 1,6-hexanediol and adipic acid.

Common polyesters based on adipic acid and ethylene glycol or ethylene glycol/1,4-butanediol provide inexpensive PUR elastomers of excellent quality. Their hydrolytic stability is lower than that of the polyesters mentioned above and can be considerably improved by appropriate stabilizers (see Section 3.5).

Polyesters have a wide molecular mass distribution and, compared with polyethers, a high viscosity; many are glassy solids (polyphthalates) or crystalline (polyadipates). Polyesters containing diethylene glycol are ether esters; their viscosities are relatively low and their stability to hydrolysis and UV radiation is moderate. The average molecular mass of polyesters is 500–4000 g/mol. For properties, see Table 3.

The polyesters always contain significant proportions of the monomeric glycol used in

Table 3. Properties of polyester polyols

Acid	Glycol	Functionality	M_r	OH no., mg KOH/g	Viscosity, mPas
Adipic acid	ethylene glycol	2	2000	56	500 – 600 (75°C)
Adipic acid	diethylene glycol/TMP	2.5	2300	60	900 – 1100 (75°C)
Phthalic anhydrid/adipic acid	ethylene glycol	2	1750	64	2200 – 3200 (75°C)
Phthalic anhydrid/adipic acid	diethylene glycol	2	560	200	6300 – 7600 (25°C)
Terephthalic acid	diethylene glycol	2	450	250	2700 – 7700 (25°C)

their synthesis, as well as short-chain oligomers and often nonfunctional cyclic esters. These substances can be removed for a limited time by thin-film distillation, but they are gradually reformed by transesterification.

3.2.3. Polycarbonate Polyols

Aliphatic carbonate esters (→ Polycarbonates) are produced by transesterification of dimethyl carbonate (DMC) or diphenyl carbonate (DPC) with glycols (e.g., 1,6-hexanediol).

Many investigations have been carried out on the production of polycarbonate by copolymerization of alkylene oxides with carbon dioxide [51–55], but this reaction is not used on an industrial scale as yet.

Polyaddition of caprolactone or pivalolactone to low molecular mass diols yields hydroxy-functional polylactones.

As the susceptibility of polyesters to hydrolysis increases with increasing dissociation constant of the acid on which they are based, the polycarbonates are highly stable against hydrolysis. In particular, polyester urethanes based on 1,6-hexanediol polycarbonate show an unmatched resistance to hydrolysis.

3.2.4. Other Polyols

Other oligomers and polymers with functional groups (OH, SH, COOH, NHR) can also be used for PUR synthesis (e.g., polythioethers, polyacetals, polyester amides, alkyd resins, and polysiloxanes).

Hydroxy-functional poly(1,3-butadienes) contain the hydroxyl function mainly in the terminal position as a result of polymerization with hydrogen peroxide. These compounds yield PURs with moderate mechanical properties but extremely high hydrolytic stability and good low-temperature behavior. Bifunctional poly(1,3-butadienes) can also be obtained by anionic polymerization.

Hydroxyl-bearing polyacrylates based on 2-hydroxyethyl acrylate or methacrylate monomers are used in the production of high-quality, two-component paints [56] (see also → Paints and Coatings, 2. Chap. 9.).

Low molecular mass polyols, in contrast to the higher molecular mass polyols mentioned above (Section 3.2.1), are used as chain extenders. In the production of PUR elastomers they are generally used in the synthesis of the "hard" segment (see Section 4.2). Important representatives are ethylene glycol, 1,4-butanediol, 1,6-hexanediol, and 1,4-bis(hydroxyethoxy)benzene.

Trimethylolpropane or glycerol is added to introduce a defined small degree of branching.

3.3. Diamines and Amino-Terminated Polyethers

Diamines are used as building blocks for PURs that contain urea groups. Due to their high reactivity the polyaddition is very fast even without a catalyst, leading to polymers with considerably improved properties compared with "standard" PURs.

For polyaddition in bulk, sterically hindered diamines (e.g., isomeric diaminodiethylmethylbenzenes 17 [57]) or other aromatic diamines with mono- or dialkyl ring substituents [58–61] are preferred.

17a   17b

These amines are liquid at room temperature and therefore suitable for processing in two-component devices (see Section 6.2).

Less reactive diamines with electron-withdrawing substituents are used for producing elastomers by the prepolymer process (see Section 5.2.2); examples include bis(4-amino-3-chlorophenyl)methane [101-14-4] (MBOCA) and 2-methylpropyl-4-chloro-3,5-diaminobenzoate [32961-44-7] [62–64]. For the toxicological and adverse health effects of MBOCA, see [65].

Amino-terminated polyethers (ATPE) are obtained primarily from the appropriate polyether polyols (see Section 3.2.1) by exchanging the hydroxyl groups with ammonia [66–68] or primary amines [69]. Aminopropoxy end groups are obtained by cyanoethylation of polyols followed by hydrogenation. The high reactivity of the amino groups can be reduced by using their adducts with carbon dioxide (carbonates, carbamates). During the polyaddition reaction the carbon dioxide is released and acts as a blowing agent.

Various processes have been proposed for producing ATPEs with aromatic amine end groups [70]. For example, NCO prepolymers can be hydrolyzed to give the corresponding amino-terminated prepolymers [71]. Other methods include the preparation of N-substituted crotonic acid esters by acetoacetylation of polyols followed by amination [72, 73] and nitrophenylation followed by hydrogenation [74–76].

The adducts of (cyclo-)aliphatic di- and polyamines with ketones are bis- and polyketimines, respectively. They function as capped amines, with which they are in equilibrium in the presence of water (see Section 3.1.3). In the tautomeric enamine form, they also react with NCO groups to form urea and amides. Combinations of ketimines and isocyanates are thus both moisture- and heat-curing [77–79].

Polyether ketimines based on aliphatic polyether amines have a rather low reactivity and therefore a very good processability [80–83].

3.4. Special Building Blocks

Aqueous PUR-ionomer dispersions are synthesized from diols or diamines that contain ionic groups or groups that can be converted to ionic groups by alkylation or neutralization with compounds such as N-methyldiethanolamine, 2-(N,N-diethylaminomethyl)-2-ethyl-1,3-propanediol, and dimethylolpropionic acid, as well as salts of 2-sulfo-1,4-butanediol, N-sulfoethylethylenediamine, N-carboxyethylethylenediamine, lysine, and glycerol 2-phosphate [84, 85].

Hydroxy-functional polyisobutylene [86, 87], siliconediamines [88–90], and chlorosulfonated polyolefins [91] can be used as oligomer building blocks leading to segmented PUR polymers.

Other special polyols can be obtained by glycolytic degradation of polymer waste [92–94] (see Section 14.3).

3.5. Catalysts and Additives

Catalysts. In order to balance the two main reactions, the isocyanate-water reaction (blowing reaction) and the isocyanate-polyol reaction (gel reaction), various catalysts are used. The reactions of the isocyanate group are extraordinarily sensitive to many different kinds of catalysts:

1. Lewis bases: 1,4-diazabicyclo[2.2.2]octane [280-57-9] (DABCO), triethylamine [121-44-8], dimethylbenzylamine [103-83-3], bis[2-(dimethylamino)ethyl] ether [3033-62-3], 1,1,3,3-tetramethylguanidine [80-70-6], 1,8-diazabicyclo[5.4.0]undec-7-ene [6674-22-2] (DBU), 2,4-bis(dimethylaminomethyl)phenol [5424-54-4]

2. Lewis acids: bis(ethylhexanoyloxy)tin [301-10-0] (tin dioctanoate), bis(dodecanoyloxy)dibutyltin [77-58-7] (dibutyltin dilaurate), dichlorodimethyltin [753-73-1]

3. Insertion catalysts: bis(dodecylthio)dibutyltin, organotin alkoxides, organotin oxides, organotin thiolates, organotin sulfides, 1,3-dicarbonyl compounds

Almost every catalyst has a specific "activity profile" [13, 95–99]. The situation is extraordinarily complex, especially since the urethane group itself exerts a catalytic effect. As hydrogen bonds and the nature of solvents used in the reaction influence the reaction rates, most kinetic studies in solvents are not applicable to solvent-free PUR production [100]. Due to the desire to reduce emanations from PUR foams less volatile or reactive catalysts have been developed [101, 102].

Metal catalysts activate the isocyanate groups by making them more electrophilic, while bases make the hydroxyl groups more nucleophilic [103].

Alkali metal salts of organic acids and phenols, as well as phenol Mannich bases catalyze not only the urethane reaction but also the trimerization of the isocyanate group. NCO prepolymers trimerize more rapidly than the corresponding diisocyanates because of the presence of urethane groups which act as cocatalysts (see Section 4.4).

Phosphines are special catalysts which dimerize the NCO group; phospholine oxides cause carbodiimidization.

Additives (see also → Plastics, Additives).

Foam Stabilizers. Besides the chemical structure and molecular weight of the polyols and isocyanates, cell size and fraction of open cells can be controlled to a certain extent by surfactants or foam stabilizers [104]. In some cases, foam formation is virtually impossible without these surface-active compounds. The majority of foam stabilizers are polysiloxane-polyether block copolymers [105–108]. Specific types, e.g., for liquid CO_2 as blowing agent [109], for flame-retarded foams [110], or for other specific foam types are available as well [111, 112].

Hydrolysis Stabilizers [113–116]. Monocarbo-Monocarbodiimides, polycarbodiimides, epoxides, cyanates, phenyliminooxazolidines.

Oxidation Stabilizers [113–115], [117–119]. These additives are used in foamed materials, especially in flexible foams, to prevent core scorching. They include sterically hindered phenols, alkylated anilines, phosphites, hydrazides, sulfides, and thioethers.

UV Stabilizers [113–115, 117–119]. Piperi-Piperidines with bulky substituents in the 2,6-positions, benzophenones, benzotriazoles, cyanoacetate ester derivatives, quenchers, Tinuvins (Ciba-Geigy).

Blowing Agents (see also → Foamed Plastics, Section 2.2.). According to the implementation of the Montreal Protocol (and the consecutive amendments) blowing agents with an ozone-depleting potential are no longer used as blowing agents for flexible foam by the member states since 1992. Several alternatives have been developed [120–127].

For flexible foams the predominant technology meanwhile is the addition of liquid or supercritical carbon dioxide (e.g., Novaflex by Hennecke, Cardio by Cannon). Another method is the application of varied ambient pressure, e.g., MegaFoam by Hennecke or VPF by Beamech (see Section 7.1.1.2) and depending on regulations of the local authorities, blowing agents like dichloromethane or acetone may be applied as well.

For rigid foams nowadays usually alkanes, namely, n-pentane, isopentane, and cyclopentane are used (see Section 7.3.1).

Flame Retardants (see also → Plastics, Additives, Chap. 6.). Various classes of chemicals are used to modify the combustion behavior of PUR flexible foams [128]. The selection of the appropriate flame retardant depends on the test that must be passed, as well as the chemistry of the foam (e.g., polyester vs polyether, HR vs. conventional foam). Most common are phosphorus- and/or nitrogen-containing compounds and halogenated phosphates such as tris(2-chloropropyl) phosphate (TCPP). As in the case of catalysts, less volatile or reactive flame-retardants have been developed in order to reduce the emissions of volatile organic compounds (VOCs) from the finished foam. The liquid flame retardants may be combined with solid ones such as melamine and ammonium polyphosphate. Brominated diphenyl ethers, which have been used for some special foam grades or in some regional markets, will be phased out due to concerns regarding their potential for bioaccumulation.

For rigid foams halogenated compounds such as TCPP, polyesters based on tetrabromophthalic acid, and special bromopolyethers are usually used. The use of polyisocyanurate rigid foams (see Section 7.3.4), which exhibit a superior flame behavior compared to PUR rigid foams, has increased. This has led to a reduction in flame retardant consumption.

Aluminum oxide trihydrate, magnesium hydroxide, and expandable graphite are also used as flame-retardants. The first two act by splitting off water vapor and the latter acts by forming protective layers.

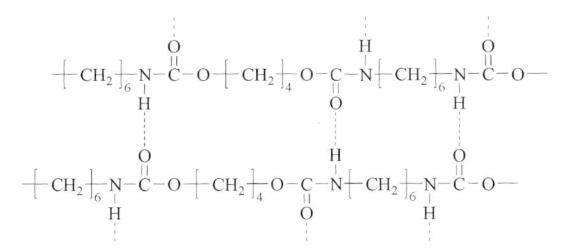

Figure 1. Hydrogen bond interactions between polymer chains of a PUR based on HDI and 1,4-butanediol

4. Structure and Morphology

Depending on their structure, PURs cover a broad range of properties. Besides the chemical composition, the chain length, and cross-linking, the properties are significantly affected by the interactions between the polymer chains (see Fig. 1), that is to say by the supramolecular structure and morphology.

4.1. Polyurethanes Without Segmented Structure

The properties of linear PURs, which are produced by polyaddition of monomeric diisocyanates and short- or long-chain diols, are similar to those of polyamides.

Short-Chain Polyols. The polyaddition of hexamethylene diisocyanate (HDI) and 1,4-butanediol leads to a very tough, crystalline material (mp 180–185 °C), which is used for manufacturing of fibers and bristles. The polyaddition of toluene diisocyanate (TDI) and butanediol or diethylene glycol yields very hard, transparent products with low heat resistance.

Long-Chain Polyols. Linear PURs based on diisocyanates (preferably MDI) and long-chain polyester or polyether diols are soft, rubbery, high molecular mass polymers with only a small proportion of urethane groups. Therefore, these materials are homogeneous (single phase), and the intermolecular physical interactions are mainly determined by the polyester and polyether segments (van der Waals forces [129]).

With increasing stiffness of the polyol chain (e.g., highly crystalline polyesters), and decreasing chain length, the resulting PUR becomes harder and less elastic.

4.2. Polyurethanes with Segmented Structure

To obtain PUR elastomers with superior properties, the polyaddition reaction mixture usually contains three components, namely, a diisocyanate, a long-chain polyether or polyester polyol, and a short-chain diol (chain extender). The resulting block copolymer type PURs are characterized by a segmented structure.

The secondary and tertiary structures, and as a consequence, the morphology of these PURs, depend on the chemical structure and the block length of the different segments. Due to the two- or multi-phase morphology, segmented PURs can exhibit extraordinary high-grade properties.

4.2.1. Hard and Soft Segments [12, 130–135]

The reaction of one equivalent of long-chain diol, one equivalent of short-chain diol, and two equivalents of diisocyanate leads to the idealized structure shown in Figure 2.

The long-chain polyols, which are flexible and are usually tangled (soft segments), alternate along the polymer chain with oligomeric rigid urethane units (hard segments).

Usually the structure of the PUR as well as the structure of the soft segment itself is determined by a Flory distribution [136, 137]. This results in a smaller number of larger hard segments than expected in theory. If polyester polyols are used, the low molecular glycols, which are usually present in considerable amounts, even enhance this effect. Further effects leading to deviations from theory include kinetic effects (if end groups with different reactivity are used) and phase transitions due

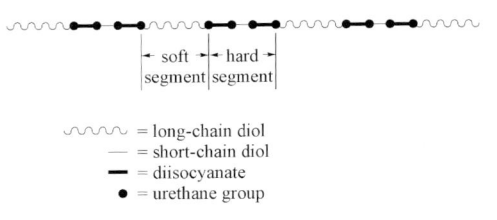

Figure 2. Schematic primary structure of a segmented PUR (idealized)

to temperature changes during the polyaddition reaction.

4.2.2. Segregation and Morphology

On cooling, the initially homogeneous PUR melt converts to a two-phase system. This is due to the incompatibility of the polar low-molecular glycols with the low-polarity polyether or polyester polyols and polyisocyanates. As the reaction progresses, the formation of polar urethane groups at the ends of the long-chain polyols leads to a certain degree of compatibilization. Nevertheless, the low miscibility of the hard and soft segments leads to segregation, which results in covalently bonded microphases in which paracrystalline oligourethane domains appear (tertiary structure, Fig. 3). On further cooling, these can aggregate and grow to form spherulitic microcrystallites (quaternary structure).

The hard segment domains form three-dimensional cross-linking centers of a noncovalent type (physical cross-linking). The formation of a large number of weak hydrogen bonds results in a high bonding strength, but allows shifts and rearrangements under load, where a covalent bond would rupture. As the hard segments are covalently bonded to the soft segments, flowability of the chains is restricted. This gives the segmented PURs their high elasticity. The hard segments are responsible for tensile strength, tear resistance, hardness, permanent elongation, and compression set, while the soft segments determine the elastic expansion and the glass transition temperature.

The interaction between different hard segment domains is mainly determined by the symmetry of the isocyanates and the nature of the chain extenders, which influence the possibility to form highly organized structures.

Most commercially produced PURs contain urea groups, which often determine the property profile more strongly than the comparatively small content of urethane groups. Since the urea group is more polar than the urethane group, the segregation of hard and soft segments is strengthened, the melting range of the hard-segment domains is higher, and the intermolecular (interchain) interaction stronger [138]. Due to these strong interactions, a single urea group

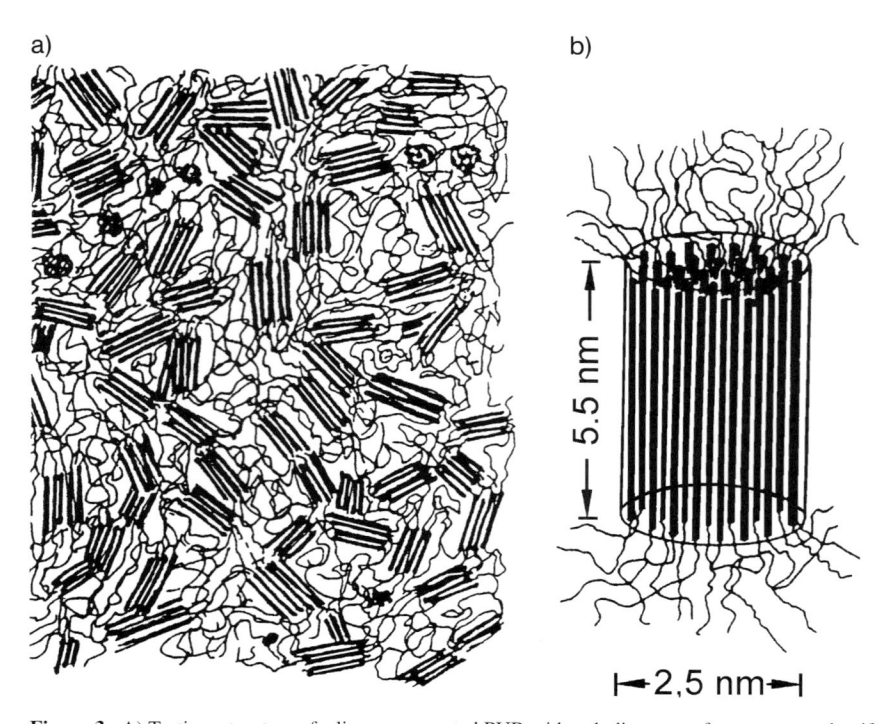

Figure 3. A) Tertiary structure of a linear segmented PUR with polydisperse soft segments and uniform (monodisperse) hard segments; B) Cylindrical model of a hard domain

shows hard-segment character. These interactions also explain the superior mechanical properties of polyureas and their low thermoplasticity.

The balance between these hard segment/hard segment interactions and the interactions between hard segments and soft segments determine the degree of segregation. Due to decreasing polarity, segregation increases from polyesters to polyethers to polybutadiene polyethers.

In the soft segments, strain-induced crystallization can be initiated by stretching or disentanglement. This strain-induced crystallization explains the high hysteresis of typical PUR elastomers [139, 140]. The mechanical properties of segmented PUR depend on the melting temperature of the crystalline domains.

Products based on 1,5-naphthalene diisocyanate (NDI) are no longer thermoplastic because the melting temperature of the hard segments is too high. They are usually produced with a small excess of isocyanate as casting elastomers (see Section 8.1). These products have extremely good mechanical properties: high tear strength, high abrasion resistance, high elasticity at high hardness, low damping, high load-bearing capacity, and excellent long-term dynamic endurance (load-bearing property) [135].

High-quality PUR thermoplastic elastomers are produced primarily from MDI. The melting temperature and the solubility of these segmented thermoplastic PURs can be influenced by modifying the chemical composition, especially of the hard segments. Use of a second short-chain diol lowers and widens the melting temperature range, and solubility is also improved.

As the properties of segmented PUR are highly determined by the phase segregation and the crystallinity of the domains, numerous production parameters [130, 131, 141–143] have tremendous influence on the behavior of thermoplastic PUR materials. Even thermoplastic PUR with monodisperse (uniform-length) N-alkylurethane hard segments, which are incapable of hydrogen bond formation show differing morphology and properties depending on their thermal pretreatment [144].

4.3. Cross-linking of Polyurethane

PUR flexible and rigid foams, cast elastomers, and most reactive systems are usually cross-linked. Even if these cross-linked PURs are not thermoplastic, the melting of the hard-segment domains is still observable, e.g., by softening (decrease in shear modulus) at elevated temperatures. With increasing cross-linking, the glass transition temperature increases, whereas the tendency to crystallize decreases.

If the cross-linking is not caused by subsequent reactions (formation of allophanate or biuret), but introduced by the use of highly functional raw materials, then the formation of segregated domains can be disturbed due to the rapid formation of a highly cross-linked PUR network, fixing the polymer chains or at least reducing their flexibility. In this case, even heating to elevated temperatures does not improve the material properties, as domain formation is not possible (e.g., in flexible foams).

4.4. Polyisocyanurates [145]

The general term PURs also covers products that contain cyclic urea structures (isocyanurates), which are formed by cyclotrimerization of isocyanate groups. The mechanism of isocyanurate formation is not completely understood. The most probable mechanism of base catalyzed trimerization [146] is shown in Figure 4.

Addition of the catalyst II to the carbonnitrogen double bond leads to an urethane anion III, which by subsequent addition of two more isocyanate molecules I is transformed via IV to the adduct V. This adduct collapses into the stable isocyanurate VI and regenerates the catalyst II.

The isocyanurates exhibit improved mechanical properties such as compression strength and dimensional stability and generally show superior fire retardancy [147]. These advantages have led to the increasing importance of isocyanurates, especially since the blowing agent for rigid foams was switched from CFC to pentane (see Section 7.3.4).

5. Production of Polyurethanes

5.1. Stoichiometry

Isocyanates (RNCO) are specified by their percentage NCO content and in some cases also by their equivalent mass (molecular mass of NCO

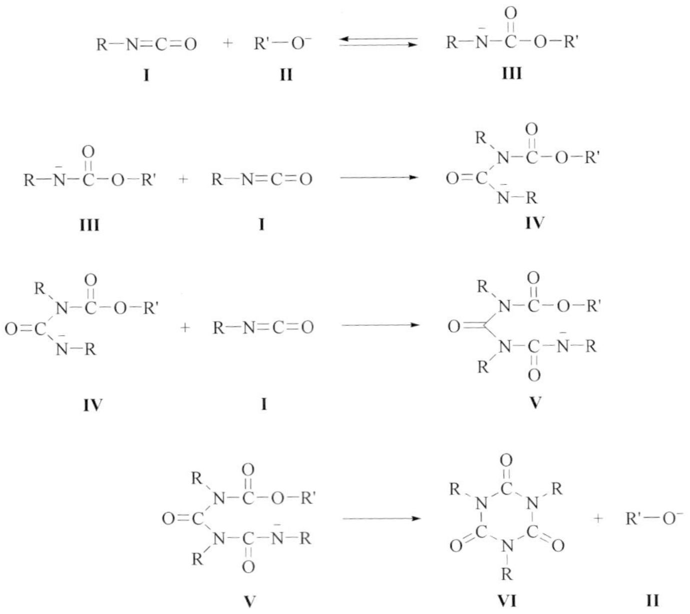

Figure 4. Mechanism of base-catalyzed isocyanate trimerization

moiety/functionality). Functionality means the number of isocyanate groups per molecule.

$$\text{Equivalent mass} = \frac{\text{molecular mass of RNCO}}{\text{functionality}} = \frac{42}{\%\text{NCO}} \cdot 100 \quad (1)$$

Polyols are specified by their hydroxyl number (OH no.) and sometimes also by their percentage OH content or their equivalent mass. The OH no. is given in mg KOH/g, owing to the method used for analyzing the OH groups: they are usually treated with acetic anhydride and, after hydrolysis with water, the excess acetic anhydride is titrated with potassium hydroxide. This means that a polyol containing 1 mmol OH groups per gram has an OH no. of 56 mg KOH/g.

$$\text{OH no.} = \%\text{OH} \cdot \frac{56 \cdot 1000}{17 \cdot 100} = \%\text{OH} \cdot 33 \quad (2)$$

$$\text{Equivalent mass} = \frac{56 \cdot 1000}{\text{OH no.}} \quad (3)$$

Water, acids (acid number AN), and secondary and tertiary amines are also most simply characterized by their equivalent mass. For mixed amines, a mean functionality must be used where necessary. For calculating quantities, 1 gram equivalent isocyanate corresponds to 1 gram equivalent Zerewitinoff-active hydrogen compound (e.g., polyol). In practice, the quantity of isocyanate is calculated almost exclusively for 100 parts by weight of the active-hydrogen compound(s). If the raw materials also contain water, then an additional quantity of isocyanate is required, since the latter reacts with water to form carbon dioxide and substituted urea.

The index (or isocyanate index) is equal to 100 (isocyanate index = 1) if the quantity of isocyanate added equals the amount necessary for reaction with all Zerewitinoff-active hydrogen atoms.

$$\text{Index} = \frac{\text{Used isocyanate quantity}}{\text{Necessary isocyanate quantity}} \cdot 100 \quad (4)$$

In practice, the index can be used to influence the properties of the PUR decisively, e.g., by "under-" or "over-cross-linking." Under-cross-linking means that free hydroxyl groups remain in the PUR; over-cross-linking means that the urethane moieties undergo consecutive reactions, i.e., the formation of allophanates. High indices of 300 or more are used to produce polyisocyanurates (PIR), in which the isocyanate groups undergo trimerization by reaction with each other (see Section 4.4).

5.2. Reaction without Solvents

The possibility of manufacturing PURs from liquid raw materials without the use of any solvent is one of the biggest advantages of PUR chemistry. As a result, most PURs are manufactured by a mass polyaddition process. This process can be a single-step reaction (one-shot process) or a two- or multiple-step reaction (prepolymer process). The reactivity is controlled by temperature or by the addition of catalysts. Due to the high exothermicity, reaction temperatures of about 100 to 180 °C are not unusual. These high temperatures guarantee good end curing of the molded PUR parts.

5.2.1. One-Shot Process

Bulk reactions without solvents are usually very fast, especially when catalysts are added. Therefore, the production of PUR foams is usually carried out by the one-shot process, in which all components are mixed directly, generally with simultaneous addition of auxiliaries such as catalysts, foam stabilizers, reinforcing agents, fillers, and fire retardants. The reaction is highly exothermic and largely completed in 0.5–30 min, depending upon the catalyst. The final properties, however, are frequently reached only after 24–48 h (post-curing).

The one-shot process requires that the reactivity of all isocyanate species is quite similar. Amines usually lead to excellent PUR properties but react much more rapidly than polyols. If diamines or polyamines are to be used it must be ensured that the urea formed at the beginning of the reaction does not precipitate and is no longer available for incorporation into the hard segments. This means that the use of polyamines is limited to amines which cannot undergo cross-linking (secondary amines), are sterically hindered [57–61, 148], or are deactivated by electron-withdrawing substituents [62–64, 149].

For easy processability all components should be liquid, of low viscosity, and miscible or at least be able to form a homogeneous reaction mixture for the very first moments of the reaction. Usually all the nucleophilic components, that is, polyols and/or polyamines, are premixed and homogenized. If these mixtures are incompatible with the short-chain diols (chain extenders), the miscibility must be enhanced by raising the temperature or adding emulsifiers.

In general, the polyaddition reaction of a multicomponent system by the one-shot process leads to a predominantly statistical structure of the polymer chains.

The term reaction injection molding (RIM) technology usually is used when highly reactive components are mixed and injected into a mold on a very short timescale (see Section 7.4). This technique allows for the handling of reactive systems with pot lives (i.e., the period in which no significant increase in viscosity is observed) of less than 1 s and reaction times of about 10–60 s.

RIM technology has initialized a development, which started with catalyzed highly reactive PUR systems [150], and, by the use of polyurethane-polyurea systems, chain extended by diamines [151–153], led to pure polyurea systems, in which the only isocyanate-reactive components are diamines [154–163].

5.2.2. Prepolymer Processes

In PUR chemistry, prepolymers are defined as intermediate reaction products of the polyaddition reaction which still contain free isocyanate or isocyanate-reactive groups. In contrast to the one-shot process, the prepolymer process allows for a highly controlled polyaddition reaction, e.g., to construct tailor-made segment structures.

In the first step of the process, polyether or polyester polyols are usually treated with excess diisocyanate to yield homogeneous reaction mixtures, which may contain considerable amounts of unconverted isocyanate groups. If diisocyanates with NCO groups of different reactivity are used (2,4-diisocyanatotoluene, isophorone diisocyanate), the resulting NCO prepolymers exhibit narrow molecular weight distribution with only minor amounts of unconverted monomeric diisocyanate.

If prepolymers free of monomeric diisocyanate are required, the latter must be removed, e. g., by thin-film distillation or extraction with a hydrocarbon solvent which does not dissolve the prepolymer. The monomer content is then lowered to below 0.1 % [164, 165].

The production of storage-stable NCO prepolymers from aromatic diisocyanates is difficult and requires special precautions (e.g., the exclusion of all traces of water and catalysts) to avoid side and consecutive reactions [166].

Low-viscosity, monomer-free NCO prepolymers can be obtained from silylated hydroxy compounds and isocyanates with a carboxylic acid chloride group [167, 168].

$$R-OH + Cl-\underset{\underset{CH_3}{|}}{\overset{\overset{CH_3}{|}}{Si}}-CH_3 \longrightarrow R-O-\underset{\underset{CH_3}{|}}{\overset{\overset{CH_3}{|}}{Si}}-CH_3$$

$$R-O-\underset{\underset{CH_3}{|}}{\overset{\overset{CH_3}{|}}{Si}}-CH_3 + Cl-\overset{\overset{O}{\|}}{C}-R'-NCO$$
$$\longrightarrow R-O-\overset{\overset{O}{\|}}{C}-R'-NCO$$

5.3. Reaction in Solvents

5.3.1. One-Component Systems

Linear or slightly branched PURs or polyurethane polyureas with molecular weights of 50 000 to 150 000 g/mol can be obtained by polyaddition reaction in solution. The PURs are usually isolated by evaporation of the solvents or by precipitation, e.g., by pouring the solution into a nonsolvent liquid.

The choice of solvent for the polyaddition is crucial, as the solvent must be able to dissolve both the soft and hard segments of the segmented PUR structure. Polar organic solvents (dimethylformamide, dimethylacetamide, 1-methyl-2-pyrrolidone, sulfolane, etc.) usually fulfill this requirement. In some cases, ketones like acetone or methyl ethyl ketone are also suitable. The ketones are especially suitable for chain extension of NCO prepolymers with aliphatic diamines as they reduce their activity by reversibly forming an adduct.

In industrial processes, linear PURs containing only urethane groups are usually produced without using solvents, but can be dissolved in a solvent if necessary for the application.

PURs containing urethane and urea groups are mainly produced by a two-step process. In the first step a prepolymer is made without the use of any solvents. In the second step this prepolymer is then treated with chain extenders such as diols or diamines. In the second reaction step, the solvent is usually added stepwise to avoid a too rapid increase in the viscosity of the solution. The obtained PUR solutions have PUR contents of ca. 20 to 30 %.

With increasing viscosity it is very difficult to achieve complete conversion of the functional groups. This problem is overcome by using NCO/OH ratios (indexes) slightly above 100 and running the reaction till the viscosity reaches a specified value, which must be determined empirically for each product. Then the reaction is stopped by adding a monofunctional stopper, e.g., dibutylamine.

If the reaction is carried out in highly polar solvents, especially in mixtures of water and acetone, methyl ethyl ketone, or DMF, even highly reactive aliphatic diamines such as ethylenediamine can be used. With increasing molecular mass the products may tend to gelation or precipitation. The solubility of the PUR can be improved by adding small portions of a second chain extender such as 1,2-diaminopropane, piperazine, or water to disturb the formation of the hard-segment domains.

5.3.2. Two-Component Systems [169]

Two-component systems are of tremendous importance for coatings and for leather and paper finishing. They usually consist of a polyhydroxy compound and an isocyanate cross-linker.

The polyhydroxy compounds are usually obtained by chain extension of common polyether or polyester polyols with diisocyanates; the isocyanate cross-linkers, which for ecological reasons must be free of isocyanate monomers, may be NCO prepolymers made by the reaction of diisocyanates with diols, diamines, or water; or they may be modified isocyanates, e.g., trimerization products of HDI (see Section 3.1.2).

To increase the amount of reactive components in the solvent-containing two-component systems, numerous improvements have been introduced to industrial processes: part of the solvent can be substituted by reactive thinners [170], which are incorporated into the cross-linked PUR. By using new catalysts an unsymmetrical trimerization product of HDI [30], with particularly low viscosity, has been developed, allowing for higher concentrations (high-solids systems) without deterioration in the processability.

5.4. Reactive One-Pack Systems

The term "one-pack system" denotes the storage-stable, ready-to-use formulation of a polymer precursor which cures without addition of further components, e.g., curing agents. Curing generally is caused by humidity, atmospheric oxygen (mainly used in paints), or by the thermal elimination of a blocking agent (see Section 3.1.3), which either vaporizes or remains in the cured polymer.

Other curing systems involve encapsulated materials with a diffusion barrier layer, which is broken, e.g., on heating. They cure without uptake or release of a substance.

Moisture Curing. Relatively low molecular mass NCO prepolymers can be dissolved in small amounts of low-polarity solvents (e.g., 75 % solution in ethyl acetate) and cured by the action of atmospheric moisture. The curing rate depends on the atmospheric humidity and temperature. Usually these solutions are used as paints or coatings, and the carbon dioxide formed as a byproduct escapes from the thin film without bubble formation.

Eliminating Blocking Agents. Mixtures of blocked NCO prepolymers with polyols are stable at storage temperatures (20–50 °C) and are cured by heating (> 120 °C) to eliminate the blocking agent (see Section 3.1.3). The reaction can be conducted in such a way that an isocyanate intermediate does not appear [39]. Appropriate hydrophilic modification can be used to produce such combinations as low-viscosity aqueous emulsions or dispersions, in which one of the components usually has a high molecular mass [38].

Removal of a Diffusion Barrier Layer. Some solid diisocyanates such as NDI, TDI-urea, and TDI-uretdione (TDI-dimer) can be stabilized by encapsulation with a diffusion barrier layer. The barrier layer is produced by dispersing the solid isocyanate in a solvent and reacting the NCO groups at the particle surface with (cyclo-)aliphatic diamines. After removal of the solvent the encapsulated isocyanates are dispersed in polyols or polyamines.

Liquid, pasty, or granular one-pack systems are produced in this way. They can be stored for several months and are made to react by heat shock, e.g., 1 min at 120 °C. Heat-curable reactive powders can also be produced on this basis [171].

5.5. Other Processes

In addition to the industrial isocyanate polyaddition processes described previously and in subsequent sections, several other synthetic methods for PURs or polyureas are of scientific interest [172].

PURs can also be obtained by the polycondensation reaction of bis-chloroformate esters and diamines. This method has the advantage over the polyaddition method that it can be carried out under very mild conditions (at ⩽20 °C).

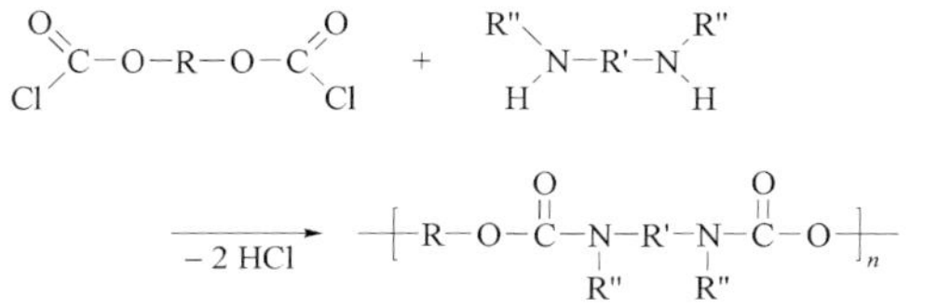

If secondary diamines such as piperazine are used in the polycondensation reaction, thermostable poly(N-alkylurethanes) are obtained. This principle has been used to produce, e.g., high-melting PURs [173] for fibers or PURs with liquid-crystalline character [174]. Liquid-crystalline PURs have also been produced from N-alkylcarbamic acid chlorides and diols.

Another method for preparing PURs is the ring-opening polyaddition of oligomers with cyclic ethylene carbonate groups and di- or polyamines, which leads to segmented PURs with free hydroxyl groups [175, 176]:

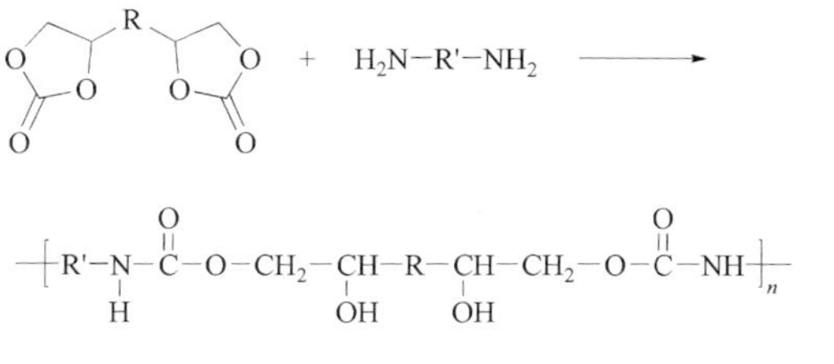

Functionalized PURs can be obtained if carbonylbiscaprolactone (CBC) is used as carbonyl source [177]. Due to its high selectivity, reactions with diamines can be carried out even in the presence of additional hydroxyl groups.

High-molecular mass α,ω-bis(amino) prepolymers for use in RIM technology (see Section 7.4) have been produced by the reaction of α,ω-bis(amino)polyethers with urea [178].

6. Processing of Polyurethanes

6.1. Supply, Storage, and Preparation of Raw Materials

The raw materials are transported in small (< 100 kg) and normal-sized drums (200–250 kg), and (road) tank trucks and trailers (ca. 23 t). The regulations applicable to the transportation of chemicals (including labeling) must be observed, as well as the transportation conditions prescribed by the manufacturer for preserving the quality of the raw materials. If raw materials are delivered by tank truck, storage tanks for polyol and isocyanate should be installed in pairs at the processing site (Fig. 5). The size of the storage tanks should be adapted to the delivery volume (frequently 15 or 30 m^3). Since the raw materials in the tank truck are usually delivered at elevated temperature, they must be cooled to the processing temperature (22–25 °C) in the storage tank. Ordinary thick plate can be used as the tank material.

6.2. Metering and Mixing Technology

In contrast to the machines used for processing thermoplastics, which consist of devices for melting, molding, and solidifying, the machines used for producing PUR moldings or semifinished parts contain all the elements of a chemical reactor.

The raw materials are conveyed from the storage tank, optionally via premixing stations for adding auxiliary materials, to the processing machine. Often the metering and mixing machine is called the wet part of the processing plant. Figure 6 schematically depicts a two-component metering and mixing machine with an impingement mixer as the mixing unit. In practice, machines using the impingement mixing principle are available for mixing up to six components. In the working tanks (a), the raw materials are prepared for processing, i. e., primarily their temperatures are adjusted precisely. Any temperature change, especially of the polyols, results in a viscosity change, which can impair later processing. The working tanks (volume 20–500 L according to the machine size) are therefore usually provided with heating and cooling jackets. Separate temperature-control circuits via heat exchangers are also common. The working tank is also equipped with an agitator for homogenization as well as an automatic refilling system and indicators for level, pressure and temperature. The conditioned raw materials are conveyed via

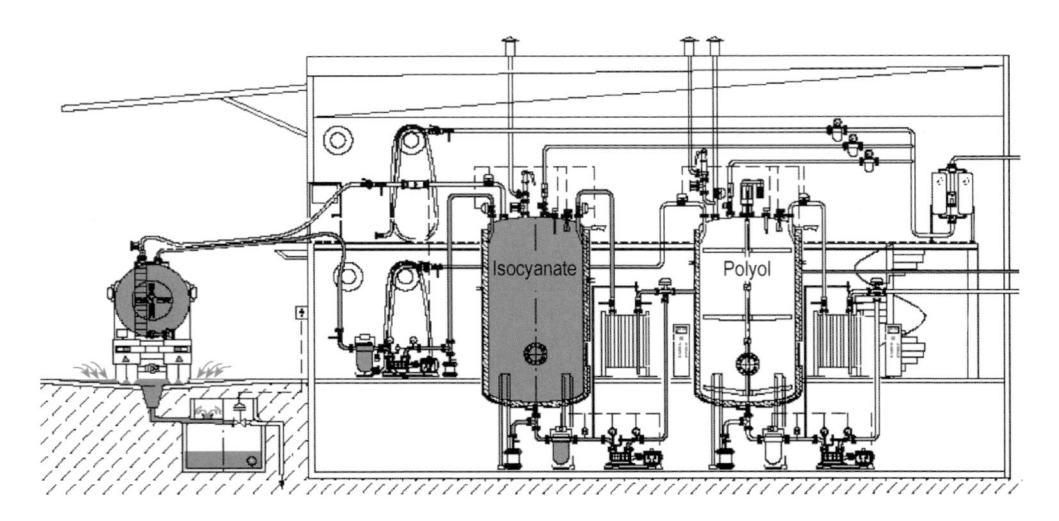

Figure 5. Tank truck unloading and storage tanks

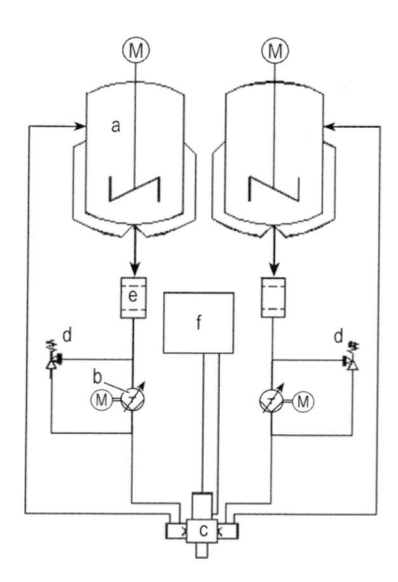

Figure 6. Principle of a high-pressure machine (circulation system)
a) Working tank; b) Metering pump; c) Mixing head; d) Safety valve; e) Filter; f) Hydraulic unit

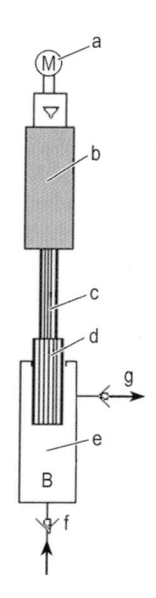

Figure 8. Plunger pump
a) Stepping motor; b) Hydraulic cylinder; c) Hydraulic piston; d) Metering piston; e) Metering cylinder; f) Stop valve: feeding side; g) Stop valve: delivery side

metering pumps (b) to the most important part of the machine, the mixing head (c), which delivers the liquid reaction mixture.

Depending on the PUR system to be processed, precision metering pumps of various outputs (up to 150 L/min) and pressure ranges (up to 250 bar) are used. The pumps must always satisfy extremely high requirements for metering accuracy in order to maintain a certain reaction mixture. In addition to conventional gear pumps and in-line piston pumps, which have been established in PUR technology for many years, axial piston pumps (Fig. 7) and, for systems containing fillers and reinforcements, slow-running single-cylinder positive-displacement pumps are used (Fig. 8).

Two types of mixing principles exist: low-pressure mixing heads with an agitator (3–40 bar) and high-pressure mixing heads (100–250 bar). In mixing heads with an agitator, the raw materials entering the upper part of the mixing chamber are intensively mixed on their way to the outlet by rotating, variable-speed agitators. Agitators include spiked, paddle, or screw agitators, propellers, or turbine impellers. After the end of the shot the agitators are cleaned with a rinsing medium. High-pressure mixing heads operate according to the countercurrent injection principle. The component streams are injected into the mixing chamber via nozzles (Fig. 9) and are mixed by virtue of the kinetic energy used and the chaotic flow behavior created [179].

In controlled-circulation mixing heads, the raw material streams are changed from the "circulation position" (Fig. 10A) (tank → pump → mixing head → tank) to the "injection position" (Fig. 10 B). After injection has occurred, a piston cleans the remaining reaction mixture from the mixing chamber and outlet pipe. These are therefore known as self-cleaning mixing heads.

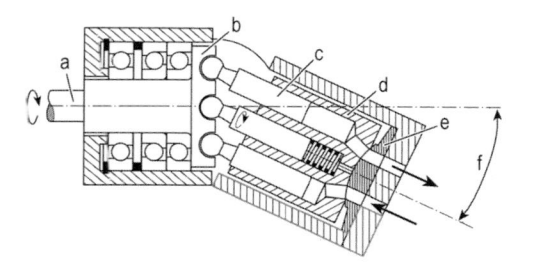

Figure 7. Axial piston pump
a) Shaft; b) Rod plate; c) Piston; d) Cylinder unit; e) Cam plate; f) Angle of inclination to adjust volume flow rate

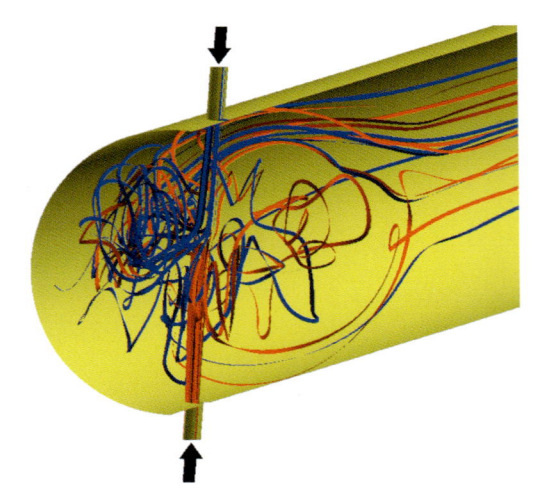

Figure 9. View into an impingement mixer during shot; visualization of streamlines

So-called L-shaped mixing heads are able to improve the homogeneity of the reaction mixture by a perpendicular deflection of the flow direction and simultaneous acceleration of the flow velocity (Fig. 11). Furthermore, the larger diameter of outlet tube compared with the dimension of the mixing tube allows the required laminar flow out of the mixing head.

6.3. Processing Plants

A two- or multicomponent metering and mixing machine is part of each processing plant. Further equipment and machines define the kind of processing plant.

In continuous plants, semifinished products are produced in the form of blocks, slabs, boards, profiles, etc. The scheme in Figure 12 depicts a plant for the continuous production of foam slabs. Four components are metered, mixed with an agitator mixing head, and poured onto the conveyor belt. The bottom and side paper are used to keep the conveyor belt clean and to avoid sticking of the foam to the conveyor belt during rising and curing. Figure 13 is a three-dimensional view of a plant for the manufacture of rectangular slab stocks.

A three-dimensional drawing of a plant for manufacturing metal sandwich panels (Fig. 14) gives an impression of how large and complex PUR plants can be. The length over all stations of processing steps in this case is ca. 350 m.

Plants for producing moldings operate batchwise. The example in Figure 15 shows a top view of a processing plant: On the left side is the wet part of the plant with a total of six components; a robot handles the high-pressure mixing head, and two clamping units are available for pouring into open moulds.

A distinction can be made between plants with movable (Fig. 16) and those with stationary molds (Fig. 17). Movable molds are supplied by a single mixing head, whereas in RIM plants, several mixing heads are used, each of them being permanently assigned to one (stationary) mold. The plants are preferably automatically controlled with electronic equipment.

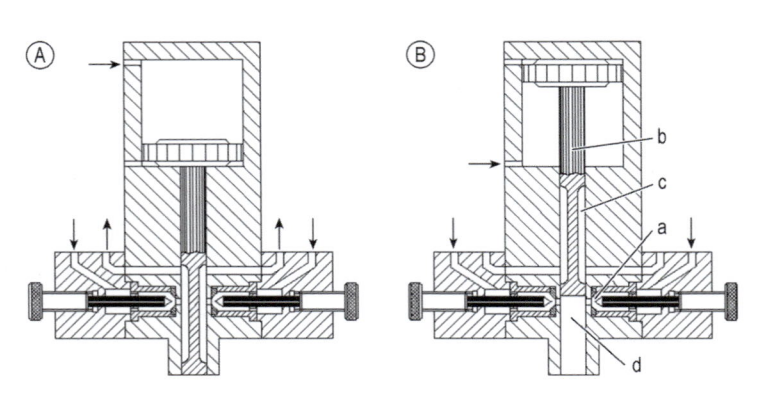

Figure 10. Circulating-groove mixing head
A) Circulation/cleaning; B) Mixing
a) Nozzle; b) Piston; c) Groove; d) Outlet tube

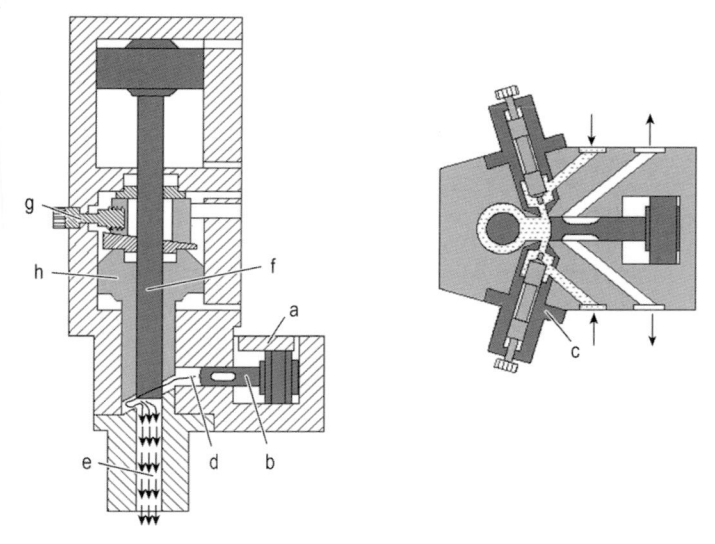

Figure 11. L-shaped mixing head with ring throttle (type Hennecke)
a) Impingement mixer; b) Actuating piston; c) Injection nozzle; d) Mixing chamber; e) Outlet tube; f) Cleaning piston; g) Adjustment of ring throttle; h) Ring throttle

7. Foams

Whilst the reaction of isocyanates with water in most PUR applications is an unwanted side reaction, for foams this reaction is of utmost importance, due to the two reaction products formed. These are CO_2, which acts as the blowing agent for foam formation, and the more or less crystalline polyurea domains, which control cell opening and act as reinforcing filler in the PUR backbone. This chemical blowing process was supplemented in the early 1960s by a physical blowing process using low-boiling liquids as blowing agents. These liquids are vaporized by the heat of reaction of the polyaddition reaction between isocyanates and polyols and/or water. The blowing agent vapor is trapped in the viscous reaction mixture and thus leads to the cellular structure of the foam. Since foam formation is a very sensitive and complex process of interdependent chemical, physical, and rheological processes, specific additives and catalysts are required (see Section 3.5).

PUR foams are generally divided into flexible, rigid, and semiflexible products. Integral skin foams, which have a densified surface layer forming an elastomerlike skin and a foamed core, are described in Section 7.4.

Polyols with a relatively high molecular mass (2000–8000 g/mol) and a nominal functionality of 2–6 yield flexible foams. Combinations of these polyols with low molecular mass glycols or amines as chain extenders and preferably polyisocyanates with a functionality of more than two yield semiflexible foams. Highly branched polyols, e.g., polyethers based on sorbitol or sucrose, with a relatively low molecular mass (<1000 g/mol) are the basis of rigid foams.

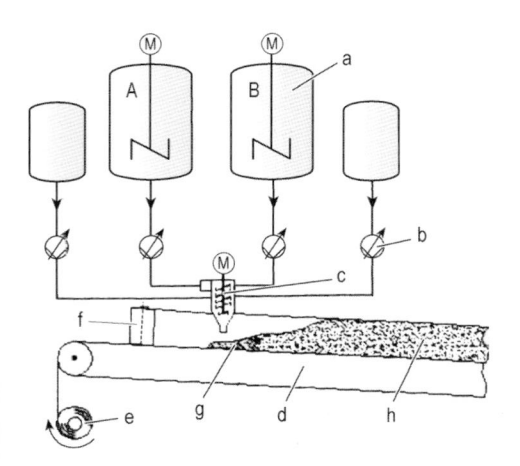

Figure 12. Principle of a plant for continuous production of foam slabs
a) Working tank; b) Metering pump; c) Agitator mixing head; d) Conveyor; e) Bottom paper; f) Side paper; g) Rising foam; h) Cured foam

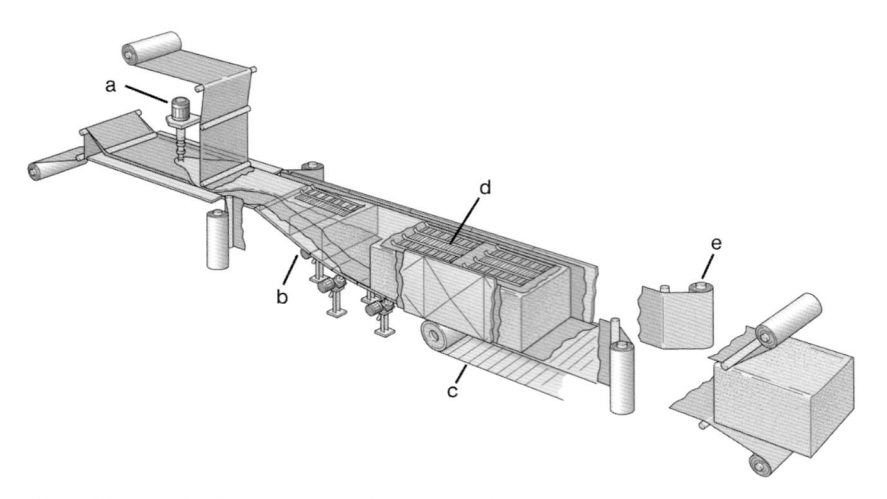

Figure 13. Plant for the manufacture of rectangular slab stocks (Quadro-Foamat-Machine: QFM, source Hennecke)
a) Agitator mixing head; b) Fall plate path adapted to foam expansion; c) Conveyor; d) Device to keep block shape rectangular; e) Paper winch

For all these classes of foams different types of manufacturing processes may be applied. An alternative way of classifying foams is therefore based upon the manufacturing process.

Depending on the processing, a distinction is made between molded foam and slabstock foam. Besides these most prominent groups of flexible foams, other foam types include cavity-filling foam (e.g., filling of cavities in car bodies for acoustical or reinforcing purposes) and spray foam (e.g., for thermal insulation of roofs). The term "filling foam" essentially refers to semiflexible foams behind appropriate overlays, e.g., flexible plastic sheets, films, etc.

7.1. Flexible Foams

Most flexible foams are produced continuously as slabs or batchwise by molding. They are

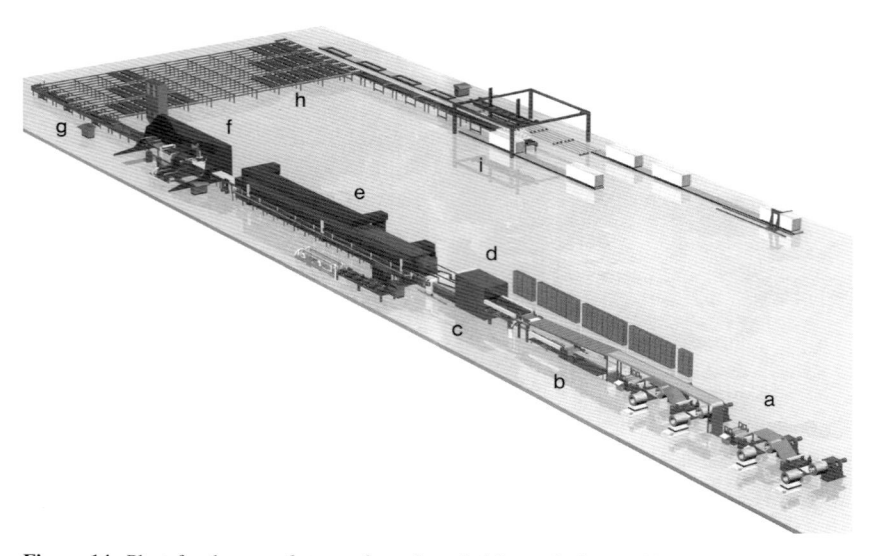

Figure 14. Plant for the manufacture of metal sandwich panels (source Hennecke)
a) Uncoiling station; b) Profiling machine; c) Heater; d) Twin-head foaming portal; e) Laminator; f) Cutting machine/bandsaw; g) Turn over system for roof elements; h) Cooling section; i) Stacker

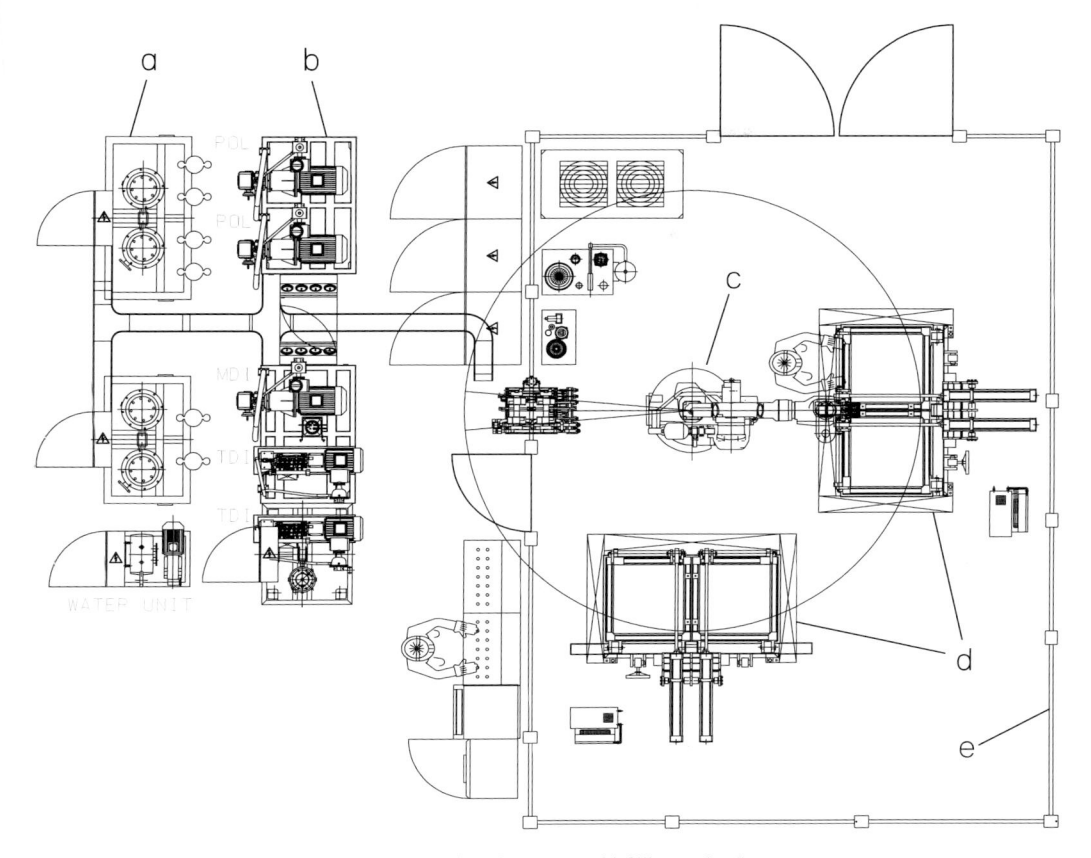

Figure 15. Layout of a plant for producing moldings by open-mold filling technology
a) Tank unit; b) Pump unit; c) Robot with mixing head; d) Clamping units; e) Protective guard

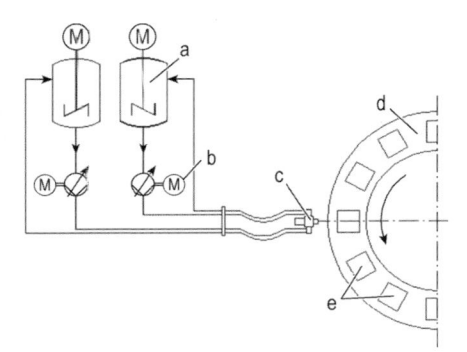

Figure 16. Principle of a plant with movable molds and stationary mixing head
a) Working tank; b) Metering pump; c) Mixing head; d) Turntable; e) Molds

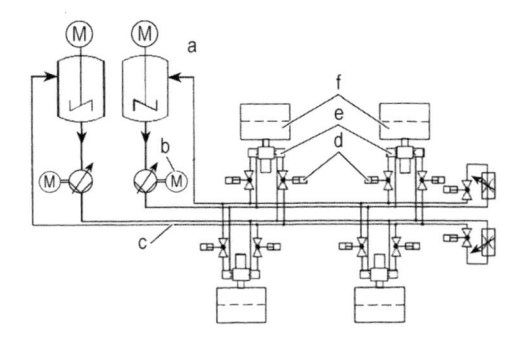

Figure 17. Principle of a RIM plant
a) Working tank; b) Metering pump; c) Circulation system; d) Selection valves; e) Mixing heads; f) Clamping units with tools

characterized by reversible deformability and an open-cell structure, associated with air permeability.

7.1.1. Flexible Slabstock Foam

7.1.1.1. Raw Materials

Based on different chemistries on the polyol side, three fundamentally different types of slabstock have to be considered: Polyester foams, made from polyester polyols, conventional polyether foam, made from polyether polyols with mainly secondary OH groups and high-resilient (HR) foamf, made from polyether foams with mainly primary OH groups.

The isocyanate used for polyester foams is usually TDI 65 (see Section 3.1.1) or blends of TDI 65 and TDI 80; the typical isocyanate for polyether foams is TDI 80. High-resilient slab stock foam is produced from polyethers and TDI 80, both optionally modified.

Main components (parts by weight) of a formulation for flexible polyester foam:

Polyester polyol (OH no. 65), e.g., Desmophen 2200	100
TDI 65 and/or TDI 80, e.g., Desmodur T65, Desmodur T80	variable
Index	95
Water	2.5–4.8
Amine catalyst	0.3–0.6
Foam stabilizer	0.7–1.3
Diethanolamine	0.8–1.5
Tin dioctanoate	0.1–0.35
Foam stabilizer	0.3–1.0

Main components (parts by weight) of a formulation for flexible polyether foam:

Polyether polyol (OH no. 48), e.g., Arcol 1108	100
TDI 80, e.g., Desmodur T80	variable
Index	98–115
Water	2.0–4.5
Amine catalyst	0.15
Tin dioctanoate	0.12–0.28
Foam stabilizer	0.6–1.5

Main components (parts by weight) of a formulation for high-resilient (HR) flexible foam:

Grafted polyol (OH no. 28), e.g., Desmophen 7653	100
TDI 80, e.g., Desmodur T80	variable
Index factor	95–113
Water	1.5–4.0
Amine catalyst	0.10–0.20

The polyesters (M_r 2000–3000 g/mol) are usually based on adipic acid and linear glycols (e.g., diethylene glycol), or triols (glycerol or trimethylolpropane) if slight branching is desired.

The polyethers used (M_r 1000–6000 g/mol) are di- and/or triols with a varying propylene oxide/ethylene oxide ratio, which is tailored to the foam's application. An important factor for polyether foaming is a well-balanced ratio of the metal and amine catalysts, which controls the simultaneous reactions of the NCO group with hydroxyl groups (polyurethane reaction) and with water (blowing reaction). If the polyurethane reaction rate predominates, the foam may shrink due to the formation of a high proportion of closed cells. If the water reaction proceeds too rapidly, the carbon dioxide formed cannot be contained in the cells; splits result and the foam can even collapse.

High-resilient foams are produced from highly reactive grafted polyethers which contain a high proportion of primary OH groups (from ethylene oxide). Grafting is achieved by in situ polymerization of styrene and acrylonitrile to form a stable SAN suspension in the polyol. Another class of grafted polyols is based on the polyaddition of TDI and hydrazine to yield PHD polyols which contain suspended polyhydrazo-dicarbonamide/polyurea particles. If highly functional (pre-cross-linked) modified TDI types are used, unmodified polyols can be applied, but due to advantages in foam processing, this approach has been generally replaced by grafted-polyol technology.

7.1.1.2. Production

The raw materials are processed in fully continuous plants to give flexible foam slabs up to 220 cm wide, 120 cm high, and of any length. Depending on the slab dimensions, internal temperatures of up to 165 °C are reached as a result of the exothermic reaction. This temperature must not be exceeded, otherwise severe scorching and even self-ignition of open-celled foam may occur. The foaming process is completed after ca. 3 min, and final curing takes

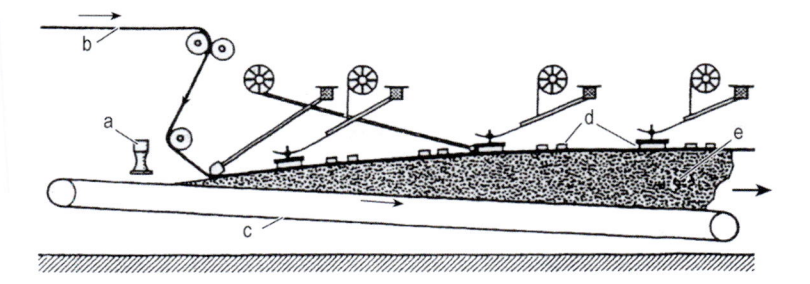

Figure 18. Principle of a plant for production of rectangular flexible foam slab stocks
a) Covering paper; b) Skids; c) Mixing head; d) Conveyor belt; e) Foam

from about 10–12 h up to 72 h, depending on the foam type. The slabs are therefore stored in curing storage facilities until they have cooled and attained their final mechanical properties, so that they can be transferred to further processing, fabrication, etc. The length of the slabs varies from 10 m (short slabs) to 120 m (long slabs). To minimize trimming losses of the foam buns, which would be dome-shaped under free-rise conditions, various technologies are applied to produce rectangular slabs.

Figure 18 shows the principle of a rectangular slabstock plant (Hennecke Planibloc process) [180]. A cover is applied to the rising foam. The pressure of the skids avoids the development of the usual slab dome without the foam becoming compacted. Besides a rectangular slab cross section, this process gives an increased yield of foam due to partial foaming of the mixture at the slab surface, which would otherwise form the surface skin.

In other processes a polyethylene film is drawn up to the slab sides in synchronization with the rising movement of the foam (Draka-Petzetakis) [181, 182]; or the reaction mixture is discharged into a trough with a sloping bottom, in which it rises and then flows off over a downward incline (fall plates) onto the conveyor (Maxfoam) [183]. A more recent development (Multiflex, Quadrofoamat, see Fig. 13) combines the fall plates with liquid laydown principles.

The application of reduced air pressure during the foaming process (variable-pressure foaming, Beamech [184]) or the introduction of carbon dioxide into the mixing chamber (e.g., Multistream (Fig. 19) and Novaflex (Fig. 20), Hennecke [185]) has allowed for the complete

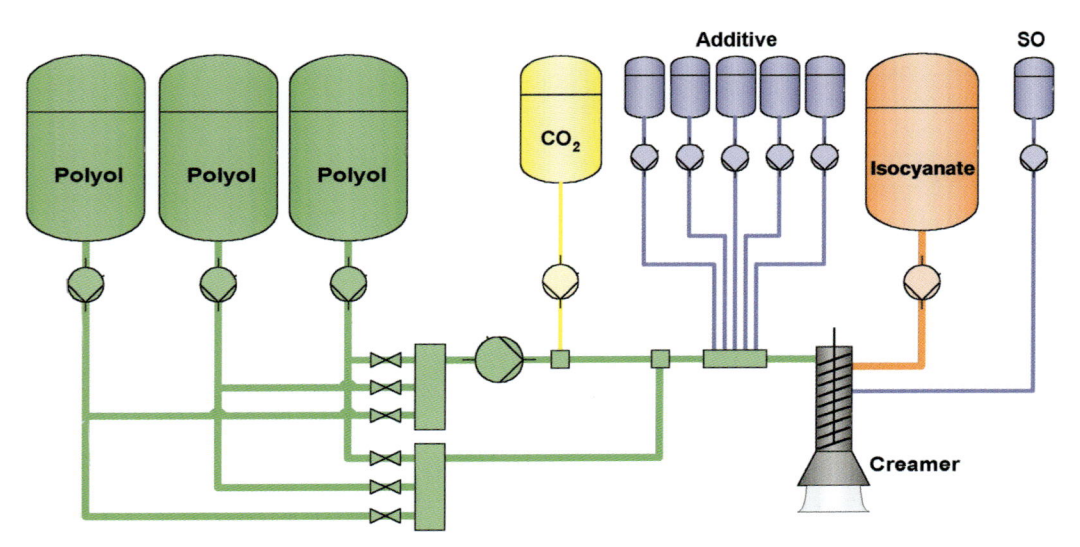

Figure 19. Multistream (Hennecke)

Figure 20. Novaflex creamer (Hennecke)

elimination of other physical blowing agents such as CFCs.

Rectangular blocks can also simply be produced batchwise by introducing the reaction mixture into metal or wooden boxes and at a given time putting a floating cover on the rising foam. In addition to primitive methods ("box foam"), mechanized box processes are also offered for this purpose with relatively low investment costs and a processing capacity of up to 1000 t/a (e.g., Hennecke Bloc-Foamat BFM 100). Special versions of these batch plants are available so that different ambient pressures can be applied during the foaming stage (e.g., Hennecke Megafoam).

In addition to the production of blocks with a rectangular cross-section, continuously produced blocks with a round cross section (diameter ca. 120 cm) have become industrially important (Reeves Brothers, General Foam [186, 187]). These buns can be peeled seamlessly.

7.1.1.3. Properties

The mechanical properties of the foam are largely determined by the raw materials and the formulation used and by its apparent density. For flexible foam slabstocks, the usual densities are 20–40 kg/m^3, but extremely light and extremely heavy types of foam (from 16 kg/m^3 and up to 130 kg/m^3) can be produced industrially for special fields. The choice of density is determined by the specific application of the foam: the higher the density, the better the properties or the lower the fatigue of the foam when in continuous use.

The hardness (measured as compressive strength at 40 % compression) is an important characteristic of PUR foams (Table 4). The hardness can be varied within wide limits by altering the formulation. Considerably softer conventional foams are obtained, e.g., by including polyethers of high ethylene oxide content and special additives. Hardness control is also possible through the index or the addition of grafted polyols.

Apart from the hardness, other characteristic values can be derived from the hysteresis loop (Fig. 21). The slope of the loading curve and the area enclosed by the loading and unloading

Table 4. Relation between apparent density and compressive strength of various grades of flexible foam slab stock

Flexible foam system	Apparent density (ISO 845), kg/m^3	Compressive strength at 40 % compression (ISO 3386), kPa
Polyester foam, based on Desmophen 2200B/ Desmodur T65	35	6.5
Polyether foam, based on Polyether 10WF15/ Desmodur T80/T65	36	6.0
Polyether foam, based on Arcol 1108/ Desmodur T80	36	4.5
HR slab stock foam, based on Desmophen 7619W /Desmodur T80	34	2.7

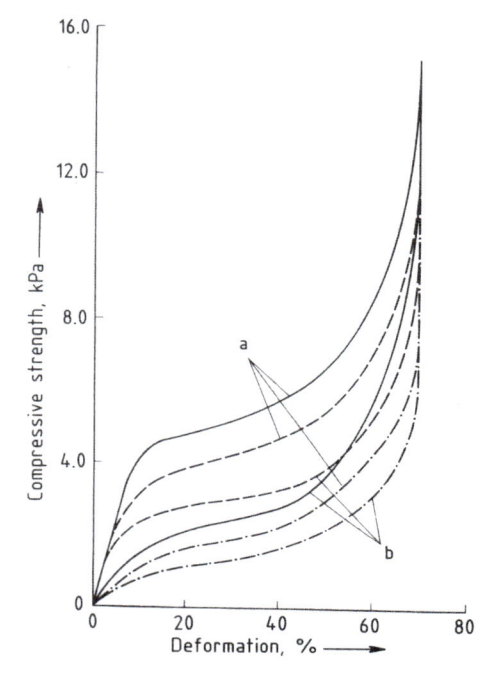

Figure 21. Hysteresis loops of different grades of flexible foam slab stock (apparent density 35 kg/m^3)
a) Loading lines; b) Unloading lines
— Polyester-based foam; – – – Polyether-based foam; –·–·– HR foam

Table 5. Mechanical properties of various grades of flexible foam slabstock

Flexible foam system	Tensile strength (ISO 1798), kPa	Elongation at break (ISO 1798), %	Compressive set (after 90% deformation; ISO 1856), %
Polyester foams	160 – 220	100 – 400	5 – 15
Polyether foams	100 – 180	100 – 400	2 – 10
HR slabstock foams	50 – 120	80 – 200	2 – 10

curves (hysteresis) are measures of the elasticity and energy absorption of the foam: the closer the curves are together, the more elastic the foam; and the larger the area, the better the energy absorption or damping properties of the foam. HR slabstock foam has the best elasticity, followed by polyether foam and polyester foam.

Some important mechanical properties are summarized in Table 5 to demonstrate their range of values. Table 6 shows the values for fatigue in continuous use under the conditions of short-duration tests.

The following list contains some additional properties of flexible PUR slabstock foams:

Thermal conductivity: 0.033–0.040 W m^{-1} K^{-1} (at 0–70 °C and densities of 20–70 kg/m^3)

Temperature stability: from −40 to +80 °C without embrittlement.

Open-cell character (see also reticulation capability): 98–100 % with respect to air permeability, not with respect to resistance to air flow.

Chemical resistance to acids and alkalis: polyether foam has a better resistance to acids and alkalis than polyester foam.

Chemical resistance to organic solvents: polyester foam shows a better resistance to organic solvents than polyether foam.

Chemical resistance to atmospheric oxygen: oxygen attacks foam slab stocks only superficially without harming the serviceability. Polyether foam is more easily oxidized than polyester foam. However, if aromatic polyisocyanates and polyether polyols are used for foam production, the foam will decompose over time if exposed to oxygen and light.

Combustibility (see Section 14.4): Flexible slabstock PURs are combustible, like all organic (carbon-containing) materials. By addition of suitable flame retardants the combustion behavior of foams can be adjusted to meet several flammability standards such as MVSS 302, BS 5852 part 2 Crib V, or even Fire Class B1 (DIN 4102). Also a post-treatment with latices containing flame retardants is possible, in particular if demanding tests have to be met. Besides flame retardants, the type of foam (polyester, conventional, or high resilient) is important if certain flammability standards have to be met.

Bactericidal and fungicidal treatment is possible by including appropriate additives in the formulation or by post-treatment.

Table 6. Change of hardness on ageing of various grades of flexible foam slab stock

Flexible foam system	Apparent density (ISO 845), kg/m^3	Decrease of indentation load deflection (ISO 3385), %	Decrease of compressive strength (5 d, 90°C, 95% rel. hum.), %
Polyester foams	25	43	35
	35	29	40
	45	26	40
Polyether foams	20	34	10
	25	29	14
	35	19	18
	50	16	16
HR slabstock foams	25	24	8
	35	18	13
	45	16	18

Reticulation capability (see also open-cell character): the flow resistance can be greatly reduced by using chemical or physical processes (Chemotronics) to remove residual cell membranes (filter foams).

Physiological aspects: According to extensive studies (both dermatological and oral toxicity in animals), flexible PUR foam of defined composition is physiologically harmless.

7.1.1.4. Trimming and Processing

Various possibilities and machines are available for processing the cured slabstock foam to finished articles. The simplest processes for foam blanks are horizontal-cutting (e.g., long-block cutting/peeling), vertical cutting, and angular cutting. These are carried out with continuously rotating knives.

Profile cutting in the form of copy cutting and punching, compression cutting, and milling is used for further forming. With profile cutting tailor-made cuts can be made by following prefabricated templates or drawings (Fig. 22).

Compression-cutting is a special variant of profile-cutting. All these processes exploit the easy deformability and excellent recovery of PUR flexible foam (Fig. 23).

Advances in foam trimming equipment such as fully automated CNC processing lines allow for economical conversion of foam buns into very sophisticated contoured parts (Fig. 24). At the same time, trim losses are minimized by specific software [188].

Form milling is used for rounding edges on seat backs and armrests as well as on seat pads. Due to the large variety of milling tools, any shape is possible (Fig. 25).

Other post-treatment methods include reticulation, impregnation, compression, welding, cementing, laminating with textiles or films

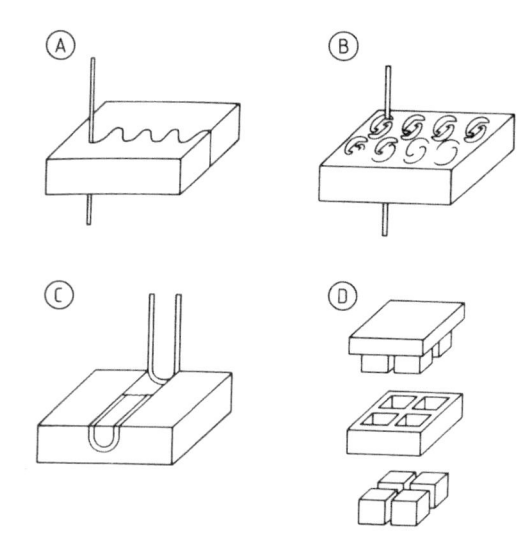

Figure 22. Cutting methods and punching
A) Hot-wire cutting; B) Copy cutting; C) Profile cutting; D) Punching

by flame lamination or adhesive bonding, quilting, and covering.

The trimmed-off foam is flaked in suitable mills. The flakes can be used as filling material or are bonded to form flake-composite blocks, referred to as rebonded foam. These composite blocks are then fabricated similarly to normal foam slab stocks (see Section 14.3).

7.1.1.5. Applications

The variety of properties and the continuous development of new foam grades and further processing techniques have made slabstock indispensable for many applications.

For furniture, the use of polyether foams extends from complete foam upholstery to the simple seat cushion. Very comfortable cushioning is achieved by combining types with different properties (e.g., density, hardness, or elasticity).

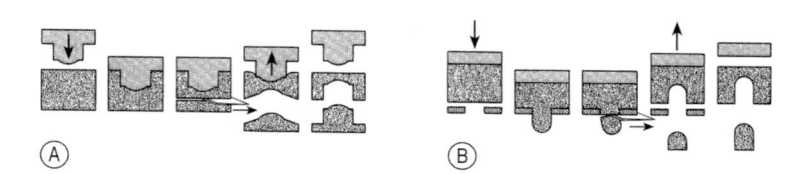

Figure 23. Compression cutting
A) Convoluting process; B) Skiving process

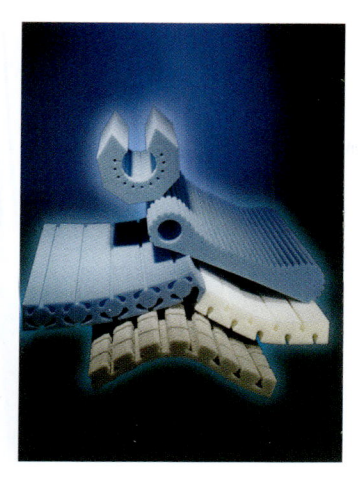

Figure 24. Contoured parts, produced by different trimming technologies

Full PUR foam mattresses at present account for approximately one-third of all mattresses sold in Europe. The relative low weight (e.g., versus latex foam) and the freedom of design by combining various foam types (high resilient, conventional, viscoelastic) and grades (density, hardness) in conjunction with advanced cutting systems have turned these mattresses into high-tech applications. All required grades of hardness are available, from the child's to the intervertebral disk mattress. Even in steel-spring mattresses, a layer of slabstock foam is used as cushioning and lining material.

An important field of application for slabstock foam is the lamination of fabrics. Polyester foams are preferred to polyether foams due to their melting properties and their specific cell structure. The foams can be bonded on one or two sides with textile widths by adhesive or flame lamination and are widely used for automotive interior trim applications such as seat covers or headliners.

Other applications in automotive industry are sun visor fillings, which can be manufactured

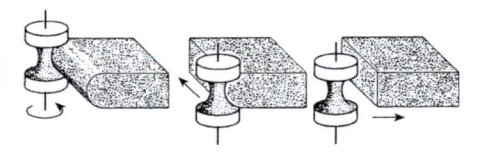

Figure 25. Shaping of flexible foam by milling

from semirigid polyester foams, and sound-absorption materials for passenger, engine, and trunk compartments. PUR foams are also used as sealing and filter materials in the ventilation system. All foams for automobiles must satisfy the fire test MVSS 302 (see Section 14.4). Depending on the density and the chemical nature of the PUR foam (HR or conventional foam) flame retardants must be added to the foam formulation to meet this requirement.

In the clothing industry the above-mentioned laminated fabrics are incorporated into jackets and coats. Other applications are paddings in ski and sports boots and underlinings in the purse-making industry.

Flexible PUR foam is particularly suitable as a packaging material for fragile goods. Because of the variety of cutting processes, extremely complicated shapes can be produced for the transportation of goods.

The main application in the household sector is sponges. They are produced from both polyether and polyester foam in very diverse colors and shapes. They can be fine- or coarse-pored and range from extremely flexible to almost rigid. The foam's hydrophilicity can be increased by using special additives. The range of uses in the household includes adhesive, sealing, and jointing tapes as well as under-carpet padding, doormats, and cleaning brushes. Special grades can be obtained by bonding with thermoplastic films by high-frequency welding.

The ability of flexible foams to absorb airborne noise efficiently is exploited for noise abatement. The low thermal conductivity of flexible block foams is used for the thermal insulation of heat accumulators and pipes. For use as filter material, e.g., in ventilation and air-conditioning systems or as dust filter in breathing masks, the flow resistance of the foam must be reduced by post-treatment. Depending on technical requirements, the foam can also be used as a filter for liquids. Since the foams do not absorb X rays, they are an ideal material for the comfortable support of parts of the body during X ray exposure.

7.1.2. Molded Flexible Foam

The foam molding process is used when it is impossible or uneconomic to produce a complex geometry by trimming slabstock foam, or when the incorporation of metal inserts (for fastening cushioning units or coverings) or even steel springs (for furniture applications) is desired. A distinction is made between hot-cure molded foam and cold-cure molded foam.

7.1.2.1. Production

To produce hot-cure foam, trifunctional polyethers with a low content of highly reactive primary OH groups in the molecular mass range of 3000–5000 g/mol are reacted with toluene diisocyanate (TDI) with ca. 80 % of the 2,4′ isomer, e.g., Desmodur T80. The blowing agent is carbon dioxide, which is formed by the water-isocyanate reaction. Polyether polysiloxanes are employed for stabilizing the foam and due to the lower reactivity of the polyols, tin dioctanoate or other tin compounds are used to catalyze the polyurethane reaction, amines are used to catalyze the blowing reaction, and rather high mold temperatures must be applied. The hardness of the foams can be adjusted by adding grafted polyols for increased load-bearing properties or by using special polyols or additives for softer grades. The resulting foam is open-celled and does not require crushing for cell opening.

Cold-cure foams are produced from trifunctional polyethers with enhanced reactivity (high proportion of primary OH groups) in the molecular mass range of 5000–6000 g/mol and grafted polyols. A variety of different isocyanates are used, usually with a functionality exceeding two (modified TDI types, blends of TDI and higher functional MDI; special MDI types). If pure TDI 80 is applied, polyethers with a functionality slightly higher than three are usually used (e.g., Hyperlite). No further blowing agents are used; the carbon dioxide formed from the reaction of water with isocyanate is the sole blowing agent. Due to the higher reactivity of the polyols no metal catalysts are required for curing, and the mold temperatures are relatively low ("cold" cure) compared to hot-cure foam. On demolding the cold-cure foam is closed-celled and must be mechanically crushed to avoid shrinkage and to achieve consistent physical properties.

7.1.2.2. Molding Process

The molding methods for hot- and cold-cure foam systems differ sharply from each other.

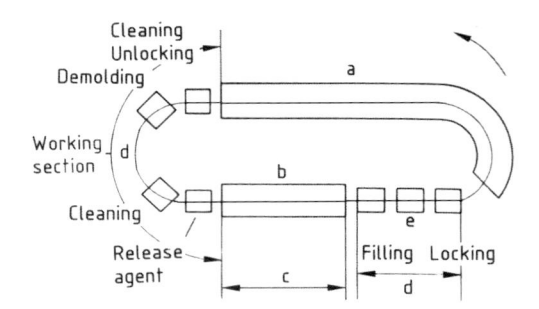

Figure 26. Plant for hot-cure moldings
a) Oven (160–250°C); b) Recirculating air/water bath; c) Cooling section; d) Working section; e) Inserts

This is mainly because external heat must be supplied to a hot-cure molded foam to cure the foam's surface fast enough.

Figure 26 shows the individual process steps of the hot-cure foam molding process. This production sequence has a cycle time of ca. 20–25 min, depending on the nature and size of the foam parts.

The molds consist of black iron plate (wall thickness 1.5–2.0 mm), aluminum sheet (wall thickness 4–6 mm), or cast aluminum (wall thickness 6–10 mm). Since the foaming process for this foam type must proceed unpressurized, good mold venting is necessary. The arrangement of the vent holes is determined experimentally. In manufacturing the molds, a volume loss during the curing process of ca. 2 % must be taken into consideration. The precise value is specific for the formulation and must be determined experimentally. Release agents are mainly aqueous wax or soap dispersions. High- or low-pressure machines, operating batchwise, are used for metering the raw materials. The molds are usually heated by hot-air ovens. The heating capacity of the ovens for the full-cure process must be adapted to the mold material and the required cycle time. After removing the molded foam parts, the molds are cooled by water or cold air to the loading temperature.

The curing process can be shortened by rapid adjustment of the internal wall temperature of the mold to ca. 120 °C after the end of the rising time. In this case, the molds are heated directly with liquid heating media to conserve energy.

Cold-cure foam systems require much lower mold temperatures. To achieve the optimum quality of the foam skin, the molds are adjusted

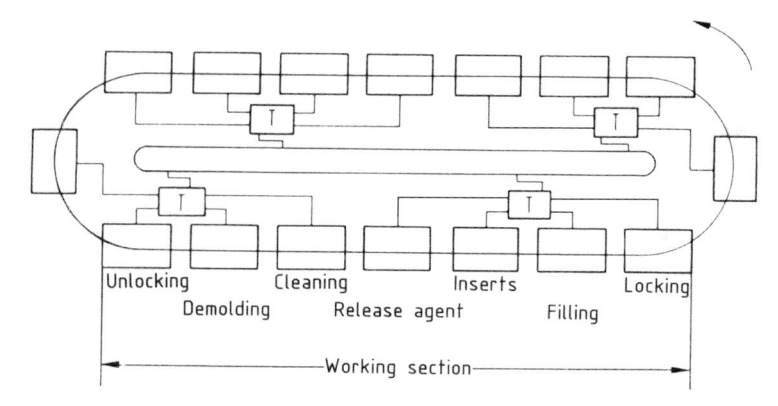

Figure 27. Plant for cold-cure moldings
T = thermostatting units

to the optimum temperature for the mold material and the release agent (between 40 and 65 °C). Cooling equipment is not required (Fig. 27). Depending on the combination of polyols, the isocyanate, the catalyst package, and the shape of the cavity, the foam can be demolded after 2–6 min.

In contrast to open-celled, hot-cure molded foam, the closed cells of cold-cure foam parts must be mechanically opened. This can be achieved by roller crushers with counter-rotating rollers, compressing the parts to a fraction of their original height, or by fast alteration of the ambient pressure in a pressure chamber.

One hour after demolding, cold-cure molded foams have reached only 50–80 % of their final hardness. Therefore, cold-cure foam parts need to be handled with care in the first few hours after production. Depending on the chemistry and the climatic conditions, final properties are achieved after 5–12 h (for comparison: a hot-cure foam will reach its final hardness within the first 2 h after demolding).

The foam molds must be designed for an internal mold overpressure of 0.3–0.6 bar. When fabricating the master patterns for the mold replica, a loss in volume of the molded part of up to 2 % must be taken into account.

Larger plants (greater space requirement, more molds) are required for the production of hot-cure foam than for cold-cure foam, owing to both the heating/cooling cycle and the longer mold holdup times of ca. 12 min compared to 4–8 min for cold-cure foam. Under ideal conditions, hot-cure molded foam has the advantage of a lower molding mass compared to cold foams at the same compression-load deflection.

7.1.2.3. Properties

Hot-cure and cold-cure foams differ fundamentally in their mechanical properties. The lowest densities can be achieved with hot-cure foam. At the same density, a hot-cure foam will usually provide a higher hardness. However, due to higher venting losses, savings in material for a hot-molded piece of foam are significantly lower than expected based on the simple comparison of apparent densities. The compression load deflection curves of molded foams can be compared with that of free-rise flexible foams (Section 7.1.1.3): the characteristics of hot-cure molded foam are similar to those of conventional slabstock foam; cold-cure molded foam can be compared with HR slabstock. The hysteresis curves show that cold-cure foam has a higher elasticity (narrower hysteresis loop) than molded parts from hot-cure foam. Table 7 gives a general overview of the properties of hot- and cold-cure foams.

With a cold-cure foam cushion, greater comfort is experienced due to its higher elasticity. However, for an automobile seat this higher elasticity means a reduced damping of the vehicle's vibrations. Since the construction of the seat (full foam or combination of foam and springs) and the damping behavior of the vehicle structure must be matched, the question whether to use hot-cure or cold-cure molded foam must be decided for each individual case. A second and more basic difference between

Table 7. Average properties for seat grades of flexible molded foams

Property	Hot-cure foam	Cold-cure foam	
		TDI based	MDI based
	HC	HR-T	HR-M
Apparent density (EN-ISO 845), kg/m^3	28 – 40	35 – 50	45 – 65
CLD-Hardness at 40 % (EN-ISO 3386-1), kPa	3 – 7	2 – 8	4 – 12
Compression set 50 %, 70 °C, 22 h (EN-ISO 1856), %	2 – 4	4 – 8	4 – 8
Elongation (EN-ISO 1798), %	120 – 160	100 – 150	90 – 120
Tensile strength (EN-ISO 1798), kPa	80 – 140	130 – 200	100 – 160
Tear strength (ASTM D 3574-F), N/m	200 – 400	200 – 450	160 – 250
Humid ageing (OEM specific)	very good	good	good – very good
Hysteresis (Energy Dissipation)	high	low	medium

hot- and cold-cure foam is their fire behavior. Cold-cure foams down to a density of ca. 35 kg/m^3 usually meet the fire standards of the automobile industry, even without added flame retardants. For hot-molded foam, flame retardants must be used.

7.1.2.4. Applications

Flexible molded foams are principally used in all kinds of seat pads, seat cushions, backrests, bench seats, headrests, etc. in passenger cars (see Fig. 35), motorcycles, and commercial vehicles, aircraft, and railways. Another important area of application in the automotive industry is sound abatement. Flexible molded foam is used either on its own for absorption of airborne noise or as a composite in conjunction with a heavy layer (EPDM, EVA, or PUR elastomers filled with, e.g., barium sulfate) or as backfilling foam on the vehicle's floor cover. These composites may act either as sound barriers or can be tailored to dampen structureborne noise. In addition, flexible molded foams are used in the furniture industry, e.g., for office chairs, armrests, and in specific combinations with steel springs, which are incorporated into the molded foam. Sealing lips for passenger car tail lamps and for filters can also be produced from molded foams.

7.2. Semirigid Foams

7.2.1. Applications

Semirigid foams are used primarily for improving inner safety and comfort in automobiles. Combined with an external layer of ABS/PVS or PUR skin and a metal or plastic carrier, they are used to produce instrument panels (dash boards), door and pillar covers, armrests, and consoles. The foam has a density range of 80–180 kg/m^3 in the finished part. Key properties are long-term adhesion to skin and carrier, a uniform, fine cell structure, low emission, and high resistance to aging and heat.

Special grades (EA foams) are optimized for energy absorption. On impact, they dissipate the kinetic energy by deformation and thus protect the passengers in a crash (Fig. 28).

7.2.2. Production

Parts made of semirigid foam are produced mostly in a two-step process. First, the preformed skin (ABS/PVC by thermoforming or PUR by spraying) is positioned in the cavity of

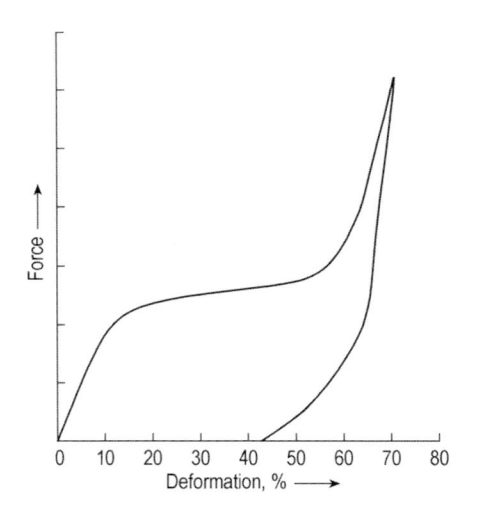

Figure 28. Dynamic force–deformation characteristics of a semi-rigid PUR foam

the mold. A plastic or metal carrier is fixed on the lid of the mold. In the second step, the mold is closed and the semirigid foam is injected. Alternately, the semirigid foam is spread in the open mold and the lid is closed before the foam expands.

The two PUR components are mixed by high-pressure machines. The molds are made from aluminum, often coated with epoxy resin. If the foam has contact to the surface of a mold, a wax-based release agent must be used. The demold times are between 1 and 6 min, depending on the size, thickness, and geometry of the parts and on the reactivity of the foam system.

7.2.3. Properties

Semirigid foams are mostly open-celled and have a higher compression load deflection and lower elasticity than flexible foams. Typical properties are:

Apparent density (ISO 845)	80–180 kg/m^3
Tensile strength (ISO 1798)	300–400 kPa
Elongation at break (ISO 1798)	30–60 %
Compression load deflection at 40 % compression (ISO 3386)	80–200 kPa
Compression set at 50 % compression (ISO 1856) 22h at 70 °C	< 10 %

The dynamic impact characteristics (Fig. 28) are important for the dampening and energy absorbing behavior.

7.3. Rigid Foams

Rigid foam is in most cases a closed-cell material that is used for thermal insulation. Besides providing thermal insulation the foam may also have to bear mechanical loads in some applications. Composite structures in which the self-adhesive effect is used to bond the foam to a substrate during the foaming process are often found in the field of rigid foam. This reinforcing sandwich-effect improves the mechanical properties significantly.

Rigid foam is used in many ways. The main applications for rigid foam are appliances (refrigerators, freezers), insulation boards (with and without facings), composite panels (mostly with metal facings), pipe insulation (as insulating shells or as composites), insulated containers, and in situ foam (spray foam).

7.3.1. Raw Materials

Rigid PURs require a highly cross-linked molecular structure. To achieve this, raw materials with high functionalities and low molecular weight are preferred on the polyol and on the isocyanate side. Therefore, short-chain polyether polyols and polymeric MDI are the materials of choice. Low molecular weight polyester polyols are also used.

Typical polyether polyols (see Section 3.2.1) are produced from a highly functional starter molecule (sucrose, sorbitol, glycerol, diethylene glycol, ethylenediamine, and others) or mixtures thereof by adding alkylene oxides, mainly propylene oxide but sometimes ethylene oxide. The resulting products are viscous liquids.

The *polyester polyols* (see Section 3.2.2) are typically based on aromatic acids like phthalic acid. (Di)ethylene glycol is mostly used as alcoholic component. Glycerol or other highly functional alcohol components may be used to increase the functionality. However, the use of starting materials with higher functionalities is limited because of the increasing viscosities of the resulting products.

The *polymeric* MDI types (see Section 3.1.1) that are used have functionalities in the range of roughly 2.5 to 3.5, corresponding to viscosities of about 100 to 1000 mPa · s at 25 °C.

Blowing agents (see Section 3.5) are not only used to create the foam. Since the blowing agent remains inside the closed cells it has a strong influence on the thermal conductivity of the resulting foam. The choice of the blowing agent depends on the most important criteria in the respective application, such as the lowest possible thermal conductivity, processability, or relationship between cost and performance. In most cases, a combination of a physical blowing agent (a low-boiling liquid or a gas) and a chemical blowing agent (mostly carbon dioxide formed by the chemical reaction of water with isocyanate) is used. While, according to the Montreal Protocol, the use of chlorine-containing halogenated hydrocarbons is still allowed in certain areas of the world, the use of these products in Western Europe is banned. Here only fluorinated hydrocarbons such as tetrafluoroethane (R 134a), pentafluoropropane (R 245fa), and pentafluorobutane (R 365mfc) may be used. Hydrocarbons such as n-, iso-, and cyclopentane or mixtures with other hydrocarbons like propane or butane are widely used, despite their flammability. Methods for a safe handling of these products are industrially available and installed. All these blowing agents have very small coefficients of diffusion; they remain within the foam during its lifetime. Where no (metallic) facings are provided, the thermal conductivity increases very slowly because of air diffusing into the foam. Therefore, in some applications, e.g., the construction industry, increments are added to the measured values of the thermal conductivity to take this ageing process into account. In this way it is ensured that a sufficient level of insulation is installed.

Catalysts (see Section 3.5), sometimes called activators, are used to control the foaming process. Products used as catalysts are tertiary amines, organometallic compounds, and alkali metal and quarternary ammonium salts of carboxylic acids. The products may be classified according to their performance-without really being able to draw sharp lines-as blowing, gel, and trimerization catalysts. A few examples taken from a wide range of products are given here: bis[2-(N,N-dimethylamino)ethyl]ether and *N*, *N*-bis(*N'*,*N'*-dimethylaminoethyl)methylamine as blowing catalysts; triethylenediamine (dissolved) and dibutyltin dilaurate as gel catalysts and potassium acetate (dissolved) and 1,3,5-tris(3-(dimethylamino)propyl)hexahydrotriazine as trimerization catalysts.

Foam stabilizers (see Section 3.5), also called surfactants, are required to prevent the small bubbles in the liquid reaction mixture from coagulating. Fine-celled foams can be prepared by using these products. To produce foams with a fine cell structure, some air (or other inert gases) may be dispersed into the polyol component or into the reaction mixture for nucleation. A fine cell structure is important for a low thermal conductivity. Chemically, most of the products used in the industry are polyether-modified polysiloxanes. The structural variations possible by using different polyethers and different polysiloxanes lead to the great number of products available on the market, optimized for various purposes. Some of these purposes are: flowability (describes the expansion of the foam into narrow cavities), adhesion to facings (the force required to remove the facing), fine cell structure (small cells of similar size), and flammability (resistivity to flames or high temperatures).

Flame retardants (see Section 3.5) are used to make the foams less flammable. The methods for testing flammability are manifold. However, official requirements regarding flammability are in place for some applications, e.g., in the construction industry. Since PUR foams are based on organic molecules, it is only possible to achieve certain flammability classifications within the group of flammable materials. Flame retardants used in the field are usually substances based on halogens and/or phosphorus. Some products have OH groups, so they can react with isocyanate and are incorporated into the macromolecule; others are nonreactive. Examples of reactive flame retardants are polyethers made from dibromobutenediol and epichlorohydrin or brominated aromatic polyester polyols. Examples of nonreactive flame-retardants are tris(2-chloroisopropyl)phosphate, diphenyl cresyl phosphate, and dimethyl propyl phosphonate. Apart from red phosphorus, which is used in the production of rigid slabstock foam by low-pressure mixing, solid flame retardants have hardly been used in the field of rigid foams. A technique has been developed for processing solid flame retardants continuously on high-pressure machines [189].

7.3.2. Processing

Mixtures of the raw materials mentioned above are processed to produce rigid foams. The foaming process is described and monitored by periods of time taken from the start of mixing:

The *cream time* is the time until the reaction mixture starts to expand. At this point in time, the exothermic reaction has heated the reaction mixture to a temperature at which the blowing agent is starting to evaporate. The gas formed remains in the reaction mixture and forms small bubbles that grow with increasing temperature and further formation of gas.

When the cross-linking of the polymer has progressed to the point where the liquid reaction mixture starts turning into a solid the *gel time* can be observed. This is mostly done by prodding the foam with a small stick. When fibers of polymer are pulled from the rising foam the gel time has been reached.

After that the foam continues to rise to its final height (*rise time*).

The *tack-free time* is the time required for the surface of the foam to lose its tackiness. The tack-free time may occur before or after the rise time.

The time required to cool down the foam depends on its thickness. For slabs 1 m in height it takes more than 24 h until ambient temperature is reached in the core. In all processes the foam applies pressure on the compartment in which it is contained. To keep the parts produced within their specified dimensions, the molds must withstand this pressure until the foam has reached a minimum strength and the part can be demolded (demold time). In continuous processes, the line speed needs to be adjusted to keep the foam in the machine for the time required to cure sufficiently. For thicker parts the demold times are longer than for thinner parts, and therefore the line speeds in a continuous production are slower for thicker parts.

Preblending some of the components-usually into the polyol side-for easier operation can be done at different levels. It can be performed by the raw material supplier or by the processor. This depends mostly on the variability required for the respective process (e.g., adjusting reactivity in continuous processes).

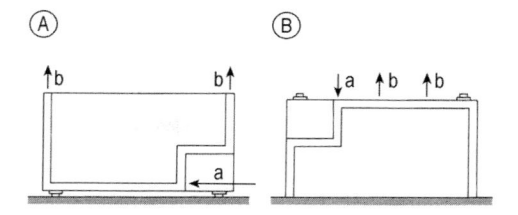

Figure 29. Principle of filling refrigerator cabinets
A) Door opening on top; B) Door opening on bottom
a) Pour-in opening; b) Ventilation openings

A discontinuous process (sometimes called a one-shot process) is used in cases where particular cavities are to be filled, such as in the production of refrigerators, freezers, insulated containers, and certain sandwich panels. Figure 29 shows the principle of filling refrigerator cabinets.

Figure 30 shows the discontinuous production of sandwich panels in a multistage press using the mixing head pull technique.

In most of these applications only one type of reaction mixture is used for the whole production range. Here all the necessary components are preblended into the polyol so that only two components, i.e, this polyol formulation and the isocyanate component, must be provided. The components are tempered and metered to the mixing head by high-pressure pumps. The mixing itself is performed by using the high-pressure impingement mixing technique. The amount of reaction mixture necessary to fill the cavity is determined by the volume of the cavity

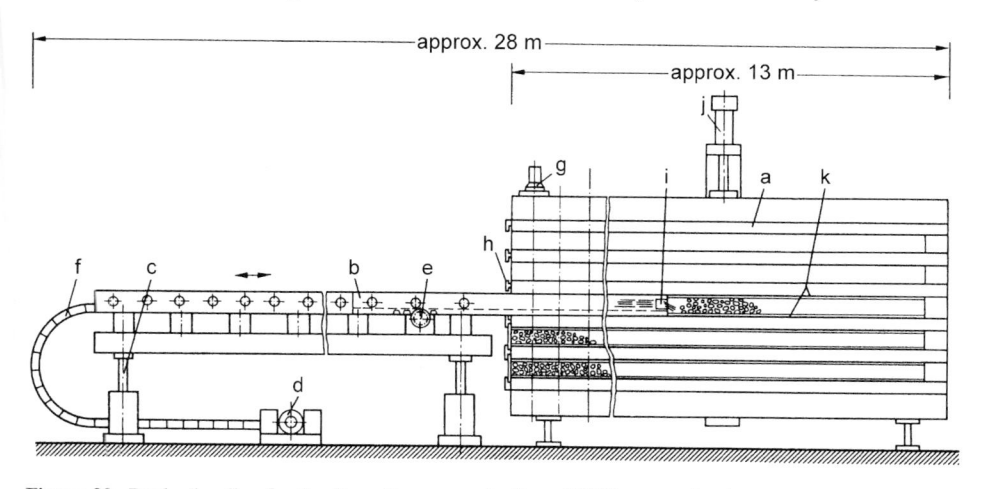

Figure 30. Production line for the discontinuous production of PUR composite elements
a) Plates of the press; b) Carrier on rolls for the hoses leading to the mixing head; c) Lifting device; d) Foam machine; e) Driving elements for horizontal movement; f) Cable holder; g) Device for closing the press; h) Closing cap for each unit; i) Mixing head with hoses; j) Opening device; k) Metal facers

and the density of the foam. The duration of the dosage is then given by the output of the pumps. This shot time must be shorter than the cream time to avoid pumping liquid reaction mixture into an already expanding foam. In so doing, the foam would be partially destroyed with negative effects on all of its properties.

In the continuous production of composite panels and insulation boards on double conveyor belts, the high-pressure impingement mixing technique is also used in most cases. The principle of this process is shown in Figure 31. Here the low weight of the mixing head and the hoses leading to it enable the oscillating of the mixing head across the moving bottom facing for an even distribution of the reaction mixture.

In the special case of the production of insulation boards on high-speed laminators at line speeds of up to 60 m/min, the high-pressure impingement mixing technique is used in most cases, but without oscillating the mixing head. The oscillation frequencies to ensure an even distribution of the reaction mixture on the bottom facing would be too high.

Here even distribution is achieved by passing the reaction mixture through a narrow gap between the top and the bottom facing (calibration technique). By this technique the small voids near the top facing that occur with the oscillating mixing head technique are eliminated. In all these applications it is necessary to adjust the reactivity of the reaction mixture according to the thickness of the board or panel. Also the amount of blowing agent must be adjusted. This means that the degree of pre-blending is lower than in a discontinuous process. Catalyst, blowing agent, and possibly other additives are metered separately into the polyol stream. Thus, only two components are finally led to the mixing head, keeping the moving weight low for the oscillation.

In the continuous production of rigid slabstock foam low-pressure mixing is used due to the high output. Here the components are metered into a mixing chamber and mixed by a mechanical stirrer. Separate metering of the components into the mixing head can be used. Preparing a polyol component containing all necessary components is also possible and allows for a certain quality control prior to foaming.

7.3.3. Properties

The most important property of rigid PUR foams is its low thermal conductivity, which is the lowest among the different insulation materials available by a significant amount. This allows for a smaller thickness of the insulating layer to achieve the same insulation value. The thermal conductivity of closed-cell PUR foams in the density range 25–60 kg/m^3 depends not so much on the apparent density (usually the thermal conductivity increases slightly with increasing density) but is dominated by the thermal conductivity of the blowing gas. The thermal conductivity can initially be near 0.020 W m^{-1} K^{-1}, and after ageing (cell gas diffusion) typically is in the range of 0.025–0.030 W m^{-1} K^{-1}.

The adhesion to substrates is important for composite constructions. The force needed to remove the foam from the substrate is the criterion used to judge adhesion.

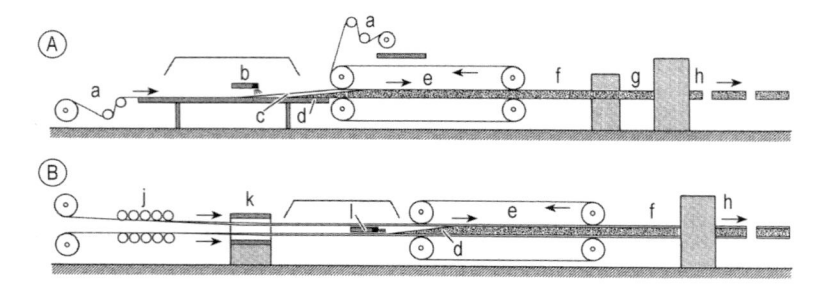

Figure 31. Principle of the continuous production of insulation board
A) With flexible facings; B) With profiled metal facings
a) Unwinding device; b) Lay-down with mixing head; c) Means for folding paper; d) Foaming reaction mixture; e) Compression; f) Cooling area; g) Longitudinal cutting; h) Transverse cutting; i) Coil unwinding; j) Corrugation rolls; k) Heating elements; l) Gate with moving high-pressure impingement mixing head

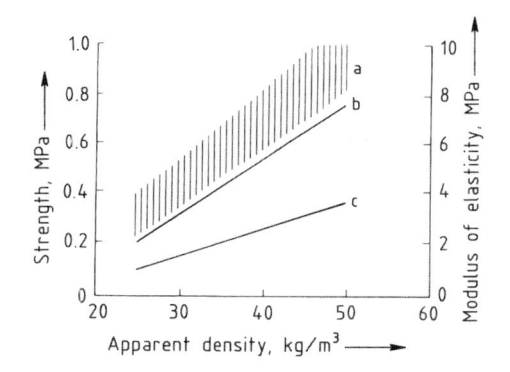

Figure 32. Relationship between typical properties and apparent density of rigid PUR foam (values at room temperature)
a) Modulus of elasticity; b) Tensile strength; c) Compression strength

Mechanical properties like compressive strength, tear strength, and shear strength are also important in load-bearing applications. Their values strongly depend on the density and the cell structure (Fig. 32).

Dimensional stability must be given under the conditions of use of the respective application. Dimensional stability is tested at low temperatures (e.g., for freezers) and at high temperatures (e.g., for hot water tanks).

Flammability is a very important property of the foam or the composite or both, depending on the test that needs to be passed. Various test methods are in place, most of them in the construction and transportation industries. For example, a harmonized test for the construction industry has been introduced in the EU. The new test procedure comprises a test on the foam (small burner) and the single burning item (SBI) test on the final product under end-use conditions. Other test procedures, e.g., those set up by insurance companies, are also in place.

For the different applications the importance of the various properties is prioritized differently.

For appliances (commercial and domestic refrigerators, freezers, hot water tanks), a low thermal conductivity is of top priority. In these applications a maximum useable space is desired at minimum or given outer dimensions. For this reason, the thickness of the insulation is limited, and only with a foam that offers low thermal conductivity can the energy consumption be minimized. Dimensional stability at low and high temperatures is the other important property here. In the low temperature applications the foam also fulfils a structural task, it forms a type of sandwich structure with the inliner and the outer facing.

Preinsulated pipes are mostly used for district heating. Here the thermal conductivity and the lifetime expectancy are the important properties. Accelerated ageing tests at elevated temperatures are performed to ensure a lifetime of, for example, at least 30 years according to EN 253.

Insulation boards, predominantly those with facings (aluminum foil, mineral fleece, papers) produced on double conveyor belts, also require low thermal conductivity and dimensional stability. On larger insulated areas, e.g., flat roofs, shrinkage would lead to areas without insulation due to gaps between the boards, while swelling might make the boards lift up and break the protective cover. For decorative boards also the appearance is important. Wrinkled facings, mostly aluminum foil, or sink marks due to voids under the facing are unacceptable here. Since most of the products are used in construction, flammability is also an important property.

Composite panels, mostly produced continuously on double conveyor belts with profiled steel sheets as facings, are mainly used in the construction of walls and roofs of large buildings like warehouses or factory buildings. The prefabricated self-supporting panels are attached to a bearing frame construction, so that the complete wall or roof can be put together very fast. The most important properties here are adhesion of the facings, dimensional stability, flammability, and thermal conductivity. The appearance is also important, especially for panels in facades. The thickness of the panels depends on the use of the building. For cold storage houses, where goods are stored at −40 °C, panels up to 24 cm thick are used. The thickness of panels for factories (from about 20 to 100 mm) depends mostly on the local climate. Discontinuously produced panels may have locks for connecting panels with one another that are incorporated into the foam during the foaming process. These panels can be used without bearing constructions, e.g., for walk-in-coolers. Another application is the use of panels in roll-up doors; large ones for truck entrances in industrial buildings, and small ones for private garages.

Slabstock foam is made for applications in very different fields. A few examples are given here: low-temperature insulation (liquid gas tanks, also in ships); refrigerated containments for lorries, where the foam is covered with glass reinforced plastic; insulation for parts that are difficult to mold such as valves and flanges (here the foam is mechanically shaped); wedge-shaped boards to give flat roofs a slope; and others. Different properties are required in these different applications, so rigid slabstock foam is a specialty field. Foams with a range of densities from about 30 to over 200 kg/m^3, different flammability classifications, foams with glass fiber reinforcements, and other variations are found here.

7.3.4. Special Types

Foam from pressurized cans, mainly one-component-foam, is used for mounting and fixing applications such as windows and doorframes. Here all the raw materials, including blowing agent and isocyanate, are blended in the can to form a prepolymer with residual NCO groups. The propellant, that is, liquid under pressure, acts first as thinner for the prepolymer. When expelled the propellant acts as blowing agent to produce the foam. The foam itself is usually an open-cell product.

A few special types of open-cell foam exist, such as thermoformable rigid foam that is used in passenger car headliners, foam for flower-arranging purposes, and packaging foam. Recently some attempts have been taken to evacuate open cell foam to reduce the thermal conductivity even further. One problem here is to find suitable facings that ensure that the vacuum is maintained during the lifetime of the product.

In certain cases the foam must be applied in situ. Two methods are predominant: spraying and layer-by-layer application. Spray foam is mostly used for the industrial insulation of storage tanks or in the insulation of buildings. While spray foam on walls is only for insulation, foam on roofs is also applied for sealing purposes. Two-component systems are used here. The components are fed through heated hoses to the mixing head, which is designed like a spray gun. The reactivity of these systems is so high that the foam cures within a few seconds. In layer-by-layer application the reaction mixture is poured between two substrates, e.g., the outside of a tank and the casing. This method is used where the application of the foam in a factory is not possible, e.g., because of the size of the tank. Here also two-component systems are used. In both cases special attention must be paid to temperature and humidity conditions. Foams should not be applied on cold or moist surfaces. Otherwise, problems with bonding the foam to the substrate may result.

Polyisocyanurate (PIR) foams [190–195], more precisely to be understood as polyisocyanurate-modified PUR foams, are products in which isocyanurate structures are incorporated into the macromolecule. These structures are formed due to the reaction of excess isocyanate with itself. Special (trimerization) catalysts are required. Three NCO groups form the cyclic isocyanurate structure, which shows a higher decomposition temperature than PUR groups. Therefore, these foams are used where flammability or heat resistance is an issue. However, this highly cross-linked structure increases the brittleness of the foams. The overall cross-linking density is controlled by the use of low-functionality polyols. In the continuous production of insulation boards on double conveyor belts, which is the major application for this type of foam, the PIR structure is usually combined with (aromatic) polyester polyols. The higher decomposition temperatures of esters compared to ethers also leads to an additional positive effect on the flammability of the foam. In the manufacturing of these foams, special attention must be paid to the temperature of the laminator because the trimerization only takes place at temperatures above 60 °C. Apart from the reduction of brittleness the polyols have the task of heating the reaction mixture by their exothermic reaction with isocyanate. When the required temperature is reached, the trimerization starts, which can often be observed as a second phase in the expansion of the foam.

7.4. Integral Skin Foams and RIM Materials [196–198]

Integral skin or self-skinning foams are products with a cellular core and a cell-free surface that are formed in a mold in a single operation (Fig. 33). Parts made by the reaction injection molding (RIM) process are microcellular materials with a high surface quality that are mostly painted.

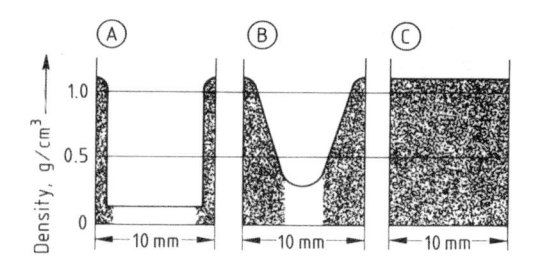

Figure 33. Principle of density distribution for integral foam boards of 10 mm thickness and various apparent densities:
A) 0.2 g/cm^3; B) 0.6 g/cm^3; C) 1.1 g/cm^3

7.4.1. Applications

Flexible self-skinning foams are used mainly in the automotive, furniture, and footwear industries. Typical automotive applications are steering wheels and gearshift knobs, but also armrests and headrests (Fig. 34). In the footwear industry, integral skin foams are used for soles of sport, leisure, and safety shoes.

RIM materials are used for exterior auto parts like fenders (Fig. 35A), panels, side protection and window encapsulation, but also for other exterior parts in agricultural machines or recreational vehicles like snowmobiles (Fig. 35B).

Rigid integral skin foams or RIM materials can be used in many different applications, mainly for structural parts. They can favorably compete with wood, thermoplastics, thermosets, and metals. They are used for furniture, skis, pump housings, window frames, decorative applications, and housings, just to name a few characteristic applications.

7.4.2. Production

Integral skin foams are mostly manufactured in metallic (aluminum) or epoxy molds. Skin formation is controlled by addition of a low-boiling solvent in amounts of about 2–10 wt %, relative to the polyol mixture. After the banning of CFCs and HCFCs, currently HFCs (hydrofluorocarbons) or alkanes are used. When using alkanes, special precautions must be taken because of their flammability.

Thermodynamic equilibrium between the liquid and gaseous phases of the blowing agent in the reaction mixture is crucial for the formation of the closed surface. The mold temperature is kept between 40 and 60 °C, and the reaction mixture in the vicinity of the mold surface has a lower temperature than the core (up to 140 °C). In the core the blowing agent vaporizes, while near the edge it condenses, with formation of the solid surface.

The quality of the surface depends very much on the quality of the mold, but precise control of the processing pressures and temperatures is also required.

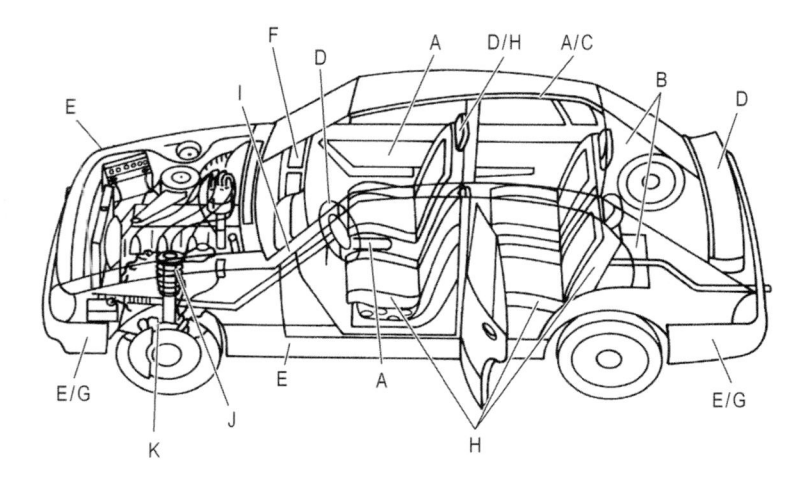

Figure 34. Use of PURs in the passenger car
A) Rigid integral foam; B) Rigid integral foam, glass-mat reinforced; C) Hot-moldable rigid foam; D) Semi-flexible integral foam; E) Semi-flexible integral foam, glass-fiber reinforced; F) Semi-flexible filling foam; G) Semi-flexible filling foam, impact-energy-absorbing ; H) Flexible molded foam; I) rigid foam; J) Hot-curing PUR elastomer; K) PUR rubber

Figure 35. Use of integral skin foams and RIM materials
A) Fender of the VW Touareg; B) Snowmobile

In special cases formation of an integral skin is possible without the addition of physical blowing agents. For this, temperature-controlled carbon dioxide formation from the NCO/H_2O reaction must be achieved.

High-pressure metering and mixing units are necessary for manufacturing of RIM parts. The processing of these materials, which are mostly of high reactivity, also requires closed, metallic molds and a clamping unit. To obtain void-free parts without porosity the reaction mixture must enter the mold cavity through a gate designed to ensure laminar flow. Also, correct venting of the mold at the highest points and nucleation of the reaction mixture by air or nitrogen are important for good quality.

After demolding, the parts must be trimmed, and gate and flash must be removed. The surface is prepared for painting by removal of the release agent by washing or sanding.

7.4.3. Properties

Depending on the field of use, soft-elastic moldings have densities of 100–300 kg/m^3, flexible molded articles 400–600 kg/m^3, and the relatively hard, microcellular elastomers 900–1100 kg/m^3.

Technical data for a typical soft integral skin foam of low density:

Density (ISO 845)	400 kg/m^3
Shore A hardness (DIN 53505)	35
Tensile strength (DIN 53504)	0.8 MPa
Elongation at break (DIN 53504)	85 %
Tear resistance (ISO 34)	1.4 kN/m
Compression set (25 %/70 h, RT; ISO 1856)	2.5 %

Mechanical properties of a flexible molding of medium density used in the footwear industry (e.g., polyester system):

Density (ISO 845)	500–600 kg/m^3
Shore A hardness (DIN 53505)	ca. 50
Tensile strength (ISO 1798)	6–8 MPa
Elongation at break (ISO 1798)	450–550 %
Tear propagation strength (ISO 34)	10–15 kN/m
Abrasion loss (DIN 53516)	50–100 mg

Mechanical properties of a microcellular elastomer molding of high density (e.g., automotive body parts):

Density (ISO 845)	1100 ± 50 kg/m^3
Shore D hardness (DIN 53505)	64 ± 3
Tensile strength (DIN 53504)	35 ± 3 MPa
Elongation at break (DIN 53504)	300 ± 30 %
Tear propagation strength with cut (ISO 34)	80 ± 10 kN/m
Modulus of elasticity according to Roelig	
at –30 °C	1400 ± 100 MPa
at + 20 °C	600 ± 50 MPa
at + 65 °C	360 ± 30 MPa

The technical data of rigid integral skin foams depend very much on the density, which can be anywhere between 150 and 1100 kg/m^3. Typical properties of a low-density rigid integral skin foam, as used for imitation wood follow:

Density (ISO 845)	400 kg/m^3
Shore D hardness (DIN 53505)	58
Flexural strength (ISO 178)	20 MPa
Flexural modulus of elasticity (ISO 178)	520 MPa
Impact strength (ISO 6603)	10 KJ/m^2
Heat deflection temperature (ISO 75)	90 °C

High-density rigid RIM materials, which are typically used at wall thickness between 3 and 5 mm, have the following characteristics:

Density (ISO 1183)	1100 kg/m^3
Shore D hardness (DIN 53505)	76
Flexural strength (ISO 178)	25–45 MPa
Flexural modulus of elasticity (ISO 178)	2000 MPa
Tensile strength (ISO 527)	53 MPa
Elongation at break (ISO 527)	14 %
Impact strength (ISO 179)	45 KJ/m^2
Heat deflection temperature (ISO 75)	115 °C

When flame retardancy tests are also considered, RIM materials are capable of passing burning tests such as UL 94 V-0 (Underwriters' Laboratories) or B2 (DIN 4102) by the addition of flame retardants.

Both flexible and rigid RIM materials can be reinforced with short or long glass or mineral fibers and also glass mats or mats made of natural fibers. These composite materials show considerably higher stiffness and especially impact strength.

8. Noncellular Polyurethanes

Although the term "noncellular" is unambiguous, the terminology of the products treated in this chapter requires explanation. "Cellular Vulkollan," well known from the early days of PURs, already breached the borderline of precise definition. In practice, this material was not considered a "foam," although it can be undoubtedly classified as a foam from a strictly physical viewpoint because of its cellular structure, however small the few cells may be. The RIM technology has made the definitions even less distinct due to the terminology of "microcellular elastomers" on the one hand and "solid integral parts" on the other. In this chapter the problem of definition is solved as follows: All products which are based on a formulation that usually leads to a foam are classified as "foams," even if cell-free solid materials can be obtained by raising the apparent density (less blowing agent). Similarly, products which are based on formulations that usually result in noncellular materials are classified as noncellular PUR, even though they may be cellular, and these materials are treated in this chapter.

There are two different processing technologies for manufacturing these noncellular PURs. Liquid, low molecular mass raw material systems are converted to high molecular mass products by the reaction of the polyol and polyisocyanate components by cold or hot curing. Linear, high-molecular-mass raw material systems are either thermoplastically processable pellets or urethane rubber are processed by vulcanization technology.

8.1. Cast Elastomers

Cast elastomers are mostly solid, noncellular materials, produced in a low-pressure casting process from ester or ether prepolymers of MDI, TDI, and NDI (see Section 3.1.1) and a cross-linker such as butanediol. If water-containing cross-linkers are used, cellular elastomers can also be obtained.

8.1.1. Applications

The highest quality elastomers are those derived from NDI [Tradenames: Vulkollan (Bayer MaterialScience), Cellasto (Elastogran)]. They are used where materials must meet extremely high requirements regarding durability and mechanical properties. Examples are high-load rolls and wheels for forklift trucks and machines, hydraulic seals, membranes, valve seats and balls, scrapers, and sieves. Cellular NDI-based elastomers are widely used in the automotive industry for bouncers and spring supports (Figs. 34 and 36).

Hot-curing elastomers based on MDI or TDI are used, e.g., for bottle stars in automatic filling machines, cutting wheels for fiber production, roll covers for the paper and printing industry, and wheels for in-line skates and skateboards.

Cold-curing PUR elastomers are used as binders for the surface of playgrounds and sport tracks. These surfaces are very durable, require little maintenance, can be used under all weather conditions, and reduce the risk of injuries. Other applications are seals for stoneware pipes, polishing disks, protective edges, and permanent fenders for boats. Cold-cure elastomers can also be sprayed on concrete or metal for corrosion protection and are used for the surfacing of bridges, parking decks, and terraces and for lining of wagons and trucks.

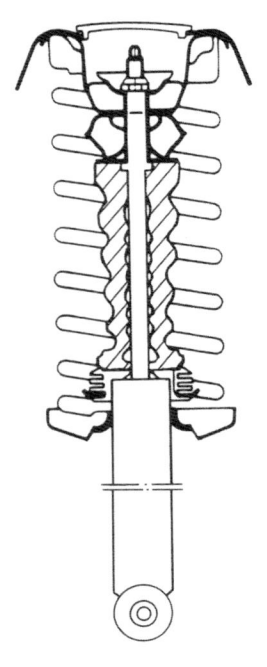

Figure 36. Use of cellular Vulkollan as supplementary spring in the passenger car

8.1.2. Production

The raw materials used are low molecular mass glycols or amines, polyester or polyether polyols, and (modified) MDI-, TDI-, or NDI-type isocyanates.

The elastomers can be manufactured in a two-step or a one-step procedure. Figure 37 shows a typical line for the production of high-quality elastomers, as is used for NDI-based high-performance materials. The system passes through several stages: dehydration of the polyester (a), production of prepolymer from polyester and diisocyanate (b), degassing of the feed component (c), admixing of cross-linking agent (d), atmospheric-pressure casting in open molds (e, f), and fully curing the molding at 100–300 °C for several hours (g). This processing technique even allows the production of voluminous and thick-walled parts in relatively simple, inexpensive molds with low energy consumption. Depending on the system to be cast, steel, aluminum, and in special cases plastics or wood can be used as the mold material.

For MDI and TDI stable prepolymers are commercially available, so that a one-step process can be used. The reactivity of the raw materials varies widely. Casting times between 2 and 60 min can be realized.

Cold-casting systems are usually processed with a casting machine. For some formulations that have a sufficiently long pot life, the starting components can also be hand mixed and cast. In

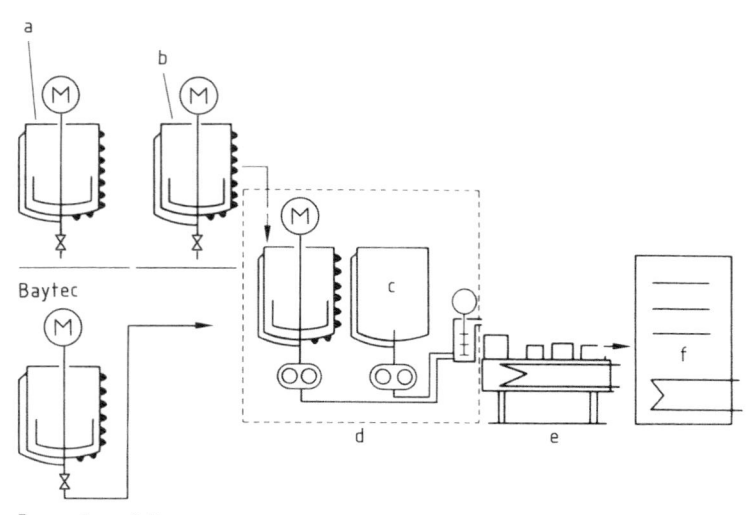

Figure 37. Flow chart for production of high-quality elastomer (e.g., Vulkollan/Baytec)
a) Dehydrogenation of the polyester; b) Prepolymer production; c) Cross-linker; d) Casting machine; e) Mold frame; f) Heating oven (thermal posttreatment)

this case the charge volumes are limited to a few liters. The reaction mixtures cure, without an external heat supply, within widely differing cure times, from less than 10 min to a maximum of 24 h. The castings reach their final strength after ca. 7 d storage at room temperature.

8.1.3. Properties

NDI-based elastomers have a temperature range for continuous use between −30 and +80 °C. At temperatures below 0 °C the material becomes increasingly hard and inelastic, though there is no danger of fracture. Embrittlement occurs only at very low temperatures (below −30 °C).

NDI elastomer (Vulkollan) is resistant to mineral oils, grease, gasoline, and most organic solvents. Gasoline and some solvents cause some swelling, but this does not impair its mechanical properties if the solvents are able to re-evaporate. Concentrated acids, alkalis, and other hydrolyzing media attack NDI elastomers. By incorporating additives or using special polyesters the hydrolytic stability can be increased so that the service life of the elastomers becomes adequate for a variety of uses even under extreme conditions. Mechanical properties of NDI elastomers are listed in Table 8.

Cellular NDI elastomers can be produced with densities of 300–700 kg/m^3 and are characterized by their high capability of absorbing energy on impact. The material can be compressed by up to 80 % of the original volume without damage. Damping units with various compression deformation characteristics can be manufactured from cellular Vulkollan by varying its apparent density, preferably in the range 300–600 kg/m^3.

Hot-casting systems with a base other than NDI are produced with Shore hardness grades from ca. 55A to 65D. The other mechanical properties and the chemical stability are basically comparable with those of NDI elastomers without reaching the same levels, especially in dynamic stability. They do not, however, reach the high values of Vulkollan in all points, as is also clear from the lower dynamic stability under load. MDI and TDI elastomers cannot replace Vulkollan but are used as supplementary products, primarily in fields where the high quality of Vulkollan is not absolutely necessary, where a system with long casting time is advantageous for producing large-volume and thick-walled moldings, and where short demolding times are desirable for producing large series.

Cold-curing products have a good elasticity, high flexibility, and a variable Shore hardness in the range of ca. 20A to 65D. They are highly resistant to aging and microorganisms and are not attacked by dilute acids and alkalis, mineral oils, soaps, or detergents. Organic solvents cause various degrees of swelling.

8.2. Thermoplastic Polyurethane Elastomers (TPU)

See also Thermoplastic Elastomers - Thermoplastic Polyurethane Elastomers, → Thermoplastic Elastomers, Chap. 2.

Production. Diisocyanate (usually MDI), polyols (polyester and polyether polyols), chain extenders (butanediol), and the necessary auxiliaries are reacted in continuous plants via mixing heads or in reaction extruders. After

Table 8. Mechanical properties of a hot-casting system (e.g., Vulkollan)[*]

Property	Vulkollan type			
	18/40	18	30	50
Shore hardness A/D (DIN 53505)	68/20	83/31	92/36	96/58
Density (ISO 1183), kg/m^3	1.25	1.25	1.26	1.26
Abrasion loss[**] (DIN 53516), mm^3	45 – 50	45 – 50	40 – 50	43 – 48
Tear propagation strength according to Graves (ISO 34), kN/m	20 – 30	40 – 55	67 – 85	85 – 96
Impact resilience (DIN 53512), %	48 – 55	45 – 50	45 – 50	45 – 50
Tensile strength (DIN 53504), MPa	42 – 49	50 – 60	37 – 45	40 – 43
Elongation at break (DIN 53504), %	600 – 700	630 – 750	550 – 650	330 – 370

[*] The properties listed here can only be regarded as approximate. For Vulkollan articles on the market, the quality characteristics specified by the manufacturer are applicable.
[**] Relative to abrasion-resistant grade of natural rubber (= 100).

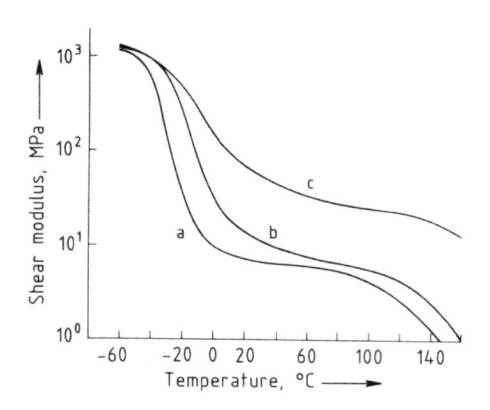

Figure 38. Shear modulus for various TPUs (e.g., Desmopan)
a) Type 385; b) Type 790; c) Type 460

solidifying, the product is processed further to pellets. The TPU can be processed further by, e.g., injection molding, extrusion, calendering, or blow molding to give finished or semi-finished products.

Properties. The principle of "tailor-made plastics" applies to the properties of thermoplastic PURs, as to all PURs. A very wide range of properties is available by careful choice of the starting materials. This is shown in Figures 38 and 39 as well as Table 9 for three selected examples.

Uses. The TPUs are valued as high-grade polymeric materials in many industrial branches for a great variety of constructional parts. Table 10 shows examples of their use in relation to their

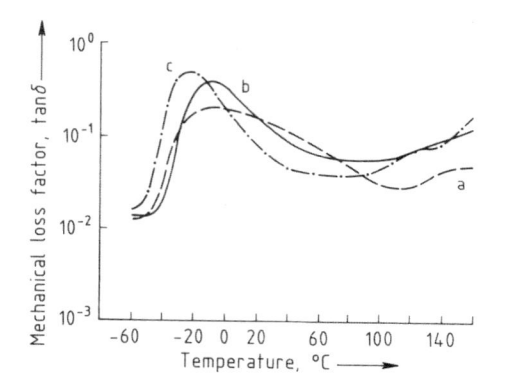

Figure 39. Mechanical loss factor for various TPUs (e.g., Desmopan)
a) Type 385; b) Type 790; c) Type 460

Table 9. Mechanical properties of different types of TPU (e.g., Desmopan)

Property	Desmopan system		
	385	460	790
Shore hardness A/D (ISO 868)	86/33	97/59	92/42
Density (ISO 1183), kg/m^3	1200	1210	1210
Tensile strength (ISO 527-1/-3), MPa	40	35	55
Elongation at break (ISO 527-1/-3), %	450	350	450
Tear propagation resistance (ISO 34-1), kN/m	80	120	85
Impact resilience (ISO 4662), %	42	35	32
Abrasion loss (ISO 4669), mm^3	30	40	30

Shore hardness. Glass-fiber reinforced types are used for bodywork units.

9. Polyurethane Coatings [199–201]

See also → Paints and Coatings, 2. Chap. 9.

PUR coatings offer a large variety of outstanding properties, which make them suitable for many high-performance applications. PUR coatings have polymeric structures with urethane, urea, biuret, or allophanate coupling groups. They are generated either from reactive systems by polyaddition of oligomeric polyisocyanates with polyol components, or from high molecular mass adducts which are further cross-linked by physical drying or other mechanisms. The nature of the isocyanate building blocks (aliphatic or aromatic) and the structure of the backbone are important. Film mechanical properties are dependent on the structure of the polymer (segmented, interpenetrated networks) and cross-link density. PUR coatings are available solvent-borne and, as a result of the trend to reduced emission systems (low VOC), as high-

Table 10. Typical applications for different grades of TPU

Hardness range		Uses
Shore A	Shore D	
75 – 90	< 35	membranes, seals, damping units, sieve units, hoses, cable sheatings
90 – 95	35 – 45	animal identification markings, roll coverings, toothed belts, sport shoe soles
> 95	45 – 55	seals, flanges, rolls, pneumatic hoses
	55 – 70	bearings, bushes, connectors, ball sockets, seals, coupling units, ski bootlegs, heel pieces

solids, waterborne, solvent-free, and powder systems. Important PUR systems are summarized below.

Two-component coatings are the most important (solvent-borne) PUR paint systems. The primary component is always polyisocyanate (mainly HDI, IPDI, or TDI derivatives). The other component consists of polyols (polyester, polyacrylic) or amines or a mixture thereof and any additional ingredients such as pigments or solvents. The reactive components must be kept separate. Mixing, preferably of an equimolar equivalent, is done immediately before application. The coatings system can be applied by all conventional coating methods, but preferably by spraying or, industrially, with automatic two-component equipment. The solids contents of the coatings are usually higher than with standard coatings and can reach 80 % (pigmented high-solids), depending on the formulation. Curing takes place at ambient temperature, but can be accelerated by heat. Outstanding film properties such as high mechanical resistance, high chemical resistance, and, with aliphatic polyisocyanates, excellent lightfastness and weather resistance, open up a broad range of uses. Major areas of application are transportation (large motor and rail vehicles, aircraft, automobile finishes and refinishes), the building sector (wood and mineral substrates), industrial paints, and steel construction.

One-component coatings include air-drying and stoving systems. The most important are solvent-borne, isocyanate-terminated prepolymers of higher molecular weight made from aromatic or aliphatic isocyanates (MDI, TDI, HDI, IPDI) with polyols. They react with atmospheric moisture to form urea linking groups. Although the final properties are comparable to those of two-component PURs, curing of these systems is slower, which limits their use mainly to wood-flooring coatings, corrosion protection, and applications on mineral surfaces.

The reaction of TDI or IPDI with polyol-modified drying or semi-drying oils generates urethane alkyds, which can be cured oxidatively (by air drying) like alkyds.

The isocyanate groups of prepolymers can be reacted with agents containing, e.g., acidic CH or NH groups to form blocked isocyanates (see Section 3.1.3). Their mixtures with polyols are stable at room temperature, but react at elevated temperatures (120–220 °C, depending on the nature of the blocking agent). The resulting films have a very high mechanical resistance. They are particularly important for automotive finishes, industrial goods and coil coating.

Powder coatings consist of a combination of solid polyols (mostly polyester or polyacrylate polyols) and solid blocked polyisocyanates as the cross-linker. ε-Caprolactam was the most commonly used blocking agent but is being replaced by other blocking agents. An alternative method, avoiding the use of an added blocking agent is the dimerization of the isocyanate groups to uretdiones. During stoving (10–25 min at 150–210 °C) isocyanate groups are liberated and undergo cross-linking. PUR powder systems are used in many industrial applications.

Waterborne coatings are growing continuously in importance, due to the requirement that emissions should be reduced. A variety of systems are available, comparable to solvent-borne coatings. One-component systems are based on self-emulsifying polyurethane or polyurethane-polyurea dispersions. They are manufactured by processes which allow the production, in contrast to solvent-borne PURs, of any molecular weight without having a significant impact on viscosity, and with only small contents of solvents or no solvent at all. These products are used in physically dried films for more flexible substrates like textile, paper, and plastics. One-component polyurethane/polyacrylate dispersions based on aliphatic or aromatic isocyanates are used for coating parquet flooring. In addition, UV-curing PUR acrylate dispersions are used in industrial wood and furniture finishes.

One-component stoving systems contain blocked polyisocyanates emulsified in OH- or NH-containing dispersions (epoxy polyester, acrylic). Important applications are automotive coatings (electrodeposition coating, primer surfacer) and other industrial coatings.

Reactive waterborne two-component systems are based on polyisocyanates (HDI, IPDI) and aqueous OH-functional dispersions (polyacrylate, polyester, polyester polyurethane). Self-emulsifying polyisocyanates reduce interfacial tension between the components and ease homogenization. The film properties are influenced by the extent of side reactions, which can occur in the presence of water. The main applications of these reactive systems are in all types of industrial, wood, and plastic coatings.

10. Polyurethane Adhesives [202–204]

See also → Adhesives, 1. General

PUR adhesives exhibit unique properties due to their wide variety of compositions, their excellent adhesion to most polar substrates (ability to form hydrogen bonds), and the possibility of chemical cross-linking.

To achieve good cohesive strength of the adhesive bond, high molecular weight polymers are necessary. On the other hand, for good adhesion to the substrate, efficient wetting of the surface is essential. This can be done either by "liquefying" the polymer by the use of solvents, aqueous dispersions, or heat or by in situ polyaddition of low-viscosity polyols and polyisocyanates to give high molecular weight products. Depending on the application, the adhesives can be differentiated as follows:

Solvent-borne adhesives are made by dissolving granules of a linear high molecular weight OH-terminated PUR in a solvent like methyl ethyl ketone, acetone, ethyl acetate, or mixtures thereof to a solids content of up to 15–20 %. The PUR building blocks are crystalline polyester diols or polycaprolactones, aromatic diisocyanates (mainly MDI or TDI), and possibly a low molecular weight chain extender like butanediol or hexanediol. After evaporation of the solvent a nonblocking adhesive film results on the substrate. Due to the crystalline nature of the polyester soft segment, this adhesive film can be thermally reactivated at temperatures between 50 and 80 °C, and in this tacky state bond formation to a second coated and activated substrate can be achieved. On cooling down again, recrystallization leads to high initial and final bond strength ("physical setting"). To improve properties such as heat resistance, resistance to solvents, water, oils, or plasticizers, two-component processing is possible. This is achieved by adding a solvent-based triisocyanate (e.g., Desmodur R-Series, Bayer MaterialScience, cf. Section 3.1.1) to the adhesive solution shortly before application, which results in chemical cross-linking by reaction with the OH groups of the base polymer and other isocyanate-reactive groups on the substrate (including water). The main application for this kind of adhesive is in the footwear industry (bonding shoe soles to uppers).

Waterborne adhesives usually consist of a 40–50 wt. % aqueous dispersion of a high molecular weight PUR made by the acetone or the prepolymer mixing process. Besides crystalline polyester diols, aliphatic isocyanates such as HDI or IPDI and ionic or hydrophilic groups containing chain extenders are used as starting materials. Properties and processing are comparable to those of solvent-borne adhesives. Cross-linking can be achieved by using solvent-free emulsifiable polyisocyanates (e.g.- Desmodur D series, Bayer MaterialScience). The main applications are in sport shoe fabrication, the furniture industry (MDF kitchen cabinet door lamination with PVC films), and the automotive industry (lamination of interior trim parts with films).

Hot-melt adhesives are available in non-reactive and reactive form. The starting materials for nonreactive systems are similar to those used for solvent-borne adhesives, but instead of dissolving the granules, these are either directly extruded onto the substrate or (co)extruded to form hot-melt films or fleeces. Another possibility for bringing the adhesive polymer into intimate contact with the substrate (without using solvents or water as a carrier) is cold grinding of the granules. The resulting powders (particle size < 600 µm) can be easily brought onto the surface by scatter coating. Common to all these nonreactive hot melt application forms is the (in this case only one-component) processing by the thermo-activation procedure described above. Application fields are textile and film lamination.

Reactive hot melts are prepolymers of a crystalline or high-T_g polyester with an excess of a diisocyanate (mostly MDI-based). They are applied at temperatures of 60–80 °C higher than the melting point of the soft segment and combine the aforementioned advantages of physical setting and chemical cross-linking (NCO/water reaction). Typical application areas are bookbinding, woodworking, packaging, and the automotive and construction industries.

Reactive adhesives can be divided into one- and two-component systems. One-component reactive adhesives liquid at room temperature are NCO-terminated prepolymers of (usually) polyether polyols and an excess of an aromatic polyisocyanate (mainly MDI derivatives). Setting occurs by reaction with humidity and therefore requires at least one porous substrate. Applications are foam rebonding, rubber crumb bonding, and in the construction and transportation fields.

Two-component reactive adhesives consist of a liquid polyether polyol (or a low-viscosity polyester polyol) and a polyisocyanate (mainly polymeric MDI). The two components are either mixed and applied immediately before use by an automatic machine or by using a handheld double cartridge/static mixer combination. A broad spectrum of polyisocyanates and polyols is available, and the mixing ratio of the two components can be varied over a wide range. Thus, it is possible to design the adhesive for the individual substrate/bond requirements, e.g., from flexible to rigid. Application fields are again transport and construction, and flexible packaging/film lamination (no porous substrate necessary).

11. Polyurethane Fibers [205]

See also → Fibers, 6. Polyurethane Fibers

Synthetic fibers containing at least 85 wt % of high molecular weight segmented PUR [206] are known as Elastane or in the USA as Spandex. The segmented structure (see Section 4.2) gives these fibers high elasticity and makes them easy to stretch. The properties are determined not only by the raw materials used, but also by the synthesis conditions, the spinning process, and the post-treatment.

Compared to rubber fibers, the PUR fibers exhibit superior properties such as the ability to form fine filaments, good colorability and chemical resistance, and high resistance to physical impact.

They are produced by a two step process. In the first step, long-chain diols are reacted with aromatic diisocyanates; in the second step, the high-melting urethane and/or urea groups which are necessary for the high-level properties are introduced by chain extension with short-chain diols or diamines (see Section 4.2) [207–211].

In industrial production, usually poly(tetraethylene glycol) [212, 213], manufactured by ring-opening polymerization of tetrahydrofuran, is used as the long-chain polyol. In some cases, copolyethers of tetrahydrofuran and 3-methyltetrahydrofuran are used [214]. If polyesters are used as diols, they are usually based on adipic acid and mixtures of ethylene glycol, propylene gylcol, butylene glycol, and/or 1,6-hexandiol.

Aromatic diisocyanates, mainly 4,4'-MDI and in some cases 2,4-TDI, are used exclusively. Due to its unsymmetrical structure, 2,4-TDI leads to a weaker segmented structure and therefore to lower thermostability. Aliphatic diisocyanates, e.g., HDI, are only used in small amounts as modifiers.

Chain extension with diamines usually gives high-performance Elastane [215–218]. Mostly primary cycloaliphatic or aliphatic diamines such as ethylenediamine and 1,2-propylenediamine are used. Modification of the hard segments is achieved by adding small amounts of, e.g., 1,3-diaminocyclohexane.

12. Polyurethanes and Isocyanates as Binders [219]

Panels and molded parts made of wood chips, fibers, and veneers are used extensively in the furniture, building, packaging, and automotive industries. These products are produced under heat and pressure by using a variety of organic binders. PURs are excellent starting materials for composite binders and are used extensively in their production. They are classified according to binder type as polyurethane and polyurea bonding agents.

Polyurethane bonding can be accomplished with one-component moisture-curing systems or two-component ambient-cure systems. The latter can be adjusted in reactivity to meet processing conditions.

PUR dispersions are also utilized in special cases. In addition to adhesion, PUR dispersions exhibit excellent film-forming properties. This is an especially important characteristic for the sizing of glass fibers. PUR dispersions meet very specific requirements regarding film formation and glass adhesion, critical for "matrix" bonding. The properties of glass-reinforced plastic depend to a high degree on the adhesion of the polymer to the surface of the glass fibers.

In polyurea bonding, polyisocyanates are transformed into polyureas at high temperatures in the presence of moisture and, if necessary, catalysts. In addition to the physical adhesion properties of polyureas, there is also the possibility of adhesion by way of chemical reactions between polyisocyanate and reactive groups of the substrate material. Compared to the PUR binders, polyureas are more rigid and hydrophobic.

Binders based on polymeric diphenylmethane diisocyanate (PMDI) have been industrially used in the wood composite industry since 1973. PMDI binders are used worldwide in the manufacture of particleboard, oriented strand board (OSB), laminated strand lumber (LSL), medium-density fiberboard (MDF), and other specially engineered wood composites, as well as panel products based on annual plants, e.g., straw and sugar cane bagasse. The manufacture of straw-based panels was indeed first made possible with PMDI binders.

The success of PMDI binders in the wood and straw composites industry has its origin in the unique properties of the polyurea bond.

Moisture resistance: polyurea is highly crosslinked, chemically stable, and hydrolytically resistant. PMDI binders are one of two binders listed in the European standard for particleboard EN 312-3/5, option 2 that require no special testing to prove hydrolytic resistance.

Dimensional stability: polyurea is also hydrophobic and lends wooden panels improved dimensional stability. This was demonstrated on particleboard during three years of outdoor exposure, where they showed the lowest moisture absorbance and weight changes. This favorable behavior was also confirmed by long-term stress bending (creep testing) under severe conditions of 40 °C and 95 % R.H.

Emission-free: PMDI wood binders contain no formaldehyde. The resultant polyurea bond in the finished panel is free from emissions.

High-strength properties: The excellent adhesion properties of polyurea and the deep anchoring of the bond in the wood structure leads to the high strengths of the different wood composite panels.

During the manufacture and processing of rigid PUR foams, cutting and milling about 5–25 % of scrap is generated. Utilization of this scrap is possible by bonding it into panels or molded parts with varying properties. This can technically be realized with the aid of the hot-press technique employing PMDI binders, analogous to the manufacture of wood composites. The physical and mechanical properties of articles made from rigid PUR foam scrap can be varied over a wide range. The properties depend upon the type and pretreatment of the raw material, the amount of bonding agent, and the density.

Cork granulates and remilled rubber waste, especially peeled tire tread from the recapping process, as well as recovered scrap from milling of frozen old tires, can be converted with PUR bonding agents to composite materials. The bonding agents are isocyanate-terminated, one-component, moisture-curing prepolymers or reactive two-component systems. They form elastic films during curing. The prepolymers are synthesized from long-chain linear or slightly branched polyethers and MDI types or toluene diisocyanate and/or from mixtures of both. The one-component prepolymers generally have an NCO content of ca. 10 % and a viscosity of 1500 to 3000 mPa · s at 23 °C.

The most important properties are elasticity, flexibility, damping capacity and void content. They can be varied over a broad range through the shape and size of the rubber particles, type and amount of binder, and degree of compaction.

13. Special Products

PURs are particularly suitable as wall materials for microcapsules (see also → Microencapsulation). Capsules can be impermeable or permeable to the core material. The capsule wall is usually produced by interfacial reaction of multifunctional isocyanates or NCO prepolymers and multifunctional amines at an oil/water interface. The microcapsules have a diameter of 1 to 5000 µm, depending on the droplet size of the emulsion used for polymer formation. The capsules are used for a wide variety of purposes, e.g., for the production of carbonless copying paper or for the delayed release of plant-protection products.

Polyoxazolidinone polyisocyanurates are produced from polyisocyanates and epoxy resins in the presence of special catalysts (see Chap. 2). They have high fire resistance, high long-term heat resistance, and excellent electrical insulation properties [220, 221].

Water-soluble PURs are of interest as retaining agents and dyeing auxiliaries, thickeners, paper-sizing agents, and other sizers.

Oligourethanes with unsaturated (e.g., acrylate) groups can be cured by UV radiation in the presence of photoinitiators [222]. Systems based on acrylate groups containing blocked NCO prepolymers have been developed. They can be cured either by heat or by radiation (dual curing [223–227]) and are used especially for automotive refinishing paint.

14. Safety and Ecology

PUR-containing products have become an essential part of modern living. For production and processing of all of these varieties of PUR, it is estimated that worldwide well over 500 000 people are handling PUR raw materials every day. For the toxicology of the raw materials, see also → Isocyanates, Organic, Chap. 10. and → Polyoxyalkylenes, Chap. 4.

Health, safety, and environmental aspects, especially for MDI and TDI (see Section 3.1.2), are described in [228].

14.1. Safety Precautions when Handling the Raw Materials

When building and operating plants for producing PUR, all laws, decrees, and regulations in force for chemical plants must be observed. Many years' experience in handling the raw and auxiliary materials for PUR has shown that these products can be handled safely. It is a prerequisite for this, however, that the persons working with them are supplied with all the necessary information and above all, that the required technical and personal protective equipment is available and used.

All isocyanates, activators, and most alkanes and haloalkanes are regarded as products having acute or chronic harmful properties.

Di- and polyisocyanates have only a moderately acute oral and cutaneous toxicity. Di- and polyisocyanates are known to have allergenic potential, and the human health risk due to the acute or chronic effect of isocyanate vapors and aerosols (of liquid and dust forms) is

Table 11. German air concentration limits (according to TRGS 900) for typical polyisocyanates (see Section 3.1)

Isocyanate	Air concentration limit	
	ppm	mg/m^3
TDI (2,4-/2,6-)	0.005	0.035
MDI	0.005	0.050
NDI	0.010	0.087
HDI	0.005	0.035
IPDI	0.010	0.092
H12-MDI	0.005	0.054

considerably higher. Therefore, it is of extreme importance that appropriate safety measures, e.g., protective clothing and ventilation, are provided.

Occupational exposure limits have been set in many countries. As examples, the currently (2003) MAK values for Germany are listed in Table 11.

The vapor pressures of the industrially most important isocyanates at their usual processing temperatures are listed in Table 12.

Many years' experience in handling the various isocyanates has shown that very effective air-extraction systems must be present at all workplaces where TDI is processed at room temperature or MDI and NDI at temperatures above 40 °C. If, for any reason, the aforementioned maximum value at the workplace is exceeded (e.g., on construction sites), the persons present in the working area must be protected against inhaling isocyanates, wherever possible by wearing suitable masks. TRGS 430 is a special German guideline on how to analyze and monitor isocyanates at workplaces where PURs are produced or processed.

More information can be obtained from literature available from the isocyanate producers, or their associations e.g.,

ISOPA: http://www.isopa.org
API: http://www.polyurethane.org/

Table 12. Vapor pressures of the industrially most important isocyanates at processing temperature

Isocyanate	Processing temp., °C	Vapor pressure, Pa	Conc. of saturated vapor, mg/m^3
TDI (2,4-/2,6-)	25	3	255.5
PMDI	25	$< 10^{-2}$	< 0.1
MDI	42	6×10^{-2}	0.65
NDI	130	ca. 120	7850

Polyols have been well studied with respect to their toxicity and, if used according to relevant regulations, do not present an industrial hygiene problem. Aminopolyethers are moderately irritating to the eyes and skin, and thus contact with them should be avoided.

Auxiliary Materials. The ready-to-use polyol preparations, which are used to a large extent nowadays, contain the auxiliary materials necessary for PUR production (especially activators, emulsifiers, and stabilizers) in quantities of up to ca. 1.5 wt %. Toxicological studies have shown that these small quantities of the auxiliary materials scarcely affect the biological behavior of the polyols.

Polyols and polyol preparations have an exceptionally low vapor pressure; a health hazard via inhalation is therefore not to be expected. If, however, they contain readily volatile aliphatic amines as activators, an unpleasant odor can be expected.

For the *tertiary amines* used as catalysts, occupational threshold values in air (MAK) defined by the German TRGS 900 (2003) are available only in some cases, e.g., triethylamine: 1 mL/m^3 (ppm) corresponding to an eight-hour average value of 4 mg/m^3. However, the intense odor of most tertiary amines gives a clear warning of excessive concentrations.

Low-boiling chlorofluoroalkanes were used as blowing agents for producing rigid and flexible foams in the past. Since these substances are suspected of damaging the ozone layer, their use was phased out in the late 1980s. These substances have therefore been replaced by alkanes and to some extent by partly halogenated alkanes, because the blowing process cannot be carried out with carbon dioxide (from the NCO/H$_2$O reaction) in all cases. Since direct contact of partly halogenated alkanes and alkanes with the skin leads to intense defatting and roughness, it is important to wear impermeable gloves and to take good care of the skin. The inhalation of vapors from these blowing agents in high concentrations can lead to intoxication.

14.2. Emissions, Accidental Release, and Waste Disposal [229]

The emission of PUR raw materials into the atmosphere during production and processing is usually small. In special cases (e.g., filling of storage tanks, flexible foam slabstock production), measures to reduce emissions are advisable or required. Since a certain concentration limit of a harmful substance must not be exceeded in the air at the workplace, the air is extracted at the processing plant, and this is one source of emissions. The volatility of the raw materials determines the extent of emission. Hence blowing agents (e.g., carbon dioxide, halogenated hydrocarbons, hydrocarbons) and tertiary amines are the main constituents, whereas isocyanates normally do not play a significant role. Since the carbon dioxide formed from the reaction of water with isocyanate is a natural constituent of air, it presents no problem. The other blowing agents are discharged into the atmosphere in varying quantities depending on the foam type. The emissions of volatile organic chemicals are being increasingly regulated in industrialized countries. National regulations must be observed.

During storage, tank-filling, and transportation of PUR raw materials, discharge due to accidents or leaks cannot be entirely ruled out. Should this happen, as a standard precaution, the accidentally released raw material should be prevented from entering soil or water by suitable measures (e.g., covering sewers). Spilled product should be covered with a liquid-binding material such as a chemical binder based on hydrated calcium silicate (e.g., Hybilat), sand, or sawdust. This can then be transferred to a waste container. In the case of spilled isocyanates, the liquid-binding material should be moist and the material in the waste container stored under moist conditions in a secure place in the open air. The lid of the container must not be tightly sealed because of danger of bursting (CO$_2$ evolution). The polluted areas must then be decontaminated (see below). In all cases, the waste collected in the containers is best disposed of by incineration in a suitable facility (see below). In the unlikely event that a large pool of escaped isocyanate has formed, all unauthorized persons should be kept away from the scene of the accident. Persons downwind should be evacuated. Further evaporation of isocyanate can be prevented very effectively by covering the pool with protein foam. Large quantities of liquids can be pumped into waste containers, before the procedure given above is applied to the remaining residues.

Nevertheless, in the case that isocyanate raw materials have leaked into the soil or water, inert polyurea is the major end product of their reaction with water. According to a three-month soil study with ^{14}C-labeled TDI and MDI, degradation products (e.g., amines) were at no time detected in the aqueous extract.

An experimental pond study indicated low environmental risk arising from spillage of polymeric MDI into a natural aquatic ecosystem. No direct ecotoxic effects and no bioaccumulation of MDI or MDA were observed. Furthermore no MDI or MDA could be detected in water.

Waste Disposal. Liquid or solid PUR production wastes which are not completely reacted are categorized as hazardous waste and should not be dumped but burned in industrial waste incineration facilities, equipped with state-of-the-art flue-gas scrubbing.

Drum decontamination. Any isocyanates residue in "empty" containers that are to be sent for reconditioning or disposal can be rendered harmless with a special neutralizing solution (for formulations, see the specialist literature). The same solution can also be used to decontaminate clothing, equipment and floors. Fully cured PUR end-product wastes, such as trimming waste from flexible foam slabs, can be recycled or recovered (see below). Wastes for which no utilization is yet possible can have their energy content recovered in appropriate incineration plants or can be disposed of in landfill sites, as a last resort (and as long as local regulations do not exclude organic materials from landfilling).

14.3. Recycling/Recovery of Polyurethanes

Because of their excellent durability, PURs in general contribute significantly to a long service life of the products containing them. This is an important contribution to waste minimization. To handle PUR wastes that cannot be prevented, technologies for recycling or recovery have been developed by many industrial associations, joint ventures, and cooperations between producers and users of raw materials throughout the world, e.g., the Polyurethane Recycle and Recovery Council (PURRC, USA), European Isocyanate and Polyol Producers Association (ISOPA, Europe), Tecpol (Germany), and the Japan Urethane Industrial Institute (JUII, Japan). The most appropriate method may vary from case to case, and depends on the properties of the PUR, the intended application, and the related capacity of the market for the recycled material, and very much on logistical, economic, and ecological factors. After all, recycling should not only serve to minimize waste; it ought to contribute also to the saving of resources and to the reduction of environmental burdens. Only then can recycling be counted as a contribution to sustainable development. Currently, the following possibilities for recycling/recovery of PUR exist.

Mechanical Recycling. Mechanical (or physical) recycling of PURs means one of the many forms of "particle recycling." The rebonding of flexible foam or adhesive pressing of rigid foam, for example, involves binding roughly 90 wt % of particles with about 10 wt % PUR binders under heat and pressure. Particle bonding uses up to 70 wt % PUR as the matrix for any kind of particles (e.g., rubber chips for sports ground surfaces). PUR powder can be incorporated into new PUR articles at a loading of round about 20 wt %. PUR powders/particles are also being reused as oil binders.

Feedstock Recycling/Chemical Recycling. Large-scale feedstock recycling processes recover oil and gas products from mixed plastic waste streams of a hundred thousand tons per year or more, of which PUR materials can be one constituent. These processes include pyrolysis, hydrogenation, synthesis gas generation, and reduction of iron ore in blast furnaces. When smaller, but pure streams of particular polymers are available, "chemical recycling" processes can be applied. It is thus possible to obtain liquid degradation products from PURs that are suitable, together with new material, for the manufacture of new PURs. For example glycolysis has been applied more often in favorable case where a suitable application for the glycolysate has been identified. Much research has resulted in several process variations [230]. Some of them include purification and chemical processing of the regenerate before use in PUR applications. Observation of appropriate industrial hygiene and safety is essential, because hazardous substances could be formed.

Energy Recovery. Combustion, or incineration with the recovery of energy, is currently the most effective way to reduce the volume of organic material which otherwise would have to be sent to landfill. Combustion is suitable for all materials for which material recycling is ruled out on ecological or economic grounds or simply because of logistical difficulties. Rotary kilns, fluidized beds, and mass burning equipment, for example, have been shown to be suitable for combustion of plastic scrap. These processes are applied, e.g., in cement production, in industrial power stations, and in municipal solid waste combustors. The heat content of PURs (lower heating value 24–30 MJ/kg) is comparable to that of coal. Numerous test runs have shown that this energy content is recovered in an environmentally sound manner in modern plants equipped with state-of-the-art flue-gas treatment facilities.

14.4. Fire Performance of Polyurethanes [128], [231]

The fire performance of a product is characterized at the various stages of a fire by the parameters ignitability, flame spread, and heat release, and by phenomena such as smoke density and the toxic potency of the combustion products. The dripping behavior, the amount of debris in general, and the corrosive effect of effluents are also determined. Fire test methods differ greatly from country to country, and also depend on the field of application. Different requirements exist, e.g., for the electrical, transportation, furniture, and building sectors. Fire testing, classification and regulatory requirements are very complex and reference must be made to summarizing literature.

In general, the fire hazards posed by a PUR foam and its fire effluents are comparable to those of natural materials like wool and wood. The chemical composition of the material is not

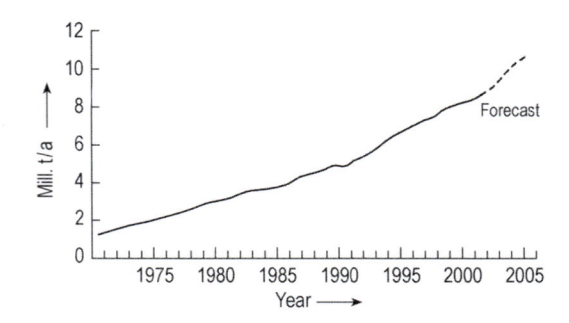

Figure 40. World PUR consumption

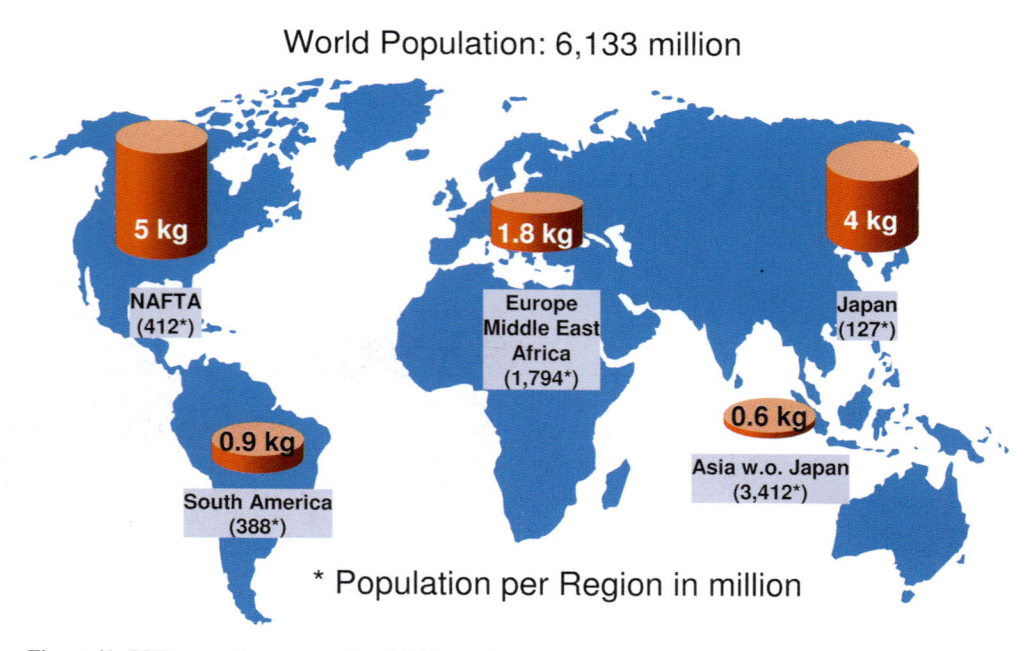

Figure 41. PUR per capita consumption 2001 by regions

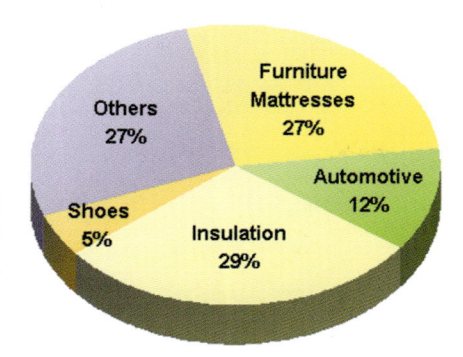

Figure 42. World PUR consumption 2004 by applications (total: 10×10^6 t/a)

the main determining factor. The behavior of the material in fires depends on, e.g., the relative surface of the material (e.g., massive wood compared to sawdust), the combination of materials, thermal conductivity, nature, duration, and intensity of ignition sources, and ventilation.

The qualitative and quantitative composition of fire effluents depends on the decomposition conditions (temperature and ventilation) as well as the amount of material involved. Comparative tests taking into account different stages of a fire have confirmed that the toxicity of fire effluents from PUR products is similar to that of other natural products like wood, cork, and wool. Fire effluents of natural and artificial organic products always pose a toxic hazard irrespective of the type of burning material.

As effluents from PUR do not differ greatly from natural materials, a similar effect on the environment is assumed. Due to their reactivity, isocyanates are rapidly decomposed photochemically or converted chemically.

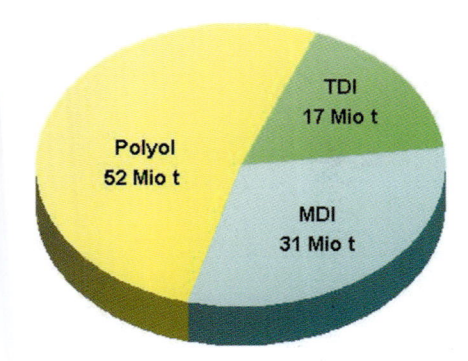

Figure 43. World PUR consumption 2004 by raw materials (total: 10×10^6 Mio t/a)

Table 13. Trade names of raw materials (exemplary)

Type	Tradename	Company
Isocyanates	Caradate	Shell
	Cosmonate	Mitsui Takeda
	Desmodur	Bayer MaterialScience
	Isonate	DOW
	Lupranate	BASF
	Mondur	Bayer MaterialScience
	Rubinate	Huntsman
	Suprasec	Huntsman
	Takenate	Mitsui Takeda
	Tedimon	DOW
	Voranate	DOW
Polyols	Acclaim	Bayer MaterialScience
	Actcol	Mitsui Takeda
	Arcol	Bayer MaterialScience
	Caradol	Shell
	Daltolac	Huntsman
	Desmophen	Bayer MaterialScience
	Isonol	DOW
	Jeffol	Huntsman
	Lupranol	BASF
	Luprapren	BASF
	Multranol	Bayer MaterialScience
	Pluracol	BASF
	Rubinol	Huntsman
	Tercarol	DOW
	Voranol	DOW
Catalysts	DABCO	Air Products
	Desmorapid	Bayer MaterialScience
	Jeffcat	Huntsman
	Niax	Witco
	Polycat	Air Products
	Toyocat	Tosoh

A special fire risk from PUR is not evident: risk of ignition and flame spread as well as emissions mainly depend on the scenario and the amount of burning material, and are not determined by the specific material involved.

15. Economic Aspects

Even though PUR makes up only for ca. 6 % of the total consumption of plastics, the annual production reached the 10×10^6 t level in 2004. The growth rate since 1970 (Fig. 40) on average exceeded 5 % and thus grew faster than GDP. The fastest growing market over the last few years was Asia, in particular China with growth rates exceeding 10 %.

Figures from 2001 show that there is still a big difference in per capita consumption between the regions (Fig. 41), which allows an optimistic outlook for further development.

Figure 42 gives an overview of the main application areas of PURs and Figure 43 the split for the basic raw materials MDI, TDI, and polyethers.

Table 13 lists trade names of raw materials.

References

General References

1. O. Bayer, *Angew. Chem.* **59** (1947) 257 – 272.
2. I. G. Farbenindustrie AG, DE 728981, 1937 (O. Bayer).
3. G. Oertel: "Polyurethane", in G. W. Becker, D. Braun (eds.): *Kunststoff-Handbuch*, 3rd ed., vol. **7**, Carl Hanser-Verlag, München 1992.
4. G. Oertel: *Polyurethane Handbook*, 2nd ed., Carl Hanser Verlag, München, 1993.
5. J. H. Saunders, K. C. Frisch: *Polyurethanes*, Interscience, New York 1962, 1964. "Chemistry," part I; "Technology," part II.
6. K. Uhlig: *Discovering Polyurethanes*, 1st ed., Carl Hanser Verlag, München, 1999.
7. R. Leppkes: "Polyurethanes" in *Die Bibliothek der Technik*, 1st ed., vol. **91**, moderne industrie Verlag, Landsberg, 2002.
8. *Houben-Weyl*, 4th ed., vol. **E20**, pp. 1561 – 1757; Science of Synthesis, vol. 18, 2005, p. 650.
9. K. N. Edwards: "Urethane Chemistry and Applications," *ACS Symp. Ser.* **172** (1981).
10. C. Hepburn: *Polyurethane Elastomers*, 2nd ed., Elsevier Applied Science Publishers, London, 1991.
11. G. Woods: *The ICI Polyurethane Book*, 2nd ed., ICI Polyurethanes and John Wiley and Sons, Chichester, New York, 1990.
12. W. Meckel, W. Goyert, W. Wieder: "Thermoplastic PUR Elastomers" in G. Holden, N. R Legge, R. Quirk, H. Schroeder (eds.): *Thermoplastic Elastomers*, 2nd ed., Hanser-Verlag, München 1996, pp. 15 – 44.
13. K. C Frisch, S. L Reegen: *Advances in Urethane Science and Technology*, vols. **1–6**, Technomic Publ., Westport, CT, 1971 – 1978. K. C Frisch, D. Klempner: *Advances in Urethane Science and Technology*, vols. **8–10**, Technomic Publ., Lancaster, PA, 1981–1987.
14. J. M Methven: "Polymeric Materials from Renewable Resources," *Rapra Rev. Rep.* **4** (1991) no. 6, Report 43, 1 – 121.
15. H. Zoebelein, *Chem unserer Zeit* **26** (1992) 27 – 34.
16. P. L Nayak, S. Lenka, S. K Panda, T. Pattnaik, *J. Appl. Polym. Sci.* **47** (1993) 1089 – 96.
17. I. Javni, Z. Petrovic, *Annu. Tech. Conf. Soc. Plast. Eng.* 1997, 791 – 795.
18. Bayer AG, EP 672694, 1995 (W. von Bonin, H.-P. Müller, M. Kapps).
19. A. Gandini, *Polym. Prepr. (Am. Chem. Soc., Div. Polym. Chem.)* **39** (1998) 69 – 70.
20. P. Ni, R. W Thring, *Int. J. Polym. Mat.* **52** (2003) 685 – 707.
21. S. Desai, I. M Thakore, B. D Sarawade, S. Devi, *Polym. Eng. Sci.* **40** (2000) 1200 – 1210.
22. Agency of Ind. Science & Technol., JP 5186586, 1993 (H. Hatakeyama *et al.*).
23. Hüls AG, EP 787754, 1996 (R. Gras, E. Wolf).
24. BASF AG, EP 716080, 1994 (W. Heider, W. Langer, H. Renz, S. Wolff).
25. Bayer AG, EP 802210, 1997 (B. Baumbach, L. Kahl, E. König, N. Yuva).
26. Bayer AG, EP 270804, 1987 (H. Heß, K. König, R. Kopp, G. Grögler).
27. R. Kopp, H. Heß, G. Grögler, Proc. SPI/FSK Polyurethane World Congress 1987, *Aachen, Technomic Publ., Lancaster*, pp. 316 – 322.
28. American Cyanamid Company, EP 234353, 1982 (B. Singh, L. W Chang, P. S Forgione).
29. V. D Arendt, R. E Logan, R. Saxon, *J. Cell. Plast.* **18** (1982) 376.
30. Bayer AG, EP 798299, 1999 (H. Mertes, F. Richter, J. Pedain, C.-G. Dieris).
31. VEBA Chemie AG, DE 2323299, 1973 (K. Schmitt, J. Disteldorf, J. Reiffer).
32. L. Born, D. Wendisch, H. Reiff, D. Dieterich, *Angew. Makromol. Chem.* **171** (1989) 213.
33. N. Bialas, H. Höcker, M. Marschner, W. Ritter, *Makromol. Chem.* **191** (1990) 1843.
34. Hüls AG, US 4912210, 1987 (J. Disteldorf, W. Hübel, K. Schmitz).
35. Mobay Corp., EP 453914, 1990 (S. D. Seneker, T. A Potter, K. L Dunlap, M. K Lowery).
36. Z. W Wicks, *Prog. Org. Coat.* **3** (1975) 73.
37. Z. W Wicks, *Prog. Org. Coat.* **9** (1981) 3.
38. J. W Rosthauser, J. L Williams, *Polym. Mater. Sci. Eng.* **50** (1984) 344.
39. V. Mirgel, K. Nachtkamp, *Farbe Lack* **89** (1983) 928.
40. B. W Peterson, T. H Austin, *Proc. SPI Annu. Polyurethane Tech./Mark. Conf.*, 1986, Toronto, Technomic Publ., Lancaster, pp. 155 – 158.
41. Th. Goldschmidt AG, US 5001189, 1986 (J. Fock, D. Schedlitzki).
42. Th. Goldschmidt AG, EP 0326890, 1989 (J. Fock, E. Esselborn, F. Nickel).
43. BASF AG, DE 4316137, 1994 (W. Franzischka, R. Becker, C. Palm, C. Sigwart).
44. W. J Pentz, R. G Krawiec, *Rubber Age (New York)* **12** (1975) 39.
45. J. R Harrison, *Plast. Ind. News* **30** (1984) 120.
46. J. L Schuchardt, S. D Harper, *Proc. 32nd Annual Polyurethane Technical Marketing Conf.*, 1989, San Francisko, pp. 360 – 364.
47. G. Wegener, M. Brandt, L. Duda, J. Hofmann, B. Kleszcewski, D. Koch, R.-J. Kumpf, H. Orzesek, H.-G. Pirkl, C. Six, C. Steinlein, M. Weisbeck, *Appl. Catal. A: Gen.* **221** (2001), 303 – 335.
48. ARCO Chemicals, EP 700 949, 1994 (B. Le-Khac).
49. E. Müller, *Angew. Makromol. Chem.* **14** (1970) 75 – 86.
50. E. Müller, *Angew. Makromol. Chem.* **16/17** (1971) 117 – 128..
51. S. Inoue, H. Koinuma und T. Tsuruta, *J. Polym. Sci. Polym. Lett.* **7** (1969) 287.
52. S. Inoue, H. Koinuma und T. Tsuruta, *Makromol. Chem.* **130** (1969) 210.
53. M. Ree, J. Y. Bae, J. H. Jung, T. J. Shin, J. Korea, *Polym. J.* **7** (1999) 333.
54. W. Kuran, S. Pasynkiewicz, J. Skupinska, A. Rokicki, *Makromol. Chem.* **177** (1976) 11.
55. T. Aida und S. Inoue, *J. Am. Chem. Soc.* **107** (1985) 1358.
56. Bayer AG, EP 68383, 1981 (J. Probst, G. Kolb, B. Riberi, P. Höhlein).
57. Bayer AG, DE 2622951, 1976 (Ch. Weber, H. Schäfer).
58. Dow Chemical, US 4477644, 1983 (T. R. Sutton, D. W Hughes).
59. BASF Wyandotte Corp., EP 107108, 1983 (J. W Light, sey, R. A Markovs, G. G Ramlow, P. T Kan).
60. Air Products, US 4816543, 1985 (W. F Burgoyne, D. D Dixon, B. Milligan, J. P Casey).

61. C. J Nalepa, A. A Eisenbraun, *Proc. S. P. I. Annu. Polyurethane Tech./Mark. Conf.*, 1986, Toronto, Technomic Publ., Lancaster, pp. 228 – 233.
62. Bayer, DE 1803635, 1968 (W. Meckel, E. Müller).
63. H. Kleimann, *Proc. SPI/FSK Polyurethane World Congress* 1987, Technomic Publ., Aachen, Lancaster, pp. 815 – 819.
64. Th. Voelker, P. Balling, *Proc. SPI Annu. Polyurethane Tech./Mark. Conf.*, 1986, Toronto, Technomic Publ., Lancaster, pp. 133 – 136.
65. U. S. Department of Health and Human Services. Hazardous Substances Data Bank, National Toxicology Information Program, National Library of Medicine, Bethesda, MD. 1993.
66. Jefferson Chem., US 3847992, 1973 (P. H Moss).
67. BP Chem. Int. Ltd, EP 356046, 1988 (I. D Dobson).
68. J.-M. Jehng, C.-M. Chen, *Catal. Lett.* **77** (2001) 147 – 154.
69. Union Carbide, EP 297536, 1988 (R. M Gerkin, D. J Schreck, D. E Smith).
70. *Houben-Weyl*, 4th ed., **E20**, 1602, 1610.
71. Bayer AG, EP 219035, 1985 (W. Raßhofer, K. König, H.-J. Meiners, G. Grögler).
72. Bayer AG, DE 1935484, 1969 (G. Oertel, G. Grögler).
73. Mobay Corp., EP 457129, 1990 (R. Mafoti, J. Sanders).
74. Dow Chemical, EP 268849, 1986 (V. R Durvasula, F. A Stuber).
75. Dow Chemical, EP 335274, 1988 (V. R Durvasula).
76. Bayer, DE 3834749, 1988 (U. Thiery, H. Reiff, D. Dieterich, G. Grögler, J. Sanders).
77. M. Bock, R. Halpaap, *Farbe Lack* **93** (1987) 264.
78. M. Bock, R. Halpaap, *Polym. Mat. Sci. Eng.* **55** (1986) 448.
79. M. Bock, R. Halpaap, *J. Coat. Technol.* **59** (1987) 131.
80. E. F. Cassidy, *Urethanes Technology* (1989),pp. 26 – 34.
81. ICI, EP 361704, 1988 (E. F Cassidy, H. R Gillis, M. Hannaby).
82. D. G Schlotterbeck, G. Matzke, P. Horn, H.-U. Schmidt, *J. Elastomers Plast.* **21** (1989) 49.
83. D. G Schlotterbeck, G. Matzke, P. Horn, H.-U. Schmidt, *Elastomerics* **121** (1989) 18.
84. *Houben-Weyl*, 4th ed., **E20**, 1603.
85. P. K. H. Lam, M. H George, J. A Barrie, *Polym. Commun.* **32** (1991) 80.
86. The University of Akron, PCTWO 9010657, 1989 (J. P Kennedy).
87. J. P Kennedy, B. Ivan, V. S. C Chang, *Adv. Urethane Sci. Technol.* **8** (1981) 245.
88. Minnesota Mining & Manufacturing, EP 380236, 1989 (C. Leir).
89. Dow Corning, EP 378079, 1989 (G. T. Decker, C.-L. Lee, G. A Gornowicz).
90. Union Carbide, EP 405494, 1989 (R. S Neale, C. L Schilling).
91. Du Pont, EP 420223, 1989 (E. G Brugel).
92. DE 3702495, 1987 (G. Bauer).
93. B.-U. Kettemann, M. Melchiorre, T. Muenzmay, W. Rasshofer, *Kunststoffe* **85** (1995) 1947 – 1950.
94. M. Kuhn, M. Kugler, *Kunststoffe* **87** (1997) 729 – 730, 732.
95. M. L. Listemann, A. L Wressell, K. R Lassila, H. C Klotz, G. L Johnson, A. C Savoca, *Proc. Polyurethanes World Congress*, 1993, Vancouver, Technomic Publ., Lancaster, pp. 595 – 608.
96. A. L Silva, J. C Bordado, *Catal. Rev.* **46** (2004) 31 – 51.
97. C F Frisch, L. P Rumao, *J. Macromol. Sci., Rev. Macromol. Chem. Phys.* **C5** (1970) 103 – 150.
98. G. Oertel: *Polyurethane Handbook*, 2nd ed., Carl Hanser Verlag, München, 1993, pp. 98 – 104.
99. D. C Fondots, *J. Cell. Plast.* **11** (1975) 250.
100. M. L Listemann, A. C Savoca, A. L Wressell, *J. Cell. Plast.* **28** (1992) 360 – 398.
101. J. G Kniss, R. Fard-Aghaie, A. R Arnold, S. Wendel, *Proc. API Polyurethanes Expo*, 2003, Orlando, pp. 361 – 366.
102. R. Hoffmann, H. Schloens, *Proc. API Polyurethane Conf.*, 2002, Salt Lake City, pp. 380 – 386.
103. L. Thiele, *Acta Polym.* **30** (1979) 323.
104. G. D Andrew, J. G Kniss, M. L Listemann, L. A Mercando, J. D Tobias, S. Wendel: "Dimensional Stabilizing Additives for Flexible Polyurethane Foams" in D. Klempner, K. C Frisch (eds.): *Advances in Urethane Science and Technology*, Rapra Technology Limited, 2001, pp. 3 – 83.
105. R. M Hill: *Silicone Surfactants*, Marcel Dekker, New York, 1999.
106. G. Rossmy, *J. Cell Plast.* **17** (1981) 319.
107. Th. Goldschmidt AG, EP 193815, 1985 (H.-J. Kollmeier, R.-D. Langenhagen).
108. B. Kanner, B. Prokai, C. S Eschbach, G. I Murphy, *J. Cell. Plast.* **15** (1979) 315.
109. O. Eyrisch, G. Burkhart, R. Borgogelli, *Proc. API Polyurethanes Expo*, 2001, Columbus, pp. 417 – 422.
110. T. Boinowitz, R. Borgogelli, W. Gower, *Proc. API Polyurethanes Expo*, 2003, Orlando, pp. 324 – 330.
111. O. Eyrisch, G. Burkhart, *Proc. API Polyurethanes Conf.*, 2002, Salt Lake City, pp 409 – 414.
112. T. Boinowitz, G. Burkhart, J. Klietsch, W. Bunting, *Proc. API Polyurethanes Expo*, 2001, Columbus, pp. 41 – 46.
113. T. Timm, *Kautsch. Gummi Kunstst.* **35** (1982) 568.
114. T. Timm, *Kautsch. Gummi Kunstst.* **36** (1983) 257.
115. T. Timm, *Kautsch. Gummi Kunstst.* **37** (1984) 933, 1021.
116. C. S Schollenberger, F. D Stewart, *Adv. Urethane Sci. Technol.* **1** (1971) 65.
117. G. Capocci, *Proc. SPI. Annu. Polyurethane Tech./Mark. Conf.*, 1986, Toronto, Technomic Publ., Lancaster, pp. 220 – 227.
118. B. P Thaspliyal et al., *Progr. Polym. Sci.* **15** (1990) 735 – 750.
119. G. Mathur, J. E Kresta, K. C Frisch, *Adv. Urethane Sci. Technol.* **6** (1978) 103.
120. R. Wiedermann, G. Heilig, *Kunststoffe* **80** (1990) 909.
121. G. Heilig, *Kunststoffe* **81** (1991) 622.
122. U. Leyrer, D. Polke, *Kunststoffe* **88** (1998) 235.
123. H. Fleurent, *Cell. Polym.* **13** (1994) 419.
124. R. D Leaversuch, *Mod. Plast. Int.* **24** (1994) 19.
125. H. Krähling, L. Zipfel, *Proc. API Polyurethanes Expo*, 2001, Columbus, pp. 333 – 338.
126. L. Zipfel, K. Boerner, P. Dournel, *Proc. API Polyurethanes Expo*, 2001, Columbus, pp. 637 – 642.
127. D. Williams, *Proc. API Polyurethanes Conf.*, 2002, Salt Lake City,pp 135 – 143.
128. J. Troitzsch, *Plastics Flammability Handbook*, 3rd ed., Carl Hanser Verlag, München 2004, pp. 78 – 83, 165 – 167.
129. G. Neumann, H. U Schimpfle, R. Becker, *Plaste Kautsch.* **24** (1977) 27.
130. G. Oertel: "Polyurethane" in G. W Becker, D. Braun (eds.): *Kunststoff-Handbuch*, 3rd ed., vol. **7**, Carl Hanser Verlag, München 1993, pp. 40 – 47.
131. W. Goyert, H. Hespe, *Kunststoffe* **68** (1978) 819.
132. C. S Schollenberger, K. Dinbergs, *Adv. Urethane Sci. Technol.* **6** (1978) 60.
133. K. Hoffmann, R. Bonart, *Makromol. Chem.* **184** (1983) 1529.
134. I. Blackwell, K. H Gardner, *Polymer* **20** (1979) 13.
135. E. C Prolingheuer, J. J Lindsey, H. Kleimann, *J. Elastomers Plast.* **21** (1989) 100.
136. M. Tanaka, *Makromol. Chem.* **187** (1986) 2345.
137. M. Tanaka, T. Nakaya, *Macromol. Sci. Chem.* **A24** (1987) 777.
138. L. Born, H. Hespe, *Colloid Polym. Sci.* **263** (1985) 335.
139. L. Morbitzer, H. Hespe, *J. Appl. Polym. Sci.* **16** (1972) 2697.

140 H. F Hespe et al., *J. Appl. Polym. Sci.* **44** (1992) 2029.
141 G. Zeitler, *Kunststoffberater* **34** (1989) 65.
142 G. Zeitler, *K-Plast. Kautsch.-Ztg.* **403** (1989) 85.
143 U. Rotermund, *J. Elastomers Plast.* **21** (1989) 122.
144 D. Boese et al., *Makromol. Chemie, Macromol. Symp.* **50** (1991) 191.
145 *Houben-Weyl*, 4th ed., **E20**, pp. 1739 – 1751.
146 S. M Clift, J. Grimminger, K. Muha, *Proc. SPI Polyurethanes Conf.*, 1994, Boston, pp. 546 – 560.
147 S. B Burns, E. L Schmidt, *Proc. Polyurethane World Congress* 1993, Vancouver, pp. 234 – 240.
148 Bayer AG, DE 3914718, 1989 (A. Ruckes, H.-J. Meiners, H. Boden, M. Schmidt).
149 J. Blahak, W. Meckel, E. Müller, *Angew. Makromol. Chem.* **26** (1972) 29.
150 Bayer AG, DE 2513817, 1975 (H. Schäfer, C. Weber).
151 Bayer AG, DE 2622951, 1976 (H. Schäfer, C. Weber).
152 D. Nissen, R. A Markovs, *J. Elastomers Plast.* **15** (1983) 96.
153 J. Blackwell, J. R Quay, R. B Turner, *Polym. Eng. Sci.* **23** (1983) 816.
154 Bayer AG, DE 3147736, 1981 (K. Seel, C. Weber, H. Wirtz).
155 Texaco Inc., EP 92672, 1982 (R. J. G Dominguez).
156 Texaco Inc., EP 93862, 1982 (R. J. G Dominguez, D. M Rice, R. F Lloyd).
157 Texaco Inc., EP 128691, 1983 (R. J. G Dominguez).
158 Texaco Inc., US 4474901, 1983 (R. J. G Dominguez).
159 Texaco Inc., US 4420570, 1982 (R. J. G Dominguez).
160 R. J. G Dominguez, *J. Cell. Plast.* **20** (1985) 433.
161 Dow, US 4495081, 1983 (J. A Vanderhider, G. M Lancaster).
162 BASF, DE 3215907, 1982 (D. Nissen, W. Schoenleben, M. Marx, K. H Illers, P. Simak).
163 BASF, DE 3215909, 1982 (D. Nissen, W. Schoenleben, M. Marx, K. H Illers, P. Simak).
164 Air Products & Chemicals, US 4892920, 1988 (J. R Quay, J. P Casey).
165 Air Products & Chemicals, EP 420026, 1989 (R. M Machado, A. J Siuta, W. E Starner, B. A Toseland).
166 *Houben-Weyl*, 4th ed., **E20**, 1613 – 1617.
167 L. Schmalstieg, L. Kahl, *Congr. FATIPEC*, 20th Nice, 1990, pp. 425 – 428.
168 Bayer AG, EP 394759, 1989 (L. Schmalstieg, K. Nachtkamp, J. Pedain).
169 C. Zwiener, L. Schmalstieg, M. Sonntag, K. Nachtkamp, J. Pedain, *farbe+lack* (1991) 1052.
170 H. Kittel, *Adhäsion* **21** (1977) 162.
171 Bayer AG, DE 3940271, 1989 (G. Grögler, H. Heß, R. Kopp, W. Raßhofer).
172 *Houben-Weyl*, 4th ed., **E20**, pp. 1703 – 1716, 1729 – 1735.
173 P. W Morgan, *J. Appl. Polym. Sci.* **40** (1990) 1771.
174 Bayer AG, DE 3808274, 1988 (H.-P. Müller, B. Jansen, W. Calaminus, R. Dhein).
175 L. Ubaghs, N. Fricke, H. Keul, H. Höcker, *Macromol. Rapid Commun.* **25** (2004) 517.
176 Eurotech ltd, US 6120905, 2000 (O. L Figovsky).
177 S. Maier, T. Loontjens, B. Scholtens, R. Mülhaupt, *Macromolecules* **36** (2003) 4727.
178 R. F Harris, C. D DePorter, R. B Potter, *Macromolecules* **24** (1991) 2973.
179 M. Bierdel: *Numerische und experimentelle Untersuchungen zum Gegenstrom-Injektions-Mischer*, Logos Verlag, Berlin 2001.
180 Maschinenfabrik Hennecke GmbH, DE 2123216, 1971 (W. Schmitz, R. Raffel, F. Althausen, F. Proksa).
181 Hollandsche Draad en Kabelfabrik (Draka), DE 1207072, 1958 (B. Hackert).
182 Hellenic Plastics & Rubber Ind., DE 1629458, 1965 (N. G. Petzetakis).
183 Unifoam AG, DE 2142450, 1970 (L. Berg).
184 Beamech Variable Pressure Foaming, Tenax Road, Trafford Park, Manchaster, M17 1JT, England.
185 NovaFlex Hennecke, Hennecke GmbH, Birlinghofener Str. 30, D-53754 Sankt-Augustin, Germany.
186 Reeves Brothers Inc., US 3325573, 1964 (J. D Boon, K. F Hager).
187 F. Buff, M. French, W. Pollock, US 3476845, 1965 (F. Buff, M. French, W. Pollock).
188 Fecken Kirfel GmbH & Co, Prager Ring 1–15, D-52070 Aachen, Germany.
189 U. Leyrer, Conf. Papers UTECH 2003, The Hague.
190 H.-U. Schmidt, E. Calgua, *Kunststoffe* **89** (1999) 104.
191 Air Products & Chemicals, US 5958990, 1980, (J. Grimminger).
192 W. W Reichmann, B. A Phillips, *J. Cell. Plast.* **24** (1988) 601.
193 J. Wu, D. Dillon, R. Crooker, *Proc. API Polyurethane Conf.*, 2002, Salt Lake City, pp. 144 – 150.
194 C. Chittolini, *Proc. Polyurethanes World Congress*, 1997, Amsterdam, pp. 655 – 659.
195 R. Franco, M. Checchin, C. Cecchini, *Proceedings of UTECH 2000*, The Hague, pp. 10 – 12.
196 V. Knipp: *Herstellung von Großteilen aus Polyurethanschaumstoffen*, Zechner & Höthig, Speyer 1974.
197 W. Schuhmacher, *Gummi Asbest Kunstst.* 1974, 698.
198 H. Piechota, H. Röhr: *Integralschaumstoffe*, Hanser Verlag, München 1975.
199 G. Oertel: "Polyurethane", in G. W Becker, D. Braun (eds.): *Kunststoff-Handbuch*, 3rd ed., vol. **7**, Carl Hanser Verlag, München 1993, pp. 599 – 642.
200 M. Bock: *Polyurethanes for Coatings*, Vincents Verlag, Hannover 2001.
201 D. Stoye, W. Freitag: *Resins For Coatings*, Hanser Gardner, Chicago 1996.
202 G. Oertel: "Polyurethane", in G. W Becker, D. Braun (eds.): *Kunststoff-Handbuch*, 3rd ed., vol. **7**, Carl Hanser Verlag, München 1993, pp. 643 – 663.
203 G. Festel, A. Proß, H. Stepanski, H. Blankenheim, R. Witkowski, *Adhäsion Kleben Dichten* 1997, 16.
204 S. Lee: *The Polyurethanes Book*, Jon Wiley & Sons, New York 2002, pp. 379 – 394.
205 G. Oertel: "Polyurethane", in G. W Becker, D. Braun (eds.): *Kunststoff-Handbuch*, 3rd ed., vol. **7**, Carl Hanser Verlag, München 1993, pp. 679 – 694.
206 DIN 60001 (1970).
207 D. C Allport, A. A Mohajer in D. C Allport, W. H Janes: *Block Copolymers*, Section 5.2.6, Applied Publishers Ltd., London, 1973.
208 H. Oertel in B. v. Falkai: *Synthesefasern-Grundlagen, Technologie, Verarbeitung und Anwendung*, Section 6.1.3, Verlag Chemie, Weinheim, Germany, 1981.
209 M. Couper, in M. Levin, J. Preston: *Handbook of Fibre Science and Technology Fibers*, vol. **3**, part A, Marcel Dekker, New York 1985, pp 51 – 85.
210 K. H Wolf, *Text.-Prax. Int.* **36** (1981) 839.
211 J. H Saunders, K. C Frisch: *Polyurethanes*, Interscience, New York 1962, 1964. "Chemistry," part I; "Technology," part II, p 129.
212 DuPont, US 2726219, 1951 (F. B Hill).
213 E. M Hicks Jr., *Am. Dyest. Reptr.* **52** (1963) 33.
214 DuPont, EP 343985, 1988 (N. E Houser, R. L Dreibelbis).
215 R. Bonart, L. Morbitzer, H. Rinke, *Kolloid. Z. Z. Polym.* **240** (1970) 807.

216 A. Blejenberg, D. Heikens, H. Meijers, H. Lampe, P. v. Reth, *Brit. Polym. J.* **4** (1972) 125.
217 H. Oertel, *Chemiefasern Textilind.* **27/79** (1977) 1090.
218 H. Oertel, *Chemiefasern Textilind.* **28/80** (1978) 43.
219 G. Oertel: "Polyurethane", in G. W Becker, D. Braun (eds.): *Kunststoff-Handbuch*, 3rd ed., vol. **7**, Carl Hanser Verlag, München 1993, pp. 665 – 678.
220 J. Franke, *Kunststoffe* **79** (1989) 999.
221 J. Franke, H.-P. Müller, *Kunststoffe* **80** (1990) 1200.
222 W. Y Chiang, S. C Chan, *Angew. Makromol. Chem.* **182** (1990) 9.
223 W. R. Grace & Co., EP 549116, 1991 (D. R Kyle).
224 W. Fischer, J. Weikard, E. Luehmann, T. Faecke, *FAPU* **14** (2002) 24.
225 H.-U. Meier-Westhues, European Coatings Conf., 2001, Zürich, Vincentz Verlag, Hannover, pp.149 – 62.
226 K. Studer, C. Decker, E. Beck, R. Schwalm, European Coatings Conf. 2004, Berlin, Vincentz Verlag, Hannover, pp. 49 – 63.
227 L. W Arndt, L. J Junker, S. P Patel, D. B Pourreau, W. Wang, *Paint Coatings Ind.* **20** (2004) 42.
228 D. C Allport, D. S Gilbert, S. M. Outterside: *MDI & TDI—Safety, Health and the Environment*, John Wiley & Sons Ltd, Chichester 2003.
229 F. K Brochhagen: "Isocyanates" in O. Hutzinger (ed.): *The Handbook of Environmental Chemistry*, vol. **3**, part G, Springer-Verlag, Berlin 1991, p. 73.
230 W. Raßhofer: *Recycling von Polyurethan-Kunststoffen*, Hüthig Verlag, Heidelberg, 1998.
231 F. H Prager: *Polyurethane and Fire*, Wiley VCH, Weinheim, Germany, 2004.

Further Reading

K. Ashida: *Polyurethane and Related Foams*, CRC Press/Taylor & Francis, Boca Raton, FL 2007.

W. Brockmann, P. L. Geiß, J. Klingenberg, B. Schröder: *Adhesive Bonding*, Wiley-VCH, Weinheim 2009.

I. Clemitson: *Castable Polyurethane Elastomers*, CRC Press, Boca Raton, FL 2008.

A. Fainleib (ed.): *Thermostable Polycyanurates*, Nova Science Publishers, Hauppauge, NY 2009.

S. Fakirov (ed.): *Handbook of Condensation Thermoplastic Elastomers*, Wiley-VCH, Weinheim 2005.

M. Ionescu: *Chemistry and Technology of Polyols for Polyurethanes*, Smithers Rapra Technology, Shawbury 2008.

P. Król: *Linear Polyurethanes*, Brill Academic Publishers, Boston, MA 2008.

E. M. Petrie: *Handbook of Adhesives and Sealants*, 2nd ed., McGraw-Hill, New York 2007.

S. Thomas, S. Ranimol: *Rubber Nanocomposites*, Wiley-Blackwell, Oxford 2009.

H. Ulrich: *Urethane Polymers*, "Kirk Othmer Encyclopedia of Chemical Technology", 5th edition, John Wiley & Sons, Hoboken, NJ, online DOI: 10.1002/0471238961.2118052021121809.a01.pub2.

C. A. Wilkie, A. B. Morgan (eds.): *Fire Retardancy of Polymeric Materials*, 2nd ed., CRC Press, Boca Raton, FL 2010.

Poly(Vinyl Chloride)

INGO FISCHER, VESTOLIT GmbH, Marl, Germany

WILHELM FRIEDRICH SCHMITT, VESTOLIT GmbH, Marl, Germany

HANS-CHRISTOPH PORTH, VESTOLIT GmbH, Marl, Germany

MICHAEL W. ALLSOPP, Independent PVC Technology Consultant, Heswall, England

GIOVANNI VIANELLO, European Vinyls Corporation (IT), Porto Maguero, Italy

1.	Introduction	1111
1.1.	Morphology	1111
1.2.	Versatility	1112
1.3.	Molecular Structure	1112
2.	Physical and Chemical Properties	1112
2.1.	Vinyl Chloride Monomer	1112
2.2.	Poly(Vinyl Chloride)	1113
3.	Resources and Raw Materials	1113
4.	Production	1114
4.1.	Suspension Polymerization of Vinyl Chloride	1114
4.1.1.	Introduction	1114
4.1.2.	Additives	1115
4.1.3.	Kinetics	1118
4.1.4.	Morphology	1119
4.1.4.1.	Control of Grain Size	1119
4.1.4.2.	Control of Porosity and Bulk Density	1119
4.1.4.3.	K-Value (Molecular Mass)	1121
4.1.4.4.	Problem Areas	1123
4.2.	Bulk or Mass Polymerization of Vinyl Chloride	1123
4.3.	Emulsion Polymerization of Vinyl Chloride	1125
4.3.1.	Introduction	1125
4.3.2.	Emulsion Polymerization Processes	1126
4.3.3.	Description of the Batch Process	1126
4.3.4.	Plastisols	1129
4.3.5.	Polymer–Plasticizer Interaction	1130
4.3.6.	Gelation and Fusion	1131
4.3.7.	Applications	1132
4.4.	Chlorinated Poly(Vinyl Chloride), CPVC	1132
5.	Environmental Protection	1133
6.	Quality Specifications and Analysis	1133
7.	Storage and Transportation	1134
8.	Processing and End Uses	1134
9.	Economic Aspects	1135
10.	Toxicology and Occupational Health	1136
	References	1139

1. Introduction

Poly(vinyl chloride), PVC, a polymer prepared from vinyl chloride monomer (VCM),

$$\underset{\text{H}}{\overset{\text{H}}{\text{C}}}=\underset{\text{Cl}}{\overset{\text{H}}{\text{C}}} \longrightarrow \left[\underset{\text{H}}{\overset{\text{H}}{\text{C}}}-\underset{\text{Cl}}{\overset{\text{H}}{\text{C}}}\right]_n$$

where $n = 700-1\,500$, holds a unique position amongst all the polymers produced today. It is relatively inexpensive and is used in such a wide range of applications that its versatility is almost unlimited. However, if it were discovered today it would probably be shelved as a somewhat intractable and thermally unstable material. How can this apparent contradiction be explained?

The uniqueness of PVC is considered in the sections: Morphology (Section 1.1), versatility (Section 1.2), and molecular structure (Section 1.3).

1.1. Morphology

As made, PVC is particulate in nature and comes in two main sizes depending on the process used. Suspension and mass polymerizations give grains (particles) of 100–180 μm in diameter, whereas the emulsion process affords a latex of particle size 0.1–3.0 µm. The latter is dried to yield friable grain-like structures of 5–50 µm.

Because of this unique particulate structure, the most frequently used word in the vocabulary of the PVC technologist is morphology. In no other polymer is it as important as it is in PVC.

In its polymerization, a growing PVC chain becomes insoluble in VCM above a chain length of about 10 units [6], PVC is essentially insoluble in its monomer, and the process is thus classified as a precipitation suspension polymerization. However, PVC is heavily swollen and partly solvated by the monomer to the extent of 27 wt% [7], and this has a major influence on the polymerization itself (see Section 4.1) as well as on the final properties and end uses of PVC. Hence, the way the PVC separates from the monomer, its future growth mechanism, and the swelling of the polymer by the monomer are critically important in its formation, handling, and subsequent processing.

1.2. Versatility

Poly(vinyl chloride) is a generic name. Each producer makes a range of PVC polymers that vary in morphology and in molecular mass, depending on the intended end use. In industry, the K-value and viscosity number are used to represent molecular mass, and producers often reflect these parameters in the grade codes used to define different products (e.g., S 68/173 refers to a suspension type material with a K-value of 68, and VY 110/57 to a resin with a viscosity number of 110). The calculation of the K-value is given in [8] and the relationship between these parameters and molecular mass in [9].

PVC with $K = 66$–68 can be processed in rigid formulations to give pipes, conduit, sheet, and window profiles; $K = 65$–71 in flexible formulations for flexible sheet, flooring, wallpapers, cable coverings, hoses, tubing, and medical products, and PVC with low K-values (55–60) in formulations for injection molding of pipe and conduit fittings, integral electrical plugs, and blow molding of bottles and other containers.

In contrast, poly(methyl methacrylate) of even very high molecular mass is still soluble in its monomer and produces only glass-like solid beads at the end of polymerization that have no internal morphology and as such are limited to a narrow range of end uses (e.g., moldings for optical applications, such as covers for car lights and illuminated signs).

1.3. Molecular Structure

Amongst the range of polymeric materials produced today PVC is unique because the bulky chlorine atom imparts a strongly polar nature to the PVC polymer chain, and the essentially syndiotactic conformation of the repeat unit in the chain leads to a limited level of crystallinity [10]. This results in good mechanical properties, particularly stiffness at low wall thickness, high melt viscosity at relatively low molecular mass, and the ability to maintain good mechanical properties even when highly plasticized. This enables a wide range of softness and flexibility to be achieved and hence leads to an even wider variety of end uses.

2. Physical and Chemical Properties

2.1. Vinyl Chloride Monomer

Vinyl chloride monomer (VCM), bp $-13.4°C$, is a gas at room temperature and pressure. Therefore, it is handled as a compressed volatile liquid in all polymerization operations. Its vapor pressure over the typical polymerization temperature range of 50 to 70°C is 800–1 250 kPa. As a result, PVC polymerization reactors are thick-walled jacketed steel vessels with a pressure rating of 1 725 kPa. VCM is slightly soluble in water (0.11 wt% at 20°C). Whereas this has some influence on the suspension polymerization process, it is critically important to the success of the emulsion polymerization process described in Section 4.3. The polymerization of VCM is strongly exothermic, and its specific heat and heat of evaporation of 1.352 kJ kg^{-1} K^{-1} and 20.6 kJ/mol, respectively, allow the use of a condenser to remove the heat of reaction as well as the more conventional jacketed vessel systems. Its explosive limits in air are 4–22 vol%, and plant design, particularly when handling unreacted VCM in the recovery system, must be designed and operated accordingly.

Table 1. Typical properties of rigid PVC (UPVC)

Property	Test	Value
Tensile strength at 23°C, MPa	ISO 527	55
Tensile modulus (1% strain, 100 s), GPa	ISO 527	2.7–3.0
Tensile modulus (1% strain, 3 years), GPa	ISO 527	1.7
Izod impact, ft lb/in	ISO 180	2 (unmodified) 10 (modified)
Specific gravity	ISO 1183	1.38–1.50
Coefficient of linear thermal expansion, K^{-1}	BS 4618 : 3.1	6×10^{-5}
Coefficient of thermal conductivity, $Wm^{-1} K^{-1}$		0.14
Flammability (oxygen index)	ASTM D2863 (Fenimore Martin)	45
Weathering resistance		very good (especially white)
Resistance to concentrated mineral acids (at 20°C)		excellent
Maximum continuous operating temperature	field experience	60°C

Further physical and chemical properties of VCM are given in → Chloroethanes and Chloroethylenes.

2.2. Poly(Vinyl Chloride)

Poly(vinyl chloride) (PVC) is always mixed with heat stabilizers, lubricants, plasticizers, fillers, and other additives to make processing possible, all of which can influence its physical and mechanical properties. Table 1 lists properties of rigid (unplasticized) PVC with a total additives content of <10%. Table 2 list properties of flexible (plasticized) PVC where the range of physical properties varies widely, depending on the plasticizer content. This can vary between 20 and 100 phr so as guide properties typical of a plasticizer content of 50 phr are given.

In addition, the K-value (molecular mass) of the PVC can also influence properties significantly (see Section 4.1.4.3).

Table 2. Typical properties of flexible PVC

Property	Test	Value
Tensile strength, MPa	ISO 527	7.5–30
Elongation at break, %	ISO 527	140–400
Shore hardness (A)	ISO 868	5–100
Specific gravity	ISO 1183	1.19–1.68
Cold flex temperature, °C	BS 2782 159B	−20 to −60
Volume resistivity at 23°C, Ω · cm	BS 2782 202A	10^{10}–10^{15}
Ageing resistance	field experience	excellent
Ozone resistance	field experience	very good

PVC has extremely good chemical resistance to all but low molecular mass chlorinated solvents. Therefore, it is widely used in the construction and lining of chemical plants.

3. Resources and Raw Materials

VCM is produced industrially by two main reactions:

1. Hydrochlorination of acetylene
2. Thermal cracking of 1,2-dichloroethane produced by direct chlorination or oxychlorination of ethylene in a balanced process. More than 90% of the VCM produced is based on this route, full details are given in → Chloroethanes and Chloroethylenes, Section 2.1.

In an ideal situation, a PVC plant is fully integrated, beginning with ethylene and chlorine (salt), but various levels of integration are employed worldwide. Stand-alone PVC plants, supplied with VCM by sea, road, or rail, are well known as are those making the bulk of their VCM requirements from scratch on site, with supplementary supplies of 1,2-dichloroethane for cracking to satisfy peak demand. Finally, there are fully integrated plants that have the benefit of uninterrupted supplies of base raw materials (ethylene and salt) and where the monomer is supplied by pipeline at a significant cost benefit. It can cost up to 65 €/t to transport VCM within continental Western Europe depending on distance and type of freight used, i.e., road, rail, or ship.

4. Production

There are three main processes used for the commercial production of PVC: Suspension (providing 80% of world production), emulsion (12%) and mass, also called bulk (8%).

4.1. Suspension Polymerization of Vinyl Chloride

4.1.1. Introduction

The layout of a typical suspension PVC plant is shown in Figure 1.

The suspension polymerization process is essentially a bulk polymerization process carried out in millions of small "reactors" (droplets). Liquid vinyl chloride under its autogenous vapor pressure is dispersed in water by vigorous stirring in a reactor (autoclave) of 25–150 m^3 capacity, fitted with a jacket and/or condenser for heat removal and baffles for optimum agitation. This results in the formation of droplets of average diameter 30–40 µm that are stabilized against coalescence by one or more protective colloids (granulating agent). The other essential ingredient is a monomer-soluble free radical initiator.

The formulations charged to the reactors (otherwise known as autoclaves, kettles, or polymerizers) are normally referred to as recipes. A basic recipe for suspension processes can be simply water, VCM, initiator, and suspension agent, for example:

VCM	100 parts
Water	90–130 parts
Protective colloid	0.05–0.15 parts
Initiator	0.03–0.08 parts

These quantities vary depending on the PVC grade, reactor size, plant type, etc.

Whereas this produces particulate PVC, it is unlikely to have the optimum morphology. This is only achieved if other additives are employed, such as oxygen, buffers, secondary or tertiary granulating agents, chain-transfer or chain-extending agents, comonomers, antioxidants, together with the right level of agitation and homogenization and the correct charging procedure and timing for each additive addition. A combination of just the right number and degree of the above parameters to achieve the optimum morphology usually determines the quality of a PVC grade.

After charging, the reactor contents are heated to the reaction temperature of 45–75°C. Reactors are 80–95% full at this stage. Heat causes some of the initiator to decompose into free radicals, and the monomer in the droplets begins to polymerize. The strongly exothermic reaction (−1 540 kJ/kg) is controlled

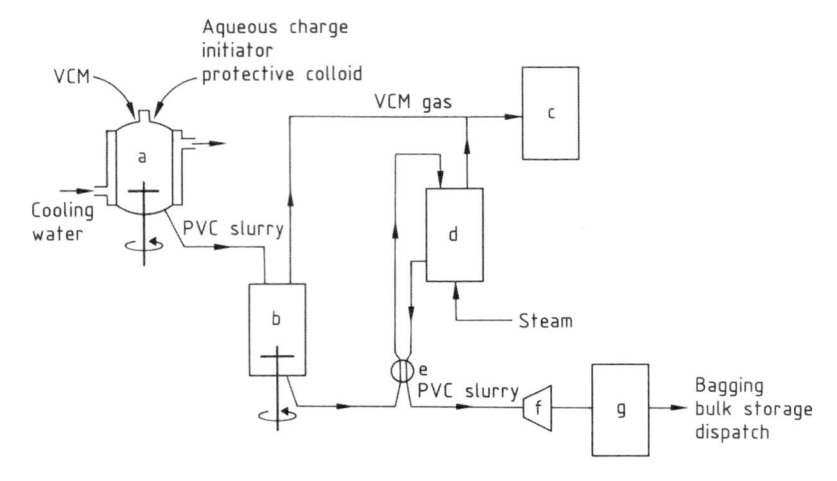

Figure 1. Suspension PVC plant
a) Reactor; b) Blowdown vessel; c) VCM recovery plant; d) Stripping column; e) Heat exchanger; f) Centrifuge; g) Driers

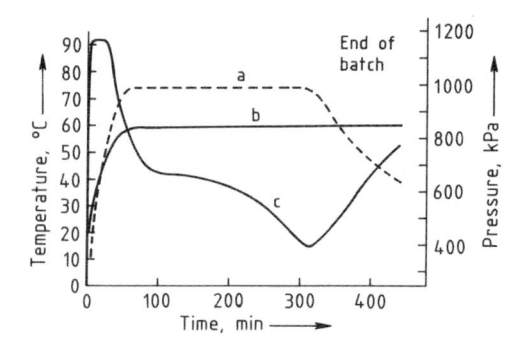

Figure 2. Reaction profile for a typical suspension polymerization
a) Pressure; b) Batch temperature; c) Jacket temperature

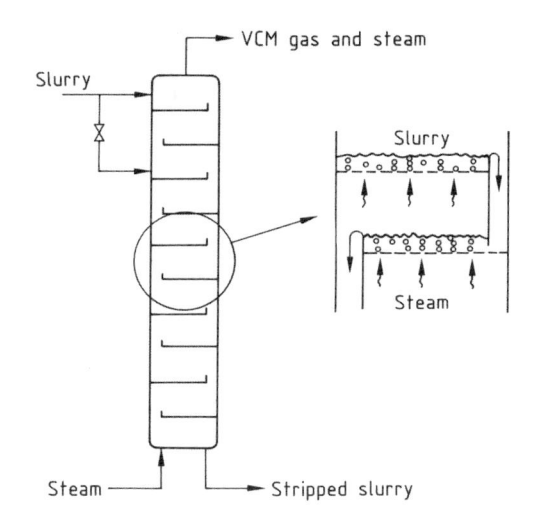

Figure 3. Stripping column

by removing heat via the jacket and/or by boil-off into a condenser from which the condensed monomer is returned to the reactor.

Progress of the reaction in a jacketed reactor can be followed by continuously monitoring the water temperature in the jacket because the flow rate is constant and a constant batch temperature is maintained by progressively reducing the cooling water temperature (see Fig. 2).

Although PVC is insoluble in its monomer (see Section 1.1) the polymer is swollen by ca. 27 wt% VCM to form a coherent gel [7]. Therefore, as the conversion increases towards 70% and beyond, the pressure in the reactor starts to fall; slowly at first, but then much more rapidly as the last of the free liquid monomer is consumed. Polymerization continues in the gel phase, very rapidly at first because chain termination is hindered by lack of mobility of the growing chain, but, as conversion increases beyond 80–85%, monomer starvation rapidly reduces the rate.

The reaction is terminated at a predetermined pressure by either adding a chain terminator and/or venting off the unreacted monomer to the recovery plant. After venting, the resin in the aqueous slurry can still contain 2–3% unreacted monomer, which is removed by stripping, either batchwise (as in older plants) or continuously in a column. The batch is discharged from the reactor to a feed vessel and then fed continuously through a stripper column (Fig. 3). The unreacted monomer is recovered, liquefied, and reused in later polymerizations. After passing through a heat exchanger, the slurry is fed to a continuous centrifuge to give a wet-cake with 20–30% moisture content. The remaining water is then removed by conventional flash and/or fluid-bed drying to give a dry, free-flowing powder with a residual VCM content below 1 ppm.

4.1.2. Additives

Water. The water used in the process should be demineralized (i.e., of low conductivity) because ionic species, especially sodium ions, can affect the performance of other additives, such as protective colloids and influence final polymer properties, such as volume resistivity.

Protective Colloids (Granulating Agents). The protective colloids used in the suspension polymerization process are of two types: Primary and secondary. The main function of the primary granulating agents is to control grain size, but they also affect porosity and other morphological properties. The demand for polymers with reduced residual VCM levels and better plasticizer adsorption led to the need for higher and more uniform porosity within each grain and this is achieved by using a secondary stabilizer in addition to the primary.

Although a wide variety of surface active materials have been or are still used, the

Table 3. Typical primary and secondary protective colloids used in the suspension polymerization of vinyl chloride

Trade name	Manufacturer	Degree of hydrolysis	Viscosity of 4% aqueous solution, mPa · s
Primaries			
Alcotex 72.5/B72	Synthomer	71.5–73.5	5.5–7.5
Alcotex 78	Synthomer	76–79	5.5–7.5
Denka W–20N	Denka	78–81	37–43
Denka MP-10	Denka	70–74	9–13
Denka HS-80	Denka	78–81	30–43
Denka B-24	Denka	86–89	40–48
Inovol PA 4	Ineos	79–81	30–6
Inovol PA 6	Ineos	70.5–72.5	4.5–6.0
Inovol PA 7	Ineos	87–89	40–50
Poval L8	Kuraray	69.5–72.5	5.0–5.8
Poval L9	Kuraray	70–72	6.0–6.5
Poval L10	Kuraray	71–73	5.5–7.5
Gohsenol KP08	Nippon Gohsei	71–75	6–9
Gohsenol KH17	Nippon Gohsei	78.5–81.5	32–38
Gohsenol KH20	Nippon Gohsei	78.5–81.5	44–52
Gohsenol GH20	Nippon Gohsei	86.5–89	40–46
Secondaries			*Solvent Base*
Alcotex 552P	Synthomer	54–57	aqueous
Polivic S202	3V-Sigma	47	methanolic
Polivic S404W	3V-Sigma	55	aqueous
Poval LM10HD	Kuraray	38–42	solid
Gohsenol LL02	Nippon Gohsei	45–51	solid
Gohsenol LW-200	Nippon Gohsei	46–53	aqueous
Inovol SA4	Ineos	54–56	aqueous
Tego SML 20 (sorbitan monolaurate)	Evonik Industries		liquid, no solvent

substances most commonly used today are cellulose ether derivatives and partially hydrolyzed poly(vinyl acetates) or poly(vinyl alcohols) (PVAs). The latter are usually preferred and are used alone or in combination with the cellulose type, which they are gradually replacing. Typical PVA protective colloids are listed in Table 3.

As PVA primary granulating agents are prepared by alkaline hydrolysis of poly(vinyl acetates), they are complex mixtures containing fractions with different degrees of hydrolysis and molecular mass and having a block distribution of the OH groups [11]. Because of this complexity suspension processes need fine-tuning around the chosen agent because decreasing the average degree of hydrolysis of the PVA increases the porosity of the PVC grains and affects the grain size, as shown in Figure 4. This behavior is used in conjunction with agitation to achieve the optimum size and porosity. Below 70% hydrolysis, the PVAs are insoluble in water and lose their ability to stabilize the VCM droplets adequately thus, the normal range of PVA granulating agents used are in the range 70 to 85%.

Many surfactants of low molecular mass have been claimed as useful secondary stabilizers but in practice, few can be used without adverse side effects. One exception is the nonionic sorbitan monolaurate. Again, PVAs are much more widely used. However, this family is prepared by acid hydrolysis of low molecular mass poly(vinyl acetates) and has a random distribution of acetate groups [11]. Initially these materials, which are insoluble in water, were prepared as methanolic solutions (e.g., Polivic S 202) but aqueous-based, low-hydrolysis PVAs (e.g., Alcotex 55, and Polivic S 404 W), which are safer to handle and are environmentally more acceptable, have been developed and are now widely used in the industry. Typical secondary protective colloids are listed in Table 3.

The use of secondary granulating agents affects the mean grain size and grain-size distribution thus the concentration of the primary must be adjusted to compensate. Because each

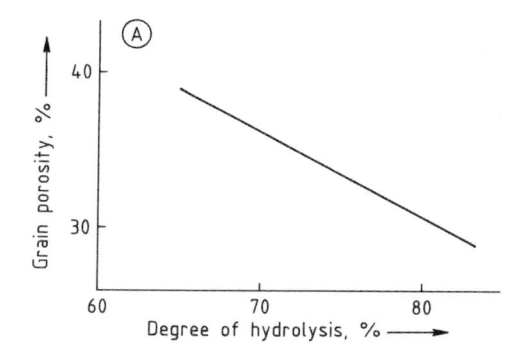

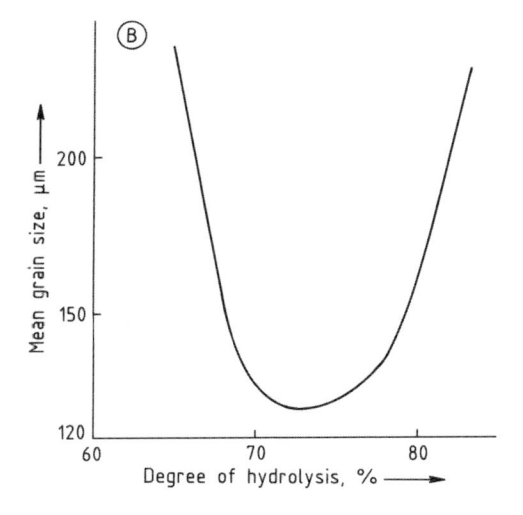

Figure 4. The effect of the degree of hydrolysis of poly(vinyl acetate) on PVC grain properties
A) Grain porosity; B) Mean grain size

producer uses autoclaves of different size and shape, equipped with different agitation and baffle systems, with or without condensers, and makes many different types of resin, no absolute formula is available that determines the magnitude of this change in primary concentration; this must be developed in-house by the suspension polymerization technologist.

Initiators. A range of initiators have been used, but relative few are widely employed. At one time benzoyl peroxide and azo initiators were typically used, but diacetyl peroxides, peroxydicarbonates (PDC), and alkyl peroxyesters are now preferred. Typical initiators are listed in Table 4.

The half-life $t_{1/2}$ at a particular temperature is the main factor that determines the choice of an initiator for a given PVC grade. The initiator should have a half-life of ca. 2 h at reaction temperature to optimize reaction time. Because the best plant output is obtained when the reaction profile is as square as possible (see Section 4.1.3) two initiators of different activity are often chosen to try to achieve this ideal. The more reactive initiator is used to start the polymerization but then as this exhausts itself the slower one takes over to completion. When preparing K 70–80 resins at temperatures at or below 52°C, a peroxydicarbonate (e.g., **4**, **5**, **6**, **7** or **8**) is used as the primary initiator, with an active initiator, such as **11** helping to bolster the slow initial rate which is obtained at these low temperatures. In the intermediate range of K 63–68 a single peroxydicarbonate (PDC) is usually sufficient, but at or below K 60 the PDC needs the extra radical flux of a slow initiator to prevent exhaustion of the reaction, so a less active diacyl peroxide, such as **3** is used as a secondary.

Three other factors are also important: Water solubility, storage and handling, and reactor fouling.

The initiators' water solubility can affect their efficiency by making them more susceptible to hydrolysis and can affect their distribution between VCM droplets; the higher the solubility the more homogeneous the distribution.

The ease of storage and handling has a major influence on the acceptability of an initiator to a PVC producer. In general the shorter the half-life of the initiator, the more unstable it is, and the lower its self-decomposition temperature. The stability also depends on the radical energy and physical form. Solid crystalline initiators, such as **7** (BCHPC) are more stable than their liquid analogues, e.g., **4** (EHPC). Many of the more active initiators must to be stored under refrigeration that tends to restrict their acceptability. Liquid initiators have the advantage that they can be automatically metered into reactors, which is essential in larger computer-controlled plants. However, the current trend is to use initiators as finely divided aqueous dispersions. This has the advantage of improving their homogeneity of distribution throughout the VCM droplets,

Table 4. Initiators for suspension or mass polymerization of VCM

No.	Name	Storage temperature[a], °C	M_r	Supplied form	$T_{1/2}$, °C[b] 10 h	$T_{1/2}$, °C[b] 1 h	$T_{1/2}$, °C[b] 6 min	$t_{1/2}$, min[c] 50°C	$t_{1/2}$, min[c] 60°C	$t_{1/2}$, min[c] 70°C
	Diacyl peroxides									
1	Dioctanoyl peroxide	10	314	solid or solution	57	75	94		700	190
2	Didecanoyl peroxide	10	343	Solid	63	80	100		700	190
3	Dilauroyl peroxide	30	398	solid	62	80	99		850	220
	Peroxydicarbonates									
4	Bis(2-ethylhexyl) peroxydicarbonate	−15	346	solution	47	64	83	250	70	21
5	Dicyclohexyl peroxydicarbonate	10	286	solid	44	59	76	250	58	16
6	Dicetyl peroxydicarbonate	20	571	solid or dispersion	48	65	84	320	70	20
7	Bis(4-*tert*-butyl cyclohexyl) peroxydicarbonate	20	398	solid or dispersion	47	64	83	250	60	18
8	Dimyristyl peroxydicarbonate	15	515	solid or suspension	48	65	84	320	70	20
	Alkyl peroxyesters									
9	*tert*-Butyl peroxyneodecanoate	−10	244	solution	46	64	84	250	70	25
10	*tert*-Amyl peroxyneodecanoate	−15	258	solution	43	61	81	560	180	55
11	Cumyl peroxyneodecanoate	−20	306	solution	38	56	75	100	30	9
12	*tert*-Butyl peroxypivalate	−5	174	solution	57	75	94	700	200	70
	Azo initiator									
13	Azobisisobutyronitrile (AIBN)	25	164	solution	64	82	101		500	170

[a] Maximum.
[b] Temperature at which $t_{1/2}$ = 10 h, 1 h, or 6 min, as indicated.
[c] Approximate.

allows them to be added at any time or temperature during the charging procedure, and renders them very safe to handle because their heat of decomposition is less than the heat of evaporation of the carrier. Dispersion of 25–40% are typical, and systems must be carefully designed to ensure complete homogenization of the dispersion is maintained and accurate metering achieved. As the water solubility of the initiator increases the tendency to cause reactor fouling also increases. This is of increasing importance as more and more processes switch from the open to the closed mode of operation (see Section 4.1.4.4).

4.1.3. Kinetics

The kinetic profile seen in Figure 2 is typical of many VCM suspension polymerizations conducted at the lower end of the temperature range and exhibits the classical acceleration with conversion as shown by the gradually decreasing jacket temperature. At higher temperatures, the reaction rate is more constant and a square-wave profile is obtained. Three different phases in the reaction can be identified:

Low Conversion (< 5%). Polymerization occurs largely in the monomer phase because the quantity of polymer produced is so small. Only this early stage follows classical kinetics where the rate of polymerization is dependent on the square root of the initiator concentration:

$$\frac{-d[M]}{dt} = \frac{k_p [M] \cdot k_d [I]^{1/2}}{k_t}$$

where

[M] monomer concentration
[I] initiator concentration
k_p propagation rate constant
k_d initiator decomposition rate constant
k_t termination rate constant

As the amount of polymer phase increases with conversion, the rate deviates more and more from this equation due to a decrease in k_t. A full description of the models used to explain the deviations from ideality is given in [12, 13].

Medium Conversion (5–65%). Polymerization continues in both the free monomer and polymer–monomer gel phases. This region shows the classical acceleration with increasing conversion [14]. Because termination occurs primarily by the diffusion together and collision of two bulky macroradicals, it is considerably slower in the viscous polymer phase. As a result, the termination rate constant is dramatically reduced and the rate of polymerization in this phase is much faster. As the amount of polymer phase increases, the rate of polymerization accelerates.

High Conversion (> 65%). No free monomer is left to maintain the autogenous pressure, which falls, and polymerization occurs in the polymer gel phase of rapidly increasing viscosity as the remaining VCM is consumed. Hence, the rate of polymerization increases still further to reach a maximum just after the pressure drop but then decreases progressively due to increasing monomer starvation. When the instantaneous polymerization rate falls below the overall rate of reaction, the batch is terminated, either by venting off unreacted VCM to the recovery plant or by adding a chain terminator and then venting. Depending on the properties required, a conversion of 70–95% is chosen. It is uneconomic to continue to very high conversion, and as properties, such as plasticizer absorption, dry up rate (i.e., the time taken to achieve a free-flowing PVC–plasticizer dry blend), and initial color–heat stability deteriorate rapidly above 85–90% conversion, most flexible resins are stopped at around 85% and rigid resins at around 90% conversion.

4.1.4. Morphology

The morphology of each PVC grain is the main parameter controlling the quality of a PVC resin. This subject covers mean grain size, grain size distribution, and, most important of all, porosity (cold plasticizer absorption) and bulk density, which are closely interrelated.

4.1.4.1. Control of Grain Size

At the beginning of the batch, the bulk vinyl chloride phase is broken down into droplets of 30–40 μm mean size under the influence of agitation and by the presence of the water-soluble protective colloid, which is adsorbed at the monomer–water interface. As polymerization proceeds, graft copolymerization of PVC onto the interfacial layer of protective colloid reduces its mobility and protective ability, and droplet coalescence begins to occur at around 4–5% conversion. As the conversion increases, changes in the surface layer of the droplet due to further polymerization suppress the droplet coalescence step, which ceases at 20% conversion, and the grain size remains largely constant for the remainder of the polymerization. Altering the stirrer speed after this point has no further effect on the mean grain size (MGS) [15] of the final product, which is controlled to between 100 and 180 μm, depending on resin type, by a combination of initial agitator speed and protective colloid(s) concentration and type.

Polymerizations with high concentrations of colloid and less agitation give spherical grains of low porosity ($\leq 10\%$) and very high bulk density (≥ 700 g/L) and are of little interest, generally due to problems of VCM removal and poor gelation. Flexible resins are made with a low MGS (100–130 μm), rigid polymers higher (150–180 μm), and low K resins with intermediate size (130–160 μm).

In extreme cases, it is possible to stabilize the droplets to such a degree that no coalescence occurs. Specialized VCM–vinyl acetate copolymers for gramophone record production fall into this category as do blending resins (30 μm MGS) for use as viscosity depressants in plastisols (see Section 4.3).

As well as achieving the desired MGS it is equally important to control the size distribution. A Gaussian distribution is normally produced and it is important to make this as narrow as possible because a coarse fraction (> 250 μm) can give rise to fisheyes (dispersion faults) and make VCM removal difficult as well as resulting in loss of yield due to screen rejects, whereas fines (< 60 μm) can be lost in aqueous and gaseous effluent or cause powder flow problems.

4.1.4.2. Control of Porosity and Bulk Density

Porosity in PVC grains is created by a complex series of interrelated steps in the formation and growth of the submicroscopic structure within each droplet and also depends on the unique nature of the VCM–PVC system.

PVC is insoluble in its monomer, and the growing oligomers start to precipitate when the chain length exceeds 10 monomer limits [6]. This initial precipitation results in the formation of microdomains about 15–20 nm in diameter

[17]. This occurs at as low as 0.001% conversion [17]. The microdomains rapidly become unstable and aggregate to form domains (primary particle nuclei) of about 0.1 µm diameter. After this stage, no further domains are formed and the number of primary particles is fixed. All new PVC radicals that are produced in the monomer phase collide with and are precipitated onto the vast existing surface area of primary particles before a new domain can be formed. Because the existing PVC phase is swollen with monomer and contains initiator, initiator radicals, and growing chains undergoing chain transfer, primary particles grow rapidly by accretion of new microradicals and microdomains as well as by gel-phase polymerization. They in turn become unstable and a final agglomeration step to form 1–2 µm primary particle aggregates takes place. By the end of the polymerization the primary particles have grown to 1 µm within the aggregates, which themselves have grown to 2–5 µm. The whole mechanism is shown in diagrammatic form in Figure 5 and discussed in detail in [18].

The arrangement of the aggregates of primary particles within each monomer droplet has a major influence on the morphology of the final PVC grain. The conversion of VCM of density 0.85 g/cm^3 at 52°C to PVC of 1.4 g/cm^3 produces a volume contraction of 39%. If the polymerizing droplets contract totally, the final

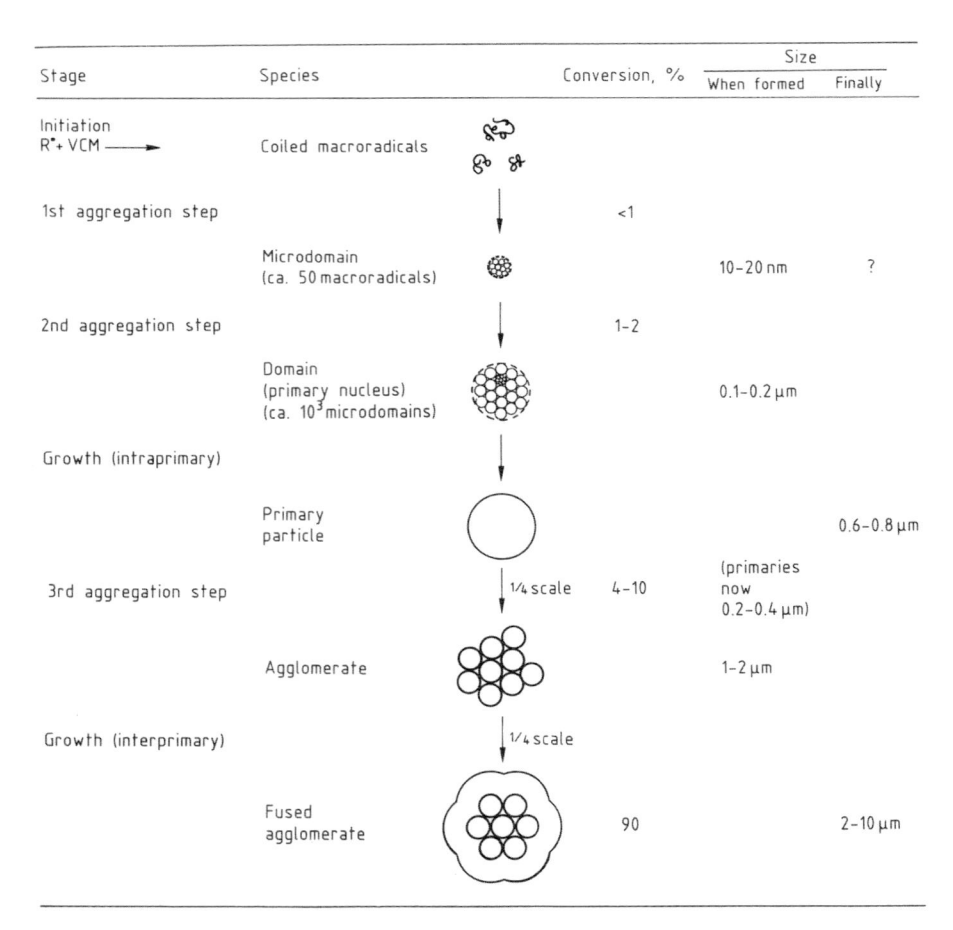

Stage	Species	Conversion, %	Size When formed	Finally
Initiation R·+ VCM ⟶	Coiled macroradicals			
1st aggregation step		<1		
	Microdomain (ca. 50 macroradicals)		10–20 nm	?
2nd aggregation step		1–2		
	Domain (primary nucleus) (ca. 10^3 microdomains)		0.1–0.2 µm	
Growth (intraprimary)				
	Primary particle			0.6–0.8 µm
3rd aggregation step		4–10		(primaries now 0.2–0.4 µm)
	Agglomerate			1–2 µm
Growth (interprimary)				
	Fused agglomerate	90		2–10 µm

Figure 5. Schematic representation of the mechanism of VCM polymerization (from [18], courtesy of Elsevier Applied Science Publishers)

grain has no porosity. If, however, the contraction is prevented totally, the final grain porosity can be as high as 39%. Therefore, the better the three-dimensional spatial arrangement of the aggregates relative to one another is, the higher the achieved porosity.

The main aim of the polymer technologist is to create the best possible spatial network of aggregates relative to one another so that a network is formed that is strong enough to resist droplet contraction and hence maximize porosity. The sooner this is achieved in the polymerization, the better the porosity attained. In practice, 34–35% is easily achievable. At ca. 65–70% conversion, all of the free monomer is consumed, and the only VCM remaining is that swelling the polymer gel (27%), in the vapor space, and dissolved in the water. At this stage, the voids in the grain (the porosity) are filled with batch water, so that the quantity of water in interstitial spaces between the grains is reduced and a significant increase in slurry viscosity is observed because less mobilizing water is available to separate the grains.

The control of bulk density or apparent density (AD) is more important than even porosity to the processor. The latter is concerned with charge size and cycle times in high-speed mixers, output rates, motor torque, and degree of gelation in extruders, and mechanical properties of the article being produced. These parameters are dependent on AD. The reason for introducing the control of porosity before control of AD is that the latter is very heavily influenced by the former. Apparent density depends on porosity, grain shape, and grain size distribution in order of priority and is best described by examining the packing fraction (PF) concept, i.e., the ability of the somewhat irregular PVC grains to fill a space to the maximum efficiency.

$$PF = \frac{(1 + 0.014 \cdot CPA)(AD \cdot 0.1)}{1.4}$$

CPA   cold plasticizer absorption (porosity)
AD   apparent density

Most extruders used in processing PVC are cf the twin-screw type, and these operate like a simple pump: Increasing the feed rate by 10% (10% higher AD) increases the discharge or output rate by 8–9% (see Chap. 8 in [14]). In practice, controlling porosity is relatively easy because a number of effective secondary stabilizers are available. Maximizing the AD is much more difficult because control of grain shape comes from a complex interaction of both agitation and choice of primary and secondary granulating agents together with in-house expertise of charging techniques and use of other additives. A range of grain shapes that can be produced is shown in Figure 6.

In practice typical property targets for porosity and AD are:

- For flexible resins (K 70): Porosity of $\geq$ 30% and AD of $\geq$ 500 g/L
- For rigid extrusion resins (K 66–68): Porosity 20% and AD 580 g/L
- For bottle resins (K 57–60): Porosity 18–20% and AD 560 g/L

4.1.4.3. K-Value (Molecular Mass)

PVC is again rather unique in its relationship between reaction temperature and K-value. With PVC the rate of polymer formation by chain–chain termination is small due to the influence of gel phase polymerization described above. Instead, because chain transfer to monomer is a predominant step in the polymerization of VCM, the length of the polymer chain (degree of polymerization, DP) is determined by the ratio of the rate of chain propagation to chain transfer:

$$DP = \frac{\text{Rate of propagation}}{\text{Rate of chain transfer}} = \frac{k_p[R_n][M]}{k_{tr}[R_n][M]} = \frac{k_p}{k_{tr}} \quad (1)$$

[M]   monomer concentration
[R_n] radical concentration
k_p propagation rate constant
k_{tr} chain-transfer rate constant

As k_p and k_{tr} depend only on temperature, the molecular mass of PVC is controlled by the reaction temperature (see Fig. 7).

If a particularly low molecular mass polymer is required or if a reactor's pressure rating is too low to operate at very high temperatures, an

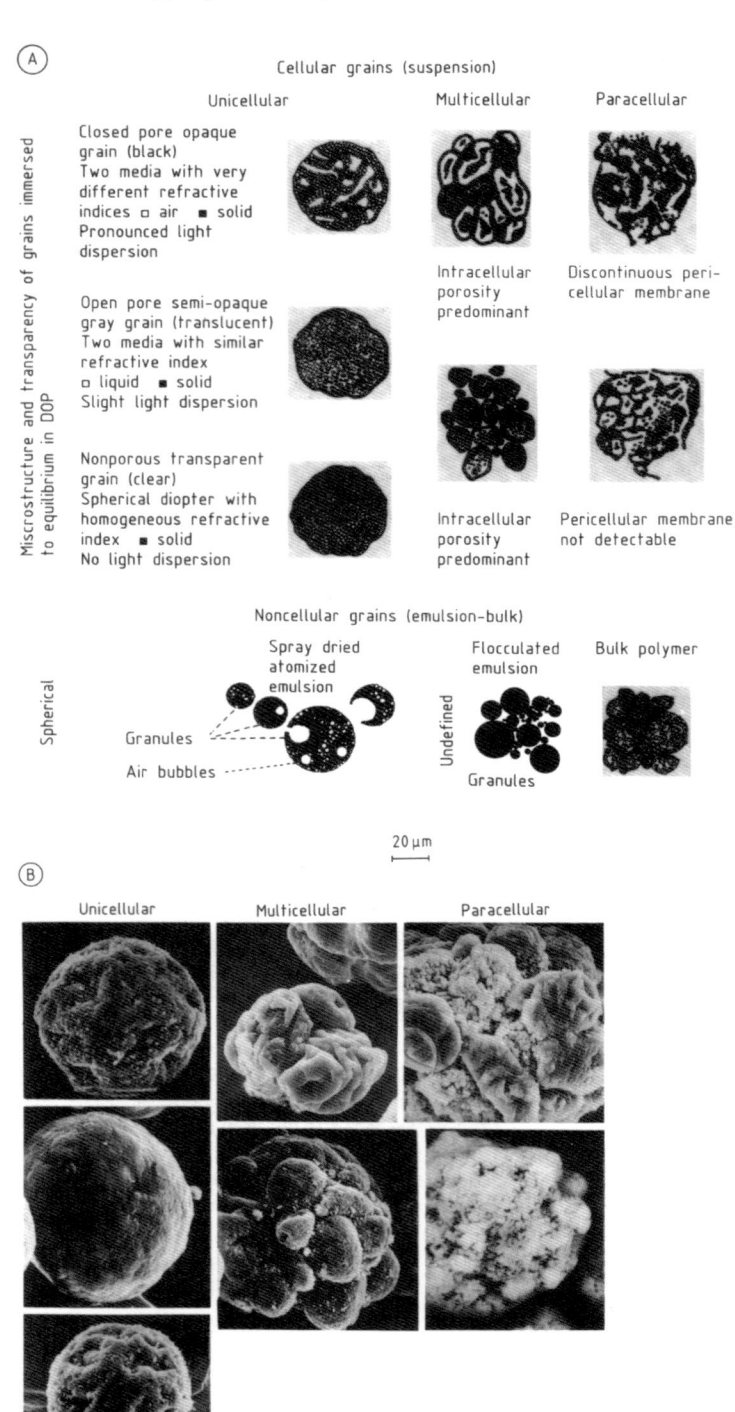

Figure 6. A) Classification of PVC grains; B) Typical cellular PVC grains — layout as in Figure 6 A From [18], courtesy of Elsevier Applied Science Publishers

For example, most bottles are made from K 57 polymers, some from K 60, but this is the limit for the application because bottles are blow molded thus restricting the melt viscosity that can be handled.

4.1.4.4. Problem Areas

There are relatively few problems in the operation of the suspension polymerization process. However, to obtain the best possible quality and consistency some points need special attention.

In the past, reactor fouling was bad and mechanical cleaning between each batch essential. With the advent of clean-wall technology, high-pressure rinse systems, and improved recipes it is now possible to operate in a closed manner, thereby reducing problems of VCM exposure (see Chap. 5 in [14]). Reactor opening for cleaning is reduced to once in several hundred batches. It is essential to ensure that the reactor is rendered completely free of polymer between batches because this is the most common cause of fisheyes. Equally important is to ensure there is no chance of cross-grade contamination in the downstream operations of slurry handling, stripping, drying, and storage. Contamination of the product by foreign substances is eliminated by good housekeeping and plant management together with the right choice of metals of construction and design. Stainless steel is widely used in many sections of PVC plants.

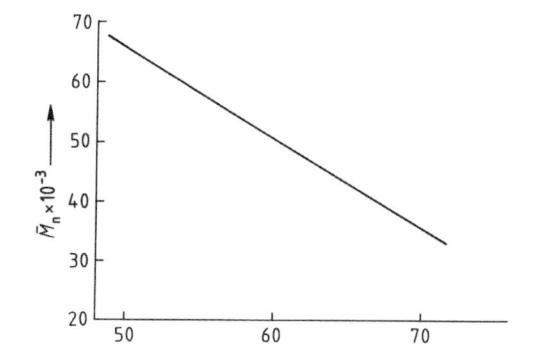

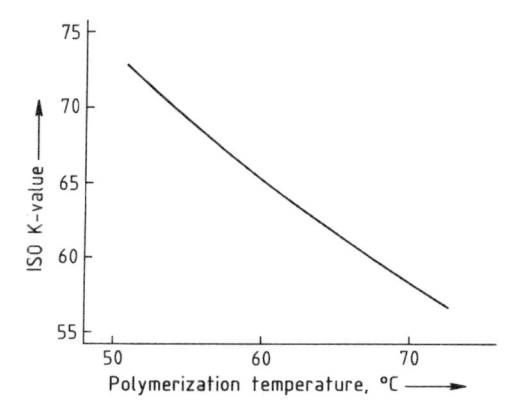

Figure 7. The effect of temperature on molecular mass and K-value

additional chain-transfer agent can be employed. This reduces the molecular mass by increasing the chain-transfer rate constant above that for monomer alone in Equation 1. As a result, DP decreases, depending on the quantity of chain transfer agent used.

The general K-value range of commercial resins is K 55 to K 75, with a few producers providing speciality resins outside this range to K 50 and K 80 and with just one or two K 90–100 polymers being available. Despite the relatively narrow K-value/molecular mass range of most commercial resins (see Table 5), the polar nature of the PVC chain gives rise to a considerable melt viscosity span. Mechanical properties increase slightly with increasing K-value so wherever possible the highest K-value polymer is used. However, the power and mechanical robustness of equipment and the complicated melt flow pattern requirements in molds often restrict the K-value that can be used.

4.2. Bulk or Mass Polymerization of Vinyl Chloride

The mass polymerization process is virtually identical to that taking place within each monomer droplet in the suspension process. As such, the mechanism and kinetics are very similar to those described in Section 4.1.3. The major difference is in the mechanical operation of the process.

About 10% of world PVC production is provided by the mass process. The process was developed by Pechiney St. Gobain (PSG). All mass producers are licensees of the PSG technology. The process is carried out in two stages:

1. In the first stage, monomer and initiator are charged and vigorously agitated in a vertical,

Table 5. Correlation of various PVC molecular mass (expressions)[a]

	European methods						Japanese method		American methods			
Viscosity number ISO/R 174 (1974)	Specific viscosity 0.5 g/100 mL cyclohexanone at 25°C	K-value 0.5 g/100 mL⁻¹ cyclohexanoe at 25°C	Relative viscosity 0.5 g/100 mL ethylene dichloride at 25°C	Specific viscosity 1.0 g/100 mL cyclohexanone at 25°C	K-value 1.0 g/100 mL cyclohexanone at 25°C	Polymerization degree JIS K 6721	Inherent viscosity ASTM D 1243-58 T (method A)	Specific viscosity ASTM D 1243-58 T (method B)	Specific viscosity 0.2 g/100 mL cyclohexanone at 30°C	Specific viscosity 0.4 g/100 mL nitrobenzene at 25°C	Approximate number-average molecular mass
70	0.35	53.9	1.304	0.70	54	560	0.62	0.239	0.13	0.24	36 000
73	0.37	55	1.316	0.74	55	600	0.65	0.25	0.14	0.25	
77	0.38	56.1	1.329	0.77	57	640	0.67	0.264	0.145	0.26	40 000
80	0.40	57.2	1.342	0.80	58	680	0.70	0.275	0.15	0.27	
83	0.42	58.3	1.355	0.83	59	720	0.73	0.285	0.155	0.29	
87	0.44	59.5	1.369	0.86	60	760	0.75	0.3	0.16	0.30	45 000
90	0.45	60.6	1.383	0.87	61	800	0.78	0.31	0.17	0.31	
94	0.47	61.9	1.397	0.92	62	840	0.80	0.32	0.175	0.32	50 000
98	0.49	62.9	1.412	0.95	63	885	0.83	0.33	0.18	0.33	
102	0.51	64	1.427	0.98	64	930	0.85	0.34	0.19	0.34	
105	0.53	65.2	1.443	1.01	65	975	0.83	0.36	0.195	0.35	55 000
109	0.55	66.3	1.458	1.04	66	1025	0.91	0.37	0.20	0.36	
113	0.57	67.4	1.474	1.07	67	1070	0.92	0.38	0.205	0.37	
117	0.59	68.5	1.491	1.10	68	1120	0.95	0.39	0.21	0.39	60 000
121	0.61	69.7	1.508	1.13	69	1175	0.99	0.40	0.22	0.39	
125	0.63	70.5	1.525	1.16	70	1230	1.01	0.41	0.225	0.41	64 000
130	0.65	71.2	1.543	1.19	70.5	1300	1.03	0.43	0.23	0.43	
134	0.67	72.2	1.562	1.22	71	1350	1.06	0.44	0.235	0.44	
138	0.69	73.0	1.581	1.25	72	1420	1.08	0.45	0.24	0.46	70 000
142	0.71	74.0	1.60	1.28	73	1490	1.11	0.46	0.25	0.48	
145	0.73	74.5	1.62	1.31	74	1570	1.13	0.47	0.255	0.49	73 000
149	0.75	75.5	1.64			1650	1.16	0.49	0.26	0.50	
153	0.77	76.3	1.661			1720	1.18	0.50	0.27	0.51	
157	0.79	77.0	1.682			1810	1.21	0.51	0.275	0.52	
161	0.81	77.5	1.704			1900	1.23	0.53	0.28	0.53	80 000
165	0.83	78.5	1.726			1980	1.26	0.54	0.29	0.54	82 000

[a] From [65], courtesy of Elsevier Applied Science Publishers.

stainless-steel autoclave (prepolymerizer) of 8–25 m³ capacity, fitted with a water-cooled jacket and condenser. Rapid polymerization takes place at 62–75°C to give ca. 100 μm aggregated spherical flocs composed of 0.1 μm primary particles, which form the basis or seeds of the final PVC grain. Conversion is taken to 7–12% in ca. 30 min, by which time the initiator used is exhausted.

2. The slurry from the prepolymerizer is discharged into the second-stage reactor, which has a volume of 12–50 m³, with fresh initiator and more VCM. Up to five second-stages can be fed from one prepolymerizer. In the second stage, the 0.1 μm primary particles making up each seed grow in size and fuse together to give the final grain of PVC of 130–160 μm diameter. As the conversion increases the physical nature changes from a wet powder at ca. 20% to a dry one at 40% conversion. The heat of polymerization is removed from the growing grains by evaporation of the VCM and condensation on the cooled wall of the reactor or in a water-cooled reflex condenser(s). The second stage takes 3–9 h, depending on K-value.

Originally, the second stage reactor was a horizontal jacketed vessel with one or more condensers, agitated by slowly rotating (6–10 rpm) ribbon blenders with some blades set very close to the reactor wall to keep it clean. However, it proved difficult to discharge and clean the reactor to a sufficiently high standard to prevent the formation of fisheyes (dispersion faults) in PVC grades intended for plasticized applications [19]. The product can contain up to 10% of coarse material that must be sieved out and recovered by grinding, milling, and reclassifying.

In 1978, this led to the development of a vertical second-stage reactor of up to 50 m³ capacity fitted with two independent agitators designed to overcome the above discharge problems. However, there has been little interest in this development and all recent investment in new PVC plants has been in the suspension process. In theory, the mass process should be cheaper because it does not need a drying stage. However, as it is a two-stage process, it is difficult to remove traces of unreacted monomer, and it gives a significant amount of oversized material, all of which offset its advantages.

Several of the smaller and older mass-polymerization plants have closed because they find it more difficult to meet increasingly stringent environmental standards and are becoming less competitive compared to suspension-polymerization plants on the basis of production rate.

The mass process is discussed in more detail in [19, 20].

4.3. Emulsion Polymerization of Vinyl Chloride

4.3.1. Introduction

The emulsion polymerization process involves the polymerization of monomer in an aqueous medium containing surfactant and a water-soluble initiator, producing PVC latices. PVC latices are colloidal dispersions of spherical particles, ranging in size between 0.1 and 3.0 μm. Most PVC latices are spray dried and then milled to obtain fine powders, made up of agglomerates of latex particles. When mixed with plasticizers they disperse readily to form stable suspensions. During mixing most of the agglomerates are broken down into the original latex particles. Such dispersion of fine particles in plasticizers are known as plastisols or pastes, and the powder is called dispersion or paste polymer. The surfactant layer around the particle surface prevents their adsorbing the plasticizer at room temperature thus they can be used as liquids and may then be spread on to fabric or other substrates, poured on molds, or deposited on formers to produce flooring, wall covering, artificial leather, balls, toys, or protective gloves. There are other grades of PVC polymers, produced by emulsion polymerization, that do not form plastisols and that are used as blends with suspension PVC grades for extrusion application or in the manufacture of battery separator plates. These so-called emulsion polymers are of only minor economic interest. Sales in latex form are very limited; latices are used in water-based paints, printing inks, and impregnated fabrics. The total

Table 6. World emulsion PVC production in 2012

Region	Production, 10^3 t	Percentage
Europe	820	32
NAFTA	220	11
Asien without China	595	22
China	580	28
South America	65	4
Rest of the world	60	3
Total capacity	2 340	100

production capacity of emulsion PVC is given in Table 6.

4.3.2. Emulsion Polymerization Processes

The ways to produce PVC latices are:

- The classic emulsion polymerization, in which the particle size and particle size distribution are controlled by a water-soluble initiator and a surfactant [21–25]. They are monodisperse with a particle size typically less than 0.1 µm.
- Seeded emulsion polymerization, in which the particle size distribution of the final latex depends on the amount and the size of the latex used as seed and on the amount of surfactant and the way in which it is added [26, 27]. A typical material has 0.2 and 1.2 µm particles (see Fig. 8).

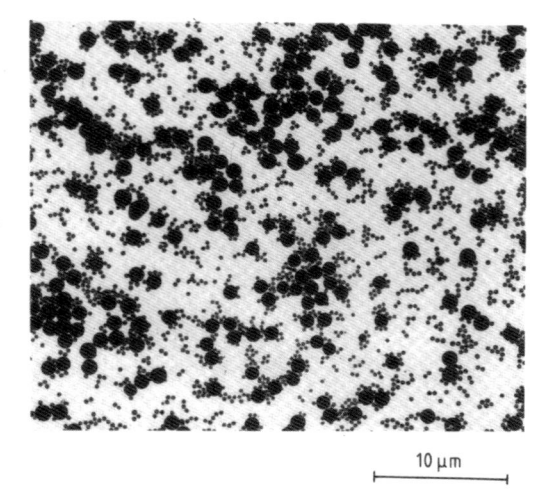

Figure 8. Electron micrograph of latex particles obtained by seeded emulsion polymerization

- Microsuspension polymerization, in which a monomer-soluble initiator is used. Polymerization takes place within the fine droplets in which the monomer is dispersed by passing a coarse emulsion of monomer, initiator, and surfactant through a mechanical homogenizer [28–30]. The resulting particle size distribution is Gaussian between 0.2 and 1.2 µm.
- Polymerization of fine emulsions achieved by using a combination of a typical surfactant, such as sodium dodecyl sulfate and a long-chain fatty alcohol. This gives rise to spontaneous emulsification of the monomer into very fine droplets. In this case, the polymerization can be carried out by using a water-soluble or a monomer-soluble initiator [31]. The latex particle size distribution is the same as for microsuspension polymerization.

The industrial polymerization techniques used for carrying out these processes are:

- Batch polymerization where all the monomer is charged at the beginning
- Semibatch or semicontinuous polymerization, with the monomer added in stages or continuously as reaction proceeds
- Continuous polymerization where monomer and other components are fed continuously into the reactor and the latex is withdrawn from the bottom of the reactor

All these processes are in current use to achieve a wide range of latex characteristics, which are related to the rheological properties of the PVC plastisols [32].

4.3.3. Description of the Batch Process

The emulsion polymerization of PVC consists of the following stages:

1. Polymerization
2. VCM removal
3. Latex storage
4. Drying
5. Milling
6. Packing and storage.

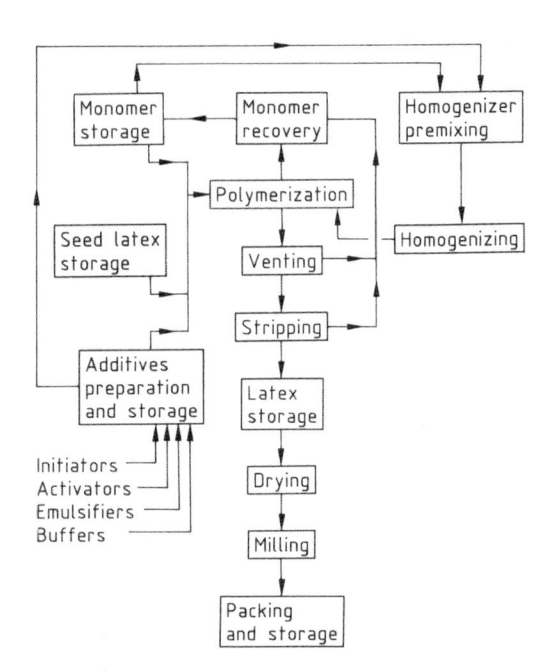

Figure 9. Schematic representation of the stages required for a plant which produces all emulsion, seeded, and microsuspension grades

Figure 9 shows a schematic representation of the stages required for a plant that produces all the emulsion, seeded, and microsuspension grades.

A recipe for a simple batch emulsion polymerization (in parts by weight) is:

Demineralized water	110–140
Vinyl chloride	100
Emulsifier	0.1–1
Initiator	0.1–0.2

The polymerization takes place in an autoclave resistant to the VCM vapor pressure. Pressures of 6.4–10.0 bar correspond to polymerization temperatures of 40–60°C, depending on the desired molecular mass. The polymerization reaction is strongly exothermic with a heat of polymerization of $-1\,534$ kJ/kg. Heat removal can be a problem as the size of the autoclave increases because the ratio between surface area and volume becomes less favorable. For this reason, industrial plants commonly use water cooling and external condensers. Modern plants use autoclaves with capacities ranging between 30 and 80 m^3. Large autoclaves can present agitation problems because low shear rates are required to maintain mechanical stability of the latex and to avoid coagulation.

Initiators. Although organic peroxides slightly soluble in water, such as methyl ethyl ketone peroxide and 1-hydroperoxy-1'-hydroxydicyclohexyl peroxide can be used, the most widely used initiators are ammonium or potassium peroxosulfate. Their decomposition in aqueous solution is first order with respect to the peroxosulfate ion concentration [33]. The rate may be increased significantly by using a reducing agent. Redox catalysis involves the use of transition metal ions, such as copper or iron [34]. The systems widely used in emulsion polymerization of vinyl chloride are potassium peroxosulfate–sodium bisulfate–copper or iron. Redox systems can also be used in microsuspension, an example is dilauroyl peroxide–copper–ascorbic acid [35].

Emulsifiers are very important in the emulsion polymerization of vinyl chloride because they determine not only the latex characteristics but also the properties of the final product [36–38]. Moreover, the nature and the concentration of the emulsifier, together with the agitation, are major parameters controlling the mechanical stability of the latex during polymerization and handling, and the major cause of reactor wall encrustation and lump formation. Because the latex is dried by spray drying, the emulsifier remains in the resin and can influence properties of the product, such as the flowability of the powder, plastisol formation, the viscosity of the plastisol, and the properties of the final article such as heat stability, water adsorption, and coat adhesion. For this reason a wide variety of emulsifiers are used. During polymerization, anionic types are usually used, whereas nonionics are generally added to the latex after polymerization. Typical emulsifiers are the sodium salts of alkyl sulfates, alkyl sulfonates, alkylbenzenesulfonates, dialkyl sulfosuccinates, alkyl ethoxysulfates, fatty acid soaps, alkyl phenol ethoxylates, and fatty acid ethoxylates.

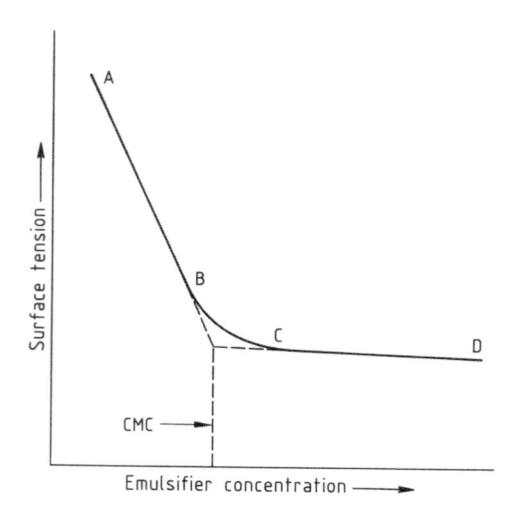

Figure 10. Dependence of surface tension on emulsifier concentration

Although VCM loss from particles by diffusion is fast, VCM removal from an emulsion is much more difficult than from suspension slurries as the surfactants lead to large amounts of foam formation. Continuous processes using a thin-film evaporator or which spray the latex into an evacuated chamber against a countercurrent stream of gas reduce considerably the problems of foaming and coagulation [41–44]. The residual vinyl chloride monomer content of the latex after stripping ranges between 200 and 2 000 ppm, depending on the type of latex and the operating conditions.

Drying and Milling. Although there are several different techniques to separate polymer from water, such as freeze drying, drum drying, or coagulation followed by centrifugation, the most widely used is spray drying because this confers a special structure to the dried particle that is important for the production of plastisols. A fine spray of latex is introduced into a hot air stream where evaporation takes place.

The spray is obtained by various types of atomizers, such as:

- Spinning disks or wheels rotating at 15 000–20 000 rpm
- Single pressure nozzles operating at 100–500 bar
- Two-fluid pressure nozzles using compressed air up to 4 bar as atomizing fluid [45]

The drying air temperature at the inlet is 150–240°C and at the outlet, 55–80°C, depending on the quality of the required product and the type of latex. By varying the drying conditions it is possible to change the resin characteristics, and hence its use, even from a typical paste grade to a general-purpose emulsion resin, e.g., for rigid extrusion. The dry powder consists of spherical aggregates, with the size usually ranging from 1 to 100 μm diameter with an average of 30–40 μm. The aggregates are called secondary particles, to distinguish them from those of the latex, which are known as primary particles. The way the aggregates form is complex and depends on many parameters, such as primary particle size, type and amount of emulsifier, mean residence time in the drying chamber, droplet size of the spray, and temperature

Particle Formation. Emulsifiers are characterized by the critical concentration at which their molecules aggregate to form micelles [39, 40]. In Figure 10 a plot of surface tension versus emulsifier concentration is shown; the point at which the curve exhibits a marked change is known as the critical micelle concentration (CMC). Conventional emulsion polymerization is carried out at a concentration above the CMC. In this way, particle formation is controlled by the micelles, which solubilize the monomer by swelling and become the loci of polymerization. The nucleation phase stops when the micelles disappear, and the emulsifier is absorbed on to the surface of the particles. Depending on the balance between the hydrophilic and hydrophobic part of the molecule, each emulsifier plays a specific role in the emulsion polymerization, as they have different CMC values, surface covering powers, and stabilizing effects.

VCM Removal. The polymerization is terminated at 90% conversion and is followed by the venting of unreacted monomer to a gas holder. After degassing, the latex contains 3% vinyl chloride and further reductions must be made to fulfill legal requirements for environmental protection. The residual VCM is removed by stripping under vacuum in a stripping vessel.

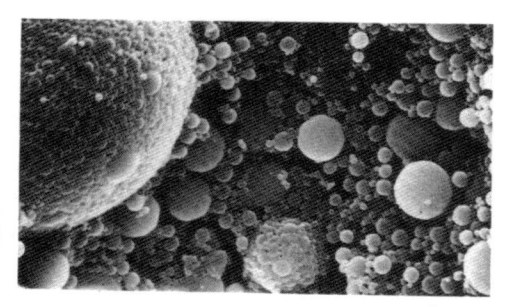

Figure 11. Scanning electron micrograph of spray-dried PVC particles produced from bimodal seeded latex

[11]. Figure 11 shows a scanning electron micrograph of spray-dried PVC particles formed from bimodal seeded latex.

Plastisol-grade resins, usually made from multimodal latex (i.e., with a Gaussian particle size distribution), are milled to reduce the size of the coarse fraction, which can cause problems in coating applications of the plastisol. Generally, the viscosity of the plastisol increases when the polymer has been milled, so it is desirable to reduce milling as far as possible and to use air classification to remove the coarse fraction (> 65 µm) under control. In industrial production, air classifiers or mills or a combination of both are used. After milling, the average particle size of the resin is in the range of 5–15 µm.

Polymers for rigid extrusion and for manufacturing plate separators in batteries are sold in unmilled form. Resins are characterized by analysis of molecular mass, emulsifier and salt content, pH of aqueous extract, granulometry, and viscosity and rheological behavior of the pastes.

4.3.4. Plastisols

Plastisols, commonly called pastes, are dispersions of PVC powders in plasticizers. A wide range of plasticizers are used in plastisol formulations, but the most widely used are C_4–C_{12} esters, such as phthalates, adipates, azelates, sebacates, trimellitates, and phosphates. Chlorinated paraffins are used as secondary plasticizers. A good plasticizer should have low volatility, low color value, neutral reaction, resistance to hydrolysis, insolubility in water, flame resistance, and nontoxicity. As no plasticizer can satisfy all these properties, several plasticizers are mixed to achieve the desired properties of the final product. Table 7 shows the influence of plasticizers on plastisol properties; Table 8, the influence of plasticizers on final product properties; and Table 9, the influence of resin characteristics on plastisol and final product properties.

The amount of plasticizer used in plastisol formulations plays a major role in determining the hardness and flexibility of the end product and can vary from 40 to 130 phr. In Figure 12, the influence of plasticizer content on plastisol viscosity is shown.

The other essential ingredients in plastisol formulations are:

- Heat stabilizers, which generally are the same compounds used for suspension polymers
- Dyes and pigments suitable for coloring plastisols; the most widely used is titanium dioxide

Table 7. Influence of plasticizers on plastisol properties

Plastisol properties	Plasticizers*
Low fusion temperature	DBP, BBP, TCP
Low viscosity	DOA, DOS, DIPA
High viscosity	BBP, TCP, polymerics
High fusion temperature	DIDA, DOZ, DTDP, DOS, polymerics

*DBP = dibutyl phthalate, BBP = butyl benzyl phthalate, TCP = Tricresyl phthalate, DOA = Dioctyl adipate, DOS = Dioctyl sebacate, DIDA = Diisodecyl adipate, DOZ = Dioctyl azelate, DTDP = Ditridecyl phthalate.

Table 8. Influence of plasticizers on final product properties

Product properties	Plasticizers
Flame retardancy	phosphates, chlorinated paraffins
Light and heat resistance	epoxy plasticizers
Low-temperature flexibility	adipates, azelates, sebacates, straight-chain phthalates
Low volatility	polymerics, trimellitates, linear phthalates, epoxy plasticizers
Stain resistance	butyl benzyl phthalate
Low migration	polymerics, trimellitates

Table 9. Influence of resin characteristics on plastisol and final product properties*

Plastisol and product properties	Effect of resin characteristics			
	Molecular mass	Particle size	Particle size distribution	Emulsifier type and quantity
Paste making		S	S	S
Plastisol viscosity		S	S	S
Air release				S
Gelation	M	M	M	S
Fusion	S			
Mechanical properties	S			
Gloss	M	S	M	
Transparency		S		S
Foaming	S	M		S
Fogging				S
Streaking flow		S		
Water resistance				S

*M = moderate effect; S = strong effect.

- Fillers, generally inorganic materials, such as calcium carbonate, barite, silicate, kaolin, or china clay, are used primarily to lower costs and also to impart special properties such as hardness, abrasion resistance, and no sticking
- Blowing agents for foamed PVC production, such as azodicarbonamide
- UV adsorbers to prevent the decomposition action of sunlight
- Esters of poly(ethylene glycol), phosphate esters, and fatty acid amines and amides are usually employed as antistatic agents
- Viscosity depressants, such as white spirit, poly(ethylene glycol) monolaurate, alkylphenols, low molecular mass paraffins
- Extender resin or filler polymer; a PVC grade of spherical form, made by the suspension process and therefore cheaper than dispersion resin, which is used to reduce cost and viscosity of the plastisol [47, 48]

The influence of plastisol ingredients on the final product characteristics is shown in Table 10.

4.3.5. Polymer–Plasticizer Interaction

Figure 13 illustrates the types of dependence of viscosity on shear rate;

- Newtonian (viscosity is independent of shear rate)
- Pseudoplastic (viscosity decreases as shear rate increases)
- Dilatant (viscosity increases with increasing shear rate)

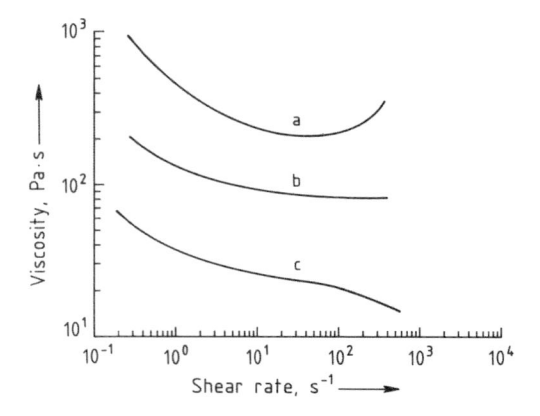

Figure 12. Influence of plasticizer content on plastisol viscosity
a) PVC/DINP = 100/40; b) PVC/DINP = 100/50; c) PVC/DINP = 100/60
DINP = diisononyl phthalate

Over a wide range of shear rate, a plastisol may exhibit all three types of behavior because the viscosity of plastisols rarely displays true Newtonian behavior. At low shear rates, they are generally pseudoplastic, at intermediate values they show dilatancy, and may sometimes become pseudoplastic again at higher shear

Table 10. Influence of plastisol ingredients on the final product characteristics[*]

Characteristics	Filler	Pigment	Stabilizers	Quantity of plasticizer	Plasticizer type	Resin
Tensile properties	M			S	M	S
Flexibility	M			S	M	
Hardness	M			S	M	
Low-temperature bending resistance	M			S	S	M
Gloss	S	S				S
Volatility				S	S	
Extraction				S	S	
Color	S	S	M			
Light resistance		S	S		M	
Cost	S		M	M	M	

[*] M = moderate effect; S = strong effect.

rates, as shown in Figure 14. There is no certainty on the exact mechanism as the rheology of PVC plastisols is complex and affected by many factors such as:

- Polymer/plasticizer mixing ratio
- Type of plasticizer and its interaction with polymer
- Particle size and their distribution of primary particles
- Structure, shape, porosity, and surface of the secondary particles
- Surfactants, added during polymerization or in the paste
- Molecular mass and molecular mass distribution
- Paste-making conditions such as mixing temperature, time, and intensity [49].

4.3.6. Gelation and Fusion

Gelation is defined as the change from liquid to semisolid state, and fusion, when an homogeneous phase at molecular level between polymer and plasticizer is reached. These processes have been examined by determining viscosity changes of the liquid system, the tensile strength, and changes in complex viscosity and viscoelastic properties over the temperature range 25–200°C.

At the beginning of heating, plasticizer viscosity decreases, as does the plastisol viscosity. At the same time, polymer particles adsorb

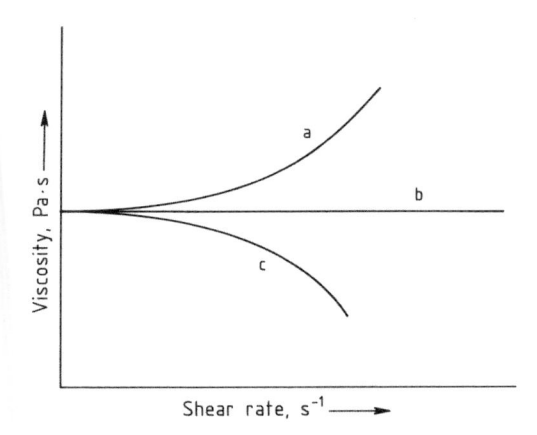

Figure 13. Effect of shear rate on dilatant (a), Newtonian (b), and pseudoplastic (c) fluids

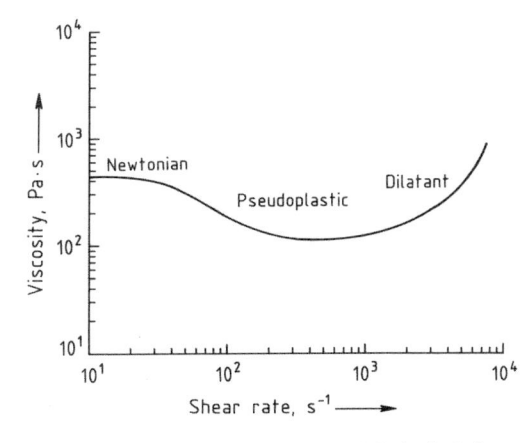

Figure 14. Effect of shear rate on the PVC plastisol viscosity (PVC/DINP = 100/50 at $T = 25°C$)

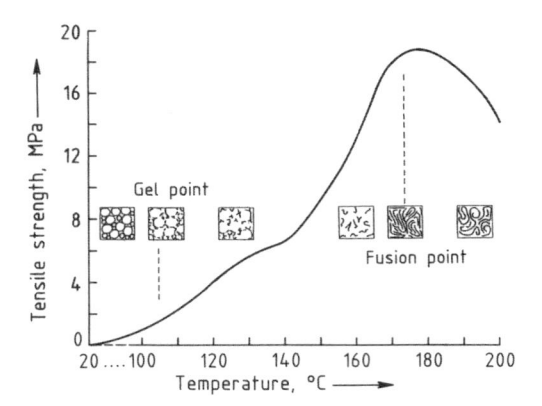

Figure 15. Changes in morphology and mechanical properties of PVC plastisol during gelation and fusion

plasticizer and swell, reducing the fraction of liquid plasticizer, and the viscosity then increases rapidly. When the liquid phase is completely absorbed by the particles, the system becomes dry. As the process continues, the interparticle boundaries disappear and the polymer chains become increasingly entangled and development of the physical properties begins. At this point, the gelation stage is complete and the fusion process starts, as illustrated in Figure 15. Further heating of plastisol makes the polymer flow into a continuous mass. The melt viscosity, after having reached a maximum, starts to decrease and the process passes through the fusion stage. Morphology of the particles, particle size, particle size distribution, and molecular mass of the resin affect the viscosity behavior. Low molecular mass resin starts the gelation phase earlier and completes the gelation process at lower temperature. The same result is obtained by using plasticizers with higher solvating action, whereas larger particle size increases gelation temperature or time because it slows the penetration of the plasticizer into the resin [50].

4.3.7. Applications

Vinyl plastisols can be applied by spread coating, knife coating, roll coating, molding, dipping, and spraying; the most widely used application is coating.

Coating. Vinyl plastisol can be used to coat paper, fabrics, metals, felt, and glass fibers to produce wallpapers, floor coverings, vinyl leather, conveyor belts, and tarpaulins. Each application needs its own formulation, in which the appropriate resin grade and plasticizer must be used to give the required rheological characteristics. In the case of fabric coating where direct spreading is employed, it is important to avoid penetration of plastisol into the fabric; therefore, the viscosity behavior of the paste must be strongly pseudoplastic at low shear rate and have moderate viscosity at high shear rate. In wallpapers, where thin coatings are applied, a low-viscosity resin with Newtonian behavior is required. In production of floor coverings, three or even four coatings are applied: An impregnation layer to saturate the fibers, one or two foam coats, and a wear layer with high mechanical strength. Each coat, which needs different grade resin to achieve the required properties, is gelled at 150°C before application of the next layer. When the last coat is applied, the entire coating is fused in an oven at 200°C. The foam layer can be chemically or mechanically foamed [51]. Usually mechanical embossing is used to create surface textures and apply colored patterns.

Dipping. The product, mainly work, household, and surgical gloves, is obtained by dipping metal or ceramic molds or an article to be coated into the plastisol and then withdrawing it, draining it, and then curing it in an oven.

Molding is used for producing hollow articles such as balls, dolls, and toys. The process requires a plastisol with low viscosity at low shear rate that has a short gelation time and is easy to deaerate. In rotational molding, the plastisol is poured into a cold mold that rotates around two perpendicular axes while entering an oven where it is heated with air to 200–250°C. After gelation and fusion, the mold is cooled in a water bath.

Further information is given in [32, 39, 40, 45, 48, 52–54].

4.4. Chlorinated Poly(Vinyl Chloride), CPVC

The softening point (heat distortion temperature or T_g) of PVC can be increased by chlorination of the polymer, either in a solvent or as a suspension. The normal chlorine content of 56.7% is raised to 63–68%, depending on type, and this increases the softening point from ca. 78–83°C (depending on formulation) to at least 120°C.

The older method involves chlorination of the PVC dissolved in a chlorinated solvent, but the process is expensive because of the need for an extensive solvent recovery system. However, it has the benefit of yielding a homogeneous material. The product is difficult to handle physically because of its very low bulk density.

The most common commercial process in use today disperses the PVC grains in water and swells them with a chlorinated hydrocarbon. The reactor is degassed to remove oxygen and heated to reaction temperature (50–60°C); chlorine is added continuously, and the system irradiated with UV light to produce chlorine radicals. Hydrogen chloride is produced as a byproduct and is removed at the end by washing with an inorganic base, after which the product is dried [55]. The degree of chlorination is variable, with the surface layers of the grains being preferentially treated.

CPVC is less thermally stable than conventional PVC and requires higher temperatures to develop good mechanical properties. Hence, it is difficult to process and quite expensive, but blends with suspension PVC can be used to achieve a specific softening point. CPVC products have not been very successful commercially but have found a specific niche in the market. They are used for hot-water systems in mobile homes in the United States and for the manufacture of fibers, for use in thermal underwear, in France.

5. Environmental Protection

Potential problem areas associated with the production, handling, and use of PVC are small if a few simple measures to avoid environmental contact and pollution are followed.

Losses of VCM from polymerization plants have been reduced to extremely low levels following conversion to closed-lid operation as a result of the toxicological problems discussed in Chapter 10 in [14]. Although modern column strippers reduce VCM levels in PVC to less than 10 ppm, a small quantity of VCM is lost from the drier section. The older batch strippers present more of a problem because they cannot achieve such low residual VCM contents.

PVC powder can be lost in centrifuge and scrubber effluent and by overfilling stock tanks and from pipeline leaks. It is important that all plant drains that are likely to contain PVC are directed to settling pits where flocculation treatment can be used so that the final discharge is free of PVC. All wet waste is collected and reclaimed as second- or third-grade material for use in noncritical applications, such as one-trip horticultural products and fencing.

Powder losses from drier stacks can be minimized by ensuring that drier cyclones are adequately sized to the air conveying rates used. Some modern plants use bag filter systems to meet current regulations that restrict the amount of PVC dust loss to 50 mg/m^3 air [56].

Compared to other polymers and typical metals PVC requires less energy per unit weight for its production [57]. Whereas plastics in general are blamed for long-lasting litter problems, it is significant that only 12% of PVC is used for consumables, whereas 64% is incorporated into products with an expected useful life of 15 years or more. Disposal is therefore less of an immediate problem than is often thought.

As landfill sites are used up more municipal waste is being incinerated. Though inherently nonflammable, PVC burns in an existing fire to release hydrogen chloride gas, which can be scrubbed out satisfactorily. Fears about the increased release of dioxins during incineration due to the presence of PVC in waste have proved to be unfounded [58].

6. Quality Specifications and Analysis

All PVC producers test a number of parameters on every batch produced as part of routine quality control operation. These include K-value, mean grain size and grain size distribution, apparent density, porosity, contamination, dispersion, color and heat stability, and residual VCM content. Most plants use statistical process control techniques to ensure that product consistency from batch to batch meets the stringent requirements of today's customers. It is difficult to give a typical testing regime because every plant makes batches of different size, has slurry-blending facilities of differing degrees of sophistication, and has hoppers of different capacity on their drying plants. For a modern plant operating large autoclaves there is considerable testing of each polymerization batch (wet stage) together with additional testing of the dried and blended material as a cross check.

In-house tests have been collected and correlated by national and international standards organizations to make them universally applicable. ISO has defined tests and standards set up by other organizations, such as the ASTM and DIN so that ISO methods are now widely accepted worldwide. ASTM D 1755 defines a system for classifying general purpose (mass and suspension) and dispersion (paste) resins [59].

7. Storage and Transportation

Most plants handle more than 80% of their production in bulk, and storage and supply in 25 kg bags is diminishing all the time as customers convert. After drying, the material is conveyed by blow-eggs (vessels in which the powder is pressurized and fluidized) of 2–15 t capacity via pipeline to the silo farm where it is stored in silos of capacity 100–500 t. PVC is discharged either directly or via a smaller loading hopper into 20 t road tankers, the most popular form of PVC transportation. Sometimes big-bags made of woven polypropylene are used for 1 t lots of PVC, and 20 t ISO metal containers are used for long distance deliveries. PVC is not a difficult material to handle and the logistics side of the operation is straightforward. It is essential to prevent cross-contamination of the PVC by other products and thorough inter-trip cleaning is essential, because many operators use their tankers for back carriage of a wide range of products, many of which are nonpolymeric.

8. Processing and End Uses

Due to its unique combination of properties, PVC is never handled on its own. Instead, a complex formulation incorporating several additives is used. A typical base formulation contains: PVC resin, heat stabilizer(s), internal lubricant(s), external lubricant(s), processing aid, and additionally, impact modifier, filler (s), pigment, UV stabilizer, as well as primary and secondary plasticizers (for flexible applications; see → Plasticizers).

PVC is intrinsically unstable because of molecular defects in some of the polymer chains [60, 61], and when subjected to heat they initiate a self-accelerating dehydrochlorination reaction. Stabilizers neutralize the HCl produced and introduce nucleophilic substitution reactions that prevent further degradation [60, 61] (see also → Plastics, Additives, Chap. 4).

The Ca–Zn stabilizer family is becoming increasingly popular as objections to the use of heavy-metal systems increases. In Europe withdrawal of cadmium stabilizers was planned for the end of 2001, and although it has been shown that use of the highly effective lead stabilizers is safe, adverse publicity will slowly erode their market position. Table 11 gives a summary of the *heat stabilizers* in regular use.

The polar, highly viscous PVC melt sticks easily to metal walls of extruder barrels, calenders, mills, etc. thus an *external lubricant* is employed to assist the smooth passage of the melt. *Internal lubricants* help to reduce melt viscosity and prevent overheating and thus help to ensure good color of the final product. *Processing aids* improve the surface appearance of extruded sections and reduce melt defects, such as screw memory, where the helical nature of the screw can be seen as regular ripples in the pipe surface. Many other additives can be used, e.g., *impact modifiers* for bottles and *UV stabilizers* as well as *fillers* (e.g., TiO_2) for house sidings and window frames to ensure the best possible in-service performance or longevity.

In all extrusion and some other conversion processes, the PVC grain is not broken down to its constituent primary particles [62], unlike emulsion PVC processing (see Section 4.3.6). Instead, the suspension grains (150 μm) gradually lose their original form by fusion and elongation under the influence of heat, pressure, and shear so that in badly processed PVC grain memory can be detected by optical microscopy of cross sections of extrudates [62]. In practice,

Table 11. Heat stabilizers for PVC

Stabilizer	Properties
Lead stabilizers	
Lead soaps	Good heat stability
Lead salts	Sulfur staining
Liquid lead complex	Lead phosphites give good light stability
Lead one-pack systems	Kickers for chemical blown foam
Tin stabilizers	
Organo sulfur tin	Excellent heat stability, good clarity; poor light stability
Organo tin carboxylates	Medium heat stability
Excellent light stability	
Barium–zinc stabilizers	
Ba–Zn liquid	Good heat and light stability on high-quality grades; general-purpose stabilizers.
Calcium–zinc stabilizers	
Ca–Zn soaps	Low toxicity
Ca–Zn liquid	Medium heat and light stability
Organic co-stabilizers	
Thiourea derivatives	Synergistic effects with epoxy plasticizer and/or Ca–Zn, Ba–Zn
Indol derivatives	
Amino crotonic acid esters	
Diketones	
Organic phosphites	
Antioxidants	
Epoxidized soybean oil (ESBO)	Synergistic effect with Ca–Zn and Ba–Zn
Epoxy esters	

converters take great care to ensure that the degree of gelation (i.e., lack of grain memory) in the final PVC article is very high. Degree of gelation can be measured by determining the degree of attack by a poor solvent, such as acetone or methylene chloride or by measurement of flow pressure to assess the strength and elasticity of partially fused material directly [63].

The enormous subject of processing, including different aspects, such as formulations, types of processing equipment, gelation, rheology, and mechanical properties is covered more widely in [62, 64–68].

The versatility of PVC can be gauged by the very broad summary of typical end uses given in Table 12.

9. Economic Aspects

Fear of litigation, the need to handle VCM (a known carcinogen), and reduced funding have all but eliminated fundamental research into

Table 12. Typical end uses for PVC

Application	Rigid PVC	Flexible PVC
Construction	window frames, gutters, pipes, cars, housesiding, ports, roofing	waterproof membranes, cable insulation, roof lining, greenhouses
Domestic	curtain rails, drawer sides, laminates, audio and video tape cases, records	flooring, wall coverings, shower curtains, leather cloth, hosepipes
Packaging	bottles, blister packs, transparent packs and punnets	cling film
Transport	car seat backs	under seal, roof linings, leather cloth upholstery, wiring insulation, window seals, decorative trim
Medical		oxygen tents, bags and tubing for blood transfusions, drips and dialysis liquids
Clothing	safety equipment	waterproofs for fishermen and emergency services, life-jackets, shoes, wellington boots, aprons and baby pants
Others	floppy-disk covers, credit cards	conveyor belts, inflatables, sports goods, toys

VCM polymerization. Work on processing of PVC continues, but at a much reduced level. Many companies have scaled down R & D and process development, and raw materials suppliers are now increasingly responsible for technological developments.

Since the early 1990s, the main trend in the PVC industry has been consolidation and amalgamation. Some well-known companies, including pioneers from the 1940s (e.g., B F Goodrich/GEON, ICI, and Chisso) have disappeared through acquisition, and many others have formed larger conglomerates to survive. The current feeling in the industry is that a minimum capacity of 500 000 t/a is required for a viable operation.

Strong investment in new plant near the peak of the economic cycle has resulted in overcapacity, weakened prices, lower company profits, and economic casualties, just as it did around 1990. Since the 2000s, the sharp increase of feedstock as well as energy costs has speeded up the process of restructuring the industry as a general trend in the more mature markets like North America and Europe, but also in Japan. Changes in the European PVC scene are summarized in Figure 16.

Table 13 lists the regional capacities (in 10^3 t/a) of the top ten producers, which represent 36% of the world capacity of 53.4×10^6 t/a. World production was 37.5×10^6 t (70% of capacity) in 2012.

Since around 1990, China has emerged as a major force on the global PVC market, and its tremendous import capacity has had a large impact on the world price/consumption relationship in the beginning. Especially since 2000, the Chinese market has been determined by a substantial increase of capacity, reaching a share of approx. 44% of the total global capacity (23.4×10^6 t/a). Of the installed capacity, approx. 81% are calcium carbide based and only 19% ethylene based.

In 2010, the Chinese production volume was approx. 10.8×10^6 t (utilization approx. 62%) and increased in 2012 to approx. 13.1×10^6 t (utilization reduced to 56%).

In 2010, there were 92 PVC producers in China, of those

- 21 sites with $> 3 \times 10^5$ t/a capacity
- 48 sites with $> 1 \times 10^5$ t/a and $< 3 \times 10^5$ t/a capacity
- 23 sites with $< 1 \times 10^5$ t/a capacity

Between 2008 and 2012 the domestic demand grew 8.5% p.a. and the domestic production output by 8.0% p.a. in China.

World PVC production was 37.5×10^6 t/a in 2012, and growth ranged from 0 (Europe) to +8% (China). Overall, long-term growth worldwide is estimated at approx. 4%. Table 14 summarizes the world supply–demand balance.

10. Toxicology and Occupational Health

Vinyl Chloride Monomer (VCM). Because vinyl chloride boils at $-13°C$, it is handled as a compressed liquid. It has a vapor density greater than air, thus, precautions are taken in design not to contain the vapor in restricted areas. Protection of reactors by relief valves or bursting disks and double-valve isolation of all lines containing VCM are basic precautions taken in the design and construction of any PVC plant vessel that contains monomer. The gas is explosive between 3.6 and 25 vol% with air and 12 vol% oxygen is required for ignition [69]. Therefore, a special series of working conditions must be satisfied in the working environment and great care is always taken to purge lines, pumps, valves, etc. before any maintenance work is started. Thorough documentation (e.g., clearance to work certificates) is always used to ensure proper isolation of plant equipment before any entry. The whole plant is contained within a "red-fence" area where smoking is not permitted and where all electrical equipment is of a gas-tight and "flame-free" construction.

VCM has a narcotic effect at 8–12 vol% concentration and can cause death at higher concentrations [69]. Therefore, early VCM and PVC plants were designed to avoid build-up of gas above 1 000 ppm in the atmosphere. VCM has a pleasant ethereal smell that is first noticed at concentrations of 500–2 000 ppm.

Exposure to VCM has been shown to lead to two distinct problems. In the mid 1960s it became clear that a number of workers involved

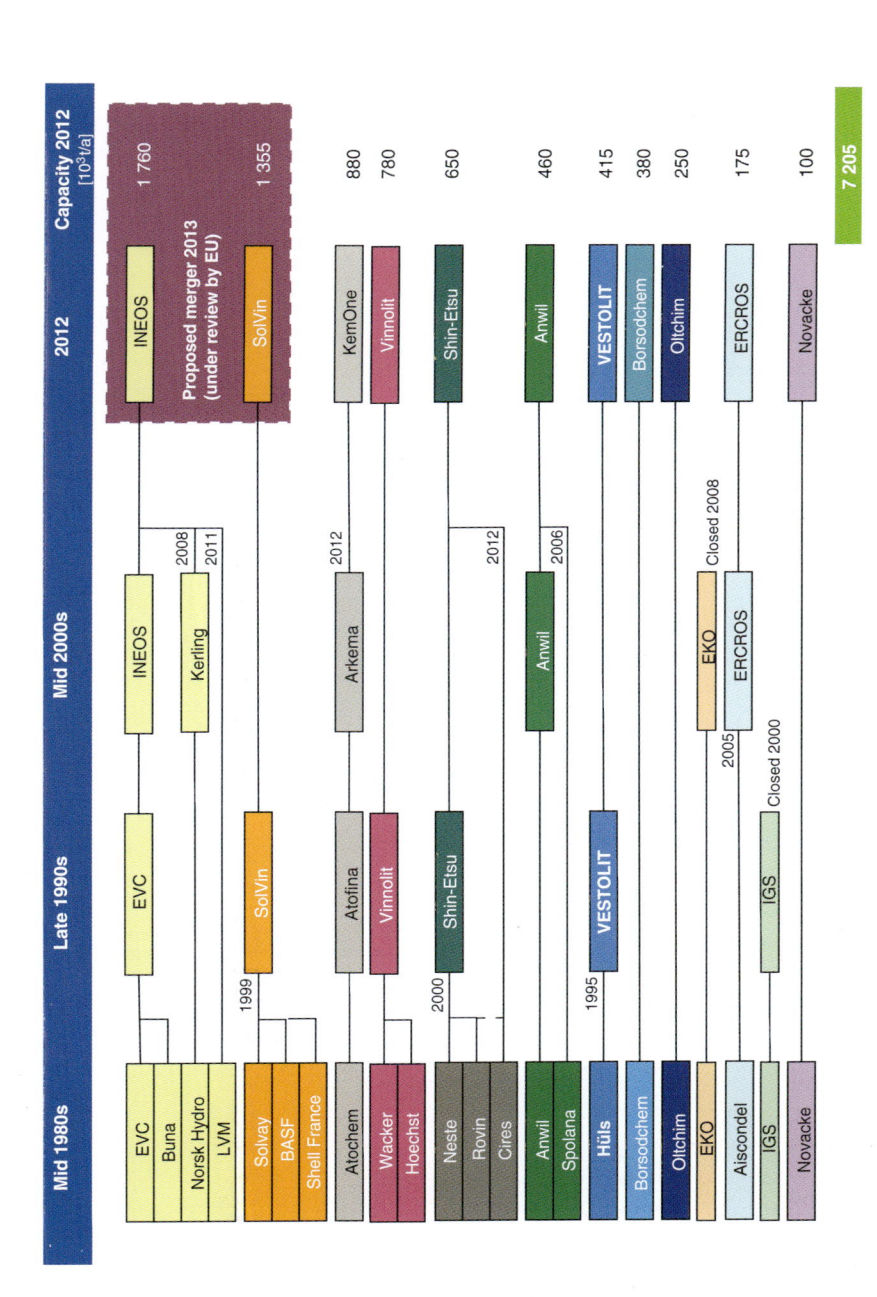

Figure 16. Evolution of the PVC industry in Western Europe

Table 13. Regional capacities (in 10^3 t/a) of the top ten PVC producers in 2012

Company	S. America	Asia	NAFTA	Europe	Total
Shin-Etsu	0	650	2 650	650	3 950
Formosa	0	1 775	1 525	0	3 300
Solvay	560	280	0	1 355	2 195
Oxy-Vinyls	0	0	1 930	0	1 930
Ineos	0	0	0	1 760	1 760
Axiall	0	0	1 330	0	1 330
LG Chem	0	1 260	0	0	1 260
Westlake	0	130	770	0	900
KemOne	0	0	0	885	885
Mexichem	390	0	480	0	880

The top 10 producers represent about 34.4% of world capacity of 53.3×10^9 t. World production in 2012 was 37.4×10^9 t (70% utilization).

PVC Production Plants:

Consumption of PVC Resin	Number of Plants
$> 3 \times 10^5$ t/a	21
$> 1 \times 10^5$ t/a	48
$< 1 \times 10^5$ t/a	23

Between 2008 and 2012
Domestic output of PVC increased 8% in yearly average.
Domestic consumption increased 15% in yearly average.
Imported quantities of PVC increased to 20% in yearly average.
PVC usage 2012 (kg/capita) is given below.

Europe (EU 28)	9.2
Eastern Europe/CIS	4.7
Turkey	11.3
Northeast Asia (without China)	12.1
Southeast Asia	2.8
China	10.2
India Subcontinent	1.5
Africa/Middle East	6.0
Oceania	5.5
North America	12.7
South America	4.2

Table 14. World supply/demand balance for PVC in 2012 (in 10^3 t/a)

	Capacity	Operating rate, %	Production	Imports	Exports	Consumption
N. America	8 040	86	6 900	90	2 520	4 470
S. America	2 350	80	1 860	950	240	2 570
European Union	7 260	80	5 780	170	1 200	4 750
Eastern Europe	950	86	820	650	20	1 450
Middle East/Turkey	1 250	65	800	1 280	70	2 010
China	23 430	56	13 120	1 130	350	13 900
Northeast Asia without China	5 630	78	4 370	110	1 660	2 820
Southeast Asia	2 320	90	2 090	320	660	1 750
Africa	625	63	400	670	70	1 000
Indian Subcontinent	1 510	88	1 330	1 210	0	2 540
Others	0	–	0	210	0	210
Total World	53 365	70	37 470	6 790	6 790	37 470

in reactor cleaning suffered from a bone condition called acroosteolysis (AOL), which affects mainly the hands and feet. Removal of the worker from exposure to VCM leads to an almost complete recovery. Since improvements to ventilating reactors and the introduction of high-pressure water cleaning the incidence of AOL has ceased [70]. A study of long-term exposure of rats to 30 000 ppm VCM did not reproduce any sign of AOL but they did develop cancers in various sites [71].

Subsequently, an extensive study with a wide range of dose conditions relating to occupational exposure was begun in 1971 at the Institute of Oncology in Bologna [72]. This showed that a very rare liver cancer, angiosarcoma could be formed in rats at levels of 250–500 ppm VCM. The first correlation between exposure to VCM and cancer in humans was made in 1973 when three workers at a plant in Louisville, Kentucky were shown to have died from angiosarcoma [73]. Since then other cases have been identified, all of which involved workers exposed to high concentrations of VCM as reactor cleaners or charging operators. As the period between first exposure to VCM and the appearance of angiosarcoma is 20 to 25 years it is likely that the incidence of the disease will soon decrease following improvements made to operation of PVC plants since 1973.

The knowledge that VCM is a human carcinogen has led to the introduction of very stringent controls by governmental and regulatory authorities to limit the exposure of workers and the general public to the monomer. In the United States, OSHA has set an 8-h TWA of 1 ppm VCM. During a shift, an employee's exposure must not exceed 5 ppm over a period of 15 min or less [74]. The European regulations, EC directive 2004/37/EC, require an annual average of < 3 ppm in the plant's airspace with alarm values of 15 ppm over 1 h, 20 ppm over 20 min, and 30 ppm over 2 min. Several Eu Member States and companies set somewhat lower levels, typically 1 ppm [75].

Poly (Vinyl Chloride). Strict limits have been set for the quantity of residual VCM in processed PVC articles intended for use in food contact applications.

In 1979, the EC set an upper limit of 1.0 ppm in the article and 10 ppb in the food. In 1986, the FDA suggested levels of ≤ 5 ppb in rigid and ≤ 10 ppb in plasticized PVC articles in food contact applications. To ensure that these limits are met PVC producers aim for these concentrations in the polymer grains themselves so that any further loss during processing ensures even greater safety.

There is little evidence that PVC powder itself causes any significant medical problems but steps are always taken to reduce powder emissions because of its nuisance value.

References
General References

1. R.H. Burgess (ed.): *Manufacturing and Processing of PVC*, Applied Science Publishers, London 1982.
2. G. Butters (ed.): *Particulate Nature of PVC*, Applied Science Publishers, London 1982.
3. L.I. Nass, C.S. Heiberger (eds.): *Encyclopaedia of PVC*, 2nd ed., vols. 1, 2, 3, Marcel Dekker, New York 1986.
4. G. Matthews: "Vinyl Chloride and Vinyl Acetate Polymers", *Vinyl and Allied Polymers*, vol. 2, Iliffe, London 1972.
5. A. Whelan, J.L. Craft (eds.): *Development in PVC Production and Processing*, Applied Science Publishers, London 1977.

Specific References

6. J.D. Cotman, M.F. Gonzales, G.C. Claver, *J. Polym. Sci. Polym. Chem. Ed.* **5** (1967) 1137–1164.
7. M.R. Meeks, *Polym. Eng. Sci.* **9** (1969) 141.
8. H. Fikentscher, *Kolloid Z.* **53** (1932) 34.
9. G.A.R. Matthews, R.B.Pearson, *Plastics (London)* **28** (1963) 98.
10. W.F. Maddams in E.D.Owen (ed.): *Degradation and Stabilisation of PVC*, Elsevier Applied Science Publishers, Barking 1984.
11. C.A. Finch (ed.): *Poly(vinyl alcohol), Properties and Applications*, John Wiley & Sons, New York 1973.
12. D.G. Kelsall, G.C. Maitland in K.H. Reichert, W. Geisseler (eds.): *Polymer Reaction Engineering, Influence of Reaction Engineering on Polymer Properties*, Hanser Publishers, München 1983.
13. G. Wieckert, G. Henschel, K.D. Weissenborn, *Angew. Makromol. Chem.* **147** (1987) 1.
14. R.H. Burgess (ed.): *Manufacturing and Processing of PVC*, Applied Science Publishers, London 1982.
15. B. Mariasi, *J. Vinyl Technol.* **8** (1986) no. 1, 20.
16. M.W. Allsopp in R.A.Burgess (ed.): *Manufacturing and Processing of PVC*, Applied Science Publishers, London 1982, p. 152.
17. J. Boissel, N. Fischer, *J. Macromol. Sci. Chem.* **A 11** (1977) no. 7, 1249.
18. M.W. Allsopp in [14] p. 181.
19. N. Fischer, *J. Vinyl Technol.* **6** (1984) no. 1, 35.
20. M.W. Allsopp in [14] chap. 2.
21. I.G. Farben, US2068424, 1931 (H.Fikentscher).
22. W.D. Harkins, *J. Am. Chem. Soc.* **69** (1947) 1428.
23. W.V. Smith, R.H. Ewart, *J. Chem. Phys.* **16** (1948) 592.

24. J. Ulgestad, P.C. Mork, J.O. Aasen, *J. Polym. Sci. Polym. Chem. Ed.* **5** (1967) 2281.
25. E. Peggion, F. Testa, G. Talamini, *Makromol. Chem.* **71** (1964) 173.
26. G. Gatta, G. Benetta, G.P. Talamini, G. Vianello: "Addition and Condensation Polymerization," in R.F. Gould (ed.): *Adv. Chem. Ser.* 1969, no. 91, 158.
27. J. Ugelstad, H. Flogstad, F.K. Hansen, T. Ellingsen, *J. Polym. Sci. Polym. Symp.* **473** (1973) no. 42, 473.
28. ICI, GB978875, 1981 (A. Frangou).
29. Rhône-Progil, GB1435425, 1976 (T. Kemp).
30. Rhône-Poulenc, GB1503247, 1977.
31. J. Ugelstad, F.K.Hansen, S. Lange, *Makromol. Chem.* **175** (1974) 507.
32. D.E.M. Evans in R.H. Burgess (ed.): *Manufacture and Processing of PVC*, Applied Science Publishers, London 1982, pp. 67–75.
33. I.M. Koltoff, I.K. Miller, *J. Am. Chem. Soc.* **73** (1951) 3055.
34. R.G. R. Bacon, *Q. Rev.* **9** (1955) 283.
35. ICI, EP38634, 1981.
36. P. Rangnes, O. Palmgren, *J. Polym. Sci., Part C* **33** (1970) 181–192.
37. Hüls, DE1964029, 1970.
38. Hüls, BE857235, 1970.
39. P. Becher: *Emulsion: Theory and Practice*, Reinhold, New York 1955, p. 173.
40. F.A. Bovey, I.M. Koltoff, A.I. Medalia, A.J. Meehan: *Emulsion Polymerisation*, Interscience, New York 1954.
41. Hoechst, DE2429776, 1980.
42. Hüls, US4233437, 1980.
43. Montedison, DE-OS3020237, 1980.
44. ICI, US4283526, 1981.
45. K. Masters: *Spray Drying Handbook*, Georg Godwin, London 1979.
46. Niro Atomizer, US4229249, 1980 (K.S.Felsvang, O.E. Hansen).
47. K. Kern Sears, J.R. Darby: *The Technology of Plasticizers*, Wiley-Interscience, New York 1982.
48. H.A. Sarvetnick: *Plastisols and Organosols*, Van Nostrand Reinhold, New York 1972.
49. H. Zecha, J. Schneider, K. Wulf, *Kunststoffe* **76** (1986) no. 1, 51.
50. N. Nakajima, E.R. Harrel, *Adv. Polym. Technol.* **6** (1986) no. 4, 409.
51. Congoleum-Nairn, US3293094, 1966.
52. F. Candau, R.H. Ottewill: *Scientific Methods for the Study of Polymer Colloids and their Application*, Kluwer, London 1990.
53. G. Matthews: *Vinyl and Applied Polymers*, vol. 2, Iliffe, London 1972.
54. L.I. Nass: *Encyclopaedia of PVC*, M. Dekker, New York 1976.
55. R.G. Parker, G.Martello in [61]2nd ed., vol. 1, 1986, p. 619.
56. TA Luft 1986.
57. British Plastics Federation, *The Energy Content of Plastics Articles*, London 1986.
58. Solvay et Cie, *PVC and The Environment*, company brochure.
59. D.J. Brandt, R.S. Guise, B.A. McCoy in L.I. Naas, C.Heiberger (eds.): *Encyclopaedia of PVC*, 2nd ed., vol. 1, Marcel Dekker, New York 1986.
60. T. Hjerberg, E.M. Sörvik in E.D. Owen (ed.): *Degradation and Stabilisation of PVC*, Elsevier Applied Science Publishers, Barking 1984.
61. D. Braun, E. Bazdadea in L.I. Nass, C.A. Heiberger (eds.): *Encyclopaedia of PVC*, 2nd ed., vol. 1, Marcel Dekker, New York 1986.
62. M.W. Allsopp in [14] chap. 8.
63. A. Gonze, *Plastica* **24** (1971) no. 2, 49.
64. D.A. Tester in [14] chap. 9 and 10.
65. G.C. Portingell in G.Butters (ed.): *Particulate Nature of PVC*, Applied Science Publishers, London 1982, pp. 135–234.
66. N.L. Perry in [61] 1st ed., vol. 2, 1976, p. 601; 2nd ed., vol. 2, 1988, p. 1.
67. W. Henschel, P. Franz in W.V. Titow (ed.): *PVC Technology*, 4th ed., Elsevier Applied Science Publishers, Barking 1984.
68. E.D. Owen: *Degradation and Stabilisation of PVC*, Elsevier Applied Science Publishers, Barking 1984.
69. J.T. Barr in [61] 2nd ed., vol. 1, chap. 5, 1986.
70. R.H. Burgess in [14] p. 101.
71. P.L.Viola, A. Bigotti, A. Caputo, *Cancer Res.* **31** (1971) 516.
72. C. Maltoni in *Vinyl Chloride Carcinogenicity; Available Scientific Evidence and Control Measures*, Association of Plastics Manufacturers in Europe, Brussels 1986.
73. J.L. Creech, M.N. Johnson, *JOM, J. Occup. Med.* **16** (1974) 150.
74. *Fed. Regist.* **39** (1974) 35890–35898.
75. www.pvc.org/en/p/vinyl-chloride-monomer (accessed: 28 March 2014).

Further Reading

R.F. Grossman (ed.): *Handbook of Vinyl Formulating*, 2nd ed., Wiley, Hoboken, NJ 2008.

C.A. Harper (ed.): *Handbook of Plastics Technologies*, McGraw-Hill, New York, NY 2006.

J.M. Margolis: *Engineering Plastics Handbook*, McGraw-Hill, New York 2006.

H.F. Mark (ed.): *Encyclopedia of Polymer Science and Technology*, 3rd ed., Wiley, Hoboken, NJ 2005.

J.W. Summers: "Vinyl Chloride Polymers", *Kirk Othmer Encyclopedia of Chemical Technology*, 5th ed., John Wiley & Sons, Hoboken, NJ.

C.E. Wilkes, J.W. Summers, C.A. Daniels, M.T. Berard (eds.): *PVC Handbook*, Hanser, München 2005.

G. Wypych: *PVC Formulary*, ChemTec Publ., Toronto 2009.

Polyvinyl Compounds, Others

MANFRED L. HALLENSLEBEN, Universität Hannover, Institut für Makromolekulare Chemie, Hannover, Germany

ROBERT FUSS, Kuraray Europe GmbH, Frankfurt am Main, Germany

FLORIAN MUMMY, Kuraray Europe GmbH, Frankfurt am Main, Germany

1.	Poly(Vinyl Alcohol)	1141
1.1.	Structure and Crystallinity	1142
1.2.	Production	1142
1.3.	Properties	1144
1.4.	Uses, Toxicology, and Economic Aspects	1145
2.	Poly(Vinyl Acetals)	1146
2.1.	General Aspects	1146
2.1.1.	Properties and Analysis	1146
2.1.2.	Synthesis and Reactions	1147
2.2.	Production	1148
2.3.	Uses	1149
2.4.	Environmental Issues, Safety, and Health	1149
2.5.	Economic Aspects	1149
2.6.	Poly(Vinyl Butyral)	1150
2.6.1.	Properties	1150
2.6.2.	Production	1151
2.6.3.	Uses	1153
2.7.	Poly(Vinyl Formal)	1156
2.7.1.	Production	1156
2.7.2.	Commercial Products	1156
2.7.3.	Properties	1156
2.7.4.	Uses	1156
2.7.5.	Toxicity	1156
2.8.	Other Poly(Vinyl Acetals)	1157
3.	Poly(N-Vinyllactams) and Poly(N-Vinylamines)	1157
3.1.	Polyvinylpyrrolidone	1157
3.2.	Polymers of other N-Vinyllactams	1159
3.3.	Polyvinylamines	1160
	References	1161

1. Poly(Vinyl Alcohol)

In 1924, W. O. HERRMANN and W. HAEHNEL [1] were the first to prepare poly(vinyl alcohol) by saponifying poly(vinyl esters) with stoichiometric amounts of caustic soda solution. At about the same time, work was resumed in the Hoechst Works of the IG Farbenindustrie on the saponification of poly(vinyl ester) derivatives such as poly(vinyl chloroacetate), which F. KLATTE and E. ZACHARIAS had begun in 1912 in the Chemische Fabrik Griesheim Elektron.

In 1932, W. O. HERRMANN, W. HAEHNEL, and H. BERG [2] discovered that poly(vinyl alcohol) can be prepared from poly(vinyl esters) by transesterification with absolute alcohols in the presence of catalytic amounts of alkali. The transesterification principle is still used today by all poly(vinyl alcohol) producers. Work on the conversion of poly(vinyl acetate) to poly(vinyl alcohol) contributed substantially to the proof of Staudinger's theory of the structure of macromolecules [3].

In the transesterification of poly(vinyl acetate) to poly(vinyl alcohol), products with different acetyl group contents can be obtained depending on production conditions. The range of commercial poly(vinyl alcohols) extends from fully saponified types to products with degrees of hydrolysis of ca. 70 mol %.

In addition to these poly(vinyl alcohols), which dominate the market, saponification products of vinyl acetate–ethylene copolymers are also commercially important [4]. Numerous saponification products of other vinyl acetate copolymers (e.g., with other vinyl esters, longer-chain α-olefins, vinyl ethers, vinyl halides, acrylate and methacrylate esters, and unsaturated acids, nitriles, and amides) are described in [5].

1.1. Structure and Crystallinity

Poly(vinyl alcohol) contains mainly 1,3-diol units. The content of 1,2-diol units in poly(vinyl alcohols) obtained by hydrolysis of poly(vinyl acetates) is <1 – 2%. The content of 1,2-diol sequences can be reduced by lowering the polymerization temperature of vinyl acetate [6] or using other vinyl esters such as vinyl formate [7] or vinyl benzoate [8]. The content of 1,2-diol units influences some properties of poly(vinyl alcohols) (e.g., the degree of swelling of films in water [9]).

Carboxyl and carbonyl end groups are more abundant than initiator end groups in poly(vinyl alcohols) [10]. Poly(vinyl alcohol) is slightly branched as a result of chain transfer reactions occurring during the polymerization of vinyl acetate [11]; → Poly(Vinyl Esters).

In partially saponified poly(vinyl alcohols) the acetyl groups can be distributed statistically or blockwise, depending on production conditions. The nature of the acetyl group distribution influences important properties of poly(vinyl alcohols), such as the melting point [11], the surface tension of aqueous solutions [12], and the emulsifying and protective-colloid properties [13].

The stereochemistry of poly(vinyl alcohols) is fixed during the polymerization of vinyl monomers. It is determined from the position and intensity in the ^{1}H NMR spectrum of the hydroxyl proton signal, which consists of three doublets assigned to syndiotactic (S), atactic (A), and isotactic (I) triads of poly(vinyl alcohol) [14, 15]. According to this, poly(vinyl alcohol) obtained by radical polymerization of vinyl acetate followed by hydrolysis is predominantly atactic in structure (24% I, 44% A, 32% S).

Stereochemically more uniform poly(vinyl alcohols) are obtained when the initial polymer is either poly(vinyl trifluoroacetate) (23% I, 36% A, 40% S) or poly(vinyl tert-butyl ether) (55% I, 32% A, 13% S). Poly(vinyl alcohols) of different tacticity have different properties. For example, in acetalization (see Chap. 2), the isotactic sequences react first, and the highest stability toward water is shown by syndiotactic poly(vinyl alcohols) [16].

Poly(vinyl alcohol) is a crystalline polymer (for X-ray structure, see [17] for the X-ray structure). The degree of crystallinity depends strongly on the structure and previous history of the poly(vinyl alcohol). Acetyl groups and other incorporated groups of any kind lower the crystallizability [18]. The degree of crystallinity of fully saponified poly(vinyl alcohols) is increased by heat treatment, which also lowers their solubility in water [19]. This effect is less marked for poly(vinyl alcohols) that contain acetyl groups.

1.2. Production

Since the monomer of poly(vinyl alcohol), vinyl alcohol, is nonexistent — the tautomeric equilibrium lies on the acetaldehyde side — only indirect methods are available for the production of poly(vinyl alcohol).

Attempts to prepare poly(vinyl alcohol) by polyaldol condensation of acetaldehyde have had little success up to now [20]. Polymerization of the metal vinyl compounds (which are relatively complicated to prepare), followed by saponification, leads in general to poly(vinyl alcohols) of low molecular mass [21, 22].

The most important manufacturing process for poly(vinyl alcohol), which is used worldwide, is the polymerization of vinyl esters or ethers, with subsequent saponification or transesterification. The preferred starting material is vinyl acetate, but derivatives of vinyl acetate (such as vinyl mono- and dichloroacetate, vinyl bromoacetate [23], and vinyl trifluoroacetate [24]), vinyl esters of other carboxylic acids (such as vinyl formate [25], vinyl butyrate [26], and vinyl benzoate [27]) and vinyl ethers (such as benzyl vinyl ether [28]) can also be used. The hydrolysis of poly(vinyl acetate) is easily controllable, and no side reactions occur.

Production Conditions. Since the properties of poly(vinyl alcohols) depend primarily on their molecular mass and residual content of acetyl groups, industrial production processes are optimized for the precise control of these two characteristics.

The preferred process for the production of poly(vinyl acetate) for further processing to poly(vinyl alcohol) is polymerization in methanol. Factors determining the molecular mass of poly(vinyl acetate) and, accordingly, of poly (vinyl alcohol) are the polymerization temperature, vinyl acetate–methanol ratio, and polymerization conversion [29]. Lowering the polymerization temperature, the proportion of methanol, and the conversion increases molecular mass.

In general, poly(vinyl acetate) should be as free as possible of monomeric vinyl acetate because the latter is cosaponified. Any acetaldehyde formed colors the poly(vinyl alcohol) yellow in alkaline saponification as a result of aldehyde resin formation and acetalizes it in acid saponification.

Conversion of poly(vinyl acetate) to poly(vinyl alcohol) can be carried out in solution, suspension, or emulsion with alkaline or acidic catalysts. The preferred process is transesterification in methanol in the presence of catalytic amounts of sodium methoxide, with formation of poly(vinyl alcohol) and methyl acetate.

Hydrolysis conditions influence the structure of the poly(vinyl alcohol) formed. By varying catalyst concentration, reaction temperature, and the reaction time, the content of residual acetyl groups can be adjusted practically at will. The nature of the distribution of the remaining acetyl groups in partially saponified poly(vinyl alcohol) is determined by the choice of catalyst and by the solvent. Thus, in alkaline saponification the residual acetyl groups are distributed largely blockwise [30], and in acidic saponification largely statistically [31].

If methyl acetate [31, 32] or benzene [18] is added to methanol in transesterification, poly(vinyl alcohols) with increased block formation by the remaining acetyl groups are obtained. In aqueous solution with acid catalysis, equilibrium conditions occur. The degree of saponification can be adjusted by the addition of different amounts of acetic acid [33]. Partly acetylated poly(vinyl alcohols) prepared by reacetylation of fully saponified products in homogeneous aqueous systems have a completely statistical distribution of the acetyl groups [31].

Poly(vinyl alcohol) is manufactured in both batch and continuous processes. For the continuous polymerization of vinyl acetate in methanol, high molecular masses can be achieved only with relatively low methanol content and low vinyl acetate conversion [34]. Part of the vinyl acetate must therefore be recycled.

In the alkaline hydrolysis of poly(vinyl acetate) in methanol, a highly swollen gel is formed initially. Then, toward the end of saponification this swollen gel contracts and releases solvent (syneresis), becoming a tough gel of poly(vinyl alcohol).

An example of a continuous process is the belt saponification of Shawinigan Chemicals (Fig. 1) [35, 36].

Suspension- or solution-polymerized poly(vinyl acetate) is mixed with methanolic alkali in a mixing vessel to give a 30% methanolic solution, which is fed to a conveyor belt, where it remains until syneresis begins. The gel is ground in a mill, neutralized with acid, washed with methanol, separated from the mother liquor, and dried. The mother liquor, consisting of excess acid, methanol, and methyl acetate, is worked up in further process stages.

Saponification in a kneader is an example of a discontinuous process (Fig. 2) [37, 38]. Poly(vinyl acetate) is dissolved in methanol in a kneader, and the amount of sodium methylate in methanol required for the desired degree of saponification is added. The transesterification process at 20–30°C usually lasts about 8 h. Then the mixture of methanol and methyl acetate is

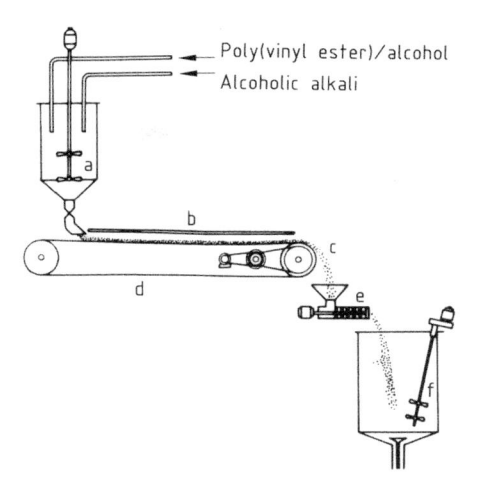

Figure 1. Belt saponification process for poly(vinyl alcohol) [38, 39]
a) Mixing vessel; b) Cover plate; c) Discharge; d) Conveyor belt; e) Mill; f) Washing vessel

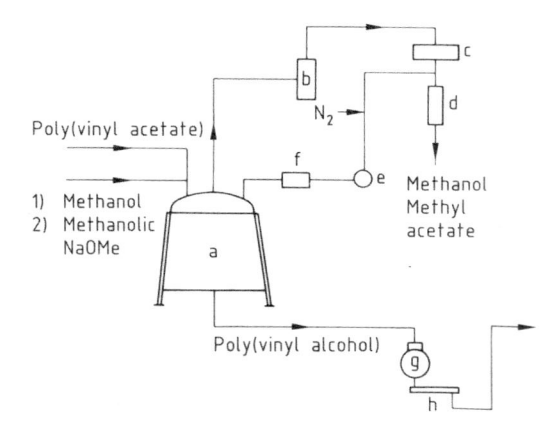

Figure 2. Manufacture of poly(vinyl alcohol) in a kneader [37, 40]
a) Kneader; b) Dust separator; c) Trap; d) Cooler; e) Blower; f) Heater; g) Mill; h) Sieve.

driven from the kneader at 60–70°C with a nitrogen stream, and poly(vinyl alcohol) is fed to subsequent processing stages.

Subsequent process stages [39, 40] include separation of vinyl acetate from partly reacted solution polymers in stripping columns, separation of the saponification mother liquor into methyl acetate and methanol, hydrolytic cleavage of methyl acetate with subsequent distillation of methanol and acetic acid, and drying of poly(vinyl alcohol), which can be carried out batchwise or continuously.

Trade Names. Poly(vinyl alcohol) is supplied by manufacturers under different trade names: Mowiol (Hoechst, Germany), Polyviol (Wacker Chemie, Germany), Rhodoviol (Rhône-Poulenc, France), Alcotex (Revertex, England), Polivinol (Rhodiatoce, Italy), Denka Poval (Denki Kagaku Kogyo, Japan), Gohsenol (Nippon Gohsei, Japan), Kurashiki Poval (Kuraray, Japan), Shinetsu Poval (Shinetsu Chem. Ind., Japan), Unitika Poval (Unitika, Japan), Elvanol (Du Pont, US), Gelvatol (Shawinigan Resins, US), Lemol (Borden, US).

Storage and transport of poly(vinyl alcohol) require no special precautions: the polymers must merely be protected from the effects of moisture and heat.

Wastewater Problems. Poly(vinyl alcohol) is degradable in sewage treatment plants with an adapted biological clarification stage. However, when discharged into receiving waters it is degraded only with difficulty, as shown by the 5-d BOD values. Tests showed that fish (guppies) are not harmed, even at a poly(vinyl alcohol) concentration of 500 mg/L of water [41].

1.3. Properties

Physical Properties. Normal commercial poly(vinyl alcohols) are white to yellowish and are supplied as powder and granules. The main factors influencing physical properties are the degrees of polymerization and hydrolysis. Most commercial products fall into two main groups: those with a degree of hydrolysis of about 98 mol% of the acetyl groups, and those with a degree of hydrolysis of 87–89 mol%. Both groups include products with degrees of polymerization from ca. 500 to 2500. Data such as the viscosity of the aqueous poly(vinyl alcohol) solutions, the molecular mass of the precursor poly(vinyl acetates), and the degree of hydrolysis or ester number are frequently used by manufacturers of poly(vinyl alcohol) to characterize their products.

The melting point and glass transition temperature depend not only on the content and distribution of acetyl groups but also on the tacticity and water content. For fully saponified poly(vinyl alcohols) a melting point of 228 °C and a glass transition temperature of 85 °C are reported [42]. Prolonged heating of dry poly(vinyl alcohol) above 110 °C can lead to evolution of gaseous cleavage products.

Poly(vinyl alcohol) Solutions. Water is the most important solvent for poly(vinyl alcohol). Certain polar solvents such as diethylenetriamine [43], dimethyl sulfoxide [44], formamide, dimethylformamide, and hexamethylphosphoric triamide [45] are relatively good solvents. The rate of dissolution in water increases with decreasing degree of polymerization and degree of hydrolysis. The degree of polymerization has a greater effect with fully saponified than with partially saponified poly(vinyl alcohols). Poly(vinyl alcohols) with degree of hydrolysis below 88% dissolve better in water at low than at elevated temperature [46].

Poly(vinyl alcohol) solutions are obtained by stirring poly(vinyl alcohol) into water at ca. 90 °C. Industrially they are produced in indirectly heated dissolving tanks or by introducing steam into poly(vinyl alcohol) suspensions. For prolonged storage, solutions must be stabilized against microbial attack by adding preservatives.

The viscosity of aqueous solutions depends on the degrees of polymerization and hydrolysis of poly(vinyl alcohols) as well as on the concentration and temperature. Fully saponified poly(vinyl alcohols) yield more viscous solutions than partially saponified poly(vinyl alcohols) at the same degree of polymerization [46]. Concentrated solutions of highly saponified poly(vinyl alcohols) undergo a viscosity increase on prolonged standing, which can be reversed by heating.

Many inorganic salts, especially sulfates and phosphates, precipitate poly(vinyl alcohols) from aqueous solutions [42]. Boric acid and borax act as thickeners [47]. The conductivity of aqueous poly(vinyl alcohol) solutions depends on their electrolyte content, which is carried over from the precipitation process: for 10% solutions it is ca. 0.1–0.7 mS/cm [48].

Chemical Properties. Poly(vinyl alcohol) undergoes the chemical reactions expected of a polyglycol with secondary hydroxyl groups. The most important are the formation of esters, ethers, and acetals [42, 49, 50]. Only a few typical reactions are mentioned here. Examples of inorganic esters are poly(vinyl nitrates), phosphates, and sulfates. Organic esters can be obtained by reaction with acid anhydrides or acid chlorides. The cinnamate esters are important for the production of light-sensitive polymers [51]. Many poly(vinyl alcohol) ethers have been described. The products of the reaction of poly(vinyl alcohol) with ethylene oxide are suitable for the production of films with good solubility in cold water. Of the reaction products of poly(vinyl alcohol) and aldehydes, poly(vinyl butyrals) have attained importance as raw materials in the manufacture of safety glass.

1.4. Uses, Toxicology, and Economic Aspects

Protective Colloid in Emulsion and Suspension Polymerization. Poly(vinyl alcohol) has proved outstandingly effective as a protective colloid for the production of plastic dispersions, especially of poly(vinyl acetate) homo- and copolymers, and as a suspension stabilizer. By selection of suitable types and combination with emulsifiers, dispersions with extremely varied properties can be produced, most of which are used in adhesives [31, 52–55].

Adhesives. Poly(vinyl alcohol) is used in aqueous solution, alone or in combination with plastic dispersions, for packaging adhesives, cigarette adhesives, and moistenable gumming [56].

Sizes. The most important field of application in the textile industry is the sizing sector. Poly(vinyl alcohol) is an excellent raw material for the sizing of stable fiber yarns and filaments [56–59].

Paper Industry. Applications in the paper industry are surface sizing and the production of coated papers and special papers with specific barrier properties. Poly(vinyl alcohol) has excellent pigment binding power and is a good carrier for optical brighteners [60–63].

Other Uses. Poly(vinyl alcohol) is used for bonding all kinds of nonwoven fabrics, especially glass fiber; as temporary bonding agents for special ceramics; in secondary brighteners in electroplating and electroforming; for the production of protective lacquers and decorator sizing and solvent-resistant dipped products, such as gloves and aprons; for the modification of surface coating formulations; as a release agent for cast resin moldings; in the production of photoresists for the printing industry and of highly absorbent sponges; and in binders and thickeners for cosmetics.

Poly(vinyl alcohol) is also very suitable for the thermoplastic production of packaging films with graded water solubility. Good plasticizers for poly(vinyl alcohol) are ethylene glycol, glycerol, trimethylolpropane, neopentyl glycol, poly(ethylene glycols) with molecular masses up to ca. 400, and ethoxylated phosphate esters. Films of plasticized poly(vinyl alcohols) are practically impermeable to nitrogen, oxygen, and carbon dioxide [64]. Saponification products of vinyl acetate copolymers (e.g., with ethylene or long-chain α-olefins) are also used in the film sector [60, 64].

Toxicology. In testing poly(vinyl alcohol) for toxicity and for compatibility with skin and mucous membranes, no negative effects were found in animals (LD_{50}; 90-d feeding test on rats; FDA patch test; dermal application) [48].

Economic Aspects. The worldwide production capacities (in 10^3 t/a) are as follows:

World	650
Japan	239
United States	127
Western Europe	60

2. Poly(Vinyl Acetals)

Robert Fuss, Florian Mummy

2.1. General Aspects

The first poly(vinyl acetal) was obtained by W. HAEHNEL and W. O. HERRMANN in 1924 by the reaction of poly(vinyl alcohol) with benzaldehyde in concentrated hydrochloric acid [65]. In the following years, more methods were developed, in which the reactions were carried out in concentrated acids and anhydrous solvents. In 1929, H. HOPF and E. KÜHN succeeded in acetalizing aqueous poly(vinyl alcohol) solutions in the presence of dilute acid [66]. Shortly afterwards, G. KRÄNZLEIN et al. reacted poly(vinyl acetate) with aldehydes with simultaneous saponification and acetalization to give poly(vinyl acetals) [67].

Poly(vinyl acetals) are terpolymers consisting of the following repeating units: vinyl acetate, vinyl alcohol, and vinyl acetals. There are poly(vinyl acetals) of different compositions, polarities, and molecular masses.

Poly(vinyl acetals) serve as very versatile and high-performance materials. They are characterized by their excellent mechanical and adhesive properties, optical clarity (of films), and good process ability.

2.1.1. Properties and Analysis

Analytical data typically quoted are:

- Grade/grade name
- Viscosity (commonly as a 10 wt.% solution in ethanol); note: different viscosity measurement systems may be utilized
- Residual hydroxyl content (also known as residual poly(vinyl alcohol)), in wt% and/or mol%
- Residual acetate content, in wt% and/or mol%
- Glass transition temperature T_g
- Solids

Solubility. Poly(vinyl acetals) are soluble in organic solvents such as alcohols, glycol ethers, ethers, Cellosolve, ketones, and others. Depending on their polarity, some grades are even soluble in hydrocarbon compounds such as toluene or xylene (highly acetalized products). Solubility in certain esters is also possible, in this case, ethyl lactate is the solvent of choice. Other solvents are terpineol, butyl carbitol, acetic acid, and dimethyl sulfoxide.

Poly(vinyl acetals) are thermoplastic polymers and may be extruded with or without plasticizers.

Particle size and density. Poly(vinyl acetals) are produced as powders ($\approx$50–$\approx$150 μm particle sizes) or compacted or granular materials and are white (colorless) to slightly yellow. Densities for powders are above 1.0 g/cm^3, but bulk densities are in the magnitude of 0.2–0.4 g/cm^3.

Molecular Mass Distribution (M_n/M_w) There is no existing standard for measuring the molecular masses of poly(vinyl acetals). The reason for this is that there is no standard poly (vinyl acetal) polymer available (different from numerous other polymers). Data quoted in some publications simply refer to PMMA standards. Thus, for the same poly(vinyl acetal) polymer, data varying up to a factor of three can be found.

Viscosities (measured as 10 wt% or 5 wt% ethanolic solutions; higher molecular mass materials are often quoted in 5 wt% solutions) range typically from 10 to 300 mPa s. Some specialty products of either higher or lower viscosity levels may also be available. At the same total solids amount, depending on the solvent used, e.g., ethanol versus isopropyl alcohol, greatly varying solutions viscosities of the same poly(vinyl acetal) are obtained. In addition, by employing the same total solids amount and the same solvent but different dissolving techniques (e.g., stirrer bar, dissolver, rotating rolls), different solution viscosities are encountered. Viscosity data of poly(vinyl acetals) may only be compared if the same solvent, concentration, temperature, solution preparation technology, and viscosity measurement method are employed.

Glass transition temperatures T_g of poly(vinyl acetals) depend on the type of aldehyde or ketone employed. Short aliphatic chains (using e.g. formaldehyde ($T_g = 108°C$), or acetaldehyde) give rise to values $T_g > 100°C$. Longer aliphatic chains (e.g., butyraldehyde) lead to T_g of approximately 63–73°C, also depending on the degree of conversion (at the same chain length of the polymer backbone). The softening point of the polymers vary in the range between 100°C and 240°C (determined by the ring-and-ball method, according to ISO 4625), mostly depending on the length of the polymer backbone.

2.1.2. Synthesis and Reactions

As most aldehydes and many ketones react with vinyl alcohol homo- and copolymers, a large number of poly(vinyl acetals) may be obtained. Poly(vinyl acetals) are specialty chemicals produced only by a limited number of producers. The chemical variety of possible poly(vinyl acetals) is rather high (numerous different poly(vinyl alcohols) are available as well as a variety of different aldehydes and ketones).

Poly(vinyl butyral) is the predominant commercial product. Of minor importance and volumes are poly(vinyl formal) and poly(vinyl acetaldehyde) as well as mixed aldehydes, the latter ones being produced by using two different aldehydes in the same reaction.

The synthesis of poly(vinyl acetals) is somewhat more demanding than that of other polymers, which are produced in very high volumes by, e.g., a single-step polymerization reaction starting from the corresponding monomer (→ Polymerization Processes, 1. Fundamentals).

Characteristic parameters used in the production of poly(vinyl acetals) are:

- Degree of polymerization
- Molecular mass distribution
- Residual hydroxyl and acetate contents
- Degree of acetalization

During the acetalization reaction starting from poly(vinyl alcohol), only the parameters residual hydroxyl content and degree of acetalization change. Degree of polymerization, molecular mass distribution, and residual acetate content remain unchanged and, hence, are determined by the choice of starting poly(vinyl alcohol). The poly(vinyl alcohols) used to produce poly(vinyl acetals) always contain residual vinyl acetate units (generally 1–8 wt%).

Poly(vinyl acetals) may differ in the degree of acetalization, which can vary within broad ranges and strongly influences the polymer properties, especially strength, solubility, and adhesiveness. According to FLORY [68], for statistical reasons, no more than 81.6% of the alcohol groups in poly(vinyl alcohol) can react with aldehydes, given the underlying poly(vinyl alcohol) also contains 1,2-glycol bonds. For acetal formation, 1,3-glycol units react leading to the characteristic cyclic 1,5-acetal. This is the case for most commercial poly(vinyl alcohols).

The acetalization reaction is assumed to be exclusively intramolecular (i.e., it proceeds without cross-linking). Therefore, only in the reaction of poly(vinyl alcohol) with formaldehyde cross-linking can be produced relatively easily, although even in this case, intramolecular acetalization is preferred. In the reaction with higher aldehydes, intermolecular acetalization is much more difficult or impossible [69].

Consequently, single aldehyde- or ketone-based poly(vinyl acetals) are terpolymers of the formula:

[structure: poly(vinyl acetal) repeating units with OAc, OH, and acetal (O–CH–O with CH$_3$ substituent) groups, indices l, m, p]

where l, m, and p can vary over wide ranges, and the alcohol, acetal (here: butyral), and acetate groups are generally distributed statistically. Although indications are that poly(vinyl acetals) with block (partially ordered) structures can be prepared, little is known about such structures.

In the reaction of formaldehyde with poly(vinyl alcohol), isotactic vinyl alcohol units, which give *cis* rings on acetalization, react more readily than syndiotactic vinyl alcohol units, which give more highly strained *trans* rings [70]. Therefore, in the acetalization of poly(vinyl alcohol), the isotactic vinyl alcohol units react first and the syndiotactic units remain. Individual acetalization processes appear to differ, however, so, e.g., poly(vinyl butyrals) of the same empirical composition can vary not only in the blocklike or statistical arrangement of their repeating units but also in the tacticity of the residual vinyl alcohol sequences. The properties of poly(vinyl acetals) are affected as a result. For example, syndiotactic vinyl alcohol polymers exhibit pronounced intermolecular hydrogen bonding, which is the reason for their high strength and poor solubility.

In poly(vinyl butyrals), the ratio between the hydrophobic and hydrophilic polymer units is variable within wide limits. The hydrophobic vinyl butyral units give the polymer good thermoplastic process ability, solubility in numerous solvents, elasticity, and toughness, as well as compatibility with many resins and plasticizers. The hydrophilic vinyl alcohol units are responsible for the high adhesion to inorganic materials such as glass and metals, high strength, cross-linking ability [71], and anticorrosive action.

Examples for cross-linking reactions include reactions with phenolic resins, epoxy resins, melamine resins, polyisocyanates, and dialdehydes.

Like all acetals, poly(vinyl acetals) are able to undergo acid-induced hydrolysis. If oxygen is also present, peroxidic oxidation can cause cleavage of the polymeric chain and increase the acidity of the polymer so that the decomposition accelerates autocatalytically.

Poly(vinyl acetals) can be stabilized effectively by the addition of alkali [72]. Poly(vinyl acetals) are also sensitive to oxidation in neutral or alkaline medium especially in the presence of UV light. In this case, stabilization with antioxidants, in addition to alkali, is advisable. Of the numerous antioxidants proposed [73], phenols and its derivatives have proved effective [74], whereas phosphites [75] or triazole derivatives [76] are effective as co-stabilizers.

Poly(vinyl butyral) powders may be stored over prolonged periods of time if kept under controlled conditions. Appropriate storage conditions consist of a dry, cool and dark environment. Manufacturers therefore recommend a storage period not exceeding 12 months.

2.2. Production

Raw Materials. The principal raw materials for poly(vinyl acetals) are poly(vinyl acetate) or poly(vinyl alcohol) and aldehydes or ketones. Most common aldehydes such as acetaldehyde or *n*-butyraldehyde are used. Basically, all types of aldehydes or ketones may be employed. However, C-1 to C-4 aldehydes have mostly been employed as they are sufficiently miscible with water (see Figs. 4 and 5). Use of formaldehyde has been greatly discontinued as a result of safety reasons (toxicity). Use of long-chain aliphatic aldehydes gives rise to remarkably different properties of the resulting products. In poly(vinyl acetals) produced from long-chain aliphatic aldehydes, compatibility with plasticizers as used for poly(vinyl acetals) film may be positively influenced. In addition, use of plasticizer and its possible migration may be reduced.

The starting materials should be pure and colorless. Commercial products may generally be used without further purification.

The poly(vinyl alcohol) must not contain insoluble components. The aqueous poly(vinyl alcohol) solution must therefore be filtered before acetalization. Although technical-grade formaldehyde can be used directly for acetalization, butyraldehyde should be purified by distillation and stored under nitrogen. It must not contain more than 2% impurities, especially other aldehydes and acids. The inorganic acids used for acetalization should be free of heavy metals.

Production Processes. Although there are a variety of potential methods for the manufacture of vinyl acetale polymers, only two different processes and their variants are commercially used [77–82]. In the *aqueous process*, the acetal is precipitated during the reaction of an aldehyde with poly(vinyl alcohol). In the *solution process*, the acetal formation takes place in an organic solvent. The subsequent reaction cascade of hydrolysis and acetalization can be performed either as a sequential or concurrent process [83].

The production processes are exemplified for poly(vinyl butyral) in Section 2.6.2.

The effect of production conditions on polymer properties is described in [84].

2.3. Uses

Owing to their transparency, elasticity, and good adhesion, poly(vinyl acetals) find various usage such as in laminated glass, as raw materials for surface coatings and primers, as a binder in ceramics and printing inks, and as adhesives. See Sections 2.6.3 and also Section 2.7.4 for a detailed list.

2.4. Environmental Issues, Safety, and Health

Poly(vinyl acetals) consist of carbon, hydrogen and oxygen. Under atmospheric combustion conditions, they burn with virtually no residue to produce carbon dioxide and water. For this reason poly(vinyl acetals) pose no disposal problems.

The U.S. Food and Drug Administration regulated poly(vinyl butyral) resins and classified them as indirect food additives [85, 86]. Poly(vinyl butyral) resin is considered nontoxic by single-dose oral ingestion (LD_{50} > 10.0 g/kg for rats) and is especially suitable for use in food packaging. The use of poly(vinyl butyral) in this area is governed by:

- EU regulation No. 0010/2011 of January 2011 on plastic materials and articles intended for contact with food (Plastic Implementation Measures, PIM Regulation),
- Council of Europe, Resolution RESAP (2004) 1, formerly AP (96) 5, on surface coatings intended for contact with foodstuffs (list of authorized monomers and starting substances in Appendix II, List 1),
- U.S. Food and Drug Administration 21 CFR
  - § 175.105 Adhesives
  - § 175.300 Resinous and polymeric coatings
  - § 176.170 Components of paper and paperboards in contact with aqueous and fatty foods.

Unformulated poly(vinyl butyral) and poly (vinyl formal) resins have flash points above 370°C. The lower explosive limit for PVB dust in air is 20 g/m^3.

2.5. Economic Aspects

Capacity. Global production capacities for poly(vinyl butyrals) are in the magnitude of 200 000 t/a (2014). Approximately 90% of the poly(vinyl butyral) production is plasticized and subsequently extruded into poly(vinyl butyral) films to be used for architectural safety glass and automotive applications. Minor use concerns applications in photovoltaics. As poly(vinyl butyrals) films contain plasticizers, the global production capacities for poly(vinyl butyral) films are in the magnitude of 225 000 t/a.

Less than 10% of the resins are used for non-interlayer applications.

Many producers are backward integrated and produce their own poly(vinyl alcohol) raw materials to be used for the production of poly(vinyl acetals).

Depending on grade in the United States and Europe, the price for nonfilm poly(vinyl butyral) resin is in range of € 5–20 per kilogram (2014).

Producers of poly(vinyl butyral) resin for nonfilm applications are Kuraray (Mowital, Pioloform) and Eastman (Butvar), as well as Chang Chun, Kingboard, and Micro Inks. Major producers of poly(vinyl butyral) interlayers are Kuraray (Trosifol, Butacite), Sekisui (S-LEC) and Eastman (Saflex) and since 2006 Chang Chun (Winlite).

The regional consumption of poly(vinyl butyral) is pictured in Figure 3. A total of 60% is utilized in Western Europe and the United States of America. In Asia Japan and China are the largest users. In the recent years China showed the largest consumption growth now reaching a total of some 12% of the global demand.

Commercial Products. Poly(vinyl butyrals) are commercially the most important poly(vinyl acetals). They are produced in numerous types, differing in molecular masses and composition. Additivation is possible but not always stated. Additives used concern oxidation and UV-light stabilization.

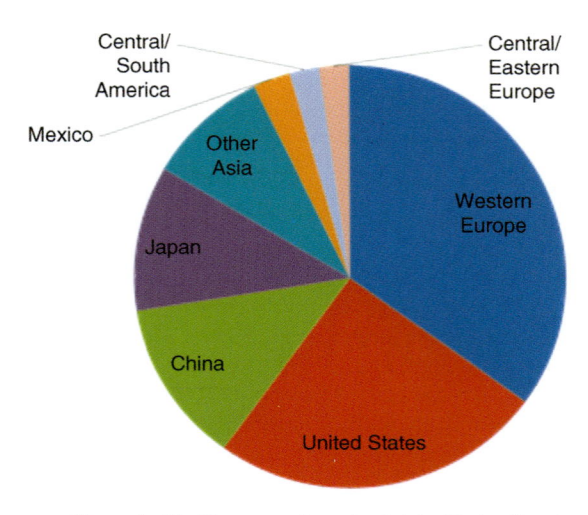

Figure 3. World consumption of poly(vinyl butyral) resin (2012)

Poly(vinyl butyral) is mostly sold as a powder. Other forms of delivery include compacted powders and granules. Next to bags (10–15 kg), deliveries in super sacks are also common. The most important manufacturers are Kuraray (Mowital, Pioloform), Eastman (Butvar), Chang Chun and Sekisui (S-lec).

As raw material for conversion into films, poly(vinyl butyral) is sold by a limited number of producers including Kuraray, DE, US, KR (Trosifol, Butacite); Eastman, US (Saflex), and Sekisui, JP (S-Lec), Chang Chun Group, TW, (Winlite) as well as Kingboard, CN.

Other forms of commercial product consist of partially acetalized poly(vinyl alcohols) that are still water-soluble. Such products are sold, e.g., from Sekisui under their trade name S-LEC K.

High-plasticizer containing poly(vinyl acetal) dispersions are also commercially available.

2.6. Poly(Vinyl Butyral)

2.6.1. Properties

Poly(vinyl butyrals) are amorphous and transparent. At low degrees of acetalization they are water-soluble. Like vinyl alcohol–vinyl acetate copolymers, they are good dispersants [87]. Also like these copolymers, their solubility in water decreases with increasing temperature [88].

Table 1 lists the composition and some average properties of various poly(vinyl butyrals), derived from product information pamphlets [85, 86, 89]. It roughly demonstrates the effects of molar mass on rheological, thermal, and mechanical properties. In detail, the softening temperature is between 105 and 155°C and increases accordingly with the molecular mass. Also the stiffness (determined by the tensile strength) increases with the chain length of the polymeric backbone. In addition, the effect of the proportion of vinyl alcohol units in the polymer is evident: with increasing vinyl alcohol content the intermolecular forces increase, and the strength and modulus of elasticity rise as do the softening point and the glass transition temperature.

Table 1. Properties of poly(vinyl butyrals) with various percentages of vinyl alcohol units and various molar masses [85, 86, 89]

Vinyl alcohol units, wt%	Molar mass, g/mol[a]	Viscosity, mPa s[b]	Water absorption in 24 h,%[c]	Tensile strength, 10^3 psi[d]	Elongation at break,%[e]	Flow temperature, °C[f]	Modulus of elasticity, 10^5 psi[g]	Glass transition temperature, °C[h]
17.5–20.0	170–250	1600–2500	0.5	7.0–8.0	70	145–155	3.3–3.4	72–78
17.5–20.0	120–150	800–1300	0.5	7.0–8.0	75	135–145	3.3–3.4	72–78
11.5–13.5	90–120	200–450	0.3	4.6–5.6	110	110–115	2.8–2.9	62–72
18.5–20.5	70–100	200–400	0.5	5.7–6.7	100	125–130	3.0–3.1	72–78
11.0–13.5	50–80	75–200	0.3	4.6–5.6	110	110–115	2.8–2.9	62–72
18.0–20.0	40–70	75–200	0.5	5.6–6.6	110	105–110	3.1–3.2	72–78

[a] Molecular mass (mass average in thousands) determined by size exclusion chromatography with low-angle laser-light scattering (SEC/LALLS) in THF.
[b] 10% by weight: 95% ethanol at 25°C using an Ostwald-Cannon-Fenske Viscometer.
[c] According to ASTM D 570-59aT.
[d] According to ASTM D 638-58T.
[e] According to ASTM D 638-58T.
[f] According to ASTM D 569-59.
[g] According to ASTM D 638-58T.
[h] Determined by differential scanning calorimetry over a range of 30–100°C on dried granular resin.

2.6.2. Production

For the commercial production of poly(vinyl butyrals), only those processes that start from poly(vinyl alcohol) isolated before acetalization are significant. Two important processes exist. One begins with aqueous poly(vinyl alcohol) solutions, which are reacted with n-butyraldehyde in the presence of mineral acid. After passage through a highly viscous intermediate stage, the polymer precipitates as fine grains. The reaction is then completed heterogeneously.

In the second process, an alcoholic poly(vinyl alcohol) suspension is acetalized. A homogeneous poly(vinyl butyral) solution is obtained that can be freed from insolubles by filtration.

Aqueous Process. In the aqueous process (Fig. 4) [77, 78], poly(vinyl alcohol) is dissolved in water and acidified with a mineral acid and reacted with butyraldehyde. As the condensation reaction proceeds and the acetals are formed, the number of hydroxyl groups present in the starting poly(vinyl alcohol) is reduced. This leads to a continuously decreasing solubility of the newly formed polymer in water. Finally the product produced becomes completely water-insoluble; the degree of acetalization at this stage is approximately 30%. However, once the product precipitates, the desired conversion is not yet accomplished. After a certain post-reaction time, sometimes in combination with heat treatment, this now heterogeneous system brings the reaction to completion. The final product is obtained as an aqueous dispersion of fine particles. This production process is characterized by a batch as well as continuous mode.

The prepared poly(vinyl alcohol) solution is charged into an acetalization vessel (up to 50 m^3 in size) and the process is batchwise. There are usually a number of such acetalization vessels running in sequences. Subsequently, the obtained acidic product slurry is worked up continuously. This work-up procedure includes thorough washing to remove the acid employed. Note: Acetals are formed under acidic conditions, however, they are not stable under acid conditions (back reaction). Being washed to neutral poly(vinyl acetals), they may be stabilized by adding an alkali, such as, e.g., sodium or potassium hydroxide. Finally the powdery material is dried and packed. Both normal bags as well as super sacks are available. In some cases, other packaging units such as full containers with polymeric inliners are used.

Solution Process. In the solution process (Fig. 5) [79–81], poly(vinyl acetate) is saponified by transesterification in the presence of ethanol and a mineral acid catalyst to obtain poly(vinyl alcohol). The ethanol and formed

Figure 4. Aqueous process
PVOH = Poly(vinyl alcohol); PVB = Poly(vinyl butyrate)

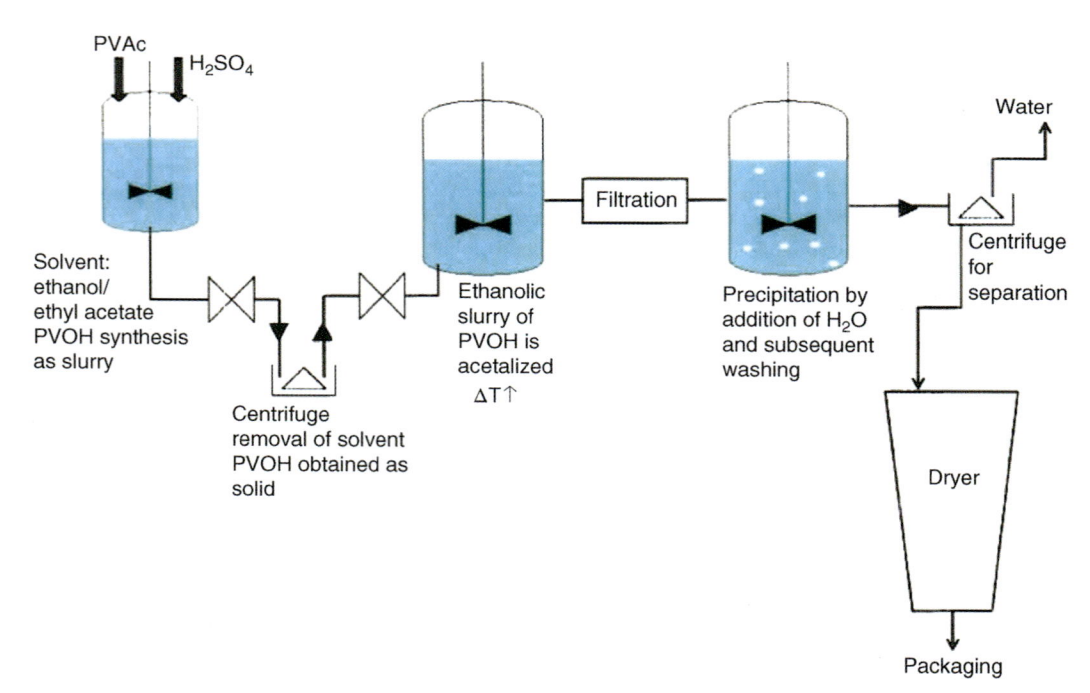

Figure 5. Solvent-based process
PVOH = Poly(vinyl alcohol); PVAc = Poly(vinyl acetate), ΔT = Rising temperature

ethyl acetate are removed from the solid by centrifugation. As the final poly(vinyl butyrate) polymer needs to have a low acetate content, the acetalization reaction cannot be performed concurrently. Poly(vinyl alcohol) is acetalized in a separate reaction unit, after being redispersed with ethanol, heated with butyraldehyde and the acid catalyst. As the acetalization proceeds, the obtained poly(vinyl acetal) becomes more and more soluble in the solvent used. Finally a homogeneous poly(vinyl butyral) solution is obtained that can be freed from insolubles by filtration. Upon completion, the obtained poly(vinyl acetal) is removed from the solvent. The major aspects of this process are the costly precipitation of the polymer and the need to recover the solvent.

In both processes the acid catalyst is neutralized upon completion of the reaction using typically sodium or potassium hydroxide, and several washing steps are necessary to remove traces of the catalyst, salts and to achieve a more alkaline resin pH to improve the resin's thermal stability [77, 90, 91].

Acetalization in Water. A small-scale synthesis may be conducted as follows:

An aqueous solution of poly(vinyl alcohol) with a polymer concentration of about 9% is heated to 95°C and slowly cooled. During cooling, butyraldehyde is added at 40°C and at a temperature of about 12°C, the hydrochloric acid (16%) is slowly added. The product starts precipitating already during acid dosage.

After 15 min, the suspension is heated to 70°C and held at this temperature for 2 h to complete the acetalization reaction. The suspension is filtered off and washed several times with deionized water at room temperature.

For stabilization, the washed polymer is suspended in water containing sodium carbonate and stirred at 70°C for 2 h. Finally, the product is rinsed several times with water and dried at 50°C overnight in a vacuum oven.

2.6.3. Uses

Laminated Glass [92]. The most important use of poly(vinyl butyral) (PVB) films is manufacture-laminated safety glass [93].

Such PVB films consist of approximately 80–70% PVB and 20–30% plasticizer. Residual hydroxyl groups mostly determine the polarity of the polymer thus resulting in adequate compatibilities of the two components. Their compatibility with plasticizers such as esters of phthalic, sebacic, ricinoleic, and citric acids is good [94]. Other important plasticizers in PVB films are ethers and esters of ethylene glycol, and esters of its oligomers with butyric or hexanoic acid. Dihexyl adipate is also used as a plasticizer for PVB films [95]. The commonly used plasticizer Hexamoll DINCH (1,2-cyclohexane dicarboxylic acid diisononyl ester) in PVB films is even suitable for food contact applications. Currently the most frequently used plasticizer in PVB films is triethylene glycol-bis-2-ethylhexanoate (3G8) [94-28-0] (suppliers: Oxea, Germany; Eastman, US).

The refractive indices of such PVB films (refractive index: 1.48) are in the same range of those of glass (1.46–1.52). Owing to the residual hydroxyl groups, adhesion to polar substrates is pronounced. Because of their flexibility, both curvatures of glass and little glass imperfections may be offset using PVB films [96].

The safety functionality derives from a well-balanced adhesion of the film to glass [97]. Upon impact dissipation of energy is achieved as the PVB film delaminates from the glass, thus reducing the velocity (energy) of the intruding object. If the sheet adheres too firmly to the glass, it is overstretched and tears when the glass breaks resulting in puncturing of the composite. Therefore, the adhesion must be adequately controlled so that when the glass breaks the sheet detaches slightly and a larger area absorbs the impact energy. The adhesion is fine-tuned by use of anti-adhesion additives. Suitable compounds are, e.g., various potassium salts such as potassium acetate [98]. However, the adhesion must remain sufficiently high to hold splinters of glass which have become completely detached from the laminate.

In automotives, next to windscreen applications sidelight as well as sun and moon roof applications are fields of use. In addition, switchable (smart) windows are on the way. In the manufacture of automobile windshields a sheet thickness of approximately 0.8 mm is generally used, whereas for structural or bulletproof glass, thinner but multiple sheets may be utilized. The surface of the sheet is structured by various techniques [99]. This reduces the tendency to stick together and facilitates deaeration during production of the glass composite. To completely prevent sticking, the sheet is either dusted with sodium hydrogen carbonate, which must be washed off prior to producing the composite, or cooled below 10°C.

In manufacturing the composite [100], sheets are laid between two or more sheets of glass and subjected in an autoclave to a special pressure and temperature program [101].

Plasticized PVB films for the production of automobile windshields and structural or bulletproof glass are produced, for example, by the following companies: Kuraray (Trosifol, Butacite), Eastman (Saflex), and Sekisui Chem. Ind. (S-lec).

Advanced-Performance Poly(Vinyl Acetal) Films. Next to safety-glass applications poly(vinyl acetal) films providing additional features or functionalities have been developed. Such additional functionalities include special reflective properties by means of doping poly(vinyl acetal) films with IR reflective nanoparticles. Whereas the optical transparency is hardly impaired, reflection of sunlight is achieved yielding less headload in, e.g., vehicles.

Single and multi-layered poly(vinyl acetal) films have been made available focusing on superior sound isolation in buildings and cars.

A further advancement is concerned with switchable windows (smart windows) [102]. Such systems are divided into active (switchable) and passive (self-switching/automatic) systems. In both systems transparent and translucent devices are possible.

In self-switching systems the window changes transparency due to increased (outside) temperatures (thermochromic windows). Tinting is completely reversible and upon cooling returns to non-tinted status (Pleotint, US).

Active systems are switchable by, e.g., change of a potential differences. Switching may be complete (full tinting) or interrupted at any intermediate state (electrochromic; suspended particle devices or gasochromic systems) (Switchmaterials, CAN; EControl-Glas GmbH & Co. KG, DE; SageGlass, US; View Inc.; US).

Poly(vinyl acetals) are used either as matrix material for such compositions (e.g., in the case of thermochromic) and are laminated between, e.g., glass. Other systems use both, poly(vinyl acetal) material as a matrix material as well as poly(vinyl acetal) film for (glass) lamination [103].

Surface Coatings and Primers. Because of their elasticity and good adhesion, PVBs are important raw materials for paints, particularly for metal coatings. They are frequently combined with phenolic, alkyd, epoxy, urea, or melamine resins. The low-acetalized and lower molecular mass types, in particular, show good compatibility with these resins. When the coatings are heat-treated ("baked"), cross-linking occurs [104].

PVBs are outstandingly effective as primers with high corrosion protection. The coatings are abrasion resistant and adhere very well. The coated plates can be welded easily [86, 105].

Wash primer formulations with PVB form a good foundation for various types of topcoats [106–108]. Wash primers provide good adhesion to metal surfaces and stabilize the metal surface by continuously releasing of anti-corrosion ions. Wash primers with phosphoric acid and zinc chromate show very good anti-corrosion performance. However, owing to toxicity concerns related to chromates, other zinc-containing pigments have been developed. Mostly a system consisting of zinc molybdate [109] and phosphate pigments [110] is recommended. The problems associated with volatile organic solvents can be eliminated, as there are waterborne PVB dispersions that can also be employed in wash primer formulations [111].

Adhesives. (→ Adhesives, 2. Applications) Poly(vinyl acetal) resins are mainly consumed in high-performance thermosetting adhesives and in thermoplastic hot-melt formulations. PVB is also found in reflective films for license plates and road signs where PVB is the matrix for the glass beads [112–114]. PVB in formulations with plasticizers, waxes, and/or combination with other co-resins is utilized to make hot-melt adhesives [115]. Structural adhesives can be obtained when PVB is cured with, e.g., epoxy resins to give a thermoset material.

Ceramics **[116].** As a temporary binder poly(vinyl acetal) is used in the manufacture of especially electronic ceramics. Next to superior dispersion, purity and binding properties a very reproducible burning out behavior (shrinkage) favors the use of poly(vinyl acetal). Poly(vinyl butyral) is completely burned out above 450°C (depending on molecular mass and atmosphere employed, N_2 vs. O_2).

Of major importance for electronic ceramic applications are sufficiently high mechanical stabilities for ever thinner ceramic layers in the green sheet stage (non-burned-out state; e.g., manufacturing of multi-layered ceramic capacitors MLCCs) as well as nonresidual (ash-free) burning out performance. Presence of foreign matters and metal ions derived from the temporary binder may render electronic ceramics unusable.

Printing Inks and Toners. PVB resin is an important binder in the formulation of printing inks [117]. Due to good solubility in mild solvents it is applied in several solvent-based formulations like for flexographic, gravure, screen and inkjet printing applications. As a binder PVB improves flexibility, adhesion, and toughness [106, 118–120]. Flexographic and gravure printing are the dominant fields in the market for printing of flexible packaging that are mainly processed in the food processing industry.

In addition to printing inks PVB is also beneficial as binder for toner formulations [106, 121]. The role of PVB is to increase viscosity, to improve the film formation in the fuser station and to prevent blocking. PVB also enhances the overall performance of the toner by minimizing the amount of fines without decreasing the flow properties.

X-Ray Imaging. Owing to their remarkable optical properties, poly(vinyl acetals) are also employed in medical devices such as X-ray imaging films. After exposure to irradiation such films are mostly developed dry. Such films consist of numerous layers where different poly(vinyl acetals) provide next to perfect optical performance additional features such as adhesion or dispersion (e.g. of silver particles). Companies producing such products include Carestream, US; Konica-Minolta, JP as well as Agfa, BE.

Poly(Vinyl Butyral) Dispersions. Poly(vinyl butyral) may be dispersed in the aqueous phase only with huge amounts of emulsifiers and dispersants. This way pollution of the environment by solvents is avoided. Nevertheless, such dispersions have minor importance because of the mentioned high emulsifier content, which impairs the adhesive properties and water resistance of coatings. Poly(vinyl butyral) dispersions can be produced by acetalization of aqueous poly(vinyl alcohol) solutions in the presence of emulsifier or by emulsification of poly(vinyl butyral). The latter process is commercially feasible. The utilization of higher molecular mass poly(vinyl acetals) is preferred, which are kneaded in the presence of high amounts of plasticizers and dispersants by shear mechanical force. Once a homogeneous mixture is obtained, careful dilution with water follows until the final dispersion is obtained. Next to virgin poly(vinyl acetal) also post-consumer poly(vinyl acetals) are utilized in the manufacture of poly(vinyl acetal) dispersions. The achieved quality levels are strongly depending on the starting materials employed. Usually the poly(vinyl acetal) content of such dispersions is in the magnitude of some 25%. High energy consumption as well as tough production conditions and equipment renders this production process to a rather niche one (producers: PPG, US; Eastman, US; Shark Solutions, DK).

In addition partly acetalized and hence water-soluble poly(vinyl acetals) are also commercially available. These waterborne products differ greatly in their performance when compared to fully, non-water-soluble poly(vinyl acetals).

Single-Serve Packaging. Unfavorable volume-to-surface ratios are obtained for small (food) packaging containers. Packaging of (oxygen) sensitive products requires gas barrier materials for sufficient product protection. Use of a combination of gas barrier providing polymers such as poly(vinyl alcohol) and poly(vinyl acetal) have been made available. Whereas the barrier material is water-sensitive, the poly(vinyl acetal)-based top coat provides sufficient moisture and mechanical protection to the container. The container materials may be based on polar (e.g., poly(ethylene terephthalate) or oriented poly(ethylene therephthalate)) or nonpolar (PE, PP) polymers. Barrier performance systems are available for various climatic conditions of use (producer: Container Corporation of Canada, CAN) [122, 123].

Other uses for which poly(vinyl butyrals) have been proposed are temporary, strippable coatings [124]; powder coatings [125]; and blends with other polymers.

2.7. Poly(Vinyl Formal)

The importance of poly(vinyl formal) is limited.

2.7.1. Production

Since the commercially important poly(vinyl formals) contain a relatively large number of vinyl acetate units, a single-stage process for manufacturing directly from poly(vinyl acetate) is particularly suitable.

In the processes developed in the United States [126, 127], 1000 parts by weight of poly(vinyl acetate) are dissolved in 2000 parts of pure acetic acid; a mixture of 875 parts of 30% formalin and 70 parts of concentrated sulfuric acid is added; and the mixture is stirred for 6 h at 85°C. After the mixture is cooled to 50°C and the sulfuric acid is neutralized, poly(vinyl formal) is precipitated with water and washed. A pale yellow powder is obtained. Instead of acetic acid, Hoechst [128] recommended the use of ethyl acetate, optionally mixed with methanol.

By altering the proportions of starting materials and varying reaction conditions, numerous types of poly(vinyl formal) can easily be produced. For the reaction mechanism, see [129, 130].

2.7.2. Commercial Products

Poly(vinyl formals) were marketed by Wacker-Chemie under the name Pioloform F and by Monsanto Chemical Company in St. Louis, Missouri under the name Formvar. The latter manufacturing unit of Monsanto was sold and Formvar is now distributed under the name "Vinylec" by SPI Supplies. The Wacker unit was purchased by Kuraray in 2008 and production of poly(vinyl formals) discontinued.

2.7.3. Properties [85, 131]

Commercial poly(vinyl formal) powders in contrast to formalized fibers from poly(vinyl alcohol), are not cross-linked. Individual types differ in molecular mass, which is usually between 15 000 and 45 000, and in the proportion of vinyl acetate units, which is usually 10 – 30 wt% of the polymer. The proportion of vinyl alcohol units is between 5 and 9 wt%.

The viscosity of polymer solutions increases not only with increasing molecular mass but also with increasing poly(vinyl alcohol) content. At the same time, the glass transition temperature increases from approximately 75 to 95°C. The softening temperature is between 140 and 170°C, depending on molecular mass. The tensile strength of the polymers is up to 70 MPa with a fracture strain of 30–70%. Depending on the type of stabilization, polymers can withstand temperatures of 120 – 150°C for up to one hour. The solubility in polar solvents increases with increasing proportion of poly(vinyl acetate). Poly(vinyl formals) are insoluble in hydrocarbons and alcohols, but soluble in mixtures of alcohols and aromatic hydrocarbons.

2.7.4. Uses

The most important use of poly(vinyl formal) is in the production of coatings for electrical wires [85], in which the poly(vinyl formal) is combined with phenolic resins. The coatings, which are cured at elevated temperatures are insoluble, tough, and resistant to abrasion and heat, and also have very good insulating properties [132, 133]. Its use in adhesive joints [91, 134] and metal coating [85] is less important.

The use of poly(vinyl formal) for foamed plastics has frequently been proposed. To produce such foamed plastics, an aqueous solution of poly(vinyl alcohol) is foamed and formalized in the presence of a water-soluble substance such as starch or sodium chloride. After washing out, an insoluble soft and spongy material is obtained [135, 136].

2.7.5. Toxicity

Poly(vinyl formals) are nontoxic provided there are no traces of residual formaldehyde.

Poly(vinyl formals) are therefore permitted for coatings that come in contact with food [85, 131].

2.8. Other Poly(Vinyl Acetals)

The numerous aldehydes that have been reacted with poly(vinyl alcohol) include furfural, benzaldehyde, crotonaldehyde, halogenated aldehydes, cyanopropionaldehyde, nitrobenzaldehyde, and glyoxylic acid. However, such poly(vinyl acetal) products have not reached economic importance. Poly(vinyl acetaldehyde acetal), on the other hand, temporarily played a role in Germany and Canada, but was displaced by poly(vinyl butyral) [137].

Studies of poly(vinyl acetals) from aliphatic C_2–C_8 aldehydes with different acetal contents in the presence and absence of plasticizers [138] have shown that the glass transition temperature decreases with increasing chain length of the aldehyde component. At the same time the poly(vinyl acetals) become softer and more hydrophobic.

Of the poly(vinyl acetals) from substituted aldehydes, the sodium salt of poly(vinyl butyral sulfonic acid) was produced commercially but has long since been abandoned (Afrilan, Hoechst). The substance was water-soluble and was employed in sizing. It was produced by passing sulfur dioxide into an aqueous solution of poly(vinyl alcohol) in the presence of crotonaldehyde at 0°C and subsequent heating the solution. Sulfurous acid was added to crotonaldehyde simultaneously with acetalization, and the sulfonic acid formed acted as the catalyst for the acetalization, so that further addition of acid was unnecessary [139].

3. Poly(N-Vinyllactams) and Poly(N-Vinylamines)

3.1. Polyvinylpyrrolidone

When *N*-vinylpyrrolidone and other *N*-vinyl compounds became commercially available by Reppe vinylation of lactams in about 1940, the polymers were soon produced on an industrial scale. Polyvinylpyrrolidone is used as an auxiliary in cosmetics and pharmaceuticals.

N-Vinylpyrrolidone

Production. Only radical polymerization is important. *N*-vinylpyrrolidone (*mp* 13.9°C, *bp* at 1.3 kPa 90 – 93°C) is usually polymerized batchwise in aqueous solution with hydrogen peroxide as initiator and regulator. The average molecular mass can be adjusted by varying the quantity of hydrogen peroxide: with 0.05% hydrogen peroxide, molecular masses of ca. 750 000 are obtained; with 3%, ca. 25 000. With increasing quantity of hydrogen peroxide the polymerization rate increases. The rate also increases with increasing concentration of *N*-vinylpyrrolidone in the aqueous solution, reaches a maximum at 40 – 60% monomer content, and then decreases. Due to the viscosity of the final polymer solution the concentration is adjusted within a range between 10% at high molecular mass and 60% at low. To control polymerization, the monomer solution must be added slowly to the polymerization mixture with cooling. Polymerization is complete after 6–12 h. The ready saponifiability of *N*-vinylpyrrolidone in aqueous acidic solution is a problem. Aqueous solutions containing monomeric *N*-vinylpyrrolidone should therefore always be neutral to basic. The pH value is best controlled with ammonia, which also activates the polymerization reaction [140].

Polyvinylpyrrolidone can be obtained as a white powder from polymer solutions by drum or spray drying. Because it is highly hygroscopic it must be protected from moisture during extended storage.

Polymerization can also be carried out in organic solvents. The initiators preferably used then are organic peroxides or azocompounds, preferably alkyl hydroperoxides, dialkyl peroxides, and peroxy esters [141]. The polymerization temperature is chosen according to the decomposition temperature of the initiator, but the polymerization rate also depends on the nature of the solvent [142, 143].

Bulk polymerization [144] and polymerization in suspension [145] have no commercial importance.

N-Vinylpyrrolidone can be copolymerized with many other vinyl compounds such as acrylic acid [144, 146, 147], methacrylates esters [140, 148], and methacrylamide [149]; for the Q and e values, see [150]. The most important are the polymers with vinyl esters [149, 151, 152], whose properties lie between those of the pure polymers.

When N-vinylpyrrolidone is heated for several hours with hydroxides and alkoxides of the alkali and alkaline earth metals [153], an insoluble polymer slightly swellable by water is formed in a spontaneous reaction. This polymer is also formed on heating N-vinylpyrrolidone with divinyl compounds in the absence of atmospheric oxygen [154] and is known as popcorn polymer [155].

Commercial Products. Polyvinylpyrrolidones with an average molecular mass of 2500 to 750 000 are on the market.

Trade names are Kollidon, Luviskol, Albigen A, and Divergan (BASF); PVP and Plasdone (General Aniline and Film Corp.); Collacral and Luviskol VA (copolymers with vinyl esters, BASF); PVP/VA (copolymers with vinyl acetate, General Aniline and Film Corp.); Polyclar and Polyplasdone XL (insoluble polyvinylpyrrolidone, General Aniline and Film Corp.).

Properties. Polyvinylpyrrolidone is a white powder. The glass transition temperature depends on the water content; extrapolation to 0% water content gives a glass transition temperature of 175°C. The mean molecular mass can be adjusted in the range of 1000 to 10^6.

Polyvinylpyrrolidone is highly hygroscopic, absorbing ca. 30% water at 60% humidity. It dissolves readily in water, and aqueous solutions are stable to electrolyte addition. The viscosity of the solution decreases with increasing temperature and with strong shearing [156].

Polyvinylpyrrolidone dissolves readily in polar solvents such as alcohols, amines, acids, and chlorinated hydrocarbons. It is insoluble in esters, ethers, ketones, and hydrocarbons. Completely anhydrous polyvinylpyrrolidone is also soluble in toluene.

Polyvinylpyrrolidone can be precipitated from aqueous solution by dropwise addition to acetone and from solution in organic solvents by addition of diethyl ether.

Polyvinylpyrrolidone has good compatibility with numerous film formers, water-soluble binders, and plasticizers. On solution casting, a clear, high-gloss, hard film is formed.

The polymer is chemically inert. The lactam group is saponified only by the action of concentrated acids, with the formation of poly[vinyl (γ-amino)butyric acid] [157].

With some substances such as iodine, polyphenols, tannin, dyes, and toxins, polyvinylpyrrolidone forms complexes [158]. Frequently, only a reversible association occurs. The protective colloid action is utilized in dispersions and in suspension polymerization.

For qualitative analysis [159], aqueous ammonium cobalt thiocyanate solutions are added to an acidic aqueous polyvinylpyrrolidone solution. The flocculation of a bright blue complex indicates polyvinylpyrrolidone. The IR spectrum is also suitable for identification. To determine concentration, the brownish red polyvinylpyrrolidone–iodine complex can be analyzed colorimetrically.

Uses. In cosmetics, polyvinylpyrrolidone and some copolymers, especially with vinyl acetate, are used as film formers in setting lotions and hairsprays. Fairly high molecular mass polyvinylpyrrolidone is used as a thickening agent and protective colloid in cosmetic emulsions. Its use as a blood plasma substitute has been abandoned.

In pharmacy, polyvinylpyrrolidone has a wide spectrum of applications: as a solubilizer, as a crystallization retarder, for detoxification, for reducing the irritant action and toxicity of certain substances, as a tablet binding and coating agent, as a suspension stabilizer, and as a dispersant for pigments in tablet-coating suspensions. The polyvinylpyrrolidone (PVP) complex with iodine, PVP-iodine, is used as a disinfectant and does not show the side effects of elementary iodine [160]. The insoluble and slightly water-swellable popcorn polymer is used as a tablet-disintegrating agent.

Polyvinylpyrrolidone and PVP-iodine are described in a series of pharmacopoeias (e.g., in the U.S.).

Since polyvinylpyrrolidone forms complexes with tannins it is used for the clarification of beer and other beverages. The cross-linked, insoluble popcorn polymer is particularly suitable [161].

Polyvinylpyrrolidone forms hard, transparent, strongly adherent films on glass, metal, plastics, and cellulose, and is used in the adhesives industry as a binder and water-soluble hotmelt adhesive.

Polyvinylpyrrolidone is also used as an auxiliary in textile finishes, as a dye acceptor for synthetic fibers, as a leveling and stripping agent for dyes, as a thickener for printing inks and latex paints, as a dispersant in laundry detergents, as a protective colloid in the emulsion and suspension polymerization of many polymers, and as a water-binding agent for the concentration of protein solutions [162].

Toxicology. Polyvinylpyrrolidone is chemically and biologically largely indifferent [163, 164] with regard to both acute toxicity and skin irritation.

In feeding tests on rats with radioactively labeled polyvinylpyrrolidone of average molecular mass 40 000, more than 99% was excreted through the intestinal tract [163]. Even when large amounts of polyvinylpyrrolidone were fed for years, no harm to experimental animals was established. Swelling of the lymph nodes, which can be interpreted as a transition stage in the excretion of polyvinylpyrrolidone, was observed only with dogs; no indications of accumulation or extensive degenerative changes to organs were detected.

Carcinogenicity of polyvinylpyrrolidone was not observed in any of the experimental procedures [165]. On the basis of extensive experience on humans as well as on the results of further experimental studies on animals with parenteral or oral administration, no evidence of carcinogenic risk to humans was found [166].

More than one million people have shown good tolerance for polyvinylpyrrolidone infusions as a blood plasma substitute without side effects. However, the high molecular mass fractions of polyvinylpyrrolidone, with molecular masses above 40 000, are excreted slowly or not at all. Polyvinylpyrrolidone is therefore no longer used for infusions because the low molecular masses types do not have a plasma-expander action. Animal experiments have shown that larger molecules are stored in the cells of the reticuloendothelial system of the spleen, the liver, and the lymph nodes, as well as in bone marrow. However, during such polyvinylpyrrolidone storage, neither morphological nor functional damage has been observed [167].

On the basis of inhalation tests on female volunteers with six different kinds of hairspray that caused no pathological changes of any kind [168], the FDA reported that the normal use of aerosol hairsprays containing PVP and PVP – vinyl alcohol as film formers is not accompanied by a health risk.

3.2. Polymers of other *N*-Vinyllactams

In the addition to polyvinylpyrrolidone, the polymers of other *N*-vinyllactams such as *N*-vinylpiperidone [162] and *N*-vinylcaprolactam [169] are known.

N-Vinylpiperidone   *N*-Vinylcaprolactam

These polymers have better solubility than polyvinylpyrrolidone in organic solvents such as aromatic hydrocarbons, ketones, and aliphatic hydrocarbons. Polyvinylcaprolactam is readily soluble in cold water: the higher its average molecular mass, the better is its solubility. In warm water (30–40°C), polyvinylcaprolactam precipitates from solution.

Vinylcaprolactam can be polymerized in bulk [170] or in solution. The polymers are used for gluing woven synthetic textiles and as components of lubricating oils [171]. The copolymers with ethylene are used as coatings for textiles and paper, and as adhesives [172].

3.3. Polyvinylamines

Poly(N-vinylcarbazole) is a transparent thermoplastic material. The glass transition temperature is 211°C and the glass point 173°C. It starts to decompose only above 300°C [173].

Vinylcarbazole

Poly(N-vinylcarbazole) dissolves in aromatic and chlorinated hydrocarbons and in tetrahydrofuran. It swells in benzene-containing fuels and is insoluble in water, dilute acid and alkali, aliphatic hydrocarbons, and alcohols.

The electrical properties are unusual. Photoconductivity depends on the molecular mass [174]; doping effects are observed upon addition of various substances such as 1,5-diaminonaphthalene, tetracyanoethylene, or anthracene [175]. The existence of traps for electronic charge carriers (electrons and defect electrons) has been demonstrated [176]. For the mechanism of conductivity, see [177].

Before polymerization, N-vinylcarbazole (mp 63–64°C, bp at 0.3 kPa 155°C) should be purified by distillation or recrystallization. When handling N-vinylcarbazole, safety instructions should be observed [178], because it has very strong sensitizing properties. No harmful side effects are known for monomer-free poly(N-vinylcarbazole).

For radical polymerization in bulk, azo initiators or peroxides are used [179, 180]. Suspension polymerization in water [181] or methanol [182] is also known. Cationic polymerization of N-vinylcarbazole proceeds easily, but whether polymerization in the presence of benzoyl peroxide occurs via a radical or a cationic route is unknown [183]. Cationic polymerization has a rate constant five times higher than radical polymerization and is used for the production of low molecular mass polymers [184].

Radical copolymerization with other vinyl compounds such as styrene, acrylonitrile, vinyl esters, and acrylate esters has been described, and a Q value of 0.28 and an e of -1.49 are given [185]. The stereospecific polymerization of N-vinylcarbazole is also known [186].

Because its softening point is ca. 200°C, polyvinylcarbazole was formerly used for moldings that are exposed to elevated temperature. Fibers and films have also been manufactured from polyvinylcarbazole [187]. Today, it is used as a heat-resistant insulator in radio and radar technology and in capacitor manufacture [188]. Because of its photoconductivity it is important for the manufacture of electrostatic photocopiers and television recording tubes [189].

Commercial Product: Luvican (BASF).

Polyvinylimidazole. Because of the second nitrogen atom, polymers of N-vinylimidazole have a basic character. The solubility therefore depends on the degree of protonation. Unquaternized polyvinylimidazole is soluble in organic solvents such as alcohol, but in water it is only swellable. It dissolves in acids, with salt formation. Quaternized polyvinylimidazole is water soluble. Depending on the length of the alkyl group used in quaternization, polymers with different hydrophobic–hydrophilic character are obtained [190].

N-Vinylimidazole (bp 190–193°C) and N-vinylimidazoles quaternized with alkyl halides or dialkyl sulfates can be polymerized by the radical mechanism using azo initiators and peroxides. Copolymerization with acrylamide, acrylates, butadiene, styrene, N-vinylpyrrolidone, and vinyl acetate has been described: values of 0.1 and 0.24 have been given for Q, and -0.91 and -1.73 for e [191].

N-Vinylimidazole

Polyvinylimidazole and copolymers are used as hairdressing preparations, antistatic agents for hydrophobic plastics, liquid detergents, and in the production of hydrogels.

Other Vinylamines. Because of their instability, aliphatic vinylamines can be prepared only with difficulty, and the corresponding polyvinylamines are obtainable only by indirect routes [192]. Cyclic amines such as carbazole and imidazole, indole, and pyrrole, on the other hand, can easily be converted to *N*-vinylamines by Reppe vinylation.

Polymers of 2-vinylpyridine and 4-vinylpyridine have also been produced. 4-Vinylpyridine can be copolymerized with other vinyl compounds such as styrene and vinyl chloride; values of 1.91 for *Q*, and −0.51 for *e* are given [193]. Polyvinylpyridines have been studied as polymeric reagents in organic synthesis, catalysts, and ion exchangers.

References

1 Chem. Forschungsgemeinschaft, DE-OS450286 (W.O. Herrmann, W. Haehnel).
2 Chem. Forschungsgemeinschaft, DE642531, 1937 (W.O. Herrmann, W. Haehnel, H. Berg); *Chem. Abstr.* **31** (1937) 59 059.
3 H. Staudinger, K. Frey, W. Starck, *Ber. Dtsch. Chem. Ges.* **60** (1927) 1782.
4 Du Pont, Elvon Hydroxyvinyl Resins, Technical Information, Wilmington, Del., 1969.
5 K. Noro: "Preparation of Modified Polyvinyl Alcohols from Copolymers," in: C. Finch (ed.): *Polyvinylalcohol*, Wiley Interscience, New York 1973, pp. 147 ff.
6 K. Noro, H. Takida, *Kobunshi Kagaku* **19** (1962) 261; *Chem. Abstr.* **58** (1963) 4650 c.
7 K. Fujii, S. Imoto, J. Ukida, M. Matsumoto, *J. Polym. Sci. Part B* **1** (1963) 497.
8 T. Ito, K. Nomo, *Kobunshi Kagaku* **15** (1968) 310; *Chem. Abstr.* **54** (1960) 8140 d.
9 J. Ukida, R. Naito, *Kogyo Kagaku Zasshi* **58** (1955) 717; *Chem. Abstr.* **50** (1956) 8245 h.
10 M. Shiraishi, *Kobunshi Kagaku* **19** (1962) 676.
11 K.R. Tubbs, *J. Polym. Sci. Polym. Chem. Ed.* **4** (1966) 623.
12 S. Hayashi, C. Nakano, T. Motoyama, *Kobunshi Kagaku* **21** (1964) 300; *Chem. Abstr.* **62** (1964) 9244 g.
13 S. Hayashi, C. Nakano, T. Motoyama, *Kobunshi Kagaku* **22** (1965) 354; *Chem. Abstr.* **63** (1965) 16476 h.
14 J.R. Member, H.C. Haas, R.L. MacDonald, *J. Polym. Sci. Part B* **10** (1972) 385.
15 T. Moritani, J. Kuruma, K. Shibatani, Y. Fujiwara, *Macromolecules* **5** (1972) 577.
16 J.F. Kenney, G.W. Willcocksen, *J. Polym. Sci. Polym. Chem. Ed.* **4** (1966) 679; H.N. Friedlander, H.E. Harris, J.G. Pritchard, **4** (1966) 649; H.E. Harris et al., **4** (1966) 665.
17 C.W. Bunn, *Nature (London)* **161** (1966) 929.
18 S. Hayashi, C. Nakano, T. Motoyama, *Kobunshi Kagaku* **20** (1963) 303; *Chem. Abstr.* **61** (1964) 5802 b.
19 I. Sakurada, Y. Nukushima, Y. Sone, *Kobunshi Kagaku* **12** (1955) 506.
20 T. Yamamoto, S. Konagaya, A. Yamamoto, *J. Polym. Sci. Polym. Lett. Ed.* **16** (1978) 7.
21 S. Nozakura, S. Kida, *J. Polym. Sci. Polym. Chem. Ed.* **12** (1974) 2337.
22 Consort. f. Ind. Elektrochem., DE1299879; FR1361830, 1964; DE-AS 1962; *Chem. Abstr.* **62** (1965) 6591 b.
23 F. Kainer, *Polyvinylalkohole*, Enke Verlag, Stuttgart 1949, p. 11.
24 C.E. Schildknecht, *Vinyl and Related Polymers*, Wiley Interscience, New York 1952, pp. 323 ff.
25 A.A. Vanscheidt, L. F. Chelpanova, *J. Gen. Chem. USSR Engl. Transl.* **20** (1950) 2261; *Chem. Abstr.* **45** (1951) 4482.
26 J.P. Flory, F.S. Leutner, *J. Polym. Sci.* **5** (1950) 267.
27 K. Imai, U. Maeda, *Kobunshi Kagaku* **16** (1959) 222.
28 Monsanto, BE636139, 1964 (R.J. Kern); *Chem. Abstr.* **61** (1964) 16187 d.
29 J. Sakurada, Y. Sakaguchi, K. Kashimoto, *Kobunshi Kagaku* **18** (1961) 694; *Chem. Abstr.* **56** (1962) 14461.
30 Y. Sakaguchi, Z. Sawada, M. Koizumi, K. Tamaki, *Kobunshi Kagaku* **23** (1966) 890; *Chem. Abstr.* **66** (1967) 65995 k.
31 K. Noro, *Br. Polym. J.* **2** (1970) 128.
32 F. Gregor, E. Engel, *Chem. Prum.* **10** (1960) 53.
33 Du Pont, US2657201, 1953 (R.W. Nebel); *Chem. Abstr.* **48** (1954) 1066 a.
34 H. Shohota, "Continuous Polymerization of Vinylacetate for Polyvinylalcohol Production," in C.A. Finch (ed.): *Properties and Applications of Polyvinyl Alcohol, SCI Monogr.* **30** (1968) 18.
35 Shawinigan Chemicals, US2643994, 1953 (L.M. Germain); *Chem. Abstr.* **47** (1953) 9058 e.
36 Eastman Kodak, US2642419, 1953 (G.P. Waugh, W.O. Kenyon); *Chem. Abstr.* **47** (1953) 7925 h.
37 M.K. Lindemann in G.E. Ham (ed.): *Vinyl Polymerization*, vol. **1**, Marcel Dekker, New York 1967, p. 252.
38 A. Hill, D.K. Hale, *Biostat Tech. Rep. (Amherst)*, no. 1418 (1946).
39 K. Noro, "Manufacturing and Engineering Aspects of the Commercial Production of Polyvinylalcohol," in [61] p. 121.
40 G. Heck, A. Schmidt, *Chem. Ing. Tech.* **47** (1975) 541.
41 T. Suziki, Y. Ichahara, M. Yamada, K. Tonumura, *Agric. Biol. Chem.* **37** (1973) 747.
42 M.K. Lindemann, *Encycl. Polym. Sci. Technol.* **14** (1971) 149.
43 H.C. Haas, A.S. Makas, *J. Polym. Sci.* **46** (1960) 524.
44 R. Naito, *Kobunshi Kagaku* **15** (1958) 597; *Chem. Abstr.* **54** (1960) 16113 i.
45 Farbwerke Hoechst, DE1111819, 1959 (G. Lohaus); *Chem. Abstr.* **56** (1962) 4973 g.
46 K. Toyoshima, "General Properties of Polyvinyl Alcohol in Relation to its Application," in [61], p. 17.
47 T. Motoyama, S. Okamura, *Kobunshi Kagaku* **11** (1954) 23; *Chem. Abstr.* **50** (1956) 2200 f.
48 Hoechst, Druckschrift über Mowiol, Frankfurt, Sept. 1976, D 1.
49 C.A. Finch, "Chemical Properties of Polyvinylalcohol," in [61] p. 183.
50 J.G. Pritchard, *Poly(vinylalcohol)*, Gordon & Breach Sci. Publ., London 1970, p. 81.
51 M. Tsuda, *J. Polym. Sci. Part B* **1** (1963) 215.
52 M. Shiraishi, *Br. Polym. J.* **2** (1970) 135.
53 H. Lamont, *Adhes. Age* **16** (1973) 24.
54 A.S. Dunn, C.J. Tonge, S.A.B. Anabtawi, *Polym. Prepr. Am. Chem. Soc. Div. Polym. Chem.* **16** (1975) 223.
55 M. Shiraishi, K. Toyoshima, *Br. Polym. J.* **5** (1973) 419.
56 K. Toyoshima, "Applications of Polyvinylalcohol in Adhesives," in [61] p. 413.
57 E.P. Czerwin, *Mod. Text. Mag.* 1966, 22.
58 C.R. Blumenstein, *Text. Ind. (Atlanta)* **130** (1966) 63.
59 V. Heap, *Text. Inst. Ind.* **11** (1973) 172.
60 H.L. Jaffe, *Pap. Trade J.* **147** (1963) no. 9, 44.

61. H. Schaefer, H.G. Oesterlin, *Wochenbl. Papierfabr.* **102** (1974) 335.
62. R. Vesanto, H. Schaefer, D. Wolf, *Pap. Puu* **57** (1975) 399; *Chem. Abstr.* **83** (1975) 117430 v.
63. H.G. Oesterlin, H. Schaefer, *Papier (Darmstadt)* **32** (1978) V 13.
64. K. Toyoshima, "Properties of Polyvinyl Alcohol Films," in [61] p. 339.
65. Consortium f. elektrochem. Industrie, DE480866, 1929 (W. Haehnel, W.O. Herrmann).
66. I. G. Farbenindustrie, DE683165, 1939 (H. Hopff, E. Kühn); *Chem. Abstr.* **36** (1942) 38799.
67. I. G. Farbenindustrie, DE692988, 1940 (A. Voss, V. Starek); *Chem. Abstr.* **45** (1941) 45203; DE737630, 1943; *Chem. Abstr.* **38** (1944) 36647.
68. P.J. Flory, *J. Am. Chem. Soc.* **61** (1938) 1518.
69. H. Kawase, *Kogyo Kagaku Zasshi* **74** (1971) 1228; *Chem. Abstr.* **75** (1971) 119111 a.
70. K. Shibatani et al., *J. Polym. Sci. Part C* **23** (1968) 647.
71. Kuraray Europe GmbH, EP1606325 B1, 2008 (B. Papenfuss, M. Steuer, M. Gutweiler).
72. S. Okamura, T. Motoyama, *Bull. Inst. Chem. Res. Kyoto Univ.* **30** (1952) 45; *Chem. Abstr.* **47** (1953) 6175 e. The Fiberloid, GB503634, 1939; *Chem. Abstr.* **33** (1939) 69927; Monsanto Chem., US2258410, 1942 (J. Dahle); *Chem. Abstr.* **36** (1942) 5922.
73. J. Schreiber, *Chemie und Technologie der künstlichen Harze*, 2nd ed., vol. **1**, Wissenschaftliche Verlagsgesell., Stuttgart 1961, p. 468.
74. General Electric, US2195122, 1940 (B.W. Nordlander); *Chem. Abstr.* **34** (1940) 52063.
75. Farbwerke Hoechst, DE-OS2208167, 1973 (E. Schmidt); *Chem. Abstr.* **80** (1974) 15739 y.
76. Monsanto, US3823113, 1974 (A.J. Reisman); *Chem. Abstr.* **83** (1975) 60333 j.
77. Shawinigan Resins Corp., US2496480, Feb. 7, 1950 (E. Lavin, A.T. Marinaro, W.R. Richard).
78. *Chem. Eng. (N.Y.)* **61** (Feb. 1954) 122, 123, 346–349.
79. E. I. du Pont de Nemours & Co., Inc., US2400957, May 28, 1946, and US2422754, June 24, 1947 (G.S. Stamatoff).
80. R.D. Dunlop, FIAT Final Report No. 1109, U.S. Government Printing Office, Washington, D.C., 1947.
81. E. I. du Pont de Nemours & Co., Inc., US3153009, Oct. 13, 1964 (L. H. Rombach).
82. Kuraray Europe GmbH, EP2730591 A1, 2014 (M. Meise, C. Lang).
83. N. Platzer, *Mod. Plast.* **28** (June 1951) 142.
84. O.V. Piastro, L.L. Ezhenkova, N.J. Tyazhlo, M.E. Rozenberg, *Plast. Massy* 1970 no. 2, 13; *Chem. Abstr.* **72** (1970) 133504 a. N.I. Tyazhlo et al., *Zh. Prikl. Khim. (Leningrad)* **47** (1974) 2285; *Chem. Abstr.* **82** (1975) 58806 b.
85. Monsanto, Butvar, Polyvinylbutyral and Formvar, Polyvinylformal, Techn. bull. 6070 and 6130, 1969.
86. Kuraray Europe GmbH, Polyvinyl butyral, Product Information, 3rd ed., Hattersheim 2013.
87. Farbwerke Hoechst, DE-OS1770580, 1968 (H. Fritz, F. Sorg); *Chem. Abstr.* **72** (1970) 133690 h.
88. L.D. Taylor, B. Biasotti, *J. Appl. Polym. Sci.* **20** (1976) 1721.
89. Sekisui Chem. Ind., S-lec B Polyvinylbutyral Resin, Technical data, 1977.
90. Monsanto Chemical Co., US2258410, Oct. 7, 1941 (J. Dahle).
91. E. I. du Pont de Nemours & Co., Inc., US2282026, May 5, 1942 (B.C. Bren, J.H. Hopkins, G.H. Wilder); US 2282057, May 5, 1942 (J.H. Hopkins, G.H. Wilder).
92. HT Troplast AG, EP1181258 B1, 2002 (U. Keller, B. Koll, H. Stenzel).
93. Hüls Troisdorf AG, EP185863 B1, 1989 (H. Pabst).
94. *Houben-Weyl, Methoden der organischen Chemie*, 4th ed., vol. **14/2**, Georg Thieme, Stuttgart 1971, p. 717.
95. Monsanto, DE-OS2429032, 1973 (R.H. Fariss, J.A. Snellgrove); *Chem. Abstr.* **82** (1975) 157268 j.
96. Dynamit Nobel AG, DE3132509 C2, 1983 (H. Brinkmann, H. Pabst).
97. Kuraray Europe GmbH, EP2548727 A1, 2013 (M. Meise, M. Frank, J. Beekhuizen, U. Keller).
98. Monsanto, DT1596895, 1965; US3271233, 1966 (E. Lavin, G. E. Mont); *Chem. Abstr.* **65** (1966) 20323 g.
99. Kuraray Europe GmbH, US8263208 B2, 2012 (H. Stenzel).
100. Kuraray Europe GmbH, Trosifol Manual, 6th ed., Troisdorf, 2012.
101. *Glass* **54** (1977) 190.
102. IDTechEx report, Smart Windows and Smart Glass 2014-2024: Technologies, Markets, Forecasts; www.IDTechEx.com/glass 30 November 2014.
103. Kuraray Europe GmbH, Gesimat GmbH, EP1647033 B1, 2003 (H. Stenzel, A. Kraft, K.-H. Heckner, M. Rottmann, B. Papenfuhs, M. Steuer).
104. E. Lavin, J.A. Snelgrove in *Adhesives*, 2nd ed., Van Nostrand-Reinhold, New York 1977, p. 507.
105. A. Steen, *Ind. Lackierbetr.* **44** (1976) 367.
106. Solutia, Inc., Butvar Polyvinyl Butyral Properties & Uses, Online Technical Bulletin 2008084D, St. Louis, Mo., 1999.
107. J.D. Scantlebury, F.H. Karman, *Corros. Sci.* **35** (1993) 1305.
108. Military Specifications, DOD-P-15328D and MIL-C-8514C (ASG), Information Handling Services (HIS), Englewood, Colo.
109. J.L. Nogueira, *Corros. Prot. Mater.* **11** (1992) 11.
110. T. Foster, G.N. Blenkinsop, P. Blattler, M. Szandorowski, *J. Coat. Technol.* **63** (1991) 91.
111. M. Gerlitz, E. Supper, *Surf. Coat. Int., Part A: Coat. J.* **84** (2001) 389.
112. 3M Co., EP223564, Aug. 28, 1991 (T.R. Bailey, R.R. Kult, L.C. Belisle).
113. 3M Co., EP360420, Mar. 27, 1996 (B.B. Wilson, R.E. Grunzinger).
114. 3M Co., US6221496, Apr. 24, 2001 (Y. Mori).
115. M. Schatz, K. Salz, J. Volek, CZ155587, Dec. 15, 2001.
116. Kuraray Europe GmbH, EP1854772 B1, 2013 (M. Frank, R. Fuss).
117. BASF, DE-OS2538097, 1977 (G. Reuss); *Chem. Abstr.* **86** (1977) 156973 f.
118. Wacker Polymer Systems, Pioloform B Polyvinyl Butyrals, Technical Bulletin 5567E, Burghausen, Germany, Jan. 2001.
119. Monsanto, US3951882, Apr. 20, 1976 (A.H. Markhart, D.R. Cahill).
120. Sandoz DE-AP2547862, May 6, 1976 (K. Taubert).
121. Sekisui Chemical Co., JP08328303, Dec. 13, 1996, and JP08292596, Nov. 5, 1996 (H. Minamino, T. Takahashi, K. Noguchi).
122. Kuraray Europe GmbH, Container Corporation of Canada, EP2431409 B1, 2014 (N.J. Gottlieb, R. Fuss).
123. Kuraray Europe GmbH, Container Corporation of Canada, EP2824133 A1, 2015 (M. Frank, R. Fuss, N.J. Gottlieb).
124. M.P. Portyanko, V.F. Yavorovskaya, A.E. Martinovich, G.M. Lishanova, *Tekhnol. Organ. Proizvod* **14** (1937) no. 45; *Chem. Abstr.* **80** (1974) 97372 k.
125. A.D. Yakovlev, I.S. Okhrimenko, *Mashinostroitel* 1973 no. 9, 20; *Chem. Abstr.* **84** (1976) 6535 k.

126 General Electric, US2085995, 1935 (W.I. Patnode, E.J. Flynn); *Chem. Abstr.* **31** (1937) 59029.
127 Shawinigan Chemicals, US2168827, 1934 (G.O. Morrison, A.F. Price); *Chem. Abstr.* **33** (1939) 94901.
128 Farbwerke Hoechst, DE878861, 1951 (H.J. Hahn); *Chem. Abstr.* **51** (1957) 2324 g; DE 889 367, 1951 (W. Fitzky).
129 G.N. Kormanovskaja, I.N. Vlodavec, *Izv. Akad. Nauk. SSSR, Ser. Khim.* **10** (1964) 1748; *Chem. Abstr.* **62** (1965) 4130 f.
130 M. Chanda, *Angew. Makromol. Chem.* **62** (1977) 229.
131 M.K. Lindemann, *Encycl. Polym. Sci. Technol.* **14** (1971) 208.
132 General Electric, US2307588, 1943 (E.H. Jackson, R.W. Hall); *Chem. Abstr.* **37** (1943) 35343.
133 A.F. Fitzhugh, E. Lavin, G.O. Morrison, *J. Electrochem. Soc.* **100** (1953) 351.
134 L. Dimter, H. Schulz, K. Thinius, *Plaste Kautsch.* **9** (1962) 318.
135 T. Goldschmidt, DE-OS2503288, 1976 (R. Mitgan, D. Schedlitzki, H. Wacker); *Chem. Abstr.* **85** (1976) 161476 e.
136 Kanebo, Tokio, DE2323968, 1972 (T. Koide et al.).
137 H. Schindler, *Kunstharz Nachr.* **9** (1975) 22.
138 *Kirk-Othmer*, 2nd ed., **21**, 304.
139 A.F. Fitzhugh, R.N. Crozier, *J. Polym. Sci.* **8** (1952) 225.
140 BASF, DE922378, 1943 (W. Reppe, K. Herrle, H. Fikentscher); *Chem. Abstr.* **52** (1958) 19252 e.
141 BASF, DE-OS2439196, 1974 (K. Herrle, W. Denzinger, K. Seelert); *Chem. Abstr.* **84** (1976) 165420 j.
142 J.W. Breitenbach, A. Schmidt, *Monatsh. Chem.* **83** (1952) 1288.
143 E. Senogles, R. Thomas, *J. Polym. Sci. Polym. Symp.* **49** (1975) 203.
144 I. G. Farbenindustrie, DE757355, 1939 (C. Schuster, R. Sauerbier, H. Fikentscher).
145 General Aniline & Film, DE-OS2602917, 1976 (D.H. Lorenz, E.P. Williams, H.S. Schultz); *Chem. Abstr.* **86** (1977) 56214 r.
146 H. Uelzmann, *J. Polym. Sci.* **33** (1958) 377.
147 S. Ponratnam, S. L. Kapur, *J. Polym. Sci. Polym. Chem. Ed.* **14** (1976) 1987.
148 BASF, DE954197, 1956 (H. Fikentscher, H. Wilhelm); *Chem. Abstr.* **53** (1959) 13674 c.
149 J.F. Bork, E. Coleman, *J. Polym. Sci.* **43** (1960) 413.
150 B. Vollmert, *Polymer Chemistry*, Springer Verlag, New York 1973, p. 139.
151 K. Hayashi, G. Smets, *J. Polym. Sci.* **27** (1958) 275.
152 D.J. Kahn, H.H. Horowitz, *J. Polym. Sci.* **54** (1961) 363.
153 General Aniline & Film, US2938017, 1956 (F. Grosser); *Chem. Abstr.* **54** (1960) 20335 f.
154 BASF, DE-OS2437629, 1976 (W. Denzinger, E. Hoffmann, K. Herrle); *Chem. Abstr.* **84** (1976) 165586 t.
155 W. Breitenbach, H.F. Kauffmann, *Angew. Makromol. Chem.* **45** (1975) 167.
156 A. Nakano, Y. Minoura, *J. Appl. Polym. Sci.* **21** (1977) 2877.
157 Chemical-Technological Institute, Moskau, SU471370, 1975 (A.S. Tevlina, V.G. Chelnokov, M.M. Mardanyan, G.V. Makarov); *Chem. Abstr.* **83** (1975) 9873 u.
158 B. Wurzschmitt, *Fresenius Z. Anal. Chem.* **130** (1949/50) 128.
159 U. Müller, *Pharm. Acta Helv.* **43** (1968) 108.
160 H.A. Shelansky, M.V. Shelansky, *J. Int. Coll. Surg.* XXC (1956) no. 6, 727.
161 BASF, DE-OS2437640, 1976 (W. Denzinger, E. Hoffmann, K. Herrle); *Chem. Abstr.* **84** (1976) 165587 u.
162 P.C. Brown, R. Cousden, *Chem. Ind. (London)* 1955, 1452.
163 BASF, Brochure B 359 d, Ludwigshafen, 1976, p. 29.
164 W. Wessel, M. Schoog, E. Winkler, *Arzneim. Forsch.* **21** (1971) 1468.
165 L.W. Burnette, *Proc. Sci. Sect. Toilet Goods Assoc.* **38** (1962) 1.
166 H. Zeller, BASF Bericht WNT, Ludwigshafen, 19.01. 1976.
167 O. Fresen, W. Weese, *Beitr. Pathol. Anat. Allg. Pathol.* **112** (1952) 47.
168 J.H. Draize, *Proc. Sci. Sect. Toilet Goods Assoc.* **31** (1959) 28.
169 N. Cobianu, S.D. Vasilescu, S. Matache, *Mater. Plast. Chem. Abstr.* **10** (1973) no. 2, 75; *Chem. Abstr.* **79** (1973) 66822 g.
170 O.F. Solomon, M. Corciovei, C. Boghina, *J. Appl. Polym. Sci.* **12** (1968) 1843.
171 General Aniline & Film, US3287272, 1966 (W. Katzenstein); *Chem. Abstr.* **66** (1967) 48127 m.
172 Bayer, DT1392354, 1964.
173 J. Pielichowski, *J. Therm. Anal.* **4** (1972) 339.
174 K. Tanikawa, *Makromol. Chem.* **176** (1975) 3025.
175 K. Okamoto, *Bull. Chem. Soc. Jpn.* **46** (1973) 2613.
176 H. Bauser, *Kunststoffe* **62** (1972) 192.
177 P.J. Rencroft, *J. Polym. Sci. Polym. Phys. Ed.* **10** (1972) 2305.
178 I.R. Tabershaw, J.B. Skinner, *J. Ind. Hyg. Toxicol.* **26** (1944) 313.
179 BASF, DE931731, 1953 (H. Fikentscher, R. Frikker); *Chem. Abstr.* **52** (1958) 12458 g; DE936421, 1955 (H. Fikentscher, R. Fricker); *Chem. Abstr.* **52** (1958) 1685 a.
180 British Oxygen, GB1034309, 1966 (L.P. Ellinger); *Chem. Abstr.* **65** (1966) 9104 f.
181 Ricoh, NL6905628, 1968.
182 General Aniline & Film, DE-OS2111293, 1971 (E.V. Hort); *Chem. Abstr.* **76** (1972) 46676 x.
183 J.C. Bevington, *Makromol. Chem.* **178** (1977) 2741.
184 General Aniline, DE-OS2111294, 1971 (E.V. Hort); *Chem. Abstr.* **76** (1972) 46678 z.
185 J. Negulesen, D. Feldmann, *Polymer* **13** (1972) 149.
186 K. Okamoto et al., *Macromolecules* **9** (1976) 645.
187 Montecatini Societa Generale per l'Industria Minerali e Chimica, GB914418, 1963; *Chem. Abstr.* **58** (1963) 9252 f.
188 Standard Telephones & Cables, GB1007040, 1965 (E.H. Hornisch); *Chem. Abstr.* **64** (1966) 836 a.
189 Kalle, DE1127218, 1962 (H. Hoegl); *Chem. Abstr.* **58** (1963) 138 c.
190 J.C. Salamone, *J. Polym. Sci. Polym. Symp.* **45** (1974) 65.
191 J.C. Salamone, P. Taylor, B. Smider, *J. Polym. Sci. Polym. Chem. Ed.* **13** (1975) 161.
192 G. Franzmann, H. Ringdorf, *Makromol. Chem.* **177** (1976) 2547.
193 K. Matsuoka, M. Otsuka, K. Takemoto, M. Imoto, *Kogyo Kagaku Zasshi* **69** (1966) 137; *Chem. Abstr.* **65** (1966) 15515 f.

Further Reading

V. Bühler, *Polyvinylpyrrolidone Excipients for Pharmaceuticals*, Springer, Berlin, Heidelberg 2010.

V. Goodship, D. Jacobs, *Polyvinyl Alcohol*, Rapra Technology, Shawbury 2006.

F.L. Marten, "Vinyl Alcohol Polymers", *Kirk Othmer Encyclopedia of Chemical Technology*, 5th ed., John Wiley & Sons, Hoboken, NJ.

E.M. Petrie, *Handbook of Adhesives and Sealants*, 2nd ed., McGraw-Hill, New York, NY, 2007.

Poly(Vinyl Esters)

HELMUT RINNO, Hoechst Aktiengesellschaft, Frankfurt/Main, Federal Republic of Germany

1. Introduction 1165
2. Raw Materials................. 1165
3. Polymer Structure 1167
4. Production 1167
4.1. Bulk Polymerization 1167
4.2. Solution Polymerization 1168
4.3. Suspension (Bead) Polymerization . 1168
4.4. Emulsion Polymerization 1168
5. Industrially Important Poly (Vinyl Esters) 1170
6. Properties 1171
7. Quality Specifications, Trade Names, Storage and Transport, Wastewater 1172
8. Uses, Toxicology, and Economic Aspects 1173
   References.................... 1174

1. Introduction

Poly(vinyl esters) are polymers with structural units of the general formula:

$$-CH_2CH{-}{\left[\,CH_2CH{-}\right]}_n{-\!-\!-\!-\!-}CH{-}CH_2{-}$$
$$\quad\;\;|\qquad\qquad|\qquad\qquad\qquad|$$
$$\quad\;\;O\qquad\qquad O\qquad\qquad\qquad O$$
$$\quad\;\;|\qquad\qquad|\qquad\qquad\qquad|$$
$$\;O{=}C{-}R\quad\;O{=}C{-}R\qquad\;O{=}C{-}R$$

They are synthesized by polymerization of vinyl ester monomers $CH_2= CH-O-CO-R$.

In addition to vinyl ester homopolymers, co- and terpolymers of mixed vinyl esters or of vinyl esters and other comonomers are also important industrially.

Poly(vinyl acetate) and vinyl acetate copolymers are the most important poly(vinyl esters).

In 1912 and 1913 KLATTE synthesized the first poly(vinyl esters): poly(vinyl chloroacetate) (R = CH_2Cl) and poly(vinyl acetate) (R = CH_3) [1].

2. Raw Materials

Vinyl Esters. Vinyl acetate is predominantly produced by acetoxylation of ethylene with acetic acid on palladium catalysts. Vinyl esters of $\geq C_3$ acids are produced by addition of acetylene to the acid, or by palladium- or mercury-catalyzed transvinylation of vinyl acetate (see → Vinyl Esters).

On an industrial scale vinyl acetate (R = CH_3), vinyl propionate (R = CH_2CH_3), vinyl pivalate [R = $C(CH_3)_3$], vinyl 2-ethylhexanoate [R = $CH(C_2H_5) - (CH_2)_3CH_3$], vinyl laurate [R = $(CH_2)_{10}CH_3$], VeoVa 10 [R = $C(CH_3)R'_2$] and VeoVa 9 [R = $C(CH_3)R'_2$] are used for the production of polymers. VeoVa 9 and VeoVa 10 are the vinyl esters of versatic acids — synthetic tertiary carboxylic acids with nine or ten carbon atoms (see → Carboxylic Acids, Aliphatic, Chap. 10.). Previously, vinyl chloroacetate, vinyl isobutyrate, vinyl caproate, vinyl isononanate, vinyl stearate, VeoVa 911, and vinyl benzoate were also used in industry for a time or were envisaged for this purpose.

Data for the industrially most important monomers are summarized in Table 1. High-purity monomers are required for the reproducible production of polymers. The content of other components can be determined by gas chromatography. The purity determination of vinyl acetate is standardized in ASTM standards [2]. Impurities which are particularly disruptive are those which severely inhibit polymerization, such as crotonaldehyde and vinylacetylene, or compounds that act as chain-transfer agents, such as acetic acid, acetaldehyde, acetone, benzene, or toluene [3], [4]. Impurities with two readily copolymerizable double bonds, such as

Ullmann's Polymers and Plastics: Products and Processes
© 2016 Wiley-VCH Verlag GmbH & Co. KGaA, Weinheim
ISBN: 978-3-527-33823-8 / DOI: 10.1002/14356007.a22_001

Table 1. Physical data for industrial vinyl esters

	Vinyl acetate	Vinyl propionate	Vinyl pivalate	Vinyl 2-ethyl-hexanoate	VeoVa 9	VeoVa 10	Vinyl laurate
CAS no.	[108-05-4]	[105-38-4]	[3377-92-2]	[94-04-2]	[54423-67-5]	[51000-52-3]	[2146-71-6]
Vinyl ester content, %	$\geq 99.9^a$	$\geq 99.9^b$	99.4^c	99.5^c	99.4	> 98.0	99.4 ± 0.1
Water (DIN 51 777), %	$\leq 0.03^a$	$\leq 0.1^b$	< 0.10	< 0.10	$< 0.1^d$	$\leq 0.1^d$	< 0.005
Acid content, %	≤ 0.0005	$\leq 0.03^b$	< 0.10	< 0.10	< 0.3	≤ 0.2	$< 0.5^e$
Color number (ISO 6271)	≤ 5	$\leq 10^b$	20	< 20	< 20	< 15	≤ 10
Inhibitor content, ppm	3 – 20	10^b	9 – 13	6 – 10	5 ± 2^d	5 ± 2^d	without^e
bp, °C	$72 – 73^a$	ca. 95	112.1	185.6	185 – 200	136 (13.3 kPa)	254
Vapor pressure, kPa							
at 20 °C	12	4.58	2.412	0.057	< 0.1	< 0.1	
at 50 °C	42.6	19.63					0.01
at 70 °C					1.2	0.7	0.38
Flash point (DIN 51 755), °C	– 8	4.5^i	14^i	65	45	75^h	136
Upper/lower explosion limit (vapor in air at 20 °C), vol %	2.6/13.4	$1.8^j – 13.5^k$	n.d.	n.d.	0.7/n.d.	n.d.	0.5/n.d.
Solubility in water at 25 °C, %	0.9^f	0.6	0.08	< 0.01	< 0.1	< 0.1	< 0.1
Solubility of water in vinyl ester at 25 °C, %	2.3^f	0.8	n.d.	n.d.	< 0.06	< 0.05	< 0.1
Heat of polymerisation, kJ/kg	1035.8	882			ca. 522	ca. 485	n.d.
T_g of homopolymer^g, °C	33	– 7 (33)	+ 70 (33)	– 36 (33)	60 (28)	– 3 (28)	$– 53^l$

^{a}Hoechst specification.
^{b}BASF specification.
^{c}Union Carbide Specification.
^{d}Shell specification.
^{e}Values from Wacker Chemie.
^{f}At 20 °C.
^{g}Values depend on measurement method; values in parentheses give T_g of poly(vinyl acetate) when same method is used.
^{h}DIN 51 758 (PM).
^{i}Tag closed cup (ASTM D56 – 61).
^{j}At 4 °C.
^{k}At 41.5 °C.
^{l}Extrapolated from copolymer data. n.d. = not determined.

vinyl crotonate, also cause problems because they can lead to cross-linking reactions and thus to the formation of insoluble polymers. Monomers used for emulsion polymerization must be completely free from previously formed polymer.

To prevent polymerization during storage and transport, 3 – 20 ppm of a phenolic compound such as hydroquinone or hydroquinone monomethyl ether is used as an inhibitor. The inhibitors generally need not to be removed before polymerization, since they can be rendered inactive during polymerization by using a sufficiently large amount of initiator.

To assess the purity and the suitability for polymer production, particularly in the case of vinyl acetate, a simple polymerization test [5] is used. A given quantity of monomer is warmed with 0.5 % benzoyl peroxide (relative to monomer) to 70 °C under defined conditions, and the time elapsing until the start of polymerization is recorded. The delay in comparison with pure, inhibitor-free monomer is a measure of the impurity content or of the amount of inhibitor added for stabilization.

Initiators. In industrial poly(vinyl ester) processes polymerization is initiated by radical initiators. Other processes, such as photopolymerization, radiation-induced polymerization, and the use of organometallic compounds as initiators have not achieved importance.

Azo compounds (e.g., azobisisobutyronitrile), diacyl peroxides (e.g., dibenzoyl peroxide and dilauryl peroxide), peroxyesters (e.g., tert-butyl peroxybenzoate and tert-butyl peroxyoctanoate), hydroperoxides (e.g., tert-butyl hydroperoxide and cumene hydroperoxide), and water-soluble inorganic peroxides (e.g., hydrogen peroxide, potassium, ammonium, or sodium peroxodisulfate) are used as radical initiators. Redox systems are also widely used, in which a peroxy compound is combined with a reducing agent (e.g., H_2O_2– ascorbic acid; H_2O_2– Fe^{2+}, ammonium peroxodisulfate – sodium sulfite, or *tert*-butyl hydroperoxide – sodium formaldehyde sulfoxalate) [6].

Initiators are chosen according to their solubility in the solvent or monomer, and the decomposition half-life of the substance or the redox combination under polymerization conditions.

Sufficient initiator must be used to oxidize any inhibitors present in the monomer, and to ensure a constant reaction rate in spite of continuous decomposition of the peroxide under the reaction conditions. High radical concentrations result in polymers of low molecular mass. Generally 0.1 – 1 % of an initiator based on the monomer is used. The initiator can be added at the start of the polymerization in one portion or added gradually during the course of the reaction.

Molecular Mass Regulators. To regulate the molecular mass of the polymer, C_1 – C_4 alcohols, C_2 – C_4 aldehydes, thiols (> C_{10}), or chlorinated hydrocarbons (e.g., carbon tetrachloride) are used; for chain-transfer constants, see [7].

3. Polymer Structure

The kinetics of homo- and copolymerization of vinyl esters has been widely studied; for a summary, see [3]. Three features are particularly characteristic of the polymerization of vinyl acetate:

1. Very high reactivity of the vinyl acetate radical
2. Relatively low reactivity of vinyl acetate monomer towards attack by radicals
3. High chain-transfer constants of vinyl acetate, poly(vinyl acetate), and other polymers relative to the vinyl acetate radical or the growing chain

The differing reactivity of the radical and the monomer, which is also the case for other vinyl esters, makes the production of uniform copolymers from vinyl esters with comonomers that behave differently (e.g., where the monomer reacts more rapidly with radicals) difficult. For copolymerization parameters r, q, and e, see [7]. Chain-transfer reactions involving the monomer or the polymer results in products with a high degree of branching being formed in the polymerization of vinyl acetate. Three types of branching are observed in the final polymer:

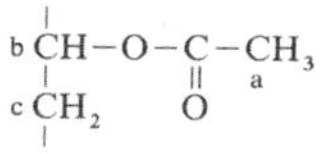

The side chains can be bound to the main chain via an ester group (a) and also directly by a C–C bond (b and c). In the case of chain-transfer in (a), the side chains can be cleaved hydrolytically. The extent of side chain formation at positions (a), (b), and (c) depends strongly on the reaction conditions. The data vary between 97 % and 75 % for (a) [8–10]. The lowering of the average degree of polymerization occurring on saponification of polyvinylacetate, the reduction in the number of side chains in the polymer, and the broadening of the molecular mass distribution, which depends on the proportion of (a) branches, must be taken into consideration in the production of downstream products of poly(vinyl acetate) (see → Polyvinyl Compounds, Others).

In poly(vinyl acetate) the monomer units are linked predominantly head-to-tail. The proportion of head-to-head segments decreases with decreasing polymerization temperature [3].

4. Production

Homo- and copolymers of vinyl esters are produced both in the homogeneous phase by bulk or solution polymerization, and in the heterogeneous phase by suspension and emulsion polymerization.

4.1. Bulk Polymerization

Bulk polymerization can be carried out discontinuously by monomer addition in stirred reactors or continuously in tubular reactors. The monomers are heated to the reaction temperature in the presence of monomer-soluble initiators. The heat of reaction can be removed for low-boiling monomers by a reflux condenser, and in all other cases through the reactor wall. The process is limited to polymers with low to medium molecular mass. Molecular mass and melt viscosity must be sufficiently low at the end of the reaction that the product can be removed from the reactor without difficulty.

Discontinuous processes for poly(vinyl acetate) with a molecular mass of $M_w = 35\,000$ have been described. The molecular mass can be controlled by addition of regulators [11]. A similar process is used for the production of graft copolymers from poly(ethylene glycol) and vinyl acetate [12].

Poly(vinyl acetate) with $M_w = 300\,000$ is obtained in tubular reactors, which give almost quantitative polymerization [11].

Terpolymers can be produced from VeoVa 10, styrene, and maleic or acrylic esters by substance polymerization at 155 – 170 °C [13]. Vinyl acetate – ethylene copolymers with an ethylene content of > 60 % are produced by using the high-pressure polyethylene process (see → Polyethylene).

4.2. Solution Polymerization

Solution polymerization allows the production of polymers with a more uniform structure and lower degree of branching in a simpler process. The removal of the heat of reaction is less problematic than in bulk polymerization, since the reaction can be carried out under reflux at the boiling point of the solvent or one of the monomers. The degree of polymerization depends on the nature and amount of solvent [14].

A process which allows the production of relatively uniform, high molecular mass poly (vinyl acetate) is solution polymerization with relatively low conversion. The unreacted vinyl acetate acts as an additional solvent. The reaction is stopped at a conversion of, for example, 50 % by adding an inhibitor, and unreacted monomer is removed.

Poly(vinyl acetate) which is to be further processed to poly(vinyl alcohol) [7], is generally produced in methanolic solution. Use of *tert*-butanol permits the production of high molecular mass solution polymers because of its low chain-transfer constants relative to the vinyl acetate radical. This is exploited industrially in the production of vinyl acetate – ethylene copolymers with an ethylene content of 50 – 70 % (see → Rubber, 6. Synthesis by Radical and Other Mechanisms). Solution polymerization can be carried out as a discontinuous process in simple stirred reactors, which operate under pressure (for vinyl acetate – ethylene) or pressureless, depending on the monomer used and the reaction conditions; or continuously in a tubular reactor or a cascade of stirred reactors, which are particularly suitable for this process.

4.3. Suspension (Bead) Polymerization

In suspension polymerization the monomer, usually vinyl acetate, is suspended in water by stirring together with a monomer-soluble initiator, and heated to the boiling point of the vinyl acetate – water azeotrope for polymerization. The polymer is produced as beads (diameter ca. 0.2 – 3 mm) and can easily be separated after cooling, and dried. Suspension polymerization combines the advantages of bulk and solution polymerizations: heat removal during polymerization is straightforward, and the product can be easily isolated and handled, and is free of solvent. Suspending agents are used to stabilize the two-phase system of water and the monomer or the solution of polymer formed in the monomer during the reaction. In the older literature a large number of active substances are described, including neutral electrolytes such as phosphates, water-soluble polymers such as modified starch, poly(vinyl alcohol), polyacrylamide, polyvinyl-pyrrolidone, copolymers with carboxyl groups such as styrene – maleic acid copolymers [11], or crotonic acid – vinyl acetate copolymers.

Apart from the nature of the suspending agent, the shape of the reactor, the stirrer speed, and the stirrer geometry significantly affect the stability of the system and the particle size distribution of the beads formed.

4.4. Emulsion Polymerization

Emulsion polymerization is the most important process for the production of vinyl ester homo- and copolymers. The emulsions, which contain water as the continuous and the polymers as the dispersed phase, are mainly used in the liquid form. Smaller amounts are converted to the polymer by spray-drying to give a powder which can be redispersed in water.

The properties of the product are determined not only by polymer composition and structure, but also by the stabilizing additives used in production, and by the particle size and particle size distribution of the dispersed polymer phase.

In the production of poly(vinyl ester) emulsions three types of stabilizer are used:

1. Water-soluble, polymeric protective colloids, such as poly(vinyl alcohol) [15], [16], [17], hydroxyethyl cellulose, polyvinylpyrrolidone [18], poly(*N*-vinylamides) [19], or starch.
2. Anionic emulsifiers that contain hydrophilic groups such as $-SO_3^-$, $-OSO_3^-$, $-(OCH_2-CH_2)_n-SO_3^-$, or $-OPO_3^{2-}$, in addition to a hydrophobic group. The hydrophobic part of the molecule can consist of $C_{12}-C_{24}$ alkyl or alkylaryl groups,

$$\left[\begin{matrix}CH_2-CHO\\ |\\ CH_3\end{matrix}\right]_n \text{ or } \begin{matrix}H_2C-COOR\\ |\\ HC-COOR\end{matrix} \text{ groups.}$$

Typical compounds of this type are dodecylbenzenesulfonate, sodium laurylsulfate, and dioctyl sulfosuccinate. Ethylene sulfonate is an anionic stabilizer [20] that leads to formation of high molecular mass polymers with $-SO_3Na$ groups by incorporation into the polymer.

3. Nonionic emulsifiers with hydrophobic groups and polyether chains as hydrophilic groups. Oxyethylation products of nonylphenol, isooctylphenol, $C_{12}-C_{18}$ fatty alcohols, or low molecular mass poly(propylene oxide) are most widely used. The degree of oxyethylation is 4 – 50.

Protective colloids usually give emulsions with an average particle size of ca. 1 – 10 µm. Ionic emulsifiers give an average particle diameter of 0.1 – 0.3 µm. With a combination of the two types the average particle diameter is ca. 0.2 – 2 µm.

The choice of emulsifiers or protective colloids depends on the desired properties of the emulsions. There is a range of known relationships between the emulsifier/protective colloid system, the amount of emulsifier or protective colloid, the polymerization reaction conditions, and the polymer properties [7], [21], [22]. However, they have not yet led to a general theory for the choice of emulsifiers and protective colloids. Therefore, for the production of new poly(vinyl ester) emulsions the most favorable emulsifiers and protective colloids and the optimal concentration must be determined experimentally.

The protective colloid most thoroughly investigated and most often used is poly(vinyl alcohol). Its emulsifying action is strongly dependent on the content of residual acetyl groups. It decreases with decreasing acetyl group content, as does the reaction rate of the emulsion polymerization [23]. Poly(vinyl alcohols) with a residual acetyl content of 0 – 12 % are generally used for the emulsion polymerization of vinyl esters. During polymerization, part of the poly(vinyl acetate) is grafted on to the poly(vinyl alcohol). In peroxodisulfate-initiated polymerization, the content of graft polymer increases with increasing molecular mass and increasing acetyl group content of the poly(vinyl alcohol) [24]. It is also strongly dependent on the initiator. With H_2O_2 and azobisisobutyronitrile as initiators significantly less graft polymer is formed.

Kinetic investigations of the emulsion polymerization of vinyl esters are mainly concerned with vinyl acetate homo- and copolymerization in the presence of ionic emulsifiers [3], [25]. Because of the considerable water solubility of the monomer, in the case of vinyl acetate a mechanism of particle formation differing from that in the emulsion polymerization of other monomers has been discussed. It is assumed that polymerization initially takes place in aqueous solution since the rate is almost independent of the emulsifier concentration [7], [25–27]. In contrast, experiments on emulsion polymerization of water-insoluble vinyl esters of long-chain carboxylic acids have shown, that here the behavior is similar to that of other hydrophobic monomers such as styrene.

For emulsion polymerization with protective colloids ideas about the stabilizing effect are mainly qualitative. It is assumed that high molecular mass hydrophilic protective colloids are adsorbed as loops on the surface of the particles and stabilize them by forming a hydrate layer.

Discontinuous Polymerization. On an industrial scale, discontinuous emulsion processes predominate. In batch processes the total quantity of monomer is heated to the polymerization temperature with the aqueous solution of the protective colloid or emulsifier and the initiator required to form the final emulsion. In the addition process the monomer or, in the production of copolymers, a monomer mixture is added to the total aqueous phase during the course of polymerization. A special case of the addition process is the preemulsion technique, in which part of the aqueous phase is charged to the reactor and a preemulsion is formed from the other part and the monomer. This preemulsion is then added during the course of polymerization.

After polymerization is complete, the emulsion formed is cooled, treated if necessary with stabilizing additives, plasticizers, or by pH adjustment, and then removed from the reactor. The treatment can also be carried out in additional stirred vessels connected to the reactor.

In the discontinuous processes stirred vessels are used. The stirrer geometry and speed have a major influence on the quality of the emulsion. Anchor stirrers, propellers, impellers, and MIG stirrers are used. In the case of vinyl acetate homo- and copolymers produced at atmospheric pressure, the heat of reaction can be removed by a reflux condenser. For other vinyl esters, and for polymerization of vinyl acetate below its boiling point, jacket cooling or the use of internal or external heat exchangers is necessary.

If low-boiling or gaseous monomers such as ethylene or vinyl chloride are copolymerized with vinyl esters, stirred autoclaves must be used. The amount of ethylene which is copolymerized mainly depends on the ethylene pressure above the reaction medium but also on other process parameters such as reaction temperature and monomer concentration during polymerization [28], [29]. Addition processes [30] and batch processes [31] at ethylene pressures up to 100 bar have been described.

Continuous Processes. For the continuous production of poly(vinyl ester) emulsions, both atmospheric-pressure, and pressurized-vessel processes (for copolymerization with ethylene [32] or vinyl chloride) have been described. The monomers, initiators, emulsifiers and protective colloids, and water are fed, either together as a preemulsion or separately as aqueous and organic phases, to a tubular reactor [33], [34], a stirred-tank cascade, a continuous stirred-tank reactor, or a loop reactor with a circulating pump [35].

Construction Materials. Vinyl esters, poly (vinyl esters), and the additives and initiators used in their production are not very corrosive. To produce poly(vinyl esters) reactors made of stainless steel or glass-lined steel are used. For parts of the plant which only come into contact with the monomers, such as pipes and heat exchangers, aluminum can also be used. Copper and copper alloys should be avoided, since copper ions can inhibit polymerization even at very low concentrations [3].

5. Industrially Important Poly (Vinyl Esters)

The only industrially important homopolymers are poly(vinyl acetate) and, to a much lesser extent, poly(vinyl propionate). All vinyl esters which are accessible on an industrial scale are also used to prepare a large number of co- and terpolymers by combination with one another and with other monomers. In this group vinyl acetate is also by far the most important vinyl ester.

In the following the most important processes for the production of the individual polymers are denoted by letters as follows:

(B)   Bulk polymerization
(S)   Solution polymerization
(BS) Bead suspension polymerization
(G)   Graft polymerization
(E)   Emulsion polymerization

Polymers are produced from the following monomers by the methods shown:

Vinyl acetate (B, S, BS, E)
Vinyl acetate – dibutyl maleate (B, E) [36]
Vinyl acetate – n-butyl acrylate (E)
Vinyl acetate – 2-ethylhexyl acrylate (E)
Vinyl acetate – n-butyl acrylate – N-hydroxymethylacrylamide (E) [37]
Vinyl acetate – crotonic acid (S, E) [38]
Vinyl acetate – VeoVa 10 (B, E) [13]
Vinyl acetate – VeoVa 10 – acrylic acid (E) [13]

Vinyl acetate – VeoVa 10 – *n*-butyl acrylate (E)

Vinyl acetate – *N*-hydroxymethylacrylamide (E)

Vinyl acetate – vinyl laurate (B, E) [39]

Vinyl acetate – vinyl laurate – vinyl chloride (E) [40]

Vinyl acetate – ethylene (B, S, E)

Vinyl acetate – ethylene – vinyl chloride (E) [41]

Vinyl acetate – ethylene – acrylic ester (E)

Vinyl acetate – ethylene – acrylamide (E) [42]

Vinyl acetate – ethylene – *N*-hydroxymethylacrylamide (E) [43]

Vinyl acetate – crotonic acid grafted on polyglycol (G) [12]

Vinyl acetate – *N*-vinylpyrrolidone (S)

Vinyl propionate (E)

Vinyl propionate – vinyl chloride (E)

Vinyl propionate – *tert*-butyl acrylate (E)

VeoVa 10 – vinyl chloride (E) [22]

VeoVa 10 – styrene – acrylic ester and/or maleate (B) [13]

VeoVa 10 – VeoVa 9 – methyl methacrylate – butyl acrylate (E)

VeoVa 10 can be completely or partly replaced by VeoVa 9 to achieve special properties.

6. Properties

Poly(Vinyl Acetate) Homopolymers. Poly(vinyl acetate) is amorphous, odorless, tasteless, and has high lightfastness, and weather resistance. The glass transition temperature (T_g = 28 °C) depends to a certain extent on the molecular mass [7]. For electrical, mechanical, and thermal properties, see [7]. The polymers are soluble in many esters, ketones, cyclic ethers, phenols, halogenated aliphatic hydrocarbons, methanol, 95 % ethanol, and 90 % isopropanol.

Commercial poly(vinyl acetate) emulsions with poly(vinyl alcohol) as the protective colloid usually have a particle size of ca.1 – 10 μm, a solids content of 40 – 65 %, and a viscosity range of 1 – 50 Pa · s. The latex particles are frequently aggregated to produce grape-like formations [44].

Emulsifier-containing homopolymer emulsions, some of which contain protective colloids such as hydroxyethyl cellulose, have lower particle size (0.1 – 2 μm).

The minimum film formation temperature of homopolymer emulsions, which is important for their applications, is 15 – 18 °C. Therefore, for many uses plasticizers, such as dibutyl phthalate, tricresyl phosphate, or auxiliary film formers, are added.

Coarse-particle emulsions stabilized with poly(vinyl alcohol) are mainly important as wood glues and adhesives. Fine- and medium-particle emulsions stabilized by emulsifiers are mainly used as binders for emulsion paints and textiles.

Vinyl Acetate Copolymers. The facile alkaline saponification of the ester groups in vinyl acetate homopolymers is a disadvantage in many applications, since the saponification products are hydrophilic and water sensitive. This low resistance to hydrolysis is one reason for combination with comonomers. The second is the glass transition temperature of 28 °C, which is too high for many uses (e.g., paints and adhesives) and renders the poly(vinyl acetate) too hard and inflexible at the usual temperatures of use. The glass transition temperature can be adjusted by the addition of plasticizers. However, for use where plasticizers are undesirable and increased resistance to saponification is required, softening comonomers are used, e.g., dibutyl maleate; *n*-butyl acrylate; 2-ethylhexyl acrylate; ethylene; or ethylene in combination with vinyl chloride, VeoVa 10, or vinyl laurate. The glass transition temperature is most effectively lowered by ethylene, vinyl laurate, and acrylic esters, while the resistance to alkali is best improved by ethylene, or ethylene combined with vinyl chloride or by VeoVa 10. Copolymerization with > 20 wt % VeoVa 9 increases the glass transition temperature and the resistance to alkali.

Since for a particular application of a copolymer, a narrow range for the glass transition temperature or for the minimum film formation temperature is required, the ratio of vinyl acetate to the comonomers can only be varied within certain limits. The additional use of VeoVa 9 in the pressureless production of emulsions or the additional use of vinyl chloride for high-pressure polymers allows the production of polymers with a desired glass transition temperature but different vinyl acetate contents. For the evaluation of a copolymer it is critical that, at a monomer ratio required by the glass transition temperature, the other properties are favorable. For example, emulsions used as binders in paints must be stable towards alkalies. LINDEMANN [28] has determined the compositions necessary to achieve glass transition temperatures of + 10 °C, 0 °C, and − 10 °C for the most important comonomers used in the production of vinyl acetate copolymers. Saponification experiments on vinyl acetate copolymers have shown that vinyl acetate – ethylene, vinyl acetate – ethylene – vinyl chloride, vinyl acetate – VeoVa 10, and vinyl acetate – vinyl laurate – vinyl chloride, co- and terpolymers with 50 – 80 % vinyl acetate have particularly favorable properties for paints [41], [45–47]. Use of N-hydroxymethylacrylamide allows the production of cross-linkable vinyl acetate copolymers, and the use of crotonic acid gives water-soluble vinyl acetate copolymers with good film-forming properties.

Vinyl Proprionate Homopolymers. Poly(vinyl propionate) differs from poly(vinyl acetate) in its lower glass transition temperature ($T_g − 7$ °C) and its higher stability towards alkalies [47]. Under the same conditions vinyl propionate polymerizes more slowly than vinyl acetate and gives products with lower molecular mass [48]. Poly(vinyl propionate) is produced industrially by emulsion polymerization.

Vinyl Propionate Copolymers. For many applications the glass transition temperature of poly(vinyl propionate) is too low. To obtain industrially more desirable polymer properties (e.g., T_g, resistance to saponification, weather resistance) it is therefore more commonly used together with comonomers that raise the glass transition temperature. Vinyl propionate – vinyl chloride and vinyl propionate – *tert*-butyl acrylate copolymers are produced by emulsion polymerization. Coatings made from these polymers are relatively resistant to saponification and have good weather resistance.

VeoVa 10 and VeoVa 9 Polymers. Homopolymers of VeoVa 10 or VeoVa 9 have not yet achieved any importance. Copolymers with vinyl acetate (see above) and vinyl acetate and acrylic esters are the most important. A range of other co- and terpolymers have been described [22], [22]. VeoVa 10 – vinyl chloride copolymers in the ratio 2 : 1 to 1 : 2 can be produced by emulsion polymerization in stirred autoclaves. The polymers are very weather resistant and have a high pigment-binding capacity.

Styrene – VeoVa 10 – acrylic ester or maleic ester terpolymers are clear, colorless, saponification resistant, and, on copolymerization with monomers with functional groups, cross-linkable [49].

7. Quality Specifications, Trade Names, Storage and Transport, Wastewater

Quality Specifications. The applications of vinyl ester homo- and copolymers are so varied that uniform, generally valid quality specifications cannot be defined.

The content of unpolymerized monomers, which is generally < 0.5 % can be determined by GC [50], [51]. Molecular mass can be determined by light scattering or gel permeation chromatography.

For emulsions the determination of dry residue (ISO 1625), sieve residue (ISO 4576), pH value (ISO 1148), viscosity (ISO 3219 and ISO 2555), freeze/thaw resistance (ISO 1147), density (ISO 8962), minimum film formation temperature, and the white point (ISO 2115) is standardized. The size of the latex particles is determined by optical or electron microscopy, by light scattering [52], by the Marshall method [53], or by aerosol spectroscopy [54]. In Germany, for polymers subject to foodstuffs regulations, the chewing gum regulation [55], the cheese regulation [56], and Recommendation XIV [57] apply.

Trade Names and Producers. Airflex (Air Products); Appretan, Mowilith, Mowiton, Imperon-Binder (Hoechst); Emultex (Harlow Chemical Comp.); Everflex (Grace); Propiofan (BASF); Rhodopas (Rhone Poulenc Polymers); Uramul (DSM); Vinamul (Vinyl Products); Ucar (UCC); Vinavil (Montedison); Vinnapas, Vipolit (Wacker).

Storage and Transport. Poly(vinyl esters) are combustible, but not readily flammable. They are generally not corrosive.

For poly(vinyl ester) emulsions, transport and storage containers must be resistant towards the aqueous phase, including low molecular mass substances dissolved therein. Irreversible damage to emulsions can arise through external effects which change the composition or the state of colloidal distribution. For example water loss can result in skin formation on the surface.

Many emulsions are frost resistant. However some products coagulate when the aqueous phase freezes. Producers generally give information on the frost resistance of emulsions.

Undesired aggregation or coagulation of the latex particles can occur on strong shearing in unsuitable pumps or in spray nozzles.

Poly(vinyl ester) emulsions must frequently be protected against attack by microorganisms [58].

Wastewater. With the exception of the less widely used crotonic acid copolymers, the polymers are insoluble in water. They can, however, enter the wastewater system as emulsions. Poly(vinyl esters) are nontoxic but are degraded extremely slowly in water. Emulsions therefore must not be discharged directly into receiving waters. Wastewater that contains emulsion can generally be treated in biological sewage treatment plants. The polymer enters the sewage sludge.

8. Uses, Toxicology, and Economic Aspects

Uses. The most important uses of poly(vinyl esters) are in emulsion paints (→ Paints and Coatings, 2. Types, Section 4.4.→ Paints and Coatings, 2. Types, Section 4.5.) and adhesives (→ Abrasives.). Other uses include antinoise compounds, chewing gum bases, concrete additives, soil stabilization, glass fiber sizing material, binders for nonwoven fabrics, fibrous leather substitute, pigment printing pastes, coatings for undersides of carpets, cheese rinds, and coatings for paper.

For some uses (building, adhesives, paints) dispersion powders, which can be redispersed in water before use, are produced from the emulsions by spray drying [59]. Here vinyl acetate homopolymers, copolymers of vinyl acetate with ethylene, VeoVa 10, or with VeoVa 10 and acrylic esters stabilized with poly(vinyl alcohol) are predominantly used. The main advantage of these products lies in their ability to be stored without any problem, and that they can be combined with hydraulic binding agents as one-component systems [60].

Toxicology. A differentiation must be made between the polymers themselves and the other substances always present in the industrial products. All investigations published to date on poly(vinyl esters) refer to industrial products; they are therefore only valid for the composition used in each case.

There are no indications that the poly(vinyl esters) investigated are toxic by oral administration; poly(vinyl acetate) see [61]. Poly(vinylesters) from C_2–C_{18} unbranched fatty acids are therefore permitted together with the comonomers listed individually in the corresponding regulations [55–57] and the accompanying substances for coating cheese and as additives for the production of chewing gum.

Economic Aspects. For the production of polymers based on vinyl esters there are no reliable statistical figures. The production capacity for vinyl acetate in 1990 (in 10^3 t/a) gives an indication of the order of magnitude [62], [63]:

	1991	1994
Western Europe	590	655
Germany	290	–
Eastern Europe	300	116
Former Soviet Union	115	–
North and South America	1480	1775
United States	1280	1500
Asia	930	953
Japan	570	–

The consumption of vinyl acetate is assumed to be ca. 800 000 t in the United States, ca. 660 000 t in Western Europe, and 550 000 t in Japan in 1990. The distribution of the consumption of vinyl acetate among the different products varies in the individual regions. In Western Europe the following quantities of vinyl acetate are consumed in the most important areas of use:

Poly(vinyl acetate) emulsions	ca. 430 000 t
Poly(vinyl alcohol)	ca. 120 000 t
Poly(vinyl acetal) and other uses	ca. 110 000 t

References

1. Dr. Klattes: Pionierpatente, Dokumente aus Hoechster Archiven, vol. 10, Hoechst, Frankfurt 1965.
2. ASTM D 2086, D 2190 D 2191 D 2193.
3. M. K. Lindemann in G. E. Ham (ed.): *Vinyl Polymerization*, vol. 1, Dekker, New York 1967, pp. 252 – 255.
4. K. K. Georgieff et al.,*J. Appl. Polym. Sci.* **8** (1964) 889 – 896.
5. K. K. Georgieff et al., *J. Appl. Polym. Sci.* **5** (1961) 212 – 217.
6. *Houben-Weyl*, **XIV/1**, 263.
7. M. K. Lindemann in H. F. Mark,N. G. Gaylord (eds.): *Encyclopedia of Polymer Science and Technology*, vol. **15**, Interscience, New York 1971, pp. 577 –703.
8. W. W. Graessley, R. D. Hartung, W. C. Uy, *J. Polym. Sci. Polym. Phys. Ed.* **7** (1969) 1919 – 1935.
9. D. J. Stein, *Makromol. Chem.* **76** (1964) 170 – 182.
10. P. Mehnert, *Kolloid Z. Z. Polym.* **251** (1973) 587 – 593.
11. J. M. de Bell, W. C. Groggin, W. E. Gloor: *German Plastics Practice*, De Bell & Richardson, Springfield, USA 1946.
12. Hoechst, DE 1 077 430, 1958 (K.-H. Kahrs et al.).
13. Shell, DE-OS 2 422 043, 1974 (A. M. C. Steenis, W. J. Westrenen).
14. Hoechst, DE 1 177 825, 1961 (W. Ehmann, K.-H. Kahrs).
15. I.G. Farbenind., DE 727 955, 1934 (W. Starck, H. Freudenberger).
16. Wacker-Chemie, DE 887 411, 1938 (H. Berg, H. Mader).
17. K. Noro, *Br. Polym. J.* **2** (1970) 128.
18. BASF, DE-AS 2 256 154, 1972 (H. Grubert, W. Druschke, Sliwka).
19. Hoechst, DE 1 206 592, 1962 (W. Bartmann, C. Beermann, W. Ehmann, D. Ulmschneider).
20. I.G. Farbenind., DE 744 318, 1940 (W. Starck).
21. Shawinigan, GB 574 863, 1942 (M. Kiar).
22. Shell Chemie: Die Herstellungvon VeoVa-Mischpolymer-Dispersionen. Technische Broschüre RES/VVx/4(G), 3rd ed., Oct. 1976.
23. T. Motoyama, S. Yamamoto, S. Okamura, *Kobunshi Kagaku* **10** (1953) 108 – 116.
24. S. Okamura, T. Yamashita, *Kobunshi Kagaku* **15** (1962) 271.
25. M. Nomura, M. Harada, W. Eguchi, S. Nagata, *ACS Symp Ser.* **24** (1976) 102 – 121.
26. F. Candau, R. H. Ottewill: *An Introduction to Polymer Colloids*, Kluwer Academic Publishes, Dordrecht 1990, p. 1, p. 159 ff.
27. M. S. El Aasser, R. M. Fitch: *Future Directions in Polymer Colloids*, Martinus Nijhoff, Dordrecht 1987.
28. M. K. Lindemann, *Paint Manuf.* **38** (1968)no. 9, 30.
29. G. Löhr, *Plast. Rubber Mater. Appl.* **4** (1979) 4, 141.
30. Hoechst, DE 1 127 085, 1960 (K.-H. Kahrs, A. Staller).
31. Air Reduction, DE-OS 1 595 402, 1966 (M. K. Lindemann, R. P. Volpe).
32. BASF, FR 1 226 382, 1959 (H. Fikentscher, E. G. Kastning, N. Rudolphi).
33. Bayer, DE-OS 2 007 793, 1970 (D. Glabisch, W.-D. Schellenberg, K. Nöthen, H. Bärtl).
34. Hoechst, AU 250 009, 1963 (D. Backmann, A. Kühlkamp, E. Paszthory, H. Schreiber).
35. Shawinigan, DE-OS 1 900 112, 1969 (R. Lanthier).
36. Deutsche Amphibolin-Werke, DE-AS 1 092 656, 1958 (G. Penell).
37. Union Oil Comp. of California, US 3 714 096, 1970 (G. Biale).
38. Henkel, DE 1 469 400, 1963 (M. Dohr, G. Tauber, J. Galinke).
39. Wacker-Chemie, DE-OS 2 229 569, 1972 (E. Bergmeister, A. Stoll).
40. Wacker-Chemie, DE 1 745 555, 1967 (E. Bergmeister, J. Heckmaier, G. P. Kirst, H. West).
41. Vinyl Products, *Pigm. Resin Technol.* **4** (1975)no. 9, 11.
42. Hoechst, US 3 870 673, 1973 (K. J. Rauterkus, J. Blazek).
43. Sumitomo Chem., DE-OS 2 727 205, 1977 (T. Oyamada et al.).
44. Hoechst: *Mowilith*, 4th. ed., 1969, p. 63.
45. G. P. Kirst, *Fette Seifen Anstrichm.* **69** (1967) 919 – 923.
46. G. Florus, *Congr. FATIPEC* **7** (1964) 149 – 153.
47. W. Sliwka, *Angew. Makromol. Chem.* **4/5** (1968) 310 –350.
48. A. J. Buselli, M. K. Lindemann, C. E. Blades, *J. Polym. Sci.* **28** (1958) 485 – 498.
49. Shell Chemicals: VeoVa 10 Mass Polymerisation. Prel. Techn. Information RES 76 : 1, 1976.
50. H. Hachenberg: *Perkin Elmer Tips* **41**, **GC**, April 1970.
51. G. Schmötzer, *Fresenius Z. Anal. Chem.* **260** (1972) 10 – 24.
52. M. Kerker: *The Scattering of Light*, Academic Press, New York 1969, pp. 311 – 413.
53. A. Kuhn: *Kolloidchemisches Taschenbuch*.Akad. Verlagsges., Leipzig 1953.
54. G. Löhr, R. Reinecke, *Angew. Makromol. Chem.* **85** (1980) 181.
55. Lebensmittelrecht, Bundesgesetze und -verordnungen über Lebensmittel und Bedarfsgegenstände, C. H. Beckssche Verlagsbuchhandlung, München, 111. Kaugummiverordnung lt. ÄndV. Dec. 20, 1977 (BGBl. I, S. 2802, 2806).
56. Lebensmittelrecht, Bundesgesetze und -verordnungen über Lebensmittel und Bedarfsgegenstände, C. H. Beckssche Verlagsbuchhandlung, München, vol. 1, Nov. 1, 1990. 64. Käseverordnung, Neufassung April 14, 1986.
57. R. Frank, H. Mühlschlegel: Kunststoffe im Lebensmittelverkehr; Empfehlungen des Bundesgesundheitsamtes. Empfehlung XIV: Kunststoff-Dispersionen, 178. Mitteilung, Heymanns Vlg., 1990.
58. K. H. Wallhäusser, W. Fink, *Farbe + Lack* **76** (1970) 471; **82** (1976) 108.
59. Hoechst, DE 2 214 410, 1972 (K. Matschke et al.).
60. J. Schulze, *Beton* **41** (1991) 232 – 237.
61. W. M. Carpenter, M. F. Grower, G. Nash, *Oral Surg. Oral Med. Oral Pathol.* **42** (1976) 461 – 469.
62. Hoechst, Zentrale Marktforschung 1991.
63. *Kirk-Othmer*, **24**, Wiley, 1997, p. 947.

Further Reading

M. Chanda, S. K. Roy: *Industrial Polymers, Specialty Polymers, and Their Applications*, CRC Press, Boca Raton 2009.

C.-S. Chern: *Principles and Applications of Emulsion Polymerization*, Wiley, Hoboken, NJ 2008.

C. F. Cordeiro, F. P. Petrocelli: *Vinyl Acetate Polymers*, Kirk Othmer Encyclopedia of Chemical Technology, 5th edition, John Wiley & Sons, Hoboken, NJ, online DOI: 10.1002/0471238961.2209142503151804.a01.pub2.

Y. H. Erbil: *Vinyl Acetate Emulsion Polymerization and Copolymerization with Acrylic Monomers*, CRC Press, Boca Raton 2000.

R. F. Grossman (ed.): *Handbook of Vinyl Formulating*, 2nd. ed., Wiley, Hoboken, NJ 2008.

H. F. Mark (ed.): *Encyclopedia of Polymer Science and Technology*, 3rd ed., Wiley, Hoboken, NJ 2005.

E. M. Petrie: *Handbook of Adhesives and Sealants*, 2nd ed., McGraw-Hill, New York 2007.

Poly(Vinyl Ethers)

GERD SCHRÖDER, BASF Aktiengesellschaft, Ludwigshafen, Federal Republic of Germany

1. Introduction 1175
2. Production 1175
3. Industrial Processes 1176
3.1. Batch Bulk Polymerization of Vinyl Methyl Ether 1176
3.2. Continuous Bulk Polymerization of Vinyl Ethyl Ether 1177
3.3. Continuous Solution Polymerization of Vinyl Isobutyl Ether 1177
3.4. Environmental Protection 1177
4. Properties and Uses 1177
5. Copolymers 1178
6. Toxicology 1178
References 1178

1. Introduction

Industrially important poly(vinyl ethers) are compounds with the general formula: $-(CH_2CHOR)_n-$ where R is an alkyl group such as methyl, ethyl, isobutyl, or octadecyl.

Poly(vinyl ethers) were first prepared by WISLICENUS, who in 1878 produced a balsamlike product from 700 mols of vinyl ethyl ether [*109-92-2*] and 1 mol of iodine [1]. This reaction was recognized as a chain reaction involving a cationic polymerization initiated by I^+ and chain growth via carbonium ions.

In the 1920s the monomeric vinyl ethers [2] became readily available by Reppe vinylation of alcohols with acetylene and grew in industrial importance. Their polymerization was researched systematically.

After 1930 vinyl ethers and poly(vinyl ethers) were produced on an industrial scale, initially at the Ludwigshafen works of the I. G. Farbenindustrie (now BASF), and since 1940 at the General Aniline and Film Corporation (GAF) in the United States, and later also by Union Carbide (UCC) in the United States.

Today (1991) poly(vinyl ether) homopolymers are only produced by BASF; UCC and GAF have ceased production. Poly(vinyl ether) homopolymers are used as raw materials for adhesives.

Copolymers of vinyl ethers with other monomers, such as maleic anhydride, vinyl chloride, or acrylic esters, are produced by GAF and BASF. They are used as auxiliary materials for adhesives and paints, and in many other industries. General information on poly(vinyl ethers) can be found in [3–5].

2. Production

Monomers. Poly(vinyl ethers) are produced exclusively by polymerization of vinyl ether monomers.

In addition to the vinylation of alcohols with acetylene, the vinylation of alcohols with ethylene also leads to alkyl vinyl ethers, according to patents of Union Oil of California [6]. However the process has not become established industrially.

The very reactive α-methylvinyl ethers [7]

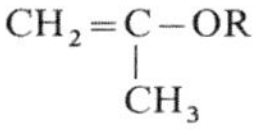

$$CH_2=C-OR$$
$$\quad\;\;|$$
$$\quad\;\;CH_3$$

are only of academic interest.

With both strong and weak initiators and independent of the polarity of the solvent used, they give polymers with syndiotactic structures [8].

The stereochemistry of the polymers from β-methylvinyl ethers

$$\begin{array}{c} CH=CH-OR \\ | \\ CH_3 \end{array}$$

has been intensively investigated [9–15]. Polymers of vinyl ethers which are substituted at the alkyl group by, for example, chlorine [16], oxyethylene groups [17], a phthalimide group [18], an acetoxy group [19], a 9-carbazolyl group [20], a trifluoroalkyl group [21], or an acrylate or sorbate group [22], also have no industrial importance.

Of the industrially important vinyl ethers (methyl [107-25-5], ethyl [109-92-2], isobutyl [109-53-5], and octadecyl [930-02-9]), vinyl methyl ether has the greatest economic growth rate.

Initiators. *Anionic initiators* are unsuitable for the polymerization of vinyl ethers.

With *radical initiators* such as azoisobutyronitrile [23], peroxides [24], or redox systems [25], vinyl ethers polymerize to give homopolymers with a low reaction rate and a low degree of polymerization. However, copolymers from vinyl alkyl ethers and maleic anhydride in a molar ratio of 1 : 1 can be produced at a high rate with radical initiation, the polymer chains being constructed from both monomer units in a strictly alternating sequence.

Copolymers of vinyl ethers and vinyl chloride or vinyl ethers and acrylic esters contain a maximum of 50 mol % vinyl ether.

Cationic initiators are very suitable for the polymerization of vinyl ethers. They require high-purity monomers and solvents and clean apparatus; the absence of water is particularly important.

Friedel – Crafts catalysts, such as BF_3, $AlCl_3$, $SnCl_4$, or their complexes (e.g., BF_3 dietherate or BF_3 dihydrate), and also acidic aluminum sulfate, are widely used [26].

Other initiators such as halogens [27], protic acids [28], halogenated metal alkyls [29], cation-forming salts (e.g., $Ph_3C^+BF_4^-$) [30], Ziegler – Natta catalysts [31], and high-energy radiation [33], [33] are of theoretical interest.

Stereoregularity. Poly(vinyl ethers) were the first group of compounds in which the phenomenon of stereoregularity was recognized [35], [35]. Depending on the choice of polymerization conditions, either an amorphous or a crystalline polymer, which were later recognized as atactic or isotactic, can be produced from vinyl isobutyl ether [36]. For further details see [4].

Living Polymers. In 1984 it was discovered that in nonpolar solvents the initiator system HI – I_2 can convert vinyl isobutyl ether to living polymers [37]. The number-average molecular mass of the poly(vinyl isobutyl ether) increases with increasing monomer conversion, but is inversely proportional to the initial concentration of initiator. The molecular mass distribution is very narrow: the quotient of the weight-average and number-average molecular masses is close to 1, whereas for conventional poly (vinyl ethers) it is ca. 3. This interesting process is, however, not used commercially.

3. Industrial Processes

3.1. Batch Bulk Polymerization of Vinyl Methyl Ether

In one process 200 mL of an initiator solution (3 % boron trifluoride dihydrate in dioxane) is charged to a stirred-tank reactor containing 300 L vinyl methyl ether (*bp* 6 °C) at 12 °C. The reactor is cooled by jacket cooling and is equipped with a reflux condenser.

After the strongly exothermic polymerization is initiated, 5300 L vinyl methyl ether (in 10 h) and 6 L initiator solution (in 16 h) are continuously pumped into the stirred reactor against a regulated internal pressure of ca. 0.3 bar. The reactor temperature is ca. 15 °C. The heat of reaction (1400 kJ/kg monomer) is removed via the reactor jacket and the reflux condenser. Three hours after addition of the monomer is complete the temperature is increased to 60 °C, and 2 h later to 90 °C. After a further 2 h the reflux condenser is removed and the automatic cooling switched off. Two hours later the stirrer is switched off.

After a further 2 hours the reactor is degassed for 4 hours. The contents of the reactor are then either emptied out directly as a hot polymer melt or transferred to a mixer where they are converted into 50 – 70 % solutions with solvents such as water, ethanol, or toluene.

About 97 % of the monomeric vinyl methyl ether is converted to a light yellow soft resin, with $\bar{M}_w \approx$ 60 000 and $\bar{M}_n \approx$ 20 000 (commercial name Lutonal M 40). Polymers from vinyl ethyl ether (bp 36 °C) and vinyl isobutyl ether (bp 83 °C) can be produced analogously.

3.2. Continuous Bulk Polymerization of Vinyl Ethyl Ether

4600 mol/h (440 L/h) of vinyl ethyl ether and 0.4 mol/h of an initiator in solution are continuously pumped into a 2.5 m^3 stirred reactor. The reaction temperature is 98 °C, and the heat of reaction (ca. 1100 kJ/kg) is removed by jacket cooling and a reflux condenser.

The liquid level in the stirred reactor is kept constant by a siphon overflow. The polymer running out is collected in a larger receiver in which unreacted monomer is polymerized. The product is homogenized and then removed in batches as a viscous liquid. The molecular mass $\bar{M}_w$ is 3300 and $\bar{M}_n$, 1400 (commercial name Lutonal A 25). A viscous polymer with $\bar{M}_w = 17 000$ can be produced analogously from vinyl isobutyl ether (commercial name Lutonal I 30).

3.3. Continuous Solution Polymerization of Vinyl Isobutyl Ether

The low-temperature polymerization of vinyl isobutyl ether in liquid propane on a conveyor belt [26] is no longer used due to poor economics.

3.4. Environmental Protection

The off-gases formed in the production of poly(vinyl ethers) are contaminated by solvent and monomer vapors; they are collected and incinerated. The off-gases from incineration are discharged to the atmosphere. In the market, solvent-containing poly(vinyl ether) products are declining in importance, while 100 % polymers and aqueous systems are exhibiting growth — a favorable tendency for environmental protection.

4. Properties and Uses

Consistency and Compatibility. Commercial poly(vinyl ethers) are highly viscous oils, soft adhesive resins, or nonadhesive elastomers, depending on the molecular mass. Poly(vinyl octadecyl ether) has a waxlike consistency.

Poly(vinyl ethers) are resistant to saponification by dilute acids and bases. Their solubility in water and organic solvents and their compatibility with other polymers depends on the nature of the alkyl group. For example, poly(vinyl methyl ether) (Lutonal M 40) is soluble in aromatic hydrocarbons, insoluble in aliphatic hydrocarbons, and soluble in alcohols and cold water. Formation of hydrogen bonds between the ether oxygen and water is responsible for the solubility in cold water. These bonds are destroyed on heating, and above 28 °C the polymer precipitates from aqueous solution.

Poly(vinyl methyl ether) forms clear films when mixed with many other polymers such as methyl cellulose, ethyl cellulose, nitrocellulose, epoxy resins, natural rubber, poly(vinyl acetate), vinyl acetate – vinyl chloride copolymers, and polystyrene. It can also be used as a plasticizer.

Poly(vinyl ethyl ether), a highly viscous oil, soft resin, or nonadhesive material, depending on the molecular mass, is soluble in ethers, esters, ketones, and hydrocarbons, but insoluble in water.

Poly(vinyl isobutyl ether) is produced as a highly viscous oil and as a soft resin, but also as a secondary emulsion in water for mixing with other aqueous polymer emulsions.

Poly(vinyl octadecyl ether) is a brittle solid which melts at 50 °C and is compatible with many waxes.

Uses. Uses of some commercial poly(vinyl ethers) are listed in Table 1.

The economically most important use of poly(vinyl ethers) is in the adhesives industry, mostly as modifying additives for other (cheaper) raw materials. Depending on the nature of the poly-(vinyl ether), adhesives for labels, for flexible tapes, or adhesives which stick well to polyethylene, on a moist

Table 1. Uses of poly(vinyl ethers)

Poly(vinyl ether)	Trade name	Consistency	Use
Methyl	Lutonal M 40	soft resin	raw material for adhesives
	ca. 50 % in water	highly viscous solution	plasticizer
			thickener
			heat sensitizer
	70 % in ethanol	highly viscous solution	raw material for adhesives
	70 % in toluene	highly viscous solution	raw material for paints
Ethyl	Lutonal A 25	highly viscous oil	raw material for paints
			auxiliary for printing inks
	Lutonal A 50	soft resin	raw material for adhesives
	70 % in naphtha	highly viscous solution	plasticizer
	70 % in ethanol	highly viscous solution	plasticizer
	70 % in toluene	highly viscous solution	plasticizer
	Lutonal A 100, ca. 25 % in hexane	highly viscous solution	raw material for adhesives
Isobutyl	Lutonal I 30	highly viscous oil	raw material for adhesives
	Lutonal I 60	soft resin	raw material for adhesives
	50 % in heavy naphtha	highly viscous solution	raw material for paints
	80 % in light naphtha	highly viscous solution	raw material for adhesives
	Lutonal I 60 D	medium-viscosity aqueous dispersion	raw material for adhesives
	Lutonal I 65 D	medium-viscosity aqueous dispersion	raw material for adhesives
Octadecyl	Luwax V	wax	polish for floors and other surfaces

background, or at low temperatures can be produced. For further information on the uses of poly(vinyl ethers) see [5], [38].

5. Copolymers

Vinyl ethers readily undergo cationic copolymerization with one another, but the resulting copolymers are unimportant industrially. Radically produced copolymers from vinyl ethers and other monomers contain a maximum of 50 mol % vinyl ether. Solutions (e.g., in ethyl acetate) or aqueous emulsions of radical copolymers containing up to 30 % vinyl isobutyl ether are used as raw materials for adhesives with special adhesive and compatibility properties. Copolymers with 75 % vinyl chloride and 25 % vinyl isobutyl ether are used as binding agents for ageing-resistant, flexible anticorrosion paints.

Copolymers of vinyl methyl ether and maleic anhydride are very versatile in their applications, being used as protective colloids, thickeners, pigment dispersants, film formers, and binding agents. Products of this type are produced by GAF (trade name Gantrez) and BASF (trade names Sokalan, Luviform, Lupasol) for use as detergent additives, in the paper and textile industries, in cosmetics, and in the production of microcapsules. For further information see [5]. GAF sells the copolymer as a white powder, which, on dissolving in water, forms a maleic acid – vinyl methyl ether copolymer and, on dissolving in an alcohol, a maleic acid half ester – vinyl methyl ether copolymer.

6. Toxicology

For the common poly(vinyl alkyl ethers) known for nearly 50 years (methyl, ethyl, isobutyl, and octadecyl), no effects which are hazardous to health have so far been found. In animal experiments they are classified as nonirritant to the skin and nonmutagenic. The oral toxicity LD_{50} (rat) is > 500 mg/kg.

The toxicity of organic poly(vinyl ether) solutions is determined essentially by the solvent. Toluene and n-hexane present in naphtha have a narcotic effect, irritate the eyes and skin, and can produce permanent damage to the nervous system. Good workplace ventilation is necessary.

References

1 I. Wislicenus, *Justus Liebigs Ann. Chem.* **192** (1878) 106.
2 *Ullmann*, 4th ed., **23**, 608 – 611.
3 J. P. Kennedy, E. Marechal: *Carbocationic Polymerization*, Wiley-Interscience, New York 1982, pp. 499 ff.

4. T. Higashimura, M. Sawamoto: "Carbocationic Polymerization: Vinyl Ethers," in G. C. Eastmond et al. (eds.): *Comprehensive Polymer Science,* vol. 3, Pergamon Press, Oxford 1989, pp. 673 – 696.
5. M. Biswas, A. Mazumdar, P. Mitra: "Vinyl Ether Polymers," *Encycl. Polym. Sci. Eng.* **17** (1989) 446 –468.
6. Union Oil of California, *US 4 057 575,* 1977 (D. L. Klass); *US 4 161 610,* 1979 (D. L. Klass).
7. M. Goodman, Y. L. Fan, *J. Am. Chem. Soc.* **86** (1964)4922, 5712.
8. K. Matsuzaki, S. Okuzono, T. Kanai, *J. Polym. Sci. Polym. Chem. Ed.* **17** (1979) 3447.
9. G. Natta, *J. Polym. Sci.* **48** (1960) 219.
10. G. Natta, M. Peraldo, M. Farina, G. Bressan, *Makromol. Chem.* **55** (1962) 139.
11. T. Higashimura, Y. Hirokawa, K. Matsuzaki, T. Uryu, *J. Polym. Sci. Polym. Chem. Ed.* **18** (1980) 1489.
12. K. Matsuzaki et al., *Makromol. Chem.* **182** (1981) 2421.
13. T. Higashimura et al., *J. Polym. Sci. Polym. Chem. Ed.* **6** (1968) 2511.
14. T. Higashimura et al., *J. Polym. Sci. Polym. Chem. Ed.* **15** (1977) 2691; *J. Polym. Sci. Polym. Chem. Ed.* **17** (1979) 1473.
15. Y. Hirokawa et al., *J. Polym. Sci. Polym. Chem. Ed.* **17** (1979) 3923; *Polym. Bull. (Berlin)* **1** (1979) 365.
16. T. Higashimura, Y. M. Law, M. Sawamoto, *Polym. J.* **16** (1984) no. 5, 401 – 406.
17. T. Nakamura, S. Aoshima, T. Higashimura, *Polym. Bull. (Berlin)* **14** (1985) no. 6, 515 – 521.
18. T. Hashimoto, H. Ibuki, M. Sawamoto, T. Higashimura, *J. Polym. Sci. Polym. Chem. Ed.* **26** (1988) no. 12, 3361 – 3374.
19. S. Aoshima, T. Nakamura, N. Vesugi, M. Sawamoto, *Macromolecules* **18** (1985) no. 11, 2097 – 2101.
20. S. A. Haque, T. Uryu, H. Ohkawa, *Makromol. Chem.* **188** (1987) no. 11, 2523 – 2533.
21. W. O. Choi, M. Sawamoto, T. Higashimura, *Polym. J.* **20** (1988) no. 3, 201 – 206.
22. S. Aoshima, O. Hasegawa, T. Higashimura, *Polym. Bull. (Berlin)* **14** (1985) 417 – 423.
23. N. M. Bortnik, S. Melamed, US 2 734 890, 1956.
24. J. F. Nelson, F. W. Banes, W. P. Fitzgerald, US 2 967 203, 1961.
25. R. R. Dreisbach, J. L. Lang, US 2 859 209, 1958.
26. *Ullmann,* 4th. ed., **19,** 382 – 385.
27. D. D. Fley, J. Saunders, *J. Chem. Soc.* 1954, 1668.
28. F. Bolza, F. E. Treloar, *Makromol. Chem.* **181** (1980) 839.
29. Y. Kishimoto, S. Aoshima, T. Higashimura, *Macromolecules* **22** (1989) no. 10, 3877 – 3882.
30. T. Kunitake, K. Takarabe, *Makromol. Chem.* **182** (1981) 817.
31. C. E. Schildknecht, I. Skeist: "Vinyl Ether Polymerizations," in: *High Polymers,* vol. 29, Wiley-Interscience, New York 1977, pp. 321 – 327.
32. Ka. Hayashi, Ko. Hayashi, S. Okamura, *J. Polym. Sci. Polym. Chem. Ed.* **9** (1971) 2305.
33. A. Deffieux et al., *Polymer* **24** (1983) 573.
34. C. E. Schildknecht, A. O. Zoss, C. McKinley, *Ind. Eng. Chem.* **39** (1947) 180.
35. C. E. Schildknecht et al., *Ind. Eng. Chem.* **40** (1948) 2104.
36. G. Natta, I. Bassi, P. Corradini, *Makromol Chem.* **18/19** (1956) 455.
37. M. Miyamoto, M. Sawamoto, T. Higashimura, *Macromolecules* **17** (1984) 265.
38. W. J. Müller, *Adhäsion* **25** (1981) no. 5, 208 – 213.

Further Reading

M. Chanda, S. K. Roy: *Industrial Polymers, Specialty Polymers, and Their Applications,* CRC Press, Boca Raton 2009.

Y. H. Erbil: *Vinyl Acetate Emulsion Polymerization and Copolymerization with Acrylic Monomers,* CRC Press, Boca Raton 2000.

R. F. Grossman (ed.): *Handbook of Vinyl Formulating,* 2nd ed., Wiley, Hoboken, NJ 2008.

R. B. Login: *Vinyl Ether Monomers and Polymers,* Kirk Othmer Encyclopedia of Chemical Technology, 5th edition, John Wiley & Sons, Hoboken, NJ,online DOI: *10.1002/0471238961. 2209142512150709.a01.*

H. F. Mark (ed.): *Encyclopedia of Polymer Science and Technology,* 3rd ed., Wiley, Hoboken, NJ 2005.

J.-P. Pascault, R. J. J. Williams (eds.): *Epoxy Polymers,* Wiley-VCH, Weinheim 2010.

Poly(Vinylidene Chloride)

JÉRÔME VINAS, Solvay, Brussels, Belgium

PIERRE EMMANUEL DUFILS, Solvay, Tavaux, France

1.	Introduction	1181
2.	Production	1182
2.1.	Emulsion Polymerization	1182
2.1.1.	Batch Emulsion Polymerization	1182
2.1.2.	Continuous-Feeding Emulsion Polymerization	1184
2.1.3.	Emulsion Polymerization for Soluble Resins	1185
2.2.	Suspension Polymerization	1185
2.3.	Comparison of the Processes	1186
2.4.	Purity and Storage of the Monomers	1187
3.	Structure and Properties	1187
3.1.	Homopolymers	1187
3.2.	Copolymers	1188
4.	Processing and Uses	1190
4.1.	Extrusion	1190
4.2.	Films and Sheets	1191
4.3.	Lacquering and Coating	1191
4.4.	Uses	1194
5.	Economic Aspects	1194
6.	Toxicology and Occupational Health	1194
	References	1195

1. Introduction

Vinylidene chloride (1,1-dichloroethylene), VDC, was first obtained by V. REGNAULT in 1838 by treating 1,1,2-trichloroethane with alcoholic caustic potash solution to produce a colorless, chemically impure liquid having a boiling point of 35–40°C, which was subsequently recognized as being impure VDC [2]. The processes used today are still based on this reaction (see → Chloroethanes and Chloroethylenes).

When VDC is stored and particularly when exposed to light (E. BAUMANN, 1872), a white precipitate is formed.

$$n\,CH_2 = CCl_2 \xrightarrow{h\nu} \left[\begin{array}{c} H \;\; Cl \\ | \;\;\; | \\ -C-C- \\ | \;\;\; | \\ H \;\; Cl \end{array} \right]_n$$

Vinylidene chloride   Poly(vinylidene chloride)

This precipitate was identified by FEISST and STAUDINGER (1930), who carried out the first systematic polymerization experiments [3, 4], as poly (vinylidene chloride) homopolymer. It was found that poly(vinylidene chloride) (PVDC) could not be processed thermoplastically because of its thermal instability. R. M. WILEY [5] succeeded in obtaining processable materials by copolymerization and plasticization. Because of the special properties of PVDC, this polymer was intensively researched and further developed in the 1930s (in the United States by Dow in particular, and in Europe by BASF). For further information on the historical development, see [6–12].

The first poly(vinylidene chlorides) were marketed more than 50 years ago. They were originally used mainly for resistant fibers and bristles. Coatings of PVDC were first applied in significant quantities around 1940 in the United States and in Germany. The first extruded shrink PVDC films for meat packaging also appeared at that time.

However, large-scale development of this class of products did not start until the 1950s when improved PVDC resins for coatings came onto the market; early examples of this class of high-grade polymers were the soluble Saran resins (F and B) and Saran dispersions of Dow and the Diofan dispersions of BASF.

All conventional commercial PVDC products are now copolymers of vinylidene chloride with small amounts of other monomers. The VDC content of these polymers is usually between 75 and 95%, and in special cases it is as low as 50% (e.g., VDC–vinyl chloride (VC) copolymers).

2. Production

Vinylidene chloride is prepared commercially by dehydrochlorination of 1,1,2-trichloroethane in the presence of lime or caustic, as described in Figure 1. Commercial VDC is then stabilized with monomethyl ether of hydroquinone (MEHQ).

This inhibitor can be removed by distillation or by washing with aqueous caustic under an inert atmosphere at low temperatures. Without inhibitor, VDC should be stored at $-10°C$ under an inert atmosphere, and used within 1 d.

Due to its asymmetric structure, VDC polymerizes readily at its double bonds; comonomers are used for modification. The main comonomers are methyl acrylate, ethyl acrylate, acrylonitrile, methyl methacrylate, butyl acrylate, and vinyl chloride.

Small amounts of unsaturated carboxylic acids or other products with adhesion-promoting groups are also used to obtain improved anchorage or other special coating properties.

Two processes are used to produce PVDC copolymers on an industrial scale, which are of comparable commercial importance: emulsion polymerization and suspension polymerization, also known as bead polymerization.

Vinylidene chloride polymerizes in the presence of ionic and radical initiators and Ziegler catalysts and under the influence of radiation. Ionic polymerization has the disadvantage that initiators such as butyllithium react with the polymer and accelerate its degradation. Therefore, only radical polymerization in aqueous systems has been of industrial interest.

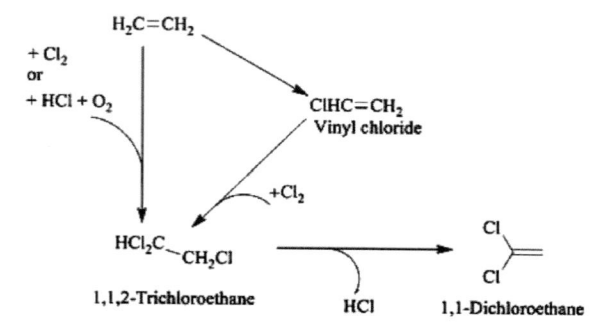

Figure 1. Synthesis of 1,1-dichloroethylene (VDC)

2.1. Emulsion Polymerization

VDC is polymerized by batch [14] or monomer-addition [15–19] processes. The mechanism of the polymerization cannot be described by the Smith–Ewart theory [20]. Typical polymerization processes require the following considerations:

Polymerization is preferably carried out at pH 1–4 to avoid attack by bases, and the operating temperature must be <80°C to suppress polymer degradation. Feed tanks, polymerization reactors, and monomer removal units are often made of glass-lined steel to avoid metallic impurities. If the reaction medium is buffered, the polymerization can also be carried out in stainless-steel reactors.

In VDC emulsion (co)polymerization, anionic emulsifiers such as sodium alkylarylsulfonates, alkyl esters of sodium sulfosuccinic acid, and fatty alcohol sulfates are used alone or in combination with nonionic surfactants such as ethoxylated alkylphenols. The usual initiators are peroxosulfates, hydrogen peroxide, organic hydroperoxides, peroxoborates, and peroxocarbonates in the presence of redox activators such as ascorbic acid, bisulfites, or formaldehyde sodium bisulfite. Redox systems are used to obtain high-molecular-mass products at low temperatures and acceptable space–time yields.

2.1.1. Batch Emulsion Polymerization [21]

Figure 2 shows a flow sheet for the production of VDC copolymers by batch emulsion polymerization.

1. VDC and comonomers are charged to a glass-lined reactor, equipped with a double jacket and containing a cold aqueous solution composed of the anionic or cationic surfactants, a water-soluble initiator, a redox activator, and different additives such as chain-transfer agents, buffer, etc. A delayed introduction of some highly reactive monomers is sometimes necessary. If the stirring is well suited, the surfactants form micelles when introduced over the critical micellar concentration and monomer droplets. By its (thermal- or redox-triggered) decomposition

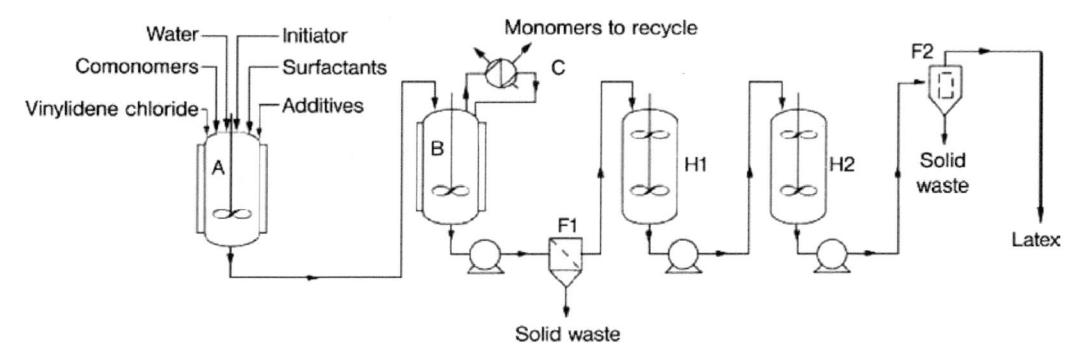

Figure 2. Flow sheet for the production of VDC copolymers by batch emulsion polymerization
A) Reactor; B) Degasser; C) Cyclone; F1, F2) Filter; H1) Buffer tank; H2) Storage tank

into radicals, the water soluble initiator provides a start of the polymerization in the aqueous phase. The newly created, water-soluble oligomers will further enter into a swollen micelle, thus creating particles, to continue the polymerization process. The gradual diffusion of monomers from the droplets to the particles, where the polymerization takes place, ensures the growth of the particles up to ca. 100–200 nm, which remain dispersed in the water phase. In some cases, it is necessary to introduce initiator gradually to keep the polymerization going. Surfactants may also be introduced throughout the polymerization in order to ensure the colloidal stability of the growing particles.

The heated double jacket allows a high increase in temperature of the media, which initiates the polymerization reaction. The exothermic nature of the latter must be carefully controlled by cold water supply in the jacket. In batch polymerization, the polymerization time can vary from 3–24 h, with polymerization temperatures from 30–80°C, leading to 95–98% final conversion [15]. Polymerization to completion of batch charges gives copolymers with broad comonomer distributions [22], except when the copolymerization parameters r_1 and r_2 equal 1, which is the case for VDC and methyl acrylate (MA) (see Table 1).

2. The copolymerization is considered as finished when the temperatures in the reactor and the double jacket are the same or when the pressure in the reactor decreases significantly, indicating the consumption of the monomers in the sky of the reactor. At the end of the polymerization, the reactor is degassed and the residual monomers are eliminated from the latex by stripping. The stripping of monomers can be done directly in the reactor or, better, in a dedicated reservoir called degasser, equipped with a double jacket and controlled in pressure. The latex is heated under reduced pressure. The combination of high temperature and low pressure causes evaporation of water which brings the unconverted monomers. The steams containing monomers are condensed, and the monomers are sent to an incinerator and the water treated before discharge.

3. The residual monomer content is strictly controlled to meet the requirements of the legislation (food contact safety) and even beyond. After the stripping, water or surfactants may be added to the latex to adjust the solid content or surface tension in order to meet sales specifications. The solid content of the latex is usually between 40 and 60%. The latex is then filtered to remove the flocs that have formed in the reactor, and then it is stored in reservoirs awaiting quality control. If the test results are positive, the product is

Table 1. Copolymerization parameters of vinylidene chloride with various comonomers

Comonomer	r_1	r_2	Temperature, °C
Styrene	0.14	2.0	60
Vinyl chloride	3.2	0.3	60
Acrylonitrile	0.37	0.91	60
Methyl acrylate	1.0	1.0	60–70
Methyl methacrylate	0.24	2.53	60
Vinyl acetate	6	0.1	68

transferred to a storage tank. Before shipping, the product is filtered again to remove any coagulated particles and analyzed (customer specifications).

2.1.2. Continuous-Feeding Emulsion Polymerization [21]

Figure 3 shows a flow sheet of the continuous-feeding emulsion polymerization process.

1. VDC, comonomers, water, surfactants, and additives are introduced in a premixing tank, which is maintained at 15°C in order to prevent thermal polymerization of the monomers. Controlled stirring of the mixture provides an emulsion of monomers in water. VDC, comonomers, and the aqueous phase can also be emulsified by using a nozzle before introduction in the reactor.
2. A given fraction of the emulsion (between 5 and 20% of the total load) is introduced in the reactor already containing a given amount of water (and, possibly, additives such as surfactants or electrolytes). A fraction of the initiator (redox system) is also introduced in the reactor, which is then heated. The polymerization starts with the fraction of monomers introduced in the reactor. This step of prepolymerization provides a seed consisting of very small particles (20–50 nm) and determines the number of particles (and thus the particle size) in the latex phase at the end of the polymerization. When the polymerization of the seed is finished, the rest of the emulsion is introduced in the reactor with a controlled flow. The monomers introduced in this way will swallow the seed particles, and the controlled introduction of the redox system (initiator and reductor) allows maintaining a high polymerization rate in the swollen seed particles. These particles will then grow progressively. However, this seed polymerization step can also be substituted by the introduction of a "ready-to-use" seed consisting of very small particles of, possibly, another polymer (e.g., polystyrene, poly (methyl methacrylate)).
3. At the end of the polymerization, the latex is transferred to a degasser. The injection of steam into the latex under a controlled reduced pressure results in the formation of foam. These foams made of latex and steam containing monomers climb in a stripping column. At the top of this column, a nozzle with a defined geometry causes the reduction of the foams in a cyclone. The latex is then redirected to the degasser and the steam containing monomers is removed by means of a vacuum circuit, condensed,

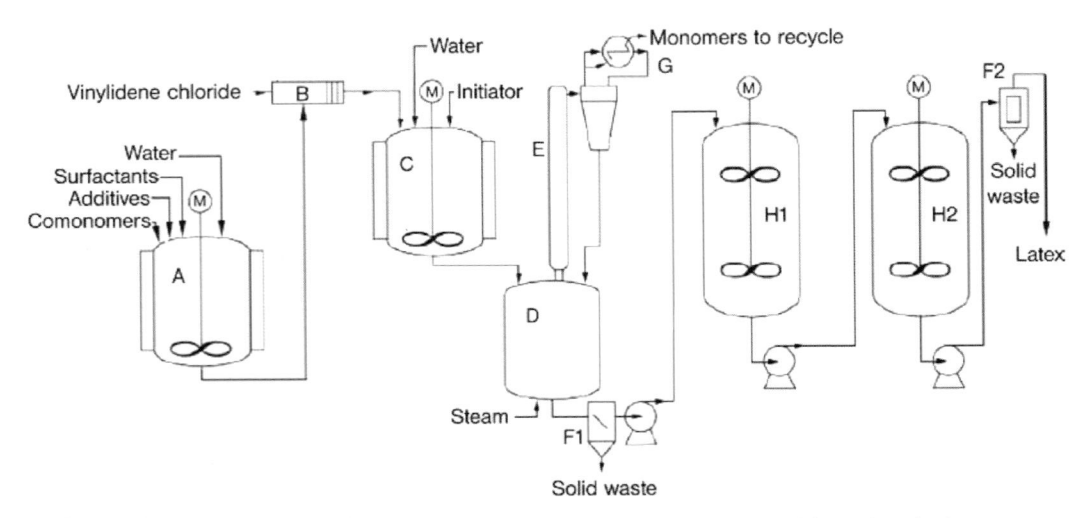

Figure 3. Flow sheet for the production of VDC copolymers by continuous-feeding emulsion polymerization
A) Premixing tank; B) Static mixer; C) Reactor; D) Stripper; E) Column; F1, F2) Filter; G) Cyclone, H1) Buffer tank; H2) Storage tank

and treated. The latex is recycled several times through the stripping column and the residual monomers are gradually eliminated. The good control of the process allows the extraction of the amount of water from the latex introduced by the steam. The latex is finally transferred to storage tanks.

Because the reaction is carried out under controlled conditions, copolymers manufactured by the continuous-addition process have uniform comonomer distribution [15–19]. The copolymers in the latex state are in an amorphous form and can be used directly as coating materials with good film-forming properties.

The principal VDC copolymers produced by emulsion polymerization are those with acrylonitrile and/or various (meth)acrylates [23–28], while VDC–vinyl chloride copolymers are produced in smaller volume [14, 29, 30]. The copolymers are used as dispersions with solids content of 50–60%.

2.1.3. Emulsion Polymerization for Soluble Resins [21]

Until the intermediate storage stage, the polymerization processes for soluble resins are similar to those for batch polymerization (see Fig. 2). The comonomers are generally acrylonitrile for soluble resins and vinyl chloride for extrudable resins.

After extraction of residual monomers, the latex is filtered and transferred to a buffer tank, which allows the feeding of a coagulation line. For coagulation, the dispersion is diluted to 30% solids content to lower its viscosity. Unlike the polymerization reaction, coagulation is a continuous process. The coagulation reactor is continuously fed from the buffer tank. In the case of latex containing anionic surfactants, salts of metallic cations are added to the latex under continuous stirring. These salts cause a significant decrease of the electric double layer which protects the particles covered by surfactants. The attractive forces between them become predominant and the latex coagulates.

The solid coagulates are successively transferred by overflowing to a series of temperature-regulated tanks. This series of stirred reactors allows for heat treatment of the grains formed. This treatment induces a partial crystallization of PVDC which limits sticky grains and facilitates further proceedings. A centrifugal dryer or a rotary filter is used to separate water from a cake of resin which was dried in a rotary dryer and overdried on a fluidized bed. Cyclones provide separation of powder and air. A vibrating screen allows the separation of large particles (which will be crushed and recycled) before storage and controls. Extrudable resins can be converted to premix by adding plasticizers and stabilizers. Since they are soluble resins, they are packed in rigid and flexible containers.

2.2. Suspension Polymerization [11, 12, 21]

Figure 4 shows a flow sheet of the suspension polymerization process.

1. The monomer mixture is dispersed in water under controlled stirring, with a small amount of dispersing agents such as poly (vinyl alcohol) or methyl cellulose added. By this, an aqueous dispersion of monomer droplets is formed, where the size of the polydispersity is governed by the stirring speed. As the dispersing agents prevent the droplets from coalescence, the stirring is the real driving force to maintain the dispersion. Indeed, the dispersion turns into a biphasic system if the stirring speed is not high enough to create shearing in the medium.

    Organic peroxides, peroxocarbonates, and azo compounds are used as initiators. Before polymerization, the initiators, which must be soluble in the monomers, must be distributed uniformly in the monomer mixture to avoid different rates of polymerization in the droplets. When the reactor temperature increases, the initiators undergo a thermal decay which generates radicals in the monomer droplets. As a consequence, each monomer droplet can be considered as bulk microreactor.

    As PVDC is not soluble in VDC, the created polymer chains precipitate in the droplets throughout the polymerization and self-assemble in a semicrystalline structure. At the end of the polymerization, the initial dispersion is converted to aqueous slurry

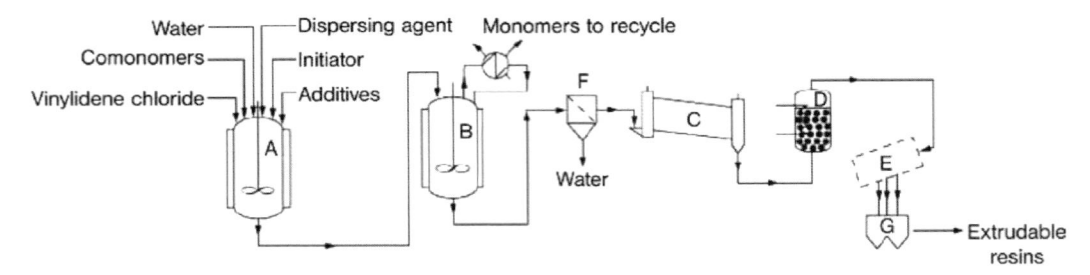

Figure 4. Flow sheet for the production of PVDC copolymers by suspension polymerization
A) Reactor; B) Stripper; C) Rotating dryer; D) Dryer with fluidized bed; E) Sieve; F) Filter; G) Storage tank

that contains hard, glassy suspended particles. These particles have diameters of 80–250 µm, and relatively narrow size polydispersity.

The polymerization temperature can vary from 60–80°C, leading to polymerization times of 8–30 h. The polymerization rate increases with increasing initiator concentration and temperature. The suspension agents influence the particle size distribution but also the porosity of the particles, and clarity and transparency of the processed products. Secondary suspension agents such as sorbitan monostearate improve the porosity and plasticizer absorption of the polymer powder.

2. After polymerization, the slurry is transferred to a stripper where residual monomers can be removed by steam treatment at reduced pressure. The monomer-free suspension enters intermediate tanks (in which several batches can be homogenized). After the transfer to theses buffer tanks, the polymer beads are separated from the aqueous phase by vacuum filtration on a rotary filter and washed on the same filter with water spray. A knife can take off the cake from the rotating filter, and the cake, of about 20% humidity, is transferred continuously to a rotating dryer.
3. Over-drying in a fluidized bed precedes the sieving. Controls and packaging in containers are made directly out of the fluidized bed.

The polymers made by suspension emulsion absorb plasticizers and stabilizers because of the highly porous internal structure of beads. They can be turned into ready-to-use premixes by the addition of various liquid or solid additives, which are needed during the extrusion process, in a dry mixer.

2.3. Comparison of the Processes [21]

On a commercial scale, heterogeneous polymerization of VDC is carried out by means of suspension and emulsion processes in the presence of radical sources.

Radiochemical and anionic initiations suffer from the impossibility of excluding polymer degradation, while cationic polymerization gives low-molecular-mass polymers. Ziegler polymerization of VDC has never been developed.

Homogeneous solution polymerization has not been described in patents. Bulk polymerization and precipitation polymerization in inert solvents such as cyclohexane or benzene cannot be controlled in practice owing to the associated engineering problems such as stirring and heat removal. In bulk polymerization, the reaction mixture thickens to a paste at 10–20% conversion. In the presence of inert solvents the reaction is too slow, only low molecular masses are attained, and the polymer must be freed of solvent.

Suspension polymerization in water can be well controlled but, especially in the case of VDC–VC copolymers, requires quite long reaction times to reach the high molecular masses attained in emulsion polymerization. Due to their low concentration of polymerization additives, suspension polymers are generally more stable and less sensitive to water.

The composition of the copolymer can be better controlled in emulsion polymerization by the metering in of monomer.

Suspension and emulsion polymerization are carried out in stirred, glass-lined or stainless-steel autoclaves. Heat is supplied or removed by jacket heating or cooling. Before polymerization, the water and empty volume of the reactor are flushed free of oxygen, which strongly inhibits the polymerization. The water added must also be free of metallic impurities, or the thermal stability of the polymers is impaired.

Reproducible setting of a defined comonomer distribution in the polymer is more difficult in suspension polymerization than in emulsion polymerization. Unmodified batch operation [31], as well as temperature-programmed polymerization and the use of initiators with different decomposition rates [32], lead to a polymer mixture with heterogeneous polymer composition. In VDC–VC copolymerization, uniform monomer incorporation is achieved by continuously removing the less reactive monomer to maintain a constant volume ratio in the nonreacted monomer mixture [33]. The polymerization rate can be increased by stepwise polymerization with more active initiators such as peroxodicarbonates [34].

Suspension polymerization is mainly used to produce VDC–VC or VDC–MA copolymers [32, 35], which are suitable for injection molding and extrusion molding. Modified VDC–alkyl acrylate [36] and VDC–methyl acrylate copolymers [37] can be processed by blow molding.

2.4. Purity and Storage of the Monomers

The VDC monomer used for polymerization contains 200 ppm p-methoxyphenol or 0.6–0.8% phenol as a stabilizer. Polymerization is practically unaffected by p-methoxyphenol, while phenol must be removed before polymerization by distillation or alkaline extraction.

Vinylidene chloride is stored under a nitrogen protective atmosphere in nickel-plated or phenol-coated vessels. The oxygen content in the stock tank must not exceed 100 ppm.

On storage VDC can form peroxides, especially with nonstabilized VDC and oxygen. These peroxides can lead to spontaneous polymerization, but are also explosive in the presence of air or oxygen, even at low temperatures ($-40°C$). The peroxides can be detected by their pungent, acrid smell or by the formation of flocculent deposits of polymer. They are deactivated on storage under water, and can be destroyed by washing with a 5% solution of methanol in tetrachloroethylene. PVDC peroxide decomposes into phosgene and formaldehyde [13].

3. Structure and Properties [10–12]

3.1. Homopolymers

PVDC does not dissolve in monomeric VDC, and polymer and monomer are practically insoluble in water. VDC polymerizes in the presence of radical sources to give a linear, symmetrical macromolecule with head-to-tail structure and the repeat unit

$$\left[\begin{array}{cc} H & Cl \\ | & | \\ C - C \\ | & | \\ H & Cl \end{array} \right]_n$$

The theoretical chlorine content of the polymer is 73.14%, although in practice chlorine contents of 72–73% are obtained.

The polymer chain is tightly packed in the solid state, and in the thermodynamically stable state it is highly crystalline (ca. 75%), indicating the absence of defects or branches. The crystallinity and rate of crystallization are influenced by the polymerization conditions, the mechanical and thermal treatment of the polymer, and, in the case of copolymers, the comonomer content.

The combination of high crystallinity and high density results in PVDC having very good barrier properties against gases and vapors and resistance to solvents.

The absolute molecular mass of PVDC cannot be determined by conventional methods due to the lack of a suitable solvent. Like other high-melting polymers, PVDC is sparingly soluble in most solvents. Only at temperatures of 130°C can 1% solutions be prepared in, e.g., 1,3-dibromopropane or bromobenzene. At room temperature PVDC dissolves only in polar solvents such as hexamethylphosphoric triamide or tetramethylene sulfoxide; amorphous PVDC dissolves in tetrahydrofuran.

The structure and solubility of PVDC are independent of the polymerization temperature.

PVDC decomposes at temperatures above 125°C, chiefly with elimination of hydrogen chloride and formation of polyene sequences. The degraded polymer is infusible, and the crystalline structure is destroyed [37]. The polymer is also degraded by the action of high-energy radiation, bases, and heavy-metal ions.

Some properties of PVDC are as follows:

Melting point	198–205°C
Glass transition temperature	– 19 to – 11°C
Amorphous density (25°C)	1.67–1.775 g/cm^3
Crystalline density (25°C)	1.80–1.97 g/cm^3
Degree of polymerization	100–10 000
M_w/M_n	1.5–2.0
Softening range	100–150°C
Ceiling temperature	350–400°C

3.2. Copolymers

Industrially important copolymers are those manufactured by radical polymerization of VDC with vinyl chloride, acrylonitrile, and (meth)acrylate esters. Table 1 lists the copolymerization parameters of VDC with the most important comonomers [10]. The Q and e values of VDC are $Q = 0.20$ and $e = 0.36$ [38].

Copolymers with high VDC content have lower melting points and, at least at low comonomer content, higher glass transition temperatures than the PVDC homopolymer. Figures 5 and 6 show the effect of the copolymer composition on the melting point and glass transition temperature.

VDC copolymers are more stable to degradation than pure PVDC. Comonomers interrupt the unstable poly(vinylidene chloride) chain. The copolymerization must therefore be conducted so that as few long PVDC sequences as possible are present in the copolymer. Furthermore, VDC–VC copolymers are more stable than PVDC since the chlorine–carbon bond is stronger in the vinyl chloride unit than in the VDC unit. Polar functional groups, however, weaken the chlorine–carbon bond, so that copolymers with polar functional groups such as the nitrile group are less stable due to neighboring group effects.

The mechanical properties of VDC copolymers depend on temperature, molecular mass, crystallinity, degree of orientation, and plasticizer content. With increasing crystallinity the strength increases, whereas toughness and elongation decrease. The above properties are improved by orientation.

Incorporation of vinyl chloride, for example, in the VDC polymer chain raises the glass

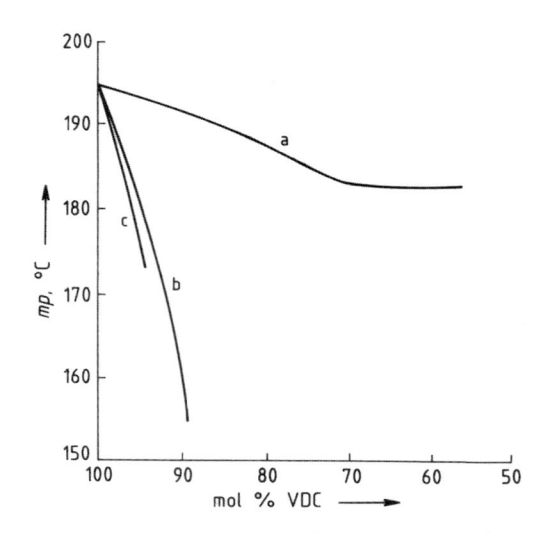

Figure 5. Dependence of melting point on copolymer composition
a) VDC–vinyl chloride; b) VDC–methyl acrylate; c) VDC–octyl acrylate

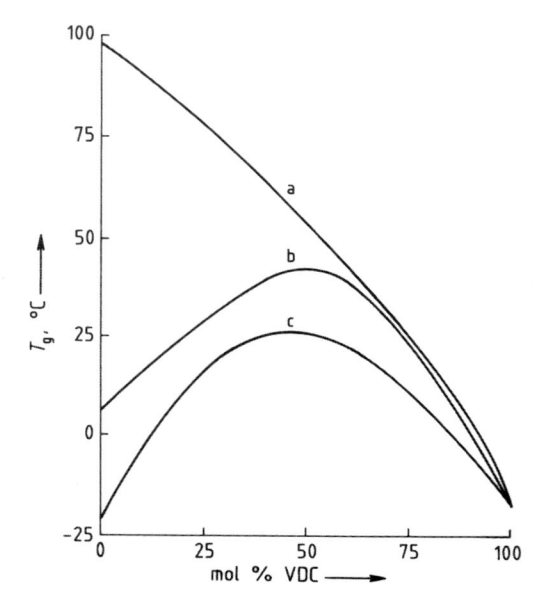

Figure 6. Dependence of glass transition temperature on copolymer composition
a) VDC–acrylonitrile; b) VDC–methyl acrylate; c) VDC–ethyl acrylate

transition temperature, which leads to less favorable low-temperature behavior (increased brittleness). The glass transition temperature can be lowered again without loss of strength properties by plasticization (2–10% plasticizer).

Crystallinity. The performance of PVDC films and coatings is greatly influenced by the high chlorine content of the polymers and especially their crystallinity.

The polymer molecule is very symmetrical if the proportion of VDC in the polymer is high, which leads to crystallization. The crystallinity is directly proportional to the mole fraction of VDC in the polymer. Thus, homopolymers of VDC are highly crystalline, i.e., they contain between 50 and 80% of crystalline material, the actual figure depending on the polymerization conditions. For this reason, they are normally unsuitable for further processing.

The symmetry of VDC polymer chains is disturbed by the incorporation of comonomers such as methyl acrylate. The crystallinity decreases steadily with decreasing mole fraction of VDC until it is completely suppressed at ca. 80 mol% VDC.

If the mole fraction of VDC in the copolymer is very high (93% VDC), the polymer particles of a PVDC dispersion may already crystallize in the aqueous phase. It is very difficult for such highly crystalline latex particles to fuse together, with the result that film formation on drying is hardly possible. For this reason, the VDC mole fraction in PVDC dispersions for general applications usually lies between 87 and 92 mol%.

The partially crystalline morphology of PVDC copolymers can best be described by a network of continuously interpenetrating crystalline and amorphous zones.

Studies on crystallization of coatings have revealed that if the mole fraction of VDC in the copolymer is ca. 88%, the amorphous state is retained for a long time (200–300 d under normal conditions) and that the crystalline phase is formed very slowly.

If the coatings contain a high proportion of VDC (91–92%), crystallization already commences after 1 d and attains a maximum within 5–20 d, depending on the product. This stage is referred to as primary crystallization. Further, secondary crystallization is slight and proceeds slowly.

Table 2 reviews the effect of increasing the VDC content in copolymers and the crystallinity on the properties of coatings and the processability of PVDC.

Permeability [39]. The most water vapor-resistant polymers are the most oxygen permeable and vice versa, with the exception of PVDC, which acts as a very good barrier to both (see → Films for comparison of permeability data). Table 3 lists the permeability of three PVDC grades.

Aromaproofness is closely related to the impermeability to water vapor or gases. It is important in practice that the contents of a pack should not lose their aroma as a result of diffusion from the inside to the outside. Likewise, their aroma must not be impaired by the diffusion of odoriferous substances from the outside to the inside.

The very good aromaproofness derives from its high resistance to diffusion and low solubility, particularly that of the crystalline sections of the polymers.

Figure 7 illustrates the good aroma barrier of PVDC-coated BOPP (biaxially oriented polypropylene) films (26 μm) in comparison with coextruded OPP and OPP coated on both sides with acrylic for limonene and menthol.

Table 2. Demands imposed on the properties of PVDC in relation to the mole fraction of VDC in the copolymers and the crystallinity

Demands on properties of PVDC	Effect of[*]	
	Increasing VDC mole fraction	Increasing crystallinity
Specific properties of PVDC impermeability to:		
water and water vapor	+	+
gases especially oxygen	+	+
aroma	+	+
resistance to solvents and chemicals	+	+
resistance to blocking	+	+
slip	+	+
flexibility	–	–
heat sealability	–	–
clarity	+	+
PVDC processability		
solubility (soluble grades)	–	–
film forming (aqueous dispersions)	–	–
schemal stability (extrudable)	–	–
film blowing aptitude (extrudable)	–	–

[*] + = improvement, – = deterioration.

Table 3. Permeability of PVDC

PVDC grade	Permeability		
	to oxygen (20–25°C)*	to carbon dioxide (20–25°C)*	to water vapor (38°C, 90% RH)**
Coating from PVDC lacquer	15–25	45–80	15–25
	10–15	30–45	10–15
Coating from PVDC dispersion	40–80	90–115	15–25
	15–40	35–90	8–15
PVDC film: plasticised slightly or unplasticised	300–1000	650–2200	50–100
	40–80	135–265	15–25

*Permeability in $cm^3 \, \mu m \, m^{-2} \, d^{-1} \, bar^{-1}$ (dry gas).
**Permeability in $g \, \mu m \, m^{-2} \, d^{-1}$.

Resistance to Oils, Fats, and Chemicals. Pinhole-free films or coatings of PVDC offer high resistance to animal and vegetable fats and oils. They are not swollen by oil or grease, nor do they permit their diffusion. Exceptions are strongly unsaturated vegetable oils, which may cause amorphous grades of PVDC to swell somewhat.

Poly(vinylidene chloride) exhibits excellent chemical resistance. The crystallinity, the large volume proportion of chlorine, as well as the high density are responsible for the high resistance to numerous solvents and chemicals.

4. Processing and Uses

4.1. Extrusion

The commercial extrusion grades of PVDC are copolymers of vinylidene chloride with comonomers such as vinyl chloride, acrylic or methacrylic acid esters, and others. They are polymerized in emulsion and coagulated, or directly polymerized in suspension; they are generally available in powder form.

Normal processing formulations of those copolymers require a plasticizer and/or a stabilizer; some include, if necessary, a lubricant for the extrusion process, a slip agent if requested by the final properties of the extrudate, pigments, light stabilizers, and impact modifiers.

A typical PVDC film formulation, for instance, contains 0–10% plasticizer (e.g., dibutyl sebacate, tributyl acetylcitrate), 0.5–5% stabilizer (e.g., epoxidized soybean oil, sodium pyrophosphate), 0–1% lubricant (e.g., fatty acids and fatty acid amines, montan wax).

The dry blend (formulated PVDC) is extruded in single-screw extruders at a compression ratio from 1/3.5 to 1/4.5. A ratio of screw length to screw diameter of 10–24 has proven to be satisfactory.

Since PVDC, even stabilized, does not tolerate long periods at high temperature, which causes the release of hydrogen chloride and leads to carbonization, all metal parts of the

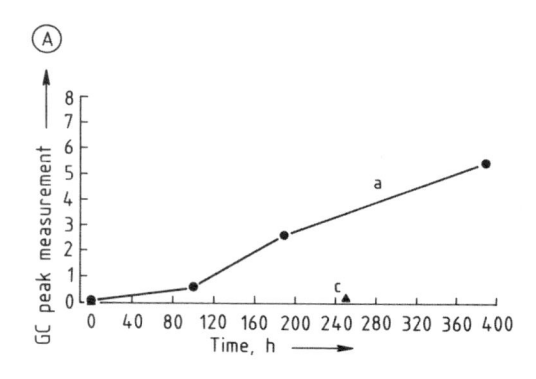

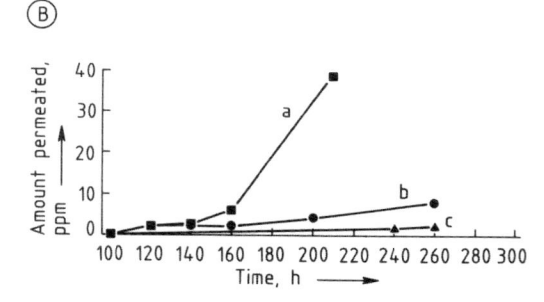

Figure 7. Aroma barrier properties of 25 μm plastic films A) Menthol permeation at 0.2 ppm concentration (GC measurements); B) Limonene permeation at 2.5 ppm a) Coextruded OPP film without barrier; b) OPP film coated with acrylic; c) OPP film coated with PVDC

extruder and dies in contact with hot molten polymer must be made of corrosion-resistant, noncatalytic metals. Recommended materials are chromium (particularly for the screw) and nickel alloys.

4.2. Films and Sheets

One of the major outputs for PVDC resins is the manufacture of films.

It is possible to process PVDC as monolayer film or as part of a coextruded sandwich, both techniques allowing the preparation of shrink or nonshrink materials.

Shrinkage can be obtained by applying tension before cooling (orientation) or by cross-linking.

Monofilm Biorientation. Usual film process for monolayer PVDC film requires the extrusion of a molten tube; the temperature of the PVDC melt is normally 150–160°C, depending on the selected additives formulation.

The shrink film bubble process can be applied to both mona and coextruded film. Immediately after the die, the molten tube passes into a cooling bath (2–10°C) which keeps the quenched cooled material in an amorphous state. The quenched, nontacky material is then slightly rewarmed in an oven and blown into a bubble. As the tube of amorphous polymer is oriented, crystallization takes place rapidly.

If a heat-stabilized film is desired, a further annealing step is applied by re-inflating the film and passing it through an infrared heating tunnel.

Coextruded Film. Since PVDC monofilm does not meet all requirements of film packaging, although formulations already allow great versatility, PVDC films are being increasingly coextruded with other materials. This technique is used to optimize the film properties by combining materials in such a way, so as to add their own advantageous properties, while minimizing their drawbacks. Furthermore, it provides better PVDC barrier properties. Since the PVDC is no longer a self-supporting material, it is possible to reduce the amount of liquid additives that adversely affect barrier properties but are necessary in monolayer extrusion to insure mechanical properties. Also, it allows lower thickness of the PVDC film to be used if required.

Two coextrusion techniques have reached a high level of development: multimanifold die coextrusion and feedblock coextrusion.

The common feature of both techniques is the ability to process the materials that will be used in the final complex, on different extruders, the difference being how the different melts are joined.

In *multimanifold coextrusion*, each molten material is brought into the die through its own flow and distribution channel.

In the *feedblock system* the materials are combined before they enter the die.

Limitations of the feedblock system are that relative viscosities and extrusion temperatures of the different melts must be similar.

Permeability data for extruded and coextruded PVDC films are summarized in Table 4.

Coextruded sheets are produced with feedblock systems in the same way as for coextruded films. The more common structures are PP–TIE–PVDC–TIE–PP or PE; PS–TIE–PVDC–TIE–PS or PE; and PVC–PVDC–PVC, where TIE denotes an adhesive (tie) layer.

4.3. Lacquering and Coating [40–43]

PVDC lacquer resins (powder form) are vinylidene chloride copolymers modified to give solvent solubility; vinylidene chloride homopolymer is almost insoluble in usual solvents. The usual comonomers are acrylonitrile, vinyl chloride, acrylates, and methacrylate esters.

Copolymers with very high vinylidene chloride contents (> 90%) are generally only soluble in THF mixtures.

Resins containing between 80 and 90% VDC are less crystalline and therefore are also soluble in methyl ethyl ketone (MEK) mixtures.

Materials with lower VDC content are soluble in common organic solvents at room temperature, such as esters, ketones, and cyclohexanone.

Coating of Regenerated Cellulose (Cellophane). A major field of application for PVDC lacquers with a high VDC content in the copolymer is the coating of cellulose film.

These tailor-made PVDC resins develop high crystallinity and are therefore soluble only in aggressive solvents such as mixtures

Table 4. Permeability of various PVDC structures

PVDC grade	Structure	Permeability to water vapor, g m^{-2} d^{-1}	Permeability to oxygen, cm^3 m^{-2} d^{-1} bar^{-1}
Aqueous dispersions	CBK + 15 g/m^2 PVDC	2.8–3.6	11–15
	CBK + LDPE + 20 g/m^2 PVDC	2–2.2	3.5–6.5
	PET (12 µm) + primer + 3–5 g/m^2 PVDC	6–10	10–15
	PA 6 (30 µm) + primer + 5–6 g/m^2 PVDC	6–11	7–10
	PVC (200 µm) + primer + 40 g/m^2 PVDC	0.6–1.1	2.6–3.3
	Cellulose film (37–48 g/m^2) + primer + 3 g/m^2 PVDC per side	5–8	3–6
Lacquer resins	Cellulose film (28–50 g/m^2) + primer + 1.5 g/m^2 PVDC per side	6–15	4–10
	PET (15 µm) + 1.5 g/m^2 PVDC per side	9–10	12–14
	BOPP (20 µm) + 3–5 g/m^2 PVDC	3–5	5–11
Extrusion grades	EVA (20 µm)/PVDC (10 µm)/EVA (20 µm) nonoriented films		
	unplasticized PDVC	1–8	5–22
	2% plasticizer	3–10	10–40
	4% plasticizer	10–28	40–140
	PVDC monolayer (30–70 µm), 6–9% plasticizer, heat shrinkable	8–15	30–200
	PVDC monolayer (15–25 µm), 4–6% plasticizer, slightly heat shrinkable	5–15	15–60

CBK = calendered bleached kraft. PET = poly(ethylene terephthalate). PA = Polyamide. BOPP = biaxially oriented polypropylene. EVA = ethylene-vinyl acetate.

of THF/toluene or MEK/toluene. These solvent mixtures (for instance, THF/toluene 70/30 to 55/45) generally allow 10–20% PVDC lacquers to be prepared. The required solution temperature is ca. 50°C; the MEK/toluene mixtures must be heated to 70–75°C to ensure complete dissolution.

Coating of cellulose film with PVDC is carried out by dipping on machines equipped with a vertical drying oven (Fig. 8).

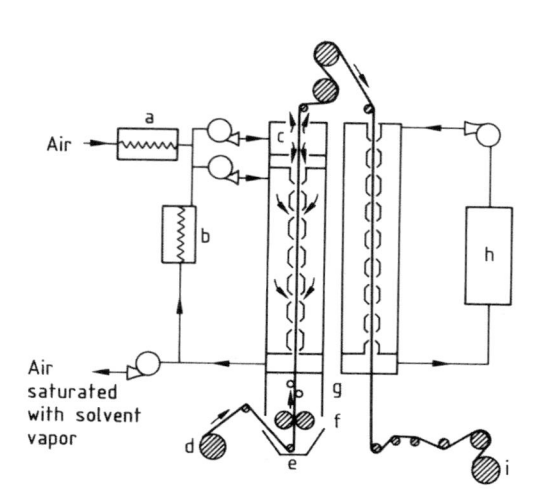

Figure 8. Cellulose film coating machine
a) Air preheater; b) Air heater; c) High-pressure chamber; d) Supply roll; e) Coating bath; f) Calibrating rolls; g) Smoothing roll; h) Rehumidifier; i) Receiving roll

To avoid blocking or sticking to the jaws of sealing machines, and to obtain the required slip properties, the following additives must be incorporated in the lacquer:

1. An additive to reduce sticking to the jaws of the sealing machine, such as fatty acids (behenic or stearic acid, for instance)
2. A natural wax (candellila, carnauba, etc.) or a synthetic wax (ketone or fatty amide) to improve blocking resistance
3. A mineral charge that roughens the surface to improve slip properties (silica, bentonite, talc, etc.)

Stabilizers are also incorporated in the solution in some cases to lower the corrosivity of the PVDC solution to metal parts of the machine.

As long as its moisture content is low, cellulose film constitutes a good barrier against gases; PVDC is thus applied to lower the water vapor permeability.

Barrier properties of cellulose film PVDC coated materials depend on the following factors:

1. Cellulose film composition
2. PVDC composition
3. PVDC coating thickness
4. Additives in the coating formula
5. Degree of crystallinity of the coating

Permeability data are listed in Table 4.

Lacquering of Paper, Foil, and Plastic Films.
Some copolymers of vinylidene chloride are tailor polymerized to obtain good solubility at room temperature in a wide range of common organic solvents such as ketones and esters.

The interest of this family of product is:

1. Easier solubility
2. Heat sealability
3. Transparency and brilliance
4. Adhesion to a wide range of substrates
5. Good barrier properties and chemical resistance

These products, although less impermeable than the PVDC for cellulose film, are used for coating paper, plastic films, foils, and cardboards to lower their permeability (see Table 4) and make them heat sealable. They are also used as protective coatings. Coatings can be made by:

1. Coating on roll machines (reverse roll coating, squeeze coating, kiss coating, gravure roll coating, etc.)
2. Dip coating
3. Spray coating
4. Use of manual rollers or brushes

Depending on the final use, plasticizers, waxes, fillers, pigments, and dyes may be added to the solution.

Vinylidene chloride copolymers for coating are aqueous dispersions of copolymers with a high content of vinylidene chloride, in the form of a milky white liquid of low viscosity.

The usual comonomers are acrylic esters, unsaturated carboxylic acids, and vinyl chloride.

The PVDC aqueous dispersions are produced by emulsion polymerization. The dry solid content is 40–60%.

They can be applied to papers and boards; plastics, films, and sheets; and aluminum. The nature and quantity of the comonomer is adjusted according to the substrate leading to the following coating grades of PVDC:

1. Grades for the production of nonblocking, sealable top coatings for paper, board, or paper-based laminates (paper–PE, paper–aluminum, etc.); these substrates may be pretreated, if necessary, with a grade of PVDC giving more flexible coatings and/or with a primer.
2. Grades for the production of sealable or nonsealable, nonblocking top coatings for plastic films (oriented poly(ethylene terephthalate), OPET, biaxially oriented polypropylene, BOPP, poly ethylene, PE, rigid PVC, polyamide 6, PA 6, oriented polyamide, OPA, etc.). The films are coated with a primer or a first layer of PVDC (flexible grade) before coating with the nonblocking top coat.
3. Grades giving coatings that are more flexible, but also slightly more blocking and more permeable, than those made from the above grades. They are used mainly to coat paper, board, and paper-based laminates (with primer if necessary), and occasionally heavy-gauge plastic films with a primer.
4. Grades giving special coatings that adhere directly to certain plastic films without a primer. Their principal application is as an undercoat for subsequent coatings of another PVDC or PE applied by extrusion coating.
5. Grades for special coating or impregnation applications.

Although PVDC dispersions are in most cases supplied ready for use, under certain circumstances the user may need to introduce one or more additives to the aqueous dispersions prior to processing to modify the properties of the coating (surface characteristics, opacity, pigmentation, etc.) or the behavior of the product when applied by a coating or impregnation machine (corrosiveness, foaming, adjustment of coating weight, etc.).

PVDC dispersions are applied in roll-to-roll operations (web coating) in widths usually not exceeding 2500 mm. Processing speed has increased considerably in the last few years to 100–400 m/min.

The barrier properties of PVDC coatings depend on

1. Nature of PVDC (comonomers)
2. Quality of the substrate (porosity)
3. Coating weight and number of layers
4. Flexibility of PVDC and the substrate
5. Drying conditions
6. Crystallinity

Permeability data are listed in Table 4.

4.4. Uses

Extrudable grades are used for:

1. Monofilaments for the production of military combat boots and insoles, draperies and curtains, and filter applications
2. Medical applications, e.g., colostomy bags are made of monolayer or multilayer structures (EVA–PVDC–EVA)
3. Food packaging:
   a. Cling films, which are very thin films, slightly adhesive, used for wrapping foodstuffs, are monolayer films
   b. Shrink multilayer films (EVA–PVDC–EVA) are used to transport and store fresh cuts of meat between the slaughterhouse and the point of sale
   c. Some synthetic sausage skins are made of monolayer PVDC
   d. Rigid sheets (PP–PVDC–PP) are stable towards sterilization process and give shelf-stable storage

Aqueous dispersions are used for:

1. Pharmaceutical packaging blister packs for drugs which are very sensitive to water are made of PVC coated with PVDC.
2. Plastic food packaging:
   Biscuits are often packed in PVDC-coated OPP. Delicate chips or peanuts are packed in PVDC-coated PET laminated with LDPE. Cellulose casing for sausages is often coated with PVDC.
3. PVDC-coated paper is often used for biscuits, cereals, and dried food.
4. Technical applications:
   PVDC is generally used for its water-resistance, fireproof or binder properties in applications such as coating carpet backings, wood chip panels, and cork coating.

Soluble grades are generally used for coating cellophane, and in some cases for plastic films (e.g., PET).

5. Economic Aspects

Poly(vinylidene chloride) copolymers, as they arise from the polymerization, are marketed by the raw material manufacturer as dispersions or powders under various names.

Known trade names and producers are:

Daran, Owensboro Specialty Polymers (aqueous dispersions)

Diofan, SolVin (aqueous dispersions)

Ixan, SolVin (extrusion and lacquer resins)

Sara; The Dow Chemical Company (extrusion resins, PVDC films)

Serfene, The Dow Chemical Company (aqueous dispersions)

Krehalon, Kureha Corporation (extrusion resins, PVDC films)

Asahi Kasei Chemicals Corporation (aqueous dispersions, lacquer resins, Barrialon PVDC films and Saran fibers)

JV Henan Shuanghui Investment & Development Co, Kureha Corp. and Toyota Tsusho Corp. (extrusion resins for PVDC casings and films)

Hainan Shiner Industrial Co. (aqueous dispersions, PVDC coated films)

Zhejiang Wildwind Plastic Co. (aqueous dispersions, PVDC coated films)

Zhejiang Juhua Co. (aqueous dispersions, extrusion resins)

Nantong Repair-Air Chemistry Bioengineering Co. (aqueous dispersions)

The PVDC world market (2004) is estimated at 160 000 t [45]:

By grades:
Aqueous dispersions	18%
Lacquer resins	9%
Extrusion grades	73%

6. Toxicology and Occupational Health [46, 47]

Monomeric VDC. VDC is a clear, colorless, volatile liquid with a sweet odor [48]. The substance is extremely flammable. In the presence of air or oxygen, 1,1-dichloroethylene can form peroxides which can be violently explosive at temperatures as low as 40°C [49]. VDC is readily absorbed following oral or inhalation exposures. VDC can be metabolized and excretion occurs via exhalation as unchanged VDC or as CO_2 or via feces or urine [50, 51]. Also dermal absorption is likely, albeit that the high volatility is probably a limiting factor [52].The substance shows no bioaccumulation potential [53]. 1,1-dichloroethylene is irritating to the eyes, but the substance is not a skin sensitizer. A single but high uptake of the substance via the oral or inhalation route can be harmful. Chronic exposure to VDC can cause liver and kidney damage. Although effects were seen in tests with cell systems, more relevant studies in living animals did not reveal a genotoxic activity of VDC [53]. A substantial number of valid carcinogenicity studies were performed which did not detect a carcinogenic potential of 1,1-dichloroethylene [54]. Nevertheless, due to its similarity to other molecules with a confirmed carcinogenic activity, a carcinogenic potential in humans by VCD can not be excluded [55]. No specific effects on the male or female fertility or on the unborn child have been detected [53].

The solubility in water is 2.5 g/L [49]. VDC is stable in water, i.e., hydrolysis is negligible, but due to its high volatility it quickly evaporates. VDC degrades rapidly by reacting with hydroxyl radicals in the atmosphere with a half-life between 11 h and 57 h. The substance does not tend to accumulate in animals or humans. It is not readily biodegradable [54]. Based on the available short-term ecotoxicology data [56], VDC should be considered as toxic towards aquatic life.

PVDC is mainly used for packaging foodstuffs or pharmaceutical products, involving both direct and indirect contact with the contents. Such packaging is governed by highly stringent regulations at an international level.

These regulations are generally based on positive lists with specific migration limits. PVDC satisfies all of these regulations.

Residual VDC monomer in commercial polymers generally ranges from indetectable up to a few ppm since all modern producers have now adapted processes to outgas PVDC. These quantities are even lower when measured on coated or extruded materials.

In the United Kingdom, information on the content of VDC in food sold in retail outlets was collected by the MAFF [57]. Of the foods analyzed, only snack foods and foods contained in VDC–VC copolymer coated packs were found to contain measurable amounts of VDC. In most other cases of food packed in barrier-coated materials (polymer films and paper), the VDC content was below the detection limit of 0.03–0.05 ppm. From all the information available, the maximum possible intake of VDC per person, when averaged over the entire U.K. population, would be no more than 1 µg/d, in the worst-case assumption that all the residual VDC in food packaging was in fact ingested [57].

References

General Reference

1 R.A. Wessling, D.S. Gibbs, P.T. Delassus: Vinylidene Chloride Polymers, *Encyclopedia of Polymer Science and Engineering*, 2nd ed., Vol. **17**, Wiley, New York 1987.

Specific References

2 V. Regnault, *Ann. Chim. Phys.* **69** (1838) 151.
3 H. Staudinger, W. Feist, *Helv. Chim. Acta* **13** (1930) 805.
4 H. Staudinger, W. Feist, *Chem. Ind.* **45** (1947) 685.
5 Dow, US 2160931, 1939 (R.M. Wiley).
6 R.C. Reinhardt, *Ind. Eng. Chem.* **35** (1943) 422.
7 J.J.P. Staudinger, *Br. Plast.* **20** (1947) 381.
8 C.E. Schildknecht: *Vinyl and Related Polymers*, J. Wiley and Sons, New York 1952, Chap. VIII.
9 J.F. Gabett, W.M. Smith in G.E. Ham (ed.): *Copolymerisation*, Interscience, New York 1964, Chap. X.
10 G. Talamini, E. Peggion in G.E. Ham (ed.): *Vinyl Polymerization*, Decker, New York 1967, Chap. V.
11 R.A. Wessling, F.G. Edwards, *Encycl. Polym. Sci. Technol.* **14** (1971) 540.
12 R.A. Wessling: *Polyvinylidene Chloride*, Gordon and Breach, New York 1977.
13 K. Mayumi et al., CA 535225, 1957.
14 Grace, US 3033812, 1962 (P.R. Isaacs).
15 Scott Bader, US 3291769, 1966 (D.M. Woodford); Scott Bader, GB 1100398, 1968 (E.B. George, B. Graham).
16 D.M. Woodford, *Chem. Ind.* (1966) 316.
17 Dow, US 3617368, 1971 (D.S. Gibbs, R.A. Wessling).
18 Grace, US 3317449, 1967 (P.R. Isaacs, D.G. Dudley).
19 Dow, US 2956047, 1960 (W.G. MacPherson, C.D. Parker).
20 R.A. Wessling, D.S. Gibbs, *J. Macromol. Sci. Chem. A* **7** (1973) no. 3, 647
21 C, Fringant: PVDC et copolymères du chlorure de vinylidène, Techniques de l'Ingenieur J6570.
22 L. Marker et al., *J. Polym. Sci.* **57** (1962) 855.
23 Dow, US 3879359, 1975 (P.E. Hinkamp, D.F. Foye).
24 ICI, GB 1338561, 1973.
25 Chemische Werke Hüls, DE 1919 705, 1970 (L. Kuhnen).
26 Eastman Kodak, US 3379665, 1968 (H.R. Lyon, H.W. Coover).
27 BASF, GB 1039383, 1966 (W. Runkel, R.W. Reffert).

28. Du Pont, US 3002956, 1961 (J.M. Perri).
29. Dow, US 3662028, 1972 (W.M. Wineland).
30. Grace, US 3088037, 1963 (C. Baum Custer).
31. Dow, US 2 968 651, 1961 (L.C. Friedrich Jr., J.W. Peters).
32. Dow, US 3642743, 1972 (J.E. Schetz, W.D. Shelburg).
33. Dow, US 2482771, 1944 (J. Heereman).
34. Kureha Chem. Ind., GB 1040826, 1966; GB 1111087, 1968.
35. Kureha, DE 2244 957, 1973 (K. Kinigo, H. Kashio).
36. ICI, GB 1171245, 1971 (B.E. Jennings, D.G.M. Wood).
37. ICI, GB 1162576, 1969 (A. Tuner–Jones, D.G.M. Wood).
38. M.P. Farr, I.R. Harrison: "A Study of Thermal Degradation Kinetics of H–H Poly(Vinylidene Chloride)," *J. Polym. Sci. Part C* **24** (1986) 257–261.
39. J. Brandrup, E.H. Immergut: *Polymer Handbook*, Interscience, New York 1966.
40. I.M. Mohr, D.R. Paul: "Comparison of Gas Permeation in Vinyl and Vinylidene Polymers," *J. Appl. Polym. Sci.* **42** (1991) 1711–1720.
41. Y. Vandendael: *Barrier Polymers: A Complex, Rapidly Developing Field with a Wealth of Possibili ties and a Promising Future*, Biotechnology International Century Press, London 1990, pp. 311–315.
42. J.P. Golstein: *New Trends in Barrier PVDC Copolymers*, Belgian Technical Seminar, Belgian Office of Foreign Trade, Tokyo, November 19, 1991.
43. Y. Vandendael: What is Barrier Packaging?, Mobil's Pack-It Basel Seminar, March 20th, 1992.
44. C. Michel: "Developments in PVDC Barrier Resin Technology," *Pira International Conference*, London, April 2nd–3rd, 1987.
45. Allied Development Estimates, 2004, www.takpi.org/content/events/07place/papers/paisley.pdf (accessed 28 June 2012).
46. European Chemical Industry Ecology and Toxicology Centre, Ecetoc Report no. 5, Vinylidene Chloride, 14 August 1985.
47. International Program on Chemical Safety (I.P.C.S.), Environmental *Health Criteria 100 Vinylidene Chloride*, World Health Organisation, Geneva 1990.
48. T.R. Torkelson: Halogenated Aliphatic Hydrocarbons containing Chlorine, Bromine and Iodine, *Patty's Industrial Hygiene and Toxicology*, 4th ed., vol. 2 part E, John Wiley & Sons, Hoboken 1994, pp. 4007–4251.
49. R.A. Wessling, D.S. Gibbs, B.E. Obi, D.E. Beyer, P.T. DeLassus, B.A. Howell: Vinylidene chloride polymers, *Kirk-Othmer Encyclopedia of Chemical Technology*. 5th ed., John Wiley & Sons, Hoboken, NJ, 1999–2011.
50. M.J. McKenna, J.A. Zempel, E.O. Madrid, W.H. Braun, P.J. Gehring: "Metabolism and pharmacokinetic profile of vinylidene chloride in rats following oral administration", *Toxicol. Appl. Pharmacol.* **45** (1978) no. 3, 821–835.
51. M.J. McKenna, J.A. Zempel, E.O. Madrid, P.J. Gehring: "The pharmacokinetics of [14C]vinylidene chloride in rats following inhalation exposure", *Toxicol. Appl. Pharmacol.* **45** (1978) no. 2, 599–610.
52. W.J. Fasano, J.N. McDougal: "In vitro dermal absorption rate testing of certain chemicals of interest to the Occupational Safety and Health Administration: ummary and evaluation of USEPA's mandated testing", *Regul. Toxicol. Pharmacol.* **51** (2008) 181–194.
53. Recommendation from the Scientific Committee on Occupational Exposure Limits for 1,1-Dichloroethene (Vinylidene Chloride), 2008, p. 15, http://ec.europa.eu/social/BlobServlet?docId=6508&langId=en (accessed 26 March 2012).
54. B. Benson, Concise International Chemical Assessment Document 51 1,1-dichloroethene (vinylidene chloride). World Health Organization, Geneva. 2003, http://www.who.int/ipcs/publications/cicad/en/cicad51.pdf (accessed 26 March 2012).
55. S.M. Roberts, K.E. Jordan, A.D. Warren, J.K. Britt, R.C. James: "Evaluation of the carcinogenicity of 1,1-dichloroethylene (vinylidene chloride)", *Regul. Toxicol. Pharmacol.* **35** (2002) 44–55.
56. Euro Chlor Risk Assessment for the Marine Environment OSPARCOM Region - North Sea 1,1-Dichloroethene, 2006, p. 27, http://www.eurochlor.org/media/49378/8-11-4-20_marine_ra_11-dichloroethene.pdf (accessed 26 March 2012).
57. Ministry for Agriculture, Fisheries and Food (MAFF), Food Surveillance Paper 3, HMSO 1980.

Further Reading

M. Chanda, S.K. Roy: *Industrial Polymers, Specialty Polymers, and Their Applications*, CRC Press, Boca Raton 2009.

C.-S. Chern: *Principles and Applications of Emulsion Polymerization*, Wiley, Hoboken, NJ 2008.

A. v. Herk (ed.): *Chemistry and Technology of Emulsion Polymerisation*, Wiley-Blackwell, Oxford 2005.

H.F. Mark (ed.): *Encyclopedia of Polymer Science and Technology*, 3rd ed., Wiley, Hoboken, NJ, 2005.

R.A. Wessling, D.S. Gibbs, B.E. Obi, D.E. Beyer, P.T. DeLassus, B.A. Howell: Vinylidene Chloride Polymers, *Kirk Othmer Encyclopedia of Chemical Technology*, 5th ed., John Wiley & Sons, Hoboken, NJ, 1999–2011.

Polymer Blends

JAN BUSSINK, GE Plastics, Bergen op Zoom, The Netherlands

HENDRIK T. VAN DE GRAMPEL, GE Plastics, Bergen op Zoom, The Netherlands

1.	Introduction	1197
2.	Market Requirements	1199
3.	Blend Technologies	1201
3.1.	Thermodynamics	1201
3.2.	Compatibility	1202
3.3.	Compounding	1203
3.4.	Reactive Compounding	1206
3.5.	Mechanical Properties	1206
3.6.	Additives	1208
4.	Property and Application Profiles of Selected Blends	1208
4.1.	Polyolefins	1210
4.2.	Poly(Vinyl Chloride)	1211
4.3.	Styrene Polymers and Copolymers	1212
4.4.	Polyamides	1214
4.5.	Poly(2,6-Dimethyl-1,4-phenylene Ether)	1215
4.6.	Polycarbonates	1217
4.7.	Polyesters	1219
4.8.	Polyetherimides	1219
4.9.	Polysulfones	1220
4.10.	Polyoxymethylenes	1221
4.11.	Polyarylates and Copolyarylate – Carbonate Blends	1221
4.12.	Poly(Phenylene Sulfide)	1222
5.	General Application Technology	1222
5.1.	Processing of Blends	1222
5.2.	Design Technology	1223
6.	Future Developments	1223
6.1.	Recycling	1223
6.2.	New Developments	1225
6.3.	Outlook	1227
7.	Acknowledgement	1227
	References	1227

1. Introduction

A continuous need for new materials is created by the increasing complexity of products that result from technological breakthroughs and the desire to obtain improved, cheaper materials for existing applications. Alongside the development of new improved base polymers and their copolymers, blend technology is one of the most widely used routes to satisfy this need. Polymer blends (also known as alloys) are combinations of two polymers with one or more distinctive properties that contribute to or are required for the proper functioning of a specific application. Their enhanced, especially adapted property profiles provide advantages to the processor and end user such as extra ease of processing, better low-temperature impact performance, higher strength, higher or lower gloss, reduced flammability, and higher transparency, often in combination with a lower price.

The first generation of blends was developed in the 1950s to improve impact performance. This impact modification normally requires the utilization of rubbery components (materials with a low glass transition temperature, T_g, and generally based on polybutadiene structures) that need a stabilization package. The first impact-modified grades were either made by special polymerization techniques (e.g., copolymerization, block and graft polymerization) or by compounding techniques. Their properties are based upon those of the main matrix polymer and are consequently limited in their scope of application. Although impact properties were improved, a number of undesirable properties were introduced (e.g., delamination, poor surface quality, and inconsistent property profiles). Better materials were made by introducing core – shell impact modifiers. A polymer with a property profile that fulfills market requirements in all areas does not exist.

Table 1. Thermoplastics

Type of plastic	Abbreviation	Name
Commodity	PE	polyethylene
	PP	polypropylene
	PS	polystyrene
	PVC	poly(vinyl chloride)
Transitional	ABS	acrylonitrile – butadiene – styrene copolymer
	SAN	styrene – acrylonitrile copolymer
	SMA copolymer	styrene – maleic anhydride copolymer acrylics
Engineering	PA	polyamide
	PBT	poly(butylene terephthalate)
	PC	polycarbonate
	PETP	poly(ethylene terephthalate)
	POM	polyoxymethylene
	PPE	poly(2,6-dimethyl-1,4-phenylene ether)
	SMA terpolymer	styrene – maleic anhydride terpolymer
High-performance	PEI	polyetherimide
	PPS	poly(phenylene sulfide)
	PSU	polysulfone

Introduced in the early 1960s, engineering polymers such as PC and PPE (Table 1) became available as new basic resins for further blend developments. These second-generation blends offer improved heat resistance and higher continuous use temperatures compared to the older generation materials.

Thermoplastics can be classified into commodity, transitional, engineering, and high-performance plastics (Table 1). Commodity or bulk plastics are made in large volumes, and generally have a low continuous use temperature. High-performance thermoplastics have a high temperature resistance but a low production volume and high price.

Engineering polymer blends are a special group of plastics, whose applications are related to and depend on their specially designed property profile. They all offer dimensional stability over a wide range of maximum temperatures in combination with a minimum continuous use temperature of 85 °C. In addition, they usually possess high impact resistance over a wide range of temperatures down to as low as − 40 °C. Their dimensional stability is determined by the component with the highest glass transition temperature or by the combination of a high crystallinity and high melting point of the highestmelting component. The latter case can be exploited best by reinforcement with inorganic fillers or fibers. The use of a polymer with a high glass transition temperature as a functional filler yields even more interesting systems because they are not restricted to a limiting volume fraction and orientation of the reinforcement. Figures 1 and 2 compare engineering blends with thermoplastics and composites in terms of continuous use temperatures and impact strengths.

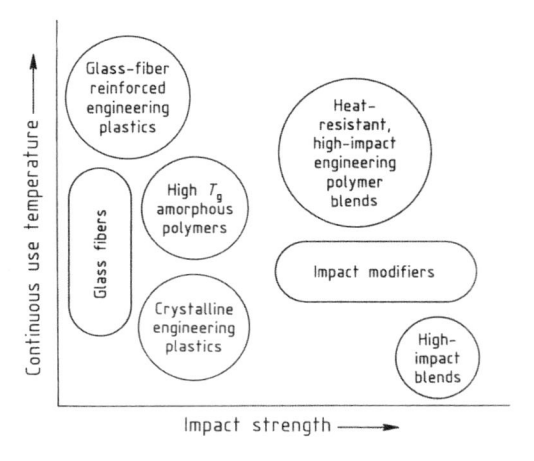

Figure 1. Second-generation polymer blends of engineering plastics

In 1990 blends accounted for 10 % of total thermoplastics production [1–7]. Annual growth figures of 6 – 8 % are predicted versus 2 –3 % for other plastics. Upgrading products based on commodity plastics by reinforcing or blending with engineering polymers creates a

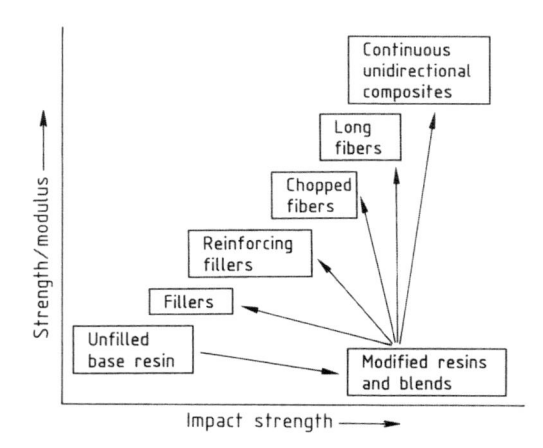

Figure 2. Relationship between mechanical properties and impact strength for various thermoplastic materials

very competitive situation for engineering plastics. It has been estimated that the U.S. market for speciality polymeric blends was $ 900×10^6$ in 1990, and was growing at an annual rate of 7 % [7]. Growth rates for individual engineering blends depend on:

1. Growth rate of major application area (e.g., automotive appliance) [8]
2. Continent of interest (U.S., Europe, Japan)
3. Position in product life cycle: e.g., PPE – HIPS blends are considered a mature blend compared with PC – ABS, and will therefore experience lower growth figure.

Companies that offer blends usually manufacture one or more of the blend components [9]. More polymer manufacturers and compounders are entering this business now because most basic patents for the production of thermoplastics have expired (e.g., PPE, PC) and blends allow facile expansion of the product range [4], [9–11].

From a technical point of view, it is important to translate the demands of a given application to the key physical and mechanical properties of the engineering plastics blends. Target values for heat and chemical resistance, high- and low-temperature impact strength, and stiffness should be assessed. The rationale used for choosing two (or more) polymers (the blend) for a specific application is important (Fig. 3). The dominant factor that determines whether a specific polymer blend is suitable for a certain application is the requirement profile (Chap. 2). The requirement can either be one demanded by the customer or end user, or one imposed by the standards of the specific industry segment (e.g., flammability standards in the aircraft industry).

Two polymers may be miscible, partially miscible, immiscible, or compatible. Consequently, the property profile resulting from blending two polymers is not a simple combination of the two individual property profiles. The relation between different blend classifications in terms of properties and morphology are discussed in Chapter 3. Techniques used to influence blend thermodynamics and ultimate property profile are also considered.

The general guidelines for the identification of a blend property profile are outlined in Chapter 4, and are subsequently applied to the most important blends. Once a blend of two polymers has been selected, qualification of the resin for a certain application depends on processing and design considerations (Chap. 5). Chapter 6 focusses on new developments in polymer blends.

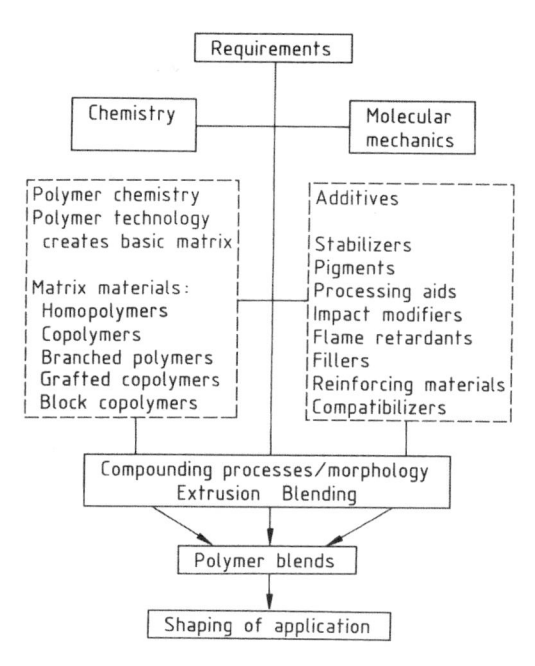

Figure 3. Schematic representation of blend development

2. Market Requirements

Specification of material requirements is determined by the needs of the end user and by the standards within an industry segment. All materials have to fulfill the requirements specified by industrial standards or mandated by law. These standards are controlled by standardized test methods and vary for different countries and industrial segments.

Blend development starts from the combination of a required property profile and an existing technology. The main industrial application areas for polymer blends are the automotive market, followed by the packaging industry, electronic and electrical industry, business equipment, and appliance sectors. Smaller application segments are construction, consumer goods, sheet manufacturers, and aerospace industries.

Automotive Applications. A number of different application areas can be discerned in an automobile. Horizontal exterior body parts demand a higher temperature resistance than vertical exterior body parts since they absorb more radiation from the sun. Both types require high surface quality and facile paintability, the coating technique being determined by the temperature resistance of the material. Mechanical properties are very important for structural body parts, but nonstructural body parts also require a high dimensional stability as characterized by a low coefficient of thermal expansion.

Bumper, or better bumper systems, have a crash-absorbing and an aesthetic function. The crash-absorbing function of bumper systems can be fulfilled by incorporating appropriate failure mechanisms. Decorative functions demand easy paintability, high UV resistance, and resistance to small impact loads. The trend is towards a separation of both functions. Facile dismantling is becoming important for recycling.

Wheel covers should have a good dimensional stability and be able to withstand the high temperatures caused by braking. Chemical resistance is usually provided by a coating.

Lighting systems consist of a light-reflecting component and a housing. The latter requires an extremely high dimensional stability and related mechanical properties. The reflector must be temperature resistant and possess sufficient reflecting capacity (surface quality and gloss).

The *passenger compartment* contains a high number of decorative applications designed for comfort. Resistance to low continuous loads (e.g., instruments in instrument panel) and glossy/nonglossy appearance are general considerations. The temperature requirement is related to exposure to the sun, but a heat distortion temperature of 120 °C is generally accepted. Key requirements for the dashboard material include a minimum heat distortion temperature of 120 °C, excellent accoustic properties (reduction of outside and engine noise), high impact resistance, and good scratch resistance. Paintability and compatibility with soft-touch overlay films are additional requirements.

Structural parts (e.g., chassis) demand such high mechanical properties that composites are the only feasible solution. *Under-the-hood applications* include cable and wiring, which demand electrical, chemical, and high-temperature resistance.

Fuel tanks place high demands on the barrier properties of the resin, in addition to its chemical resistance. PE in fuel tanks has been subjected to (sulfo)fluorination to increase surface polarity and improve barrier properties. Other techniques are employed in utilizing PA – PE or PA – PVAL and PVAL – PA blends [PVA = poly(vinyl alcohol)] for fuel tanks where a special layered morphology ensures impenetrability.

Applications in the *motor compartment* have to withstand harsh environmental conditions, and thus require a high-temperature and environmental (including chemical) resistance. The most suitable products are semicrystalline polymers.

Electronic and Electrical Applications. The dominating requirements in this industry are high heat and chemical resistance and high dimensional stability. A key requirement is a low, preferably temperature-independent, conductivity. The trend towards miniaturization demands better flowing resins coupled with a high heat resistance. Higher heat materials are also required for the introduction of the surface-mount circuitboard assembly [11] in which thermoplastics are combined with metals. The materials must be able to withstand chemical cleaning baths and temperatures over 260 °C for more than 1 min resulting from high-temperature soldering via wave, vapor-phase, IR, or laser methods. Dimensional stability should be maintained during these operations and under operating conditions. Electrical (office) equipment must fulfill stringent flammability requirements, particularly when used in restricted areas.

Appliance Sector. Basic requirements for household appliances are water resistance, chemical resistance, appearance (glossy etc.), and temperature resistance. Electrical properties (e.g., conductivity) and absence of toxicity risks must conform to consumer and government standards.

Building and Construction Applications. Material requirements differ from country to country. Public buildings are subject to the most severe rulings. The greatest drawback of plastics is their intrinsic flammability resulting from their organic nature. The fire tests mandated for plastics are thus difficult to pass. To reduce the risk of fire, only relatively small amounts of plastics can be used in low-risk areas (e.g., outside sidings). The most important flammability characteristics are ignition and flame propagation behavior. Materials may have horizontal, vertical, or no fire propagation. Assessment of flammability characteristics is extremely complex, since the behavior of materials under fire conditions depends not only on the nature of the material, but also on its thickness and the configuration of the part.

Mechanical requirements are much easier to fulfill and depend largely upon the temperature resistance of the material.

Packaging Applications and Pipes. Basic requirements for packaging materials are barrier properties for water, oxygen, and carbon dioxide as well as excellent extrudability resulting from a high melt strength. Permeability requirements are usually satisfied by multilayer systems in which polymers are alternated with tie layers. Properties can be optimized by bidirectional orientation. These multilayer systems can be substituted by a single, special two-phase blend material.

Pipes for water distribution should have a long guaranteed life time (> 100 years), and prevent migration of low molecular mass compounds into the water.

3. Blend Technologies

From a thermodynamic perspective, polymer blends may be completely miscible, partially miscible, or immiscible (Fig. 4). An almost fluid transition is possible between these systems; by modifying the base polymers each blend can be optimally positioned for the foreseen application.

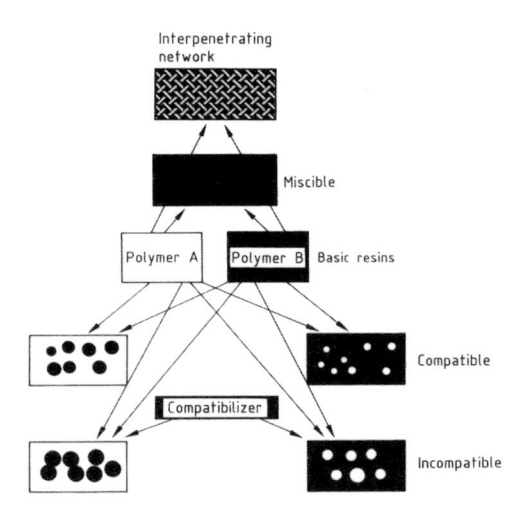

Figure 4. Possibilities of blend morphology

3.1. Thermodynamics

The thermodynamic blend requirements have been reviewed elsewhere [12–15]. Blend polymers interact with each other in their amorphous state, either in a melt or dissolved in a common solvent. A thermodynamically stable mixture of two polymers is obtained if the free energy of the mixture is lower than the combined values of the individual components, i.e., the energy of mixing ΔG_{mix} should be negative. The free energy of mixing is composed of an enthalpy and entropy term [12–15]

$$\Delta G_{mix} = \Delta H_{mix} - T\Delta S_{mix} \tag{1}$$

A homogeneous blend over the complete composition range of two components 1 and 2 is only obtained if

$$(\partial^2 G/\partial x^2) > 0 \tag{2}$$

in which x denotes the volume fraction of component 2. The entropy contribution to the overall free energy of the system is usually negligible, and is only important in special cases (e.g., blending with relatively low molecular mass components such as plasticizers).

The enthalpy resulting from the interaction energy between the two polymers determines whether they are completely or partially miscible. Complete mutual solubility results in a single-phase system. Partial solubility provides a two-phase system with a very low interfacial energy allowing the production of very finely dispersed blend systems without the need for a third component such as a compatibilizing block copolymer. Normally, even when very similar components, such as polyolefins are blended, the enthalpy term is positive and leads to non-miscibility. Phase separation can only be prevented if interaction between the different chemical groups in the two polymers is possible (e.g., dipole–dipole interaction, charge transfer complex formation, hydrogen bonding). A negative enthalpy value leads to a completely or partially miscible blend system. The miscibility of a two-component system can be improved by increasing the interaction energy between the two components or reducing the interaction energy within each component. The latter effect is achieved by using random copolymers made from monomers that, by themselves, form immiscible homopolymers. The internal repulsion energy in these copolymers can thus improve miscibility [16]. The interaction energy between two polymers is the main factor that determines the final blend result.

The morphology of incompletely miscible systems can be structured. In a two-phase system with a dispersed phase and a continuous phase, the polymer constituting the more important continuous phase can be selected. The second possibility consists of an interpenetrating, co-continuous network formed from both polymers (Fig. 4). Morphological control depends on chemical interaction and rheological factors such as the melt viscosity ratio of the two polymers and the dependence of melt viscosity on shear rate. The compounding conditions used during melt blending are therefore of major importance for the final blend property profile (see Section 3.2 and Fig. 3).

The influence of temperature on blend thermodynamics is important because it determines phase separation behavior and morphology (see Section 5.1). A typical example is shown in Figure 5. Three regions can be discerned: (1) a thermodynamically stable region (miscibility), (2) a metastable region where the system is stable only towards small concentration fluctuations, and (3) a region where spinodal decomposition yields a two-phase system. The binodal

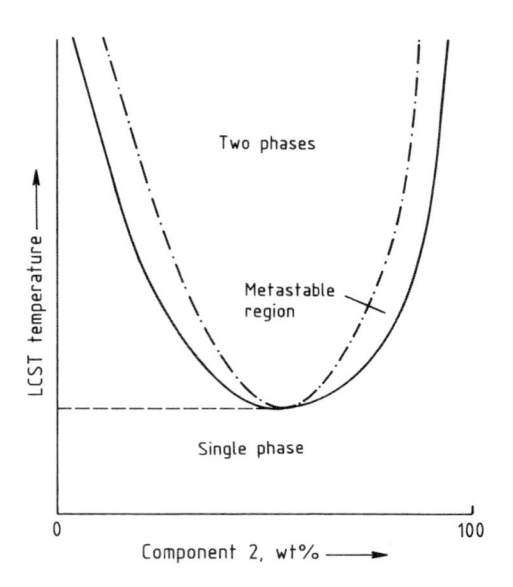

Figure 5. A schematic phase diagram for a polymer–polymer blend with a lower critical solution temperature (LCST)

(solid line) corresponds to $\Delta G_{mix} = 0$, the spinodal (dashed-dotted line) is defined by $(\partial^2 G/\partial x^2) = 0$.

3.2. Compatibility

The term compatibility describes all useful (scientific) means of producing stable blend microstructures that provide technologically important properties for the performance of the part. In the absence of interactions between two blend components, the blend morphology is largely determined by the interfacial tension. The interface between two polymers in the melt is characterized by the interfacial tension and the domain adhesion [17]. The interfacial tension γ_{12} between polymers 1 and 2 is defined as the work required to increase the interface

$$\gamma_{12} = (\delta G/\delta A)_{P,T,n} \qquad (3)$$

where A is the surface area between polymers 1 and 2. The magnitude of the interfacial tension is determined primarily by the polarity difference between the two phases [17]. It is also related to the surface tensions of the constituent polymers which are defined as the work required to increase the surface of a polymer by 1 cm². Increase in the surface area of a polymer involves the transport of polymer segments from the bulk to the surface which deprives them of the small forces acting in the bulk (e.g., van der Waals forces). In total,

$$\gamma_{12} = \gamma_1 + \gamma_2 - W_a \qquad (4)$$

where γ_1 and γ_2 are the surface tensions of polymers 1 and 2, and W_a is the work of adhesion (equal to the decrease in Gibbs free energy on forming an interface between polymers 1 and 2). In practical terms this means that the work of adhesion increases as the interfacial attraction increases, corresponding to a decrease in interfacial tension. If two immiscible polymers are mixed, more interface is created and the free energy of the system increases. The system is thermodynamically unstable and strives toward a reduction of the total interface surface. Consequently, if two individually dispersed polymer droplets encounter each other, they agglomerate immediately. This leads to coarsening of the phase morphology which can only be prevented by the use of compatibilizers. A compatibilizer reduces the interfacial tension, improves the interfacial adhesion, permits finer dispersion during mixing, and stabilizes morphology during processing.

The most common types of compatibilizers are AB block copolymers for a blend of polymers A and B (Fig. 6) that are added prior to compounding or synthesized in situ [18], [19]. The block copolymer decreases the interfacial tension by dissolution of the copolymer block segments within their respective homopolymer phases, leading to covalent bonds being positioned across the interface. Physical anchoring of both phases contributes to the stability of the resulting morphology and mechanical strength. In commercial blends, 5 – 15 wt % compatibilizer is used.

Several other routes are available for compatibilizing phases in commercial multicomponent polymer blends. Large-scale phase separation can be prevented by promoting cocrystallization, by cross-linking the dispersed phase, or by the formation of interpenetrating networks. Ionic attraction between blend components can also be used to stabilize the phase structure. High stress shearing conditions may be applied to mechanically interlock both phases, but this technique is seldom practised.

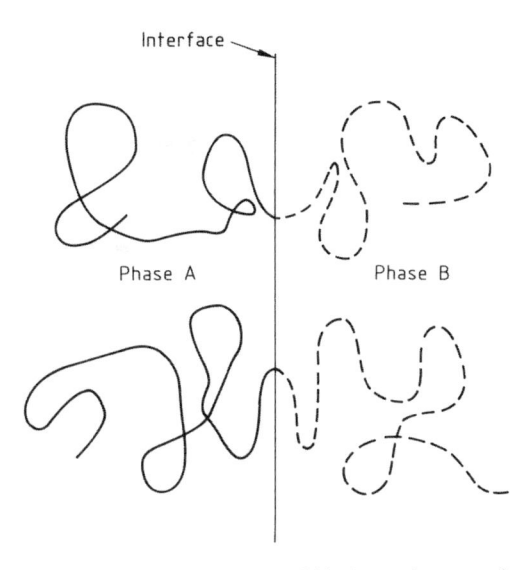

Figure 6. Ideal configuration of block copolymer at the interface between polymer phases A and B

3.3. Compounding

For scientific and screening purposes, limited amounts of blends are prepared by combining solutions of the blend components from the same solvent or mutually miscible solvents. Blends of miscible polymers can then be obtained by sol-vent evaporation or coprecipitation. Care must be taken not to heat a blend beyond a potential LCST temperature in a subsequent melt processing step. In dealing with partially miscible or compatible systems, coprecipitation followed by a melt cycle is usually employed [20], [21].

The main method of blend production is batch or continuous melt blending in kneaders or compounding extruders [20], [21]. The machines used are corotating twin-screw compounders with a mixing section based on the Erdmenger principle, Busch co-kneaders, and Banbury kneaders.

The technology involved in this method of manufacturing polymer blends is determined by the thermodynamics of the blend system (Section 3.1) and by the rheological conditions during melt blending. The final morphology of the blend in the compounding step is of prime importance for its resulting property profile.

Compounding machines are not simple extruders which merely melt and transport the resin [15], [22], [23]. Intensive mixing of the melt is achieved by the dispersive or distributive action of the screw element. Depending on the thermodynamics of the mixture of molten components, a dispersion of component 1 in 2 or 2 in 1 is obtained. The criteria determining the continuous phase are the volume fractions used in the melt-blending process, the ratio of the individual melt viscosities, and the surface tension (the polymer with the lowest surface tension usually becomes the continuous phase).

Two basic microrheological processes can be discerned in the blending of polymers: distributive and dispersive mixing. Distributive mixing determines the homogeneous distribution of the second phase throughout the matrix, whereas dispersive mixing causes large particles to break up into smaller ones. During actual mixing both processes occur simultaneously.

Distributive Mixing. At the start of the blending process, the droplet size of the second phase is relatively large. Two counteracting stresses are in effect: the deforming shear stress τ and the interfacial stress σ/R. Their ratio is usually called the capillary number Ca.

$$C_a = \tau R/\sigma = \eta_m \gamma R/\sigma \qquad (5)$$

where

γ = shear rate (s^{-1})
η_m = viscosity of matrix (Pa · s)
σ = interfacial tension (N/m)
R = radius of particle (m)

Typical shear stresses in polymer melts are in the order of 10^4 N/m^2. The local radius of a dispersed droplet is approximately 10^{-3} m in the early stages of the mixing process, and σ is typically 10^{-2} N/m. An affine deformation is the result.

The total interface is a measure of the effect of distributive mixing and, in the absence of reorientation, is directly proportional to the total shear γ.

$$A = A_0 \gamma \qquad (6)$$

where A is the interfacial area and A_0 the initial interfacial area.

If the layers are oriented along the shearing direction, such as in corotating twin-screw extruders, distributive mixing becomes more efficient.

When two polymers dissolve in each other, they do so by diffusion. The time required for complete dissolution is greatly reduced by decreasing the droplet size of the starting dispersion; this is facilitated by minimal surface tension between the two initially molten phases. The distributive mixing action is produced by the shearing and elongating forces that act upon the continuous phase and are transferred to the dispersed phase [15], [22], [23]. Under stagnant conditions, the dispersed phase consists of molten droplets which are deformed into threadlike structures. The length – diameter ratio of these structures depends on the interfacial tension constraints of the dispersed phase, and on the effectiveness of the transfer of the deformation forces from the continuous phase to the dispersed phase. The latter is maximized when the ratio of the apparent melt viscosities of the continuous and dispersed phases is close to one. Equation (7) describes the equilibrium conditions for droplet formation [24], [25]:

$$\eta_1/\eta_2 = \Phi_1/\Phi_2 \qquad (7)$$

where η is the melt viscosity and Φ the partial volume.

Dispersive Mixing. The dispersed phase deforms affinely with the fluid motion, resulting in elongated droplets. This process continues until the capillary number approaches unity. This limit is reached if the local radii are in the order of 10^{-6} m, resulting in $\sigma/R = 10^4$ N/m^2. The elongated droplets become unstable due to interfacial tension driven Rayleigh distortions, and droplets are formed, which are again subjected to shear stress.

This process of elongating and breaking up repeats itself until the droplet size reaches an equilibrium dimension determined by the above-mentioned factors. These process steps are shown in Figure 7.

Figure 8 illustrates the difference between the effect of shear and elongational flow on particle size. For high viscosity ratios (η_1/

process. The interfacial tension has to be very low(< 5 mN/m) to obtain a dispersed phase with dimensions smaller than 1 µm. In cases when the two blended polymers have a higher interfacial tension, compatibilizers are necessary to obtain the necessary dispersion quality. These compatibilizers are generally block copolymers consisting of monomeric units that are (preferably) miscible or at least highly compatible with the two phases of the blend system. During the melt-mixing process, they position themselves at the interface between the two miscible components, and thus overcome the repulsive effects of interfacial tension by forming chemical bonds between the two blocks. In this way, addition of a compatibilizer leads to a decrease in interfacial tension, yielding smaller particle sizes at the same viscosity ratio and shear stresses. It also provides the necessary morphological stability in this thermodynamically unstable system. Furthermore, mechanical properties are optimized if compatibilizer blocks of sufficient molecular mass are entangled in the relevant polymer phase.

The basic blend of two polymers described above can be regarded as the base matrix, which has to be further adjusted to the needs of the application by addition of additives (e.g., impact modifiers). The location of the impact modifier can be predetermined by chemical bonding with the desired phase or by specific surface interfacial effects.

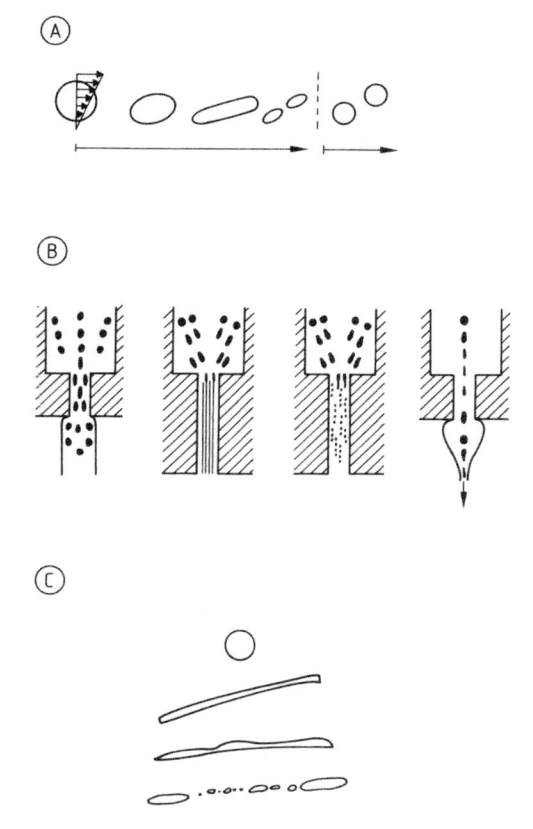

Figure 7. Droplet deformation A) Deformation caused by shear flow; B) Deformation caused by elongational flow; C) Actual droplet formation (combination of A and B)

$\eta_2 > 4$), dispersed particles do not break up in shear flow because they rotate around themselves. Elongational flows are irrotational, and dispersive mixing is possible. The final quality of the blend dispersion depends entirely on this

Predicting Morphology. Why does one of the dispersed resin phases encapsulate the other? Under equilibrium conditions such effects are probably due to interfacial energy differences between the blend components. The tendency of a liquid to spontaneously spread across a solid or liquid substrate can be expressed in terms of the surface and interfacial tensions of the components using Harkin's equation

$$\lambda_{ij} = \gamma_j - \gamma_i - \gamma_{ji} \tag{8}$$

where γ_j, γ_i, and γ_{ji} are the surface and interfacial tensions of the solid, liquid, and solid/liquid respectively, and λ_{ij} is defined as the spreading coefficient. Equation (8) can be approximated by

$$\lambda_{12} = \gamma_1 - \gamma_2 - \frac{4\gamma_1^d \gamma_2^d}{\gamma_1^d + \gamma_2^d} - \frac{4\gamma_1^p \gamma_2^p}{\gamma_1^p + \gamma_2^p} \tag{9}$$

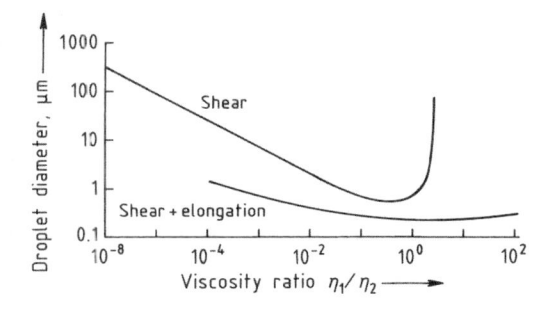

Figure 8. Influence of deformation forces on droplet diameter as a function of viscosity ratio η_1/η_2 with η_1 the dispersed and η_2 the matrix viscosity

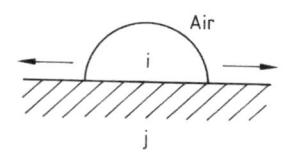

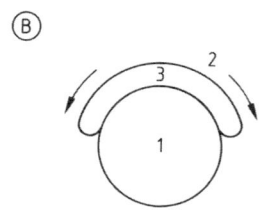

Figure 9. Schematic diagram showing spreading behavior of liquid on solid substrate (A) and one polymer phase on another within a third (B)

where γ^d and γ^p constitute the distributions of the contribution to the surface tension caused by dispersive and dipole interactions, respectively.

Spreading is predicted to occur only for positive values of λ_{12}. Melt blending is usually carried out at elevated temperature, interfacial tensions should therefore be known at processing temperatures but variation in surface tension with temperature is comparable for most polymers (0.06 – 0.08 dyne · cm/ °C).

With three or more different components a number of interfacial tension combinations result. This can create a competitive situation leading to a number of options. If the spreading coefficient λ_{12} is positive, component 1 will encapsulate component 2 and as such will be dispersed in 3 (Fig. 9) [26]. A three-phase system develops in those cases where none of the possible three-phase combinations has a positive spreading coefficient. All these morphologies are thermodynamically controlled.

One of the most severe problems encountered in blends of two immiscible components with an insufficiently low interfacial tension is loss of morphology or dispersion through agglomerization. This leads to uncontrollable thermodynamically driven, phase separation and delamination with total loss of mechanical properties.

3.4. Reactive Compounding

Reactive compounding is used in those cases where the above-mentioned agglomeration problem has to be overcome by creating chemical bonds across the interface [27]. This technique can be applied with a single-phase matrix and a two-phase blend matrix. An example is phase-selective impact modification where the modifier is chemically bonded to one phase component of the blend matrix. This is possible through the use of specially functionalized components containing reactive groups. One component may be functionalized with a group that can react with an existing reactive site in the second component. Alternatively, both components may be functionalized with mutually reactive sites.

The most commonly used chemical reactions are those between carboxylic acids or anhydrides with amines or epoxy groups, and between amino and epoxy groups. Reactive extrusion involves the introduction of these reactive groups into a nonreactive polymer by means of radically induced graft reactions with unsaturated monomers or by replacing available reactive groups with more or different reactive groups.

The term in situ compatilization denotes those cases where a functional group is introduced to establish chemical bonds between the two phases or to generate a compatibilizer in situ. An example of this technology is found in the manufacture of the completely immiscible blend of PPE and PA with the assistance the addition of, for example, maleic anhydride (see Section 4.5, PPE – PA blends).

3.5. 3.5Mechanical Properties [28]

Nearly all thermoplastics are subjected to mechanical loading during their lifetime. Mechanical properties are therefore often used as the basis for material selection. Engineering data sheets supplied by the manufacturer contain information on mechanical properties, thermal properties, electrical properties, flammability behavior, and chemical resistance. These properties are measured by standardized test procedures (e.g., ASTM, DIN, ISO), see → Plastics, Properties and

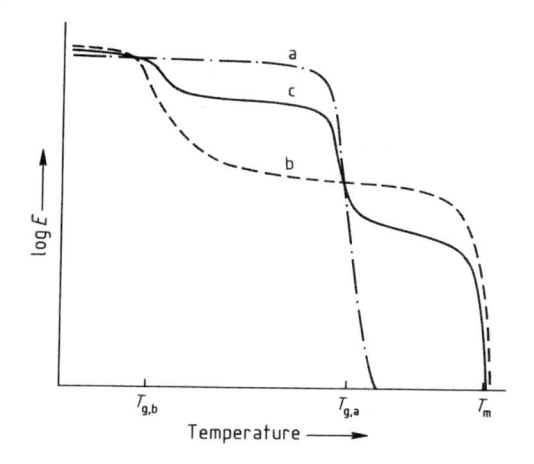

Figure 10. Generalized modulus – temperature diagram for two polymers and their blend a) Amorphous polymer with high T_g; b) Crystalline polymer (e.g., polyamide, polyester) with low T_g; c) Blend of a and b

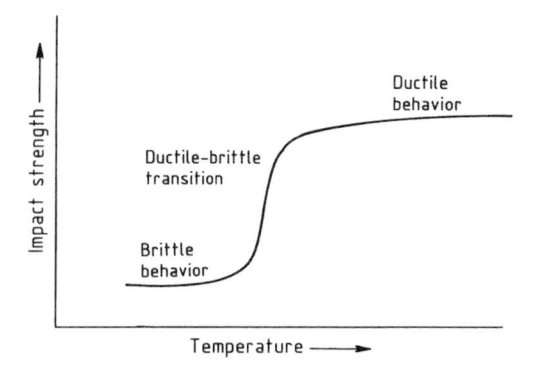

Figure 11. Typical impact behavior as a function of temperature

Testing. Important mechanical properties for engineering blends are discussed briefly in this section [28].

Characterization of a thermoplastic (blend) usually starts with a curve of tensile modulus (E) plotted against temperature (Fig. 10).

The *glass transition temperature* T_g (1 Hz) is the temperature at which the larger chain segments either gain or lose rotational freedom. Its location shifts with the frequency of the test method, i.e., the performance of a material is difficult to predict based on merely T_g values. Secondary transitions also play an important role in determining impact resistance, especially if the T_g (1 Hz) lies close to the use (or room) temperature.

Impact strength is the ability of a material to withstand impulsive loadings. Tests are performed using notched and unnotched samples. The material fracture mode is important, being either ductile or brittle and depends on temperature (Fig. 11). The location of the ductile – brittle transition is usually reported.

The *toughness* (ability to absorb mechanical energy without fracture) is measured as the area under standard (tensile) stress – strain curves. For many (engineering) thermoplastics it is usually improved by the incorporation of a low-modulus component (rubber) as a separate phase [29], [30]. When highly dispersed the rubber acts as an effective stress concentrator and initiates both crazing and shear-yielding phenomena in the matrix. Since both these processes can absorb large amounts of energy, impact-modified materials exhibit superior resistance towards crack propagation under impact conditions. The low-modulus phase may be incorporated by grafting, copolymerization, or melt blending. High stress concentration is reached when the ratio of the moduli of the rubber phase and the matrix is high. This value is increased when the rubber particle cavitates under the onset of the three-dimensional stress in the part.

Enhancement of the intrinsic toughness of the blend system increases impact resistance. The most important factor in optimizing the impact performance of a polymer resin is identifying the preferred deformation mechanism of the matrix polymer — crazing or shear band formation — because each mechanism requires a specific type of impact modification. In all cases the predominant deformation mechanism depends not only on the modified matrix system and its morphology, but also on variations in test conditions.

Two important phenomena are usually considered in *long-term behavior*: creep (the increase in strain for a material subjected to a constant load) and *stress relaxation* (reduction of stress induced in a material subjected to constant strain). Both processes provide estimates of the materials long-term dimensional stability and are important in selecting resins and design parameters for a certain application (see Section 5.2).

The *ultimate use temperature* is the maximum allowed temperature at which the material will provide the required dimensional stability under the foreseen conditions. The *continuous use temperature* is related to the oxidative stability or intrinsic stability of the blend system. This is usually determined by the weakest blend component, whereas the ultimate use temperature of the most thermally stable blend component determines the thermal resistance.

The two most commonly employed ultimate use temperatures are the Vicat softening temperature (Vicat A or B) and the heat distortion (deflection) temperature (HDT). Typical HDT values for completely amorphous polymers are 20 °C lower than the T_g values. The Vicat values are somewhere inbetween. For typical semicrystalline polymers, large differences are found between the high- and low-load HDT values. The Vicat values are usually close to the low-load HDT values.

3.6. Additives

The role of additives in modifying polymer properties should not be underestimated. See also, → Plastics, Additives.

The distribution of stabilizers between the various blend components (including impact modifiers) is important. Redistribution of stabilizers initially present in the weakest component, may yield surprising effects. For instance the UV resistance of PC – ABS blends is markedly higher than expected for an ABS system due to the presence of UV stabilizers in the continuous PC phase. Redistribution processes can be prevented by using chemically bonded additives.

Oxidation at moderate temperatures (thermal oxidation) and under the influence of UV radiation (photo-oxidation) can lead to cross-linking, lowering of molecular mass, and deterioration of mechanical properties. Unsaturated rubbery polymers, such as those used for impact modification, are particularly sensitive to attack by oxygen and heat. *Heat stabilizers* and *antioxidants* are added to suppress these reactions.

Flame retardants are applied to allow a polymer to conform to flammability requirements. A large variety of compounds (e.g., aluminum trihydrate, organic phosphates, antimony oxide in combination with various halogen compounds) are used for this purpose. They increase the amount of solid (nonvolatile) material by carbonization or cross-linking, and/or interfere in the oxidation processes taking place in the flame (flame poisoning).

Plasticizers are used to reduce the modulus of the resin (i.e., to increase flexibility) over the expected use temperature range. They are also effective as flow-improving additives and modify the continuous phase. They have to be selected on the basis of compatibility parameters. They are sometimes moderate molecular mass polymers with a low T_g, but more commonly oligomers with a molecular mass of 3000 – 1000. The use of polymeric plasticizers versus the traditional organic materials is growing due to environmental concerns (nonvolatile and less migrating) and also due to practical performance reasons.

Release agents used to provide easy mold release are selected according to their volatility and thermal stability in relation to the processing temperatures.

A number of types of *fillers* exist. Mica and lime are used to increase volume and reduce the material price. Other particulate fillers such as graphite, carbon black, and aluminum flakes are used to enhance the material's conductivity for a variety of purposes. Reinforcing fillers (e.g., glass fibers, continuous glass mats, or glass fabrics) are used chiefly to improve mechanical properties such as modulus and strength and increase the heat distortion temperature (Fig. 2).

Transesterification catalysts in blends of two polyesters may catalyze copolymer forming transesterification reactions during (melt) processing.

Catalyst complexing agents are added to prevent these unwanted side reactions. Compatibilizers have been discussed in Section 3.2. The role of *impact modifiers* has been discussed in Section 3.5.

4. Property and Application Profiles of Selected Blends

Thermoplastics are crystalline or amorphous. Crystalline resins (PA, POM, PP, PSU,

Table 2. Property profile of engineering thermoplastics

Resin	Advantages	Disadvantages
ABS	impact strength, processibility, surface appearance	HDT, flammability
ASA	weathering performance, ductility	HDT
HIPS	processibility, impact strength	chemical resistance, HDT
PA	processibility, impact strength if saturated with water, toughness, chemical resistance (crystalline PA)	water adsorption leads to general change in mechanical properties and dimensional stability, HDT
PBT	chemical and solvent resistance	processibility, shrinkage
PC	low-temperature toughness, HDT, constant modulus, high T_g	stress-crack sensitivity, solvent/chemical resistance, impact strength under certain conditions
PETP	chemical and solvent resistance	shrinkage, low-temperature toughness, crystallization speed
POM	tensile strength, modulus	impact strength
PP	chemical resistance, low price, high flammability	HDT, oxidation sensitive, brittle
PPE	HDT, rigidity, flame retardancy, impact strength	processibility, oxidative stability (UV)
PS	processibility	brittleness
PSU	excellent thermal and oxidative stability	expensive, notch sensitive

polyesters) have superior resistance to organic solvents, a low melt viscosity, and are often not transparent. They also exhibit considerable mold shrinkage and warpage. High dimensional stability is only achieved below the glass transition temperature. Above this temperature, reinforcing fillers (e.g., glass fibers) are required to provide the desired dimensional stability and strength; they also enhance the tensile modulus. Furthermore, the crystallization rate and the history determine whether a material acts according to the characteristics described in Table 2.

Blending with engineering polymers is needed to improve their intrinsic property profile and expand their application range [31]. The amorphous polymers (ABS, PC) have excellent dimensional stability up to 20 °C below their T_g; maintain their properties at elevated temperature; and show good impact behavior, high transparency, and low warpage. Amorphous polymers with a very high T_g are very difficult to process and/or very brittle. Also their application area is limited due to lack of chemical resistance. These issues had to be addressed before they were accepted by the market. Semicrystalline engineering polymers frequently exhibit a too slow or too rapid crystallization with a poorly controlled crystalline phase morphology. This causes a processing-dependent property profile, leading to unpredictable impact behavior. Furthermore their mechanical properties vary widely depending whether they are used below or above their glass transition temperature.

The advantages and disadvantages of the most important thermoplastics are listed in Table 2. The key to the development of a successful blend is to maintain the strengths of both polymers while removing their disadvantages. The heat distortion temperatures of these polymers and some engineering blends follow:

ABS	80 – 100
PBT	70
PC	135
PEI	170 – 240
PP	70
PPE	200
PS	80
SMA	105
ABS – PC	110
ABS – SMA	95
PA – PPE	170 – 190
PC – ABS	100 – 125
PC – PBT	120 – 140
PC – SMA	115
PEI – PC	170
PPE – HIPS	80 – 175

In the following sections selected commercially important blends and engineering blends are discussed. The treatment of the blends is based upon the main component (defined as being present in an amount of more than 50 wt %), and not the (for mechanical properties very important) continuous phase. Each section is concluded with a short list of applications that may be similar because different blends are targeted for the same application.

4.1. Polyolefins

Polyethylene. Polyethylene [9002-88-4] (PE) is classified according to density which is controlled by the attainable degree of crystallization and the density of the amorphous phase [32]. Densities range from 0.8 to 0.95 g/cm^3 and are primarily determined by their underlaying molecular structure which results from the way these polymers are synthesized. See also → Polyolefins, Chap. 1..

High-density PE (HDPE) is produced by comparatively low-pressure polymerization with Ziegler or complex oxide catalysts; density 0.95 g/cm^3, melting temperature 130 – 135 °C.

Low-density PE (LDPE) is traditionally made by a high-pressure route; density 0.92 g/cm^3, melting temperature 115 °C, but lower crystallinity than HDPE due to presence of branching structures.

Linear low-density PE (LLDPE) is comparable in structure and general properties with LDPE but lacks long-chain branching. It is produced by copolymerizing ethylene with butene, hexene, octene, or 4-methylpentene at low pressure.

Very low-density PE (VLDPE) is an extreme version of LLDPE; density may be as low as 0.88 g/cm^3.

Ethylene copolymers are typically polymers of ethylene with polar monomers such as acrylates.

Ultimate rheological properties can be adjusted by blending to suit the processing techniques used. Density can also be adjusted by blending because it simply follows linear additivity of the components used. From a practical point of view the polyolefin systems are completely compatible due to their very low interfacial tension.

Vinyl, acrylic, and even carbon monoxide comonomers are used to modify polyethylene. The possibilities of these modifications are, however, restricted because crystallinity and melt temperature rapidly decrease with comonomer content, resulting in products with a very low modulus of elasticity. Rubbery materials are produced in this way and are frequently used as impact modifiers for blends. Although chemical modification of polymers is preferred (primarily due to cost reasons) some property profiles can only be obtained through blending.

Polyethylenes are widely used as blend components because they improve a number of important properties such as flow, impact resistance, chemical resistance, and environmental stress cracking sensitivity. Cost – price reductions and volume – price improvements are an additional advantage.

One of the most important application areas of PE blends is the packaging industry. They display barrier properties in particular toward oxygen and water vapor, and chemical resistance toward aqueous systems, edible fats, and oils. In blends of polyamides with slightly functionalized PE, the desired blend morphology is best described by the term controlled delamination (i.e., agglomeration of the dispersed PE phase leads to a two-phase layer structure but does not reach an undesirably high level that would cause delamination). This can only be achieved through extrusion shaping [33]. The inherent incompatibility in PE – PA blends is overcome by chemical bond formation between carboxylic or maleic anhydride groups of the modified PE phase and amine end groups of the PA phase. The following types of compatibilizers have been used: maleic-anhydride or chlorine-modified polyolefins [34], ethylene – methacrylic acid copolymers, ionomers (maleic-anhydride-grafted PE), and epoxy compounds. Increasing the degree of PE modification with polar groups improves the adhesion of these intrinsically nonadhering polymers to more polar polymer blend components to give extremely tough materials [33], [36], [36]. Applications include automotive parts (mainly injection-molded compounds), power tools, farm and garden equipment, as well as sport and recreational markets.

PE – isobutene films have improved properties and toughness compared to HDPE films. Typical applications are single-layer extrusion films, and heavy-duty bags. HDPE, LDPE, and LLDPE polymers are blended in different ratios to obtain films in a customer-preferred density. Uses include films for bags.

Polypropylene. Polypropylene (PP) [9003-07-0] (→ Polyolefins, Chap. 2.) is used as homopolymers, random copolymers and block copolymers. Polypropylene is one of the most widely used commodity plastics. It has a good chemical resistance, excellent thermal stability, a low price, and high flowability. The polymer has three drawbacks: (1) it is sensitive to oxidation during processing and use, (2) it has a poor resistance to UV-induced degradation, and (3) it is rather brittle even though its glass transition temperature (-10 °C) is below room temperature.

Thermal oxidation problems have been solved by use of specially developed additive packages. Impact performance can be improved by chemical modification (copolymerization) or blending. Comonomers are mainly ethylene and higher olefins. Cheap rubbers such as EPDM (ethylene – propylene – diene rubber) are frequently used for blending [37]. The miscibility between polyolefins is extremely limited in spite of their great chemical similarity (Fig. 7). However, due to the small interfacial tension the dispersibility is acceptable. Nevertheless, when blends of PP and other polyolefins are being made, they often require a compatibilizing block copolymer to provide the necessary degree of dispersion and to maintain morphological control during processing. The latter is essential for obtaining the desired mechanical properties of the product. Morphological stability can be further enhanced by in situ cross-linking of the dispersed phase (dynamic vulcanization) [38]. The main applications of PP and its blends are in bumper systems and dashboards.

Blending with incompatible polymers can be achieved via utilization of a functionalized PP (e.g., by grafting with maleic anhydride). The inherent interfacial tension differences are thus overcome by means of chemical bonds instead of physical adhesion. Blends of PP and PA based on this technique are becoming available [37].

4.2. Poly(Vinyl Chloride)

Poly(vinyl chloride) [9002-86-2] (PVC) is the oldest and one of the most important plastics, see → Poly(Vinyl Chloride). Worldwide sales of PVC-based plastics were expected to exceed 13×10^6 t/a in 1991 [39].

PVC has a number of important properties. Its amorphous character and T_g of 75 °C provide an excellent basis for good dimensional stability under load up to 60 °C. Its high polarity gives chemical resistance towards a broad range of solvents and an excellent flame retardancy. However, pure PVC is rather brittle under notched impact conditions and has a limited thermal stability. This last property has successfully been addressed by stabilization techniques. In combination with extensive utilization of blend technology, all conventional processing techniques are now possible for PVC. The abundant presence of strong polar C – Cl bonds make PVC highly suitable for blending. Depending on the concentration of strongly interacting dipoles in the blend components, completely miscible or compatible blends can easily be created. The resulting two-phase morphologies either have a particulate or a controlled delaminated character which controls the final deformation mechanism and its performance in blends of PVC.

Chlorinated polyethylene (CPE) is being used as a blend component to improve the impact and flow behavior of PVC [40]. Compatibilizers are not necessary because the chlorine content can be adjusted to the required compatibility level (35 wt %).

Similar effects are obtained for copolymers of ethylene and vinyl acetate (EVA) and/or ethylene and ethyl acrylate (EEA) [40]. The ester groups interact with the H – C – Cl groups of the PVC and provide the required compatibility if the comonomer content is in the range 30 –40 wt %. Afterchlorination of PVC yields CPVC which has a higher heat distortion than PVC. The T_g increase is directly related to the chlorine content. CPVC can be used directly or in completely miscible blends with PVC to obtain products having higher T_g s.

Heat-resistant blends can also be obtained by blending PVC with α-methylstyrene – acrylonitrile copolymer (see Section 4.3) which is completely miscible with PVC. Blends are also available with a low T_g. An important example is the blend of PVC with NBR rubber; depending on the nitrile content (critical level, 32 wt % acrylonitrile) a single-phase soft blend or a two-phase impact-modified PVC can be made.

Impact performance can be improved by blending with two-phase impact-modified materials such as ABS and ASA (acrylonitrile – styrene grafted onto acrylic rubber). ASA does not impair weatherability as is the case for ABS. Blends with poly(methyl methacrylate) (PMMA) also show good weatherability performance. Blends of PVC with ABS, ASA, SMA, and PMMA are mainly applied as structural panels in the building industry. The blends are extruded in sheets which are shaped by vacuum forming and press molding. The PVC provides mechanical strength and flame resistance. The second component should provide the required heat distortion temperature and surface properties. The impact modification of PVC and its blends has been increased by using shell – core impact modifiers [41]. These modifiers consist of a very small rubber core, usually a butadiene homo- or copolymer or a poly(butyl acrylate) copolymer, and a "hard shell" made of PMMA or SAN. The shell polymers should have optimal interaction with the matrix PVC. These impact modifiers are not only very effective but also allow high shear processing without deterioration of surface appearance.

Since 1980, PVC and its blends have been able to penetrate the injection-molding market. This was made possible by (1) improving the flowability of PVC by an order of magnitude via reduction of its molecular mass, (2) a greatly improved stabilization package, and (3) improved thermal control in modern molding machines to prevent overheating.

The construction and pipe markets are supplied by PVC – chlorinated polyethylene blends (preferred in the United States) and PVC/EVA blends (preferred in Europe). Both blends have better impact properties than PVC. EVA blends have a slightly better UV stability.

PVC – polyurethane blends are typically used in shoe soles, boots, oil-resistant articles, and calendered coatings by virtue of their UV stability, better thermal stability, and low temperature flexibility.

The additions of small amounts of methyl methacrylate (MMA) copolymers to PVC improves its hot melt strength and reduces melt fracture during processing. Toughness and impact performance can be further improved by blending with MMA-containing impact modifiers. Applications include clear films and window profiles.

4.3. Styrene Polymers and Copolymers

See also → Polystyrene and Styrene Copolymers.

Polystyrene. Polystyrene [9003-53-6] (PS) is a very brittle polymer (T_g 100 °C) due to a low entanglement density and lack of secondary transitions. Its usefulness has therefore been limited severely. Poor impact performance has been a major concern and was originally addressed by blending PS with styrene – butadiene (SBR) rubber and subsequently by the now generally established chemical processes for making high-impact PS (HIPS). The block copolymer styrene – butadiene (SB) and terpolymer styrene – butadiene – styrene (SBS) as well as their hydrogenated derivatives, are used as blend components to improve impact properties [36]. The resulting deterioration of the continuous use temperature (T_g reduction by at least 10 °C) can be solved by either making copolymers of styrene with stiffer or more polar comonomers such as α-methylstyrene, maleic anhydride, acrylonitrile, or by blending with a miscible high-T_g polymer such as poly(tetramethylbisphenol A carbonate) or PPE (see Section 4.5). Blending relatively small percentages of PPE (< 30 %) with PS can be important because it increases the heat distortion temperature by about 1 °C per percent PPE. The mechanical deformation behavior of the blends is also greatly improved by the strong increase in the overall entanglement density. The overall property profile of pure PS is not good enough to make it an interesting blend component for a two-phase system.

PS – polyolefin blends have been developed for packaging purposes bringing the combination of printability, fat resistance, and barrier properties. They again have a controlled delaminated morphology and are typically extrusion materials. However, the more polar and functionalized PS copolymers offer far better blend possibilities and are widely used and discussed below.

ABS. See also → Polystyrene and Styrene Copolymers. Most commercial acrylonitrile – styrene – butadiene polymers [9003-56-9] (ABS) are blends of rubber-impact-modified styrene – acrylonitrile (SAN) copolymer with

SAN or other SAN-miscible copolymers. Blending can take place in a suspension or emulsion by coagulation or melt blending during the compounding phase. ABS can be produced by three different techniques: emulsion, suspension, and mass (bulk) polymerization [43], [43]. Each technique delivers ABS with slightly different properties.

The main properties of ABS resins are their good to very high impact strength and low price, coupled with very acceptable mechanical (engineering) properties up to 90 °C. They have an excellent surface appearance and good chemical resistance. In addition, they possess the excellent injection-molding behavior of PS, good melt flow, low mold shrinkage, and an excellent surface finish. Their main disadvantage is their poor resistance toward prolonged exposure to light and heat due to the unsaturated bonds in the butadiene rubber impact modifier systems. UV resistance can be improved by incorporation of acrylic rubbers (ASA). Principal advantages of ASA (SAN grafted onto acrylic rubber) over ABS are high outdoor stability, high resistance to yellowing on exposure to UV light, and good resistance to environmental stress cracking [43]. Other essentials of ABS are its composition-dependent T_g (100 – 120 °C) and a continuous use temperature between 80 and 100 °C. The actual use temperatures for ABS can be enhanced up to 118 °C through the incorporation of α-methylstyrene. Its high flammability can be improved by addition of 10 – 15 % bromine-containing compounds and 3 – 5 % antimony trioxide. This results in a higher density, reduced color intensity, lower surface properties, reduced impact performance and processability, and corrosivity. Blending is a preferred method used to increase the heat distortion temperature, to improve weatherability (unsaturated impact modifiers), and overcome flammability concerns.

ABS – PC and ASA – PC Blends. These blends offer a somewhat higher heat distortion temperature than ABS and good processing characteristics [43–45]. The Vicat softening temperature of 100 °C for ABS can be raised to 130 °C by blending with PC. The carbonate group of the PC and the nitrile group of the SAN matrix interact [46], providing just enough compatability for a useful blend. Optimum interaction

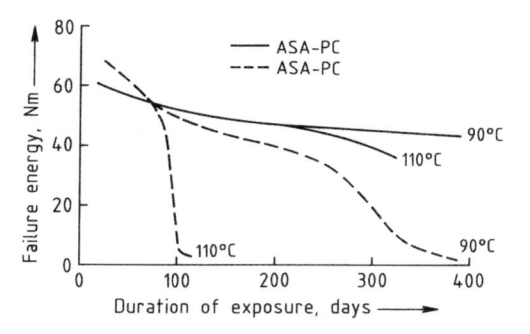

Figure 12. Failure energy of ASA – PC and ABS – PC as a function of the duration of exposure to heat

between the polymers is obtained for a SAN containing 30 wt % acrylonitrile [46]. Intrinsic compatibility can be enhanced through addition of a third component containing ester groups (e.g., PMMA or PBT) [47], an effect that can be explained in terms of mutual miscibility.

ASA – PC blends offer a better resistance towards UV light and heat aging compared to ABS/PC (Fig. 12). Applications include instrument panels, distribution boxes, and kitchen exhaust hoods [48].

ABS – PVC Blends. ABS has excellent surface properties with high gloss, but burns readily and completely. PVC (T_g 85 – 95 °C) on the other hand lacks the good impact behavior and processibility of the styrenics. An excellent interaction exists between PVC and ABS, removing the need for compatibilizers. ABS – PVC blends are used in low-cost electrical panels such as covers and enclosures [7].

SMA – ABS Blends. Copolymers of styrene and maleic anhydride (SMA) have glass transition temperatures ranging from 100 to 140 °C, depending on the maleic anhydride content. The materials are rather brittle, but impact-modified resins are known. Impact performance can be improved by blending with ABS. Both resins are compatible and have a good processibility (styrenics). The miscibility window is remarkably small considering the similarity in structures [49]. Instrument panel frames is a large automotive application.

ABS – PA Blends. Both chemical resistance and heat distortion temperature can be improved by blending ABS with semicrystalline polyamides. Typical applications are automotive body panels, connectors, and under-the-hood components [7], [48]. See also Section 4.4.

4.4. Polyamides

Polyamides (PA) are either semicrystalline or amorphous. The polymer chain can consist of aliphatic, cycloaliphatic, and aromatic components. More than 60 different commercial polyamides are known. For further details, see → Polyamides.

Examples are included of the two most important general-purpose polyamides: (1) PA 6 [25038-54-4] made by ring opening polymerization of ϵ- caprolactam, and (2) PA 66 [32131-17-2] made from hexamethylenediamine and adipic acid. These aliphatic PAs are mostly used for their combination of wear resistance strength, chemical resistance, creep resistance, and low coefficient of friction. Good dimensional stability can only be obtained for (glass) filled materials which have the disadvantages of warpage and machine wear. Polyamides have a good strength and fatigue resistance. Tensile modulus values of the matrix can be increased by incorporation of aromatic monomers in the main chain. The fully aromatic PAs have a high intrinsic flame retardance. Decomposition temperatures for most aliphatic and aliphatic – aromatic PAs, are 280 – 300 °C, limiting the processing temperature.

The properties of PAs are highly dependent on the interactions of the polar amide linkage. Higher-melting PAs can be obtained by increasing the frequency of amide linkages [50]. The amide linkage interaction between the polymer chains in the crystal structure via hydrogen bonding is particularly strong in the amorphous phase. It also causes water absorption which is related to the methylene – amide ratio. Water absorption decreases the dimensional stability of the polymer, reduces the T_g (plasticizing effect), but increases the toughness. The magnitude of the T_g reduction with respect to relative humidity changes depends on the amount of amide linkages and stiffness of the chain. A major drop in tensile modulus at room temperature results when the T_g drops from 50 to below 0 °C. Other physical properties (e.g., dielectric constant, water permeability) also change considerably.

There are three major reasons for blending PAs:

1. To increase their heat distortion temperature (in a conditioned state)
2. To reduce their moisture sensitivity and thus enhance dimensional stability
3. To improve their dry impact strength (at low temperatures)

Polyamides are very suitable for in situ compatibilization techniques (reactive extrusion) due to the presence of reactive amine or carboxyl end groups. Carbonamide groups are also available for hydrogen bonding.

PA – PA Blends. Melting points of semicrystalline PAs are sometimes very close to their degradation temperatures, creating either a very small or no processing window. Blending aromatic semicrystalline PAs with aliphatic PAs improves their processibility. Aliphatic PAs generally have lower T_gs, but a higher water absorption value which can lower the T_g in the dry state significantly. Aromatic PAs possess more free volume, and have a lower dimensional growth with the same water absorption.

Blends of amorphous transparent PAs have been reported [51]. Predictions of miscibility windows between PAs using binary interaction and intramolecular repulsion models proved that the amide concentration was more important than compositional differences [52–54].

PA – ABS Blends. Blends of PA 6 and PA 66 with ABS combine chemical and abrasion resistance with high impact resistance. The ABS phase provides good dimensional stability up to 90 °C that can be further improved by substitution of styrene in the ABS with α-methylstyrene [55]. These blends are also more warp resistant due to the reduced (semi)crystalline volume fraction. Compatibilization is achieved through hydrogen bonding between amide NH and nitrile CN groups. Anhydride-containing polymers such as SMA have been used to improve compatibilization by covalent bonding and to obtain a smaller particle size [56].

Triax 1000, (Monsanto) is an example of a PA – ABS blend. It is used for a variety of markets such as functional (mainly under-the-hood) automotive components, power tools, appliances, and sports goods. PA – ABS blends have excellent chemical resistance, stress resistance, and toughness due to their co-continuous network.

PA – PC Blends. Blends of PA 6 and PC have been investigated over the full range of compositions, and proven to be immiscible [57]. Chemical reactions are claimed to produce low molecular mass compounds that both plasticize and slightly compatibilize the blend. No commercial PA – PC blends have yet been developed.

PA – PP Blends. Blending PAs with PP raises impact strength, solvent and moisture resistance, and improves strength [34]. The addition of compatibilizers and/or glass fibers is required to obtain a blend with a stable morphology. Compatibilizers are formed in situ by using maleic anhydride-functionalized styrene – ethylene – butadiene – styrene (SEBS) triblock copolymers [58], [59], ethylene – propene copolymers, PP, HDPE, or glycidyl methacrylate copolymers [60] in concentrations of 1 – 30 wt %. Maleic-anhydride-modified compatibilizer intermediates yield a better dispersion (with particle diameters in the range 0.1 – 0.3 μm) and high impact values. Interestingly, material toughness is not enhanced by using smaller rubber particles [59].

PP is usually processed at 200 °C, but PA – PP blends are processed at 250 °C — the presence of PP leads to a readily flowing resin. These resins are used for decorative applications where dimensional stability is much more important than mechanical properties. Examples include automotive applications, sprinkler housings, and housings for industrial tools.

Impact-Modified PA. The notch sensitivity of most PAs can be reduced by blending with 20 – 30 % functionalized impact modifier such as maleic-anhydride-functionalized HDPE [61], [62], or maleic-acid-modified EPDM rubber. PA 66 has an intermediate level of ductility, characteristic of a semicrystalline resin. Chain mobility is reduced by the presence of crystalline regions and relaxation times of polymers are increased. Impact modifiers are used to create overlapping stress fields (optimum conditions at 30 vol % impact modifiers) which enhance ductility [61]. Good adhesion is provided by the maleic anhydride functionalities on the impact modifiers.

The materials are used for all applications where ordinary PAs fall short of required impact strength and toughness: oil pans, fan blades, connectors, and power tool housings.

4.5. Poly(2,6-Dimethyl-1,4-phenylene Ether)

Poly(2,6-dimethyl-1,4-phenylene ether) [25134-01-4] (PPE) and its blends have been extensively reviewed elsewhere, → Poly(Phenylene Oxides), but will be briefly summarized. PPE is an amorphous resin (T_g 215 °C) with a number of attractive properties [63]. It has a serious drawback however: its (melt) processing temperature is higher than temperatures at which severe side reactions take place, thus altering the polymer property profile.

PPE – HIPS Blends. The full miscibility of PPE with PS, discovered in the early 1960s [64], [65], led to the development of blends with lower (and thus acceptable) processing temperatures. Polystyrene has a low T_g and good melt processibility.

PPE – PS blends are fully miscible over the complete composition range as illustrated by a single T_g for each composition. This provides the end user with a tremendous flexibility in achieving required maximum use temperatures. Resulting heat distortion and Vicat B temperatures are discussed in [63], → Poly(Phenylene Oxides).

The miscibility of PPE with styrene copolymers and polymers of styrene derivatives, e.g., poly(α-methylstyrene) [64–66] has been studied. Many of these blends also demonstrate full miscibility, probably due to the interaction of the electron-deficient methyl groups of PPE with the electron-rich phenyl rings of PS [63], [67]. The reduction of impact performance of the ductile PPE by blending with brittle PS (Fig. 13) can be attributed to a reduction of the entanglement density and some loss of free volume [63], [67]. Impact performance is

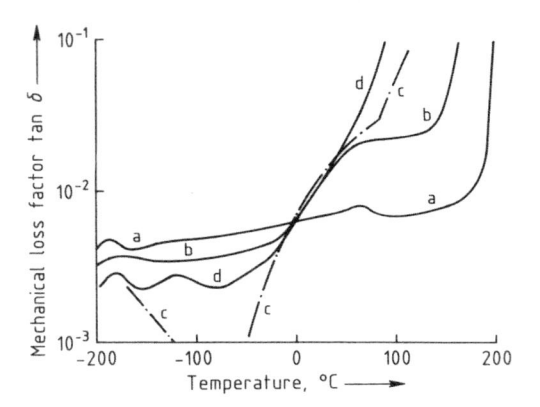

Figure 13. Impact strength of PPE – PS blends as a function of temperature a) PPE; b) PPE – PS (70:30); c) PPE – PS (30:70); d) PS

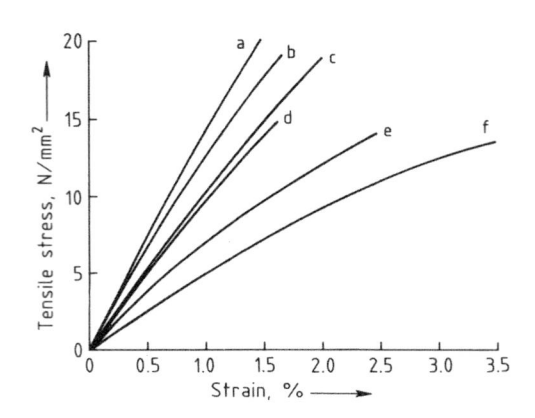

Figure 14. Creep behavior of PPE – PA 66 (Noryl GTX 900) at 23 °C and 50 % RH a) PPE – PA 66, 1 h; b) 100 h; c) 5000 h; d) PA 66 1 h; e) 100 h; f) 5000 h

greatly improved by incorporation of impact modifiers. Thermograms of commercial blends have two glass transition temperatures, with one of them being attributable to the rubber impact modifier [68].

Impact strength increases with PPE molecular mass. Commercial PPEs have weight-average molecular masses in the range 30 000 – 40 000. The rubber particle size is also important. Since their commercial introduction in 1964 by General Electric, PPE – HIPS blends have become the engineering polymer blends with the largest annual sales. They are used for applications ranging from automotive interior parts including dashboards and instrument panels, to business machine housings and chassis; from electrical conduits and lighting covers to gears and sprockets.

PPE – PA Blends. PPE – PA 6 and PPE – PA 66 blends have been extensively discussed elsewhere, see → Poly(Phenylene Oxides). In these blends, the finely dispersed PPE phase raises the continuous use temperature. The tensile modulus is maintained up to almost 205 °C with only a minor drop over the temperature range from 0 – 50 °C due to the T_g of the amorphous PA phase. The moisture sensitivity of the system is also reduced. The PA 6 and PA 66 phases provide easy processibility and chemical resistance. These PAs can be easily replaced by PAs with higher glass transition and melting temperatures [69–71]. Rubbery impact modifiers have been added to improve low-temperature impact performance [69–71]. These blends exhibit mechanical properties which are less moisture-sensitive than PA 66 as shown by the stress – strain curves (Fig. 14). At a stress level of 7 MPa, PA 66 exhibits three times as much creep as the PPE – PA 66 blend.

Blends of PPE and PA have been specifically developed for exterior automotive parts with a Class A surface quality (e.g., fenders of the Renault Clio 16 S). Other applications are rear quarter panels, and hood materials for lawn and garden tractors.

PPE – Polyolefin Blends. The usefulness of unmodified PPE is limited by its notch sensitivity and difficult melt processability. Addition of polyolefins improves flow by modifying the melt rheology of the system but sometimes lowers the heat distortion temperature. New blends have been developed in which PPE is combined with special rubbers (e.g., high molecular mass polyoctanomer [72], [73]) that have a high impact strength, a high softening temperature up to 205 °C, and facile processibility. These PPE – polyolefin blends have a high gloss and are used in metallizable head lamp reflectors. Other application areas are dimensionally stable exterior automotive body panels, and major appliances.

The intrinsic flame retardancy of PPE has been enhanced with additives resulting in a new, halogen-free, flame-retardant material that meets the most stringent low smoke [73], low toxicity, and low corrosivity standards. The key factor is increased char formation under fire conditions which reduces the amount of volatiles [73]. Potential applications of this material (Noryl Xtra LS) are in office buildings, trains, and tunnels.

4.6. Polycarbonates

The polycarbonate of bisphenol A [25037-45-0] (PC) is a linear polymer with a low rate of crystallization. See also → Polycarbonates. Thus, it is regarded as an amorphous polymer (T_g ca. 145 °C). Its maximum use temperature is approximately 130 °C and, because of its excellent thermal oxidative stability, this almost coincides with its continuous use temperature (125 °C). PC is noted for its high impact resistance, especially at low temperature. This is due to the high segmental mobility of the polymer chain (gamma transition at −100 °C [74]). Polycarbonate suffers from stress – crack sensitivity coupled with poor chemical resistance. The latter prevents PC from being used in automotive applications where frequent contact with gasoline and oils would ruin its material properties. Chemical resistance can be improved by blending with semicrystalline polyester resins. The stress cracking sensitivity of PC can be overcome by blending with ABS or ASA.

PC – Polyester Blends. The advantages resulting from blends between PC and PBT are summarized in Figure 15. Blends of PC with polyesters are based on a partial miscibility of both components in the amorphous phase

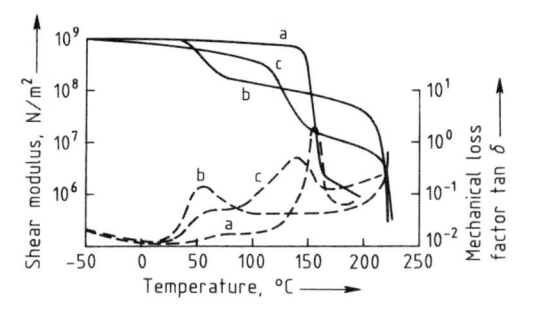

Figure 16. Dynamic mechanical properties of PC, PBT, and PC – PBT blends : shear modulus (————) and mechanical loss factor (– – –) as a function of temperature a) PC; b) PBT; c) PC – PBT (55 : 45)

[75–77]. Thermal and dynamic mechanical analysis shows a sharp T_g transition for PC at 150 °C (Fig. 16), and a sharp melting point at 225 °C with a broad glass transition for PBT due to the spectrum of relaxation times of the crystal – amorphous polymer interphase. Partial miscibility of PC and PBT is indicated by a shift of the PC T_g to 141 °C in a PC – PBT blend.

Blending both resins results in a complex co-continuous morphology. Resistance towards chemicals (e.g., oils and lubricants) is strongly enhanced as compared to PC. However, the impact properties of these blends are still insufficient for automotive applications. Impact modification with a rubber compound (frequently core – shell impact modifiers) solves this problem [78], [79].

The key to morphology control and thus mechanical properties is control of transesterification. Care should be taken to prevent these reactions (Figs. 17 and 18) catalyzed by residual

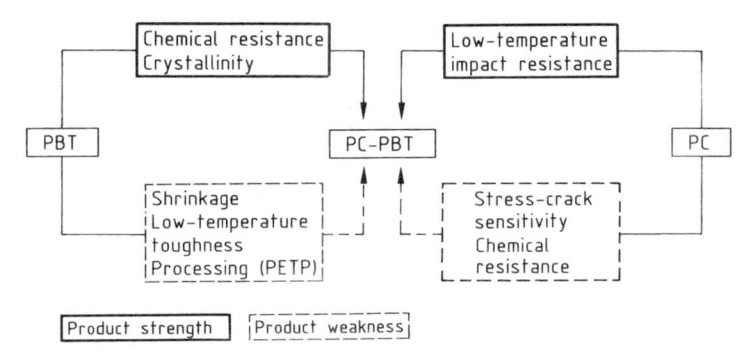

Figure 15. Blend motives for PC and PBT

Figure 17. Ester interchange between PC and PBT

polycondensation catalysts present in the PBT [79–81]. Addition of proprietary additive packages is commonly practised. In addition, processing temperatures of 270 – 290 °C should not be exceeded.

One of the most successful applications of the blend of PC, PBT, and impact modifier (Xenoy, GE Plastics) is a car bumper, developed for the Ford Sierra in Europe, and transferred to the Ford Escort in the United States. This bumper withstands 5 km/h impact at high and low temperature.

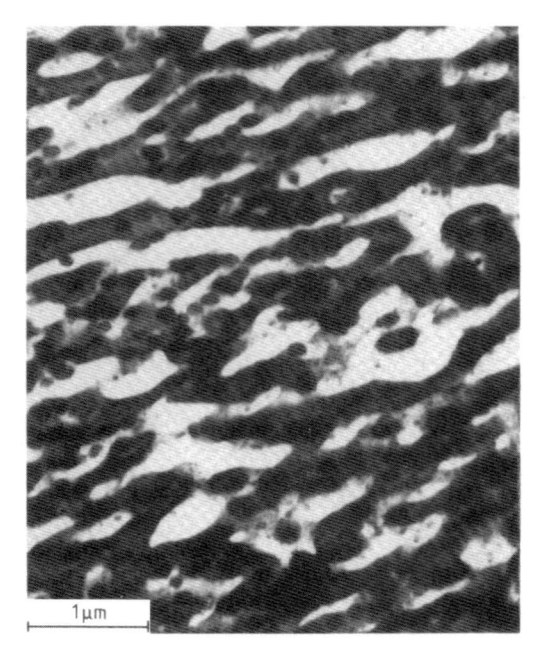

Figure 18. Morphology of impact-modified PC – PBT blend (Xenoy, GE Plastics)
The darker areas are PC, the black dots are impacted modifiers, the remainder is PBT.

PC – PETP blends have similar properties to their PBT analogs [7], [82], [83]. The lower crystallization rate of PETP allows the development of transparent materials for use in medical applications by virtue of their good dimensional stability, clarity, and low water absorption. The products have a better chemical resistance than PC and are impact modified to give acceptable low-temperature impact performance. PC –PETP blends flow better than PC, and are injection-moldable. Unreinforced PETP cannot be injection molded due to its low rate of crystallization.

PC – Polyolefin Blends. Despite its good toughness, PC has a distinct ductile to brittle transition which is dependent on temperature, thickness, test speed, stress configuration, and molecular mass. The dependency is reduced if PC is blended with low levels (4 – 8 %) of LDPE. This shifts the ductile – brittle transition of PC from about 0 °C to below − 20 °C. Moreover, the effect of part thickness on notch Izod impact strength is reduced [84].

PC – ABS Blends. Polycarbonate blends with ABS resins have been successfully used in automotive interior and exterior applications [85], [85]. The acrylonitrile content of the ABS plays a key role in determining the miscibility of the SAN phase with the PC phase. ABS has no low-temperature impact performance, whereas PC does through its gamma transition at − 100 °C and ample free volume [74]. The impact values of PC – ABS blends are directly correlated with their heat distortion temperatures. The presence of the rubbery impact modifier slightly reduces the tensile modulus. The HDT values are hardly affected at modulus values above 3 GPa. These are important values because they are required for high-speed impact tests for automotive applications. In addition, PC – ABS blends provide paintability and better UV stability if PC is the continuous phase [86–89]. The addition of UV absorbers to the (transparent) PC phase improves the protection of ABS against the damaging effects of UV light (see Section 4.3). PC –ASA blends have superior weather resistance, however.

PC – ABS swells in aromatics, ketones, esters, and hydrocarbons. It has good electrical insulating capability. Major applications are instrument panels [7] and business equipment housings [90].

4.7. Polyesters

Thermoplastic polyesters have a low moisture absorption and are chemically resistant apart from being susceptible to hydrolysis at high temperature in alkaline environments. They have a low dimensional stability due to their low T_g; glass-filled materials are required to achieve acceptable use temperatures. Impact performance for these highly filled materials is poor, but can be improved by blending with impact modifiers. The most important polyesters are poly(ethylene terephthalate) [25038-59-9] (PETP) and poly(butylene terephthalate) [24968-12-5] (PBT). See also, → Polyesters PBT crystallizes within 1 – 3 s whereas PETP crystallizes within 15 – 20 s. PBT's higher melt flow and faster crystallization lead to a shorter cycle time. Special processing conditions are required if (partially) crystalline PETP is to be obtained: high mold temperature (120 – 140 °C) and long cycle times (2 min). The advantages of PETP over PBT are its higher melt temperature (256 vs. 226 °C) and strength.

PBT – PETP blends have enhanced surface quality and reduced warpage when reinforced with glass fibers. Both qualities are attributable to the lower rate of PBT crystallization caused by the addition of PETP. The weight ratios used in PBT – PETP blends are usually 85 : 15 to 75 : 25. HDT values are not affected, but costs are reduced due to the lower price of PETP. Typical ap-plications include hot-appliance housings (e.g., toasters, irons, and frying pans).

PBT – PC Blends. Addition of PC to glass-filled PBT grades increases impact strength values and improves thermal expansion, flow, and flatness (reduced warpage). The chemical resistance of these blends is slightly lower than that of a PBT compound. Core – shell impact modifiers are used for toughening [78], [79], [91]. The importance of interfacial adhesion with the matrix has been proven [91], [92]. Applications include grill doors, air cleaners, and appliance components where improved dimensional stability over PBT is required.

Miscelleneous Polyester Blends. Noteworthy are blends of PBT with copolyester elastomers containing short crystallizable PBT blocks [93]. The tensile strength and modulus of these compounds are increased upon blending with PBT while retaining its flexibility. Excellent surface quality and improved paint adhesion qualify these resins for bumper fascias. Similar uses hold for PETP – acrylic and PBT – acrylic materials. Toughened PBT and PETP blends are used to extend existing product lines by varying the amount and type of impact modifier. Ignition compounds, housings, and windshield wipers are typical automotive applications; flame-retardant grades are used for connectors and switches.

4.8. Polyetherimides

Polyimides are condensation polymers derived from bifunctional carboxylic anhydrides and primary diamines. Polyetherimides [61128-24-3] (PEI) contain an ether group in either or both the aromatic dianhydride or aromatic diamine building block. Ultem polyetherimide (GE Plastics) produced from bisphenol A bisether dianhydride is the only commercially available thermoplastic polyetherimide [94] (see also → Polymers, High-Temperature). It has a glass transition temperature of 217 °C, and can be subjected to conventional melt-processing techniques. Its thermal stability is maintained up to ca. 500 °C under nitrogen [95].

In commercial PEI grades, additives (e.g., reinforcing fibers) and modified base resins are used to increase the glass transition temperature and chemical resistance. Primary reasons for blending are therefore to improve processibility, impact strength, and chemical resistance. PEI is used mainly in under-the-hood automotive applications, head-lamp reflectors, injection-molded printed circuit boards, trays for microwave ovens, and high-performance composite sheets with PEI as the matrix.

PEI – PC Blends. PEI – PC blends combine the high temperature resistance and intrinsic flame retardance of PEI with the good impact performance and ease of processing of PC. Interactions are strong enough to form an excellent blend without using compatibilizers. The high continuous use temperature of PEI is not decreased because PC is stable to thermal oxidation up to ca. 175 °C. Blending with PC reduces the price of the material and its notch sensitivity.

Two grades of PEI – PC thermoformable sheets exist, one is targeted to household applications (e.g., Tupperware microwave cookware), the other to aircraft interior components where the material satisfies all regulations pertaining to flammability, smoke, and toxicity. PEI – PC is used for aircraft window housings and seat backs.

Thermoformable PEI – PC products are also used as high-temperature paint masks in the automotive industry to prevent overspray. The high heat resistance and excellent dimensional stability enable the material to withstand the temperature conditions in the painting operation.

PEI – PEEK Blends. PEI is an amorphous resin. Its chemical resistance can therefore be improved by blending with a semicrystalline resin. PEI is miscible with the amorphous phases of polyetherketone [27380-27-4] (PEK) and polyetheretherketone [31694-16-3] (PEEK) over the entire composition range. Crystallinity and crystallization are not affected by incorporation of up to 40 wt % PEI [96]. The product range of these materials can be further enlarged by copolymerization of the PEI structure leading to enhanced inherent properties.

Markets for PEI – PEK and PEI – PEEK blends have not yet been fully exploited [97]. A 70:30 PEI – PEEK grade developed by Amoco Performance Polymers was not commercialized due to competition with the liquid crystalline polymer Xydar [7].

4.9. Polysulfones

Polysulfones [25135-51-7] (PSU) and polyethersulfones [25667-42-9] (PES) are amorphous high-T_g polymers with excellent thermal and oxidative stability (see also, → Polymers, High-Temperature). Their environmental stress-cracking resistance against most liquids encountered in the food and automotive industries is good because of the polar sulfone group.

Polysulfones are, however, expensive, demonstrate a severe notch sensitivity, and have a limited chemical resistance resulting in stress cracking sensitivity under certain conditions. Both problems can be overcome by blending with ABS (Fig. 19), PAs, PETP, and PBT.

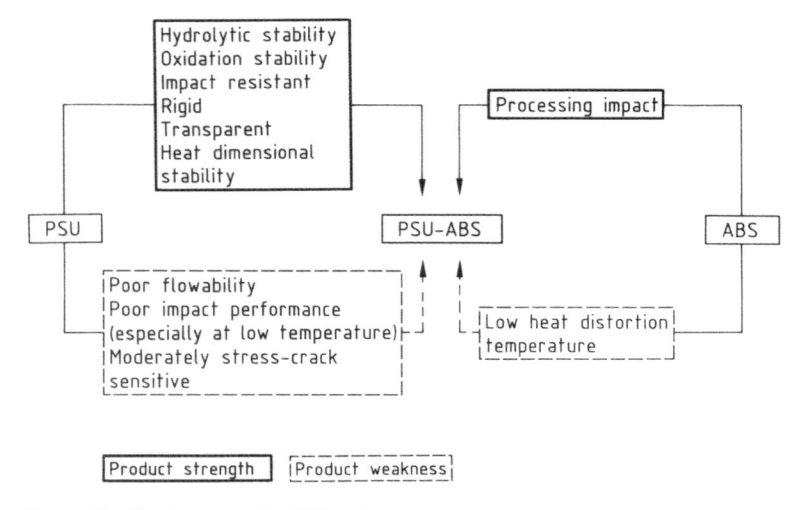

Figure 19. Blend motives for PSU and ABS

Chemical resistance is strongly improved if the semicrystalline resin constitutes the continuous phase. The principles and objectives followed for developing these blends are the same as those for PC – ABS, PC – PETP, and PPE – PA blends. Polar interactions are possible between the carboxyl groups of PSU and the nitrile groups of ABS.

Applications include household water faucets, electric, and electronic applications. PSU – PBT replaces PBT in selected connectors and switches due to its higher strength [7], [48].

4.10. Polyoxymethylenes

Polyoxymethylene [9002-81-7] homo- and copolymers are highly crystalline materials well known for their hardness, low friction coefficient, chemical stability, and relative brittleness; see also → Polyoxymethylenes. They are highly flammable (LOI 16) and their thermal stability is limited to 210 °C. In view of their melting temperature of 182 °C (homopolymer), careful processing is required.

Blends of polyoxymethylene homo- and copolymers are known. Silicone oils and polytetrafluoroethylene have been added to POM to enhance the high slip characteristics. Impact-resistant grades of POM use polybutadiene, acrylic rubber, grafted synthetic rubber, polyolefins, and polyurethane [98].

Impact-resistant POMs, comparable to super-tough nylon, have been produced by blending with compatible linear thermoplastic polyurethanes. They demonstrate a special morphology since the polyurethane phase forms an interpenetrating network structure in the POM matrix.

Applications are hinges and connectors that demand improved low-temperature flexibility and impact strength.

4.11. Polyarylates and Copolyarylate – Carbonate Blends

In principle the polyarylates (PAr) comprise all fully aromatic polyesters. In practice this term denotes polyesters derived from bisphenol A and mixtures of isophthalic and terephthalic acids in ratios between 70 : 30 and 30 : 70, but most frequently 50 : 50 [99], [100]. Due to their fully aromatic structure, the PArs are amorphous, transparent, practically colorless, and heat resistant. The T_g of the commonly used PAr containing bisphenol A and isophthalic/terephthalic acid in a weight ratio of 50 : 50 is 184 °C. This polymer is soluble in chlorinated hydrocarbon solvents, has a reasonable hydrolytic stability, good mechanical properties, strength, modulus, impact resistance, and excellent weatherability (despite yellowing). The polyarylates are intrinsically self extinguishing in flammability tests [101], [102]. Processibility is, however, marginal in view of the high processing temperatures (320 – 360 °C) required. Copolyester carbonates (PPCs) have therefore been developed with T_g's between those of PC (145 °C) and PAr (184 °C). PPCs are aromatic copolyesters in which part of the ester groups are substituted with carbonates. This group of polymers fills the T_g gap which existed between PC and the polysulfones [103].

Polyarylates are mainly used in applications demanding high heat stability and transparency such as lenses and medical equipment (high-pressure steam and radiation resistance needed for sterilization). Since the ultimate thermal dimensional stability of polymers under loading conditions is determined by the T_g of the matrix polymer, polyarylates and copolyarylates contribute significantly to the engineering property profile of blends.

Many PAr blends, especially those with semicrystalline aliphatic – aromatic polyesters such as PETP, PBT, and poly(cyclohexane terephthalate) (PCHT) have been made but very few have been successful. These polyester – PAr blends show the same features as the polyester – PC blends. PETP and PBT blends are partly miscible with PAr, whereas PCHT is completely miscible over the whole composition range. Furthermore, under melt-processing conditions, transesterification reactions catalyzed by polymerization catalysts rapidly transform the physical blend into a homogeneous copolymer with an averaged T_g value. Only by adding phosphites can the catalytic system be deactivated, yielding a usable blend with different characteristics from the existing PPCs. This holds particularly for the PETP and PBT blends. The ester interchange reaction does not need to be stopped in PAr – PCHT blends due to its complete miscibility [104–107].

PETP – PAr blends are used in the automotive industry for their improved chemical resistance and heat distortion temperature over PC and PC – PETP blends. They are also used in the lighting industry for reflectors. PCHT – PAr blends are used in medical applications because they have a very good resistance to yellowing when exposed to high-energy radiation for sterilization [103].

4.12. Poly(Phenylene Sulfide)

Poly(phenylene sulfide) [9016-75-5] (PPS) is a versatile thermoplastic engineering resin (see → Polymers, High-Temperature, Chap. 5.). It possesses high temperature resistance, inherent flame resistance, good mechanical properties, and high chemical resistance. The mechanical properties and dimensional stability of parts are determined by the crystallinity and orientation of the polymer during processing. Unfortunately, PPS is very brittle resulting from stiff spherulites and weak interspherulitic boundaries. Research is therefore directed at improving ductility while maintaining overall crystallinity levels.

Two types of PPS are commercially available: (glass) filled grades and blends with polytetrafluoroethylene (to improve moldability and crack resistance). A large number of other blends have been reported (e.g., PPS – PEI [108] and PPS – PETP [109]) but have not been introduced commercially.

5. General Application Technology

5.1. Processing of Blends

The utilization of thermoplastics, including blends, involves processing and adjustment of the bulk rheological properties of each commercial plastic grade to the requirements of the processing technique (e.g., injection molding, extrusion, blow molding). Adjustment is only possible if an acceptable processing window exists within the molecular mass constraints of the polymers which are needed to obtain the desired mechanical properties. The processing window is determined by the thermal stability and the processibility (viscosity) of the melt.

In blends, the maximum melt temperature is determined by the stability of the least stable component.

The rheology of a blend system is determined by its components. Blend processability can only be predicted to a limited extent. The rule of logarithmic additivity only applies to completely miscible systems of homologous polymers such as mixtures of the same polymer with different molecular masses (Eqs. 10 and 11), all functions having a molecular mass > entanglement molecular mass.

$$log\eta_b = \Phi_1 log\eta_1 + \Phi_2 log\eta_2 \qquad (10)$$

$$\eta_b^{1/3.4} = \Phi_1 \eta_1^{1/3.4} + \Phi_2 \eta_2^{1/3.4} \qquad (11)$$

where η is the viscosity, Φ is the partial volume, and the subscripts b, 1, and 2 denote the blend, component 1, and component 2, respectively.

Even for a completely miscible blend system (e.g., PPE – PS), corrections have to be made for the "positive deviation" which is associated with the mixing enthalpy (Eq. 12):

$$log\eta_b = \Phi_1 log\eta_1 + \Phi_2 log\eta_2 - \frac{\Phi_1 \Phi_2 \Delta H_{mixing}}{2.45RT} \qquad (12)$$

This enthalpy factor depends on the temperature (LCST behavior, Section 3.1). In miscible or partly miscible blends with processing temperatures close to the LCST temperature this can have extremely important effects, such as shear-induced phase separation (temperature rises due to the introduction of mechanical energy via the processing equipment) leading to complete loss of morphology and mechanical properties. Furthermore, the elastic and viscous parameters that depend on temperature and shear rate also contribute to the rheology characteristics of a polymer, and are different for each polymer.

In impact-modified multiphase blend systems, the prediction of the rheology characteristics by means of (empirical) rules becomes highly unsatisfactory. Simple approaches based on the Einstein relation for dispersions of solid particles in a liquid are followed by assuming that two-phase systems behave as a dispersion of fixed particles of different deformability (from solid to rubbery). In more complex two-phase blends where the morphology of the dispersed phase is not fixed and delamination is possible, a combination of parallel and series model of rheological effects is used with an interfacial layer correction.

The impact modifiers interact strongly with the matrix to create a thixotropic melt behavior in which the viscosity value at zero shear rate is much higher than in the unmodified systems. These characteristics can be very useful in creating the higher melt strength needed for extrusion and blow molding. Their presence may, however, also lead to poor surface appearance and knitline problems under excessive shearing forces.

In conclusion, the rheology of polymer blends is extremely complicated to predict, but commercial products are tuned to the requirements of most standard processing methods. The tolerance level for deviation from the recommended processing conditions is not unlimited.

5.2. Design Technology

The design considerations of a thermoplastic part involves functional factors related to production and assembly, and material factors related to the performance of a material in service.

Design procedure can be subdivided into the selection of the production process (e.g., injection molding, extrusion), resin characteristics (mechanical and electrical properties), flow analysis, stress analysis, long-term considerations (creep, fatigue, stress relaxation), appearance of a part, its assembly (including automation), and recycling. Computer programs aid the designer in developing a part capable of meeting all the mechanical requirements, while maintaining its production feasibility [28], [110].

The current emphasis on recycling demands that produced parts can be easily disassembled from final applications. In the case of cars, routes for facile dismantling and separation should be conceived in the designing step (see Section 6.1). Product designers and engineers must therefore consider all aspects of the primary and secondary life cycles.

Blends require special consideration of flow (viscosity), weld lines, processing temperature versus phase behavior (see Section 3.1), flow versus continuous phase, and high shear conditions. Thermodynamically controlled, shear-induced phase separation is possible leading to delamination and morphology changes (Section 5.1). Care should be taken to avoid high shear conditions.

The dependence of viscosity and shear rate on temperature is also very important for blends because the ratios between the blend components influence blend morphology and determination of the continuous phase (see Section 5.1). Prediction of rheology and flow behavior is more complicated than for single thermoplastic resins. Weld lines are an issue at positions where two flows encounter each other [111]. Design should result in placing weld lines in non-stressed regions.

6. Future Developments

6.1. Recycling

Recycling of plastics and polymer blends has drawn public attention, see also → Plastics, Recycling. Concern regarding plastic waste has risen because of the large volume of plastics (20 – 30 %) in municipal waste. It corresponds, however, to a weight fraction of only 8 % for which packaging materials are mainly responsible. The volume of waste must be reduced, particularly in densely populated areas. Thermoplastic blend producers and end users are faced with finding commercially effective recycling strategies. This is particularly acute for the automotive industry in Germany, where recycling laws demand the recycleability of 25 % of plastic car parts by the end of 1994.

The use of smaller quantities of better, more durable blend products may or may not have a positive overall environmental effect. One of the solutions to the waste problem practised in Japan is energy recycling by burning (70 %) and deposition of solid waste (30 %). This simple solution will soon disappear in the western world for political and economic reasons — waste disposal will become very expensive due to increasing tariffs. Other options (i.e., reuse of plastics in other applications) must therefore be considered.

The most important factor in plastic recycling is the purity and compatibility of the waste. Minimum waste is generated in plastic manufacturing and compounding facilities due to economic reasons; reworking of plastic waste is a standard procedure. Separation of the different plastics used facilitates backfeed and rework programs. Multilayer coathanger extrusion and two-component molding facilitate the usage of off-color and slightly off-grade materials.

The real problem starts with postconsumer plastic waste where purity and compatibility requirements are no longer fulfilled. Separation, sorting, and cleaning of the plastic waste from the bulk waste are necessary. This is very expensive and only economically feasible for a small number of cases. Any recycled material derived from plastics with different colors and grades therefore represents a downgrade product. In addition, this approach will only succeed if the recycle concept is economically viable, e.g., in the case of high-value engineering plastics.

Reusing the chemicals from which the polymers were made is another route. Thermal, pyrolytic, hydrolytic, or glycolytic processes are used to degrade the polymers [112]. This approach is only applicable in a few cases and has been proven technically feasible with PETP bottles [112]. However, the process economics are very limited.

Finally, the fuel value of the plastics should be considered. Polymers have a considerable caloric value being (mainly) organic materials. The energy in the material can be reclaimed by burning and then converted to electrical energy or steam.

Already 20 % of all plastics used can be considered blends. Blend thermodynamics, morphology, and careful processing are critical for obtaining the required/desired mechanical properties (Chap. 3). Lower-grade applications should be found for recycled blends. Typical examples are blends of postconsumer PE with PA 6 [113], scrap rubber [114], and recycled PETP with PE [115], [116] and PC [117].

Two important principles govern the role of engineering blends in recycling strategies:

1. Designing and developing better, more durable products via blend technology.

2. Combining compatible resin systems (i.e., minimization of the number of different resins in a module for various applications). The target is to obtain materials from one product family which are mutually compatible (Section 3.2). Only these compatible materials can be combined to yield economically viable, useful materials (Fig. 20).

An example of a module is a dashboard structure made of a glass-filled, injection-molded PPE – HIPS blend, which could include, for example, a large proportion of recycle material from dashboards. The air ducting is made of compatible, blow-molded PPE – HIPS and is mounted on the carrier unit by means of PPE – HIPS clip fasteners. The complete unit is overmolded in PPE – PS foam to form the outer surfaces and to provide sound insulation. A soft-feel foil is used to give a "luxury" look.

This principle influences the material selection step of the design process. Functionality (i.e., good performance) and cost are no longer the most important selection criteria because they may lead to a combination of incompatible materials. Instead compatibility considerations dominate. The flexibility of blend technology allows the creation of materials that satisfy compatibility and use criteria. It is conceivable that concessions may be made regarding functionality, i.e., although not all requirements are met, certain materials are chosen to give the cheapest module costs.

Recycling materials after their life cycle involves several steps [118]: dismantling, transportation of parts to general assembly, cleaning, processing of reground material into an application. The most important issues are identification of the numerous blends in similar applications, definition of separation procedures in which contaminants are removed, and facile disassembly steps during designing. Costs involved with collection, separation, cleaning, and reprocessing materials mean that this is only a viable concept for blends of engineering plastics with their high property level and high profit margin.

The cascade model (GE Plastics) envisages a cycle of applications for a material in keeping

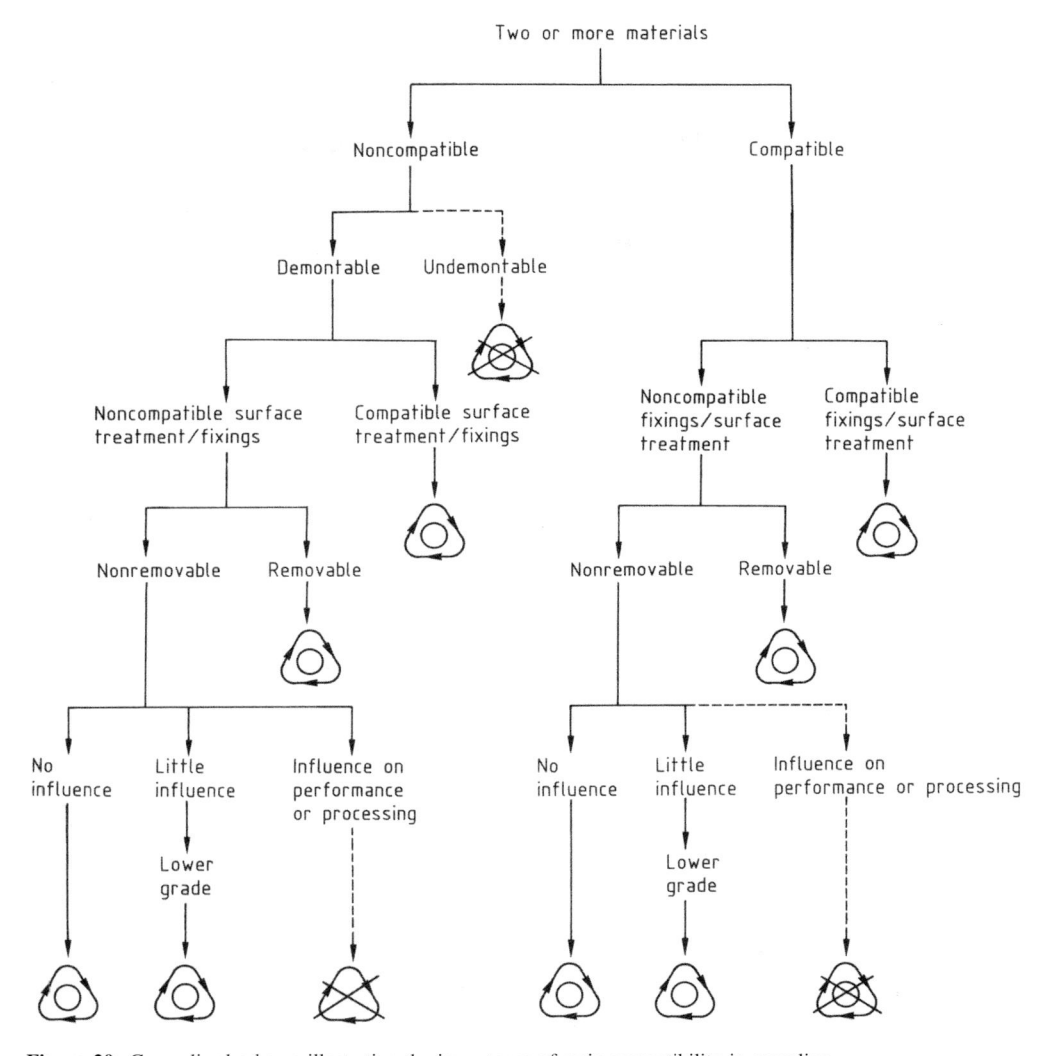

Figure 20. Generalized scheme illustrating the importance of resin compatibility in recycling

with the deterioration of mechanical and other properties.

Developments are under way to compatibilize the coatings, paints, and other materials used for aesthetic purposes with the main component.

Examples of cascade models for thermoplastics employed by Opel combine PP from battery cases and bumpers to make new fender liners, bumper carriers, and air-filter housings, and convert PC – PBT from old painted bumpers into new spoiler supports [119].

6.2. New Developments

Several promising new developments can be mentioned. PPE – PS blends are used to make expandable beads, which give strong low-density (down to 20 g/dm^3) foams with a heat distortion temperature > 100 °C, i.e., a sterilizable, re-usable material [120].

The use of PPE as an organic, heat-resistant, flame-retardant filter that does not affect mechanical properties at high filling ratios has been demonstrated for PA – PPE blends, and can be applied to other blend systems. These new blends include a PPE – PBT blend under development (trade name Gemax, GE Plastics), which has an even lower moisture absorption and thus higher dimensional stability than PA – PPE [121]. The material is targeted for automotive and appliance markets. Blends of chemically modified PPEs will also find increasing use.

Hybrid systems are related blends and consist of thermoset networks filled with heat-resistant engineering thermoplastics [122]. High-performance thermoset materials are used in structural applications where dimensional stability is of the utmost importance. Their temperature stability is determined by the glass transition temperature, whereas their mechanical stability above this temperature is controlled by the network density. Stiffer, more heat-resistant materials can be obtained by using stiffer building blocks or increasing the network density. However, this also increases the brittleness.

Incorporation of heat-resistant engineering thermoplastics improves impact behavior while maintaining dimensional stability at high temperatures. Dissolution of these resins in epoxy systems allows easy processing. The thermodynamically controlled phase separation that takes place during the curing of the epoxy resin leads to a high flexibility in controlling the ultimate network density and thus mechanical properties. Commercially available materials include PEI – epoxy systems (Brochier, France) [123–125] intended for metal-laminating structures in the aircraft industry, and PPE – epoxy systems (General Electric, USA) [126], [127]. The latter systems provides another solution to the poor melt processability of PPE (Section 4.5); it combines the low viscosity and easy processibility of thermosets with the toughness of thermoplastics. A commercial product (Getek, GE Electromaterials) is used as a circuit board material, the presence of PPE allows compliance with highly demanding techniques used for surface mounting and impedance control. A commercially available PPE has been functionalized with head and tail groups to provide optimal adhesion between the PPE phase and final cured epoxy resin. The PPE volume fraction determines whether the product is an epoxy-filled PPE matrix or a continuous epoxy matrix with a highly dispersed PPE phase.

A new reactive polymer system is based on oxazolines. Copolymers of styrene with a functional oxazoline monomer can react with carboxyl-, amino-, or anhydride-functional polymers to produce graft copolymers [128]. This reactive PS has been used to compatibilize the combination of PPE and an ethylene – acrylic acid copolymer, producing a high-impact engineering blend with excellent strength and processing properties.

Also new are blends of thermoplastics with 5 – 25 % liquid crystalline polymers. The resulting "molecular composite" can be regarded as a material in which the rigid crystalline polymers are reinforced at a molecular level. Liquid crystalline polymer blends have been reported for PC [129], [130]; PEI [129]; PP [131]; PBT [132]; PPE – PS [133]; ABS [134]; PSU [135]; PETP [136], [137]; and PA 6 [138]. These blends can be processed by conventional techniques such as extrusion, injection molding, and fiber spinning. Nearly all combinations of liquid crystalline polymers lead to an immiscible polymer blend. Consequently the mechanical properties of the blends are determined by the morphology of the liquid crystalline polymer domains in the thermoplastics. This morphology depends on composition, processing conditions, viscosity ratio of the component polymers, and the rheological characteristics.

Efforts are being made to use more biodegradable plastics for packaging. Commercial blends of PE with starch have been introduced by Feruzzi. The biodegradability of blends of polycaprolactone and LDPE or PP with cornstarch has been reported [139].

The development of thermoplastic blends for high-temperature composites (at > 200 °C) to be used in structural applications is described in [140]. Polymers with high temperature stability tend to be intractable or brittle, whereas easily processible polymers usually fall short of the projected temperature requirement. Luckily, many high-temperature, aromatic and heterocyclic polymers form miscible blends with each other which permits balancing of thermal response and processibility. Poly(2,2'-metaphenylene-5,5'-bibenzimidazole) (PBI), for example, is miscible with

PEI and with PAr; PEI is miscible with PEEK and PEKK.

Blending is also applied to improve the processibility of conductive polymers. Examples are Versicon (Zipperling) intended for transparent electrostatic coatings and corrosion prevention; electrically conductive blend or polyaniline and PVC with EMI properties; and conductive blends of poly(*p*-phenylene vinylene) and polyacrylamide [141]. Blends of (conductive) polyaniline in a matrix of elongated/stretched ultrahigh molecular mass PE are being commercialized by Neste.

6.3. Outlook

Blend technology will accelerate penetration of thermoplastics in markets where its presence is still limited [142]. Designers will have to be made aware of the advantages of working with thermoplastics instead of conventional materials such as wood and metal: their superior performance in a large number of applications, their versatile property profiles, the design freedom, and recycle options. The construction industry is an area with enormous potential for thermoplastic blends and composites. Short-term growth will be driven by applications such as windows, electrical systems, and skylights, followed by the replacement of wood, steel, and concrete. Advantages are lighter constructions, labor savings, and low maintenance costs. A number of possibilities are demonstrated in GE Plastics "Living Environment" concept house and a similar principle developed by Neste. They show the possibilities of thermoplastics and their blends, as well as the use of recycle materials in novel design concepts.

A breakthrough in the automotive market would be large exterior parts and structural components [143]. Two factors limit the use of thermoplastics in body panels. Firstly, cars will continue to feature steel – plastic body construction throughout the 1990s, requiring plastics to withstand vehicle assembly processes designed for steel. Secondly, car makers must compare the overall cost of plastics to steel including tooling, assembly, and customer value rather than on a raw material basis.

Miniaturization is the main driving force for the electronics expanded use of synthetic polymers in the electronics industry. With the development of new molding techniques for high-temperature resins such as polyarylsulfones and poly(phenyletheretherketone), new applications can be realized.

7. Acknowledgement

Acknowledgement. The authors are deeply indebted to Veronique Hopmans for her contributions.

References

1 H. Rudolph, *Makromol. Chem. Macromol. Symp.* **16** (1988) 57.
2 A. Moro, A. Chiolle, L. Credall, G. Foschini, *Makromol. Chem. Macromol. Symp.* **16** (1988) 137.
3 J. Levy, *Perform. Chem.* **6** (1991) no. 2, 36.
4 L. M. Rossi, R. S. Stevenson, *Compalloy '90* (1990) 357.
5 D. T. Wark, *ANTEC '90* (1990) 1156.
6 S. Izawa, *Int. Polym. Sci. Tech.* **17** (1990) 35.
7 *Specialty Polymeric Blends and Alloys*, Skeist Incorporated, Whippany, N.J. 1991.
8 *Macplas*, November 1991, 65.
9 J. B. Crossfield, *Des. Eng. (N.Y.)*, Nov 1990, 45.
10 J. Bussink in P. J. Lemstra, L. A. Kleintjes (eds.): *Integration of Fundamental Polymer Science and Technology*, vol. 3, Elsevier Applied Science, New York 1989.
11 *European Plastic News* **18** (1991) no. 10, 42.
12 O. Olabisi, L. M. Robeson, M. T. Shaw: *Polymer – Polymer Miscibility*, Academic Press, New York 1979.
13 W. J. Macknight, F. E. Karasz, in D. R. Pual, S. Newman (eds.): *Polymer Blends*, vol. **1**, Academic Press, New York 1978. Chapter 4, 111.
14 W. J. Macknight, F. E. Karasz, J. R. Fried, in D. R. Pual, S. Newman (eds.): *Polymer Blends*, vol. 1, Academic Press, New York 1978.
15 L. A. Utracki: *Polymer Alloys and Blends*, Carl Hanser Verlag, München 1989.
16 M. Bosma, G. ten Brinke, T. S. Ellis, *Macromolecules* **21** (1988) 1465.
17 S. Wu: *Polymer Interface and Adhesion*, Marcel Dekker, New York 1982.
18 N. G. Gaylord, *J. Macromol. Sci. Chem.* **A 26** (1989) 1211.
19 L. Leibler, *Makromol. Chem. Macromol. Symp.* **16** (1988) 1.
20 C. H. Han: *Multiphase Flow in Polymer Processing*, Academic Press, New York 1981.
21 R. M. Ottenbrite, L. A. Utracki, S. Inoue: *Current Topics in Polymer Science*, vol. 2, Hanser Verlag, München 1987.
22 J. G. M. van Gisbergen: "Electron Beam Irradiation of Polymer Blends", Ph. D. Thesis, Eindhoven University 1991.
23 J. J. Elmendorp: "A Study on Polymer Blending Microrheology," Ph. D. Thesis, Delft University 1986.
24 *Aufbereiten von Polymerblends*, VDI Verlag, Düsseldorf 1989.
25 H. P. Grace, *Chem. Eng. Commun.* **14** (1982) 225.
26 S. Y. Hobs, M. E. J. Dekkers, V. H. Watkins, *Polymer* **29** (1988) 1598.
27 M. Xanthos, S. S. Dagli, *Compalloy '90* (1990) 277.
28 GE Plastics, Engineering Materials Design Guide, Bergen op Zoom 1991.

29. C. B. Bucknall: *Toughened Plastics*, Applied Science Publishers, London 1977.
30. H. Kim, H. Keskkula, D. R. Paul, *Polymer* **32** (1991) 2372.
31. H. T. van de Grampel: *RAPRA Review Report Series 49*, Pergamon Press, Oxford 1992.
32. G. Allen, *Polymer J.* **19** (1987) 1. L. A. Utracki in L. A. Utracki, R. A. Weiss (eds.): *Multiphase Polymeric Materials*, ACS Books, Washington, DC. 1989.
33. P. M. Subramanian in W. J. Koros (ed.): "Barrier Polymers and Structures," *ACS Symp. Series* **423** (1990).
34. A. D. Abshire: "Blends Containing Polyolefins", *SRI report* **175 A** (1990).
35. G. Serpe, J. Jarrin, F. Dawans, *Polym. Eng. Sci.* **30** (1990) 553; R. Bell, *ANTEC '90*(1990) 671.
36. J. Willis, B. D. Favis, *Compalloy '89*(1989) 173.
37. A. K. Bhowmik, H. L. Stevens(eds.): *Handbook of Elastomers*, Marcel Dekker, New York 1988.
38. A. J. Kinioch, R. J. Young: *Fracture Behavior of Polymers*, Applied Science Publishers, London 1983.
39. PVC Blends, Alloys and Graft Polymers, Conference Proceedings, Atlanta, Oct. 1990, SPE Vinyl Div.
40. D. Braun, B. Boehringer, J. Herth, *Makromol. Chem. Macromol. Symp.* **29** (1989) 227.
41. A. D. Roberts (ed.): *Natural Rubber Science and Technology*, Oxford University Press, Oxford 1988.
42. F. Mark et al. (ed.): *Encycl. Polym. Sci. Technol.*, J. Wiley, New York 1991, vol. 9, p. 769.
43. G. Lindenschmidt, R. Theysohn, *Engineering Plastics* **3** (1990) 1.
44. W. K. Chin, L. J. Hwang, *ANTEC '87*(1987) 1379.
45. D. J. Stein et al., *Angew. Makromol. Chem.* **36** (1974) 89.
46. J. Huang, M. Wang, *Adv. Polym. Technol.* **9** (1989) 293.
47. J. M. Wefer, US 4 895 921, 1985.
48. H. F. Giles, *Modern Plastics Encycl.*, 1987 – 1988.
49. J. Maruta, T. Ougizawa, T. Inoue, *Polymer* **29** (1988) 2056.
50. J. G. Dolden, *Polymer* **17** (1976) 875.
51. P. Maskus, *ANTEC '91*(1991) 1385.
52. T. S. Ellis, *Polym. Commun.* **32** (1991) 489.
53. T. S. Ellis, *Macromolecules* **22** (1989) 742.
54. T. S. Ellis, *Macromolecules* **24** (1991) 3845.
55. V. J. Triacca et al., *Polymer* **32** (1991) 1401.
56. R. E. Lavengood et al., *SPE Tech. Pap.* **33** (1987) 1369.
57. E. Gattiglia et al., *J. Appl. Polym. Sci.* **38** (1989) 1807; **41** (1990) 1411.
58. S. S. Dagli et al., *ANTEC '90*(1990) 1924.
59. M. J. Modic, L. A. Pottick, *Plast. Eng.* **49**, July 1991, 37.
60. M. Wajs, M. Glotin, *Br. Plast. Rubber* 1989 March, 15.
61. S. Y. Hobbs, R. C. Bopp, V. H. Watkins, *Polym. Eng. Sci.* **23** (1983) 380.
62. B. N. Epstein, J. A. Rakeshaw, E. A. Flexman, D. D. Huang, *Polym. Prepr. (Am. Chem. Soc. Div. Polym. Chem.)* **32** (1991) no. 2, 42.
63. J. Bussink, W. Minderhout, W. Sederel in print (ed. Bottenbruch), Carl Hanser Verlag, München 1992.
64. M. B. Djordjevic, R. S. Porter, *Polym. Eng. Sci.* **23** (1983) 650.
65. General Electric, US 3 383 435, 1968 (E. P. Cizek).
66. V. T. Bui, Polymer Blends, Symposium Proceedings, Cambridge 1990, C 15/1–4.6125.
67. A. F. Yee, *Polym. Eng. Sci.* **17** (1977) 213.
68. T. Szabados, E. Galambos, *Int. Polym. Sci. Techn.* **13** (1986) 42.
69. S. Y. Hobbs, M. Dekkers, V. H. Watkins, *J. Mater. Sci.* **24** (1989) 2025.
70. H. J. Sue, A. F. Yee, *J. Mater. Sci.* **26** (1991) 3449.
71. J. R. Campbell, S. Y. Hobbs, T. J. Shea, V. H. Watkins, *Polym. Eng. Sci.* **30** (1990) 1056.
72. Chemische Werke Hüls, US 4 656 220, 1987 (H. Jadamus et al.).
73. C. Bailly, *Compalloy '90*(1990) 375.
74. D. Freitag, G. Fengler, L. Morbitzer, *Angew. Chem. Int. Ed. Engl.* **30** (1991) 1598.
75. G. J. Pratt, M. J. A. Smith, *Polymer* **30** (1989) 1113.
76. W. N. Kim, C. M. Burns, *J. Appl. Polym. Sci.* **41** (1990) 1575.
77. D. C. Wahrmund, D. R. Paul, J. W. Barlow, *J. Appl. Polym. Sci.* **22** (1978) 2155.
78. M. E. J. Dekkers, S. Y. Hobbs, V. H. Watkins, *J. Mater. Sci.* **23** (1988) 1225.
79. S. Y. Hobbs, M. E. J. Dekkers, V. H. Watkins, *J. Mater Sci.* **23** (1988) 1219.
80. A. W. Birley, X. Y. Chen, *Br. Polym. J.* **16** (1984) 77.
81. D. Delimoy et al., *Polym. Eng. Sci.* **28** (1988) 104.
82. W. N. Kim, C. M. Burns, *J. Appl. Polym. Sci.* **41** (1990) 1409.
83. R. W. Avakian, R. B. Allen, *Polym. Eng. Sci.* **25** (1985) 462.
84. General Electric, US 4 122 131, 1978.
85. K. W. McLaughlin, *Polym. Eng. Sci.* **22** (1989) 1550.
86. K. W. McLaughlin, *SAE Proceedings*, 1988.
87. D. Quintens, G. Groeninckx, M. Quest, L. Aerts, *Polym. Eng. Sci.* **30** (1990) 1474; **22** (1990) 1484.
88. H. Takahashi et al., *J. Appl. Polym. Sci.* **36** (1988) 1821.
89. W. Chiang, D. Hwung, *ANTEC '86*, 492.
90. H. Kress, D. Folajtan, N. Lazear, *ANTEC '88*, 1834.
91. M. E. J. Dekkers, S. Y. Hobbs, V. H. Watkins, *Polymer* **32** (1991) 2150.
92. D. J. Hourston et al., *Polymer* **32** (1991) 1140.
93. M. Gaztelumendi et al., *Makromol. Chem. Macromol. Symp.* **20** (1988) 269.
94. General Electric, US 3 730 946, 1973 (G. Wirth, D. R. Heath).
95. H. Farong, W. Xueqiu, L. Shijin, *Polym. Degrad. Stab.* **18** (1987) 247.
96. G. Crevecoeur, G. Groeninckx, *Macromolecules* **24** (1991) 1190.
97. J. E. Harris, L. M. Robeson, *J. Appl. Polym. Sci.* **35** (1988) 1877.
98. Delrin General Guide to Products and Properties, E. I. du Pont de Nemours & Co., Wilmington, Del. 1991. Celcon Acetal Copolymer Properties, Celanese Corp., New York 1991.
99. G. Bier, *Polymer* **15** (1974) 527.
100. D. Freitag, K. Reinking, *Kunststoffe* **71** (1981)no. 1, 46.
101. M. Kimura, G. Salee, R. S. Porter, *J. Appl. Polym. Sci.* **29** (1984) 1629.
102. A. Golovoy, M.-F. Cheung, *Polym. Eng. Sci.* **29** (1989) no. 2, 85.
103. Bayer; Unikka; GE Plastics; Amoco; product brochures.
104. L. M. Robeson, *J. Appl. Polym. Sci.* **30** (1985) 4081.
105. A. Golovoy, M.-F. Cheung, K. R. Carduner, M. J. Rokosz, *Polym. Eng. Sci.* **29** (1989) 1226.
106. A. Ausin et al., *Polym. Eng. Sci.* **27** (1987) 529.
107. C. E. Desper, M. Kimura, R. S. Porter, *J. Polym. Sci. Polym. Phys. Ed.* **22** (1984) 1193.
108. S. Akhtar, J. L. White, *Polym. Eng. Sci.* **31** (1991) 84.
109. V. L. Shingankuli, P. Jog, V. M. Nadkarni, *J. Appl. Polym. Sci.* **36** (1988) 335.
110. D. V. Rosato, D. P. Di Mattia, D. V. Rosato: *Designing with Plastics and Composites: A Handbook*, Van Nostrand Reinhold, New York 1991.
111. G. Menges, *Makromol. Chem. Macromol. Chem.* **23** (1989) 13.
112. *Modern Plastics*, July 1991.
113. F. P. La Mantia, D. Curto, *Polym. Degrad. Stab.* **36** (1992) 131.

114 J. R. M. Duhaime, W. E. Baker, *Plast. Rubber Process Appl.* **15** (1991) 87.
115 E. Wissler, *ANTEC '90*(1990) 1434.
116 I. M. Chen, C. M. Shiah, *ANTEC '89*(1989) 1802.
117 A. G. Staniulis, *ANTEC '90*(1990) 794.
118 GE Plastics, Design for Recycling, Bergen op Zoom 1992.
119 N. H. Naitove, R. Monks. *Plast. Technol.,* **38** (1992) 101.
120 L. M. Martynowicz, *J. Cell. Plast.* **26** (1990) 423.
121 A. McHale, P. Y. Liu, SPE Conference Proceedings, Rosemont II, 1987, 147.
122 J. L. Hedrick *et al.*,*Polymer* **32** (1991) 2020.
123 C. B. Bucknall, A. H. Gilbert, *Polymer* **30** (1989) 213.
124 A. H. Gilbert, C. B. Bucknall, *Makromol. Chem. Macromol. Symp.* **45** (1991) 289.
125 D. J. Hourston, J. M. Lane, N. A. Macbeath, *Polymer Int.* **26** (1991) 17.
126 S. J. Kubisen, P. C. Long, Proceedings Electronic Materials and Processing Conference, A.S.M. International, Chicago Ill., Sept. 1988.
127 S. J. Kubisen *et al.*: *A New Performance Material*, GE Laminated Products, Coshocton, Ohio 1991.
128 G. Sinai-Zingde *et al.*,*Makromol. Chem. Macromol. Symp.* **42/43** (1991) 329.
129 K. G. Blizard, C. Federici, O. Federico, L. L. Chapoy, *Polym. Eng. Sci.* **30** (1990) 1442.
130 T. M. Malik *et al.*,*Polym. Eng. Sci.* **29** (1989) 60.
131 H. J. O'Donell *et al.*,*Polym. Prepr. (Am. Chem. Soc. Div. Polym. Chem.)* **33** (1992)no. 1, 376.
132 M. Paci *et al.*,*J. Polym. Sci. Polym. Phys.* **25** (1987) 1595.
133 G. Crevecoeur, G. Groeninckx, *Polym. Eng. Sci.* **30** (1990) 532.
134 M. Takayanagi, K. Goto, *J. Appl. Polym. Sci.* **29** (1984) 2547.
135 V. G. Kulichikhin *et al.*,*J. Appl. Polym. Sci.* **42** (1991) 363.
136 A. M. Shukhadia, D. Done, D. G. Baird, *Polym. Eng. Sci.* **30** (1990) 519.
137 M. Kyotani *et al.*,*Polym. Prepr. Jpn. (Engl. Ed.)* **39** (1990) E 1423.
138 F. P. La Mantia, A. Valenza, M. Paci, P. L. Magagnini, *Polym. Eng. Sci.* **30** (1990) 7.
139 Y. Tokiwa, A. Iwamoto, M. Koyama, *Polym. Mater. Sci. Eng.* **63** (1990) 742.
140 M. Jaffe *et al.*,*Makromol. Chem. Macromol. Chem.* **53** (1992) 163.
141 J. Machado, F. E. Karasz, R. W. Lenz, *Polymer* **29** (1988) 1412.
142 J. Alper, G. L. Neldon: *Polymeric Materials: Chemistry for the Future*, ACS, Washington, DC, 1989.
143 *Plast. World*, September 1991, 55.

Further Reading

A. I. Isayev (ed.): *Encyclopedia of Polymer Blends*, Wiley-VCH, Weinheim 2010.

V. Mittal (ed.): *Optimization of Polymer Nanocomposite Properties*, Wiley-VCH, Weinheim 2010.

L. M. Robeson: *Polymer Blends*, Hanser, München 2007.

M. Xanthos (ed.): *Functional Fillers for Plastics*, 2nd ed., Wiley-VCH, Weinheim 2010.

Polymers, Biodegradable

MICHAEL BREULMANN, BASF SE, Ludwigshafen, Germany
ANDREAS KÜNKEL, BASF SE, Ludwigshafen, Germany
SABINE PHILIPP, BASF SE, Ludwigshafen, Germany
VALENTINE REIMER, BASF SE, Ludwigshafen, Germany
KAI O. SIEGENTHALER, BASF SE, Ludwigshafen, Germany
GABRIEL SKUPIN, BASF SE, Ludwigshafen, Germany
MOTONORI YAMAMOTO, BASF SE, Ludwigshafen, Germany

1.	Glossary and Abbreviations	1231
2.	Introduction	1232
3.	Biodegradability and Toxicology	1232
3.1.	Mechanism of Biodegradation	1232
3.2.	Toxicity of Biodegradable Polymers	1233
3.3.	Standards for Biodegradable Polymers	1234
3.3.1.	Controlled Composting Test	1234
3.3.2.	Standard Assessment of Biodegradable Polymers	1235
4.	Biodegradable Polymers: Synthesis, Properties, and Suppliers	1236
4.1.	Raw Materials	1236
4.2.	Starch	1237
4.3.	Poly(lactic acid) (Polylactide, PLA)	1238
4.4.	Polyhydroxyalkanoates (PHA)	1239
4.5.	Synthetic Polyesters	1240
4.5.1.	Aliphatic Polyesters	1240
4.5.2.	Aliphatic/Aromatic Polyesters	1242
4.6.	Other Synthetic Polymers	1242
4.7.	Compounds	1243
4.7.1.	Compounds of Starch and Biodegradable Polyesters	1243
4.7.2.	Compounds of PLA and Biodegradable Polyesters	1245
4.8.	Cellulose Derivatives	1246
5.	Processing and Additives	1246
5.1.	Introduction	1246
5.2.	Extrusion	1247
5.3.	Blown-Film Extrusion	1247
5.4.	Cast-Film Extrusion	1248
5.5.	Modification of Biodegradable Polyester Films	1249
5.5.1.	Additives	1249
5.5.2.	Printing	1249
5.5.3.	Metallization	1249
5.5.4.	Multilayer Films	1250
5.6.	Extrusion Coating	1250
5.7.	Sheet Extrusion and Thermoforming	1250
5.8.	Injection Molding	1250
5.9.	Blow Molding	1251
6.	Market Overview and Growth Drivers	1251
7.	Value of Biodegradability — Life-Cycle Assessment (LCA)	1252
8.	Applications	1253
8.1.	Organic Waste Bags	1253
8.2.	Shopping (Carrier) Bags	1255
8.3.	Mulch Film	1255
8.4.	Horticulture	1255
8.5.	Packaging	1255
8.5.1.	Food Packaging	1255
8.5.2.	Nonfood Packaging	1256
9.	Production Capacity	1257
10.	Outlook	1257
	References	1258

1. Glossary and Abbreviations

Glossary: Terms for Carbon Origin and Degradation

Biobased	Refers to the origin of the raw materials. Only significant if the carbon content of annually renewable raw materials is given as the share of the total carbon content.
Renewable	Refers to renewable (annually or otherwise renewable) raw materials such as corn, wheat, paper, wood, etc.
Fossil (nonrenewable) or fossil-based	Refers to raw materials not considered to be renewable, such as gas, crude oil, coal, which are organic in origin.

Fragmentable (degradable)	Certain materials containing special metal additives as decomposition aids to obtain fragments under the effect of time, heat, and stress. Not fully biodegradable according to current scientific standards.
Biodegradable	Microorganisms such as bacteria, fungi, or algae metabolize these materials completely, giving off CO_2, water, energy, and biomass (aerobic process). In anaerobic biodegradation, also methane is produced.
Compostable	Fully biodegradable under composting conditions, as defined by current standards, i.e., ISO 17088, EN 13432, ASTM D 6400, Japanese GreenPla.

2. Introduction

Biodegradable polymers have been around on an industrial scale since the end of the 1990s. In contrast to biopolymers which are by definition made partially or completely of renewable raw materials, biodegradable polymers are not defined in terms of their raw material basis. After all, biodegradability means that a given substance can be completely converted into water, CO_2, and biomass through the action of microorganisms such as fungi and bacteria. This property is not dependent on the origin of the raw materials, but only on the chemical composition of the polymers. As will be shown, biodegradable polymers can be made either of renewable or of fossil raw materials [1].

Biodegradability is primarily a specific functionality which, like other properties, gives a product additional value and thus optimizes it for certain areas of application. Biodegradability is the possibility of disposing packaging material along with organic waste or of tilling agricultural films into the ground, with all the advantages this entails from the standpoint of hygiene, disposal logistics, and system costs.

A distinction has to be made between degradability (mechanical disintegration) and *bio*degradability (metabolism): not every polymer that can no longer be seen after a few weeks because it has disintegrated into small pieces has actually been biologically degraded. Biodegradability is a certified performance characteristic (ISO 17088, EN 13432 in EU, ASTM D 6400 in North America, GreenPla in Japan). To fulfill, e.g., the EN 13432 norm, the polymer has to be converted to CO_2 by over 90% within 180 d under defined conditions of humidity, temperature, and oxygen level.

The following chapters will describe in more detail the key prerequisite for biodegradability followed by a description of the synthesis properties and processing of the various biodegradable polymers. The value of biodegradability including life-cycle assessment (LCA) will be addressed in Chapter 6. The significant growth of the market, the various market applications and the corresponding development of production capacities are reviewed in Chapter 5.

3. Biodegradability and Toxicology

The property of biodegradability is not dependent on the origin of the raw materials but only on the chemical composition of the polymers. A polymer is biodegradable if it is metabolized by microorganisms (bacteria, yeasts, algae, fungi) in their natural environment (e.g., surface or sea water, soil or compost) to energy, biomass, water, and carbon dioxide or methane within a given time period (e.g., in composting standards: six months).

Biodegradation is related to either aerobic or anaerobic processes. In an aerobic process, organic matter is transformed to carbon dioxide, water, energy, and biomass by microorganisms in the presence of oxygen. Aerobic biodegradation of polymers occurs in soil or in compost with a variety of microorganisms [2]. In anaerobic biodegradation, organic matter is metabolized by microorganisms in the absence of oxygen. In this process, organic matter is converted to methane, carbon dioxide, water, energy, and biomass.

3.1. Mechanism of Biodegradation

Most of the biodegradable polymers are insoluble in water. They are too big to pass the cellular membrane of microorganisms. Therefore, biodegradation of these polymers begins with chain cleavage by enzymes (e.g., lipases and esterases) secreted by the microorganism in the natural environment [3]. The polymer chains are cleaved to oligomers which can be absorbed and metabolized by microorganisms. The cells of the microorganism gain energy from the mineralization process producing water, biomass, carbon dioxide, or methane. Biodegradation is a surface erosion process because enzymes are too large to diffuse into the biodegradable polymer in bulk [4]. The biodegradation process is influenced by humidity, oxygen level, light, and temperature.

The enzymatic degradation of a polymer chain depends on the following factors:

- Within the semicrystalline polymer, amorphous regions with flexible chains are present
- These polymer chains are flexible enough to enter the catalytic center of the enzyme
- Chemical bonds (e.g., ester bonds) can be cleaved by the enzyme

The process of biodegradation of polymers not only depends on the activity of enzymes but also on the temperature, humidity, nutrients, and gases (e.g., oxygen in aerobic biodegradation). Therefore, it is important to test the combination of the different factors to define the biodegradability of a specific polymer in a natural environment (e.g., as mulch film). Detailed biodegradation processes of the individual polymers are described in Chapter 4.

As an example for the class of biodegradable polymers, the aliphatic aromatic polyester poly [(butylene adipate)-co-(butylene terephtalate)] (PBAT, Ecoflex), will be used to describe the different test procedures and standards. Laboratory tests have been performed with PBAT to monitor the biodegradation of polymers with respect to the formation of intermediates. It has been shown by interrupting the biodegradation process that Ecoflex was degraded to oligomers and to the monomers used to synthesize the polymer. The residual polymer, oligomers, and monomers could at any stage be metabolized by > 99 % in an inoculum prepared from compost (Table 1) [5]. No accumulation of residues has been found.

3.2. Toxicity of Biodegradable Polymers

To prevent any negative impact of polymers which are designed for applications ending up in the environment, it is important to provide adequate information for comprehensive assessment of the environmental and toxicological safety of such a plastic. In general, this may be achieved by characterizing the raw materials, e.g., the polymer itself, and potentially resistant residues after biodegradation with respect to ecotoxicological and toxicological properties. As an example, results from laboratory tests are summarized in the following, which have been conducted with the polyester Ecoflex in order to identify toxic effects (see also → Ecotoxicology).

Water-Soluble Intermediates—Daphnia Test. In the toxicity tests, the toxicity of the water-soluble intermediates is particularly important because they can easily enter groundwater or be more readily absorbed by the organism.

Testing was carried out in accordance with DIN 38412 Part 30. In this test, the pollutant-dependent immobilization of the daphnia in solution of different concentrations (series of dilutions) is used. The control solution contains microorganisms that biodegrade Ecoflex enzymatically. The stock solution at the end of the test also contains the degradation intermediates of Ecoflex.

Table 1. Fragments of Ecoflex after degradation with isolated pure strain or pure strain and mixed cultures from compost (x = detected; n.d. = not detected)

	Monomers[*]			Aliphatic oligomers[*]		Aromatic oligomers[*]	
	B	A	T	BA	ABA	BT	BTB
Test 1	x	x	x	x	x	x	x
Test 2	x	x	x	x	x	n.d.	n.d.
Test 3	x	x	x	n.d.	n.d.	n.d.	n.d.
Test 4	n.d.	n.d.	n.d.	n.d.	n.d.	n.d.	n.d.

[*] A = adipic acid; B = 1,4-butanediol; T = terephthalic acid; BA and ABA = co-monomers of butane diol and adipic acid; BT and BTB = co-monomers of butane diol and terephthalic acid. Test 1: 1750 mg polyester in 80 mL media. Intermediates from isolated pure culture after 21 d. Enzyme activity stopped by pH shift (in situ building of high amounts of acids during degradation). Test 2: 350 mg polyester in 80 mL media. Intermediates from isolated pure culture after 7 d. Test 3: 350 mg polyester in 80 mL media. Intermediates from isolated pure culture after 21 d. Test 4: Residues after a 7-d inoculation by isolated pure culture and a 14-d inoculation with compost eluate. No intermediates were detected by GC analysis.

The polymer was successively diluted and for each concentration 10 daphnia were placed in the test solution (20 °C, pH 7.0). After 24 h, the number of daphnia still swimming was counted. Even with a low dilution (Stage 2) as in the control solution, there were still nine daphnia swimming. The test was therefore passed.

Plant Growth Test. The ecotoxicity of composted biodegradable polyester (Ecoflex) was studied in a plant growth test following the EN 13432, Annex E, which is based on the OECD guideline 208. In this test, effects on seedling emergence and early plant growth are investigated with different higher plant species exposed to treated compost. Seeds are placed in contact with treated and control soil. Four plant species covering the three categories outlined in OECD guidance 208 were introduced in the test, namely wheat (*Triticum sativum*), summer barley (*Hordeum vulgare*), mustard (*Sinapis alba/Brassica alba*), and the mung bean (*Phaseolus aureus*).

The following samples were prepared and used for testing:

- Mixture of reference soil and 25 % compost with addition of Ecoflex after 12 weeks composting
- Mixture of reference soil and 50 % compost with addition of Ecoflex after 12 weeks composting.

The evaluation of the test results on seedling emergence and biomass showed no significant effects when treated soil was compared to control soil (Fig. 1). With all four plant species, both parameters reached at least 90 % of the control level regardless of the test concentration.

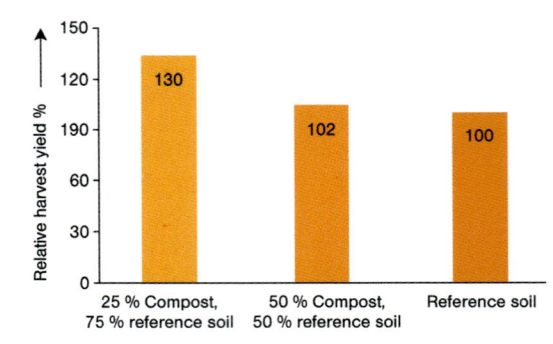

Figure 1. Result of plant growth test for PBAT (Ecoflex) according to OECD 208

An overview of the described and other tests is shown in Table 2 [6]. Biodegradation of PBAT (Ecoflex) causes no accumulation of environmentally dangerous compounds either in organisms or in the ecosystem.

3.3. Standards for Biodegradable Polymers

3.3.1. Controlled Composting Test

Biodegradability of polymers is appropriately described by test and assessment methods in international standards. Industrial composting is today a well-established disposal process for organic waste materials. In order to adapt this industrial composting process, laboratory tests have been developed and certified in standards.

The most important proof of aerobic biodegradability is the controlled composting test according to ISO 14855, which is a central part of every standard for biodegradable polymers. In a composting vessel, the test polymer is

Table 2. Ecotoxicological tests of PBAT (Ecoflex)

Test	Result
Acute toxicity to daphnia DIN 38412 Part 30, fishes	passed
Terrestrial plant toxicity OECD 208	no effects at the highest concentration
Earth worm toxicity OECD 207	no effects at the highest concentration
Primary skin irritation rabbit OECD 404	nonirritant
Primary irritations of the mucus membrane rabbit OECD 405	nonirritant
Guinea pig OECD 406 (modified Buehler test)	nonsensitizing
LD_{50} rat (oral) OECD 423	> 4000 mg/kg, virtually nontoxic after a single ingestion
Ames test OECD 471	substance was not mutagenic

incubated in a batch process—mixed with a defined, mature compost quality. The vessel is aerated continuously with carbon dioxide free air at 58 °C and at defined humidity. Thus, the composting process runs under optimum moisture and oxygen conditions. Simultaneously, a blank compost inoculum without additional carbon source is tested under the same conditions. The carbon dioxide content of the exhaust air of both vessels is compared. After subtracting the carbon dioxide evolution of the blank inoculum, the carbon dioxide evolution related to the test polymer is monitored as the biodegradation curve. Figure 2 shows that in a controlled composting test of Ecoflex more than 90 % of the polymer was converted to carbon dioxide after 80 d.

The biodegradation curve shows the relative carbon dioxide evolution of the test inoculum with test polymer related to the theoretical carbon dioxide evolution of the test polymer. According to the ISO standard compostable polymers, ISO 17088, 90 % of the theoretical carbon dioxide evolution of the test polymer has to be detected after 180 days. The activity of the compost inoculum in the controlled composting test is validated using a cellulose reference instead of the polymer. The strict requirements on the test result ensure that only compost with a defined degree of maturity is used to test the polymer.

3.3.2. Standard Assessment of Biodegradable Polymers

Like the ISO 17088, the European standard 13432 and the American standard ASTM D 6400 define basic requirements for packaging

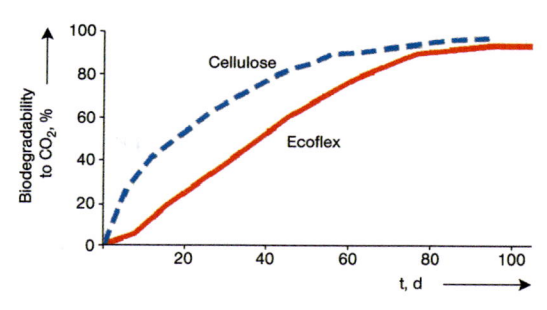

Figure 2. Controlled composting test of PBAT (Ecoflex) and cellulose

and packaging materials to be considered as biodegradable and compostable in industrial composting facilities by addressing the scheme summarized in Table 3. The test results have to be certified by registered test institutes (e.g., DIN-Certco, D; Vincotte, B, Biodegradable Products Institute, USA).

The new ISO standard 17088 became effective in 2008 and is mainly based on EN 13432 and ASTM D 6400. This standard can be used worldwide. Additional tests may be necessary to obtain the product registration according to ISO 17088 based on a regional standard.

Registered biodegradable polymers are used to produce biodegradable finished products, e.g., bags, packaging, cutlery, plates, which have to be certified in the next step. This procedure at a certified test institute shall prevent the use of nonconforming additives, colors, and packaged goods. The certificate entitles the holder to use the regional symbol for biodegradable plastics on his packaging or article in combination with his certification number (Table 4):

Table 3. Registration scheme for biodegradable polymers

Test	Assessment
Declaration of polymer composition	volatile solids > 50 %; heavy metal content < limits in EN 13432
Laboratory test of biodegradability (ISO 14855, controlled composting test or equivalent)	EN 13432: > 90 % of theoretical CO_2 evolution after 180 d ASTM D 6400-04 (homopolymers): > 60 % of theoretical CO_2 evolution after 180 d ASTM D 6400-04 (heteropolymers): > 90 % of theoretical CO_2 evolution after 180 d
Disintegration in pilot-scale composting test using specimen of maximal thickness	< 10 % of the weight of the specimen shall fail to pass through a > 2 mm fraction sieve
Analysis of compost quality	density; dry solids; volatile solids; salt content; pH; content of elemental N, P, Mg, Ca
Ecotoxicity of compost using minimum 2 species of plants	rate of germination and biomass > 90 % of blind value of compost without polymer

Table 4. Logos of biodegradable polymers according to the standards [7]

Organisation	DIN-Certco	Vincotte	Jäitelaito-syhdistys	BPI/USCC	BPS
Location	Germany	Belgium	Finland	North America	Japan
Logo					
Standard	EN 13432 ASTM D 6400	EN 13432	EN 13432	ASTM D 6400	GreenPLA certification scheme

4. Biodegradable Polymers: Synthesis, Properties, and Suppliers

4.1. Raw Materials

Biodegradability is a functional performance characteristic of the polymer backbone, which can be achieved with renewable as well as fossil-based raw materials.

Fossil-Based Raw Materials. Major fossil-based monomers are butanediol (BDO) and the dicarboxylic acids adipic acid, terephthalic acid, and succinic acid. BDO is produced in world-scale plants with different processes and feedstocks (acetylene by Reppe process, butane, butadiene). Compared to BDO, succinic acid is produced in smaller scale based on maleic acid anhydride (MSA). Adipic acid is based on cyclohexane derived from benzene, and terephthalic acid is based on p-xylol.

Renewable raw materials are used as

- feedstock for monomer production (e.g., lactic acid) or polymer production (polyhydroxyalkanoates, PHA)
- structural material (e.g., cellulose) and as a compound ingredient (e.g., starch) for biodegradable polyesters

Feedstocks for Monomer or Polymer Production. Starch (from corn, wheat, etc.) → Starch, is transformed to glucose, which is further used as fermentation feedstock for monomers (e.g., lactic acid) or PHA production. Other feedstocks for fermentative monomer production are sucrose, glycerol, or plant oils [8]. Intensive work is underway to use cellulose raw materials (C_6 and C_5 sugars) not only for bioethanol production but as well for different chemical intermediates and monomers like lactic and succinic acid [9].

Besides the indirect use of feedstocks for monomer production via fermentation, the direct access to monomers via extraction and processing of renewable feedstocks (e.g., sebacic acid, biorefinery) → Biorefineries – Industrial Processes and Products is intensively elaborated and being commercialized. For example, Novamont plans to install a biorefinery process for the production of biodegradable polyesters using local feedstocks [10].

The most established monomer production via fermentation for biodegradable polymers is the production of lactic acid [11] whereas commercial succinic acid production is underway [12].

Structural Materials and Compound Ingredients. Cellulose can be used as a basic material for coating and/or lamination with biodegradable polyesters. Starch (from corn, wheat, potato) is used as a compound ingredient for biodegradable polyesters to produce film materials (see Section 4.2). The biodegradable polyester is needed as an "enabler" to improve the properties and processability of the starch. Furthermore, starch is used as loose fill material (not further described in this article).

4.2. Starch

→ Starch

Native starch occurs in the organs of plants as discrete particles (starch granules). Starch molecules are polysaccharide compounds composed of anhydroglucose units. Starch serves as a food reserve and energy source of plants.

Starch granules consist of a multilayer structure from two polymers: amylose and amylopectin. Amylose is a predominantly linear 1,4-α-D-anhydroglucose polymer. Potato starch contains about 21 % of amylose at an average degree of polymerization M_n of 4900 (Fig. 3). Corn starch contains about 29 % of amylose at a much lower M_n of about 900 [13].

Amylopectin forms 70 – 85 % of regular starches at a very high M_n of approximately 2 000 000. Amylopectin consists of anhydroglucose units connected by 1,4-α-linkages. Branch points are generated by 1,6-α-linkages at selected sites (Fig. 4). Amylopectin and amylose together form grain structures which are partially crystalline because of hydrogen bonds.

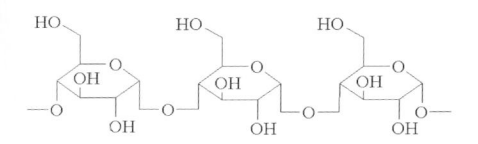

Figure 3. Representative structure of linear amylose

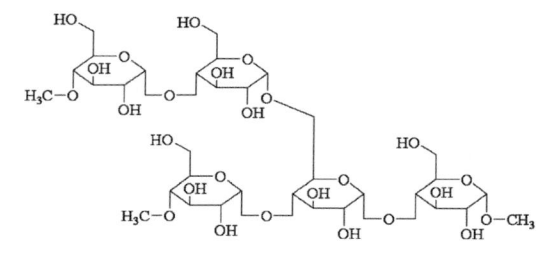

Figure 4. Representative structure of amylopectin including 1,6-α-D branch point

The particle size distribution depends on the type of plant (Table 5). The size of the potato starch granules is considerably higher than the size of corn and tapioca starch granules. Also the moisture absorption of potato starch is much higher than for corn and tapioca starch. Thus, processing of tapioca starch in technical processes like compounding is similar to that of corn starch [14].

Pure starch can be successfully formulated only for a limited number of plastic applications such as foam, e.g., for loose fill. Pure starch films and sheets are brittle and moisture-sensitive. They even disintegrate in the presence of water. Additionally, the temperature window for processing is very small because of the limited temperature stability of natural starch of about 170 – 180 °C. Chemical modification of starch by partial substitution of hydroxyl groups (e.g., with esters or ethers) can significantly improve hydrophobicity and the rheological properties. Cross-linking of the starch chains improves stability against acids, heat treatment, and shear forces [15]. However, the high requirements of film applications are not fulfilled by modified starch alone. Compounds with biodegradable polyesters as "enablers" are the appropriate

Table 5. Particle size distribution and moisture of starch from different plants

	Potato starch	Corn starch	Tapioca starch
Diameter (average), μm	5 – 100 (23)	2 – 30 (10)	4 – 35 (10)
Number of granules per gram	100×10^6	1300×10^6	500×10^6
Moisture absorption at 65 % relative humidity, 20, °C %	19	13	13

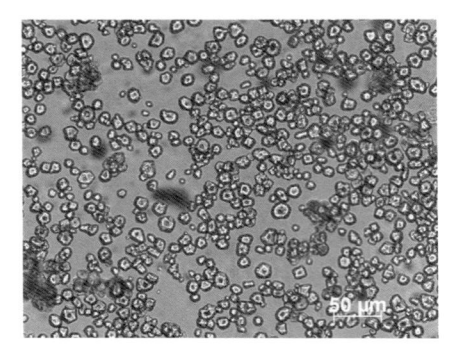

Figure 5. Light micrograph of granular corn starch dispersed in PBAT as continuous phase

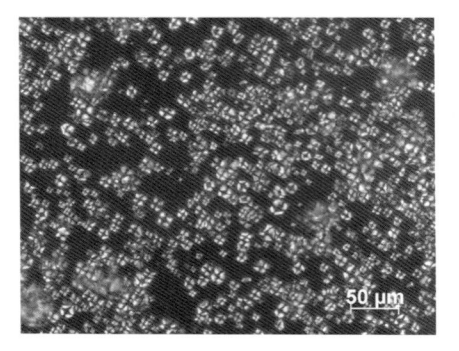

Figure 6. Light micrograph of granular corn starch dispersed in PBAT in polarized light to demonstrate cristallinity

solution (see Section 4.7). Figures 5 and 6 show microscopic pictures of starch compounds.

4.3. Poly(lactic acid) (Polylactide, PLA)

→ Lactic Acid

Synthesis. PLA is a melt-processible thermoplastic polymer completely based on renewable resources. The manufacture of PLA includes one fermentative step followed by several chemical transformations. The typical annually renewable raw material source is corn starch which is broken down to unrefined dextrose. This sugar is then subjected to a fermentative transformation to lactic acid (LA). Direct polycondensation of LA is possible but usually LA is first chemically converted to lactide, a cyclic dimer of LA, via a PLA prepolymer. Finally, after purification, lactide is converted via a ring-opening polymerization (ROP) to PLA [16–20]:

Properties and Applications. Due to its stereogenic center, lactic acid exists in two enantiomeric forms (D- and L-lactic acid) leading to three different lactide stereoisomers (D-, L-, and *meso*-lactide, Fig. 7). Depending on the relative amounts of the different stereoisomers in the final polyester, the crystallinity of the resulting PLA is heavily influenced and in this way the properties of the polymer can be adjusted to satisfy the needs of different applications [17, 21, 22].

PLA shows properties similar to polystyrene. Its high stiffness and transparency makes it a suitable material for applications such as plastic bottles, cups for cold drinks, stiff packages like clamshells as well as degradable films. PLA is completely nontoxic and classified in the USA as generally recognized as safe (GRAS) [23].

Figure 7. Lactic acid and lactides

PLA can be processed on standard equipment and used in injection molding, injection stretch blow molding, and for the production of blown and cast films, biaxially oriented films, or transformed to extrusion paper-coating compounds, etc. Although PLA is a compostable material, it is also more and more used in durable applications because of its biogenic origin and its sustainable image.

Limitations and Technical Developments Underway. Nevertheless, PLA still suffers from several drawbacks like its low impact strength, its poor barrier properties and especially its low heat resistance. A lot of research activities are currently in process to overcome these drawbacks. More and more companies offer solutions to the problems based on masterbatches of PLA copolymers or compounds.

Biodegradation. Chemical hydrolysis is considered to be the main degradation route for PLA [17]. This hydrolysis process takes place at high humidity and at elevated temperatures, which, for example, are found in industrial composting facilities. The fragments that result from the hydrolysis process, i.e., short oligomers and monomers, can then be consumed by microorganisms. Therefore, PLA articles can be certified to be compostable under these conditions but will not degrade in home composting piles because of the fact that temperatures are typically much lower. Apart from composting (or incineration), PLA can be chemically recycled by conversion of the used polymer back to the monomers followed by repolymerization.

Producers. The leading producer of PLA is NatureWorks LLC, a joint venture of Cargill Dow LLC and Teijin, with a nominal production capacity of about 140 000 t/a (see Chap. 9) [24]. NatureWorks LLC sells PLA resins for packaging and fiber applications under the trade name Ingeo. In Japan, NatureWorks PLA polymer is sold by Mitsui Chemicals under the trade name Lacea [25].

4.4. Polyhydroxyalkanoates (PHA)

Introduction. Structurally related to PLA are the polyhydroxyalkanoates (PHA), a class of polyesters derived from hydroxyalkanoic acids which can differ in chain lengths and in the positions of their hydroxyl groups (Fig. 8). The most common representatives among this class of biodegradable polymers, which are available so far in pilot-scale amounts, are poly(3-hydroxybutyrate) (P(3-HB)), poly(4-hydroxybutyrate) (P(4-HB)), poly(hydroxybutyrate-*co*-valerate) (PHBV), and poly(hydroxybutyrate-*co*-hexanoate) (PHBH).

Synthesis. Like PLA, PHAs are derived from annually renewable resources like sugars, starch, or fatty acids. Moreover, hydrocarbons like methane have recently been described to be used for the fermentative synthesis of PHB by

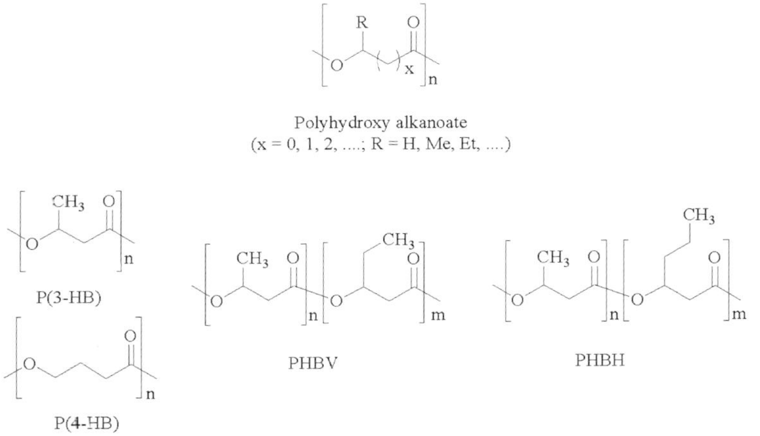

Figure 8. Polyhydroxyalkanoates, general structure and most important derivatives

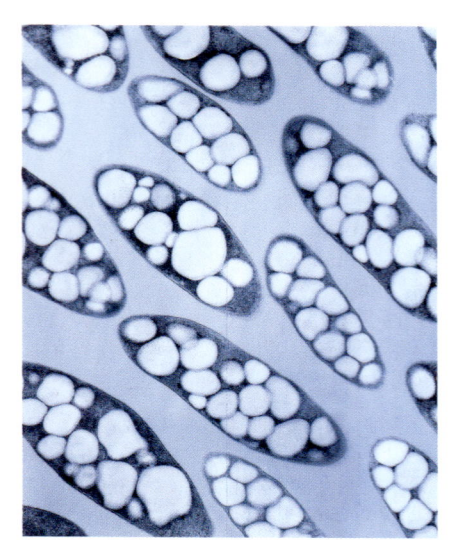

Figure 9. Polyhydroxyalkanoate particles in microorganism (photograph provided by Telles, USA)

methanotrophic bacteria [26]. In contrast to the PLA production, for PHAs is no chemical transformation needed. The polymer itself is made by various microorganisms as a carbon- and energy-storing substance (Fig. 9) [27]. The mass of the polymer in the microorganism can reach up to 90 % of the dry weight of the cell mass. The chemical structure of the resulting PHA depends on the bacterial species and the composition of the carbon source on which the microorganisms are grown [28]. After fermentation and cell lysis, high-purity PHA can be isolated by chemical extraction. Companies are currently working on the optimization process by designing genetically modified microorganisms, and recently transgenic plants that produce PHAs in their cell membranes have become objects of research [27, 29]. However, bacterial fermentation is still the most commercially viable way to produce PHAs in industrial scale [30].

Properties and Applications. The great structural variety of PHAs (see Fig. 8) gives rise to a multitude of different property profiles and different possible applications including injection-molded articles for bioresorbable medical and nonmedical applications such as fishing nets, degradable films, etc. [27, 31, 32].

Biodegradation. As PHAs are direct metabolites of microorganisms found in natural environments, they are accepted as food sources by many abundant microbe species which break the polymer down by extracellular poly(hydroxyalkanoic acid) depolymerases [33]. This is the reason why PHAs are quickly biodegraded both under aerobic conditions (e.g., composting) and in environments where anaerobic degradation takes place (e.g., marine environments, landfills).

Producers Although fermentative PHB has been known since 1926, commercial interest occurred not earlier than in the fourth quarter of the 20th century [34]. The major player today is Telles (joint venture of Metabolix and ADM) with the trade names Mirel, Biopol. Other companies active in the area of PHAs are Kaneka (Nodax), Mitsubishi Gas Chemicals (Biogreen), PHB Industrial (Biocycle) and Tianan (Enmat). The production capacities of PHAs are still modest and only pilot-scale quantities are found in the market today. But it can be expected that the capacities will be increasing significantly in the next years: Telles plans to produce Mirel in a capacity of 50 000 t/a by the end of the second quarter of 2009 [35] (see Chap. 9).

4.5. Synthetic Polyesters

4.5.1. Aliphatic Polyesters

Introduction. It is not the origin of the raw materials used in the manufacture of a polymeric material that renders it biodegradable: It is the chemical structure which allows enzymatic attack and enzymatic hydrolysis of the polymer chain followed by complete digestion of the fragments by microorganisms (see Chap. 3) [36]. With this knowledge, polymers can be specifically designed and produced from existing fossil-based monomeric building blocks in existing production plants. These polymers are then not only biodegradable but also exhibit beneficial mechanical properties. Aliphatic polyesters were the first completely fossil-based polymers which combined both the biodegradability of polymers found in nature as well as the useful mechanical properties of modern plastics.

Structure and Synthesis. There are two fundamentally different structures of synthetic polyesters. One of them is closely related to fermentative PHAs (Section 4.4) and is derived from hydroxyalkanoic acids or their intramolecular, cyclic condensates, lactones. The most common derivative is poly-ε-caprolactone and, as the name suggests, this polyester is produced in a ROP from ε-caprolactone [37, 38]:

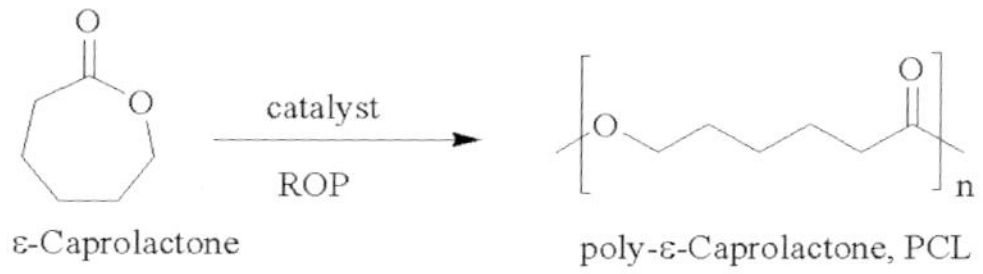

ε-Caprolactone →(catalyst, ROP)→ poly-ε-Caprolactone, PCL

The other type of polyester is based on a direct condensation reaction of aliphatic dicarboxylic acids and aliphatic glycols [39]. The most important derivatives are polycondensates from 1,2-ethylene glycol or 1,4-butanediol with succinic acid, adipic acid, or combinations of both (PES, PBS, and PBSA, respectively, Fig. 10).

Properties and Applications. Due to the broad spectrum of different monomers that can be used for the synthesis of these polymers, the properties can be adjusted to a certain degree. While PBS is a relatively stiff polyester, other combinations with longer dicarboxylic acids lead to more flexible materials. Generally, aliphatic polyesters are low-melting, flexible plastic materials which are best suited for and usually found in film applications like organic waste bags or mulch films and monofilament fibers. However, they are also found in foams and injection-molded parts. Processing on conventional polyolefin equipment is typically possible [39]. A drawback of aliphatic polyesters is the fact that they are prone to chemical hydrolysis leading to limitations in shelf life and transportation.

Biodegradation. In microbially active environments like soil and compost, aliphatic polyesters are degraded in a combination of chemical and enzymatic hydrolysis and the fragments are ultimately degraded to carbon dioxide and water [39].

Producers. PCL and biodegradable derivatives are sold by Solvay (Capa), Daicel (CellGreen), and by Dow under the trade name Tone. Due to the low melting point and low modulus of PCL, this biodegradable polymer is mostly used as a component in blends with biopolymers (see below). The Japanese company Showa Highpolymers produces aliphatic polyesters of the latter type in amounts of several thousand tons per year and sells them under the tradename Bionolle [39]. IRE Chemicals, a Korean company, sells polyesters of the same structure under the tradename EnPol, and Mitsubishi

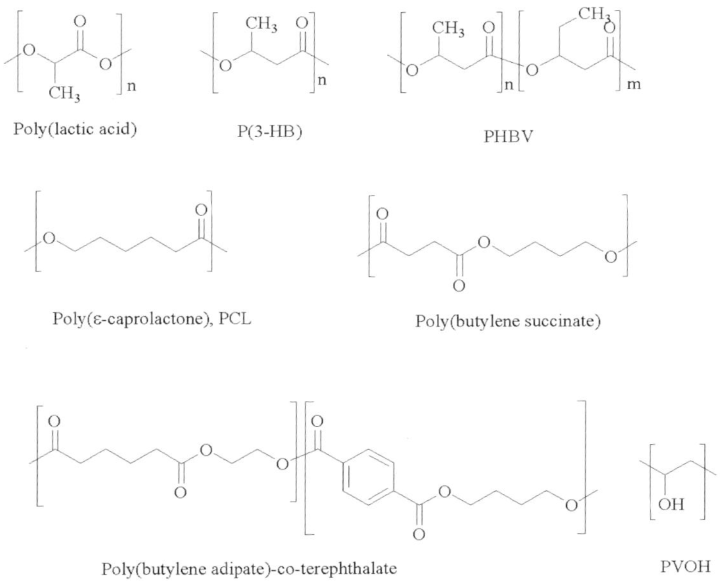

Poly(lactic acid)   P(3-HB)   PHBV

Poly(ε-caprolactone), PCL   Poly(butylene succinate)

Poly(butylene adipate)-co-terephthalate   PVOH

Figure 10. Commercially available biodegradable polymers

Chemicals has a biodegradable, aliphatic copolyester with a small amount of PLA (GS-PLA) in its portfolio.

4.5.2. Aliphatic/Aromatic Polyesters

Synthesis. The good biodegradability of aliphatic polyesters can be combined with the good mechanical properties of aromatic polyesters to some extent by copolyesterification of aliphatic and aromatic monomers [40]. Aliphatic/aromatic polyesters are synthesized in melt polycondensation processes from aliphatic acids (typically adipic acid) as well as aromatic dicarboxylic acids (typically terephthalic acid) with aliphatic glycols (typically 1,4-butanediol). The best known aliphatic/aromatic polyester is Ecoflex by BASF (Fig. 11).

Properties and Applications. Ecoflex is designed to be a strong and flexible material with mechanical properties similar to PE [41]. Therefore, Ecoflex can be melt-processed on standard polyolefin equipment and is mainly used in film applications like organic waste bags, mulch films, shopping bags, cling films, etc. Ecoflex is also used as a replacement for PE in paperboard coating, e.g., to yield completely biodegradable paper cups (see Chap. 9).

Due to its beneficial properties and its complete biodegradability as well as to a clear trend to renewable raw materials, Ecoflex is used as an "enabler" for renewable biopolymers like starch, PLA, PHAs, lignin, and cellulose. Ecoflex/starch as well as Ecoflex/PLA blends (e.g., Ecovio by BASF) are the most common renewable and biodegradable plastic materials for film applications in the market.

Biodegradation. Although pure aromatic polyesters like poly(butylene terephthalate) (PBT) are not prone to microbial attack, copolyesters of aliphatic and aromatic monomers are completely biodegradable if the aromatic content does not exceed a certain limit [40–43]. Similar to aliphatic polyesters, these mixed structures also degrade in microbially active environments mainly by enzymatic hydrolysis followed by complete mineralization of the fragments [5].

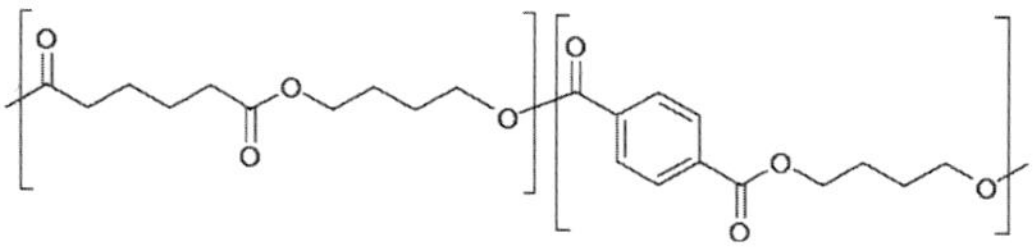

Producers. Ecoflex is the most important aliphatic/aromatic polyester in the market. Ecoflex is a product by BASF and current production capacities amount to 14 000 t. Recently, BASF announced to increase capacities by another 60 000 t by the end of 2010. Another supplier of aliphatic aromatic polyesters is DuPont (Biomax).

4.6. Other Synthetic Polymers

Another class of synthetic polycondensates produced in industrial scale are polyamides. Due to their ability to hydrogen bonding, polyamides generally show high melting points and have a highly suppressed mobility of the polymeric chains. This is the reason why polyamides have a good profile of mechanical properties but are—apart from naturally occurring homopolyamides and some materials of academic interest (e.g., PA4)—not biodegradable [44, 45]. The homopeptide polyaspartate → Polyaspartates and Polysuccinimide is a synthetic analogue to naturally occurring peptides. It is synthesized in a polycondensation reaction of aspartic acid via polysuccinimide, which is hydrolyzed to polyaspartate. This water-soluble and biodegradable polyamide (not discussed here in detail) is commercialized by Lanxess under the tradename Baypure and used, e.g., as an additive for detergents [46].

Poly(ethylene glycol) (PEG) is a water-soluble polyether mainly used in detergents and shampoos. PEG is both aerobically and anaerobically biodegradable if the molecular weight is not too high [47–50].

The only existing biodegradable polymer with an all-carbon backbone is poly(vinyl alcohol) (PVOH). Poly(vinyl alcohol) is synthesized by (partial) hydrolysis of poly(vinyl acetate) and shows a solubility in water which depends on the degree of hydrolysis. PVOH can be processed into films by solution casting or by thermoplastic melt processing [51]. The films are biodegraded in wastewater treatment or in composting facilities. Main application of PVOH as biodegradable polymer is in flushable items.

Figure 11. Aliphatic/aromatic polyester, Ecoflex

The mechanism of biodegradation is different from the ones discussed for polycondensates: it includes oxidation of the alcohol functionalities followed by enzymatic chain scission to smaller fragments which can finally be mineralized [52]. Major suppliers of PVOH are Kuraray (Poval, Mowiol), Nippon Gosei (Hi-Selon) and Indroplast (Hydrolene).

4.7. Compounds

Biodegradable polymers have defined property profiles (Table 6) as explained in the previous chapters, which limit their application range to a certain extent. If the market demands biodegradable polymer solutions using renewable resources, a compound can be developed by using a combination of rigid polymers like PLA or PHA and soft and impact-resistant synthetic biodegradable polymers like PBAT. Such a material responds precisely to the application requirements (Fig. 12). One example is the PBAT/PLA compound (Ecovio L BX 8145), which enables the production of film products similar to PE-HD.

4.7.1. Compounds of Starch and Biodegradable Polyesters

Based on the experience in extruder cooking of starch for products like peanut flips and dry pet food, compounding equipment for compounds of biodegradable plastics has been developed. Corotating, intermeshing, and self-cleaning twin-screw extruder concepts are available from machinery suppliers like Coperion, Clextral, Extricom, Leistritz, Krauss Maffei Berstorff, and others. These extruder concepts prevent caking and charring of the dew during starch cooking.

Starch compounds with polyesters are used to enhance hydrophobicity as well as mechanical and thermal properties of compounded products. The crystalline structure of starch granules has to be disintegrated prior to its use in applications (e.g., production of thin films and bags) because starch granules as large as the film thickness reduce the mechanical properties. High shear forces, heat, and/or plasticizers (e.g., moisture, glycerol) are used in a separate compounding step to turn granular starch into thermoplastic starch.

Compounding of starch and biodegradable polyesters can be efficiently performed in a four-stage process:

1. Starch is destructurized by temperature, pressure, and plasticizer (e.g., water, glycerol).
2. The biodegradable polyester (e.g., PCL, PBAT, PVOH) is added.
3. The melt has to be degassed by an effective vacuum, leaving only a few percent of moisture in the compound.
4. The compound is granulated under water or air and subsequently dried in a vacuum drier.

Film products with good mechanical characteristics are only obtained if thermoplastic, plasticized starch has been used. Another

Table 6. Properties of biodegradable polyesters and reference polymers[*]

Test	Standard	PET	PLA	PHB	PBAT	PBS	PE-LD
Transparency		transparent	transparent	opaque	translucent	translucent	transparent
Density, g/cm^3	ISO 1183	1.35	1.25	1.25	1.25	1.26	0.92
Melting point, °C	DSC	245	140–175	175	115–120	115	105–115
Glass transition temp. T_g, °C	DSC	75	58	−4	−33	−32	−100
Vicat A, °C	ISO 306	75	56	96	91	93	90
Mechanics MD/CD[**]							
Module of elasticity, MPa	ISO 527	3100	3600	1900	80	645	250
Tensile stress, MPa	ISO 527	58	70	27	23	35	20
Ultimate elongation, %	ISO 527	50	3	8	470	47	500
Barrier properties							
Oxygen, mL m^{-2} d^{-1} bar^{-1}	ASTM D 3985	15		45	600	100	1500
Water absorption, %		0.15	0.3	0.5	0.5	0.4	
Food contact	2002/72/EC	o.k.	o.k.	o.k.	o.k.	no	o.k.
Biodegradability	EN 13432	no	yes	yes	yes	yes	no

[*] Injection-molded specimen, 4 mm, barrier values: blown film: 100 μm.
[**] MD = machine direction = direction of extrusion of blown film, CD = cross direction = perpendicular to direction of extrusion of blown film.

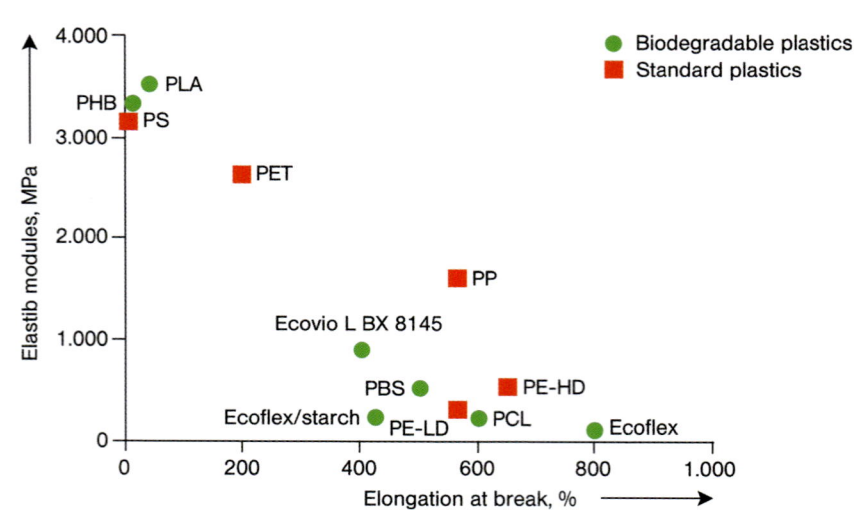

Figure 12. Stiffness versus flexibility for typical standard and biodegradable plastics, ISO 527
Ecoflex = PBAT; Ecovio LBX 8145 = compound containing 45 % PLA, about 55 % Ecoflex, and compatibilizers

condition is that the biodegradable "enabling polyester" forms the coherent phase [53]. If no plasticizer is used, biodegradable polymers with merely limited mechanical properties will be obtained. In this case, coextrusion with the base polyester or another polymer is the way to upgrade mechanical and/or moisture barrier properties.

In Table 7, most of the mechanical properties of average PE-LD carrier bags from blown film, 30 µm, are close to the PBAT/starch compounds containing thermoplastic starch. The reduction in stiffness by 20 – 25 % compared to that of PE-LD is lower than for PBAT/starch compounds with granular starch, which amounts to 50 – 55 %. The reduction in tensile strength is similar for both starch compounds: 10 – 35 %. If granular starch is used, the film samples exhibit a rough surface. Thus printability and mechanical properties (stiffness and puncture resistance) of PBAT compounds with granular starch are inferior to PBAT/starch compounds with thermoplastic starch by about 30 – 55 %.

In comparison to PE-LD, compounds of PBAT/thermoplastic starch are stiffer than pure PBAT. PBAT/thermoplastic starch compounds are used because they combine biodegradability and renewable resources.

Table 7. Examples for blown films, 30 µm

Test	Standard	PE-LD carrier bags (mean)	PBAT	PBAT+granular starch compound	PBAT+thermoplastic starch compound
Transparency		opaque	translucent	opaque	opaque
Printability		8 colors flexoprint	8 colors flexoprint	poor printability	8 colors flexoprint
Mechanics MD/CD[*]					
Modul. of elasticity, MPa	ISO 527	330/270	110/100	150/140	270/205
Tensile stress, MPa	ISO 527	32/25	35/40	23/22	21/20
Ultimate elongation, %	ISO 527	460/640	640/750	390/590	490/540
Puncture resistance, J/mm	DIN 53373	17	26	9	19
Barrier properties					
Oxygen, mL m^{-2} d^{-1} bar^{-1}	ASTM D 3985	4800	2000		
Water vapor, g m^{-2} d^{-1}	ASTM F 1249	3	240		
Food contact	2002/72/EC	not limited	not limited	dry food	dry food
Biodegradability	EN 13432	no	yes	yes	yes

[*] MD = machine direction = direction of extrusion of blown film, CD = cross direction = perpendicular to direction of extrusion of blown film.

4.7.2. Compounds of PLA and Biodegradable Polyesters

Polylactic acid is a transparent, stiff polyester (see Section 4.3). It can also be transformed into transparent flexible films with properties close to cellophane using biaxial orientation technology and into fibers using standard production process of fiber spinning and subsequent orientation and fixation. However, for most of the flexible-film applications the stiffness of PLA (3600 MPa) is too high. Therefore, compounds of PLA and soft biodegradable polyesters like PBAT are used to reduce the stiffness efficiently [54].

The addition of 20 % PBAT (e.g., Ecoflex) in PLA already reduces the stiffness of PLA by 25 % (Fig. 13), keeping the impact strength (e.g., Charpy, unnotched, according to ISO 179/1eU at −20 °C) at 22 kJ/m^2, which is above the level of high-impact polystyrene. Containers produced from this compound are resistant even to sudden impact—they deform without brittle failure at room temperature (23 °C).

Flexible-film applications are much more demanding regarding the stiffness/toughness ratio of the PBAT/PLA compound. Usually, PP with a stiffness of 1600 MPa is the stiffest product to be used in blown-film extrusion. But most blown-film lines for stiff products can only handle conventional PE-HD with a stiffness of 600 – 1200 MPa. Moreover, high requirements have to be fulfilled concerning the bubble stability at lower film thicknesses, e.g., at 10 – 15 μm or below.

Corotating, intermeshing, and self-cleaning twin-screw extruder concepts (e.g., from machinery suppliers like Coperion, Clextral, Leistritz, Kraus Maffei Berstdorf, Extricom as being employed for starch compounds can be modified for the compounding of PBAT and PLA. The volume elements for deaeration of starch powder have to be changed to elements for conveying and melting of granulates. Adapted shear mixing devices are used to blend the melts at low shear energy to avoid high temperatures causing pronounced hydrolysis of the PLA phase. A thorough venting in one or two steps as needed for starch compounds is not necessary for PBAT/PLA compounds. It is frequently possible to achieve higher output rates with PBAT/PLA compounds than with PE-LD.

Good performance characteristics are obtained for PBAT/PLA compounds with PBAT as the continuous phase and PLA as the discontinuous phase. Examples are given in Table 8 for a blend with 45 % PLA (PLA as discontinuous phase, PBAT as continuous phase) and a blend with 60 % PLA (PLA as continuous phase, PBAT as discontinuous phase), compared to PE-HD.

Table 8 shows that a compound with 45 % PLA and PBAT can achieve about 70 – 75 % of the mechanical properties of an average PE-HD carrier bag at 30 μm thickness. The values of a PE-LD compound (see Table 7) are reached or exceeded. The 45 % PLA compound (Ecovio L BX 8145) yields a biodegradable film that comes closest to the properties of PE-HD films on the market.

The flexibility of PBAT/PLA compounds with PBAT as continuous phase can always be enhanced by using PBAT in the dryblend.

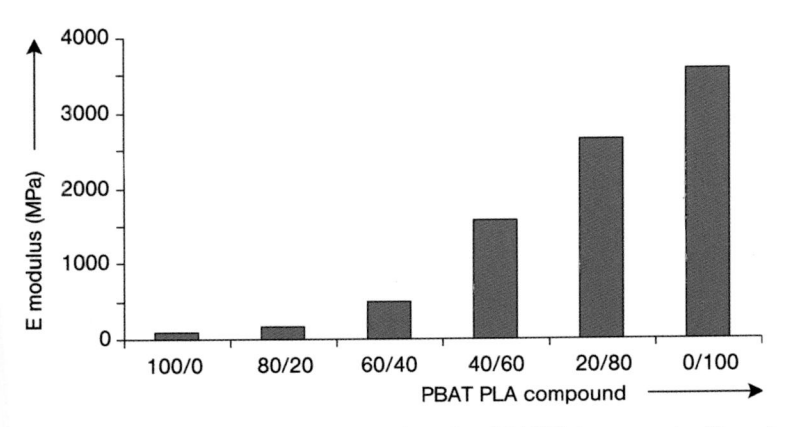

Figure 13. Stiffness of injection-molded specimen from PBAT/PLA compounds with varying PLA content according to ISO 527

Table 8. Characteristics of blown films, 30 μm, from PE-HD and compounds with PBAT

Test	Standard	PE-HD[a]	PBAT+PLA 55/45 dryblend[b]	PBAT+PLA 55/45 compound[c]	PBAT+PLA 40/60 compound[d]
Transparency		opaque	opaque	opaque	translucent
Printability		8 colors flexoprint	8 colors flexoprint	8 colors flexoprint	8 colors flexoprint
Mechanics MD/CD[e]					
Modul. of elasticity, MPa	ISO 527	650/630	1180/490	1020/440	1560/1080
Tensile stress, MPa	ISO 527	45/42	39/21	50/32	47/30
Ultimate elongation, %	ISO 527	640/520	360/170	430/360	160/160
Puncture resistance, J/mm	DIN 53373	42	19	31	31
Barrier properties					
Oxygen, mL m^{-2} d^{-1} bar^{-1}	ASTM D 3985	2000		1400	
Water vapor, g m^{-2} d^{-1}	ASTM F 1249	1.3		160	
Food contact	2002/72/EC	not limited	not limited	not limited	not limited
Biodegradability	EN 13432	no	yes	yes	yes

[a] Average values for carrier bags.
[b] Dryblend of 55 % PBAT and 45 % PLA.
[c] Compound of approx. 55 % PBAT and 45 % PLA.
[d] Compound approx. 40 % PBAT and 60 % PLA.
[e] MD = machine direction = direction of extrusion of blown film, CD = cross direction = perpendicular to direction of extrusion of blown film.

Thus, PE-HD-like or PE-LD-like properties can be achieved by using an appropriate combination of PBAT and PBAT/PLA compound [55].

Another important aspect of biodegradable PBAT/PLA blends is their shelf life under regular storage conditions. Because standard climate is used to condition plastic specimens prior to test, standard room climate (23 °C, 50 % relative humidity) has been used to store film samples from PBAT and PBAT/PLA compounds. After three years of storage under these conditions, biodegradable films from PBAT (Ecoflex F BX 7011) achieve or exceed the mechanical properties of PE-LD carrier bags except for stiffness. Films from PBAT/PLA compounds (e.g., Ecovio L BX 8145) maintain the property level of PE-LD carrier bags during two years of storage for stiffness, tensile strength, and puncture resistance. The ultimate elongation can be adjusted by using PBAT (e.g., Ecoflex F BX 7011) [55].

4.8. Cellulose Derivatives

Cellulose esters (e.g., with acetic acid or propanoic acid) are well-known amorphous plastics which are used for injection molding of transparent articles. Thus, there are compounds for injection-molded products on the market, e.g., for candle housings in cemeteries or as disposable cutlery. However, only cellulose acetate with a degree of substitution of not more than 2.5 is biodegradable. Biodegradability certificates according to EN 13432 are difficult to achieve because the biodegradation of cellulose derivates is a very slow process.

Cellulose films (e.g., cellophane) are very transparent and exhibit an excellent barrier against oxygen and aroma if the moisture level is kept below 50 % relative humidity. When a coating based on natural waxes is used, a high water-vapor barrier can be achieved, which makes these films applicable for packaging of food, e.g., cereals, pasta, cheese, fruit, and vegetables.

5. Processing and Additives

(→ Plastics, Processing; → Films, Section 6.10.)

5.1. Introduction

Biodegradable polyesters are designed for processing on regular equipment for standard polymers. Nevertheless, the equipment has been originally designed according to the property profiles of standard polymers. Therefore, the following limitations have to be taken into account if a conversion process is selected for a biodegradable polyester.

1. Limited compatibility to standard polymers. Biodegradable polyesters are incompatible to standard polymers like polyolefins, PS, and PVC, forming large domains in blends with standard polymers.

Table 9. Sensitivity of polymers to moisture, thermal melt instability, and crystallization rate

Polymer	Moisture tolerance	Thermal melt instability	Crystallization rate
Polyolefins	no	> 320 °C	low – high
PVC	no	> 180 °C	no
PET	< 100 ppm	> 300 °C	low
PS	no	> 270 °C	no
PLA	< 200 ppm	> 270 °C	no – high
PBAT	< 1000 ppm	> 270 °C	low
PBS	< 200 ppm	> 270 °C	high
PHA	low	< 190 °C	high

2. Sensitivity to moisture. Chain cleavage caused by hydrolysis depends on the type of polyester, the moisture level, and the processing temperature and time [56–58].
3. Reduced thermostability at higher melt temperatures. Natural polymers like starch or cellulose become instable at processing temperatures above 170-180 °C, depending on the processing time. Decomposition and charring occur under these conditions. Synthetic biopolymers like PBAT, PCL, and PBS are stable without significant change of viscosity until 200 °C. A significant change in melt viscosity is observed at a melt temperature above 230 °C for PBAT [59].
4. Low crystallization rate. Biodegradability is related to a low degree of crystallinity [58, 60]. Thus a low crystallization rate enables the converter to control crystallinity. This is important because a high degree of crystallinity correlates to high mechanical properties.

The sensitivity of biodegradable polyesters to these limitations depends on their chemical composition (Table 9).

5.2. Extrusion

Extrusion is a continuous conversion process by melting and subsequent transformation of granules or powders of biodegradable polymers and additives into semifinished or finished articles like sheet, profiles, tubes, bottles, films, tapes, or monofilaments (e.g., lines, yarns). The term "extrusion" also refers to compounding processes, because the same definition applies to the production of granules by either cutting strands or directly by underwater pelletizing.

Single-screw extrusion is the most common conversion process for biodegradable plastics. Only injection-molded and injection stretch blow molded rigid containers like closures and bottles are not produced by a continuous extrusion process.

The common element of extrusion processes is the extruder itself. Thermoplastic biopolymer granules are introduced into the hopper by means of automatic dosing equipment. The granules enter the screw at the cooled barrel inlet. In the screw the material is conveyed, densified, deaerated, and eventually melted by means of energy dissipation and external barrel heating. Shear and mixing elements on the screw allow for a good and even temperature distribution in the screw channel. At the end of the screw, the back pressure from the die needs to be generated for proper flow control [61].

Basically, biodegradable polyesters are designed to run on existing extruders for polyolefins, PS, PVC, or PET. The limitations of biodegradable polyesters have to be addressed by the selection of an appropriate extruder.

Because of the incompatibility of standard polymers to biodegradable polyesters, an appropriate purging procedure has to be developed.

Because of their sensitivity to moisture, biodegradable polymers need to be processed either in a dried form or on extruders which extract volatile components [62].

Usually high shear forces and stagnation zones should be avoided when processing biodegradable polyesters, e.g., starch compounds because they are sensitive to thermal degradation [63].

5.3. Blown-Film Extrusion

In general, mechanical properties of polymers are improved by orientation processes [54, 56, 58]. Because of the biaxial orientation, films from blown-film lines exhibit a high puncture resistance. Blown-film lines are designed to produce flexible PE-LD or PE-HD films ranging from 5 – 200 µm.

The extrusion process requires the following steps:

1. The biodegradable polyester is first processed by the extruder forming a melt stream with constant melt temperature and output

rate. The melt is transformed into a tube in the die head using spiral mandrel-type dies.
2. The tube leaves the die at or close to melt temperature. It is cooled down quickly by the air ring and in some cases the internal bubble cooling.
3. The sizing unit allows for an even cooling process of the film as well as a fixation of the film bubble. The ratio of the maximum bubble diameter to the diameter of the die exit yields the blow-up ratio — an important measure for the biaxial orientation of the film.
4. After the sizing basket, the film thickness is determined by the thickness measuring unit.
5. The film bubble is collapsed to a flat tube in the collapsing frame, which is typically equipped with aluminum rolls, wooden boards called slats, or an airboard.
6. A reversing unit changes the position of the film in the cross direction by turning the film continuously.
7. The web is split and wound in two flat film rolls.
8. The production process can be controlled by automatically keeping constant meter weight, thickness distribution, and film width. This device enables an automatic quality control of the production process.

Specific requirements for biodegradable polyesters are:

- Biodegradable polyesters like PBAT and their compounds, PBS, PBSA, and PCL run well on extrusion lines for PE-LD. Branching and chain extension, e.g., of PBAT lead to a good bubble stability even at low thickness of ≤ 15 µm.
- Low melt temperatures, e.g., 140 – 170 °C for PBAT and PBAT/starch compounds and 165 – 190 °C for PBAT/PLA compounds can be achieved, which are beneficial for blown-film stability and thermal stability of biodegradable polymers.

Welding. PCL, PBAT, and its compounds are weldable on existing bag-making machines for PE-LD und PE-HD—but at lower welding temperatures of, e.g., 90 – 100 °C for PBAT. As the crystallization process is slower than with PE-LD, the welding line can stick to the surface, if direct surface contact is possible. Therefore, the welding machine has to allow for extra cooling of the weld lines.

5.4. Cast-Film Extrusion

Cast-film extrusion of polyolefins has been developed to obtain flexible films with a high level of transparency by freezing the amorphous polymer structure of the melt on a chill roll. Cast films are mono-oriented in the extrusion direction.

The process is divided into the following steps:

1. The biodegradable polyester is first processed by the extruder, which forms a melt stream with constant melt temperature and output rate.
2. In the forming section, the melt streams of several extruders can be merged into one using an adapter feed block. This device controls the flow of each stream to obtain an even layer distribution [64].
3. Then, the melt stream is transformed into a flat film in the film die. The thickness distribution is controlled automatically by means of expansion bolts. The melt film leaves the die in a defined angle to obtain a tangential contact with the chill roll, maximizing the contact angle with the chill roll and thus the cooling capacity. Between die exit and the contact line on the chill roll, the film is stretched by a factor of 10 – 50 within a fraction of a second. The film is positioned on the chill roll by static discharge units. After the subsequent chill rolls, the film thickness distribution is determined by a thickness measuring unit. The film can be corona-treated using an electrical discharge process to increase surface tension and thus facilitate printing. In most cases, this procedure is not mandatory for biodegradable polymers because their surface tension is > 38 dyn without corona treatment.
4. The film is wound into rolls using a contact/surface winder with an option for automatic roll change.
5. The production process can be controlled automatically, keeping meter weight and thickness distribution constant. This device enables an automatic quality control in the production process.

Specific requirements for biodegradable polyesters are:

- Biodegradable polymers are processed at low melt temperatures, e.g., 170 °C for PBAT and 170 – 190 °C for PBAT/PLA compounds.
- A mat chill roll with a high surface roughness should be used to minimize sticking problems of biodegradable polyesters. The chill-roll temperature should be kept at about 30 – 40 °C for PBAT/PLA compounds. Pure PBAT can be run at lower chill-roll temperatures.
- Because of the limited thermal stability of, e.g., PBAT/starch compounds, dead spots in the feed block adapter or the film die have to be avoided. Thus, a reduction of the film width by closing the die gap with metal bars (deckling) is not recommended for PBAT/starch compounds.

5.5. Modification of Biodegradable Polyester Films

5.5.1. Additives

To improve processing and performance of films made of biodegradable polyesters, film qualities can be modified by using additives in the form of masterbatches:

- Slip agents are used to reduce the coefficient of friction of the final film as well as the adhesion of the film to metal parts or to itself during processing. Biodegradable amides of fatty acids, e.g., oleamide, erucamide, ethylene-*bis*-stearylamide, fatty acid esters like glycerol oleates or glycerol stearates, as well as saponified fatty acids, e.g., stearates are typically used as slip agents for biodegradable polyesters.
- Antiblock agents like talc, chalk, or silica can be used in form of masterbatches, e.g., based on PBAT [59, 67].
- Pigment masterbatches can be used to achieve specific colors in film applications. The use of pigment masterbatches is limited by the requirements on heavy metal content by the standards for biodegradability, e.g., EN 13432 [65]. Examples for heavy metal free pigments are carbon black and titanium dioxide. Carbon black pigment is used to make black films, which are applied as mulch film to increase the soil temperature in spring. Coated titanium dioxide pigments allow the production of white films, e.g., for carrier bags or for white mulch films, which reflect infrared radiation to reduce the soil temperature. The maximum loading with titanium dioxide is limited by the standards that have to be met.
- Inorganic nucleation agents like talc or calcium carbonate and organic nucleation agents can be used to improve the clarity of pure biodegradable polyesters: With their help, PBAT, PHA, or PLA films of thicknesses > 20 μm achieve the clarity level of PE-LD.
- Antifog agents like fatty acid esters are used to avoid the formation of water droplets on the film under condensation conditions, e.g., in cling film applications at the transition from room temperature to a refrigerated warehouse. The antifog agent is hydrophilic. It reduces the surface tension of droplets so that a uniform water film does not impair the clarity of the cling film [67].

5.5.2. Printing

In general, PBAT and PBAT/PLA compounds can be printed and welded on standard equipment for PE-LD. Both alcohol-based or water-based inks can be used after testing. Prior to printing, the material has to be corona-treated if the surface tension is < 38 dyn. The drying temperatures should be kept below PE-LD conditions, depending on the content of PLA in the PBAT/PLA compound. As drying conditions depend very much on the machine design, they should be determined during production trials [67].

5.5.3. Metallization

Depositing of a thin metal layer under a high vacuum using a plasma process is one of the most efficient production processes for high-barrier films. Slip and antiblock agents in the film have to be avoided, because surface defects reduce the barrier properties of the coating. Biodegradable polyesters like PLA, PBAT/PLA compounds, and cellulose derivates like cellophane can be metallized on standard equipment [66].

5.5.4. Multilayer Films

Film properties of different polymers can also be combined to meet the requirements of specific applications.

In the lamination process, films of different materials are bonded using heat and/or biodegradable adhesives [68]. A metallized film, e.g., from PLA with high barrier and stiffness can also be laminated onto a PBAT/starch compound with good welding performance to produce a stand-up pouch for detergents.

A reason for the use of multilayer films is to minimize the use of additives. Surface active additives like slip, antiblock, or antifog agents are used only in the outer layers of a multilayer film. The migration of the organic additives and their solubility in the polymers has to be considered.

5.6. Extrusion Coating

→ Plastics, Processing, 1. Processing of Thermoplastics

Extrusion coating has been developed to form thin polymer layers on flexible substrates like paper, paper board, and metal foils in a high-speed process [68–70]. In combination with paper and paper board, the polymer protects the substrate from moisture, oil and fat. The polymer provides the welding properties to transform, e.g., the paper board/polyethylene substrate into a paper cup, a liquid carton or a box for frozen food.

Biodegradable polyesters like PBAT and PBAT compounds have a good adhesion to paper and cardboard. PBAT compounds can be processed on existing extrusion coating lines designed for PE.

5.7. Sheet Extrusion and Thermoforming

Sheet extrusion lines produce foils and sheet from 200 µm to > 10 mm. Packaging applications usually require thicknesses < 2.5 mm, which can be provided in rolls for the subsequent thermoforming process. Sheet extrusion lines have originally been designed for PP, PS, PS copolymers, PET, and PC [70, 71].

The extruder provides a continuous, thermally homogeneous melt stream which is transformed into a sheet by a die. The sheet thickness is controlled by the die and the line pressure distribution in the roll stack. The sheet has to cool down slowly to avoid the build up of high internal stresses which influence the subsequent thermoforming process.

During the thermoforming process the sheet is heated above the glass transition temperature and below the melting point of the crystalline phase. Afterwards, the hot sheet is formed into a chilled mold using vacuum, pressure, and/or mechanical force. After a cooling step, the thermoformed containers are trimmed and ejected. The skeleton — some 30 – 70 % of the total volume — is recycled in the same application [72].

Inline extrusion/thermoforming machines integrate the processes of sheet extrusion and thermoforming. After sheet extrusion, the hot sheet is transferred and processed in the thermoforming unit.

Biodegradable polyesters have to be dried before sheet extrusion. Melt temperatures of 180 – 220 °C can be used for PLA and PLA/PBAT compounds. Reprocessing of about 30 – 70 % of regrind from the subsequent thermoforming process requires an acceptable processing stability of the biodegradable polyester for 5 – 10 processing cycles. Inline extrusion/thermoforming machines provide the best recycling performance because the contact of the scrap to air moisture is minimized.

5.8. Injection Molding

The advantage of the injection molding process is the accurate reproduction of complex three-dimensional shapes [73, 74].

The melt is plastified into a reservoir in front of the screw using an extruder unit as described in Section 4.2. In contrast to this extruder, the screw of the plastification unit of an injection-molding machine is pushed back while the melt for the next shot is produced. The next shot starts with the injection of the melt through a melt distribution system of runners and sprues which provide an even filling of the mold.

Injection pressure is maintained on the injected part to refill the loss of volume of the polymer during the cooling phase until the sprues are frozen. After the cooling period, the part is ejected. Hot runner sprue systems are constantly kept at melt temperature to avoid scrap from sprues in cold runner systems.

Biodegradable polyesters can be injection-molded as well. But because of the slow crystallization of most biodegradable polyesters the cooling times have to be prolonged, thus increasing production costs. In the production of injection-molded packaging like dairy cups, lids or caps, and closures, hot-runner systems are standard to enable a scrap-free production. To be used in these hot-runner systems, the thermostability of the biodegradable polyester must be at a high level. Currently, only a few applications have been realized using PLA, PBS, and cellulose acetates in injection-molding processes.

5.9. Blow Molding

Blow molding is the main process for the production of hollow articles like bottles, jerry cans, drums, vats, fuel tanks, and integrated bulk containers. Originally the process has been designed for PE-HD, but today a variety of different plastics can be used in blow molding [75, 77]. Blow molding can be performed in a continuous process (extrusion blow molding) and in a discontinuous process (injection stretch blow molding).

Extrusion Blow Molding. In order to achieve a high melt strength, a parison is extruded at low melt temperature (e.g., PBAT/PLA compound: 170 °C; PE-HD: 180 – 220 °C) using an extruder as described in Section 4.1 and 4.2. The parison is clamped between the two parts of a mold and inflated into the mold. After a cooling step, the hollow article is ejected. Scrap (30 – 70 %) is trimmed off and recycled into the same application [75, 76].

In extrusion blow molding, biodegradable polyesters have to fulfill basically the same requirements as for the sheet extrusion/thermoforming process.

Injection Stretch Blow Molding. The injection stretch blow molding (ISBM) process has been designed to improve the performance of PET bottles by using a biaxial stretching process at temperatures above glass transition temperature to maximize crystallization, clarity, barrier, and mechanical properties without scrap. Thus, less material is needed than in extrusion blow molding. The process can also be used well with PP or PS [78]:

A preform of the shape of a test tube with a defined thickness distribution is injection-molded in one unit (Section 5.8). The preform is transferred to the stretch-blowing unit and heated above glass transition temperature. Then, it is stretched by using a metal device and inflated by pressure. After cooling, the hollow article is ejected.

Biodegradable polyesters like PLA and PBAT/PLA are suitable for ISBM because they can be oriented sufficiently without failure in the biaxial stretching process.

The advantage of the ISBM is the scrap-free production, thus one is free from the thermostability requirements due to the recycling process. For hollow biodegradable articles, it is a challenge to keep the balance between good mechanical properties—requiring a high level of crystallinity—and good biodegradability—requiring a low level of crystallinity.

6. Market Overview and Growth Drivers

Compared to the global market for PE film applications (approx. $30\,000 \times 10^3$ t in 2007), the market for biodegradable polymers (65×10^3 t in 2007; excl. loose fill) represents a small niche segment, which has been established over the last decade. The competitive advantages and market drivers of biodegradable polymers in specific applications are based on

- superior life-cycle eco-efficiency (e.g., for garden waste bags)
- changing consumer behavior based on higher sensitivity for environmental issues and an increased interest in environmentally friendly products
- increased interest of retailers to differentiate in the market
- support by municipal authorities (providing a composting infrastructure)
- legislative frameworks to enhance the use of biodegradable products (e.g., in France, Spain, Italy)

- technology progress including access to new applications
- larger production plants and increasing production capacities

The authors expect the market for biodegradable polymers to grow from 65×10^3 t (2007) to 400×10^3 t (2015), which means an average annual growth rate of 25 %. As this forecast takes into account not only the potential but also the risks of new technologies/markets, it is quite conservative compared to other market studies ranging from 480×10^3 t to 6350×10^3 t in 2015 [79]. Depending on the development of the above-mentioned factors and technology progress, the market growth for biodegradable polymers may be influenced significantly.

Biodegradable Polymers versus Feeding the World. In contrast to fuels and energy, the much lower amount of renewable resources needed for the production of biodegradable polymers does not lead to a competition to global food production. As mentioned, according to the authors, the market for biodegradable polymers will grow to 400×10^3 t in 2015. The volume of corn needed for the production of this amount of biodegradable polymers would equal approx. 0.1 % of the total $700\,000 \times 10^3$ t of annual global corn harvest [80]. This amount does not influence the overall volume balance of corn, especially when considering that other raw materials like sucrose (from sugar cane or beet) or potato starch, etc. can be used, too. With the long-term perspective of making cellulosic materials available for monomers (e.g., lactic acid) and biodegradable polymers (e.g., PLA, PHA), no correlation to food production is going to exist.

7. Value of Biodegradability — Life-Cycle Assessment (LCA)

Biodegradable plastics are developed for special applications where a possibility for a safe release into the environment is required. They are particularly attractive when economic and/or ecological benefits can be gained by leaving plastic products in the soil or organic waste stream.

To define the value of biodegradable polymers, the overall system costs and the environmental impact of individual products in their respective target applications have to be considered. To this end, comprehensive life-cycle assessments (LCA) are an appropriate tool, especially when accompanied by costs evaluations that cover all phases from the cradle to the grave.

When looking at the life cycle of biodegradable plastics, two aspects are of particular importance: the end-of-life options and the use of renewable resources in the material production (the major part of the currently available biodegradable plastic products are made of blends of fossil-based polymers and polymers derived from biomass).

By using renewable carbon from biomass, an improvement in the CO_2 balance can be achieved. However, significant effects beyond the impacts on greenhouse gas emissions are possible, e.g., soil modification, eutrophication, impact on biodiversity, land requirements, and water consumption. These aspects depend on different factors like feedstock type, scale of production, cultivation and land-management practices, location, and downstream processing routes. Especially the environmental implications of agriculture are sometimes difficult to assess by the LCA methodology and require further research.

With regard to the end-of-life phase, there are different disposal routes for biodegradable plastics.

Organic Waste. The waste in the EU contains approx. 30 – 40 % of organic waste. For the disposal of this organic waste, different end-of-life options are possible: landfilling, incineration, and biological treatment.

Landfilling. This traditional way is still the one disposal method used most widely in the EU. In landfills, biodegradable waste decomposes to produce landfill gas and leachate. The landfill gas consists mainly of methane and, if not captured, it contributes considerably to the greenhouse effect. For this reason, the diversion from landfill is an important part of the European Waste Framework Directive.

This directive (amended in 2008) stimulates a more intensive utilization of organics waste, the diversion from landfill, separate collection of organic waste, and treatment in a way that fulfills a high level of environmental protection. It supports the use of environmentally safe materials produced from organic waste. The member states are requested to amend their national waste laws within two years after its coming into force.

Incineration. Organics waste is usually incinerated as part of mixed municipal waste stream. Depending on the facility and the energy use, this process can be regarded as energy recovery or as disposal. As the moisture content in organics waste is mostly very high (about 60 %), the efficiency of such processes is quite poor.

Biological Treatment. Composting is the most common biological treatment option. It may be considered as a recycling option when the resulting compost is used as fertilizer or soil improver. Effects like increased water retention capacity and improved soil structure are ecological benefits which are of particular importance when soil erosion is a serious problem (for example in some Southern European countries).

Anaerobic digestion is a method of producing biogas from organic waste for energy purposes and, thus, could be seen as energy recovery. This waste management option is especially suitable for treating wet organics waste. In addition, the processes of digestion and composting are usually combined as successive steps at one facility.

Direct Release in Agricultural Applications. A nonbiodegradable agricultural film has to be recollected after use (labor costs) and disposed (or recycled), which generates additional costs within the system. In contrast to PE film, the biodegradable agricultural film can be simply tilled into the ground.

8. Applications

The following paragraphs describe the major applications today and to come (loose-fill applications excluded). An overview of applications and volumes for 2007 and 2015 is shown in Table 10. Volumes are expected to grow from 65×10^3 t in 2007 to 400×10^3 t in 2015. Compounded annual growth rate (CAGR) is 25 % Table 11 shows an overview of the material properties with respect to the applications and the standard polymers.

8.1. Organic Waste Bags

As described above, composting is the most favored method of recovery of post consumer organic waste. Composting is already well-established in some European countries, and is being established in others. The Netherlands and Germany are the leading counties in the development of a composting infrastructure. In these two countries, 95 % and 60 % of all households, respectively, have access to industrial composting plants. In the EU, organic matter accounts for 30 – 40 % of total domestic refuse. With the expectation of continuous expansion of the composting infrastructure, a significant growth of organic waste bags is expected.

From a technical point of view, compost bags have to be biodegradable according to EN 13432. Beyond the mechanical properties, it is necessary to down-gage the bags at 15 – 30 µm to achieve a good LCA. The temperature resistance should allow transport and storage at 60 °C. Breathability, i.e., low-barrier properties for water and gases are an advantage. The proposed biodegradable polymers fulfill these basic requirements for compost bags. Only the moderate stiffness of PBAT and the low heat resistance of PCL (40 °C) limit the use of these materials for compost bags.

Paper is a biodegradable alternative to plastics. Compost bags made of paper are rather heavy. Pure Kraft paper exhibits a limited resistance to oil and fat and the rough surface limits the printability. But the acceptance on customer level is high because paper is perceived as an environmentally friendly material.

Table 10. Development of applications for biodegradable polymers

Application	Volume 2007, 10^3 t	Volume 2015, 10^3 t	Comment
Organic waste bags and shopping/carrier bags	16	131	most established segment
Packaging including foam	42	248	food and nonfood packaging
Mulch film and horticulture	7	21	
Sum	65	400	

Table 11. Applications of biodegradable polymers and compounds

Application	Reference polymers	PLA	Starch+PBAT	PLA+PBAT	PHB	PBS/PBSA	PCL	Biodegradable alternative	Comment
Basic: flexible film Compost bags	PE	○	●	●	○	●	◕	paper	material selection depends on desired property profile, e.g., PE-LD- or PE-HD-like
Mulch film	PE	○	◕	◕	○	◕	◐	paper	biodegradability depending on local climate is key issue
Development flexible film Carrier bags	PE	○	●	●	○	●	◕	paper, cotton	material selection depends on desired property profile, e.g., PE-LD- or PE-HD-like; political application
Shrink film	PE	○	●	●	?	?	?		development status
Hygiene film	PE	○	●	●	◐	◐	◐		development status
Knitted nets	PE-HD	◐	◕	◕	◐	◕	◐	cotton	development status, biodegradability depends on stress induced crystallization.
Rigid container Extrusion + thermoforming	PET, PP, PS, PVC	◕	○	◕	◐	◐	◐	cardboard	extrusion-coated cardbord is an alternative for nontransparent market
Injection molding, ISBM	PP, PS, PET	◐	◐	◐	◕	◐	◐		development status, biodegradability of bottles to be enhanced

● Application requirements fulfilled as described
◕ 1 Property < 70%
◐ 2-3 Properties < 70%
○ > 3 Properties < 70%

8.2. Shopping (Carrier) Bags

Changing consumer behavior based on higher sensitivity for environmental issues and an increased interest in environmental friendly products accompanied by an increased interest of retailers to differentiate in the market are the major drivers for the market growth of biodegradable shopping bags. After shopping, these bags can be applied for the disposal of organic waste resulting in a double use of the material.

Thin biodegradable carrier bags have a property profile which is similar to that of compost bags:

- Good mechanical properties for loads of about 1000 times of their own weight
- Good puncture resistance, e.g., for liquid beverage cartons
- Down-gaging to 10 – 20 µm
- Good printability (8 color flexo printing) for superior presentation
- Good welding performance for high-speed bag making
- Usefulness as compost bags after several services as a carrier bag

Except for the optical presentation and the resulting need for printability, biodegradable carrier bags have to fulfill the same requirements as biodegradable compost bags. Thus the material options are also the same.

Because of the higher prices of high-end carrier and shopping bags, paper bags are more often used as shopping than as compost bags. However, paper bags with high-quality printing and coatings are generally not biodegradable because of the coating materials applied.

8.3. Mulch Film

Agricultural films are a well-established application in Japan and the EU. This application is very cost-sensitive because of the subsidy structure of the agricultural sector of the EU. However, the regulations for waste disposal of mulch film require either recycling or adequate treatment (e.g., incineration). If thin mulch films are concerned, the recovery of the film in the field is crop-dependent and difficult. Cleaning of the soil residues is cost intensive. In this case it can be more cost-effective to use biodegradable mulch film, which is adapted to the climate and the fruit application.

Cucumber, e.g., is harvested up to 25 times per growing season using heavy machinery. At the end of the season, the mulch film is difficult to recover. For this application the use of biodegradable mulch film is of interest.

Because of the limited UV resistance of PBAT, it is difficult to use this polymer and its compounds for mulch-film applications: In Europe, more than eight weeks of intact performance of the mulch film is required. PCL does not exhibit the necessary temperature resistance for this application. PBS and PCL biodegrade too quickly for most mulch-film applications—they need to be formulated with a plastic of slower biodegradability like PBAT.

Since the most important property of mulch film is the appropriate biodegradability, extensive experiments in each region are indispensable.

8.4. Horticulture

Different horticultural items made of biodegradable polymers such as plant pots, seed/fertilizer tape and binding materials, foams, and nets for erosion control offer reduction of system complexity by reducing the number of work steps.

8.5. Packaging

The packaging market, especially the food packaging market, offers large opportunities for biodegradable polymers. In the following section, the focus will be on flexible food packaging, rigid food packaging, and paper coating (see also → Foods, 4. Food Packaging). For nonfood packaging, the example of hygiene film applications will be described in more detail.

8.5.1. Food Packaging

Shrink films are used to combine several sales items in one packaging — e.g., six bottles in a six-pack. The bottles are packed in a piece of film which is heated and shrunk in a heating tunnel above the melting point of the film for a short time. The shrink forces after relaxation have to stay high enough to store and carry the packaged goods along the logistic chain: Thus the requirements are:

- High shrink values in extrusion direction MD > 60 %
- Low shrink values in cross direction CD < 20 %
- High shrink rate during heating in the oven
- Welding of the loose end during shrink process
- High shrink forces after relaxation

Although the market is large, biodegradable shrink films are still under development. PBAT/PLA compounds have a good strength and an adequate heat resistance. Films thereof exhibit the necessary shrink values. These prerequisites are promising—but still the materials need to be tested in more detail.

Knitted nets are produced from blown film or specialized extrusion lines with corotating dies. The film rolls are slit into small tapes of 2–5 mm width. Then the tapes are stretched on heated rolls (galettes) in several steps to achieve and maintain a high level of orientation. Typical stretching factors are 3–5 times. Thus, the materiel undergoes a strain-induced crystallization which maximizes the crystallinity level in the tape.

Based on the processing, polymers like PCL, PBS, and PHA obtain a high crystallinity. High crystallinity levels correlate to low biodegradability speed. Therefore, amorphous polymers like PBAT compounds are more suitable. However, pure PBAT is too soft and elastic for orientation.

Rigid Packaging Containers. Transparent rigid containers account for about 60 % of the European market for rigid packaging. They are best produced using PLA in an extrusion/thermoforming, blow-molding or injection blow-molding process. If biodegradability is an issue for the application, it needs to be checked under composting conditions.

Nontransparent containers (e.g., dairy cups) account for the remaining 40 % of the European market. If toughness requirements are not met by PLA, PBAT/PLA blends can be used for the extrusion of rigid containers. They have an adequate viscosity and they can be recycled if the moisture problem is solved by drying the recycled material (like virgin PLA: to a moisture level below 200 ppm).

PBAT/PHA compounds have a high temperature resistance and a good stiffness/toughness ratio. However, up to now they are too sensitive to thermal decomposition in processing. PBS and PCL are too soft for rigid containers. Moreover, PCL shows a limited thermal stability and PBS still has to prove its suitability for food contact in the EU.

Paper Coating. Coated or laminated paper products represent another potential market for biodegradable polymers. At present, paper packaging such as disposable cups are extrusion coated with PE-LD, which restricts the biodegradation of the paper substrate since it acts as an impervious barrier. By using biodegradable polymers, the synergies with the biodegradable paper can be fully explored. Synthetic biodegradable polymers, such as PBAT or PBAT compounds, thus take over the function of the PE-LD and are used to toughen and to protect against fat, moisture, and temperature variations.

8.5.2. Nonfood Packaging

Hygiene films have to be soft, permeable, strong and thin:

- Soft films comply with the haptics of, e.g., diapers
- Only water-vapor-permeable films are comfortable to wear
- Strong films fulfill the requirements for high-speed processing during assembly of hygiene products
- Thin films comply with the requirements of the LCA as well as reduce cost
- The film has to comply with various regulations concerning skin contact.

Currently only PBAT compounds fulfill all the requirements, because the ratio of stiffness/flexibility can be adjusted by the PBAT content. Pure PBAT is too soft, PBS is too stiff and PCL too temperature-sensitive for hygiene applications. The materials still have to pass the relevant tests for skin contact.

Foam. For loose-fill applications, foamed biodegradable polymers based on starch have been

Table 12. Expected development of production capacities of biodegradable polyesters and compounds

	Manufacturers	Product	Capacity 2007, t/a	Capacity 2010, t/a	Country
PLA	Cargill	NatureWorks	70 000	140 000	USA
	Pyramid Bioplastics Guben GmbH	PLA	0	60 000	Germany
Starch-based compounds					
	Novamont	Mater-Bi	20 000	60 000	Italy
	Biotec	Bioflex	10 000	20 000	Germany
PHA					
	Tianan	Enmat	1000	10 000	China
	Metabolix/ADM	Mirel	5000	50 000	USA
Aliphatic/aromatic polyesters					
	BASF	Ecoflex	14 000	74 000	Germany
Aliphatic polyesters					
	Showa H.P.	Bionolle	3000	3000	Japan
	Daicel	Celgreen	1000	1000	Japan
	IRE Chemicals	EnPol	3000	3000	Korea
	Mitsubishi Chemicals	GS-PLA	6000	6000	Japan

used for more than one decade (not addressed in detail in the text and excluded in Table 10). A new generation of biodegradable foams for food packaging applications has been introduced to the market (e.g., by NatureWorks) or is under development (BASF, consortium PURAC/Synbra/Sulzer). The foams have the potential to replace PS foams currently used for food and protective packaging.

Others: Medical Applications. Historically, one of the first applications has been the use of biodegradable polymers for medical purposes. Quantitatively, these applications play a minor role in market size, but a major role in health and safety of medical procedures.

Nonbiodegradable Applications — Durable Use. In the last years, biodegradable polymers like PLA have been established in different nonbiodegradable applications. PLA is processed into fibers (e.g., for textiles) and is used for durable parts in electronics. The driver for these applications is the content of renewable materials and not the polymer's biodegradability.

9. Production Capacity

After one decade of market development, production capacities will be increased significantly in the next years. Next to a planned increase of the NatureWorks PLA capacity to 140 000 t/a, further suppliers of PLA will enter the market. In 2009, Telles (Metabolix/ADM joint venture) will start up its 50 000 t/a PHA capacity plant and BASF has announced the increase of its Ecoflex capacity by 60 000 t/a in 2010. A summary of existing and planned capacities is shown in Table 12.

10. Outlook

Biodegradable polymers are specialty plastics for selected applications where biodegradability adds value. The market for biodegradable polymers (65×10^3 t in 2007; excl. loose fill) represents a small niche segment compared to the global market for PE film (approx. $30 000 \times 10^3$ t in 2007) or to the global polymer market (approx. $250 000 \times 10^3$ t in 2007). Based on different market drivers (changing consumer behavior, improved composting infrastructure, technology, application, and capacity development), the expected growth rate of approximately 25 % p.a. can be influenced significantly. Biodegradable polymers combine different chemical and biotechnological steps in their synthesis and degradation and thus are an excellent example of a symbiosis between these disciplines.

Abbreviations

BDO	butanediol
LA	lactic acid
LCA	Life-cycle assessment
MSA	maleic acid anhydride
PBAT	poly[(butylene adipate)-*co*-(butylene terephthalate)]
PBS	poly(butylene succinate)

PBSA	poly[(butylene succinate)-*co*-(butylene adipate)]
PBT	poly(butylene terephthalate)
PC	polycarbonate
PCL	polycaprolactone
PE	polyethylene
PE-HD	high-density polyethylene
PE-LD	low-density polyethylene
PEG	poly(ethylene glycol) (also poly(ethylene oxide), PEO)
PES	poly(ethylene succinate)
PET	poly(ethylene terephthalate)
PHA	poly(hydroxyalkanoate)
PHB	poly(hydroxybutyrate)
PHBH	poly(hydroxybutyrate-*co*-hexanoate)
PHBV	poly(hydroxybutyrate-*co*-valerate)
P(3-HB)	poly(3-hydroxybutyrate)
P(4-HB)	poly(4-hydroxybutyrate)
PLA	poly(lactic acid)
ROP	ring-opening polymerization
PP	polypropylene
PS	polystyrene
PVC	poly(vinyl chloride)
PVOH	poly(vinyl alcohol)

References

1. C. Bastioli (ed.): *Handbook of Biodegradable Polymers*, Rapra 2006; A. Steinbüchel, Y. Doi (eds.): *Biopolymers*, vol. 4, Wiley-VCH, Weinheim 2002.
2. I. Kleeberg, C. Hetz, R. M. Kroppenstedt, R.-M. Müller, W.-D. Deckwer, *Appl. Environ. Microbiology* **64** (1998) 1731.
3. I. Kleeberg, K. Wetzel, J. VandenHeuvel, R.-J. Müller, W.-D. Deckwer, *Biomacromolecules* **6** (2005) 262.
4. R. J. Müller, *Biol. unserer Zeit* **31** (2000) 215.
5. U. Witt, T. Einig, M. Yamamoto, I. Kleeberg, W. Deckwer, R. J. Müller, *Chemosphere* **44** (2001) 289.
6. O. Fränzle, M. Straškraba, S. E. Jørgensen: Ecology and Ecotoxicology, *Ullmann's Encyclopedia of Industrial Chemistry*, 6th ed., Wiley-VCH, Weinheim 2002.
7. B. De Wilde, "Comparison of Certification Systems", Lecture on 30.09.2008.
8. S. S. Yazdani, R. Gonzalez, *Current Opinion in Biotechnology*, **18** (2007) 213.
9. B. Kamm, M. Kamm, *Adv. Biochem. Eng. Biotechnol.* **105** (2007) 175.
10. C. Bastioli, *Lectio Magistralis* 2008, University of Genua.
11. M. Malveda, M. Blagoev, A. Kishi, Lactic Acid, Its Salts and Esters, *Chemical Economics Handbook 2006*, SRI consulting, Menlo Park 2006.
12. J. B. McKinlay, C. Vieille, J. G. Zeikus, *Appl. Microbiol. Biotechnol.*, **76** (2006) 727.
13. J. J. M Swinkels: *Starch Terminology*, Ref. no. 109202/1095, AVEBE, Veendam, Netherlands, 1995, p. 4.
14. J. J. M Swinkels: *Differences between commercial starches*, Ref. no. 05.00.01.012.EF, AVEBE, Veendam, Netherlands, 1988, p. 4.
15. J. J. M Swinkels: *Industrial Starch Chemistry*; Ref. no. 109341/0399/vG/1000, AVEBE, Veendam, Netherlands, 1999, p. 28.
16. D. Bendix, *Polym. Degr. Stabil.* **59** (1998) 129.
17. J. Lunt, *Polym. Degr. Stabil.* **59** (1998) 145.
18. D. Garlotta, *J. Polym. Envir.* **9** (2001) 63 – 84.
19. Cargill, US 5357035, 1994 (P. R. Gruber et al.).
20. Mitsui Toatsu Chemicals, US 5310865, 1994 (K. Enomoto, M. Ajioka, A. Yamaguchi).
21. M. Spinu, C. Jackson, M. Y. Keating, K. H. Gardner, *J. Macromol. Sci.* **A33** (1996) 1497.
22. J. Kolstad, *J. Appl. Polym. Sci.* **62** (1996) 1079.
23. R. E. Conn, J. J. Kolstad, J. F. Borzelleca, D. S. Dixler, L. J. Filer, B. N. LaDu, M. W. Pariza, *Food Chem. Toxicol.* **33** (1995) 273.
24. NatureWorks LLC, www.natureworksllc.com (accessed 29 October 2008).
25. Cargill Dow LLC, Mitsui Chemicals, http://www.mitsuichem.com/release/2001/pdf/010926e.pdf (accessed 26 September 2001).
26. K.-D. Wendlandt, M. Jechorek, J. Helm, U. Stottmeister, *Polym. Degr. Stabil.* **59** (1998) 191.
27. S. Y. Lee, *Biotechnol. Bioeng.* **49** (1996) 1.
28. A. Steinbüchel, H. E. Valentin, *FEMS Microbiol. Lett.* **128** (1995) 219.
29. S. F. Williams, O. P. Peoples, *CHEMTECH* **26** (1996) 38.
30. S. Y. Lee, S. J. Park, J. P. Park, Y. Lee, S. H. Lee: Economic Aspects of Biopolymer Production, in A. Steinbüchel (ed.): *Biopolymers*, vol. 10, Wiley-VCH, Weinheim 2003, p. 307.
31. J. Asrar, K. J. Gruys: Biodegradable Polymer (Biopol), in Y. Doi, A. Steinbüchel (eds.): *Biopolymers*, vol. 4, Wiley-VCH, Weinheim 2002, p. 53.
32. S. F. Williams, D. P. Martin: Applications of PHAs in Medicine and Pharmacy, in Y. Doi, A. Steinbüchel (eds.): *Biopolymers*, vol. 4, Wiley-VCH, Weinheim 2002, p. 91.
33. D. Jendrossek, *Polym. Degr. Stabil.* **59** (1998) 317.
34. M. Lemoigne, *Bull. Soc. Chim. Biol.* **8** (1926) 770.
35. Mirel, www.mirelplastics.com/about/telles.html (accessed 23 December 2008).
36. U. Witt, M. Yamamoto, U. Seeliger, R.-J. Müller, V. Warzelhan, *Angew. Chem. Int. Ed.* **38** (1999) 1438.
37. A. Löfgren, A.-C. Albertson, *Rev. Macromol. Chem. Phys.* **C35** (1995) 379.
38. D. Mercerreyes, R. Jérôme, *Macromol. Chem. Phys.* **200** (1999) 2581.
39. T. Fujimaki, *Polym. Degr. Stabil.* **59** (1998) 209.
40. R.-J. Müller, U. Witt, E. Rantze, W.-D. Deckwer, *Polym. Degr. Stabil.* **59** (1998) 203.
41. M. Yamamoto, U. Witt, G. Skupin, D. Beimborn, R.-J. Müller: Biodegradable Aliphatic-Aromatic Polyesters: Ecoflex®, in Y. Doi, A. Steinbüchel (eds.): *Biopolymers*, vol. 4, Wiley-VCH, Weinheim 2002, p. 299.
42. U. Witt, R.-J. Müller, W.-D. Deckwer, *J. Environ. Polym. Degr.* **3** (1995) 215.
43. U. Witt, R.-J. Müller, W.-D. Deckwer, *J. Environ. Polym. Degr.* **4** (1996) 9.
44. F. B. Oppermann, S. Pickartz, A. Steinbüchel, *Polym. Degr. Stabil.* **59** (1998) 337.
45. K. Hashimoto, T. Hamano, M. Okada, *J. Appl. Polym. Sci.* **54** (1994) 1579.
46. M. Schwamborn, *Polym. Degr. Stabil.* **59** (1998) 39.
47. F. Kawai: Biodegradation of Polyethers (Polyethylene Glycol, Polypropylene Glycol, Polytetramethylene Glycol and Others), in S. Matsumura, A. Steinbüchel (eds.): *Biopolymers*, vol. 9, Wiley-VCH, Weinheim 2002, p. 267.
48. K. Ogata, F. Kawai, M. Fukaya, Y. Tani, *J. Ferment. Technol.* **53** (1975) 757.
49. E. L. Fincher, W. J. Payne, *Appl. Microbiol.* **10** (1962) 542.
50. F. Kawai, M. Fukaya, Y. Tani, K. Ogata, *J. Ferment. Technol.* **55** (1977) 429.
51. Mowiol® Polyvinyl Alcohol, Kuraray Product Brochure, http://www.kuraray-am.com/pvoh-pvb/downloads/Mowiol_brochure_en_KSE.pdf, (accessed 08 January 2009).

52. S. Matsumura: Biodegradation of Poly(vinyl alcohol) and its Copolymers, in S. Matsumura, A. Steinbüchel (eds.): *Biopolymers*, vol. **9**, Wiley-VCH, Weinheim 2002, p. 329.
53. W. Wiedmann: *Maschinenkonzepte für biologisch abbaubare Werkstoffe*, VDI Aufbereitungstechnik, VDI-Verlag, Düsseldorf 2004, p. 10.
54. L. Jiang, M. P. Wolcott, J. Zhang: "Study on Polylactide/Poly (butylene adipate)-*co*-terephtalate) Blends", *Biomacromulecules* **7** (2006) 199.
55. G. Skupin: "Ecovio® - a modular system", in Ecoflex, Ecovio, BASF Brochure, Ludwigshafen, Germany, 2008, p. 16.
56. G. W. Becker, D. Braun, L. Bottenbruch (eds.): Polycarbonate, Polyacetale, Polyester, Celluloseester, *Kunststoff-Handbuch*, vol 3/1, Hanser Verlag, München 1992.
57. H. J. Saechtling: *Kunststoff Taschenbuch*, Hanser Verlag, München 2007, p. 539.
58. R. Ishioka, E. Kitaguni, Y. Ichikawa: Aliphatic Polyesters: Bionolle, in Y. Doi, A. Steinbüchel (eds), *Biopolymers*, vol. 4, Wiley-VCH, Weinheim 2002, p. 10.
59. M. Yamamoto, U. Witt, G. Skupin, D. Beimborn, R. J. Müller: Biodegradable aliphatic-aromatic Polyesters: Ecoflex, in Y. Doi, A. Steinbüchel (eds): *Biopolymers*, vol. 4, Wiley-VCH, Weinheim 2002, p. 11.
60. Z. Gan, K. Kuwabara, M. Yamamoto, H. Abe, Y. Doi, "Solid-state structures and thermal properties of aliphatic-aromatic poly[(butylene adipate)-*co*-(butylene terephthalate)]", *Polym. Degrad. Stab.* **83** (2004) 289.
61. C. Rauwendaal: *Understanding Extrusion*, Hanser Verlag, München 2004, p. 75.
62. C. Rauwendaal: *Understanding Extrusion*, Hanser Verlag, München 2004, p. 97.
63. W. Wiedmann, *Maschinenkonzepte für biologisch abbaubare Werkstoffe*, VDI Aufbereitungstechnik, VDI-Verlag, Düsseldorf 2004, p. 3.
64. *Polyethylene film*, Technical Brochure Basell, Hoofddorp 2005, p. 27.
65. Requirements for packaging recoverable through composting and biodegradation, EN 13432:2000, Beuth Verlag, Berlin 2000.
66. H. G. Lotz, R. Kukla, P. Sauer, G. Steiniger, "Latest developments in sputtering thin films for transparent barrier films", *Fraunhofer Institut für Verfahrenstechnik und Verpackung, Konferenz ICE 2007*, München, Germany, 2007.
67. G. Skupin, "Processing of Ecoflex® and Ecovio®" in Ecoflex, Ecovio, BASF Brochure, Ludwigshafen, Germany 2008, p. 18.
68. J. Nentwig: *Kunststofffolien*, Hanser Verlag, München 2006, p. 195.
69. H. J. Saechtling: *Kunststoff Taschenbuch*, Hanser Verlag, München 2007, p. 256.
70. C. Rauwendaal: L*Polymer Extrusion*, Hanser Verlag, München 2004, p. 547.
71. F. Johannaber: *Kunststoff Maschinenführer*, Hanser Verlag, München 2004, p. 321.
72. J. L. Thorne: *Technology of Thermoforming*, Hanser Verlag, München, 2004, p. 16.
73. F. Johannaber: *Kunststoff Maschinenführer*, Hanser Verlag, München 2004, p. 11.
74. G. Pötsch, W. Michaeli: *Injection Molding*, Hanser Verlag, München 2007, p. 29.
75. D. V. Rosato: *Blow Molding Handbook*, Hanser Verlag, München 2004, p. 75.
76. *A Guide to Polyolefine Blow Molding*, Technical Brochure Basell, Hoofdorp 2003, p. 11.
77. M. Thielen: *Extrusion Blow Molding*, Hanser Verlag, München 2004, p. 5.
78. D. V. Rosato: *Blow Molding Handbook*, Hanser Verlag, München 2004, p. 129.
79. D. K. Platt, *Biodegradable Polymers, RAPRA Market Report, 2006*, Rapra technology 2006; Institut National de la Statistique et des Etudes Economiques, Les exploitations agricoles europeennees; cited in Club — Bioplastique, 2007.
80. V. Reimer, S. Philipp, A. Künkel, *Kunststoffe Int.* **8** (2008) 12.

Further Reading

C. Bastioli: *Handbook of Biodegradable Polymers*, Rapra Technology Ltd., Shawbury 2005.
E. Chiellini, R. Solaro (eds.): *Biodegradable Polymers and Plastics*, Kluwer Academic, New York 2003.
S. W. Shalaby, K. J. L. Burg: *Absorbable and Biodegradable Polymers*, CRC Press, Boca Raton 2004.
R. Smith (ed.): *Biodegradable Polymers for Industrial Applications*, CRC Press, Boca Raton 2005.
L. Yu: *Biodegradable Polymer Blends and Composites from Renewable Resources*, Wiley, Hoboken, NJ 2009.

Polymers, Electrically Conducting

HERBERT NAARMANN, BASF Aktiengesellschaft, Ludwigshafen, Federal Republic of Germany

1.	Introduction	1261	5.4.	Polythiophene	1269
2.	Synthetic Routes	1262	5.5.	Polyphenylene	1270
3.	Principles of Electrical Conduction	1263	5.6.	Poly(Phenylene Sulfide)	1272
4.	Orientation Processes	1264	5.7.	Poly(Phenylene Vinylene)	1272
5.	Types of Electrically Conducting Organic Materials	1265	5.8.	Polyaniline	1273
			5.9.	Miscellaneous Polymers	1274
5.1.	Polyacetylene	1265	6.	Uses	1277
5.2.	Polydiacetylenes	1267		References	1278
5.3.	Polypyrrole	1267			

1. Introduction

Electrically conducting polymers (ECPs) are materials with an extended system of C=C conjugated bonds. They are obtained by reduction or oxidation reactions (called doping), giving materials with electrical conductivities up to 10^5 S/cm. These materials differ from polymers filled with carbon black or metals because the latter are only conductive if the individual conductive particles are mutually in contact and form a coherent phase.

This review concerns the synthesis routes, polymerization techniques, doping, orientation, and development of well-defined, highly conducting polymeric materials. Their wide range of potential uses from electrodes in rechargeable batteries to organic transistors is limited by their vulnerability to air and moisture due to their highly conjugated structures and the doping agents. Electrically conducting materials are compiled, their specific properties and potential applications are described.

Numerous attempts have been made to synthesize "conductive organic materials". The first was the synthesis of polyaniline by F. GOPPELSROEDER in 1891 [1]. After decades interest grew in organic polymers as insulators, but not as electrical conductors.

In the late 1950s organic semiconductors became the focus of investigations. Preliminary studies in this field up until the mid 1960s are reviewed in [2]. The semiconducting polymers were termed "covalent organic polymers", "charge-transfer complexes", "organometallic polymers", "hydrogen-bonded polymers", and "mixed polymers". Highest conductivity values reached about 10^{-3} S/cm. In 1964 LITTLE theoretically evaluated the possibility of superconductivity in polymers and suggested a model, consisting of a polyene chain with cyanine, dyelike substituents [3]. In the same year systematic studies were presented based on aromatic and heterocyclic compounds exhibiting electrical conductivities of 0.5 S/cm [4], followed by studies correlating doping, pressure, irradiation, and chain length to conductivity, with values up to 100 S/cm [5].

Interest heightened and became acute from 1975 when IBM scientists showed that poly (sulfur nitride), $(SN)_n$, was superconductive [6] and MACDIARMID's group reported [7] the doping of polyacetylene films prepared by SHIRAKAVA [8] reaching conductivity values of 38 S/cm. Since then many expectations have been raised, but scientific progress and practical applications have been limited; they depend on the reproducible production of well-defined specimens, the determination of synthesis conditions, and the laws relating these conditions to product properties. Synthetic methods are improving; more easily processable, soluble, flexible materials are now being produced.

2. Synthetic Routes

The synthesis of electrically conducting polymers with conjugated −HC=CH− bonds requires the controlled coupling of a large number of monomers (unsaturated compounds or compounds with difunctional groups). Alternatively, pendant groups attached to existing polymers can be cyclized to give conjugated systems.

Polymerization of unsaturated monomers gives polymers with long chains that are often branched and cross-linked and therefore insoluble. Alkynes polymerize as follows:

1. n HC≡CH → CH=CH$_n$ insoluble polyacetylenes [9]
2. n HC≡C–CH=CH$_2$ → CH=CH–CH=CH$_n$ insoluble polyacetylenes [9]
3. n RC≡CH → CR=CH$_n$ soluble polyacetylenes where R denotes alkyl or aryl groups [10]
4. Alkyne or allene → 460 °C thermal polymerization, insoluble polyenes and aromatics [11]

Stepwise coupling of monomers with difunctional groups runs in a more controlled fashion than polymerization of unsaturated monomers but problems are caused by low conversion rates and purification of the starting monomers. This method includes Wittig, MacMurry, retro Diels – Alder, and elimination reactions. It is accessible to a wide variety of soluble precursors and produces defined polymers. Examples follow:

1. Oxidative coupling [12] can be applied to pyrrole, thiophene, aromatic, and heteroaromatic systems (0 – 350 °C)

2. Grignard coupling [13]:

3. Boronic acid coupling [14]:

4. Stille reaction [15]:

5. Elimination reactions [16]:

6. Feast reaction (retro-cycloaddition) [17]:

7. Grubb's method: polybenzvalene is isomerized to polyacetylene in the presence of HgCl$_2$ [18]:

Cyclooctatetraene is also polymerized to give soluble polyenes [19]:

Polycyclization starts with polymers whose pendant side groups are cyclized to give condensed systems, often under pyrolytic conditions. Defects cannot be excluded.

These methods lead to the basic structures of electrically conducting organic materials (ECOMs) and are reviewed in [9]. Examples follow [12]:

Analogous reactions occur with −CH=CH$_2$ pendant groups.

Monomers may also be cyclized [10], [20]:

3. Principles of Electrical Conduction

The electrical properties of materials are determined by their electronic structure (Fig. 1). The band theory accounts for the different behaviors of metals, semiconductors, and insulators.

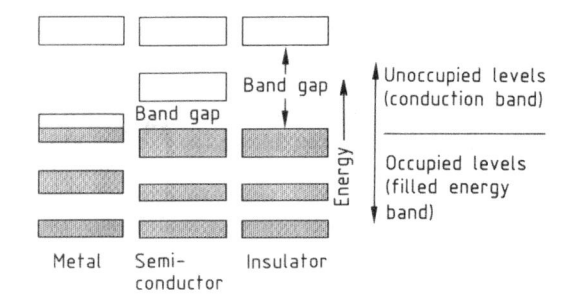

Figure 1. Model of band structure

The band gap is the energy spacing between the highest occupied energy level (valence band) and the lowest unoccupied energy level (conduction band). Metals have a zero band gap which means that they have a high electron mobility, i.e., conductivity. Semiconductors have a narrow band gap (ca. 2.5 – 1.5 eV), conductivity only occurs on excitation of electrons from the valence band to the conduction band (e.g., by heating). If the band gap is larger (> 3 eV), electron excitation is difficult; electrons are unable to cross the gap and the material is an insulator [21]. Electrically conducting organic materials such as polyphenylene, polyacetylene, or polypyrrole are, however, peculiar in that the band theory cannot explain why the charge-carrying species (electrons or holes) are spinless. Conduction by polarons and bipolarons is now thought to be the dominant mechanism of charge transport in organic materials [22]. This concept also explains the drastic deepening of color changes produced by doping. A polaron (a term used in solid-state physics) is a radical cation that is partially delocalized over several monomer units (e.g., in a polymer segment). The bipolaron is a diradical dication. Low doping levels give rise to polarons, whereas higher doping levels produce bipolarons. Both polarons and bipolarons are mobile and can move along the polymer chain. See also, → Plastics, General Survey.

Doping. The process that transforms insulating polymers (e.g., polyacetylene, conductivity 0.1 S/cm) to excellent conductors (Fig. 2) is the formation of change-transfer complexes by electron donors such as sodium or potassium (n doping, reduction) or by electron acceptors such as I$_2$, AsF$_5$, or FeCl$_3$ (p doping, oxidation). The doped polymer backbone becomes

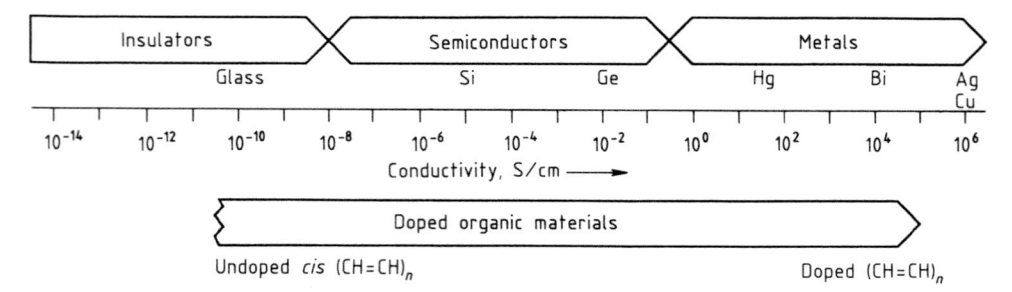

Figure 2. Comparison of the electrical conductivity (300 K) of organic and inorganic materials and the effect of doping

positively or negatively charged [23] with the dopant forming oppositely charged ions (Na^+, K^+, I_3^-, I_5^-, AsF_6^-, $FeCl_4^-$). The polymer can be switched between the doped, conductive state and the undoped, insulating state by applying an electric potential that makes the counterions move in and out. This switching corresponds to charging and discharging when these materials are used as electrodes in rechargeable batteries.

Measurement. Electrical conductance is a measure of the flow of current through a material for a given applied voltage [24].

The electrical conductivity σ in reciprocal ohms or siemens per centimeter ($\Omega^{-1}\,cm^{-1}$ or S/cm) denotes the current flow through a body of length 1 cm and with a cross-sectional area of 1 cm². The conductance σ is usually measured via its inverse, the electrical resistance R ($\sigma = 1/R$).

The resistivity is the inverse of the conductivity

$$\sigma = \frac{d}{F}\frac{1}{R}$$

(R in Ω, σ in $\Omega^{-1}\,cm^{-1}$, d is the thickness of the specimen in cm, and F its surface area in cm²).

Ohm's law ($R = V/I$) is valid for electrical conductors (metals) and semiconductors and allows R to be determined by measuring the applied voltage V in volts and the current I in amperes with a simple measuring system (Fig. 3) or with a Wheatstone bridge.

With high resistances (i.e., $\sigma = 10^{-4}$ to 10^{-14} S/cm), two-point measurement is enough. Higher conductivities, as in electrically conducting polymers, require four-point measurements because the contact resistances in the circuit are frequently as great or greater than that of the material under investigation.

Contactless measurements of the conductivity are possible in a microwave field [24].

4. Orientation Processes

Orientation processes are powerful methods that are used to improve conductivity and other material properties (e.g., transparency, anisotropy). Orientation can be achieved in several ways, including stretching.

Mechanical stretching can be performed after polymerization, e.g., in noncross-linked polymers [25]. In the case of polyacetylenes prepared with aged Ziegler–Natta catalysts [26], [27], stretching increases conductivity from 2500 S/cm to values as high as 10^5 S/cm.

Continuous electrochemical polymerization (e.g., of polypyrrole) on the surface of a rotating drum permits simultaneous peeling off,

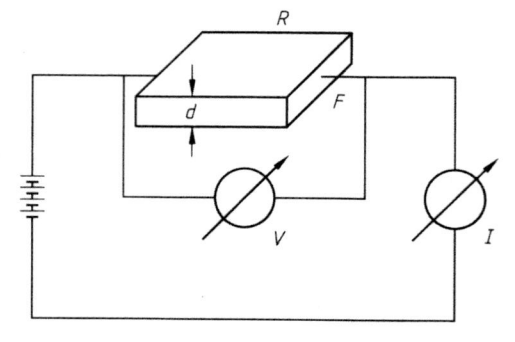

Figure 3. Conductivity measurement equipment

mechanical stretching, and orientation $\sigma \leq 200$ S/cm [25]. Greater stretching rates and therefore greater conductivities are reported in poly(pyrrole perchlorate) films (σ up to 10^3 S/cm) [28]. Biaxially stretched films yielded conductivities of 800 S/cm parallel to the stretching direction and 290 S/cm in the cross direction. Stretched poly(phenyl vinylenes) and poly(thienyl vinylenes) yielded conductivities of ca. 10^3 S/cm [29].

Orientation can also occur during polymerization or by performing polymerization in an oriented matrix consisting of liquid crystals and using magnetic fields [30]. Variants are the use of liquid crystal matrices during the electro chemical synthesis of polyheterocycles [31] and the synthesis of polymers (e.g., substituted thiophenes) with liquid crystal side chains that contain sulfonate groups [32]. The sulfonate groups act as "self dopants" and the liquid crystal side chains are responsible for orientation.

Polymerization of extremely thin polyacetylene films (< 1 μm) on crystal surfaces by epitaxial growth (e.g., on frozen benzene) [33] also induces orientation in the deposited polymer layer. Substituted polypyrrole films with a high anisotropy can be produced by the Langmuir – Blodgett technique [34–36].

5. Types of Electrically Conducting Organic Materials

Electrically conducting organic materials can be divided into three main groups: polyenes, polyheterocycles, and polyaminoaromatic compounds. Polyacetylene, polypyrrole, and polyaniline are the most intensively studied polymers.

5.1. Polyacetylene

Polyacetylene (PAC) exists in various isomeric forms:

cis-transoid (cis-polyacetylene)
[25786-70-1], [74373-36-7]

trans-cisoid
[73589-68-1]

trans-transoid (trans-polyacetylene)
[25768-71-2]

cis-Polyacetylene is relatively unstable and reverts to the thermodynamically stable trans-polyacetylene via the metastable trans-cisoid form. cis-cisoid PAC has not yet been prepared in pure form, it has a cyclic, helical structure [37].

The term "polyacetylene" was first mentioned in 1866 [38]. In 1898, E. ERDMANN and P. KÖTHNER [39] described a "high-molecular-weight copper acetylene" that was formed when acetylene gas was passed over copper powder below 250 °C. In 1948 REPPE [40] prepared Cuprene film with a metallic luster. In 1961 HATANO reported the polymerization of acetylene with a $AlEt_3/Ti(OBu)_4$ catalyst to give polymers with conductivities up to 0.001 S/cm [41]. Since then intensive work has been carried out on the various polymer types, reviews are given in [42]. Some important syntheses are listed in Table 1. Table 2 shows the conductivities of various types

The polyacetylenes grow primarily in the form of fibrils with an internal surface area of up to 200 m^2/g [45]. The fibrils are insoluble and can be rendered electrically conductive by incorporation of n-type or p-type dopants.

The polyacetylenes are chemically unstable and react, for example, with atmospheric oxygen to form epoxy, carbonyl, or carboxyl groups. N–(CH=CH)$_n$ compounds are more stable, probably due to the absence of defect sites at high crystallinity. Conductivity increases as a function of crystallinity and decreases as a function of defects (sp^3-bonded carbon atoms, $-CH_2CH_2-$) [46].

Figure 4 illustrates two types of polyacetylene. The standard Shirakawa type (Fig. 4 A) is crosslinked and contains ca. 2 % sp^3 fractions. The new BASF technique involves polymerization at room temperature (instead of − 76 °C) and the use of a tempered catalyst. The stretched polyacetylene product has parallel fibrils (Fig. 4 B). It is linear (no sp^3 fractions), is highly orientable (can be stretched by up to 660 %), and has a conductivity exceeding 10^5 S/cm. A convincing demonstration of the high anisotropy (1: 100) in the stretched samples is the fact that when strips of such a stretched polymer are laid across each other, polarized light (sunlight) is extinguished in the region of overlap [47].

Substituted Polyacetylenes. The synthesis of polymers from substituted acetylene monomers

Table 1. Preparation and properties of polyacetylenes

Name of method	Catalyst	Polymerization-temperature, °C	cis/trans ratio	Defects (sp^3 fraction), %	Conductivity, S/cm (iodine doped)	References
Reppe	Ni acetylacetonate	100	5/95	15	10^{-3}	[43] (Table 1, no. 2)
Luttinger – Green	$NaBH_4$/CO (II)/Ni (II)	– 76	50/50	8	10^{-3}	[43] (Table 1, no. 6)
Hatano	$AlEt_3$/Ti(OBu)$_4$	100	20/80	10	10^{-5}	[43] (Table 2, no. 8)
Shirakawa	$AlEt_3$/Ti(OBu)$_4$ in toluene	– 76	90/10	3	520	[43] (Table 1, no. 16)
Naarmann (BASF)	1) $AlEt_3$/Ti(OBu)$_4$ in silicone oil	20	85/15	0	500	[44]
	2) as 1), but with an aged (> 100 °C) catalyst	20	85/15	0	2 000	[44]
	3) as 2) but oriented	20	85/15	0	$> 10^5$	[44], [45]
Feast	thermal cleavage of tetracyclobis (trifluoromethyl) decatriene	80	75/25		100	[17]
Grubbs	isomerization of polybenzvalene,	20	40/60		1	[18]
	cyclooctatetraene ring opening by carbenes	20	40/60		350	[19]

is directed toward the preparation of substituted, conjugated chains which ameliorate the negative properties of polyacetylenes (e.g., sensitivity to air, insolubility, and infusibility) while maintaining the desired electrical properties of acetylene's conjugated backbone [10], [48]. Alkyl- and aryl-substituted polymers result. They are soluble (e.g., in toluene and cyclohexane) and processible, but have low conductivities (< 0.1 S/cm) compared with the unsubstituted polyacetylene.

Modification of Polyacetylenes. The controlled modification of the polyene chain by radical reactions and Diels – Alder addition leads to functionalized systems with slightly improved atmospheric stability but diminished conjugation length and electrical conductivity. Examples

Table 2. Conductivities of some metals and different types of polyacetylene

Metal or polyacetylene	Conductivity (volume), S/cm	Density[d], g/cm^3	Conductivity (weight), S cm^{-2} g^{-1}
Hg	10 365	13.54	767
Pt	101 522	21.45	4 733
Cr	77 519	7.19	10 781
Fe	102 986	7.87	13 085
Au	470 588	19.3	24 382
Ag	671 140	10.49	63 972
Cu	645 000	8.94	72 147
Shirakawa polyacetylene [8]	897[a]		813
	30[b]	1.23 (0.40)	22
Shirakawa polyacetylene, oriented in liquid-crystal matrix [30]	1 600[b]	1.10 (0.50)	1 750
BASF polyacetylene, oriented [44]	23 000[c]	1.12 (0.85)	20 000
BASF polyacetylene with additional reducing agent, oriented [44]	120 000[c]	1.15 (0.90)	104 347

[a] Doped with $FeCl_3$.
[b] Doped with I_2 vapor.
[c] Doped with a saturated solution of I_2 in CCl_4.
[d] Densities for polymers denote values after doping; values before doping are given in parentheses.

of polyacetylenes and compares them with those of metals.

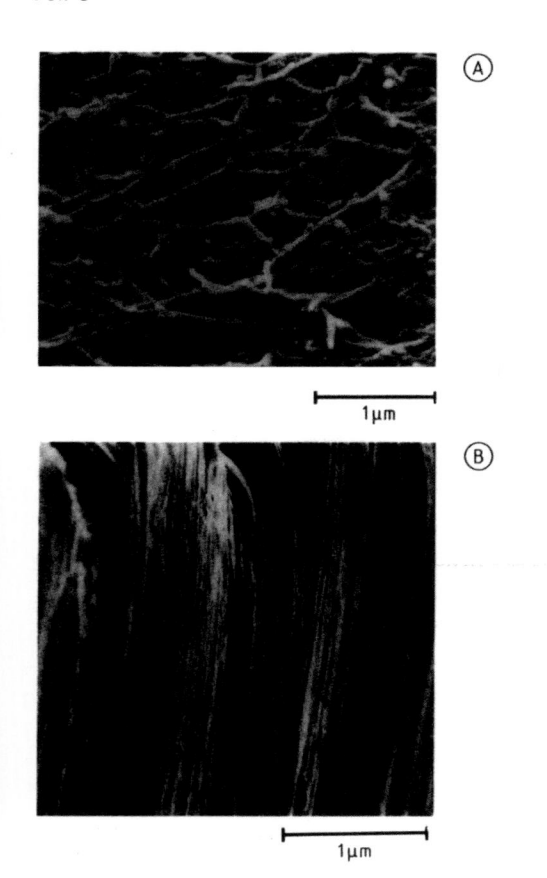

Figure 4. Scanning electron micrographs of iodine-doped polyacetylene a) Standard, cross-linked polyacetylene (conductivity ca. 500 S/cm); B) Stretched polyacetylene (conductivity > 100 000 S/cm)

include reaction with tosylmethyl isocyanide (TosCH$_2$–N=C) to give polyenes with pyrrole units [49], with nitrogen-substituted spin-labeled compounds to give spin labels (e.g., maleimido peroxyl groups) [50], with chlorosulfonyl isocyanate to give –SO$_3$ and –COO$^-$ groups [47], with tetrachloromethane to give –Cl and –CCl$_3$ groups [27], and with 3-chloroperbenzoic acid to give epoxy groups [27]. Cycloaddition with 3,4-dichloromaleic anhydride leads to a fusible polyacetylene (165 – 180 °C) [47].

5.2. Polydiacetylenes

In contrast to polyacetylene, polydiacetylenes [27987-87-7] have limited electrical conductivity (≤ 0.1 S/cm), but can be obtained as large, single crystals.

The polymerization of diacetylene

$R^1 = R^2 = -CH_2-O-\text{（}NO_2\text{）}-NO_2$ mp 172 °C

$R^1 = R^2 = -CH_2-SO_2-\text{（）}-F$ mp 193 °C

$R^1 = R^2 = -CH_2-CONH-\text{（）}-CH_3$ mp 96 °C

$R^1 = -CH_2-O-SO_2-\text{（）}-CF_3$
$R^2 = -CH_2-O-SO_2-\text{（）}-F$ mp 61 °C

is an example of a topochemical polymerization in which 1,4-addition of 1,3-diyne units takes place in the crystalline state. The reaction does not require a catalyst and is performed by irradiation of the diacetylene crystals with visible or UV light, X-rays, or γ-rays, or by annealing the crystals below their melting point. The unreacted monomer is then extracted with a suitable solvent (e.g., hexane, toluene), leaving a single, dark red crystal of polydiacetylene. This unique polymerization process was already observed in 1882 [51] and has now been intensively studied [52]. A correct interpretation of the phenomena was first provided in the early 1970s [53].

5.3. Polypyrrole

Polypyrrole [30604-81-0] is a polysalt that can be produced in the form of powders, coatings, or films; it is intrinsically conductive.

Synthesis. The preparation of polypyrrole (pyrrole red and pyrrole black) by oxidation of pyrrole dates back to 1888 [54] and by electrochemical polymerization to 1957 [55]. A fairly

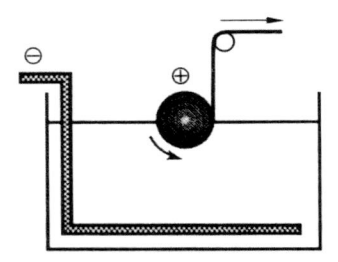

Figure 5. Continuous electrochemical process (BASF) for producing polypyrrole on a rotating drum electrode

long period elapsed before this organic π-system attracted general interest and was found to be electrically conductive [4].

Conductive polypyrrole films are obtained directly by anodic polymerization of pyrrole in aqueous or organic electrolytes (acetonitrile) [56]. They are black and, under suitable reaction conditions, can be detached from the anode in the form of self-supporting films (minimum thickness ca. 30 µm). Some of the conducting salt used in the electrolyte solution is incorporated in the film as a counterion.

In contrast to polyacetylene, polypyrrole has a high mechanical and chemical stability and can be produced continuously as flexible film (thickness 80 µm; trade name: Lutamer, BASF) by electrochemical techniques (Fig. 5) [57].

Other electrochemically polymerizable heterocycles are thiophene, furan, and their substituted and oligomeric derivatives.

The polymerization reaction is very complex. The main reaction steps are shown in Figure 6 [58]. A radical cation is formed in the initial oxidation step. This is followed by a coupling reaction, deprotonation, and a one-electron oxidation which regenerates the radical cation system. Also important is the reaction introducing oxygen with formation of a labile −CONH group in the ring system.

The quality of the polymers is greatly influenced by many factors, e.g., impurities, electrode material, pressure, concentrations, temperature, and comonomers. The most decisive, however, are the current density and the electrolyte, particularly the conducting anion X^- [59] because it is incorporated into the polymer as a counterion.

The properties of the counterion (e.g., its size, geometry, charge) influence the properties of the polymer. The amount of counterion (anion) incorporated depends on the reaction conditions. In general, one anion is incorporated for every three pyrrole units. Exceptions are pyrrole- or thiophenesulfonic acids where the counterion is coupled directly to the monomer (self doping) [60]. Some typical conducting anions are fluoroborate, perchlorate, aromatic sulfonic acids, penicillin, n-dodecyl sulfate [59], [61], phthalocyanine sulfonic acid, poly(styrene sulfonic acid), camphor sulfonic acid, styrene sulfonic acid, and heparin [52].

By changing reaction conditions, polymers with different surface morphologies, (e.g., an open porous structure) can be obtained.

The anion X^- can also be released, e.g., by applying a negative potential. Release can be specifically controlled, offering interesting possibilities for active counterions of medical interest (e.g., heparin and monobactam) that are incorporated into polypyrrole.

Variation of the monomers and their substituents yields polymers with conductivities between 10^2 and 10^{-4} S/cm [59]. Alkyl substituents also increase the solubility of the polymers with the result that electrically conducting polymers can be applied as coatings from solutions. This also applies to polymers derived from bridged pyrroles. Examples follow [62]:

Figure 6. The oxidative coupling of pyrrole [58]

An interesting variant is the chemical oxidation of heterocycles (e.g., thiophene or pyrrole

[58]) dissolved in an organic solvent (e.g., ethanol) on the surface of various materials. Conductive coatings (thickness 0.01 µm) can be produced on films of poly(phenylene sulfone), block copolymers of butadiene and styrene (Ultrason, Styrolux), poly(vinyl chloride), or other polymer films to give transparent, antistatic films with conductivities of about 0.001 S/cm. Ceramics and glass can also be coated in this way. Porous material (e.g., wood, fabrics, and open-celled foams) or fibers (e.g., polyamide, glass, or carbon fibers) can also be modified and rendered antistatic by this method.

Conductive powders (e.g., polypyrrole) with particle sizes of about 0.1 µm and conductivities of up to 10 S/cm can also be produced by chemical oxidation and can be incorporated as fillers in thermoplastics. These materials can be used for chip carriers.

Potential Uses. The conductivity of polypyrrole film suggests applications such as flexible conductive paths in printed circuits, heating films, and film keyboards.

Polypyrrole films show good electromagnetic shielding effects of about 40 dB over a wide range of frequencies (0 – 1500 MHz).

Although the new systems appear to be promising, stability problems may be encountered. High mechanical stability and/or stable conductivity are required under a variety of conditions (e.g., presence of air and moisture, temperature, and duration of exposure). Polypyrroles in which the counterion is 4-hydroxyphenyl sulfonic acid do not undergo any change in conductivity if they are exposed to nitrogen for two months at 140 °C. No change in conductivity was observed when these polymers were stored in the laboratory for three years at room temperature and 55 % R. H. Polypyrrole is sensitive to moisture because this leads to leaching of the counterion and thus to a decrease in conductivity. This can be avoided by use of appropriate hydrophobic or polymeric counterions [e.g., camphor sulfonic acid or poly(styrene sulfonic acid)] or by incorporating hydrophilic compounds. Polypyrroles obtained by synthesis in aqueous electrolytes maintain their conductivity at a level of about 20 S/cm. Polypyrroles with perchlorate as a counterion are unstable under atmospheric conditions but can be used as electrodes in rechargeable batteries [63].

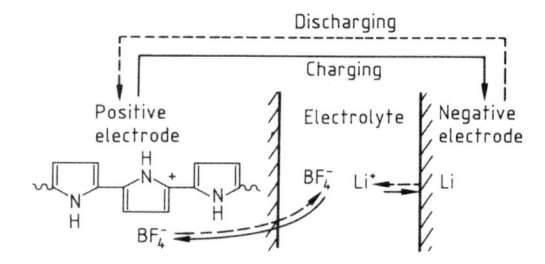

Figure 7. Electron reactions of a rechargeable polypyrrole – lithium battery
During discharge electrons flow from the negative electrode to the positive electrode and reduce the polypyrrole from the doped to the neutral state. During recharging an opposite potential is applied to the electrodes. The polymer takes up BF_4^- ions from the electrolyte and is oxidized. Lithium ions are also deposited at the other electrode.

The redox character of polypyrrole was first studied in 1975 [64]. It is a suitable electrode material for rechargeable electrochemical cells. The advantage of polymer electrodes is that they can be easily shaped, allowing novel battery designs (e.g., for the electronics sector) and less expensive production methods. Polymer cells with polypyrrole and lithium electrodes have been developed (Fig. 7) [63]. In the flat cell, the polypyrrole and lithium films are sandwiched together; in the cylindrical cell, the two films are wound concentrically. Their energy per unit mass and their discharge characteristics are similar to those of the nickel – cadmium cells now on the market. More than 500 charging and recharging cycles have been achieved with laboratory cells. Applications include dictaphones and pocket radios [63].

5.4. Polythiophene

Since the first report of thiophene polymerization in 1883 [65] decades passed until thiophene was considered to be an attractive monomer for conducting polymers [66]. Interest is still growing because thiophene is easier to handle than pyrrole (less sensitive to oxygen) and allows the simple preparation of substituted monomers leading to soluble polymers. Three methods have been used to produce polythiophene [25233-34-5]: chemical oxidation (Lewis acid), coupling with organometallic agents or by the Grignard reaction, and electrochemical oxidation (anodic). The preparation techniques for films, powders, and coatings are similar to those of pyrroles but thiophene needs stronger oxidants for its polymerization than pyrrole or other heterocycles. Standard oxidation potentials follow:

Pyrrole	+0.8 V
Indole	+0.9 V
Azulene	+0.9 V
Thiophene	+1.6 V
Furan	+1.85 V

The oxidation potential of the monomers correlates to the ease of polymerization.

Figure 8 illustrates the correlation between UV – VIS absorption and the oxidation potential of thiophene oligomers [67]. The oxidation potential of the series (thiophene)$_n$: $n = 1 - 6$ decreases with the increasing number of thiophene units.

Substituted Polythiophenes. Alkyl-substituted thiophenes with alkyl chains up to C_{18} yield polymers that are soluble in common organic solvents with electrical conductivities up to 100 S/cm. Due to their good processability [68] and high chemical and electrochemical stabilities in air and moisture, polythiophene and its derivatives are excellent candidates for practical applications (e.g., display devices, electrochromic devices [69], energy storage, field transistors [70], solar energy cells, and overchargeable batteries [66].

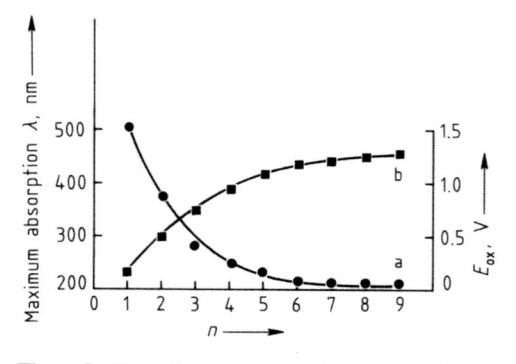

Figure 8. Correlation between maximum absorption wavelength (a) and oxidation potential (b) of thiophene oligomers Oxidation potential measured in 0.5 mol/L Et$_4$NBF$_4$ in acetonitrile versus saturated calomel electrode.

5.5. Polyphenylene

Synthesis. First attempts to prepare polyphenylene [9033-83-4], [24991-24-0], [25190-62-9], [26499-97-8] date back to 1842 [71]. RIESE describes a process in which polyphenylene ($n=13$) is synthesized from 1,4-dibromobenzene and sodium. Further methods are the Ullmann reaction, thermal decomposition of diazonium salts, coupling of phenylene dihalogenide – Grignard compounds [72], and the oxidative coupling of benzene with AlCl$_3$ and CuCl$_2$ (Kovacic method [73]). In 1963 the Kovacic method was systematically extended by varying reaction conditions (temperature, catalysts, Lewis acid, and oxidants) and starting materials [4]. Many side reactions occur leading to cross-linked graphite-like lattices.

Other aromatic compounds, substitution products, condensed systems (e.g., diphenylene, terphenylene, pyrene, anthracene, chrysene), unsaturated cyclic compounds (e.g., cyclopentadiene, cyclooctatetraene), heterocyclic compounds (e.g., thiophene, pyrrole, quinoline), and metal complexes (e.g., phthalocyanines, ferrocene) were investigated.

All Friedel – Crafts catalysts and the usual dehydrogenation agents can be used for oxidative coupling. Oxidizing Friedel – Crafts agents, such as FeCl$_3$, assume both functions. The degree of suitability has been tested on benzene/CuCl$_2$ (Eq. 1) and on benzene/AlCl$_3$ (Eq. 2).

$$AlBr_3 > AlCl_3 > FeCl_3 > MoCl_6 > NbCl_5 > TiCl_4 > SnCl_4 > BF_3 > ZnCl_2 \quad (1)$$

$$Pd(II) > Co(III) > Mn(III) > Cu(II) > Fe(III) > chloranil > V(III) > I_2 \quad (2)$$

The oligomeric and polymeric reaction products were also characterized in terms of thermoelectric power, dark conductivity, and photoconductivity. They showed electrical conductivities up to 0.1 S/cm, the highest value obtained at that time [4]. The polymers were obtained as powders and were processed by press sintering. Later synthesis by Diels – Alder [74] or boronic acid coupling [75] lead to soluble or meltable materials.

A novel route to high molecular mass poly(p-phenylene) is based on the polymerization of derivatives of 5,6-dihydroxycyclo-1,3-hexadiene, a compound that is easily prepared by the bacterial oxidation of benzene [76].

R = methyl, methoxy, phenyl   Soluble precursor

The advantage of this process is that it involves a soluble precursor polymer, which means that films can first be cast, objects coated, or fibers spun from solution; poly(p-phenylene) is then formed on heating.

The Marvel method is similar to the Kovacic method [77]:

Properties. The properties of polyphenylenes depend mainly on their linkage (m-, p-), but also on the method used to prepare them. o-Polyphenylenes are unknown: m- and p-polyphenylenes are crystalline (> 50%), intractable, and insoluble. p-Polyphenylenes are stable up to ca. 500 °C in air and up to 650 °C under nitrogen (10% weight loss); the stability of m-polyphenylenes is approximately 200 °C lower [78]. Figure 9 demonstrates the drastic differences between m- and p-substituted poly- phenylenes. The *para* derivatives are soluble in toluene up to n = 6, the *meta* derivatives are soluble up to n = 16. Polymer powders can be processed by press sintering (400 °C, > 100 bar).

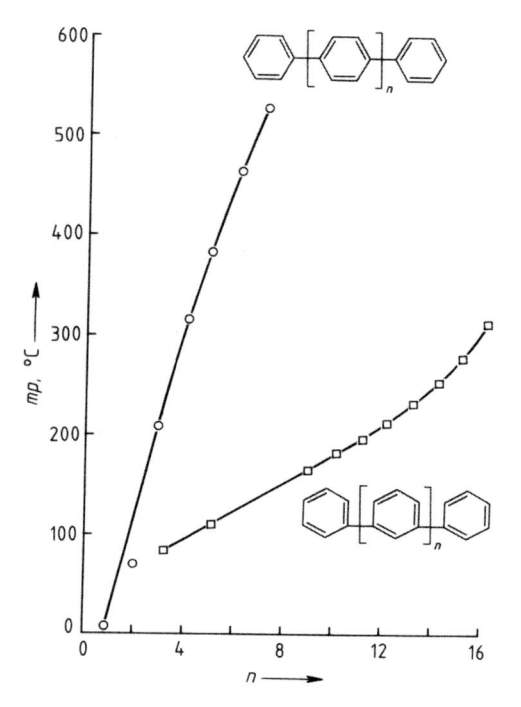

Figure 9. Correlation of the melting points and the degree of condensation (n) of m- and p-polyphenylenes

Poly(p-phenylene) can be n-doped with electron donors or p-doped with electron acceptors, electrical conductivities reach 10^2 S/cm. Donor doping is best accomplished with the aid of an electron-transfer agent (e.g., by adding naphthalene or α-methylstyrene in tetrahydrofuran) [79], [80]. Acceptor doping is usually done by exposing the powders to AsF_5 gas or heating with nitromethane solutions of SbF_5. Electrochemical doping (anodic oxidation or cathodic reduction) is performed in suitable electrolytes (e.g., acetonitrile), the electrodes being separated by diaphragms.

Anodic oxidation:

$p\text{-PP} + AgBF_4 \rightarrow p\text{-PP}^+BF_4^- + Ag^+$

Cathodic reduction:

$p\text{-PP} + LiClO_4 \xrightarrow{+e^-} p\text{-PP}^-Li^+ + ClO_4^-$

PP = polyphenylene

Oxidation potentials, conductivities, and spectra of polyphenylene are cited in [79].

Soluble, substituted polyphenylenes and polyaromatic systems are discussed in [81].

Potential Uses. Poly(*p*-phenylene) can be used as electrodes for rechargeable batteries.

5.6. Poly(Phenylene Sulfide)

Poly(*p*-phenylene sulfide) [9016-75-5], PPS, (white powder, T_g 92 °C, *mp* 270 – 290 °C, 65 % crystallinity) was the first melt-processible polymer to be doped with strong electron acceptors (e.g., AsF_5) to yield highly conductive products [82]. The first laboratory synthesis of PPS was reported by MACALLUM [83] and involved the melt reaction of 1,4-dichlorobenzene, sulfur, and sodium carbonate. A commercially product has been available as powder, film, or fiber since 1973 from Phillips Petroleum under the trade name Ryton; it is produced from 1,4-dichlorobenzene and sodium sulfide (high-pressure process) in a polar solvent (*N*-methylpyrrolidone) [84].

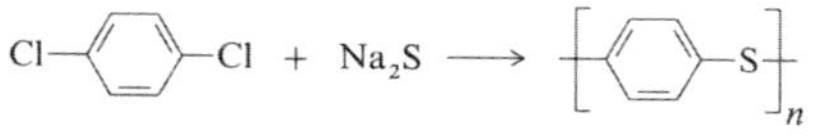

$$Cl-\langle\ \rangle-Cl\ +\ Na_2S\ \longrightarrow\ [\langle\ \rangle-S]_n$$

Another process involves self condensation of a halogenated thiophenol resulting in "head-to-tail" polymerization [85]:

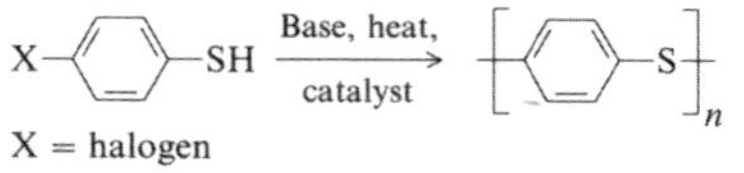

$$X-\langle\ \rangle-SH\ \xrightarrow[\text{catalyst}]{\text{Base, heat,}}\ [\langle\ \rangle-S]_n$$

X = halogen

Poly(*m*-phenylene sulfide) and substituted (methyl-, fluoro-) poly(*p*-phenylene sulfide) are prepared by analogous reactions from appropriate monomers.

Doping of PPS with AsF_5 produces conductivities up to 2×10^{-2} S/cm [86]. Attempts to make *n*-doped PPS have been unsuccessful. In addition to AsF_5, doping with $FeCl_3$, H_2SO_4, $HClO_4$, FSO_3H, CF_3SO_3H, $AlCl_3$, TaF_5 has been investigated [87]. Heavily doped PPS undergoes intramolecular cross-linking to form benzothiophene rings [79]:

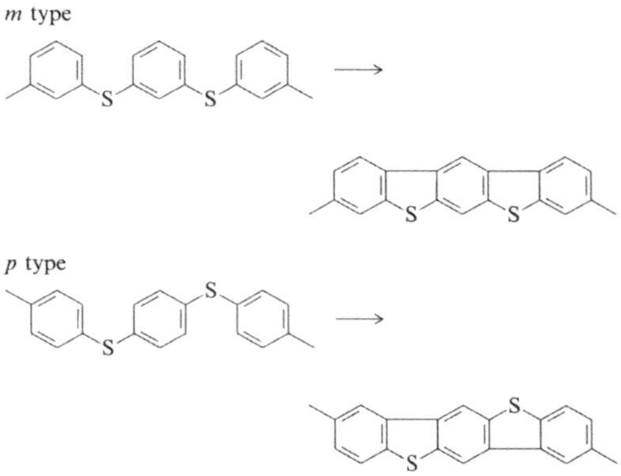

m type

p type

Poly(phenylene sulfides) are used where resistance to heat and chemicals is required, for example in the chemical process industry (pump housings, impellers, valves, metering devices, tanks, coil bobbins). See also, → Polymers, High-Temperature, Chap. 5..

Poly(*p*-phenylene selenide) [52410-66-9] and poly(*p*-phenylene telluride) [84174-18-5] have been prepared by techniques similar to those used to prepare PPS [79]. Conductivities of AsF_5-doped samples reach 0.01 S/cm.

5.7. Poly(Phenylene Vinylene)

Poly(*p*-phenylene vinylene) [88] [96638-49-2], [87092-55-5], [26009-24-5] was first synthesized by the Wittig condensation of terephthaldehyde and *p*-xylene bis(triphenylphosphonium) chloride. The method which is now mainly employed starts from a soluble polysulfonium salt [89], for example:

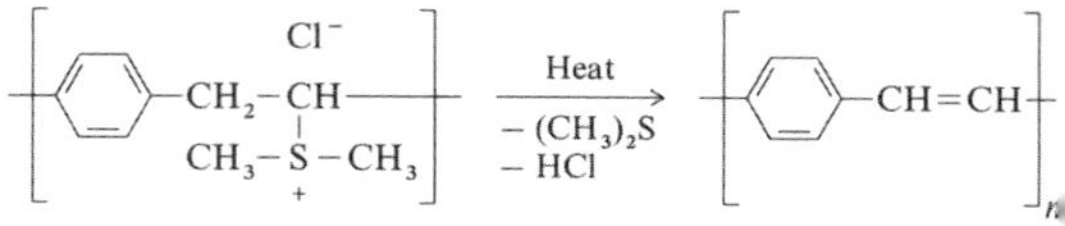

$$\left[\langle\ \rangle-CH_2-\overset{Cl^-}{\underset{CH_3-\overset{+}{S}-CH_3}{CH}}\right]\ \xrightarrow[-HCl]{\text{Heat}\atop -(CH_3)_2S}\ [\langle\ \rangle-CH=CH]_n$$

The polysulfonium salt is obtained by the reaction of α,α'-dichloro-*p*-xylene with excess dimethyl sulfide (50 °C, 20 h), and polymerization with sodium hydroxide (0 °C, 1 h). The first doped poly(phenylene vinylene) was reported by KARASZ using AsF_5 as dopant [90]; its conductivity was 3 S/cm. Stretching the precursor film at high temperature and doping with

AsF$_5$ and SO$_3$ yielded conductivities up to 2800 S/cm [36], [91].

Substituted derivatives [92], copolymers [93], and blends [94] show better thermal stability. Highly conductive graphite films have been prepared by pyrolysis of poly(phenylene vinylene) (> 3000 °C), stretched samples doped with SO$_3$ had conductivities of 10^5 S/cm [95]. Poly(phenylene vinylene) can be used as tunable polymer diodes or luminescent electrodes [96].

5.8. Polyaniline

The synthesis of polyaniline [25233-30-1] is remarkably simple and is summarized in Figure 10 [97], [98]. Aniline can be chemically (1863, LIGHTFOOT) or electrochemically (1865, LETHEBY) oxidized and dehydrogenated to form a quinone diimine.

Oxidation of the primary aromatic amine usually does not stop at the monomeric quinone imines but gives rise to more complex quinone imine products. The type of product depends largely on the oxidants and the reaction conditions. 1,4-Benzoquinone-4-imino-1-phenylimine is an unstable intermediate which reacts with further aniline molecules and oligomerizes to give the final product, polyaniline.

Studies on polyaniline suggest that the polymer can exist in a wide range of structures [99], which can be regarded as copolymers of reduced (amine) and oxidized (imine) units of the form:

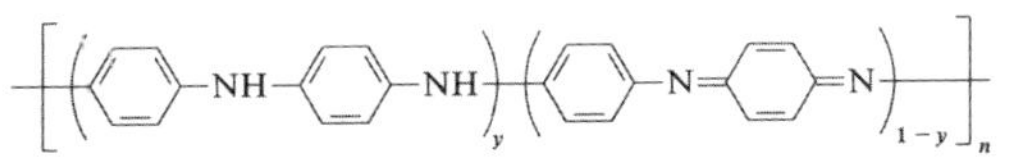

$$\left[\left(\!\!\left\langle\bigcirc\right\rangle\!\!-\!NH\!-\!\!\left\langle\bigcirc\right\rangle\!\!-\!NH\right)_y\!\!\left(\!\!\left\langle\bigcirc\right\rangle\!\!-\!N\!=\!\!\left\langle\bigcirc\right\rangle\!\!=\!N\right)_{1-y}\right]_n$$

When $0 < y < 1$ these structures are poly(p-phenylene amine imines), in which the oxidation state of the polymer increases with increasing content of the imine form. The fully reduced form ($y = 1$) is leucoemeraldine, the fully oxidized form ($y = 0$) is pernigraniline, and the 50%-oxidized structure ($y = 0.5$) is emeraldine. Each structure can exist as the base or as its salt, formed by protonation. Four repeating units can therefore be envisaged in the polymer chain in amounts which depend on the degree of oxidation and protonation of the structure. The "metallic state" is

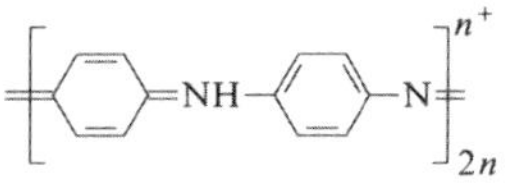

$$\left[\left(\!\!\left\langle\bigcirc\right\rangle\!\!=\!NH\!-\!\!\left\langle\bigcirc\right\rangle\!\!-\!N\right)_{2n}\right]^{n+}$$

The reaction of aniline to "aniline black" proceeds through a number of intermediates [98]:

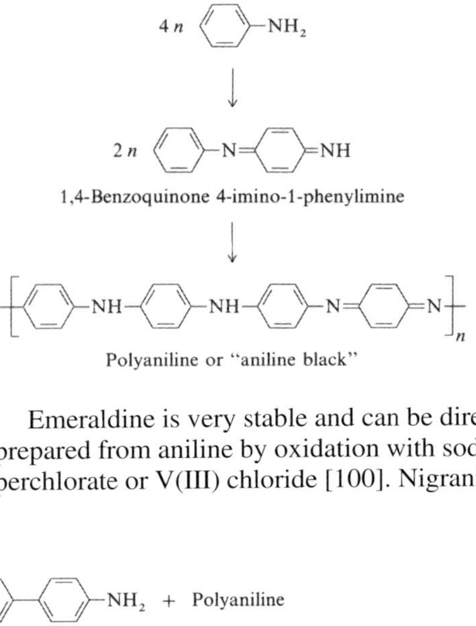

$$4n \;\left\langle\bigcirc\right\rangle\!-\!NH_2$$

$$\downarrow$$

$$2n \;\left\langle\bigcirc\right\rangle\!-\!N\!=\!\left\langle\bigcirc\right\rangle\!=\!NH$$

1,4-Benzoquinone 4-imino-1-phenylimine

$$\downarrow$$

$$\left[\left\langle\bigcirc\right\rangle\!-\!NH\!-\!\left\langle\bigcirc\right\rangle\!-\!NH\!-\!\left\langle\bigcirc\right\rangle\!-\!N\!=\!\left\langle\bigcirc\right\rangle\!=\!N\right]_n$$

Polyaniline or "aniline black"

Emeraldine is very stable and can be directly prepared from aniline by oxidation with sodium perchlorate or V(III) chloride [100]. Nigraniline

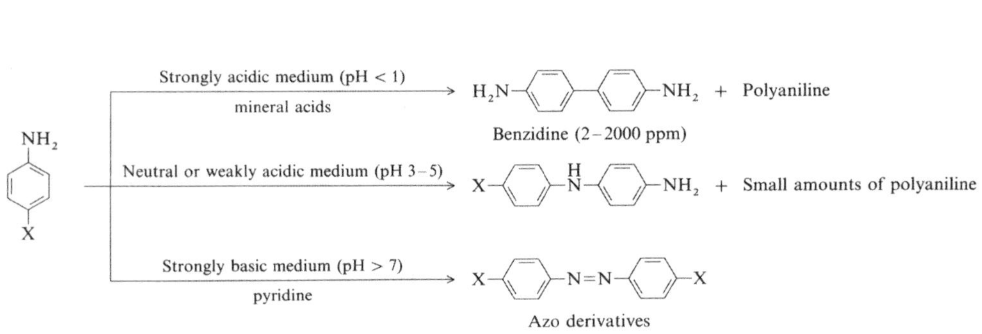

Figure 10. Products of aniline oxidation formed at different pH values

and pernigraniline are further oxidation products of emeraldine, both are very unstable. The structure of emeraldine and related molecules is described in [101].

Nitrogen- or ring-substituted homologues of aniline can apparently be oxidized similarly, yielding corresponding "aniline blacks".

Oxidants used for preparing oligomers and polymers of aniline include $Cr_2O_7^{2-}$, $S_2O_8^{2-}$, Ce^{4+}, and Pb^{4+} ions. Aniline can also be polymerized by anodic oxidation.

The reaction conditions (temperature, duration, pH, oxidants, and catalysts) determine the yield and type of oxidation product [97].

A decisive handicap in the production of polyaniline is the formation of small quantities of carcinogenic benzidene (Fig. 10) [102].

HONZL synthesized benzidene-free polyanilines by avoiding the use of aniline as the starting monomer [103]:

Similar structures can be obtained by polycondensation of quinones and aromatic amines [104].

Properties. The conductivity of polyaniline is 10 S/cm up to pH 4, but decreases to 10^{-10} S/cm above pH 4. The polymers are stable up to 250 °C (undoped) and up to ca. 150 °C (doped).

Conduction in polyaniline can be accounted for in terms of electrons hopping under the assistance of proton transfer, for which the presence of water plays an essential role [105].

In contrast to other electrically conducting polymers, polyanilines may be doped with protons. Polyaniline is available as powder, film, or fibrils (Versicon, Allied Signal) [106], [107].

The electrochemistry of polyanilines was first investigated at the end of the nineteenth century. In the mid 1960s, interest in this topic was renewed. The use of polyaniline as a battery electrode on account of its redox and proton transfer behavior was described in 1968 [108]. In 1986, polyaniline was presented as a "novel conducting polymer"; its preparation and redox behavior were described [109]. The phenomena of anodic oxidation and discharge by means of reversed polarity have been known since 1891 [1].

Uses. Polyaniline is primarily of interest as an electrode material. Its discharge capacity is greater than that of polypyrrole(+)/perchlorate (−), its self-discharge is higher than that of polythiophene or Ni/Cd battery systems [66]. GENIES has announced a rechargeable battery based on polyaniline (anode), propylene carbonate – $LiClO_4$ (electrolyte), and Li – Al (cathode) [110], [111]. This polyaniline is made by oxidizing aniline with ammonium persulfate in ammonia/2.3 mol/L hydrogen fluoride as solvent. The discharge capacity of the polymer is 100 A · h/kg at 25 °C and 140 A · h/kg at 40 °C for current densities of 0.5 mA/cm^2 and for an amount of material giving a capacity of 10 mA · h. The voltage in open circuit for the fully charged battery developed by GENIES is 3.6 V. The average utilizable potential is 2.8 – 3 V. The energy density for the polymer lies between 280 and 420 W · h/kg. Its behavior with regard to self-discharge and to constant applied voltage (floating life) is excellent [111]. Like polypyrroles, polyanilines show excellent antistatic behavior and have a high shielding efficiency for electromagnetic interference.

5.9. Miscellaneous Polymers

There has been a flood of literature concerning new electrically conducting polymers, examples are listed below.

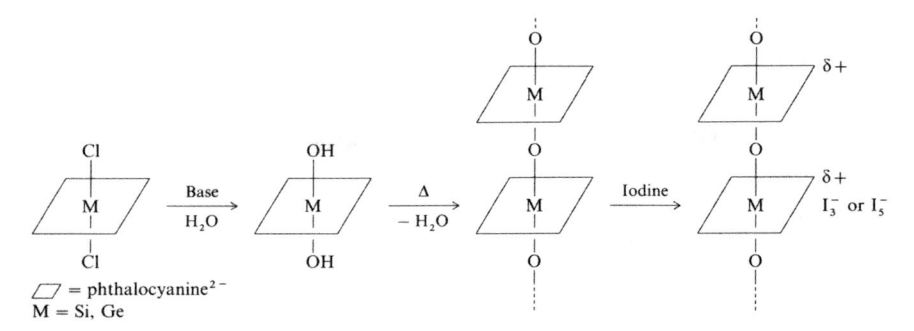

Figure 11. Synthesis and doping of bridged phthalocyanine polymers

Bridged macrocyclic complexes [81] are mainly derivatives of tetraazaporphyrin or phthalocyanine:

Tetraazaporphyrin (M = H_2)

Phthalocyanine (M = H_2)

The synthesis and doping of bridged phthalocyanine polymers is illustrated in Figure 11. The bridging ligand can be linked by two σ bonds, by two coordinate bonds, or by one σ and one coordinate bond (Fig. 12).

After doping with iodine, conductivities up to 10.1 S/cm are obtained. Substituted phthalocyanines are soluble and can be cast as films or handled by the Langmuir – Blodgett technique to give ultrathin, well-defined molecular layers. The insoluble powders can only be processed by press sintering.

Polyazulene [82451-56-7] [112], [113] is synthesized by electrochemical polymerization with ClO_4^- as counterion (similar to that of polypyrrole) yielding amorphous polymer films which can be peeled from the anode. Conductivity is 0.01 S/cm. The films may be electrochemically and reversibly discharged to the nonconducting form.

Polycarbazole [51555-21-6] [113]. Solutions of carbazole in acetonitrile may be electrochemically oxidized (counterion ClO_4^- or

Figure 12. Construction of bridged macrocyclic transition-metal polymers

BF_4^-) at a platinum anode to give electrically conductive films with poor mechanical stability and conductivities up to 0.001 S/cm. Polycarbazole has also been obtained by vacuum evaporation of carbazole and by chemical condensation [114].

Polyindole [82451-55-6] [48]. Electrochemical oxidation of indole (counterion ClO_4^- or BF_4^-) yields brittle films with conductivities of ca. 0.01 S/cm. Polyindole has been employed as a polymer coating in a glucose sensor [113].

Polypyrene [41496-25-7] [113]. Electrochemical oxidation of pyrene (counterion BF_4^-, ClO_4^- or AsF_6^-) yields insoluble, brittle films with conductivities up to 1 S/cm.

Polyazepine [115483-32-4] [115]. Phenyl azide can be photopolymerized in the gas phase to yield films of poly-1,2-azepine with a conductivity of 10^{-2} S/cm (after I_2 doping).

Polyfulvenes [107889-59-8] [116]. Polyfulvenes are formed by cationic polymerization of 6,6-dimethylfulvene followed by chemical or electrochemical dehydrogenation, conductivity 10^2 S/cm (after I_2 doping).

Low-band-gap aromatic polymers [117]. An important prerequisite for obtaining low-band-gap polymers is that the ground state of the polymer has a quinonoid contribution. These polymers have the general formula:

X, Y = SO_2, SO, S, N−R, CH=CH

Benzenoid and anthracenoid precursors are:

Low- or narrow-band-gap systems (gap < 0.5 eV) already contain a high proportion of thermally excited electrons at room temperature. These materials can be oxidized electrochemically at a low oxidation potential to give conducting polymers. Cross coupling with organometallic intermediates affords the undoped polymers. Conductivities up to 10 S/cm are reached (after iodine or electrochemical doping).

Polyindophenines [138309-48-5] [118] are prepared by reaction of isatin with thiophene and have conductivities up to 10^{-2} S/cm without additional dopant. The polymers are soluble in dimethyl sulfoxide and can be cast as films; the material is thermostable up to 230 °C.

Charge-transfer complexes [119] are combinations of electron donor compounds with electron acceptor compounds. They are synthesized, for example, by electrochemical charging of aromatic hydrocarbons with ClO_4^-. The charge-transfer complexes are assembled in defined stacks and have conductivities up to 10^3 S/cm, in some cases with superconductivity. Examples are given below [120], [121]:

Bisethylene dithiotetrathiofulvalene – I_3 complex
T_c 8.1 K

Dicyanodiimine quinone–copper complex
$\sigma = 10^3$ S/cm

Finally it should be remembered that all electrically conducting organic materials are due to the various dopant charge-transfer complexes [123].

Buckminsterfullerene [99685-96-8] [122] is an allotrope of carbon (Fig. 13). When doped with potassium, it reaches conductivities up to 500 S/cm; cooling to 18 K makes it superconducting.

6. Uses

Few of the organic materials discussed above have been demonstrated to be sufficiently stable for critical applications.

Stability is a relative property and is the "Achilles heel" of electrically conducting polymers. For example, polypyrroles with BF_4^- as counterion are unstable under atmospheric conditions but are as stable as Ni/Cd electrodes in rechargeable batteries. It was hoped that electrically conducting organic materials could combine conductivities comparable to those of metals with the processibility of plastics. Work over the last decades has gone a long way in this direction but there are still many obstacles to be mastered before these materials become widely used.

Polyene chains with extended double bond systems are unstable against oxygen, forming epoxy, carbonyl, and carboxyl groups. They also undergo cross-linking and substitution reactions (Diels – Alder and radical reactions). Aromatic and heterocyclic structures are much more stable against environmental conditions (e.g., temperature, irradiation).

Doped materials (n or p doped) are saltlike compounds and are therefore unstable against humidity, oxygen, and temperature. Conductivity is lost as a result of interruption of the conjugation system, this destroys the charge-transfer complexes within a few days.

Defect-free oriented polyacetylenes are more stable, they do not loose any conductivity for 3 months [24]. Polypyrroles and polythiophenes show unchanged conductivities over several years (Sections 5.3, 5.4). Polyindophenines are "intramolecularly doped", containing covalently bonded electron donor and acceptor groups within the same molecule (Section 5.9).

How many uses do conducting polymers have? The short answer is "scarcely any" [123], but they do have some applicability [80], [124], [125] for:

1. Rechargeable batteries [120], [126] (Sections 5.3 and 5.4)
2. Coatings for electromagnetic radiation shielding or electrochemical metallization [63], [127]
3. Semiconducting devices and photovoltaics [128]
4. Catalysis [129]
5. Sensors and electrochromic displays [130], e.g., for gas detection (NH_3, CO_2)
6. Solar energy conversion, e.g., electroluminescence [131]

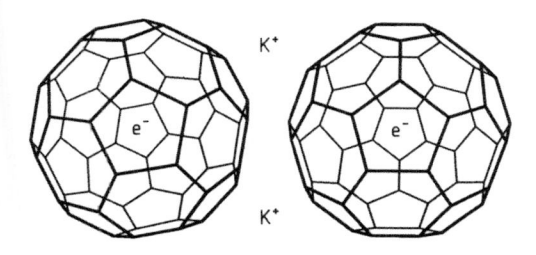

Figure 13. Potassium-doped C_{60} molecules

It seems that this class of electrically conducting materials will form the basis for a new generation of "intelligent" material systems [132].

References

1. F. Goppelsroeder, *Die Internationale Elektrotechnische Ausstellung* **18** (1891) 978; **19** (1891) 1047.
2. H. A. Pohl: "Electronic Behavior of Organic Macromolecular Solids," B. A. Bolto: "Semiconducting Organic Polymers Containing Metal Groups," D. D. Eley: "Semiconducting Biological Polymers," in J. E. Katon (ed.): *Organic Semiconducting Polymers*, Marcel Dekker, New York 1968.
3. W. A. Little, *Phys. Rev.* **134** (1964) A 1416.
4. H. Naarmann, F. Beck: "Neuartige Polymerisate aus aromatischen und heterocyclischen Verbindungen und ihre elektrophysikalischen Eigenschaften," GDCh Meeting, Munich, Oct. 12, 1964. BASF, DE 1 178 529, 1964 (H. Naarmann, F. Beck, E. G. Kastning). BASF, DE 1 195 497, 1963 (H. Naarmann, F. Beck, E. G. Kastning). BASF, DE 1 092 137, 1964 (H. Naarmann, F. Beck, E. G. Kastning).
5. W. Slough et al. in [2] pp. 55, 118 – 163. H. Naarmann, *Angew. Chem. Int. Ed. Engl.* **8** (1969) 915. BASF, DE-OS 1 953 898, 1969 (H. Willersinn, H. Naarmann, K. Schneider).
6. R. Greene, G. B. Street, L. J. Süter, *Phys. Rev. Lett.* **34** (1975) 577.
7. A. J. Heeger et al.,*J. Chem. Soc. Chem. Commun.* 1977, 578 – 580.
8. H. Shirakawa, S. Ikeda, *Polym. J. (Tokyo)* **2** (1971). T. Ito, H. Shirakawa, S. Ikeda, *J. Polym. Sci. Polym. Chem. Ed.* **12**, (1974) 11.
9. H. Naarmann: "Synthese elektrisch leitfähiger Polymere," *Angew. Makromol. Chem.* **109/110** (1982) 295 – 338. W. J. Feast, "Synthesis of Conducting Polymers," in T. A. Skotheim (ed.): Handbook of Conducting Polymers. Marcel Dekker, New York 1986, pp. 1 – 44. N. C. Billingham, P. D. Calvert: "Electrically Conducting Polymers," *Adv. Polym. Sci.* **80** (1989) 1 – 90.
10. H. W. Gibson: "Substituted Polyacetylenes," in T. A. Skotheim (ed.): *Handbook of Conducting Polymers*, Marcel Dekker, New York 1986, pp. 405 –435.
11. H. Hopf, O. Kretschmer, H. Naarmann, *Adv. Mater.* **1** (1989) no. 12, 445.
12. H. Naarmann, *Adv. Mater.* **2** (1990) no. 8, 345 – 348. W. R. Sorenson, T. W. Campbell: *Präparative Methoden der Polymeren Chemie*, Verlag Chemie, Weinheim, Germany 1962, p. 167.
13. T. Yamanato, Y. Hayashi, A. Yamamoto, *Bull. Chem. Soc. Jpn.* **51** (1978) 2091.
14. N. Miyaura, T. Yagani, A. Suzuki, *Synth. Commun.* **11** (1981) 513 – 519.
15. K. J. Stille, *Makromol. Chem.* **154** (1972) 49.
16. M. Kanabe, M. Okawara, *J. Polym. Sci. Polym. Chem. Ed.* **6** (1968) 1058.
17. H. H. Edwards, W. J. Feast, *Polymer* **21** (1980) 595.
18. T. M. Swager, D. A. Dougherty, R. H. Grubbs, *J. Am. Chem. Soc.* **111** (1989) 4413.
19. E. J. Ginsburg et al.: "Conjugated Polymeric Materials," in J. L. Bredes, R. R. Chance (eds.): *NATO ASI Ser.* **182** (1990) 65.
20. A. W. Snow, *Nature (London)* **292** (1981) no. 40.
21. K. Menke, S. Roth, *Chem. Unserer Zeit* **20** (1986) no. 2, 33, 53.
22. R. R. Chance, D. S. Boudreaux, J. L. Bredas: "Solitons, Polarons, Bipolarons" in T. A. S. Skotheim (ed.): *Handbook of Conducting Polymers*, vol. 2, Marcel Dekker, New York 1986, pp. 825 – 857. P. Bätz, Dissertation, Universität Tübingen, Tübingen 1991.
23. H. Schäfer, Dissertation, Universität Göttingen, Göttingen 1983.
24. S. Pekker, A. Janossy in T. A. Skotheim (ed.): *Handbook of Conducting Polymers*, vol. 1, M. Dekker, New York 1986, p. 45.
25. G. Lugli, U. Pedretti, G. Perego, *J. Polym. Sci. Polym. Lett. Ed.* **23** (1985) 129.
26. BASF, BMFT-Forschungsbericht: Entwicklung von elektrisch leitfähigen Alternativ-Polymeren, 03 C 134–0, chap. II, Ludwigshafen, Aug. 1, 1982 – July 31, 1985, pp. 1 – 75.
27. H. Naarmann, *Synth. Met.* **17** (1987) 223. H. Naarmann, N. Theophilou, *Synth. Met.* **22** (1987) 1. T. Schimmel, M. Schwoerer, *Solid State Commun.* **65** (1988) no. 11, 1311. J. Tsukamoto, A. Takahashi, K. Kawasaki, *Jpn. J. Appl. Phys. Part* **29/1** (1990) 125.
28. M. Yamaura et al.,*Synth. Met.* **28** (1989) C 157 –164.
29. D. Gagnon, J. Capistran, F. Karasz, R. Lenz, *Polym. Prepr. (Am. Chem. Soc. Dir. Polym. Chem.)* **25** (1984) 284 – 285. D. Gangon, F. Karasz, E. Thomas, R. Lenz, *Synth. Met.* **20** (1987) 85 – 95.
30. H. Shirakawa, Y.-C. Chen, K. Akagi, T. Norahara, *Synth. Met.* **14** (1986)173, 199. M. Aldissi, *J. Polym. Sci. Polym. Lett. Ed.* **23** (1985) 167. H. Shirakawa, K. Akagi, S. Katayama, *J. Macromol. Sci. Chem.* **A25** (1988) no. 5 – 7, 643. H. Shirakawa, K. Akagi, M. Suezaki, *Synth. Met.* **28** (1989) D 1. A. Moutaner et al.,*Synth. Met.* **28** (1989) D 19.
31. BASF, DE-OS 3 533 252 Al, 1985 (H. Naarmann, M. Portugall).
32. H. Naarmann, *Synth. Met.* **41 – 43** (1991) 1 – 6.
33. A. G. MacDiarmid, T. Woerner, A. G. Heeger, A. Feldblum, *J. Polym. Sci., Polym. Lett. Ed.* **20** (1982) 305; **22** (1984) 119.
34. T. Shimidzu, *Langmuir* **3** (1987) 1169.
35. H. Nakahara, *Thin Solid Films* **60** (1988)87, 153. J. Watanabe, K. Hong, M. F. Rubner, *Synth. Met.* **28** (1989) C 473.
36. M. Matsumoto et al.,*Synth. Met.* **27** (1988) B 601.
37. I. Bozovic, *Mod. Phys. Lett. B.* **1/3** (1987) 81.
38. H. Wiu, *Jahresber. Fortschr. Chem.*(1866) 516.
39. H. Erdmann, P. Köthner, *Z. Anorg. Allgem. Chem.* **18** (1898) 48.
40. W. Reppe, *Justus Liebigs Ann. Chem.* **560** (1948) no. 1, 140.
41. M. Hatano, S. Kambara, S. Okamoto, *J. Polym. Sci.* **61** (1961) 26.
42. Houben-Weyl, 4th ed., vol. **20/2**,p. 1212; *Science of Synthesis*, vol. 45, 2008, p. 1421.
43. H. Naarmann, *Angew. Makromol. Chem.* **109/110** (1982) 295.
44. H. Naarmann, N. Theophilou, *Synth. Met.* **22** (1988) 1.
45. T. Schimmel et al., *Bayreuther Polym. Symp.*,April 12, 1989; *Synth. Met.* **37** (1990) 1 – 6; *Makromol. Chem.* **190** (1989) 3217.
46. H. Haberkorn et al.,*Synth. Met.* **5** (1982) 51.
47. H. Naarmann in J. L. Bredas,R. R. Chance (eds.): *Conjugated Polymeric Materials*, Klüver Acad. Publ., Dordrecht 1990, pp. 11 – 51.
48. H. Naarmann, P. Strohriegel: "Conducting and Photoconducting Polymers," in H. Kricheldorf (ed.): *Handbook of Polymer Synthesis*, part 13, Marcel Dekker, New York 1992, pp. 1353 – 1425.
49. BASF, DE-OS 3 118 630, 1981 (H. Naarmann, G. Köhler).

50. H. Winter, G. Sachs, E. Dormann, H. Naarmann, *Synth. Met.* **36** (1990) 353.
51. A. Bayer, L. Landsberg, *Ber. Dtsch. Chem. Ges.* **15** (1882) 52. F. Bohlmann, *Angew. Chem.* **69** (1957) 82.
52. G. Wegner in E. W. Halfield (ed.): *Molecular Metals*, chap. 4.0, Plenum Publ., New York 1979, p. 209.
53. E. Hädicke et al., *Angew. Chem.* **83** (1971) no. 7, 253.
54. M. Dennstedt, J. Zimmermann, *Ber. Dtsch. Chem. Ges.* **21** (1888) 1478. A. Grossauer: *Die Chemie der Pyrrole*, Springer-Verlag, Berlin 1974, p. 149.
55. H. Lund, *Acta Chem. Scand.* **11** (1957) 1323. A. Stanienda, *Z. Naturforsch.* **228** (1967) 1107.
56. K. K. Kanazawa et al., *J. Chem. Soc. Chem. Commun.* 1979, 854. A. F. Diaz, J. Bargon: "Electrochemical Synthesis of Conducting Polymers," in T. A. Skotheim (ed.): *Handbook of Conducting Polymers*, vol. 1, Marcel Dekker, New York 1986, p. 82. G. B. Street: "From Powder to Plastics," in T. A. Skotheim (ed.): *Handbook of Conducting Polymers*, vol. 1, Marcel Dekker, New York 1986, p. 266.
57. BASF, US 4 468 291, 1982 (H. Naarmann, G. Köhler, J. Schlag).
58. E. M. Genies, G. Bidan, A. F. Diaz, *J. Electrochem. Soc.* **149** (1983) 101; **129** (1982) 1685.
59. H. Naarmann, *Macromol. Chem. Macromol. Symp.* **8** (1987) 1.
60. BASF, DE-OS 3 425 511, 1968 (H. Naarmann, G. Köhler). A. O. Patil, Y. Ikenoue, F. Wudl, A. Heeger, *J. Am. Chem. Soc.* **109** (1987) 1858.
61. BASF, EP 129 070, 1984 (G. Wegner, W. Wernet).
62. M.-A. Sato, S. Tanaka, K. Kaeriyama, *Synth. Met.* **18** (1987) 229. B. F. Goodrich, US 4 487 667, 1984 (L. Traynor). G. Schiavon, G. Zotti, *Synth. Met.* **28** (1989) C 199. N. Oyama, T. Ohsaka, H. Miyamoto, *Synth. Met.* **28** (1989) C 193. T. Inagaki et al., *Synth. Met.* **18** (1989) C 245. R. Bittihn, *Kunststoffe* **79** (1989) no. 6, 530
63. BASF, EP 0 206 133 Al, 1986 (H. Naarmann, W. Heckmann). BASF Plastics, Research and Development KVX 8611, Ludwigshafen, Oct. 1986. R. Bittihn, *Makromol. Chem. Symp.* **8** (1987) 51.
64. Agence National de Valorisation, FR 7 518 383, 1975 (R. Buvet, R. Vallot, I. Gal, L. T. Yu).
65. V. Meyer, *Ber. Dtsch. Chem. Ges.* **16** (1883) 1465.
66. G. Tourillon: "Polythiophene and its Derivatives," in T. A. Skotheim (ed.): *Handbook of Conducting Polymers*, vol. I, Marcel Dekker, New York 1986, p. 293.
67. F. Martinez, R. Voelkel, D. Naegele, H. Naarmann, *Mol. Cryst. Liq. Cryst.* **167** (1989) 227.
68. G. Gustafsson et al. in J. L. Bredas, R. Silbey (eds): *Conjugated Polymers*, Klüver Acad. Publ., Dordrecht 1991, p. 315.
69. M. Gazard: "Electrochromic Display Device," in T. A. Skotheim (ed.): *Handbook of Conducting Polymers*, Marcel Dekker, New York 1986, p. 673.
70. F. Garnier, G. Horowitz, X. Peng, D. Fichou, *Adv. Mater.* **2** (1990) no. 12, 592.
71. F. Riese, *Justus Liebigs Ann. Chem.* **164** (1872) 161.
72. H. Naarmann, *Angew. Makromol. Chem.* **109/110** (1982) 295.
73. P. Kovacic, A. Kyriakis, *Tetrahedron Lett.* **11** (1962) 467.
74. J. K. Stille, *Makromol. Chem.* **154** (1972) 49.
75. M. Miyaura, T. Yanagi, A. Suzuki, *Synth. Commun.* **11** (1981) 513.
76. D. G. H. Ballard, A. Courtis, I. M. Shirley, S. C. Taylor, *J. Chem. Soc. Chem. Commun.* 1983, 954.
77. C. S. Marvel, *J. Am. Chem. Soc.* **81** (1959) 448.
78. P. E. Cassidy: *Thermally Stable Polymers*, M. Dekker, New York 1980, p. 34. M. Busch, W. Weber, *J. Prakt. Chem.* **146** (1936) 1–55.
79. R. L. Elsenbaumer, L. W. Schacklette in T. A. Skotheim (ed.): *Handbook of Conducting Polymers*, Marcel Dekker, New York 1986, p. 214.
80. R. H. Baughman, L. W. Shacklette in W. R. Salaneck, D. T. Clark, E. J. Samuelsen (eds): *Science and Applications of Conducting Polymers*, A. Hilger, New York 1990, p. 47.
81. H. Naarmann et al., Entwicklung neuer Polymerer mit konjugierten Bindungssystemen und definierten zweidimensionalen Strukturen, BMFT Report 03M4019, part 1, Feb. 1991.
82. L. W. Shacklette, R. L. Elsenbaumer, R. Chance, H. Eckhardt, *J. Chem. Phys.* **75** (1981) 1919.
83. A. Macallum, *J. Org. Chem.* **13** (1948) 154.
84. Phillips Petroleum, US 3 607 843, 1971 (S. Edmonds, H. Hill, jr.).
85. R. W. Lenz, C. E. Handlorits, H. A. Smith, *J. Polym. Sci.* **58** (1962) 351.
86. J. Rabolt et al., *J. Chem. Soc. Chem. Commun.* 1980, 347.
87. J. Tsukamoto, K. Matsumura, *Jpn. J. Appl. Phys.* **23** (1984) 584.
88. M. MacDonald, T. Campbell, *J. Am. Chem. Soc.* **82** (1960) 4669.
89. M. Kanabe, M. Okawava, *J. Polym. Sci. Polym. Chem. Ed.* **6** (1968) 1058; *Chem. Abstr.* **68** (1968) 105721t. R. Wessling, H. Zimmermann, US 3 706 677, 1972; *Chem. Abstr.* **78** (1973) 85306 n.
90. G. Wnek, J. Chien, F. Karasz, C. Lillya, *Polymer* **20** (1979) 1441.
91. J. Murase et al., *J. Mol. Cryst. Liq. Cryst.* **118** (1985) 335.
92. H. H. Hörhold et al., *Materials Sciences Forum*, vols. **62–64**, (1990) 411–417. J. Murase, T. Onishi, T. Nogushi, M. Hirooku, *Synth. Met.* **17** (1987) 639.
93. M. Lux et al., *Proc. Macromol. Colloq.*, Freiburg, FRG 1989, p. 44.
94. J. Schlenhoff, J. Machado, P. Gletkowki, F. Karasz, *J. Polym. Sci. Polym. Phys. Ed.* **26** (1988) 2247.
95. F. Ohnishi, J. Murase, N. Takanobu, M. Hirooka, *Synth. Met.* **14** (1986) 207.
96. R. H. Friend et al., *Nature (London)* **356** (1992) 47.
97. *Houben Weyl*, 4th ed., **VII/3b**, 242.
98. R. Willstätter, C. W. Moore, *Ber. Dtsch. Chem. Ges.* **40** (1907) 2665.
99. A. G. Diarmid et al., *Synth. Met.* **18** (1987) 285.
100. K. Venkataraman: *The Chemistry of Synthetic Dyes*, vol. **II**, Academic Press, New York 1952, p. 776.
101. C. B. Duke et al., *Synth. Met.* **21** (1987) 143.
102. E. Genies, *New J. Chem.* **12** (1988) no. 4, 184.
103. J. Honzl, M. Metalova, *Tetrahedron* **25** (1969) 3641.
104. A. Everaets, S. Roberts, H. K. Hall, *J. Polym. Sci. Polym. Chem. Ed.* **24** (1968) 1703.
105. J. P. Travers, M. Nechtschein, *Synth. Met.* **21** (1987) 135.
106. M. Kryszewski: *Semiconducting Polymers PWN*, Polish Scientific Publ., Warzawa 1980, p. 76.
107. A. J. Heeger (pp. 1 – 12, 105 – 115), E. Genies (pp. 93 – 104), A. G. MacDiarmid, A. J. Epstein (pp. 117 –127) in W. R. Salaneck, D. T. Clark, E. J. Samuelsen (eds.): *Science and Application of Conducting Polymers*, Adam Hilger, New York 1990.
108. R. Buret et al., *Electrochim. Acta* **13** (1968) no. 2, 1441, 1451.
109. W. S. Huang, B. D. Humphrey, A. G. MacDiarmid, *J. Chem. Soc. Faraday Trans. 1* **82** (1986) 2385.
110. N. C. Billingham, P. O. Calvert: "Electrically Conducting Polymers," *Adv. Polym. Sci.* **80** (1989) 85.

111. E. Genies, P. Hany, C. Santier, *J. Appl. Electrochem.* **18** (1988) 751; *Synth. Met.* **28** (1989) C 647.
112. J. Bargon, S. Mohmand, R. J. Waltman, *Mol. Cryst. Liq. Cryst.* **93** (1983) 279.
113. P. C. Pandey, *J. Chem. Soc. Faraday Trans.* **1** (1988) no. 84, 2259.
114. S. A. Jenekhe, S. Reed, *J. Mol. Cryst. Liq. Cryst.* **105** (1984) 175; **106** (1984) 289.
115. A. W. Meijer, S. Nijhius, F. Van Vroonhoven, E. Havinga in L. J. Bredes,R. R. Chance (eds.): *Conjugated Polymeric Materials*, vol. **182**, Klüver Acad. Publ., Dordrecht 1989, p. 115.
116. H. Naarmann, *Synth. Met.* **17** (1987) 223.
117. M. Hanack, G. Hieber, G. Dewald, H. Ritter in W. R. Salaneck, D. T. Clark,E. J. Samuelsen (eds.): *Science and Application of Conducting Polymers*, Adam Hilger, New York 1990, p. 153.
118. H. Naarmann in: Entwicklung neuer Polymerer mit konjugierten Doppelbindungssystemen und definierten zweidimensionalen Strukturen, BMFT Report 03M 40197, Feb. 1990, p. 8. G. Koßmehl in T. A. Skotheim (ed.): *Handbook of Conducting Polymers*, M. Oelsler, New York 1986, vol. 1, p. 394.
119. E. P. Goddings, *Endeavour* **34** (1976) 125; *Synth. Met.* **27** (1988) 1 – 4.
120. S. Hünig et al.,*Z. Naturforsch. A Phys. Phys. Chem. Kosmophys.* **44 A** (1989) 825; *Synth. Met.* **27** (1988) B 181.
121. K. Bechard, D. Jerome, *Spektrum Wissensch.* **9** (1982) 38. D. Jerome, *Synth. Met.* **21** (1988) A183.
122. A. W. Sleight: "Sooty Superconductors," *Nature (London)* **350** (1991) 557.
123. N. C. Billingham, P. D. Calvert, *Adv. Polym. Sci.* **90** (1989) 4.
124. T. A. Skotheim (ed.): *Handbook of Conducting Polymers*, vol. 1, Marcel Dekker, New York 1986.
125. J. R. Ellis: "Commercial Applications of Intrinsically Conducting Polymers," in [124] p. 489.
126. A. G. MacDiarmid, R. B. Kaner: "Electrochemistry of Polyacetylene, Application to Rechargeable Batteries," in [124] p. 689.
127. H. J. Mair,S. Roth (eds.): *Elektrisch leitende Kunststoffe*, Hanser Verlag, München 1989, pp. 59, 89, 105, 137, 149, 237.
128. J. Kanicki: "Polymeric Semiconductor Contacts and Photovoltaic Applications," in [124] p. 543.
129. K. Soga, S. Ikeda: "Catalytic Properties of Conducting Polymers," in [124] p. 661.
130. G. Weddingen, H. Suhr in [127] p. 545.
131. *Nobel Symp.* NS 81, June 12 – 18, 1991, Lulea, Sweden.
132. D. T. Clark in [107] p. 161.

Further Reading

P. J. S. Foot, A. B. Kaiser: *Conducting Polymers*, Kirk Othmer Encyclopedia of Chemical Technology, 5th edition, John Wiley & Sons, Hoboken, NJ,online DOI: 10.1002/0471238961.0512050318052514.a01.pub2.

M. S. Freund, B. A. Deore: *Self-Doped Conducting Polymers*, Wiley, Chichester 2007.

G. Inzelt: *Conducting Polymers*, Springer, Heidelberg 2008.

T. A. Skotheim, J. R. Reynolds: *Conjugated Polymers*, 3rd ed., CRC Press, Boca Raton, FL 2007.

T. A. Skotheim, J. R. Reynolds: *Handbook of Conducting Polymers*, 3rd ed., CRC Press, Boca Raton, FL 2007.

M. Wan: *Conducting Polymers with Micro or Nanometer Structure*, Springer, Heidelberg 2008.

Polymers, High-Temperature

DAVID PARKER, ICI Advanced Materials, Wilton, Middlesborough, United Kingdom

JAN BUSSINK, GE Plastics Europe, Bergen op Zoom, The Netherlands

HENDRIK T. VAN DE GRAMPEL, GE Plastics Europe, Bergen op Zoom, The Netherlands

GARY W. WHEATLEY, Humberside Polytechnic, Hull, United Kingdom

ERNST-ULRICH DORF, Bayer AG, Leverkusen, Germany

EDGAR OSTLINNING, Bayer AG, Leverkusen, Germany

KLAUS REINKING, Bayer AG, Leverkusen, Germany

FRANK SCHUBERT, Evonik Industries AG, Marl, Germany

OLIVER JÜNGER, Ticona GmbH, Frankfurt, Germany

REINHARD WAGENER, Fresenius University of Applied Sciences, Idstein, Germany

1.	Introduction	1281
2.	Polyetherimides	1283
2.1.	Production	1283
2.2.	Properties	1285
2.3.	Processing	1289
2.4.	Uses	1289
2.5.	Economic Aspects	1290
3.	Polyaryletherketones	1291
3.1.	Production	1291
3.2.	Properties	1293
3.3.	Processing	1294
3.4.	Uses	1295
4.	Polysulfones	1295
4.1.	Production	1296
4.2.	Properties	1297
4.3.	Processing	1298
4.4.	Uses	1298
5.	Poly(Phenylene Sulfide)	1298
5.1.	Production	1299
5.2.	Properties	1301
5.2.1.	Material Properties	1301
5.2.2.	Properties of Nonreinforced Moldings	1302
5.2.3.	Properties of Reinforced Moldings	1303
5.3.	Processing	1305
5.3.1.	Injection Molding	1305
5.3.2.	Other Molding Methods	1305
5.3.3.	Finishing of Shaped Articles	1305
5.3.4.	Extrusion	1306
5.3.5.	Semifinished Products	1306
5.4.	Uses	1306
5.5.	Economic Aspects	1307
6.	Liquid Crystal Polymers	1307
6.1.	Introduction	1307
6.2.	Physical Properties	1311
6.2.1.	Mechanical Properties	1311
6.2.2.	Thermal Properties	1315
6.2.3.	Behavior on Exposure to Flame	1317
6.2.4.	Electrical Properties	1317
6.2.5.	Rheological Properties	1317
6.3.	Chemical Properties	1318
6.3.1.	Chemical Resistance	1318
6.3.2.	Hydrolysis Resistance	1318
6.4.	Production	1319
6.5.	Uses	1320
6.6.	Trade Names (Examples)	1321
6.7.	Economic Aspects	1321
	References	1321

1. Introduction

Definition. Any definition of a *high-temperature polymer* is necessarily arbitrary. Today's conventional engineering plastics [e.g., nylon, poly(ethylene terephthalate)] were yesterday's high-temperature polymers, and the quest for improved temperature performance is ongoing. In the context of this review, thermoplastic polymers exhibiting useful mechanical properties at temperatures in excess of 150 °C would seem an appropriate description.

Various other synonyms such as *high-performance polymers* or *advanced engineering polymers* have been employed. These materials are characterized by relatively low-volume sales compared with conventional engineering polymers and a relatively high selling price. Thus the ratio of the sales price for nylon to the sales price for the high-temperature polymers considered here varies from about 1 : 3 to 1 : 20. However, these polymers enjoy applications in varying market sectors such as the electrical–electronic, automotive, aerospace, and chemical process industries because they offer good performance to cost ratio.

The fact that the polymers can be used at high temperatures should not disguise that they possess many other useful and interesting properties. The crystalline polymers, polyetheretherketone and poly(phenylene sulfide), for example, find many room-temperature applications as a result of their good environmental resistance, particularly to solvents.

Classification. As with all polymers, high-temperature polymers fall into two classes, amorphous and (semi)crystalline. Polysulfone (PSU), poly(ethersulfone) (PES), and polyetherimide (PEI) are amorphous whereas poly-(phenylene sulfide) (PPS), polyetheretherketone (PEEK), and polyetherketone (PEK) are semi-crystalline.

Despite the high linearity of the sulfone polymers (100% 4,4'-specificity), the differences in the conformations of the phenyls bordering the sulfone and ether links preclude close packing of chains whereas the regular conformations of the phenyls in chains of PPS, PEEK, and PEK lend themselves to crystal formation. Interestingly, the ether and carbonyl moieties in PEEK are crystallographically equivalent.

Crystalline polymers (particularly filled grades–glass, carbon, mineral) retain useful mechanical properties above their glass transition temperature (T_g). Thus, although PEEK has a T_g of 143 °C its continuous service temperature (CST) is 250 °C. A further advantage of the crystalline range is inertness to aggressive media: PEEK shows outstanding resistance to aqueous acids, alkali, and organic solvents.

History [1]. Two synthetic routes lend themselves to the production of high-temperature polyaromatics: polycondensation (nucleophilic) or electrophilic processes. All of the polymers considered here may be made by polycondensation, and are currently manufactured by this process. Initially, electrophilic routes to PES (ICI) and various polyketones were developed but were quickly superseded by the polycondensation process, largely due to the lack of selectivity for linear polymers and the corrosive reactants used in the electrophilic process.

Raychem in 1972 launched their version of PEK (Stilan) produced via electrophilic chemistry, but this material was short-lived in the marketplace. Development of electrophilic processes continues, with recent announcements of polyketones by Du Pont (PEKK), and BASF (Ultra PEK).

Routes to PPS (Philips), PES (ICI), and PSU (Union Carbide, now produced by Amoco) were developed during the 1960s with commercial launch in the early 1970s. Similarly, polycondensation routes to PEEK (ICI), PEK (ICI), and PEI (General Electric, GE) were developed in the 1970s with launches in the 1980s.

To put the timescale of one of these developments into perspective, ICI research for routes to polyketones started in 1972. The first 20 g sample of PEEK was made in ICI laboratories at Welwyn Garden City in 1977, followed quickly by scaleup to semitechnical quantities. PEEK manufacture moved to full production scale at Hillhouse in late 1978 and became commercially available in 1981. A dedicated plant for PEEK production was brought on-line at Hillhouse in 1987.

One of the keys for success for polymers made by polycondensation is the use of high-purity monomers to guarantee the high degree of reactivity required (> 99%) for useful molecular masses, and the high degree of stereospecificity for properties such as toughness. Indeed, the history of the development of high-temperature polymers is mirrored by the development of their component monomers.

Stability. High-temperature thermoplastic polymers which have found commercial application have done so because they can be processed using conventional techniques such as extrusion or injection molding. However, for these materials to stand the processing temperatures employed they must exhibit stability against degradation. Typically an amorphous polymer is processed at 130–150 °C above its

T_g and a crystalline polymer at 30–50 °C above its crystalline melting temperature (T_m). Thus, PES (T_g 225 °C) and PEI (T_g 217 °C) are processed at 350–370 °C while PEEK (T_m 343 °C) is processed at 370–390 °C. Filled and high-viscosity grades may require even higher temperatures.

High temperature stability is attributable to high melting or softening points, high bond dissociation energies, structures not susceptible to chain (unzipping) degradation reactions, lack of inter- or intrabond formation, and low chemical (particularly to oxygen) reactivity. Resistance to chain degradation is perhaps the most important and is most easily provided by replacing aliphatic with aromatic units. Thus although aliphatic polymers have high bond dissociation energies and the first step in degradation is a high-energy process, it yields highly reactive, free-radical species which "unzip" the chain. Unzipping is precluded in aromatic polymers because any free radicals generated are stabilized by delocalization via the π system. Furthermore, volatile fragments cannot be readily formed.

Clearly, the totally aromatic polyparaphenylene should therefore have optimum stability. It is indeed stable to above 500 °C. However, extreme chain stiffness results in an intractable material. In order to take advantage of aromatic stability but provide processability, flexibilizing link units are incorporated into the chain [O,S,C(CH$_3$)$_2$]. Other, relatively inert moieties may be used to replace the aromatic ring (e.g., SO$_2$, CO). The balancing of these groupings gives the polymers their characteristic properties.

2. Polyetherimides

In 1982 General Electric Plastics introduced the first and still only thermally processible polyetherimide (PEI) resin under the trade name Ultem [61128-24-3]. It was the result of a ten-year development effort headed by J. WIRTH of General Electric's Center of Research and Development in Schenectady (USA). The idea behind this development was to maintain the already known thermal and chemical stability associated with the aromatic ether and imide structures while building a polymer chain flexible enough to provide melt processability within the limits of the melt stability.

Polyimides are condensation polymers derived from bifunctional carboxylic anhydrides and primary diamines. Polyetherimides form a special class within these polyimides as they contain an ether group in either or both the aromatic dianhydride or aromatic diamine building block. A number of possible monomeric units are shown in Table 1. To date, Ultem produced from bisphenol A bisether dianhydride is the only commercially available thermoplastic polyetherimide. The synthesis thereof forms the basis for the development of these PEI resins [2, 3].

2.1. Production

Polyetherimides are synthesized by melt polycondensation of bisphenol A dianhydride with a diamine, usually m-phenylenediamine [4]. The starting dianhydride for PEI is produced as follows. First, an N-alkyl phthalimide is nitrated:

The NO$_2$ groups are then nucleophilically displaced by bisphenolate A (BPA) ions:

Bisphenol A (BPA) salt

BPA-diimide

The resulting BPA diimide can be converted into the desired BPA dianhydride by an

Table 1. Possible monomers in polyetherimides

Name	Structure
Dianhydrides	
Hydroquinone bis(3,4-dicarboxyphenyl)ether dianhydride	
Bis(3,4-dicarboxyphenyl)ether dianhydride	
Bisphenol A bis(3,4-dicarboxyphenyl)ether dianhydride	
Diamines	
4,4′Diaminodiphenyl ether	
1,4-Bis(4-aminophenoxy)benzene	
1,3-Bis(4-aminophenoxy)benzene	

exchange reaction with phthalic anhydride:

BPA-diimide + 2 Phthalic anhydride → BPA-dianhydride

The final polycondensation of the BPA dianhydride with an aromatic diamine (e.g., m-phenylenediamine) can be performed in a number of ways, ultimately resulting in the formation of PEI:

BPA-dianhydride + H_2N—⬡—NH_2 → Polyetherimide

m-Phenylenediamine

PEI resins with a number-average molecular mass of 15 000–17 000 are usually produced.

The number-average molecular mass can be adjusted with monofunctional phthalic anhydride [5, 6]. Commercial PEI grades are tuned to the specific needs of an application and are obtained in three ways:

1. Use of additives, such as reinforcing fibers, mineral fillers, conductive fibers, pigments, polytetrafluoroethylene (to reduce frictional resistance), release agents, and flow improvers.
2. Modification of the base resin by copolycondensation with other monomers such as *p*-phenylenediamine and pyromellitic anhydride. This allows for higher glass transition temperatures and/or changes in chemical resistance.
3. Blending with other polymers (e.g., polycarbonate) improves processability and impact resistance.

2.2. Properties

The structures of Ultem and Kapton, a polyetherimide produced by Du Pont (available as "Kapton polyimide film"), are shown in Figure 1 [7–9]. The segmental mainchain rotational freedom in Ultem (shown by arrows) provides this PEI molecule with many more configurational possibilities than the more symmetrical Kapton. Ultem therefore has a lower glass transition temperature (217 °C) than Kapton (360 °C) and can be more readily melt processed. The thermal stability of both polymers is almost the same (± 500 °C under nitrogen) [10]. Properties of some Ultem grades are listed in Table 2.

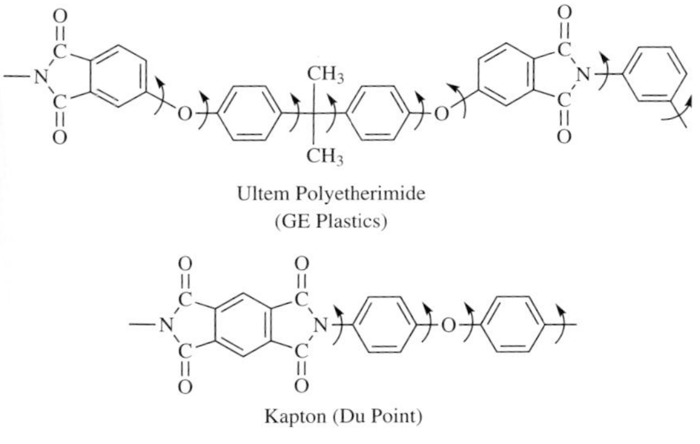

Mechanical Properties. Ultem has a very high tensile strength of about 110 MPa compared to, for example, polycarbonate (60 MPa), as well as a long-term, high-load, high-temperature dimensional stability. This is the combined effect of a high glass transition temperature, a high tensile modulus, a low thermal expansion coefficient, low creep, and a low water absorption. Stress–strain behavior is shown in Figure 2.

The relatively constant mechanical property profile of the PEI grades as a function of glass fiber content over a wide temperature range (up to 175 °C) is a direct result of the high glass transition temperature (217 °C) of this amorphous resin (Fig. 3). The excellent adhesion of PEI resins to glass fibers gives a linear relationship between the flexural modulus or tensile strength and glass fiber content up to 30 wt% [11].

Fatigue properties are excellent. The endurance (stress) limit at 10^6–10^8 cycles for PEI reinforced with 30% glass fibers is 30 MPa. Isochronous stress–strain diagrams from which the creep performance up to 100 000 h can be extrapolated, indicates a good resistance of PEI against creep (Fig. 4).

Aging characteristics (UL 94 temperature index) indicate a continuous service life up to 100 000 h at 175 °C.

Impact Behavior. The unfilled resin has excellent ductility due to the large number of possible segmental rotations and the high entanglement density that results from the kinked structure of the macromolecular chain. Its rather high notch sensitivity can be overcome by making sure that parts made of PEI have wellrounded edges.

Chemical Resistance. Environmental stress crazing and cracking are typical problems of amorphous engineering plastics, including to a lesser extent also PEI resins. Contact and subsequent interaction between the polymer and a liquid may result in dissolution or swelling, depending on the strength of the molecular interaction forces between the polymer and the liquid molecules. These forces are

Figure 1. Polyetherimide (PEI) structures

Table 2. Properties of Ultem PEI

Property	Standard	Ultem grade (x_1, x_2, x_3, x_4)*					
		1000	1010	2100, 2110	2200, 2210	2300, 2310	2400, 2410
Mechanical							
Tensile strength, yield, MPa	ISO 527	105	105	120			186
Tensile strength, break, MPa	ISO 527	90	90	115	140	160	
Tensile elongation, yield, %	ISO 527	8	8	5			
Tensile elongation, break, %	ISO 527	60	60	6	3	3	2
Tensile modulus, GPa	ISO 527	3.0	3.0	4.5	6.9	9.0	11.7
Flexural strength, yield, MPa	ISO 178	145	145	200	210	230	250
Flexural strength, break, MPa	ISO 178			4.5	6.2	9.0	11.7
Flexural modulus, GPa	ISO 178	3.3	3.3	M 109	M 114	M 114	M 125
Hardness, Rockwell		M 109	M 109				
Izod impact strength notched, J/m	ISO 180/1A	50	30	60	90	100	105
Thermal							
Vicat softening temperature, rate B, °C	ISO 306	219	219	223	226	228	234
Vicat VST/B/120, °C	ISO 306	215		222	224	226	
DTUL** (0.45 MPa), °C	ISO 75	210	207	210	210	212	216
DTUL** (1.82 MPa), °C	ISO 75	200	197	207	209	210	213
Thermal conductivity, W m^{-1} °C^{-1}	DIN 52 612	0.22	0.22				
Coefficient of linear expansion, flow, mm·mm^{-1} °C^{-1}	DIN 53 752	5.6×10^{-5}	5.6×10^{-5}	3.2×10^{-5}	2.5×10^{-5}	2.0×10^{-5}	1.4×10^{-5}
Physical							
Density, g/cm^3	ISO 1183	1.27	1.27	1.34	1.42	1.51	1.61
Water absorption, 24 h, %	DIN 53 495	0.25	0.25	0.28	0.26	0.18	0.12
Water absorption, equilibrium, %	DIN 53 495	1.25	1.25	1.00	1.00	0.90	0.90
Mold shrinkage, flow, %	ASTM D955	0.7	0.7	0.5–0.6	0.3–0.6	0.2–0.4	0.1–0.3
Electrical							
Volume resistivity, Ω·m	IEC 93	6.7×10^{15}	6.7×10^{15}	1.0×10^{15}	7.0×10^{14}	3.0×10^{14}	
Dissipation factor (2450 MHz)	IEC 250	0.0025	0.0025	0.0046	0.0049	0.0053	
Comparative tracking index, V	IEC 112	100–175	100–175	100–175	100–175	100–175	100–175
Flame characteristics							
Oxygen index, %	ISO 4589	47	44	47	50	50	54
Flame retardancy (mm thickness)	UL 94	V-0 (1.60) 5 V (1.90)	V-0 (0.71)	V-0 (0.41)	V-0 (0.41)	V-0 (0.25)	V-0 (0.25)

*The first three digits (x_1, x_2, x_3) have the following meanings: $x_1=1$ unfilled resin; $x_1=2$ glass-fiber-reinforced resin; $x_2=1$, $x_2=2$, $x_2=3$, glass fiber content=10, 20, and 30%, respectively; $x_3=0$ standard resin; $x_3=1$ easy flow. Ultem is a registered trade mark of GE Plastics.

**DTUL=deformation temperature ultimate limit.

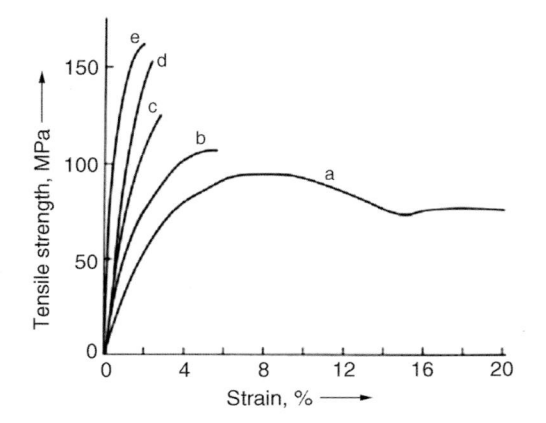

Figure 2. Stress–strain curves of unmodified PEI and glass reinforced PEI
a) Unmodified PEI; b) 10% glass-reinforced PEI; c) 20% glass-reinforced PEI; d) 30% glass-reinforced PEI; e) 40% glass-reinforced PEI

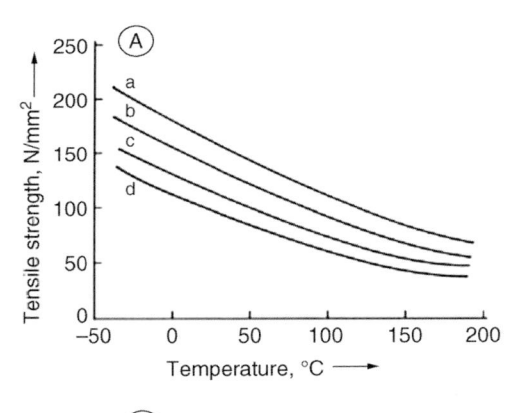

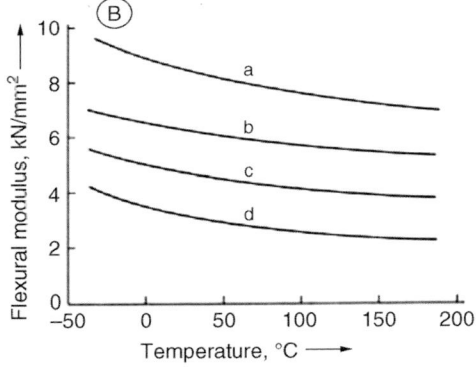

Figure 3. Tensile strength (A) and flexural modulus (B) of nonreinforced and glass-reinforced PEI
a) Ultem 2300 (30% glass-reinforced); b) Ultem 2200 (20% glass-reinforced); c) Ultem 2100 (10% glass-reinforced); d) Ultem 1000 (nonreinforced)

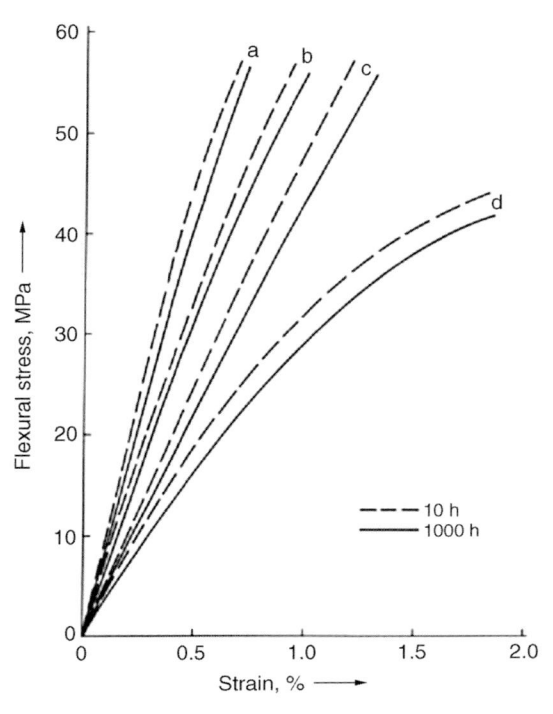

Figure 4. Isochronous stress–strain diagram of PEI (Ultem) at 23 °C
a) 30% glass-reinforced; b) 20% glass-reinforced; c) 10% glass-reinforced; d) Unfilled

characteristic for each combination of liquid and amorphous polymer and are determined by the differences in solubility parameter δ, defined as

$$\Delta = \sqrt{(\text{cohesive energy}/\text{molar volume})}$$
$$= \sqrt{\text{cohesive energy density}}$$

For PEI δ is 9.3–11.7 $(\text{cal/cm}^3)^{0.5}$ [12]. Its behavior toward contact with liquids of different solubility parameters is shown in Figure 5. If the difference between the solubility parameters of PEI and the liquid is larger than 5, i.e. $|\delta_{\text{PEI}} - \delta_{\text{liquid}}| > 5$ $(\text{cal/cm}^3)^{0.5}$, then no stress cracking problems are to be expected. However, if $|\delta_{\text{PEI}} - \delta_{\text{liquid}}| < 5$ $(\text{cal/cm}^3)^{0.5}$, swelling of PEI is possible and stress cracking may result from load, leading to actual strain which in time may result in crack problems. Moreover, if $\delta_{\text{PEI}} \approx \delta_{\text{liquid}}$, the

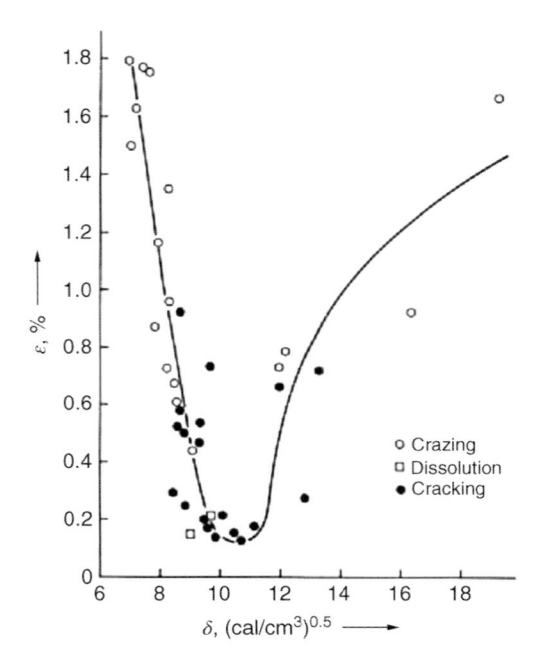

Figure 5. Dependence of critical strain ϵ_c for crazing, cracking, and dissolution of PEI versus liquid solubility parameter δ

polymer dissolves and the part disintegrates. The chemical resistance of PEI towards several solvents is summarized in Figure 6.

Flammability. The flammability behavior of Ultem as characterized by a high limiting oxygen index (LOI) of 47 results from the highly aromatic structure with a hydrogen/carbon molar ratio of 0.64. Selfsustained burning requires an oxygen rich atmosphere of over 45 vol%, making Ultem a selfextinguishing engineering plastic [13]. Therefore PEI resins fulfil the most severe flammability requirements stipulated by various countries for different industrial segments and part thicknesses:

1. Electrical and electronics industry: Rating V-0 at 0.64 mm wall thickness according to UL 94 Rating 5 V at 1.5 mm wall thickness according to UL 94
2. Building and construction industry: Class B-1 at 1.5 mm according to DIN 4102 (Germany) M 1 rating at 1.5 mm according to NFP 92–501 (France) Class 0 at 1.5 mm according to BS 476, parts 6 and 7 (UK)
3. Aircraft interior construction Passed FAR 25.853 at 1.5 mm (Federal Aviation Regulations, USA)

Requirements with respect to heat, smoke, and toxic gas generation during burning can also be met. Both the heat release rate criterion of ≤ 65 W at a 65 W radiation energy input and the NBS smoke density criteria (measured according to ASTM D 662) can easily be complied with.

Electrical properties of unfilled PEI and PEI reinforced with 30% glass fiber are presented in Table 3. The dielectric constant ϵ has a nearly constant value over the entire service temperature and frequency range. The temperature and frequency dependence of the dissipation factor (Fig. 7) are much lower for PEI than for other engineering resins. The values at 915 and 2450 MHz are of particular interest since these are the most well-developed microwave bands for power applications. The low value (in an

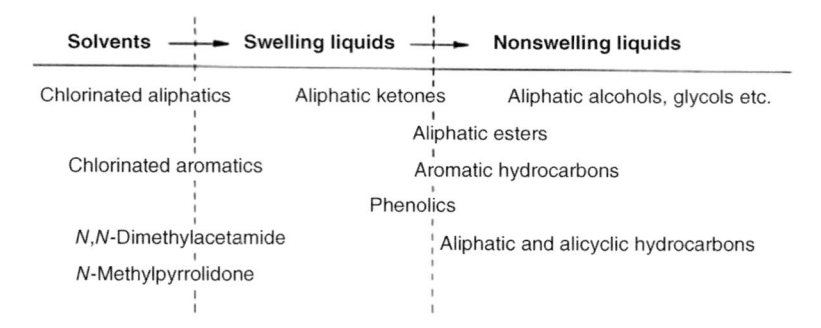

Solvents	Swelling liquids	Nonswelling liquids
Chlorinated aliphatics	Aliphatic ketones	Aliphatic alcohols, glycols etc.
		Aliphatic esters
Chlorinated aromatics		Aromatic hydrocarbons
	Phenolics	
N,N-Dimethylacetamide		Aliphatic and alicyclic hydrocarbons
N-Methylpyrrolidone		

Figure 6. Chemical resistance of polyetherimides
Neutral basic, acidic and/or oxidative aqueous systems have no effect with the exception of aqueous amine and ammonia which cause hydrolytic breakdown of PEI resins.

Table 3. Electrical properties of PEI

Property	Test	Unfilled	30% reinforced with glass fiber
Dielectric strength, V/mm	D 149	2840	2520
Dielectric constant ϵ (1 Hz)	D 150	3.15	3.7
Dissipation factor (1 kHz, 50% R.H.)	D 150	0.0013	0.0015
Volume resistivity, $\Omega \cdot$ cm	D 257	6.7×10^{17}	3×10^{16}

absolute sense) ensures that dielectric losses are minimal, especially at high temperatures.

Weather and radiation resistance of PEI parts is excellent. A 1000-h weatherometer test had no effect on mechanical properties.

2.3. Processing

The rheological characteristics of PEI resins allow them to be subjected to all thermoplastic processing techniques, i.e., injection (blow) molding, extrusion, structural foam molding, and compression molding [14]. Processing should be based on guidelines applicable to amorphous engineering plastics, but with appropriate temperature adjustments. The high glass transition temperature (217 °C) requires a melt temperature typically in the range 360–425 °C. Clean, well-processed regrind can be recycled in the feed in amounts of 20–50%. PEI resins absorb water and have to be extensively pre-dried before they are melt processed to prevent severe surface splay upon molding or extrusion.

Heating for at least 4 h at 150 °C is required to lower the water content to the desired level of 0.03%.

Although processing fumes are negligible, proper ventilation is advised. Purging of the processing equipment can be achieved with high molecular mass polyethylene in combination with a gradual lowering of the melt temperature from 400 to 200 °C.

2.4. Uses

Commercial PEI grades are used in several industrial segments, of which the most important are reviewed below [15].

The *automotive industry* uses PEI mainly in under-the-hood applications, where long-term mechanical strength and dimensional stability are required at high service temperatures in combination with resistance to oil, grease, brake- and cooling fluids. Examples are sensor housings, oil pump parts, ball bearing cages, and bearings.

PEI can easily be metallized and is therefore used in head-lamp reflectors, especially in combination with energy-intensive halogen lamps.

The *lighting industry* has similar requirements to those mentioned above for head-lamp reflectors. A special feature of PEI is its IR transparency. Thus unfilled, unpigmented PEI reflectors made by the recently developed dichroic layer technology can produce "cool" light. Heat is not concentrated by the reflector but diffuses through it. This principle is of importance in lighting systems employed in operating theaters, art exhibitions, etc. The dichroic coating technique uses the different refraction of various wavelengths at interfaces and its dependence on layer thickness to separate the IR and the visible part of the emitted light. The visible part of the spectrum is

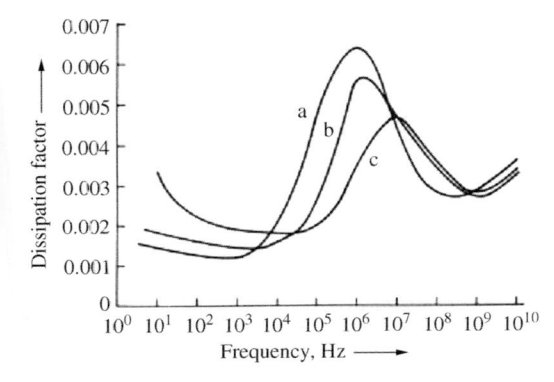

Figure 7. Dissipation factor of PEI unmodified resin versus frequency at 50% relative humidity (ASTM D 150)
a) 23 °C; b) 49 °C; c) 82 °C

reflected, whereas the IR part diffuses through the PEI.

The *electrical and electronic industries* make use of PEI's intrinsic flame resistance and its excellent dielectric properties. Notable is the development of two- and three-dimensional injection-molded, printed circuit board structures where the design allows integration of mechanical and circuitry functionality. A special, newly developed, metal-plating technique for PEI makes this even more interesting [16]. Other applications include circuit boards, connectors, switches, and controls. PEI's low tracking resistance (comparative tracking index=145 V, according to IEC standards) makes it unsuitable for high-voltage parts that carry direct current.

Polyetherimides are also used in the *food processing and beverage industries* on account of their high heat resistance, favorable mechanical properties, good chemical resistance to and cleanability (sterilization) of contact liquids, as well as lack of toxicity and disadvantageous effects in taste and odor. Specially developed PEI grades fulfil these requirements and have been formally approved by most European health agencies (e.g., Bundesgesundheitsamt in Germany) and the FDA in the United States. Polyetherimides are used in liquid-handling equipment, pumps, high-pressure dispensers, coffee makers, kitchen utensils, microwave trays, dishes and cups, cheese molds, and lightweight unbreakable baking forms. The food processing industry is becoming one of the most important volume markets for PEI.

PEI parts are also utilized in *household appliances*, e.g., water reservoirs for steam irons, hair dryers, doors of microwave ovens, and coffee-maker reservoirs.

Medical equipment often has to be sterilized and must be inert to body fluids. Here, parts made of PEI have a good performance and can be sterilized by all conventional methods (e.g., use of steam, irradiation, and chemicals). The intrinsic transparency of the material is also advantageous. Examples of applications in the medical equipment sector include surgical staplers and sterilization tanks.

The *aircraft industry* is also interested in materials that offer cost and weight savings without sacrificing safety (especially in relation to flammability issues). Components made of PEI with a high volume fraction of glass fibers or other high modulus fiber provide an excellent substitute for metal (especially aluminum) parts. The injection moldability of PEI resins allows the economic production of complex parts and multifunctional integration. High-performance composite sheets with PEI as the binding resin are used in monolithic and laminated structural parts, such as interior partition walls and floor elements. Ducts and casings are also made from PEI composites. The composites have a thermoplastic matrix and can easily be thermoformed into complex shapes.

2.5. Economic Aspects

The commercial development of PEI has been very successful. Since the introduction of the first commercial Ultem grade in 1982, annual worldwide sales volume exceeded 5000 t in 1990. Future growth is expected to increase because applications are expanding and further market development activities are underway. Potential large-volume markets in composites and in household and industrial appliances are emerging. Extra market growth may be expected from developments in high-speed trains, where the same composite materials can be used as in the aircraft industry due to the emphasis on fire safety issues. Possible markets for the compatible PEI–polycarbonate blends and the completely miscible PEI–polyetherketone or PEI–polyetheretherketone blends have not yet been fully exploited [17]. Moreover, the product range of these materials will be further enlarged by simple modification of the PEI structure through copolymerization, leading to enhanced inherent properties. An example of the flexibility offered by chemical modification of the polymer main chain is siloxane–etherimide block copolymers (PSEI) [18]. Siloxane segments provide not only flexibility, toughness, and reduced moisture absorption, but also specific surface properties and lower dielectric constant. The length of the siloxane blocks is the key to these properties [18]. PSEI copolymers have recently been commercialized by GE Plastics under the trade name Siltem as high-performance coating materials for electrical wires. They are also used with Ultem in aircraft interior applications where they improved the impact performance of PEI, while maintaining its flame retardancy and heat resistance. Applications of PEI do not pose any environmental issues. It is possible to recycle PEI, provided the material has been well processed and does not contain impurities.

3. Polyaryletherketones

The first reports of syntheses of polyaryletherketones (PAEKs) appeared in the early 1960s [19, 20], and the syntheses were based on electrophilic substitution reactions, essentially Friedel–Crafts acylations. The problem of limited solubility of the crystalline polymers was solved by conducting the polymerization in strongly acidic media, initially polyphosphoric acid [21] and later HF/BF$_3$ [22], which dissolved the polymers by protonation of their carbonyl groups. Despite the fact that high-molecular-mass polymers with useful mechanical properties could be produced, production problems associated with the highly aggressive reaction medium and problems of obtaining pure, stable polymers retarded the full commercialization of PAEKs.

Poly(oxy-1,4-phenylenecarbonyl-1,4-phenylene) [27380-27-4] (polyetherketone, PEK), synthesized by polyacylation in HF/BF$_3$, was available on a limited scale from Raychem for a short period in the mid 1970s as "Stilan".

Polyetherketone PEK

The first synthesis of PAEKs by nucleophilic displacement was reported by JOHNSON et al. [23] (Union Carbide) in the late 1960s, but the molecular masses of the polymers were low because of crystallization resulting in premature precipitation; mechanical properties were consequently poor.

The key breakthrough in the synthesis of PAEKs by nucleophilic displacement was achieved when ROSE et al. [24, 25] (ICI) identified diphenyl sulfone as a suitable solvent for the potentially highly crystalline, high-molecular-mass polymers. Nucleophilic displacement polymerization in this solvent formed the basis of the large-scale industrial route to poly(oxy-1,4-phenyleneoxy-1,4-phenylenecarbonyl-1,4-phenylene) [31694-16-3] (polyetheretherketone, PEEK), the first PAEK to be fully commercialized. PEEK has been commercially available from ICI since the early 1980s and PEK, synthesized by a similar nucleophilic route, since 1987.

Polyethetherketone PEEK

In the 1980s and 1990s, Hoechst tried to enter the PAEK market with poly(oxy-1,4-phenyleneoxy-1,4-phenylenecarbonyl-1,4-phenylenecarbonyl-1,4-phenylene) (polyetheretherketone ketone, PEEKK, Hostatec) produced by the nucleophilic route, and BASF with polyetherketone ether ketone ketone (PEKEKK, Ultrapek) produced by the electrophilic route, but they failed to gain substantial market share and, therefore, left the field of PAEK.

In 1993, ICI sold its Victrex PEEK business to a management buyout group, which established Victrex plc. Since then, Victrex has become the major manufacturer of PAEK resins, producing Victrex PEEK, Victrex HT (PEK), and Victrex ST (PEKEKK).

In 2005, Degussa (after 2007 Evonik Degussa) formed a joint venture with Jilin University in China for the production of PEEK (VESTAKEEP). Gharda Chemicals in India produced PEEK, based on a novel electrophilic route. This business was acquired by Solvay in 2005, which built a plant in India for the production of PEEK by the nucleophilic route (KetaSpire).

Rallis in India is currently (2010) the main manufacturer of PEKK and supplies material to Cytec for aerospace applications and thermoplastic composites. PEKK, being used for various industrial and medical applications, is sold by Oxford Performance Materials, which was acquired by Arkema in early 2009.

3.1. Production

The vast majority of PAEKs is produced by methods based on nucleophilic displacement reactions.

PEEK. The commercially significant PAEK, PEEK, is prepared by reacting 1,4-benzenediol (hydroquinone) with 4,4′-difluorobenzophenone in diphenyl sulfone in the presence of alkali-metal carbonates under an inert atmosphere at temperatures approaching the melting point of the polymer (> 300 °C) [26].

$$n \text{ HO-C}_6\text{H}_4\text{-OH} + n \text{ F-C}_6\text{H}_4\text{-CO-C}_6\text{H}_4\text{-F} \xrightarrow[\text{Diphenyl sulfone}]{n \text{ Na}_2\text{CO}_3} [\text{-O-C}_6\text{H}_4\text{-O-C}_6\text{H}_4\text{-CO-C}_6\text{H}_4\text{-}]_n + n \text{ CO}_2 + n \text{ H}_2\text{O} + 2n \text{ NaF}$$

The hydroquinone is deprotonated in situ by the alkali-metal carbonate to give a hydroquinone salt. The hydroquinone anion then attacks the 4,4′-difluorobenzophenone in a nucleophilic manner, replacing the fluorine atom and leading to the formation of an ether bond.

The polymer is isolated from the diphenyl sulfone and inorganic impurities, which can destabilize the polymer melt, by solvent extraction (leaching). A slight excess of 4,4′-difluorobenzophenone is normally employed to prevent the formation of phenoxide chain ends, which can destabilize the polymer. The water formed as a result of polycondensation could potentially hydrolyze the fluoro groups and prevent formation of the high-molecular-mass polymer but is lost because of the high reaction temperatures and the hydrophobicity of the solvent.

The expensive difluoro monomer is employed in the reaction rather than the less expensive dichloro analog because the weakly electron-withdrawing carbonyl group does not strongly activate the halogen for a nucleophilic displacement reaction and, consequently, a good leaving group is necessary to make reaction efficient. Nevertheless, some attempts have been made to use the less expensive chloro monomers for the production of PEEK [27, 28] but a relevant industrial process using the chloro monomers has not been established so far.

An electrophilic route for the production of PEEK was developed by Gharda [29]. The process is based on the reaction of phenoxy phenoxy benzoic acid in alkane sulfonic acid (e.
g., methane sulfonic acid) with a condensating agent like thionyl chloride.

PEK can be produced by two nucleophilic routes [25] which are essentially similar to the route to PEEK. The two-monomer route employs 4,4′-dihydroxybenzophenone in place of 1,4-benzenediol. The single-monomer route employs the potassium salt of 4-fluoro-4′-hydroxybenzophenone. The latter route has the greatest tendency to exhibit side reactions leading to branching (via carbanion formation due to abstraction of a proton *ortho* to the fluoro group of the monomer), which results in polymers with inferior mechanical properties. The two-monomer route is, therefore, usually preferred.

PEK has been produced by electrophilic acylation [30]. The most convenient synthesis involves the Friedel–Crafts polycondensation of 4-phenoxybenzoyl chloride. The preferred reaction medium of HF/BF$_3$ necessitates the use of special polytetrafluoroethylene reaction vessels. This electrophilic route employs relatively cheap nonfluoro substituted aromatic monomers. If the problems associated with the aggressive reaction medium can be solved, this route may prove to be the cheapest route to PEK.

Gharda Chemicals developed a route for production of PEK using the low-cost chloro monomer 4-chloro-4′-hydroxybenzophenone in the presence of potassium carbonate [31]. The monomer is prepared by reaction of 4-chlorobenzoylchloride with phenol in the presence of AlCl$_3$ as a catalyst.

PEKEKK is produced similar to PEEK and PEK (two-monomer route) by a nucleophilic displacement reaction of 1,4-bis(4-fluorobenzoyl)benzene with 4,4′-dihydroxybenzophenone.

PEKK is produced by the electrophilic route using diphenyl ether and terephthaloyl chloride and isophthaloyl chloride, and AlCl$_3$ as a catalyst [32]; the incorporation of the isophthaloyl groups reduces the crystallinity.

Although the nucleophilic route is also possible here, it is not the preferred one because of the complexity of the required monomers, 1,4-bis(4-fluorobenzoyl)benzene and 1,4-bis(4-hydroxybenzoyl)benzene.

PAEKs with high glass-transition temperatures. Since the mid 2000s there has been increased interest in PAEKs with glass-transition temperatures (T_g) higher than those of PEKK due to the increasing challenges of oil and gas extraction. The exploitation of offshore oil and gas reserves requires the use of materials that have outstanding mechanical and chemical properties at high pressures (>150 MPa), high temperatures (>200 °C), and chemically aggressive environments (hydrogen sulfide, brine, crude oil).

The glass-transition temperature can be increased by the incorporation of sulfone groups into the backbone of the polymer chain. Various sulfone-containing monomers can be used, ranging from dichlorodiphenyl sulfone to monomers especially designed to maximize the increase in T_g while minimizing the effect on crystallinity [33].

The reaction of 4,4′-difluorobenzophenone with 4-(4-hydroxyphenyl)phthalazinone in a polar solvent in the presence of potassium carbonate yields an amorphous PAEK with a high T_g [34]. This attempt can be further extended by copolymerizing the aforementioned monomers with 4,4′-biphenol, which results in a melt-processible semicrystalline PAEK with T_g ranging from 180–240 °C and melting temperatures (T_m) ranging from 310–376 °C, depending on the monomer ratios [35].

3.2. Properties

Physicochemical Properties. Most wholly aromatic PAEKs are semicrystalline because of the strong interchain forces between the polar groups in the backbone and the crystallographic near-equivalence of the ether and carbonyl units [36]. Many of the desirable properties of commercial PAEKs are related to their tendency to develop a high degree of crystallinity under normal processing conditions.

The semicrystalline morphology of PAEKs is largely responsible for their excellent resistance to solvent attack. For example, PEEK is insoluble in all common solvents. It is soluble at room temperature only in strongly acidic media (e.g., concentrated sulfuric acid, hydrofluoric acid, and trifluoromethanesulfonic acid) that can protonate the carbonyl groups. At temperatures well above the glass-transition temperature the polymer is soluble in a few unusual media (e.g., diphenyl sulfone, some high-boiling-point esters, benzophenone, and 1-chloronaphthalene). PEEK absorbs small amounts of water [37] but can absorb significant amounts of dichloromethane and related solvents that contain a hydrogen atom activated by neighboring electronegative groups [37, 38].

The insolubility of PAEKs causes problems in the molecular-mass characterization of the polymers. Viscometry and light-scattering methods using concentrated sulfuric acid and a high-temperature gel-permeation-chromatography method have been developed [39]. The number-average molecular mass of commercial PEEK grades is typically in the order of tens of thousands.

The essentially linear aromatic backbone of PEEK results in excellent resistance to chemical attack and ionizing radiation. The only common reagents that degrade PEEK are chlorine, bromine, and concentrated nitric acid [40]. Dissolution of PAEKs in concentrated sulfuric acid results in monosulfonation of the aromatic rings substituted by two ether linkages [41].

PAEKs display limited resistance to degradation by UV light, as is the case for all linear polyaromatics, but the problem is only serious in high-insulation environments (i.e., prolonged exposure to intense sunlight) and is much reduced by incorporation of a suitable pigment (e.g., carbon black).

PAEKs exhibit excellent resistance to gamma radiation due to the delocalized bonding electrons in the aromatic rings.

Thermal Properties. Differential scanning calorimetry studies of quenched amorphous PEEK indicate a T_g at 143 °C and a melting endotherm with a maximum at 334 °C [42]. Increasing the ratio of carbonyl-to-ether linkages in polyarylether backbones increases both T_g and T_m; PEK, therefore, has a T_g at 154 °C and T_m at

367 °C, and PEKK has a T_g at 165 °C and T_m at 386 °C [43, 44].

Isothermal crystallization studies indicate that PEEK crystallizes very rapidly [42] and normal processing conditions generally result in degrees of crystallinity of 25–35%.

The thermal stability of PAEKs is excellent. PEEK melts are stable for > 1 h at 400 °C. Thermoxidative stability is lower with prolonged heating above 300 °C, which results in cross-linking. The presence of copper impurities significantly accelerates the thermal degradation of PEEK melts [45].

Mechanical Properties. The high crystallinity rapidly developed by PEEK and PEK results in very high flexural moduli and tensile strengths. These properties decrease substantially above the T_g but useful residual values are maintained up to temperatures over 250 °C, which is close to the melting temperature of the crystalline regions, this is reflected by the maximum service temperature of the polymers. PAEKs with higher T_g and T_m show improved high-temperature retention of their properties.

PAEKs are notch-sensitive like many thermoplastics, i.e., the presence of a stress-raising feature lowers the otherwise acceptable impact strength.

Selected mechanical properties of various PAEKs are given in Table 4.

Flammability. PAEKs have low hydrogen-to-carbon ratios and do not produce large amounts of combustible volatiles; therefore, they are difficult to ignite. Moreover, PAEKs have very low smoke emissions and burn in excess oxygen producing only nontoxic gases.

Electrical Properties. PAEKs have a high volume resistivity (i.e., they are good insulators) and a low dissipation factor for electromagnetic radiation. Both properties are stable over a wide temperature range up to the T_g.

Selected electrical properties of various PAEKs are listed in Table 4.

3.3. Processing

The thermal stability of commercial PAEKs allows melt processing in conventional equipment [50] capable of reaching the required processing temperatures (approx. 40–60 °C above T_m). This stability also allows reworking and recycling of PAEKs.

PAEKs can be injection- and compression-molded; extruded into amorphous or crystalline film, sheet, and mono- or multifilament fiber; powder- and dispersion-coated; blow-molded; used for thermoplastic composites; welded; metallized, adhesively bonded; and machined. A recent processing technology to PAEKs is laser sintering [51], in which a focused laser beam melts PAEK powder layer by layer to form a solid object. Starting from construction data (e.g., computer-aided design data), a highly accurate manufacturing of objects having very complex shapes can be achieved.

Table 4. Selected properties of various PAEKs [46–49]

Property[*]	PEEK (VESTAKEEP 4000G)	PEK (Victrex HT G45)	PEKEKK (Victrex ST G45)	PEKK (OXPEKK-C)[*]
Tensile modulus, GPa (ISO 527)	3.5	3.7	4.3	4.4
Tensile strength, MPa (ISO 527)	96	115	115	110
Tensile elongation,% (ISO 527)	30	30	20	12
Charpy notched impact strength, kJ/m² (ISO 179)	7		4	
Izod notched impact strength, kJ/m² (ISO 180)		7		7
Heat deflection temperature A, °C, (1.8 MPa, ISO 75)	155	163	172	175
Dielectric strength, kV/mm (IEC 60243-1)	16	17	21	23
Volume resistivity, $\Omega \cdot cm$ (IEC 60093)	10^{15}	10^{16}	10^{16}	10^{16}
Dissipation factor (1 kHz, IEC 60250)	0.003			0.004

[*] ASTM data

3.4. Uses

PAEKs can be found in almost all industry segments and their use is generally connected with their outstanding high-temperature performance and chemical resistance.

Compounds and Blends. The properties of PAEKs can be modified by adding various high-temperature-stable additives or fillers (→ Plastics, Additives, Chap. 7). The most common fillers are glass fibers and carbon fibers for reinforcement. Other fillers such as graphite and polytetrafluoroethylene for lubrication, hard inorganic fillers for the improvement of the abrasion resistance, talc and mica for the reduction of shrinkage and warpage, and conductive fillers for the preparation of highly conductive grades or grades with electrostatic dissipation properties are used.

PEEK can be blended with various other polymers to improve some of the properties (→ Polymer Blends). Some common blend partners are polyarylethersulfones [52] for the reduction of warpage and cost, polyetherimide [53] for the increase of T_g and the reduction of the crystallization rate, polyphenylene sulfide [54] for cost reduction and the improvement of melt flow, and polybenzimidazoles [55] for the improvement of wear resistance at high temperatures.

Stock shapes (e.g., rods, sheets, tubes) are manufactured by extrusion or compression molding of powder and used in various industry segments for prototyping, small production runs, and large-volume applications.

Films (→ *Films*) can be produced by means of a conventional extruder; the crystallinity of the film is controlled by the temperature of the casting rolls.

PAEK films are used for different applications, e.g., seals, bearings, liners, heat exchangers, speakers, and membranes.

Fibers (→ *High-Performance FibersI*). PAEK monofilaments and hollow monofilaments, multifilaments, as well as staple fibers are available. The fibers are produced by extrusion and orientation and have high strength.

PAEK fibers are used for open-mesh conveyor systems, textile printing, food processing, aerospace components, thermoplastic composites, dry filtration, chemical separation, sports strings, braids, brushes, and cords.

Coatings. Cable coatings are obtained by extrusion of PAEK. Another coating process is the use of fine powders for powder coating applied by electrostatic powder spraying, flame spraying, or dispersion coating applied by dispersion spraying techniques.

PAEK coatings are applied where thin layers are sufficient, e.g., for the protection of other materials, especially metals.

Composites (→ *Composite Materials*). PAEK composites are mainly continuous carbon fiber reinforced and produced by proprietary impregnation processes. Fine powders with particle sizes down to 10 μm are available for the impregnation process.

PAEK composites are mainly used in the aerospace industry for bolts, brackets, fasteners, inserts, and nuts, as well as for structural objects.

Medical Applications (→ *Surgical Materials*). Medical devices like surgical instruments have been manufactured from PAEK for many years. In 1999 implantable PEEK grades were launched by Invibio Ltd.. Ten years later, PEEK has become one of the major polymers used for medical devices and implants, and implantable PEEK grades are available from all major PEEK producers.

Compared to metals used in implantology, PEEK offers various advantages: the tensile modulus is close to the one of bones (reduced stress shielding), metal ions are not released, artifacts in X-ray imaging do not occur, parts can be sterilized by gamma radiation, steam, and ethylene oxide, and complex shapes can be easily formed.

4. Polysulfones

Polysulfone is the generic term for all sulfone-containing polymers. Polysulfones are an important class of engineering thermoplastics. The polymer backbone consists of *para*-linked aromatic groups connected by ether, sulfone, and in some cases alkyl groups.

Development work on polysulfones was active by the early 1960s. Early syntheses were electrophilic, being in effect polysulfonylations [56]. This type of reaction did not generally produce a linear *para*-linked chain, and the introduction of *ortho* linkages and branch-points resulted in polymers with poor mechanical properties [57].

An alternative strategy for the synthesis of polysulfones based on nucleophilic aromatic substitution reactions [58, 59] proved more successful for the production of *para*-linked, essentially linear polymers with good mechanical and thermal properties. This route was subsequently employed in all large-scale commercial routes to polysulfones. By the late 1960s the first commercial polysulfone, poly(oxy-1,4-phenylenesulfonyl-1,4-phenyleneoxy-1,4-phenylene(1-methylethylidene)-1,4-phenylene) or bisphenol A polysulfone [25135-51-7] (Udel, Union Carbide) was available.

Udel

This was followed by the introduction in the early 1970s of the polyethersulfone poly(oxy-1,4-phenylenesulfonyl-1,4-phenylene) [25667-42-9] (Victrex, PES) by ICI and Astrel 360 by 3 M, a biphenyl containing polysulfone with a high glass transition temperature but poor processability.

Victrex PES

More recent developments include the introduction of polysulfones containing biphenyl groups in the repeat unit (Radel R polyphenylsulfone [68518-59-2], Union Carbide; Victrex HTA [121763-41-5], ICI) which have higher T_g values but retain melt processibility.

Radel R

HTA

The major manufacturers of polysulfone resins are Amoco, which acquired the Union Carbide polysulfone business (Udel, Radel A and R), and ICI (Victrex PES and Victrex HTA). BASF have recently marketed comparable polysulfones to the ICI and Amoco polymers but on a smaller scale (Ultrason S Polysulfone, Ultrason E Polyethersulfone).

Since drafts were first submitted for publication, ICI has announced that it is to withdraw from the manufacture and sale of Victrex PES and HTA. BASF have announced a capacity increase for their Ultrason products.

4.1. Production

Bisphenol A polysulfone (Udel) is synthesized by nucleophilic substitution of 4,4′-dichlorodiphenyl sulfone (DCDPS) by 4,4′-(1-methylethylidene)bisphenol (bisphenol A) in a dipolar aprotic solvent such as dimethyl sulfoxide (DMSO) [60]. The reaction is carried out in two stages. Firstly addition of two equivalents of aqueous NaOH or KOH to a solution of the bisphenol in a mixture of DMSO and azeotroping solvent (e.g., chlorobenzene) generates the more nucleophilic bisphenate salt. Excess water is then removed by azeotropic distillation at 120–140 °C before the dichloro monomer, which could be hydrolyzed by water and lead to a low molecular mass product, is added and polymerization proceeds. Polymerization is carried out at 130–160 °C under an inert atmosphere to prevent oxidation of the bisphenate salt. The dipolar aprotic solvent promotes very rapid rates of nucleophilic substitution and the molecular mass of the polymer can become excessively high unless it is controlled by the addition of monofunctional halides or phenols.

Radel R and Radel A (structure not disclosed) are probably also synthesized by this type of route but employing the appropriate bisphenol monomers.

The synthesis of Victrex PES is similar to the synthetic route to Udel with 4,4′-sulfonylbisphenol (bisphenol S) being condensed with DCDPS [61]. The lower solubility and reactivity of the bisphenol (nucleophilicity reduced by electron-withdrawing sulfone group para to the hydroxy group) necessitates a significantly higher reaction temperature (up to 285 °C). Consequently, the dipolar aprotic solvent used is diphenyl sulfone, phenate salts being notionally generated "in situ" by the action of K_2CO_3 or Na_2CO_3. The high reaction temperature and the hydrophobicity of the solvent result in the efficient removal of water that might otherwise hydrolyze the dichloro monomer. Azeotropic distillation is not necessary.

Victrex HTA is produced by an analogous route but with the DCDPS being replaced by bis (4-chlorophenylsulfonyl)biphenyl.

4.2. Properties

Physicochemical Properties. Key properties of Udel polysulfone and Victrex PES are listed in Table 5.

All commercial polysulfones are essentially amorphous as utilized. The unfilled polymers display good clarity.

The polysulfones are relatively polar and their solubility characteristics reflect this. Polysulfones are soluble to some extent in a variety of dipolar aprotic solvents (e.g., dimethyl sulfoxide, dimethylformamide, N-methylpyrrolidone) and certain chlorinated hydrocarbons (e.g., dichloromethane, chlorobenzene). Udel polysulfone is significantly soluble in low molecular mass ketones and some aromatic hydrocarbons [62]. Victrex PES and biphenyl-containing polysulfones are more solvent resistant than Udel polysulfone [63]. The polysulfones are crazed but not dissolved by solvents with polarities similar to that of the polymer and this tendency increases with stressed samples.

Polysulfones can absorb small amounts of water but this has a minimal adverse effect on dimensions and mechanical properties. Polysulfones are very resistant to hydrolysis, even by hot water or steam. They are stable to dilute acid and caustic but are degraded by concentrated nitric and sulfuric acids. Chlorine and bromine also degrade polysulfones.

Polysulfones show limited stability to UV light and the materials must be filled for outdoor applications. Resistance of the wholly aromatic polysulfones to ionizing radiation is excellent, Udel polysulfone has a lower resistance due to the presence of aliphatic groups.

The ready solubility of polysulfones allows the determination of average molecular masses and distributions by a variety of common techniques including gel permeation chromatography. Number-average molecular masses of commercial polysulfones are typically of the order of tens of thousands.

Thermal Properties. Commercial polysulfones cannot be crystallized from the melt or by annealing and are amorphous as utilized. The T_g tends therefore to be responsible for short-term temperature ceilings (e.g., heat deflection temperature) for applications of the polymers. Aromatic rings and sulfone groups introduce chain rigidity whereas ether links increase flexibility. Hence increasing the ratio of sulfone to ether links in the repeat unit increases the T_g. The presence of biphenyl groups (e.g., HTA and Radel R) also increases chain rigidity and hence T_g.

The thermal stability of polysulfones is excellent due to the strong bonds in the backbone and results in stable melts during processing below 400 °C. The aliphatic isopropylidene groups in Udel polysulfone significantly reduce the thermooxidative stability relative to the

Table 5. Properties of Udel polysulfone and PES

Properties[*]	Udel polysulfone P 1700	Victrex PES 4100 G
T_g, °C	185	223
Heat distortion temperature, °C (D 648)	174	203
Water absorption, % (D 570)	0.22	0.66
Tensile strength, MPa (D 638)	70	84
Flexural modulus, GPa (D 790)	2.7	2.6
Izod impact strength, J/m (D 256)		
Unnotched	no break	no break
Notched	69	83

[*] Appropriate ASTM test method given in parentheses.

wholly aromatic polysulfones and prevent prolonged high-temperature use close to T_g. Melt processing on conventional equipment is still possible however.

Commercial polysulfones possess selfextinguishing behavior. Despite the presence of sulfur-containing groups, they exhibit relatively low levels of smoke and toxic gas emission.

Mechanical Properties. The polar groups in the polysulfone chain result in a very high flexural modulus. Modulus decreases rapidly as T_g is approached. The impact strength of the unnotched polymers is also high, but the presence of molded or accidental notches dramatically reduces the impact strength. The notched impact strength of the biphenyl-containing polysulfones is significantly higher than that of other polysulfones.

Filling of the polymers with glass or carbon fiber substantially increases the modulus of the polysulfones but generally only increases the heat deflection temperature slightly.

Electrical Properties. The polysulfones have high resistivities and low dielectric constants and dissipation factors for electromagnetic radiation which are substantially retained up to the T_g of the polymer.

4.3. Processing [62, 64]

Commercial polysulfones are all melt processible on conventional equipment, the required processing temperatures increasing with the T_g of the polymer. The polysulfones are dried before processing to prevent the evolution of adsorbed moisture from lowering the quality of moldings. Annealing at a temperature close to the T_g of the polymer is recommended to prevent stress cracking of moldings.

4.4. Uses

The mechanical properties of Udel polysulfone and Victrex PES are similar. If enhanced high-temperature performance or chemical resistance is required Victrex PES is preferred. The biphenyl-containing polysulfones are significantly more expensive than Udel and Victrex PES, their use is limited to situations where extreme heat or impact resistance is required. Their use as films and foamed plastics are described elsewhere, see → Films, Section 6.10., → Films, Section 6.11., → Films, Section 6.12., → Films, Section 6.13. and → Foamed Plastics.

Udel and Victrex PES have been widely used to produce injection-molded articles, replacing metals in low- to medium-temperature applications where weight saving is important — particularly aerospace applications. Notch sensitivity limits the application of the polymers in critical load-bearing components.

The high-temperature hydrolytic stability of polysulfones has permitted widespread use in fluid-handling components and steam-sterilizable biomedical moldings. Chemical resistance enables a range of chemical process and automotive applications.

Good insulating properties up to high temperatures favor polysulfones as insulation for electrical cables and as a circuit board material (→ Insulation, Electric, → Insulation, Electric).

5. Poly(Phenylene Sulfide)

Poly(phenylene sulfide) [9016-75-5] (PPS), poly(thio-1,4-phenylene), is a high-temperature-stable, thermoplastic accessible from very cheap monomers. It has therefore seen continuously high growth rates since its commercialization in 1972. PPS consists of repeating aromatic monomer units joined by sulfur bridges:

$$\{Ar-S\}_n$$

Ar = aryl (—⟨phenyl⟩— for PPS)

Reactions between sulfur and aromatic hydrocarbons carried out by FRIEDEL and CRAFTS [66, 67] in 1888 and by GENVRESSE in 1897 [68, 69] resulted in cross-linked resins and low-molar-mass products but not well-defined linear polymers. In the late 1940s, MACALLUM developed a process for preparing poly(arylene sulfides), including PPS, by reacting dichlorinated aromatic compounds with sulfur. He also obtained partially cross-linked products [70].

In the early 1960s a process was developed by Dow Chemicals which involved copper-induced self-condensation of aromatic halothiophenols and resulted in an experimental product [71, 72].

In 1963 EDMONDS and HILL developed a process, eventually commercialized in 1972 by Phillips Petroleum, in which PPS was prepared from *para*-dichlorobenzene (*p*-DCB) with sodium sulfide in *N*-methylpyrrolidone (NMP) as a suitable polar aprotic solvent [73].

$$n\,Cl-\langle\rangle-Cl + n\,Na_2S \longrightarrow [S-\langle\rangle] + 2n\,NaCl \quad (1)$$

It appears that all technical PPS polymerization processes make use of this polymerization reaction. Reaction kinetics investigations [74] demonstrated that NMP not only acts as a mere solvent but undergoes a ring-opening hydrolysis to sodium *N*-methyl-4-aminobutyrate (SMAB) before the polymerization starts:

$$Na_2S + H_2O \rightleftharpoons NaOH + NaSH \quad (2)$$

$$NaOH + \text{(N-methylpyrrolidone)} \rightleftharpoons CH_3NH(CH_2)_3COONa \quad (3)$$

$$CH_3NH(CH_2)_3COONa + NaSH \rightleftharpoons \text{"SMAB-NaSH"} \quad (4)$$

SMAB forms a truly crystallizable stoichiometric 1:1 complex with NaSH, which is considered to be the nucleophile attacking the chloroaromatics. During the polymerization reaction, HCl is formally set free, which subsequently reacts with SMAB to regenerate NMP with NaCl and water as byproducts. Because of the reaction sequence (2)–(4), the monohydrate appears to be the optimum sodium sulfide form for PPS polymerization. Less water would not allow the SMAB-NaSH to be formed, more water would slow down the polymerization at least in the early phase.

Although Reaction resembles a step-growth polycondensation of the A–A plus B–B type following CAROTHER'S law, the actual polymerization kinetics are different. Oligomers of higher molar masses are already formed early in the polymerization process, i.e., at a low conversion of $X \approx 20\%$. This observation has triggered a scientific discussion on the polymerization mechanism, including the proposal of a chain polymerization carried out by activated species [75, 76]. Researchers at Phillips Petroleum finally demonstrated [74] that the PPS reaction is a polycondensation reaction but with monomers (in particular, *p*-DCB) being about a factor of ten less reactive than growing-chain end groups. Because this low monomer reactivity would eventually lead to polydispersities of $D \approx 20$ or higher, a cleavage reaction of already formed polymer chains by Na_2S has been included in the polymerization model to explain the experimentally observed polydispersities of $D \approx 2$–5 [77].

5.1. Production

The key challenge in PPS polymerization engineering is the achievement of a sufficiently high polymerization degree. At high monomer conversions of $X > 0.9$ the polymerization degree P_n approximately follows CAROTHER'S law:

$$P_n = \frac{1}{(1-X)}$$

Therefore, an exact stoichiometric match between the monomers Na_2S and *p*-DCB is required to achieve sufficient molar masses. Typically the reaction is carried out with a slight excess (1–3 mol%) of *p*-DCB over Na_2S because

- *p*-DCB is much less reactive than chlorine-terminated polymer chains and, therefore, reacts only very slowly with sulfur-terminated polymer chains
- an excess of Na_2S, on the other hand, cleaves already formed polymer chains in a very fast reaction
- an excess of Na_2S undergoes complex side reactions to form bad smelling organosulfur compounds

Achievement of an exact stoichiometric match is further complicated by the fact that sulfur equivalents charged to the reactor might be lost from the reaction mixture as H_2S over the vapor phase or coprecipitate as NaSH with the byproduct NaCl [77].

In the initial process variant the reaction product is a suspension that contains comparatively low-molar-mass PPS and sodium chloride as the solid phase. The suspension can be freed and p-DCB by processes such as flash evaporation. The resulting powder consists of polymer and salt. It is washed with water at various pressures and temperatures to remove the salt and then dried [78, 79].

The PPS powder is then cured to achieve the necessary molar mass. Curing means chain extension and branching through thermooxidation, in the simplest form with air [73, 80]. The melt flow index (MFI) can be reduced from about 6000 g/10 min to 50–700 g/10 min (ASTM D 1238–70, 316 °C, 5 kg). Similar products of increased melt viscosity but branched chain architecture may be obtained by adding small amounts of trifunctional aromatics to the reaction mixture [73, 81].

Because a branched PPS polymer cannot be processed into useful films or fibers, several processes were developed to obtain higher molar masses and linear PPS molecules. In general the processes are based on two concepts:

1. Phase separation
2. Molar mass build-up in a homogeneous reaction

Concept 1 involves the addition of "catalyst" substances such as sodium or lithium acetate to the reaction mixture [81, 82]. The reaction mixture then partially separates into a solvent- and water-rich phase of high ionic strength and a polymer-enriched phase where the further condensation of low-molar-mass chains continues at increased end-group population density. Phase separation has also been induced by adding water or low-volatile ethers to the reaction mixture to recover the polymer in granular form directly [83].

In a further development of the phase-separation concept, a well-defined amount of water is added to the reaction mixture at low conversions, i.e. the low-molar-mass, prepolymerized PPS solution. The reaction mixture then undergoes complete phase separation and linear PPS polymers of high molar masses are formed as fine granules in a separate molten phase. After cooling down the reaction mixture, the granules may be separated by mechanical screening from the liquid or fine-particle phase of the reaction mixture. They are subsequently washed, dried, and directly melt-processed into specific compounds. Under these reaction conditions, a low-molar-mass polymer fraction of up to 15% of the total PPS remains dissolved in the solvent phase, which is removed as a fine-particle fraction from the main product in the screening step [84]. This technology has been extended to a continuous work-up process [85].

In a process development following concept 2, lithium sulfide is added instead of sodium sulfide. Li_2S employed in the polymerization reaction leads to formation of LiCl as a byproduct, which, in contrast to NaCl, is soluble in NMP. Thus complications during the polymerization reaction due to reaction mixture inhomogeneity and salt byproduct precipitation can be avoided. As a result, high-molar-mass PPS is formed under well-defined, homogeneous-phase reaction conditions which may enable a continuous polymerization process [86]. However, the presence of lithium requires additional downstream separation and cleaning steps [87].

Almost homogeneous reaction conditions are also achieved in a two-step process using conventional Na_2S. In the first step, p-DCB is employed in stoichiometric excess to form easily crystallizable, chlorine-terminated prepolymers. A major part of NaCl is formed in this step and removed. Subsequent reaction of the prepolymers with further amounts of Na_2S produces only very little salt and proceeds in an essentially homogeneous reaction mixture [88]. A similar approach comprises firstly the preparation of cyclic oligo(phenylene sulfide) in dilute solution and, in a second step, a ring-opening polymerization, optionally in the presence of some linear polymer. This step might be carried out as a solvent-free process in the melt [89]. Though effective in providing linear, high-molar-mass PPS of narrow molar-mass distribution, such two-step process schemes add complexity to the overall process.

The PPS polymerization reaction approximately follows second-order reaction kinetics [74, 75, 77] and shows very low reaction rates at high conversions. Temperatures of up to 290 °C have been combined with cycle times of 10 h and beyond in PPS batch reactions [73, 78–84]. Moreover, the PPS formation reaction is strongly exothermic [75, 90], because the lattice energy of the insoluble NaCl is set free. A number of attempts have thus been undertaken to improve reaction control by employng continuous or semibatch reactors. A cascade of continuously stirred tank reactors has been designed [91] for a fully continuous PPS reaction.

Semibatch reaction control schemes enable high polymer concentrations in the reaction mixture, because the immiscible monomers Na_2S and p-DCB do not have to be compatibilized at their respective maximum concentrations. In one of the semibatch reaction processes, the p-DCB is charged to the reactor and the sulfide component metered under reaction conditions. A eutectic mixture of Na_2S and NaSH is added as a melt. This approach is combined with the use of N-methyl--caprolactam as a solvent, which allows water to be removed in an azeotropic distillation near atmospheric pressure around 212 °C. The polymerization was thus carried out in a water-free mode [92]. In another semibatch reaction approach, the sulfur component is added as a SMAB-NaSH complex to the reactor already charged with some p-DCB. The two monomers are then added in parallel. An improved control of H_2S loss is achieved by counter-currently absorbing the H_2S in the highly alkaline sulfide stream and simultaneously distilling off the water [93].

Generally, the synthesis of PPS according to Reaction has a poor efficiency. For each equivalent of PPS (108 g) two equivalents of byproduct NaCl (117 g) are formed. A higher efficiency would be achieved by directly producing PPS from benzene and sulfur, which are even less costly compounds than Na_2S and p-DCB. Attempts using p-diiodobenzene (p-DIB) made from benzene as an intermediate, which subsequently reacts with elemental sulfur in the melt, have been published [94–96]. The key technical hurdles in this approach are the formation of di- or polysulfide bridges in the polymer chain, iodine recovery, and the iodine content in the polymer product.

5.2. Properties

5.2.1. Material Properties

The properties of PPS may vary to a limited extent, depending on the production process and workup.

Various methods have been described for characterizing the melt behavior and estimating the molecular mass of PPS. A commonly employed method is the measurement of a melt index: (melt flow index (MFI) in g/10 min, or melt volume index (MVI) in mL/10 min). In many publications the ASTM D 1238–70 standard with a modified application force (5 kg) and melt temperature (316 °C) is recommended [80–82]. The method determines only one parameter, however, as a single-point measurement on a shear-dependent melt curve. Furthermore, the standard does not permit measurement of higher values (i.e., high flowability). Measurement of the melt viscosity as a function of the shear force in a cone–plate or capillary viscosimeter is more useful [99].

Solution viscosity and light scattering measurements in combination with size-exclusion chromatography (gel permeation chromatography) have also been used to determine molecular masses [100]. Industrially produced poly(arylene sulfides) generally have broad distribution curves, which become wider after curing (see above). For further details of measurement of molecular mass, see → Plastics, Analysis.

PPS can be characterized by IR or nuclear magnetic resonance spectroscopy [101]. The thermal degradation of PPS has been investigated by direct pyrolysis in the ionization source with the aid of mass spectroscopy. In the presence of oxygen, aromatic ether bonds are formed by oxidation, and condensed aromatic compounds are produced by elimination of hydrogen sulfide. Sulfur dioxide is another inorganic gaseous product [102].

Surface oxidation produces sulfoxide and sulfone groups that can be detected by electron spectroscopy for chemical analysis (ESCA) methods [103]. Thermogravimetric measurements have shown that the high thermal stability of PPS compared to other plastics is due to cross-linking reactions [104].

Chloroaryl, thiol, or thiolate groups may be expected as end groups for poly(arylene sulfides) with a regular structure. The chlorine end group content can be determined by elemental trace analysis (e.g., Wickbold combustion) [105]. Determination of thiol groups is more difficult, a wet chemical method is described in [106].

The content of ionic impurities may be critical in certain applications, for example in the

Table 6. Mechanical properties of nonreinforced PPS

Property	Standard	Melt viscosity[*], Pa·s		
		14	66	168
Impact strength, kJ/m^2	ISO 180/1 C	14	26	34
Tensile strength, MPa	DIN 53 455	40	67	59
Elongation at break, %	DIN 53 455	1.1	2.3	10.9
Modulus of elasticity (tension), MPa	DIN 53 457	3940	3560	3460

[*] Measured in a rotation viscosimeter at 306 °C and a shear rate of 100 s^{-1} (measurements made at Bayer).

electronics sector. The polymer must be brought into solution to release and determine its salt content [107]. A physical method for determining chloride ions has also been published [108].

5.2.2. Properties of Nonreinforced Moldings

PPS is a partially crystalline material. Moldings thus consist of an amorphous phase and a crystalline phase. The amorphous phase has a glass transition temperature of 90 °C. The thermodynamic melting point of the crystalline phase for "ideal" crystals is between 300 and 310 °C [109]. The platelet-shaped, orthorhombic crystals (length and width ≤ 1000 nm, thickness ≤ 10 nm) in moldings have a melting point between 280 and 285 °C. The PPS crystallites form spherical spherulites from still melts, and stacks with a hard-fiber morphology from flowing melts [110].

During injection molding or extrusion, PPS crystallizes when the melt is cooled. Differential scanning calorimetry (DSC) shows that crystallization begins at about 250 °C. The final degree of crystallinity depends on whether the melt is cooled above or below the crystallization temperature of ca. 127 °C, and how quickly this takes place.

It is often impossible to obtain the maximum degree of crystallization during injection molding. Such moldings may become distorted due to post-shrinkage during subsequent use under thermal stress. In order to ensure that PPS moldings are dimensionally stable during subsequent use, it may therefore be necessary to temper them beforehand (i.e., to store them at elevated temperature, e.g., 220 °C) to complete their crystallization under controlled conditions.

Injection moldings of nonreinforced PPS have high tensile and flexural strengths, as well as high moduli of elasticity (Table 6). They have a high rigidity and a low flexibility, with the result that they may fracture even under very small loads. The impact strength and notched impact strength are also extremely low. Whereas moldings of low-viscosity PPS exhibit brittle fracture even under slight deformation, moldings of high-viscosity PPS can be deformed before they fracture in a ductile manner [111].

Thermal and electrical properties of nonreinforced PPS are listed in Table 7. Moldings of nonreinforced PPS that are not mechanically stressed retain their shape up to the melting temperature. Under mechanical loads, however, they become deformed above ca. 90 °C, i.e., the glass transition temperature. For a partially crystalline thermoplastic the coefficient of linear thermal expansion is extremely low. The thermal conductivity of PPS is also very low [112].

On the basis of its dielectric values and resistance values PPS is a slightly polar insulator.

PPS is an inherently flame-resistant thermoplastics (oxygen index 0.44). It is classified as a

Table 7. Thermal and electrical properties of nonreinforced PPS (Ryton R 6)

	Standard	Value
Thermal properties		
Vicat softening temperature (VST/B$_{120}$), °C	DIN 53 660	≥ 240
Heat deflection temperature (HDTA), °C	DIN 53 641	117
Linear coefficient of expansion, K^{-1}	DIN 53 752	
at 20–90 °C		60×10^{-6}
at 100–180 °C		110×10^{-6}
Electrical properties		
Dielectric strength, kV/mm	ASTM D 149–64	15
Volume resistivity, Ω·cm	ASTM D 257–66	4.5×10^{16}
Dielectric constant	ASTM D 150–70	
at 1 kHz		3.1
at 1 MHz		3.1
Dissipation factor	ASTM D 150–70	
at 1 kHz		0.0005
at 1 MHz		0.0009

Table 8. Chemical resistance of PPS on storage at 93 °C

	Weight increase after 24 h, %	Tensile strength after 24 h, MPa	Tensile strength after 3 months, MPa
	0	74	74
Gasoline benzene	0.07	71	64
Trichloroethylene	6.5	51	
Butanol	0.05	72	74
Butanone	1.02	77	52
Dibutyl ether	0	79	74
Amyl acetate	0.14	92	74
Butylamine	1.52	49	0
Hydrochloric acid (37%)	0.57	75	41
Sulfuric acid (30%)	0.14	81	74
Nitric acid (10%)	0.32	83	23
Sodium hydroxide (30%)	0.07	69	74

V-0 material (UL 94) at a thickness of 0.8 mm (→ Flame Retardants, Chap. 3.) [112].

At low temperature PPS is resistant to all organic solvents and chemicals. PPS moldings swell, however, under the action of some organic compounds on heating, particularly after prolonged exposure (Table 8). PPS is extremely resistant to acids and alkalis but oxidizing agents, in particular nitric acid, attack PPS on heating and destroy the moldings. PPS is moderately permeable to diffusion of gases and water vapor [113].

5.2.3. Properties of Reinforced Moldings

Reinforcement of PPS with fibers and/or mineral fillers increases its mechanical strength and thereby compensates for its low impact strength, and also improves its dimensional stability at high temperature. Fibrous reinforcing materials are usually made of glass but carbon fibers and aramid fibers are also employed on a small scale. Mineral fillers are generally used in combination with glass fibers and include calcium carbonate, kaolin, calcium sulfate, mica, talc, and quartz. These fillers and reinforcing materials can be incorporated into PPS in high concentrations (up to about 70 wt%) by using conventional compounding equipment [113–116].

Injection moldings made of glass-fiber-reinforced PPS have a high hardness, ultimate tensile strength, flexural strength, and modulus of elasticity, but a very low formability and ductility and only moderate impact strength and notched impact strength. Such moldings can permanently withstand high vibrational stress. The slip properties and abrasion resistance of glass-fiber-reinforced PPS molding compounds are inferior to those of other partially crystalline thermoplastics reinforced with glass fibers [115].

Glass fiber–mineral-reinforced PPS injection molding compounds have relatively high hardness and rigidity, and relatively low impact strength and percentage elongation at break (Table 9). The properties may be altered within wide limits by varying the weight ratio of glass fibers to mineral fillers [113, 114].

Reinforcing PPS with mineral fillers also considerably improves its heat resistance, hardness, and rigidity (Table 10). It also results in moldings with very smooth surfaces and high dimensional stability. Omission of fibrous reinforcement means, however, that the flexural

Table 9. Mechanical properties of PPS molding compounds reinforced with 40 wt% glass fiber

Property	Standard	Craston XMB 3100 (Ciba-Geigy)	Fortron 1140 A 4 (Hoechst Celanese)	Ryton R 4 (Phillips Petroleum)	Supec G 401 (GE Plastics)	Tedur KU 1–9510–1 (Bayer)
Tensile strength, MPa	DIN 53 455	175	159	116	170	190
Elongation at break, %	DIN 53 455	1.6	1.4	0.9	1.0	1.6
Modulus of elasticity (tension), GPa	DIN 53 457	17.5	14.2		14.5	15.0
Flexural strength, MPa	DIN 53 452	250	240	190	240	300
Modulus of elasticity (bending), GPa	DIN 53 457	15.5	12.4	13.0	14.0	13.0
Impact strength (Charpy), kJ/m^2	DIN 53 453	20	29	11	25	35
Notched tensile impact strength (Charpy), kJ/m^2	DIN 53 453	8	9	6	8	9
Ball indentation hardness, MPa	DIN 53 456	300	305	300	300	300

Table 10. Mechanical properties of PPS molding compounds reinforced with glass fibers and mineral fillers

Property	Standard	Craston X MB 3110	Fortron 6165 B 4	Ryton R 10	Tedur KU 1–9521
Density, g/cm^3	ASTM D 1505–68	1.93	2.1	2.0	1.9
Glass fiber/mineral content, wt%		60	65	70	60
Tensile strength, MPa	DIN 53 455	110	111	86	110
Elongation, %	DIN 53 455	1	1	0.7	1
Modulus of elasticity (tension), GPa	DIN 53 457	21.5	18.2	18.0	20
Flexural strength, MPa	DIN 53 452	165	190	140	180
Modulus of elasticity (bending), GPa	DIN 53 453	20.0	16.7	15.0	19
Impact strength (Charpy), kJ/m^2	DIN 53 453	11	19	6	10
Notched tensile impact strength (Charpy), kJ/m^2	DIN 53 453	6	7	4	6
Ball indentation hardness, MPa	DIN 53 456	370	430	350	300

and tensile strengths of the moldings are very low: they can be deformed only very slightly before they fracture [113, 114].

Nonreinforced PPS exhibits partial softening at 90 °C and moldings cannot be mechanically stressed above this temperature. Reinforcement with glass fibers or minerals improves the strength to such an extent that moldings can withstand mechanical stress at elevated temperature. This is reflected in the heat resistance values according to ISO 75, i.e., the heat deflection temperature (HDT), which consistently lies between 260 and 270 °C [113, 114].

Maximum service temperature is characterized by relative temperature indices which are the temperatures at which 50% loss of initial mechanical properties occurs after an exposure of 25 000 h. The relative temperature indices of reinforced and filled PPS molding materials are 200–240 °C. The thermal conductivity of reinforced PPS injection molding compounds is primarily determined by their glass fiber content and the fiber orientation in the molding. In moldings fabricated from materials reinforced with glass fibers and mineral fillers, the anisotropy of the thermal expansion is less. Reinforced PPS injection molding compounds are recommended as excellent electrical insulators on account of their high volume resistivity and surface resistivity (Table 11) [113, 114].

Reinforced PPS molding materials are extremely flame-resistant, with a very low fume density. In addition they pass the Ohio State University (OSU) test [113, 114].

Moldings are extremely resistant to most organic compounds, even at elevated temperatures. However, on prolonged storage at elevated temperature in water or in water vapor they lose up to 60% of their initial strength [116]. Water absorption is extremely low, even in boiling water, and accordingly dimensional changes due to water absorption do not occur. Under prolonged storage in hot air, above ca. 120 °C PPS injection moldings turn brown. However, this is not associated with reduction of mechanical strength (maximum use temperature: 200–240 °C) [113, 114].

PPS undergoes photooxidation in the light, and prolonged weathering results in yellowing and moldings with a dull, rough surface. Their mechanical strength decreases only very slowly however. PPS is resistant to high-energy radiation [113, 114].

Repeated processing of PPS injection molding compounds does not result in degradation of the PPS macromolecules, as is confirmed by the constant melt viscosity. Recycling of regenerated material from sprues and defective parts is therefore acceptable [113, 114].

Table 11. Electrical properties of reinforced PPS molding compounds (measurements made at Bayer)

Property	KU 1–9511	KU 1–9521
Specific volume resistance, $\Omega \cdot cm$ (VDE 0303, part 1)		
4 d, 23 °C, 50% R.H.	7.2×10^{15}	2.5×10^{15}
4 d, 23 °C, 92% R.H.	6.7×10^{14}	9.2×10^{14}
Surface resistivity, $\Omega \cdot cm$ (VDE 0303, part 3)		
4 d, 23 °C, 50% R.H.	6.4×10^{15}	2.7×10^{15}
4 d, 23 °C, 92% R.H.	9.6×10^{13}	3.4×10^{13}

5.3. Processing

5.3.1. Injection Molding [117]

PPS injection molding materials containing glass fibers and/or minerals have an extremely good processability on account of their low melt viscosity.

Injection Conditions. It has proved advantageous to store the molding powder for 3–6 h at ca. 150 °C prior to injection molding to remove any volatile constituents. Materials that contain hydrophilic fillers should always be dried beforehand under these conditions.

PPS injection molding compounds are processed at melt temperatures of 350–370 °C, the most common range is 320–340 °C. PPS only crystallizes extremely slowly and incompletely below 120–130 °C, the temperature of the molds must therefore be above 130 °C and, if possible, between 150 and 170 °C to obtain moldings with smooth, uniform surfaces.

Solidification during filling results in surface damage and weak spots on the moldings. The injection rate should therefore be as high as possible and is usually performed at 75–150 MPa. High injection pressures are advantageous since they ensure uniform moldings, good surfaces, and favorable mechanical properties. Overinjection can easily occur however, with the formation of mold marks. After-pressures of 50–70% of the injection pressure are normally employed.

PPS molding compounds can be processed in all conventional screw-injection molding devices, triple-zone screw arrangements with compression ratios of 1 : 2 to 1 : 3 are often used. PPS materials generally contain glass fibers or mineral fillers and thus, wear down screws and cylinders, particularly in the feed section. Reinforced (armor-plated) screws and cylinders are therefore recommended, with shutoff nozzles instead of open nozzles. Heated channel nozzles may also be used.

Molds. Since the PPS melt has a low viscosity, it may penetrate between the mold parting surfaces, and thus cause formation of burrs and marks. High clamping pressures are used to prevent this.

PPS moldings reinforced with glass fibers shrink anisotropically. Nonwarping moldings can, however, be obtained by modifying the mold shape, injection rate, mold temperature, and the temperature of the melt. Shrinkage is also lower in PPS materials that contain both mineral fillers and glass fibers. Such materials are consequently better suited to the production of nonwarping, dimensionally stable moldings.

The high content of glass fibers and fillers in most PPS injection molding compounds can lead to considerable abrasion of the molds, and wear-resistant mold steels are therefore used. Steels with Rockwell hardnesses of 59 HRC to 62 HRC are preferred.

The mold cavities should be vented sufficiently rapidly, normally through venting slits in the mold parting surfaces. The slits should not be wider than 0.01 mm to avoid the formation of burrs and marks on the moldings.

5.3.2. Other Molding Methods

PPS moldings can be produced not only by injection molding (Section), but also by compression or by sintering PPS powders or mixtures of PPS powders and glass fibers, fillers, and/or additives (e.g., molybdenum sulfide or polytetrafluoroethylene).

The property spectrum of PPS pressed moldings is substantially inferior to that of injection moldings. The mechanical strength of sintered parts roughly corresponds to that of pressed parts of the same composition. An advantage over compression molding is that hot (360 °C) molds do not need to be handled [118].

Low-viscosity PPS powders are used to coat metallic parts (coating from aqueous dispersions, electrostatic powder coating, fluidized-bed coating). Coatings of oxidatively cross-linked PPS are rigid, flexible, and wear-resistant, and can withstand high thermal stress over prolonged periods. The chemical resistance of PPS coatings is decisive for their use in protecting metal parts against corrosion [119].

5.3.3. Finishing of Shaped Articles

Most PPS materials contain a high concentration of glass fibers and mineral fillers. The use of hard-metal drills, milling machines, and saws to finish PPS moldings is therefore recommended.

PPS moldings can be strongly bonded to one another by ultrasonic welding, hot-plate welding, and with two-component adhesives, in particular epoxy and polyurethane adhesives.

Coatings on PPS moldings can be obtained only if a polyurethane primer is applied.

5.3.4. Extrusion

Films [120]. See also → Films, Section 6.12.4..

Films, particularly biaxially stretched films, constitute one of the main uses of unfilled PPS, and are mainly used in capacitors. Biaxially stretched PPS films have good mechanical properties, their outstanding property being their high thermal resistance. If ignited, they rapidly cease burning after the flame has been removed. PPS films are resistant to bases, acids, solvents, and other chemicals. However they are rapidly destroyed by concentrated sulfuric or nitric acid at room temperature.

PPS films can be coated and printed after corona or plasma pretreatment. The very low polarity of the PPS molecules accounts for the low dielectric constant and the low dielectric loss factor.

To produce biaxially stretched PPS films the filtered PPS melt is extruded at ca. 320 °C onto cooled rollers. It has also proved advantageous to remove volatile constituents by degassing the melt before the extrusion, this is achieved by releasing the pressure and applying a vacuum. The amorphous film is then stretched at 90–120 °C successively or simultaneously in the longitudinal and transverse directions.

Fibers [121]. See also → High-Performance Fibers, Section 2.3.2.

Fibers constitute the second largest area of use of unfilled PPS, and are employed particularly in the form of fabrics and nonwovens for filtering hot and/or aggressive media.

PPS fibers have a high mechanical strength, and can be used for short periods up to about 230 °C, whereby they retain 40% of their tensile strength at 23 °C. The longterm service temperature is 190 °C. If fabrics made from PPS are ignited, they rapidly cease burning once the source of ignition is removed.

PPS fibers have a high resistance to acids, alkalis, and most organic solvents, even at elevated temperature. Oxidizing agents attack PPS fibers however. In boiling water PPS fibers may shrink by up to 10%.

Extrusion of the raw fibers is followed by two-stage stretching and heat setting, in which the fibers are heated under tension at 265–270 °C for 1 min to ensure satisfactory crystallization of the PPS.

5.3.5. Semifinished Products

Reinforced PPS materials are processed into continuous profiled sections, rods, pipes, etc. On account of its high resistance to heat and solvents, poly(phenylene sulfide) is also an excellent matrix for high-performance composite materials [122].

5.4. Uses

The high price of PPS injection molding materials limits their uses and applications. They are therefore mainly used where technical and economic requirements cannot be satisfied by other industrial thermoplastics.

The applications of PPS molding materials are determined by their unique combination of technical advantages, the most important being:

1. High dimensional stability even at high temperature
2. Very high rigidity and hardness
3. Very good long-term static and dynamic loadability
4. Low static and dynamic coefficients of friction
5. Extremely high resistance to almost all chemicals
6. Low flammability
7. Good processability by injection molding
8. Machinability and weldability
9. Inscribability of the moldings by printing or laser inscription

The main areas of use for PPS molding materials include the electrical/electronics, automotive, and lighting sectors. PPS moldings are also used in many other market sectors, for example in household appliances, chemical plant construction, and mechanical engineering.

In the *electronics sector*, special, highly filled PPS molding compounds with particularly low melt viscosities have to some extent replaced epoxy compositions in the housing of electronics components such as transistors, coils, and capacitors. Glass fiber-reinforced PPS is also used for integrated circuit housings (chips). Connectors (e.g., high-density connectors) made of PPS reinforced with glass fibers (and minerals) are produced on a large scale in the United States. The high mechanical strength of moldings made from glass fiber-reinforced PPS accounts for its use in components of electric motors (e.g., brush holders and wound cores).

PPS molding compounds are also being used increasingly in the *automotive sector*. For example pumps in petrol tanks are fabricated from PPS reinforced with glass fiber and minerals. In pump impellers of vacuum pumps used in steering and braking systems, the PPS moldings have to satisfy extremely stringent requirements as regards dimensional accuracy, dimensional stability, mechanical strength, dynamic longterm stress, and abrasion resistance.

In the *lighting sector* the trend towards small automobile headlamps and high light intensities means that a great deal of heat is generated in a confined space. Housings are therefore made of special, highly-filled PPS molding compounds. Caps and holders of halogen lamps are also produced from glass fiber-reinforced PPS.

PPS molding compounds are employed in *chemical plant construction* (e.g., for valves, column packings, and distillation columns). In feed pumps the danger of spark formation as a result of buildup of electrostatic charge can be excluded by using PPS molding materials whose electrical conductivity has been increased by incorporating carbon black.

On account of its resistance to chemicals, PPS is suitable for *applications involving contact with food*. Certain PPS compounds have been approved by the BGA and FDA for use as a material for kitchen utensils [123].

5.5. Economic Aspects

Despite the fact that PPS has been commercially available since the mid 1970s, it is a very "young" product. On account of its outstanding property profile it has a broad, constantly increasing range of applications. Newly developed copolymers and poly(arylene sulfides) based on dihalogenated aromatics with tailor-made property profiles are reinforcing this development [124].

Table 12. PPS turnover forecast [65]

Region	Turnover, t		Average annual growth, %
	1989	1995	
North America	5 500	10 500	11.4
Far East (including Japan)	4 000	7 700	11.5
Western Europe	2 500	4 800	11.5
Total	12 000	23 000	11.5

PPS compounds are used in strongly expanding sectors. In parallel with growing awareness among customers, further processors are beginning to use PPS which provides a good precondition for high growth rates for PPS compounds. Average annual growth rates exceeding 11% are expected in North America, the Far East (including Japan), and Western Europe (see Table 12) [65, 125].

Trade Names. In the mid 1970s only Phillips Petroleum was represented on the market, with its Ryton types. Following the expiry of the basic patents other companies are becoming increasingly active both as regards resin production and compounding [125].

Trade names of companies with their own primary resin production include Ryton (Phillips); Tedur, Tregalon (Bayer); Fortron (Kureha); Tonen PPS (Tohpren); Susteel (Tosoh Susteel); Torelina (Toray); (Sung Yong)

Trade names of companies without their own primary resin production include Supec (General Electric, United States); Tedur (Miles, United States); OFL and OCL (LNP, United States); Fortron (Hoechst Celanese, United States, Western Europe); Craston (Ciba-Geigy, Western Europe); Primef (Solvay, Western Europe), Larton (Lati; Western Europe); DIC-PPS (Dainippon Ink, Japan); Sumicon (Sumitomo Bakelite, Japan); Nova PPS (Mitsubishi Kasei, Japan); Fortron (Polyplastics, Japan); Denka PPS (Denka, Japan); Toyobo PPS (Toyobo, Japan); PPS (Ube, Japan); Petchem Idemitsu PPS (Idemitsu, Japan).

6. Liquid Crystal Polymers

6.1. Introduction

After first approaches in the late 1960s, the development of liquid-crystalline polymers (LCPs) has accelerated as a direct result of the increased knowledge of structure–property relationships. Highly oriented materials have been made by designing rodlike LCPs [126].

The family of LCPs can be divided into two main classes:

1. Thermotropic LCPs. The liquid crystalline phase is formed when the pure component is heated.
2. Lyotropic LCPs [127]. The liquid crystalline phase is formed when the polymer is mixed with a solvent. These polymers are mainly used in fiber application.

The liquid crystalline state is a highly ordered state where the thermal energy has exceeded the intermolecular forces (molten, liquid state) while still maintaining the ordered state of a crystal structure. Molecules that form liquid crystalline phases have either rigid, long lathlike shapes with a high aspect ratio (unidirectional) or disklike molecular structures (bidirectional) The behavior of LCPs is determined by the fact that parts of the polymer, or the whole material, cannot be bended or entangled, showing a high level of stiffness.

The stiff elements in the polymer chain are called mesogens. The mesogens can be connected in three different ways:

- Linking the mesogens to form a rigid main-chain LCP, e.g., by means of a polycondensation
- Attaching to or grafting the mesogens onto a flexible polymer backbone to form a side-chain LCP with a comblike structure
- Combination of both concepts [127]

The geometric arrangements of main-chain LCPs and side-chain LCPs are shown in Figure 8.

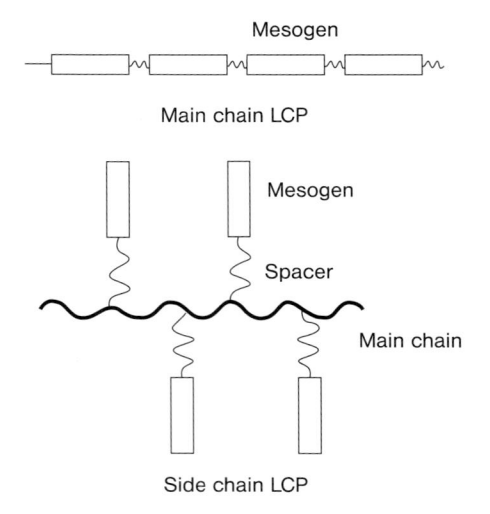

Figure 8. Schematic representation of the two geometric arrangements of main-chain and side-chain LCPs [128]

Processing the thermotropic LCPs yields highly oriented fiberlike materials with anisotropic properties. These materials show extraordinary stiffness and rigidity and typically low coefficients of linear thermal expansion in the direction of fiber orientation. Perpendicular to this direction, these properties are less pronounced. The anisotropic behavior can be reduced by implementing flexible segments into the polymer or using suitable fillers like, e.g., glass fibers that disturb the orientation of the material and its crystalline phases [129].

Liquid crystals exhibit phases of one- or two-dimensional order, which is between that of regular crystals (three-dimensional) and amorphous phases (no order). These phases are called mesophases. Figure 9 shows the orientation and order of molecules in smectic (sm), nematic (nem), and cholesteric (chol) mesophases. For detailed information on liquid crystal structure and phases, → Liquid Crystals.

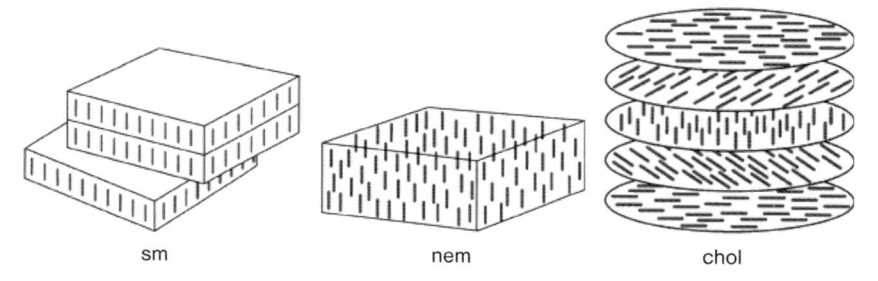

Figure 9. Schematic representation of the molecule ordering in smectic (sm), nematic (nem), and cholesteric (chol) mesophases [130]

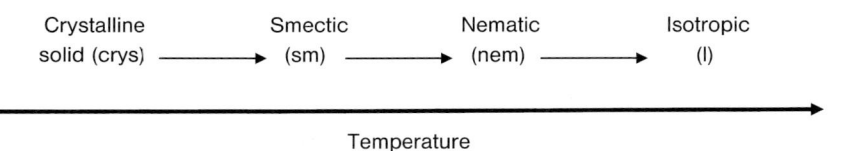

Figure 10. Sequence of the phases by heating crystalline LCP [126]

Crystalline solid (crys) → Smectic (sm) → Nematic (nem) → Isotropic (l)

Temperature

In smectic phases, the molecules are structured in layers. The axis of the molecules is perpendicular to the layers, i.e., parallel to the layer normal. In the nematic state, the molecules are oriented parallel to each other but they do not form distinguished layers. The cholesteric state lies in between the smectic and nematic state, i.e., the molecules are ordered in layers but the molecular axis is tilted with respect to the layer normal, and they are ordered in a spiral fashion [131].

For thermotropic LPCs, increasing the temperature leads to a transition from the crystalline solid state (crys) to the smectic (sm), nematic (nem), and, finally, to the isotropic state (i) (Fig. 10). Not all possible phases may appear, and many types of smectic crystals may be present simultaneously. Additionally, the liquid crystalline phases may appear rather upon cooling than upon heating [126].

In 1956, FLORY predicted the behavior of rigid rodlike polymers. He postulated that this kind of material would form ordered structures in solution at some critical concentration. This lyotropic behavior was first reported in 1937 for solutions of tobacco mosaic virus. FLORY offered an explanation for the lyotropicity, which was verified by systematic experiments on concentrated solutions of poly(γ-methyl glutamate) and poly(γ-benzyl glutamate) [126, 127, 132].

In the late 1960s, also thermotropic liquid crystalline polymers were synthesized, which opened a market for thermoplastic applications. Since then, this class of materials has seen a rapid development. Carborundum developed poly(p-hydroxybenzoic acid), a homopolyester based on p-oxybenzoyl repeating units that was first developed in 1955 by AELONY and RENFREY [133]. Poly(p-hydroxybenzoic acid) was the first commercial LCP and was commercialized under the trade name Ekonol. Saint-Gobain acquired the Ekonol business in the mid-1990s. Poly(p-hydroxybenzoic acid) is a rigid linear polymer with extremely high crystallinity but no melt transition up to 600 °C. Flow and creep are virtually nonexistent below the melting point. Because this melting point is significantly above the degradation temperature, thermoplastic processing is not realizable without damaging the polymer matrix. Additionally, the material is insoluble in all known solvents and does not allow any casting processing technologies, which makes analysis and characterization difficult. However, the extraordinary thermal stability of poly-(p-hydroxybenzoic acid) makes this material very interesting for high-temperature applications. Because of its insolubility and the lack of a melt transition, Ekonol has to be processed by hammering, plasma spraying, or sintering [126, 131, 134–136].

The structural principles of thermotropic LCPs are exemplified by poly-(p-hydroxybenzoic acid):

The high melting point of the polymer is a result of the high symmetry in the polymer chain. Breaking the molecular symmetry leads to a reduction of the melt transition temperature below the degradation temperature. This can be done by several methods [127] (Fig. 11):

- Copolymerization of mesogenes of different size or functionality
- Incorporation of flexible mesogene units
- Introduction of lateral substituents
- Synthesis of chains with kinks or branches

Changing the bridging group of the mesogens is one option to introduce either flexibility

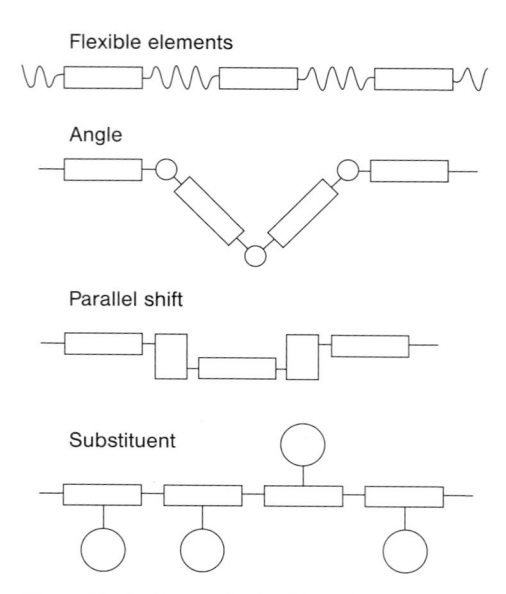

Figure 11. Various methods of breaking the symmetry of LCPs to control the transition temperature [128]

or stiffness. Table 13 gives examples of bridging groups which can be used alternatively to the ester function.

Both the nature and the orientation of the bridging groups influence the flexibility and the crystallization properties and, hence, the transition temperature of the LCP. For example, substitution of p-hydroxybenzoic acid (bridging group type A–B) by a hydroquinone (A–A) and terephthalic acid (B–B) lowers the crystallinity and reduces the flexibility of the chain. Although the ester groups cannot be oriented absolutely parallel, their conjugation leads to stiffening. On the other hand, addition of biphenylic groups to the backbone (no bridging group) introduces a twist at the C-C bond between the two non-conjugated aromatic groups. Linking groups and polymerization mechanism have to be selected in a way that the tacticity of the polymer will be maintained [127].

Another option for varying the mesogen is changing the aromatic structure that defines the major amount of stiffness of the backbone. For example, naphthalene derivatives can be used besides benzoic systems. Although higher conjugated systems are theoretically also possible, they are not commercially relevant. The position of the substituents on the aromatic systems defines the orientation of the polymer chain and influences crystallinity and flexibility of the chain significantly (Table 14) [127, 137–140].

Another approach is the introduction of aliphatic segments into the polymer chain. After the commercialization of poly(p-hydroxybenzoic acid) by Carborundum (Ekonol) and Dartco (Xydar), the next series of LCPs entered the market, Tennessee Eastman's X7G, a modification of poly(ethylene terephthalate) (PET). Here, the mesogenic units are implemented into regular PET by a transesterification of PET with p-acetoxybenzoic acid, which produces a thermotropic LCP composed of p-oxybenzoic (60%) and ethylene terephthalate units (40%). The use of this material is limited because of its reduced thermal stability, which is due to the content of aliphatic segments [126, 127, 141–143].

Table 13. Examples of bridging groups as building blocks in LCPs

Esters	—C(=O)—O—
Amides	—C(=O)—NH—
Ethers	—O—
Urethanes	—NH—C(=O)—O—
Alkenes	CH=CH
Substituted alkenes	CH=C(CH$_3$)
Azo compounds	—N=N—
Aldimines	—CH=N—

Table 14. Examples of possible aromatic groups as building blocks in LCPs

Phenylidene	(1,4-, 1,3-, 1,2-phenylene)
Naphthylidene	(naphthalene isomers)
Biphenylidene	(biphenylene)

Most of the commercial concepts for fully aromatic thermotropic LCPs use poly-(p-hydroxybenzoic acid) that employs chain-disrupting comonomers to obtain a better processability. Comonomers are, e.g., isophthalic acid, hydroxynaphthoic acid, hydroquinone, terephthalic acid, and p, p'-diphenyl ether groups.

In the early 1980s, Vectra LCP was introduced by Ticona, followed by the market entry of Polyplastics Co. as a second commercial manufacturer of Vectra LCP in 1996 and by other manufactures such as DuPont (Zenite), Sumitomo (SumikaSuper), Solvay (Xydar), and Toray (Siveras). The materials comprise a family of copolyesters and copolyesteramides, based on p-hydroxybenzoic acid copolymerized with hydroxynaphthoic acid and, optionally, a selection of other suitable monomers. These monomers provide the building blocks given in Table 13 and Table 14 [127, 139–142].

6.2. Physical Properties

Typically, LCPs exhibit a high mechanical strength even at high temperatures, inherent flame retardancy, and good weather resistance. LCPs have a high z-axis coefficient of thermal expansion. The polymers show extreme chemical resistance and resist stress cracking at elevated temperatures in the presence of most chemicals, including aromatic or halogenated hydrocarbons, strong acids, bases, ketones, and other aggressive industrial substances. The hydrolytic stability in boiling water is excellent. Environments that deteriorate the polymers are mainly high-temperature steam, concentrated sulfuric acid, and caustic materials [144].

6.2.1. Mechanical Properties

The properties of LCPs are influenced to a high degree by its liquid crystal structure. The rod-shaped molecules are highly oriented in the flow direction during injection molding or extrusion and remain rod-shaped even in the melt phase (Fig. 12).

Due to the highly ordered nature of LCPs, mechanical properties of its parts, shrinkage, and other characteristics depend on the flow pattern in the specific part. During mold filling, for example, the molecules are oriented in the flow direction. In the finished part, these molecules are ultimately aligned on the surface, where they form a skin that is highly oriented in the flow direction. The skin makes up to 15–30% of the total thickness of the part (Fig. 13).

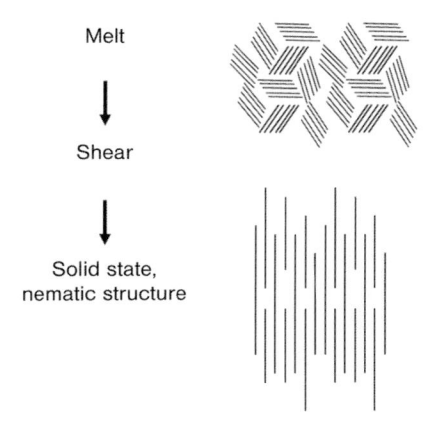

Figure 12. Orientation change of rod-shaped polymers due to the influence of shear [144]

This molecular orientation causes a self-reinforcement effect giving exceptional flexural and impact strength as well as good tensile performance. The increase in relative strength values is directly correlated to a decrease in wall thickness.

These properties depend primarily on the type of filler or reinforcement used. Glass fibers impart increased stiffness, tensile strength, and heat deflection temperature. Carbon fibers give the highest stiffness. Mineral fillers might also improve stiffness and provide increased toughness and a smoother surface compared to glass

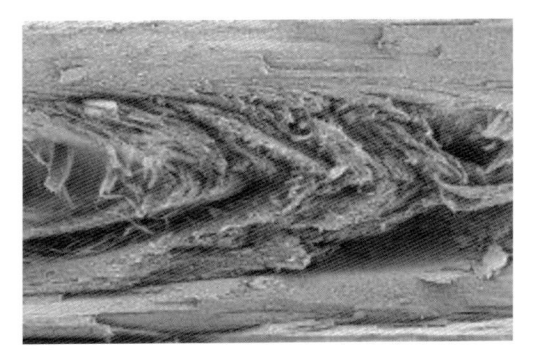

Figure 13. Layered structure of injection-molded LCP [144]

fiber reinforcement. Graphite improves elongation at break and adds lubricity. Polytetrafluoroethylene (PTFE)-modified grades have excellent sliding and wear properties. The impact strength of unfilled Vectra polymers is reduced by the addition of fillers and reinforcing materials but still remains high [144].

Anisotropy and Wall Thickness. LCPs are well known to have anisotropic properties when molded into parts. Unlike other engineering polymers, LCPs become significantly less anisotropic if formulated with mineral fillers or, to a lesser extent, with glass-fiber reinforcement. A comparison of the anisotropy of an LCP test specimen with that of a test specimen produced from a conventional engineering plastic, poly(butylene terephthalate) (PBT) with and without glass-fiber reinforcement is shown in Figure 14.

The anisotropy of 30% glass-fiber-reinforced Vectra and 30% glass-fiber-reinforced PBT is very similar. In grades with a high mineral filler loading, the anisotropy ratio can even be reduced to 1 (isotropic) [144]. One explanation for this effect is the orientation of the LCP fibrils in the specimen: Without solid fillers, the fibrils follow the shear profile within the flowing melt. If solid fillers are added, the polymer melt tends to flow faster and the fibrils flow around the fillers. This gives the fibrils a partial orientation in any direction of the part, even perpendicular to the flow direction. The total distribution of the fibril direction becomes more statistical and, consequently, the microscopic properties of the material become more isotropic.

Another factor causing anisotropy is the wall thickness of the part: as the wall, film, or sheet thickness decreases, the highly oriented outer layer accounts for a higher proportion of the total wall thickness. This higher percentage of oriented surface layer results in greater strength and modulus in the thinner sections [144].

Short-Term Mechanical Stress. Tensile stress–strain curves show that the ultimate stress of LCPs is nearly untouched by the addition of fillers or reinforcing additives [128]. The tensile stress–strain curves of Vectra with different reinforcements and fillers shown in Figure 15 are representative for other LCPs. The stress as break differs considerably with the base polymers as can be shown by the comparison of Vectra A130 with Vectra E130i (both 30% glass-fiber-reinforced). If differs less within the polymer series (compare Vectra S135 (35% glass-fiber-reinforced) to Vectra S540 (40% mineral-filled)).

As with any thermoplastic, the stiffness and strength of Vectra decreases with both increasing temperature and increasing wall thickness. The influence of the wall thickness on the

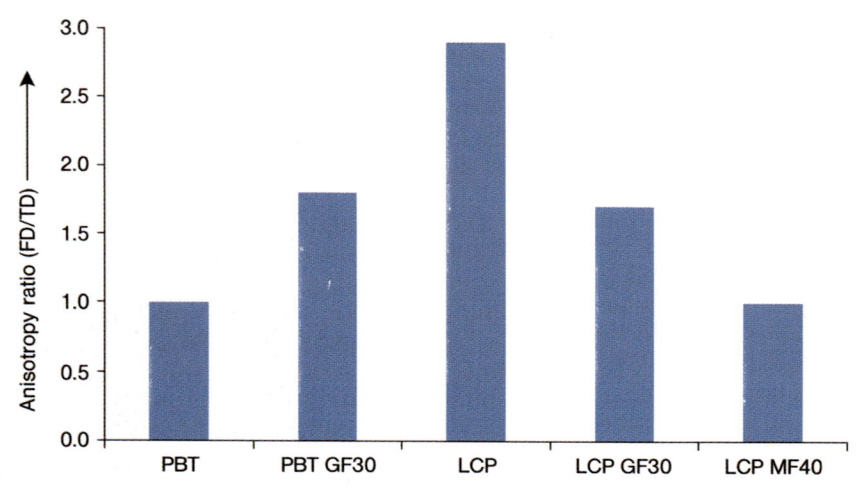

Figure 14. Comparison of the anisotropy ratios of LCP grades versus those of PBT grades (ISO test specimen) [144]
FD/TD = Tensile strength in flow direction/tensile strength in transverse direction
GF30 = 30% glass-fiber-reinforced
MF40 = 40% mineral filled

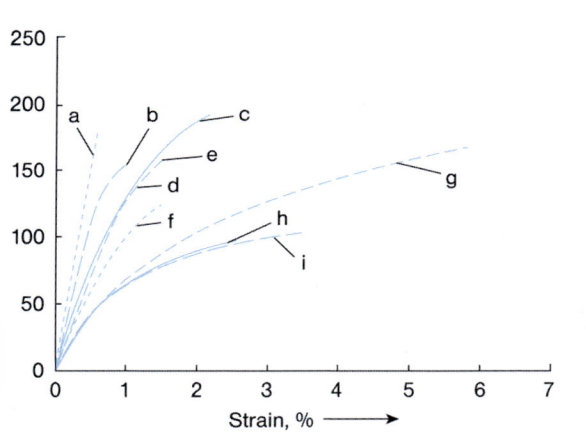

Figure 15. Stress–strain curves of Vectra with various fillers and reinforcements (tensile test at 23 °C) [144]
a) B230 (30% carbon-fiber-reinforced); b) A230D-3 (30% carbon-fiber-reinforced); c) A130 (30% glass-filled); d) S135 (35% glass-fiber-reinforced); e) E130i (30% glass-filled); f) S471 (45% glass/mineral-filled); g) A430 (25% PTFE filled); h) S540 (40% mineral-filled); i) E540i, (45% glass/mineral-filled)

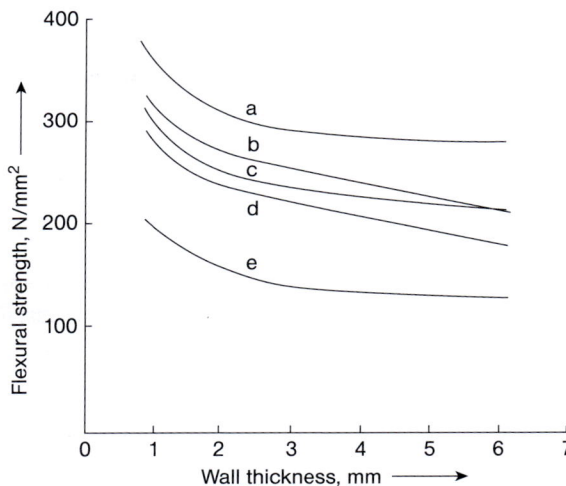

Figure 16. Flexural strength of Vectra with various fillers and reinforcements in relation to the wall thickness [128]
a) B130 (30% glass-filled); b) A130 (30% glass-filled); c) A230 (30% carbon-fiber-reinforced); d) C130 (30% glass-filled); e) A 625 (25% graphite-filled)

flexural strength is shown in Figure 16, that of the temperature in Figure 17 [128].

Long-Term Mechanical Stress. LCPs exhibit a good creep resistance. In particular, the creep resistance above 200 °C is remarkable compared to that of other thermoplastics. Figure 18 shows the flexural creep modulus of Vectra A130 between 23 °C and 220 °C for various stress levels. The maximum exposure time was 10 000 h. The stress levels were 30% of the short-term failure stress. Creep rupture was not observed at stress levels below 30% [128, 144].

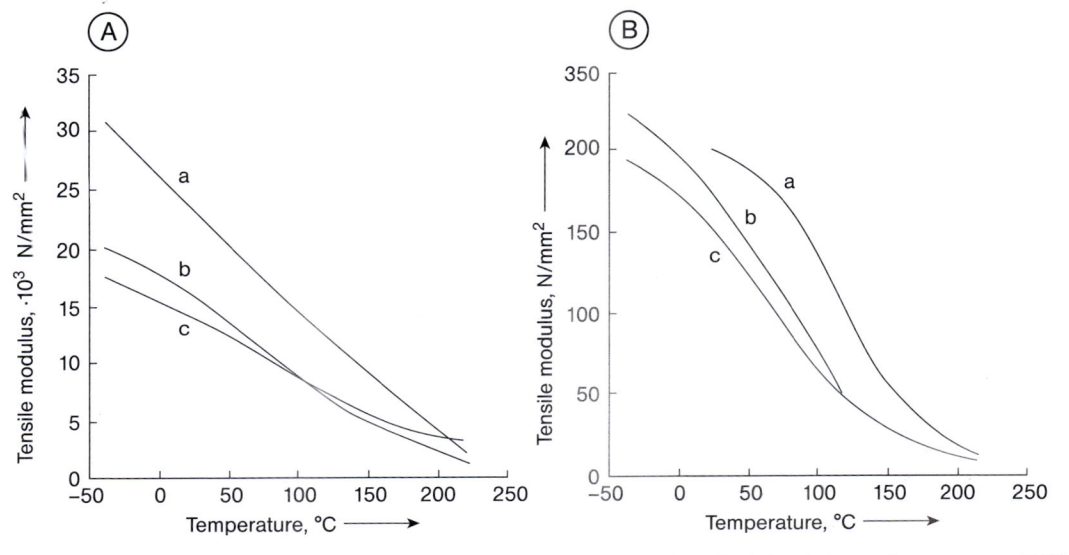

Figure 17. Tensile modulus (A) and strength (B) of Vectra types (all 30% glass-filled) in relation to the temperature [128]
a) B130; b) A130; c) C130

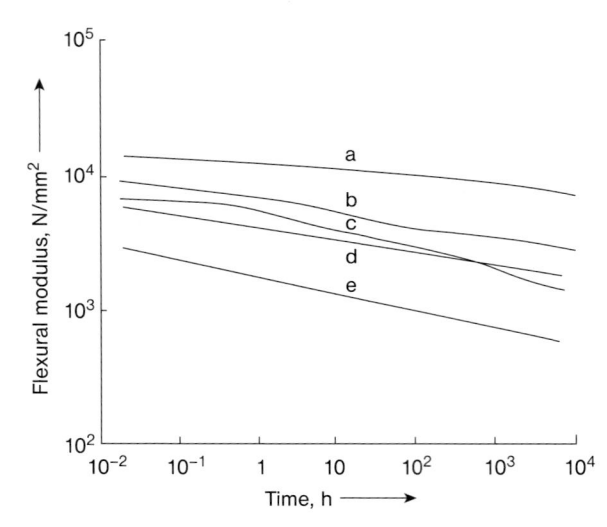

Figure 18. Flexural creep modulus of Vectra A130 (LCP GF30) at various temperatures and under loads from 4 to 80 N/mm² [128]
a) 23 °C, 80 N/mm²; b) 80 °C, 40 N/mm²; c) 120 °C, 20 N/mm²; d) 180 °C, 8 N/mm²; e) 220 °C, 4 N/mm²

Impact Stress. LCPs have very high notched and unnotched impact strength because of their woodlike fibrous structure. The process-determined orientation of the fibers parallel to the surface and, hence, perpendicular to the impact results in a high level of impact resistance. Again, the high performance of LCPs at elevated temperatures is remarkable (Fig. 19).

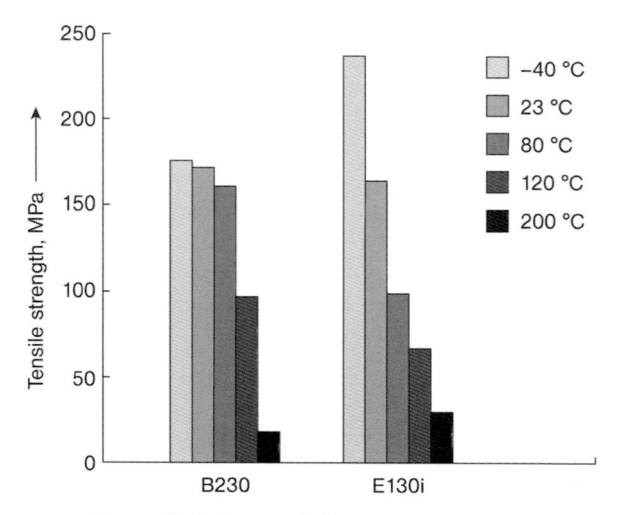

Figure 19. Influence of the temperature on the tensile strength of Vectra (B230, 30% carbon-fiber-filled; E130i, 30% glass-fiber-reinforced) [144]

If the fibrous structure is cut by notching, such as in a notched Izod or Charpy specimen, this resistance is significantly reduced. Although the energy to break drops, it remains still high compared to that of other glass-reinforced thermoplastics [128, 144].

Tribological Properties. In general, LCPs show good friction and wear properties under low loads. However, the friction and wear characteristics are very specific to the kind of application. Typical slip- and wear-modified grades contain PTFE, carbon fibers, graphite, or a combination of these with other fillers and reinforcing materials. Dynamic coefficients of friction, μ, typically range from 0.1 to 0.2 (also depending on the friction partner). More specific data are obtained from standardized tests (Table 15).

Figure 20 compares the dynamic coefficient of friction, μ, of a number of Vectra LCP grades with that of standard polyoxymethylene (POM). The test specimens were in unlubricated sliding contact with a rotating steel shaft under low load. Conditions were as follows: For friction testing, a polymer plate run against a steel ball (diameter = 13 mm, roughness height = 0.1 µm), load F_N = 6 N, sliding speed v = 60 cm/min. Wear testing was performed on a rotating shaft with a peripheral speed v = 136 m/min, load F_N = 3 N, for 60 h [144].

Although LCPs exhibit a moderate surface hardness, the wear tends to be low in the direction of the LCP and filler fibers. Perpendicular to this direction, the wear is significantly higher because the inner adhesion of the material is significantly lower in this direction than in the direction of fiber orientation. Again, LCPs exhibit a high performance at elevated temperatures, also in terms of low friction. This performance is used, for example, in

Table 15. Coefficient of friction of Vectra with various fillers (according to ASTM D1894) [144]

Description	Vectra LCP grade	Coefficient of friction μ (in flow direction)	
		Static	Dynamic
Glass-fiber-reinforced	A115	0.11	0.11
	A130	0.14	0.14
Carbon-fiber-reinforced	A230D-3	0.19	0.12
PTFE modified	A430	0.11	0.11
	A435 FDA	0.16	0.18
Graphite	A625	0.21	0.15

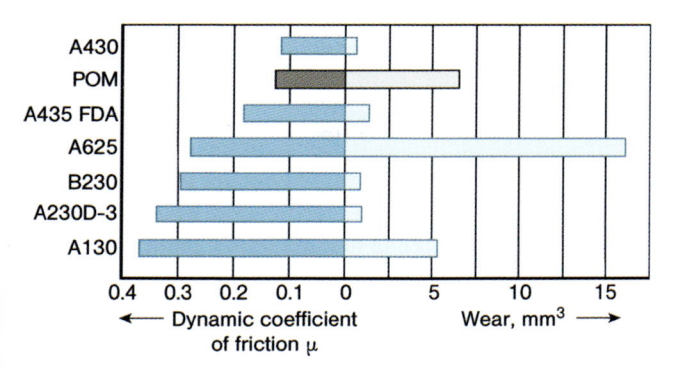

Figure 20. Coefficient of friction μ and wear (average from longitudinal and transverse to flow direction) of several modified LCP grades and standard POM [144]

moving parts in automotive lighting applications where the temperatures actually go up to 200 °C [145].

6.2.2. Thermal Properties

Thermodynamics. LCPs have a significantly lower heat of fusion than semicrystalline thermoplastics. Figure 21 shows the differential scanning calorimetry (DSC) curves for Vectra A130 (LCP GF30) and two other thermoplastics.

The curve for the LCP resembles more that of an amorphous thermoplastic ($\rightarrow$ Plastics, Properties and Testing). This is attributed to the liquid crystalline structure of LCPs. In LCPs, the transition from the solid to the melt phase is associated with only a minor order change because the melt maintains the high orientation of the solid. There is nearly no entropy increase when melting an LCP.

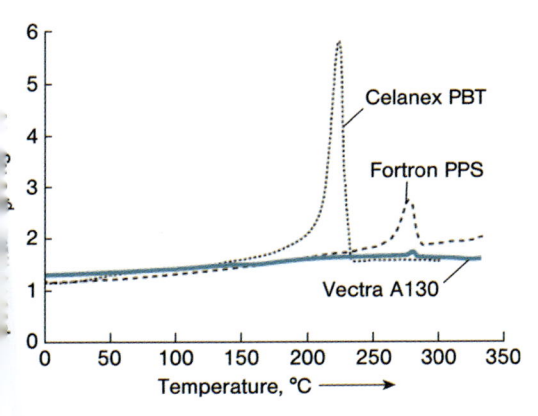

Figure 21. DSC curves of Vectra A130 (LCP GF30), poly(butylene terephthalate) (PBT), and poly(phenylene sulfide) (PPS) [144]

Because of the high order of the melt state and its ability to solidify with minimal change in structure, the transition energy during melting or freezing of LCPs is one to two orders of magnitude lower than that of semicrystalline thermoplastics (Fig. 22) [144, 145].

Dynamic Mechanical Analysis (DMA). Because LCPs are anisotropic, both the orientation of the test specimen and the direction of the mechanical stress have a significant influence on the test results. Dynamic mechanical analysis (DMA) of two or more LCP grades can be used to compare their retention of stiffness at increasing temperatures (Fig. 23). The higher the stiffness at any temperature, the more creep-resistant the LCP will be at this temperature [144].

Table 16 gives the temperatures at which the moduli G′ fall to 50% of the ambient-temperature moduli for various commercial LCP grades.

In general, the higher the temperature $T_{1/2G'}$, the more creep-resistant the grade will be at elevated temperatures. For example, Vectra A130 ($T_{1/2G'}$ = 88 °C) is more creep-resistant than Vectra A625 ($T_{1/2G'}$ = 83 °C) in the temperature range 23–90 °C. Likewise, Vectra E130i ($T_{1/2G'}$ = 119 °C) is more creep-resistant than Vectra A130 ($T_{1/2G'}$ = 88 °C) in the temperature range 80–120 °C [144].

Similarly, peaks in the damping factor curve indicate transitions and temperature ranges where the polymer will be more energy-dissipating. Typically, LCPs have two damping peaks, one at the α-transition temperature (T_g) and the other at the lower α-transition temperature. The damping-peak temperatures are listed in Table 16. Glass transitions usually occur in

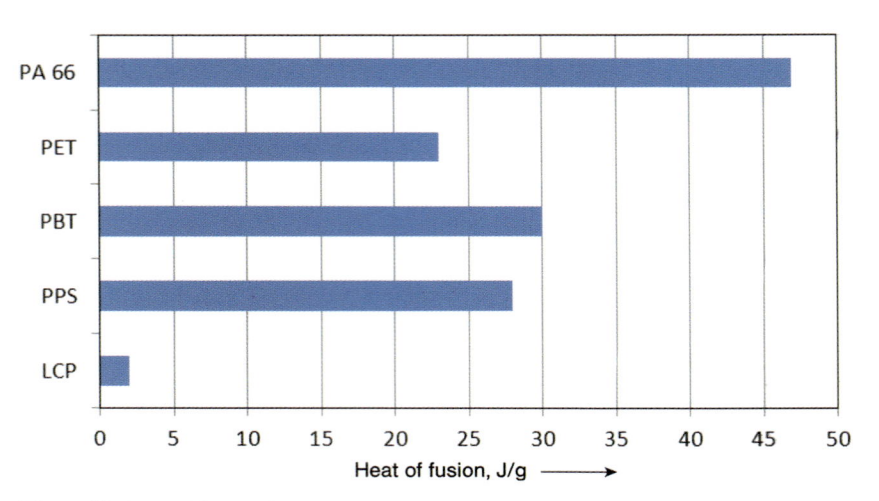

Figure 22. Heat of fusion of an LCP and other engineering thermoplastics [144]

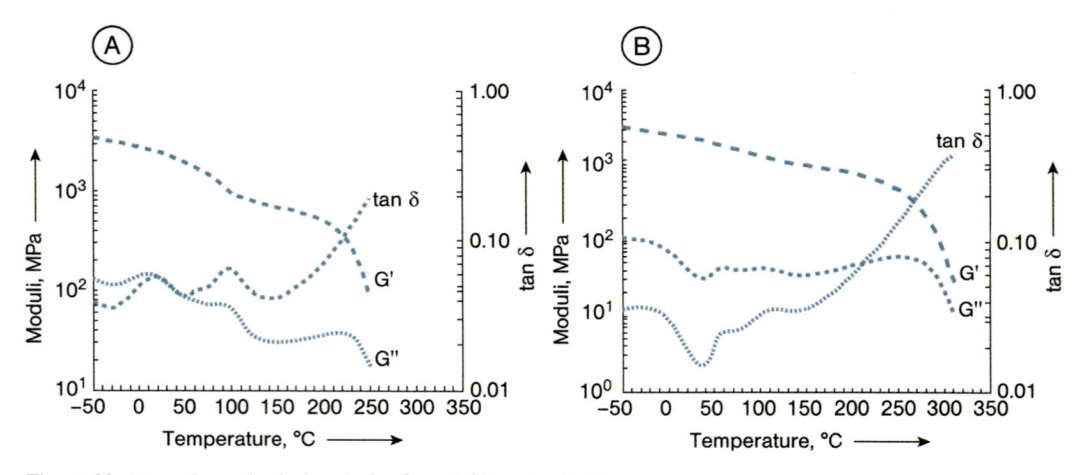

Figure 23. Dynamic mechanical analysis of two LCP grades [144]
G' = shear storage modulus; G'' = shear loss modulus; $\tan\delta$ = damping factor
A) Vectra A130 (GF30, T_m = 280 °C); B) Vectra E130i (GF30, T_m = 335 °C)

Table 16. DMA data of various commercial Vectra LCP grades [144]

		α-Transition temperature, T_g, °C	β-Transition temperature, °C	Modulus G' at 23 °C, MPa	Half-modulus temperature, $T_{1/2G'}$, °C
A130	30% GF	97	18	2400	88
A230D-3	30% CF	103	21	3200	95
A625	25% graphite	98	22	3000	83
E130i	30% GF	113	59	2300	119
E471i	35% GF/Min	115	57	2500	119
E540i	40% Min	117	61	3200	113
S135	35% GF	128	35	2400	121
S471	45% GF/Min	131	39	3600	119

the 120–155 °C range while the low-temperature secondary loss peaks are at 10–80 °C. In general, a damping-peak temperature just above the ambient temperature makes the LCPs a good sound absorber. When struck, they do not "ring", but they "clunk" or sound "dead" [144].

Coefficient of Linear Thermal Expansion. The coefficient of linear thermal expansion of LCPs is very low compared to that of other thermoplastics. While the coefficient of linear thermal expansion depends on the nature of resin, it is also strongly influenced by the nature and amount of fillers. Most commercial LCPs have coefficients of linear thermal expansion in flow direction of about 10^{-6} K^{-1}. This meets the values of glass, steel, ceramic, or glass fiber/epoxy substrates. The coefficient of linear thermal expansion is up to ten times higher perpendicular to the flow [128, 144].

The low coefficients of linear thermal expansion result in a high dimensional stability. In combination with the high thermal stability (decomposition temperatures >370 °C), this makes LCP suitable, for example, for all common soldering technologies, including vapor-phase (wave) and IR (reflow/SMD/through-hole) soldering [128, 144] (→ Soldering and Brazing).

6.2.3. Behavior on Exposure to Flame

Most LCPs are inherently flame-resistant and self-extinguishing. On exposure to very high flame temperatures, the fully aromatic LCPs form a carbon char layer, which retards the development of flammable gases. Some LCPs have a self-ignition temperature of >540 °C [144, 146].

6.2.4. Electrical Properties

Without any conductive additive, LCPs are, like most thermoplastics, electrical insulators. Surface resistivity typically exceeds 10^{13} Ω m. Volume resistivity is typically in the range of 10^{14} Ω.

Conductive LCPs are usually modified with carbon fibers or conductive carbon black. This modification results in conductive properties comparable to those of other conductive-modified thermoplastics. Surface resistivity can drop down to 10^0 Ω m, volume resistivity can be reduced down to 10^1 Ω [144].

6.2.5. Rheological Properties

Solid LCPs have a nematic liquid crystal structure. The melt viscosity decreases continually with increasing deformation (shear) rate. At the shear rates that normally occur during injection molding, the melt viscosity of Vectra is lower than that of conventional filled or reinforced semicrystalline thermoplastics (Fig. 24).

From a technical point of view, it is important that this drop in melt viscosity with increasing shear rates is both continuous and reversible, i.e., not caused by degradation of the polymer backbone. The nonlinear behavior that some semicrystalline polymers show is either caused by nonlinear (or turbulent) flow effects within the measurement or by polymer degradation. The latter can be proved by retesting at lower shear rates.

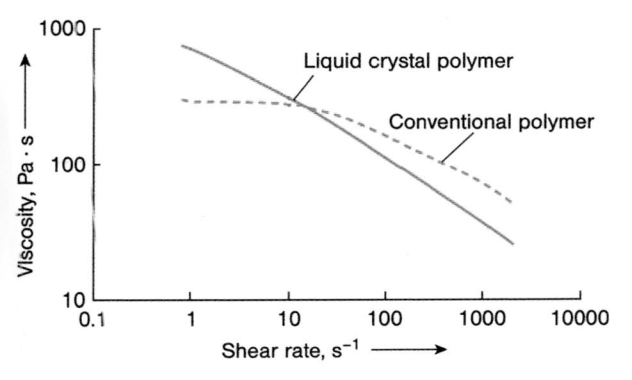

Figure 24. Comparison of the melt viscosity of LCPs and conventional semicrystalline polymers [144]

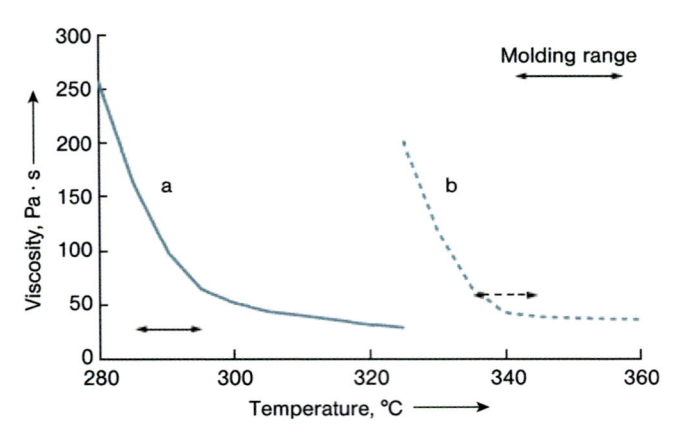

Figure 25. The melt viscosity of two Vectra grades in relation to the temperature [144]
a) A130 (LCP GF30, $T_m = 280\,°C$); b) E130i (LCP GF30, $T_m = 335\,°C$)

This rheological behavior allows LCPs to be processed to thin-walled parts (>0.2 mm wall thickness) or to complicated part structures with high flow lengths. In contrast to other thermoplastics, the flow length can be increased by increasing the shear rate, either by higher injection speeds or (which is unusual for other polymers) by reducing the injection gate diameter [128, 144].

Above the melt-transition temperature (which is in fact the transition temperature from the nematic to the isotropic state), the melt temperature has only a minor influence on the melt viscosity (Fig. 25).

This means that filling problems are rather to be solved by adjusting tools or the injection speed and pressure, and not the temperature.

6.3. Chemical Properties

6.3.1. Chemical Resistance

LCP show a very good resistance to chemicals, particularly to organic solvents (even at high temperatures). Their resistance to concentrated mineral acids and bases (both inorganic and organic) at elevated temperature is reduced. In general, caustic components in the presence of moisture or alcohols are problematic; acids become more critical with increasing acidic and oxidation strength.

The resistance of LCPs to methanol and methanol-containing solvents or fuels depends on the temperature and time of contact. In permanent contact with methanol-containing liquids the temperature is limited to max. 70 °C [144].

Two degradation mechanisms are involved. First, LCPs, similar to any polyester, are generally sensitive to hydrolysis, especially alcoholysis and aminolysis. This sensitivity depends on the reactivity and the size of the groups. In general, primary alcohols are more critical for LCPs than secondary alcohols, while tertiary alcohols are usually uncritical, which might be due to steric effects [127, 147]. Alcoholysis is catalyzed mainly by caustic components. In the case of amines, which are often present in solvents or fuels, primary amines are more critical than secondary amines. Tertiary amines and hindered amine light stabilizers (HALS) are not critical for LCPs.

The other degradation mechanism is oxidation of either the aromatic sections of the backbone (mainly leading to discoloration) or the ester bond.

Discoloration and degradation can also be caused by the presence of iron ions.

6.3.2. Hydrolysis Resistance

LCPs show a higher resistance to hydrolysis than other polyesters. Figure 26 shows the results of immersion tests in hot water.

However, prolonged exposure at high temperatures leads to gradual hydrolytic degradation. Similarly to other polyesters, LCPs have an increased degradation rate in the presence of caustic components and at increasing temperatures.

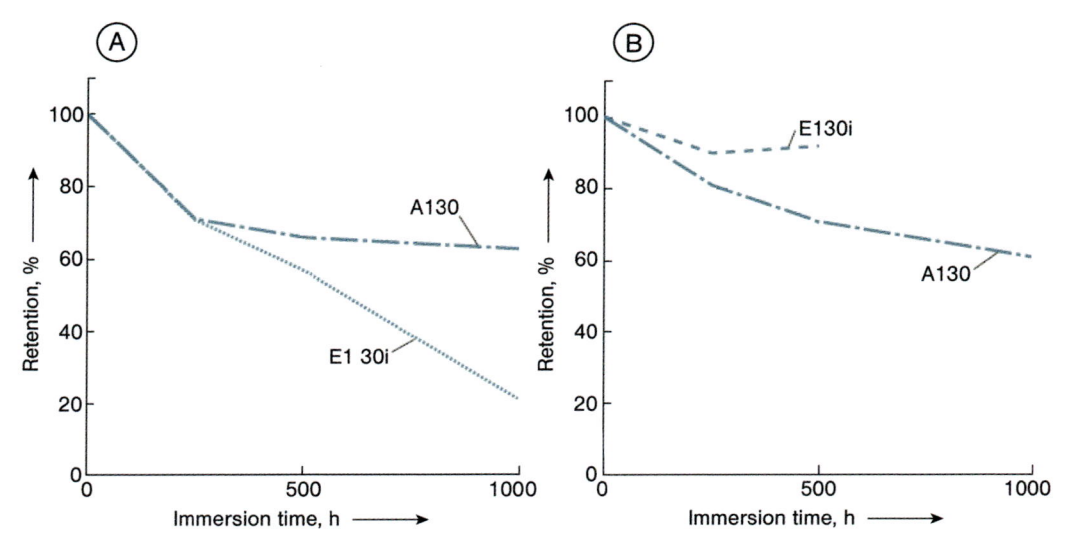

Figure 26. Hydrolysis resistance of Vectra A130 (LCP GF30) and E130i (LCP GF30) tested by immersion in hot water (120 °C, 2 bar) [144]
A) Tensile strength retention; B) Tensile modulus retention

6.4. Production

In general, the fully aromatic LCPs consist of (co) polyesters or polyesteramide copolymers and all common mechanisms for the polymerization of polyesters can be applied to produce LCPs (→ Polyesters, Section 3.2). Technically, all manufacturers use the polycondensation reaction of mesogens with suitable functionalities.

Aromatic hydroxycarboxylic acids are suitable mesogens, but aromatic diols (e.g., hydroquinone) combined with aromatic diacids (e.g., 4,4′-dihydroxydiphenyl (4,4′-biphenol) and terephthalic acid) or combination of hydroxycarboxylic acids with diols and diacids are also possible.

Besides the nonthermoplastic poly(p-hydroxybenzoic acid), another simple structure is the copolymer of p-hydroxybenzoic acid (HBA) and hydroxynaphthoic acid (HNA), Vectra A950:

In this material, the mesogen HBA builds a rigid, stiff polymer chain. HNA is employed to the backbone as a chain-disrupting comonomer. The addition of a "crankshaft" into the polymer chain reduces the melting point and allows melt processing steps.

A huge number of comonomers for controlling the thermal and mechanical properties of LCPs are available. Most commercial LCPs are based on a selection of HBA, HNA, isophthalic acid (IA), terephthalic acid (TA), hydroquinone (HQ), bisphenol (BP) and naphthoic diacid (NDA). Patents give a broad overview on the comonomers used [135, 137–140].

Three different methods are employed for the polycondensation on a commercial scale.

1. Bulk polymerization in the melt
2. Bulk polymerization in the solid state
3. Suspension polymerization in nonaqueous media

All three methods share a common reaction scheme comprising three steps. The first step involves a complete esterification of the hydroxy groups followed by the formation of a low-molecular-mass prepolymer. In the last step, the molecular mass of the polymer is increased by continuing the polymerization under vacuum.

The polymerization reaction is usually catalyzed by, e.g., sodium alkoxides, titanium alkoxides, lithium hydroxide, or *p*-toluenesulfonic acid, but also shows good results without a catalyst [148].

In the first step, all hydroxy functions are transferred into acetates with the help of acetic acid anhydride (Fig. 27). This step is required both to reduce the thermodynamic barrier of the polycondensation reaction (reduction of the activation energy) and to achieve milder process conditions because of the lower melting points of the corresponding acetates [127, 147].

For esterification, an excess of 5–20 mol% of acetic anhydride is usually added [149]. Acetylization is performed under reflux at ca. 150 °C for 2–4 h. Then, the mixture is quickly heated up to 240 °C and then heated to 300 °C at a slower rate. The mixture is kept for several hours at this temperature, while the acetic acid and excess acetic anhydride are removed by destillation. This process produces low-molecular-mass polymers [150, 151].

In the suspension polymerization process, an inert solvent remains in the reaction mixture and the polymer will precipitate from the suspension as a powdered material [148].

In the melt polymerization process, the temperatur is increased further above the melting point of the resulting polymer, up to more than 370 °C. The polycondensation then continues in the melt state; the degree of polymerization can be measured by the amount of acetic acid formed and the viscosity of the melt. At the end of the reaction, the product is driven off the reactor and granuled for further use [151].

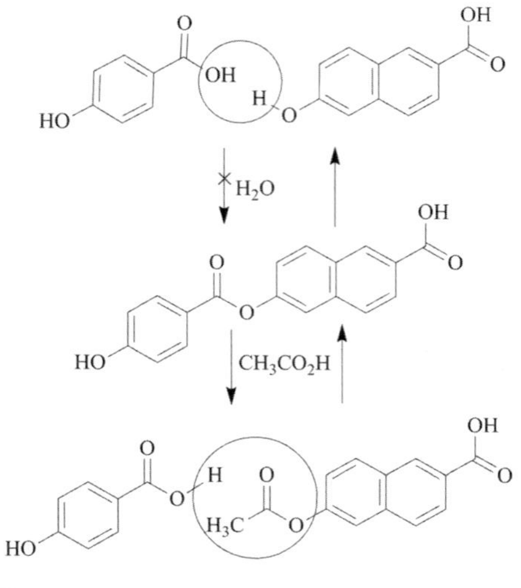

In the solid-state polymerization method, the product is driven off the reactor at a lower degree of polymerization and chopped to pellets. This allows a faster and more effective exhaustion of the reactor [151].

Both the powdery products of the suspension polymerization and the prepolymer of the bulk polymerization methods need a post-polymerization step to yield high-molecular-mass polymers. Therefore, the powder or pellets are heated under vacuum to a temperature just below the melting point. Again, acetic acid is formed and removed by vacuum. While the higher surface area of the powder allows shorter reaction times in the post-polymerization step, the occurrence of powder-clotting during solidification is more likely in this process [135].

6.5. Uses

LCPs are obtained in a variety of forms, from sinterable high-temperature compounds to injection-moldable compounds. LCPs can be welded, though the lines created by welding are a weak point in the resulting product.

For the thermoset poly-(*p*-hydroxybenzoic acid) polymers, the fabrication is accomplished by metallurgical techniques which are not normally used for polymers, for example, compression sintering at 420 °C and ca. 70 MPa [134].

Thermoplastic LCPs are usually processed by injection molding or extrusion techniques. The flowability and mechanical stability of the thermoplastic LPCs at thin wall dimensions, combined with soldering resistance and good electrical properties, offers a broad use in the electric and electronics market. Many applications as connectors and microconnectors (e.g., in play consoles, computer and laptops, mobile phones, television, and automotive), switches, and electronic components with small wall thicknesses are reported. But also the automotive market and the application in lightings (e.g., sensors, actuators, lamp sockets, LED sockets, heat shielding in automotive application) offer a broad use [145, 152–154].

Additional properties, e.g., no embrittlement at low temperatures, are relevant for automotive,

Figure 27. Esterification step in the synthesis of LCP [135]

aerospace, sporting goods (e.g., rovings and sails). Furthermore, the excellent heat resistance which is interesting for applications in, e.g., protection clothes, bakeware (heat resistance combined with a low surface energy gives non-staining and nonsticking surfaces) offers huge opportunities besides the electric and electronics market [145, 155].

6.6. Trade Names (Examples)

Ekonol (Saint Gobain)
Laxtar (Lati)
Selcion (Samsung)
Siveras (Toray)
Sumikasuper (Sumitomo)
Vectra LCP (Ticona, Polyplastics)
Xydar (Solvay)
Zenite (Ticona)

6.7. Economic Aspects

The ratio for the sales price for nylons to the sales price for the high-temperature LPCs varies from 1:5 to 1:30. However, the mechanical performance of LPCs in combination with their flow performance and the ability to be processed into finest structures allows a miniaturization of parts, which is one of the drivers for the use of LCPs in electronic applications such as cellphone components. This miniaturization means a significant saving in the consumption of polymers for the production of parts.

Another economic aspect is the high tolerance of most LCP to the use of regrind in injection-molding processes. Typically, amounts up to 50% of regrind can be tolerated in molding processes without a significant drop of material performance [156].

References

1 R.B. Seymour, G.S. Kirshenbawm: *High Performance Polymers: Their Origin and Development*, Elsevier, Amsterdam 1986.
2 J.G. Wirth, D.R. Heath, US 3730946, 1973.
3 T. Takekoshi, J.E. Kochanowsky, J.S. Manello, M.J. Webber, *Polym. Prepr. (Am. Chem. Soc. Div. Polym. Chem.)* **20** (1979) no. 1, 179.
4 T. Takekoshi, J.E. Kochanowsky, J.S. Manello, M.J. Webber, *J. Polym. Sci. Polym. Chem. Ed.* **23** (1985) 1759–1769.
5 T. Takekoshi, J.E. Kochanowsky, J.S. Manello, M.J. Webber, *J. Polym. Sci. Polym. Symp.* **74** (1986) 93–108.
6 L.R. Schmidt, E.M. Lovgren, P.G. Meissner, *Int. Polym. Proc.* **IX 4** (1989) 270–276.
7 J.W. Verbicky in H.F. Mark, N.M. Bikales, C.G. Overberger, G. Menges (eds.): *Encyclopedia of Polymer Science and Engineering*, vol. **12**, John Wiley, New York 1988, 364–382.
8 *Kirk-Othmer Encyclopedia of Chemical Technology*, 3rd ed., vol. **18**, J. Wiley & Sons, New York, pp. 611–615.
9 I.W. Serfati, *Proceedings of the First Technical Conference on Polyetherimides*, New York 1982, 149–161.
10 H. Farong, W. Xueqiu, L. Shijin, *Polym. Degrad. Stab.* **18** (1987) 247–259.
11 R.O. Johnson, E.O. Teutsch, *Polym. Compos.* **4** (1983) 162.
12 S.A. White, S.R. Weissman, R.P. Kambour, *J. Appl. Polym. Chem.* **27** (1982) 2675–2682.
13 D.E. Floryan, G.L. Nelson, *J. Fire Flammability* **11** (1980) 284.
14 H.D. Burks, T.L. St. Clair, *J. Appl. Polym. Sci.* **30** (1985) 2401–2411.
15 R.O. John, H.S. Burklis, *J. Polym. Sci. Polym. Symp.* **70** (1983) 129.
16 General Electric, US 4959121, 1990 (W.V. Dumas, D.F. Foust).
17 J.E. Harris, L.M. Robeson, *J. Appl. Polym. Sci.* **35** (1988) 1877.
18 C.A. Arnold et al., *Polymer. Prepr.* **28** (1987) 217; **29** (1988) 349.
19 Du Pont, US 3065205, 1962 (W.H. Bonner).
20 ICI, GB 971227, 1964 (I.E. Goodman et al.).
21 Y. Iwakura et al., *J. Polym. Sci. Polym. Chem. Ed.* **6** (1968) 3345.
22 Du Pont, US 3441538, 1969 (B.M. Marks).
23 R.N. Johnson et al., *J. Polym. Sci. Polym. Chem. Ed.* **5** (1967) 2375.
24 ICI, GB 1414421 and GB 1414422, 1975 (J.B. Rose, P.A. Staniland).
25 T.E. Attwood et al., *Polymer* **22** (1981) 1096.
26 ICI, EP 0001879, 1977 (J.B. Rose and P.A. Staniland).
27 Mitsubishi Gas Chemical Company, EP 0306051, 1988 (S. Ebata, Y. Higuchi).
28 ICI, EP 0182648, 1985 (J.A. Daniels).
29 Gharda Chemicals Ltd., EP 1170318, 2001 (K.H. Gharda et al.).
30 Raychem, US 3953400, 1976 (K.J. Dahl).
31 Gharda Chemicals Ltd., GB 2446397, 2007 (K.H. Gharda et al.).
32 Du Pont, EP 0225144, 1986 (E.G. Brugel).
33 ICI, EP 0397356, 1990 (D.J. Kemmish et al.).
34 Hay, US 5254663, 1993 (A.S. Hay).
35 Polymics, WO 2010062361, 2009 (Y.-F. Wang et al.).
36 P.C. Dawson, D.J. Blundell, *Polymer* **21** (1980) 577.
37 E.J. Stober et al., *Polymer* **25** (1984) 1845.
38 J.N. Hay, D.J. Kemmish, *Polymer* **29** (1988) 613.
39 J. Devaux et al., *Polymer* **26** (1985) 1994.
40 ICI Advanced Materials, Victrex PEEK, Lit. Ref. VK 2/0988, Wilton.
41 X. Jin et al., *Br. Polym. J.* **17** (1985) 4.
42 D.J. Blundell, B.N. Osborn, *Polymer* **24** (1983) 953.
43 T.E. Attwood, P.C. Dawson, J.L. Freeman, L.R.J. Hoy, J.B. Rose, P.A. Staniland, *Polymer* **22** (1981) 1096.
44 M. Shibata, R. Yosomiya, J. Wang, Y. Zheng, W. Zhang, Z. Wu, *Macromolecular Rapid Communications*, **18** (1997) 99.
45 R.B. Prime, J.C. Seferis, *J. Polym. Sci. Polym. Symp.* **24** (1986) 641.
46 Evonik Degussa GmbH, VESTAKEEP® 4000G technical data sheet.
47 Victrex plc, Victrex® HT™ G45 technical data sheet.

48. Victrex plc, Victrex® ST™ G45 technical data sheet.
49. Oxford Performance Materials, OXPEKK® technical data sheet.
50. ICI Advanced Materials, Victrex PEEK, Lit. Ref. VK 3/0988, Wilton 1988.
51. EOS, US 7847057, 2008 (F. Müller et al.).
52. Amoco, EP 0176988, 1985 (L.M. Robeson, J.E. Harris).
53. Amoco, US 5079309, 1989 (J.E. Harris et al.).
54. Ticona, WO 2009128825, 2008 (M. Ajbani et al.).
55. Hoechst-Celanese, EP 0392855, 1990 (E. Alvarez et al.).
56. 3 M, GB 1060546, 1963 (H.A. Vogel).
57. J.B. Rose, *Polymer* **15** (1974) 456.
58. ICI, GB 1153035, 1965 (D.A. Barr, J.B. Rose).
59. R.N. Johnson et al., *J. Polym. Sci. Polym. Chem. Ed.* **5** (1967) 2375.
60. Union Carbide, GB 1078234, 1963 (A.G. Farnham, R.N. Johnson).
61. T.E. Attwood et al., *Polymer* **18** (1977) 259.
62. Union Carbide, Udel Polysulphone: An Outstanding Engineering Polymer for Moulding and Extrusion, 1978.
63. ICI Advanced Materials, Victrex PES: Chemical Resistance Data, Lit. Ref. VS 7/0689, Wilton 1989.
64. ICI Advanced Materials, Victrex PES: Properties and Processing, Lit. Ref. VS 10/1089, Wilton 1989.
65. Nippon Chemtec Consulting Inc.: "PPS & Competitive Resins," *Evaluation of High Performance Engineering Resins*, part 2, R & D Evaluation Rep. Ser. No. 50, Osaka 1990.
66. C. Friedel, J.M. Crafts, *Ann. Chim. Phys.* **14** (1888) 433.
67. J.W. Cleary, *Polym. Sci. Technol. (Plenum)* **31** (1985) 159.
68. P. Genvresse, *Bull. Soc. Chim.* **17** (1897) 599.
69. J.W. Cleary, *Polym. Sci. Technol. (Plenum)* **31** (1985) 173.
70. 52. A.D. Macallum, *J. Org. Chem.* **13** (1948) 154; US 2538941, 1951.
71. R.W. Lenz, C.E. Handlovits, *J. Polym. Sci.* **43** (1960) 167.
72. R.W. Lenz, C.E. Handlovits, H.A. Smith, *J. Polym. Sci.* **58** (1962) 351. Dow Chemical, US 3274165, 1965.
73. Phillips Petroleum, US 3354129, 1963.
74. D.R. Fahey, C.E. Ash, *Macromolecules* **24** (1991) 4242; D.R Fahey, H.D. Hensley, C.E. Ash, D.R. Senn, *Macromolecules* **30** (1997) 387.
75. I. Koschinski, K.-H. Reichert, H.-J. Traenckner, *Angew. Makromol. Chem.* **11** (1992) 201; M. Dauben, K. Platkowski, K.-H. Reichert, *Angew. Makromol. Chem.* **234** (1996) 177.
76. W. Koch, W. Heitz, *Makromol. Chem.* **184** (1983) 779; W. Koch, W. Risse, W. Heitz, *Makromol. Chem., Suppl.* **12** (1985) 105.
77. M. Haubs, R. Wagener, *e-polymers* 2002, no. 56.
78. Phillips Petroleum, US 3707528, 1971.
79. Phillips Petroleum, US 4526684, 1985; EP-A 103261, 1984.
80. Phillips Petroleum, US 4383080, 1982; R.T. Hawkins, *Macromolecules* **9** (1976) 189.
81. Phillips Petroleum, US 4038261, 1977; DE 3030488, 1979.
82. Phillips Petroleum, US 3919177, 1975; US 4038260, 1977; US 4039518, 1977; Idemitsu PetrochemicalsEP 409105, 1991.
83. Phillips Petroleum, US 4415729, 1982; Dainippon, US 4490522, 1984.
84. Kureha, EP-A 166368, 1984; EP-A-244187, 1987; EP-A 256757, 1986, EP-A 259984, 1986.
85. TiconaWO 2001043846, 2001; WO 2002047795, 2002; Kureha, WO 2012008340, 2012; EP 1788010, 2007.
86. Idemitsu Petrochemicals, US 5756654, 1998; US 5898061, 1999; US-A-2005118093, 2005; US 5344973, 1994.
87. Idemitsu Petrochemicals, US-A-2003027943, 2003; EP 1106643, 2001; EP 477964, 1992.
88. Ticona, EP 737705, WO 9738040, WO 9814503.
89. Toray, WO 2008105438, 2008; WO 2007034800, 2007.
90. Phillips Petroleum, US 3645697, 1972.
91. Phillips Petroleum, US 4056515, 1977; US 4060520, 1977, US 4066632, 1978.
92. Bayer, EP 501217, 1992; EP 499929, 1992; EP 372252, 1990; EP 0374462, 1990; EP 0126369, 1984.
93. Ticona, WO 99/45057, 1999.
94. M. Rule, D.R. Fagerburg, J.J. Watkins, P.B. Lawrence, *Makromol. Chem., Rapid Commun.* **12** (1991) 221; D.R. Fagerburg, D.E. van Sickle, *J. Appl. Pol. Sci.* **51** (1994) 989; D.R. Fagerburg, J.J. Watkins, P.B. Lawrence, *Macromolecules* **26** (1993) 114; D.R. Fagerburg, J.J. Watkins, P.B. Lawrence, *J. Appl. Pol. Sci.* **50** (1993) 1903.
95. Eastman Kodak, US 4746758, 1988; US 4786713, 1988; US 4792600, 1988; US 4855393, 1989.
96. SK Chemicals, WO 2009054555, 2009; WO 2009078667, 2009, US 20100022743, 2010.
97. Bayer, DE-OS 3318401, 1983; DE-OS 3338501, 1983; DE-OS 3243189, 1983.
98. Bayer, EP-A 142024, 1985; DE-OS 3339233, 1985; EP-A 162210, 1985; EP-A 171021, 1986; EP-A 175968; DE-OS 3713669, 1988; DE-OS 3723071, 1989.
99. J. Schurz: *Viskositätsmessung an Hochpolymeren*, Berliner Union Verlag, Stuttgart 1972; Struktur-Rheologie, Berliner Union Verlag, Stuttgart 1974.
100. C.J. Stacy, *J. Appl. Polym. Sci.* **32** (1986) 3959. T. Housaki, K. Satoh, *Polym. J. (Tokyo)* **20** (1988) 1163 BayerDE-OS 3909599, 1990.
101. D.O. Hummel: *Atlas der Polymer- und Kunststoffanalyse* 2nd ed., Hanser Verlag, München 1978. C.E. Brown, I. Khoury, M. D. Bezoari, J. Kovacic. *J. Polym. Sci. Polym. Chem. Ed.* **20** (1982) 1697.
102. G. Montaudo, M. Przybilski, H. Ringsdorf, *Makromol. Chem.* **176** (1975) 1763. G. Montaudo, C. Puglisi, E. Scamporrino, D. Vitalini, *Macromolecules* **19** (1986) 2157. F. Quella, *Kunststoffe* **71** (1981) 386.
103. A. Kaul, *Polym. Prep. Am. Chem. Soc. Div. Polym. Chem.* **28** (1987) 229.
104. P.A. Lowell, R.H. Still, *Polym. Degrad. Stab.* **18** (1987) 33. A. Kinugawa, *Kobunshi Ronbunshu* **44** (1987) 139.
105. F. Ehrenberger, S. Gorbach: *Methoden der organischen Elementar- und Spurenanalyse*, Verlag Chemie, Weinheim, Germany 1973. T.S. Ma, R.C. Rittner: *Modern Organic Elemental Analysis*, Marcel Dekker, New York 1979.
106. Kureha, EP-A 238193, 1986.
107. Bayer, DE-OS 3828056, 1990.
108. Toyo Soda, Hodogaya, EP-A 225 471, 1987.
109. L.C. Lopez, G.L. Wilkes, *Polymer* **30** (1989) 147.
110. B.J. Tabor, E.P. Magré, J. Boon, *Eur. Polym. J.* **7** (1971) 7. J. Garbarczyk, *Polym. Commun.* **27** (1986) 335.
111. Phillips Petroleum, Ryton PPS-Information no. 2, Bartlesville 1986. Celanese Eng. Resins, Fortron Polyphenylene Sulfide Introduction, Chatham 1987. R.S. Shue, *Plastics Eng.* **4** (1983) 37.
112. Phillips Petroleum; Ryton PPS, Polyphenylene Sulfide Resins, Bartlesville1986.
113. Phillips Petroleum, Ryton PPS Tesins, Technical Service Memorandum, TSM-275, Bartlesville 1976; TSM-266, Bartlesville 1983.
114. Ciba Geigy, Craston Polyphenylene Sulfide Moulding Compounds, Cambridge 1988. Celanese Eng. Resins, Fortron Polyphenylene Sulfide (PPS), Chatham1987. Susteel PPS, *Plas. Ind. News* **33** (1987) 4. Bayer, Polyphenylensulfid Tedur, Leverkusen 1990.

115 LNP Eng. Plastics, Selbstschmierende verstärkte Thermoplaste, LNP-Bulletin 254–983, Ramsdonkveer 1983.
116 *Mod. Plast. Int.* **5** (1974) 86. M.W. Woods, *1st Int. Ryder Conf. Special Performance Plast. Markets* **2** (1987).
117 M.W. Hill, *Polym. Eng. Sci.* **16** (1976) 831. C.W. Osborn, *Plas. Eng.* **33** (1977) 25. Phillips Petroleum, Ryton PPS, Spritzgießverarbeitung, Bartlesville 1979.
118 Phillips Petroleum, Plastics Techn.Center Report *SSL-243*,Bartlesville 1974. Phillips Petroleum, Ryton PPS Resins, Technical Service Memorandum, TSM-260, Bartlesville 1974, 1976.
119 E. Heitz, *Gummi, Asbest, Kunstst.* **28** (1975) 578. W.D. Powell, *31st Ann. Conf. Exhibition Soc. Plast. Ind. Canada*, Vancouver 1973. Phillips Petroleum, Ryton PPS Resins, Technical Service Memorandum, TSM-275, Bartlesville 1976, TSM-278, Bartlesville 1977.
120 Toray, PPS-Film Torelina, Tokyo 1988. M. Ito, *J. Polym. Sci. Polym. Chem. Ed.* **23** (1985) 245.
121 Phillips Petroleum, Preliminary Bulletin, Bartlesville1983.
122 F.N. Cogswell, *Int. Polym. Process.* **1** (1987) 157. H. Gupta, *Gummi, Fasern, Kunststoffe* **39** (1986) 288. J.B. Cattanach, G. Guff, F.N. Cogswell, *J. Polym. Eng.* **6** (1986) 348. W. Soll, T.G. Gutowski, *SAMPE J.* **24** (1988) 15. H. Heißler, D. Muser, H. Wurtinger: *Kunststoffeinsatz im Automobilbau*, VDI-Verlag, Düsseldorf 1984, pp. 167–177.
123 FDA Regulation 177.2490, Title 21, Code of Federal Regulations.
124 D.G. Brady, J.F. Geibel, *ACS-Meeting, Division Polym. Chem.*, Reno 1987, p. 7. M. Wood, J. Geibel, V. Vives, *ANTEC 1988*, p. 1673.
125 A.S. Wood, *Mod. Plast. Int.* 1988 April, 34.
126 L.C. Sawyer, D.T. Grubb: *Polymer Microscopy*, Chapman and Hall, London 1987, pp. 239–254.
127 J.MG. Cowie in *Polymers: Chemistry and Physics of Modern Materials*, 2nd ed., Blackie Academic & Professional, London 1991, pp. 362–398.
128 H. Domininghaus, P. Elsner, P. Eyerer, T. Hirth: *Kunststoffe, Eigenschaften und Anwendungen*, 7th ed., Springer, Heidelberg 2008, pp. 893–910.
129 W. Kaiser: *Kunststoffchemie für Ingenieure*, Carl Hanser, München 2006, pp. 434–457.
130 http://de.wikipedia.org/wiki/Nematisch#Nematische_Phase (19.08.2011).
131 H.-G. Elias: Chemical Structures and Syntheses, *Macromolecules*, vol. 1, Wiley-VCH, Weinheim 2005, pp.
132 P.J. Flory: Molecular Theories of Liquid Crystals, inA. Cifferi, W.R. Krigbaum, R.B. Meyer (eds.): *Polymer Liquid Crystals*, Academic Press, New York 1982.
133 Gen Mills, US 2728747, 1955 (D. Aelony, M.M. Renfrew).
134 Saint Gobain Ceramic Materials, http://www.coatingsolutions.saint-gobain.com/ekonol.aspx (19.08.2011).
135 O. Juenger in W. Keim (ed.): *Kunststoffe – Synthese, Herstellungsverfahren, Apparaturen*, Wiley-VCH, Weinheim 2006, pp. 264–267.
136 Sumitomo Chemical, B1, 1990 (H. Sugimoto, Y. Ohbe, K. Hayatsu).
137 Amoco Corp., EP 0347228 A2, 1989 (R. Layton, J.W. Cleary).
138 Sumitomo Chemical, JP 56057821 A, 1979 (Y. Kato, H. Sugimoto, M. Hanabata).
139 Du Pont, EP 0778867 B1, 1996 (M.G. Waggoner).
140 Amoco Corp., WO 9004002, 1990 (P.J. Huspeni, B.A. Stern, P. D. Frayer).
141 W.J. Jackson Jr., H.F. Kuhfuss, *J. Polym. Sci, Polym. Phys. Edn.* **14** (1976) 2043.
142 W.C. Wooten Jr., F.E. McFarlane, T.F. Gray Jr., W.J. Jackson Jr. in A. Cifferi, I.M. Ward (eds.): *Ultra-high Modulus Polymers*, Applied Science, London 1979, p. 227.
143 J. Economy, W. Volksen in A. Zachariades, R.S. Porter (eds.): *The Strength and Stiffness of Polymers*, Marcel Dekker, New York 1983, p. 293.
144 Ticona GmbH, *Vectra(r) – Liquid Crystalline Polymers (LCP)*, company brochure, Frankfurt, a.H. 2007.
145 F. Johänning, *Kunststoffe* **10** (2008) 196–199.
146 J. Hein, *Kunststoffe* **9** (2010) 114–115.
147 P. Sykes: *Reaktionsmechanismen der Organischen Chemie. Eine Einführung*, 9th ed., Wiley-VCH, Weinheim 1988.
148 Dart Industries, DE 2025971 C2, 1970 (S.G. Cottis, J. Economy, B.E. Nowak).
149 Amoco Corp., EP 0390915 B1, 1990 (B.A. Stern, M. Matzner, R. Layton).
150 Du Pont, EP 0639204 B1, 1993 (G.R. Alms, M.R. Samuels, M. G. Waggoner, M.G.).
151 Polyplastic Co, JP 2000248056 A, 2000 (T. Shiwaku, Y. Fukute).
152 J. Hein, *Kunststoffe* **12** (2010) 98–100.
153 H. Käll, *Plastverarbeiter* **61** (2010) no. 7, 48.
154 W. Leonhard, *Plastverarbeiter* **53** (2002) no. 8, 52–53.
155 Vectran, http://www.vectranfiber.com/ (accessed 19.08.2011).
156 Underwriter Laboratories, http://data.ul.com .

Reinforced Plastics

GOTTFRIED W. EHRENSTEIN, Universität Erlangen-Nürnberg, Erlangen-Tennenlohe, Federal Republic of Germany

JOSEF KABELKA, Universität Erlangen-Nürnberg, Erlangen-Tennenlohe, Federal Republic of Germany

1.	Introduction, Classification	1325
1.1.	Fiber-Reinforced Polymers	1326
1.2.	Particulate Polymer Composites	1328
1.3.	Laminar Polymer Composites	1328
2.	Reinforced Plastic Components	1329
2.1.	Reinforcing Fibers	1329
2.1.1.	Glass Fibers	1330
2.1.2.	Graphite (Carbon) Fibers	1330
2.1.3.	Aramid Fibers	1331
2.1.4.	Other Continuous Fibers	1332
2.1.5.	Other Discontinuous Fibers	1332
2.2.	Fillers	1332
2.3.	Polymeric Matrices	1332
3.	Fabrication of Polymer Composites	1333
3.1.	Hand Lay-up Technique	1333
3.2.	Spray-up	1334
3.3.	Filament Winding	1334
3.4.	Pultrusion	1335
3.5.	Bag Molding	1335
3.6.	Sheet Molding Compounds (SMC)	1335
3.7.	Thick Molding Compounds (TMC)	1336
3.8.	Bulk (or Dough) Molding Compounds (BMC, DMC)	1336
3.9.	Prepregs	1337
3.10.	Reinforced Reaction Injection Molding (RRIM)	1337
3.11.	Resin Transfer Molding (RTM)	1337
4.	Physical and Chemical Properties	1337
5.	Quality Specifications	1339
6.	Economic Aspects	1340
	References	1340

1. Introduction, Classification

Reinforced plastics, also called polymer composites, are now an important class of engineering materials. They are distinguished by attractive mechanical properties and corrosion resistance, unique flexibility in design capabilities, and ease of fabrication. The word "composite" means "consisting of two or more distinct parts." Polymer composites consist of one or more discontinuous phases embedded in a continuous-phase polymer matrix.

The discontinuous phase is usually harder and stronger than the continuous phase, and is called the reinforcement. The matrix can be classified as thermoplastic (capable of being separately hardened and softened by increases and decreases, respectively, in temperature) or thermoset (changing into a substantially infusible and insoluble material when cured by the application of heat, or through chemical means).

The properties of composites are strongly influenced by the properties of their constituents and the distribution and interactions among them. The constituents usually interact in a synergistic way, providing properties that are not accounted for by a simple volume-fraction sum of the components. Along with the volume fraction and the distribution of discrete units in the discontinuous phase, the interfacial area plays an important role in determining the extent of interaction between the reinforcement and the matrix and — in this way — the final properties of the composite. The fabrication technology of composites, as well as some of their physical properties, is dominated by the chemistry and rheology of the matrix resin and by the type and physical form of the reinforcement.

The use of reinforcing agents makes it possible for any thermoset – or thermoplastic – matrix property to be improved or changed to meet varying requirements. Most polymer

Ullmann's Polymers and Plastics: Products and Processes
© 2016 Wiley-VCH Verlag GmbH & Co. KGaA, Weinheim
ISBN: 978-3-527-33823-8 / DOI: 10.1002/14356007.a23_049

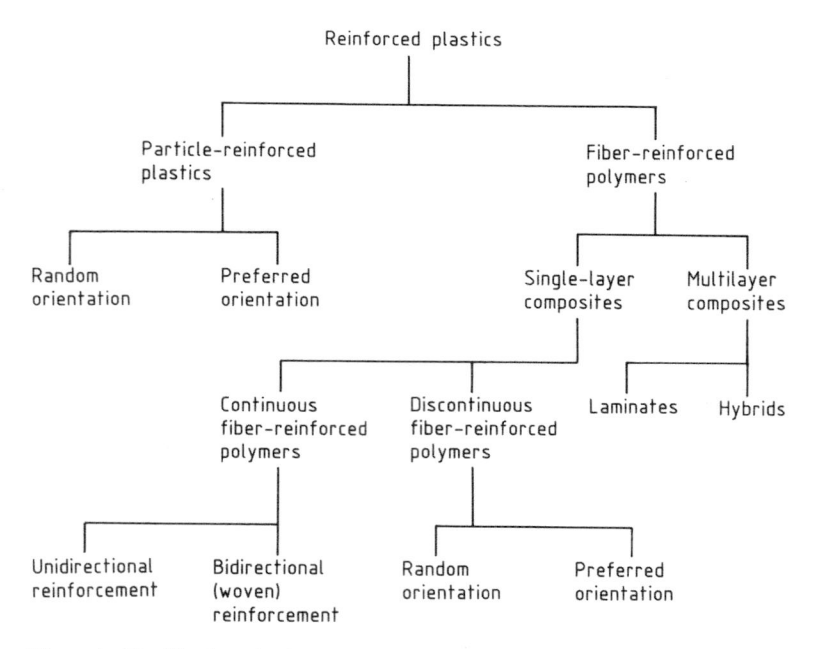

Figure 1. Classification of polymer composites [9]
The single-layer composites include materials composed of identical layers having the same orientation.

composites developed thus far have been fabricated to improve mechanical properties such as strength, stiffness, or toughness. The strengthening efficiency of the discontinuous phase plays the most important role in these products, and the strengthening mechanism depends strongly on the geometry of the reinforcements. Therefore, polymer composites can be classified according to reinforcement geometry [9]. The three major classes of polymer composites are fibrous, laminar, and particulate. The commonly accepted classification scheme for polymer composites is presented in Figure 1.

1.1. Fiber-Reinforced Polymers

Fibrous composites consist of reinforcing fibers embedded in a matrix. All fibers are characterized by a high aspect (length-to-diameter) ratio. This results in a high surface – volume ratio, permitting good anchorage of fibers within a matrix. Fiber reinforcement can be either continuous or discontinuous. Fibers are generally much stronger and stiffer than the surrounding polymer matrix, but they are small in diameter, and bend easily when pushed axially. Although such composites may have outstanding tensile strength, support is essential to prevent individual fibers from bending and buckling. Thus, the principal purpose of the matrix is to function not as a load-bearing constituent but essentially to bind the fibers together and protect them. Fiber-strengthening efficiency is highest when the direction of loading corresponds to the orientation of the fibers. In the transverse direction, properties of the matrix and the quality of the fiber – matrix interface predominate. The combination of matrix and fibrous elements results in a composite structure that may assume different structural forms depending on the forming and processing methods used.

Because the fibers are oriented in a particular direction, material properties may vary with direction. Such a material is referred to as "anisotropic." Many types of anisotropic materials exist, with special designations such as orthotropic, transversely isotropic, and cubic. The fundamental building block for fiber-reinforced plastics is an element with a unidirectional (UD) but transversely random array of fibers embedded in a surrounding matrix, as illustrated in Figure 2. Thus, a unidirectional composite is transversely isotropic (axes O_T and O_P can be selected arbitrarily). In the $O_L - O_T$ plane, however, the properties of a UD lamina

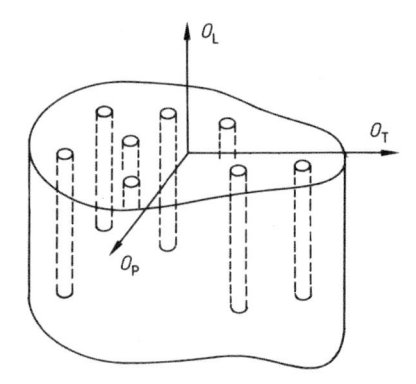

Figure 2. Basic element of fiber-reinforced plastics; coordinate system
O_L = longitudinal; O_T = transverse; O_P = perpendicular

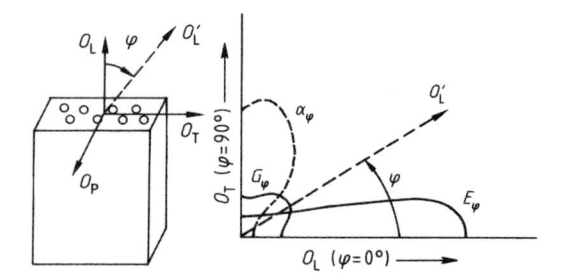

Figure 3. The directional nature of UD lamina properties (E, G, α — Young's modulus, shear modulus, and coefficient of thermal expansion)

are strongly dependent on direction (Fig. 3). The longitudinal properties of UD composites are controlled by the fiber, whereas transverse properties are matrix-dominated. The transverse properties are unsatisfactory in most engineering applications, a problem which is overcome by bonding together two or more laminae to act as an integral structural element displaying the desired properties in all directions. Combining layers with different properties or different orientations gives the designer the ability to tailor a laminate (provide the optimal configuration) for a particular set of loading conditions. However, in applications where the stress may not be predictable, or where stresses are likely to be approximately equal in all directions, the composites themselves must have approximately equal strength in all directions.

An effective way of producing such an isotropic composite material is to use randomly oriented short fibers as reinforcement. Molding compounds consisting of short fibers are usually capable of producing isotropic composites. In short-fiber composites, the so called fiber-end effect plays an important role, and composite properties become a function of fiber length. Loads are not applied directly to the fibers in a composite, but are transferred instead by the matrix to the fiber surface through nonuniform shear stress acting along the fiber – matrix interface (Fig. 4). The tensile stress in the fiber reaches a maximum at the fiber midpoint. The *critical length* of a fiber can be defined as the minimum length at which the ultimate strength of that fiber is reached. Interfacial strength is influenced strongly by the effectiveness of bonding between the fiber and the matrix via coupling agents. The higher the interfacial strength, the shorter is the length that will ensure full exploitation of the fiber's strength.

In the case of short-fiber injection-molded composites, some fibers will have aspect ratios lower than the critical value. These fibers cannot contribute effectively to overall composite strength. Compounding and molding cause fibers to undergo attrition, resulting in a broad distribution in length and a significant fraction of fibers falling below the critical length. In long-fiber injection-molded composites, the mean fiber length is typically 10 times as great as in short-fiber composites. Thus, a higher proportion of the fibers remains effective in load-bearing

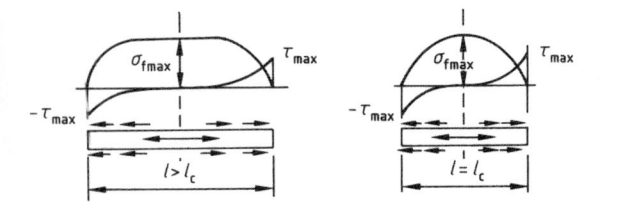

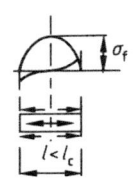

Figure 4. Distribution of tensile stress σ and shear stress τ in short fibers aligned with the load (l = length, l_c = critical fiber length)

situations, and the composites themselves are stronger than their short-fiber counterparts. Figure 4 shows that a small region adjoining the fiber end is subject to less than the maximum fiber stress. This in turn affects the elastic modulus. When the fiber length is much greater than the critical length, composite behavior approaches that of a continous-fiber-reinforced polymer. The tensile strength of a composite is more strongly dependent on interfacial bond strength than is its tensile modulus. This is evident in particulate fillers, where the aspect ratio approaches 1. Here the addition of filler contributes to an increase in modulus but generally to a decrease in the strength of the composite.

In general, the incorporation of reinforcing fibers into a polymeric matrix increases short-term stiffness and strength, increases creep resistance and fatigue endurance, and reduces the influence of temperature on all of these properties. Thermal expansion can be reduced as a way of improving dimensional stability.

Impact strength, which is related to the toughness of a material, is reduced in some polymer composites, a circumstance that can be influenced by several factors. The loading of fibers may transform the failure mode from ductile to brittle, which can in turn lower impact strength. However, this reduction can be counteracted, at least to some extent, if interfacial adhesion between the polymer matrix and the reinforcement permits the absorption of impact energy through debonding. Nevertheless, in systems where matrix-governed ductile failure occurs, increased interfacial adhesion generally improves impact strength.

1.2. Particulate Polymer Composites

By definition, a particle is nonfibrous; it generally lacks a long dimension, and is not effective in improving fracture resistance when added to a polymer matrix (with the exception of rubberlike particles in brittle polymer matrices, which contribute to an improvement in fracture resistance by promoting and then arresting crazing). Despite this definition, polymers containing very short fibers are generally classified as particulate composites. The introduction of hard particles into a polymer matrix leads to a reduction in strength due to stress concentration in the adjacent matrix material.

The two basic reasons for adding particulate filler or reinforcement to a polymer are (1) to fill the system (i.e., replace some of the more expensive polymer with less expensive filler) and (2) to modify properties of the matrix material (e.g., thermal or electrical conductivity), improve processibility, reduce friction or shrinkage, improve performance at elevated temperature, increase wear or abrasion resistance, improve fire retardancy, provide color, and improve appearance.

Platelet-shaped particles, such as flakes, can also serve as reinforcing agents. They offer two-dimensional geometry and, when placed parallel to each other, can be packed more closely than fibers or spherical particles. However, such possibilities have not been fully exploited due to fabrication difficulties. In terms of reinforcement, the longer the particles, the better is the reinforcement. Most particle-filled polymers are manufactured by injection molding, however, so a balance must be struck in terms of a particle length that gives both optimum properties and good processibility. If relatively long fiberlike particles are used as filling, their distribution and orientation may change within the molding. The flow of fiber-filled plastics into a mold during injection molding is a very complex process, and ensuring flow-induced orientation in the optimum direction is difficult. On the other hand, sometimes the challenge is to minimize undesirable effects of anisotropy (e.g., warping and postmolding shrinkage), which may be accentuated by the flow-orientation of fibers. In these cases particulate fillers may be preferable to fibers.

1.3. Laminar Polymer Composites

Laminar composites consist of at least two different layers (laminae) bonded together into one composite piece. Such layers may differ in material, form, or orientation. Combinations of matrix, fibrous mat, aligned fibers, cloth, paper, and other materials can be laminated together in either continuous or discontinuous processing.

Because of the nearly unlimited variety of possible combinations of layers, material forms, and orientations, generalization regarding laminar design is difficult. This great variety offers the designer considerable freedom in both

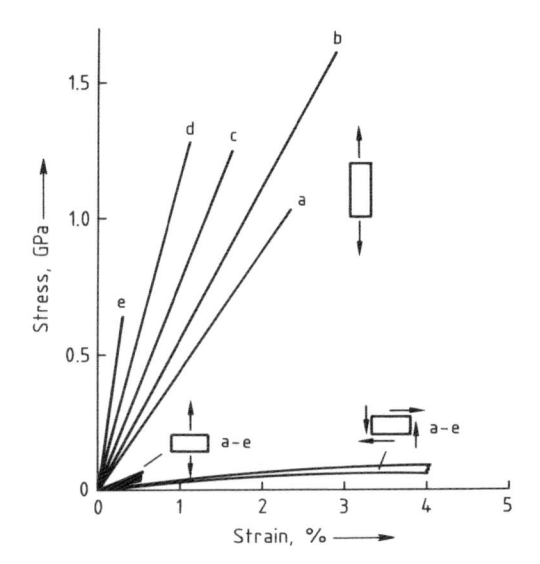

Figure 5. Stress – strain relationships for UD composites under loading in different directions
Matrix = epoxy resin; fiber volume fraction = 0.6
Fiber reinforcement: a) E-glass; b) R-glass; c) Aramid; d) HS graphite; e) UHM graphite

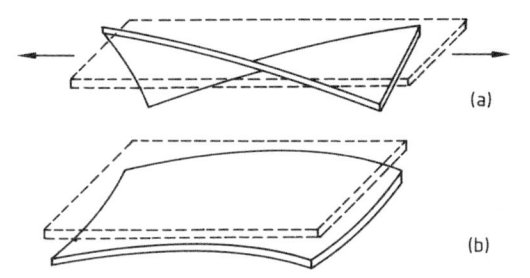

Figure 6. Distortion of an unsymmetrical laminate due to (a) load and (b) temperature change

laminate composition and product form. The unique properties obtained with layered construction encourage the designer to develop composite materials that meet new and innovative design parameters.

The principal design criterion is appropriate fiber orientation within each layer. Polymeric composite structures that incorporate continuously oriented fibers laid in plies can be radically anisotropic in nature. Strength, stiffness, and coefficients of thermal or moisture expansion can all vary more then tenfold in different directions (cf. Fig. 5). Whereas the in-plane strength of such materials may be comparable to the strength of metals, for example, strength in the through-thickness direction can be lower than that of pure matrix. Differing properties of the different layers (e.g., thermal expansion, stiffness) may lead to considerable residual stress when a polymer matrix is cured at elevated temperature. Interlaminar stresses induced through external loading may also be significant, and can cause distortion when the layer-stacking sequence is not symmetric and situated symmetrically to the midplane (Fig. 6).

The theory of laminates has been elaborated [9–17], and many computer programs are now available to help determine the optimal configuration of a laminate under specific conditions; [e.g., LAMINATE (Engineering Software, Dallas, Texas), COMPCAL and CYCLAN (American Society for Composites, Dayton, Ohio), MIC-MAC (Think Composites, Dayton, Ohio), and CADFIBER (IKV, Aachen, Germany)]. In addition to these general programs, many special software packages are being developed for special tasks (e.g., computation of fatigue and crack propagation).

2. Reinforced Plastic Components

2.1. Reinforcing Fibers

As a rule, fibers used for reinforcement purposes are rather fine, ranging from 2 to 20 µm in diameter. The main exception is boron fiber, available in diameters up to 200 µm. Fibers are typically stronger and stiffer than a corresponding bulk material. Fiber strength generally decreases with increasing diameter and length

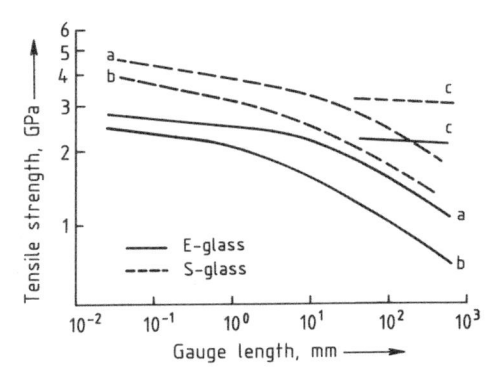

Figure 7. Dependence of fiber strength on gage length
a) Mean filament strength; b) Strength of fiber roving; c) Strength of fiber roving saturated with polyester resin

(Fig. 7). A high fiber aspect ratio permits an effective transfer of load to the fibers via the matrix, which makes such fibers an excellent reinforcing agent.

The major categories of synthetic reinforcing fibers are glass, graphite (carbon), boron, organic, metallic, and ceramic. Natural fibers (asbestos, jute, and sisal), formerly used for economic reasons, have now been generally supplanted by synthetics (in the case of asbestos, because of adverse health effects).

Synthetic fibers are usually prepared in continuous processes and then converted into a form suitable for their intended composite application. The major finished forms are continuous roving, woven roving, fiber mat, chopped strand, and yarn for textile applications.

The form of reinforcement varies with the fabrication process. Three categories can be distinguished: In the first, fibers are placed directly on the component being manufactured. Examples are the filament winding (Section 3.3) and spray-up contact (Section 3.2) molding processes. The second group includes those processes in which a sheet, web, or preform of fibers is prepared and placed in position on a mold where it comes in contact with a resin (e. g., resin-transfer molding, Section 3.11). The third category consists of processes involving the preparation of a compound of fiber and plastic that can flow during molding. Sheet molding compounds (Section 3.6), bulk molding compounds (Section 3.8), and short-fiber-reinforced thermosets and thermoplastics are typical examples within this category.

2.1.1. Glass Fibers (→ Composite Materials; → Fibers, 12. Glass Fibers)

Glass fibers constitute the most widely used form of synthetic composite reinforcement. Chemically, glass is an amorphous material consisting essentially of a silica network. Other chemical components are added to decrease its viscosity to levels suitable for melting and fiber formation. Both the physical and the chemical properties of the resultant glass are altered to varying degrees by the type and amount of modifiers.

The four main glass compositions used to create continuous fibers for reinforcement are high-alkali glass (A-glass), low-alkali or electrical-grade glass (E-glass), chemically resistant glass (ECR-glass),

Table 1. Properties of glass fibers

Property	A-glass	ECR-glass	E-glass	S-glass
Density, kg/m^3	2500	2490	2540	2480
Tensile strength, GPa	3.0	3.0	3.5	4.6
Elasticity modulus, GPa		69	72	85
Coefficient of thermal expansion, 10^{-6} K^{-1}	8.6	7.2	5.0	5.6
Tensile elongation, %		2.5	2.4	2.9

and high-strength glass (S-glass). Properties of glass fibers are listed in Table 1.

Glass fibers are supplied in several basic forms, including rovings, chopped strands, mats, yarns, woven fabrics, woven rovings, nonwoven fabrics, tapes, and other special configurations.

The principal advantages of glass fibers are low cost and high strength. The disadvantages are a low modulus, poor abrasion resistance, and poor adhesion to the polymer matrix. Fiber *sizing* (containing a coupling agent) is used to improve fiber – matrix adhesion; its basic function is to form a chemical link between the glass surface and the matrix. Typically, a coupling agent consists of a silane that is adsorbed onto the glass and is capable of reactions with functional groups in the resin (→ Fibers, 11. Inorganic Fibers, Survey, Section 1.5.3.). The precise mechanism of action is not known, but silane sizing performs a useful role in glass-reinforced systems involving unsaturated polyester, epoxy, vinyl ester, or thermoplastic matrices.

Glass fibers generally have good chemical resistance, are noncombustible, and do not absorb water. Substantial reduction of thermal dilatation can be achieved by the addition of glass fibers because their thermal expansion coefficient is an order of magnitude lower than that of most plastics. The useful temperature range is quite large: glass does not soften significantly until a temperature > 400 °C is reached. Glass fibers are isotropic.

2.1.2. Graphite (Carbon) Fibers (→ Composite Materials, Section 4.1.1.; → Fibers, 15. Carbon Fibers)

Graphite and carbon fibers are generally prepared by the thermal decomposition of three fibrous organic precursors: polyacrylonitrile (PAN), rayon (cellulose), or pitch. They are available in tows containing typically 2000 – 10 000

Table 2. Properties of graphite (C) and aramid fibers

Property	C, high strength	C, high modulus	C, ultrahigh modulus	Aramid
Density, kg/m^3	1800	1900	2000	1440
Tensile strength, GPa	2.5	1.8	1.0 – 1.3	2.3
Young's modulus, GPa				
Longitudinal E_L	230	390	520 – 620	124
Transverse E_T	14	8	6	5
Shear, G_{LT}	10	8	7	2.6
Poisson ratio [1]				
Longitudinal ν_{LT}	0.25	0.2	0.2	0.35
Transverse ν_{TT}	0.35	0.3	0.25	0.35
Coefficient of thermal expansion, 10^{-6} K^{-1}				
Longitudinal	– 1.0	– 1.2	– 1.5	– 4.0
Transverse	10	12	15	52
Elongation, %	0.9	0.3	0.2	1.6

untwisted individual filaments. Fiber cross sections may be circular, dog bone, or irregular, with a diameter of ca. 6 – 10 µm. The primary advantages of graphite over glass fibers are higher modulus, improved creep-rupture resistance, better fatigue endurance, and lower density. However, they are also characterized by a lower limit of elongation, which results in a relatively low fracture energy. Therefore, impact resistance for graphite-fiber composites is lower than for glass-fiber composites. The high strength and stiffness of graphite fibers result from a high degree of structural orientation. They are strongly anisotropic, with transverse tensile and shear moduli generally an order of magnitude lower than the corresponding axial modulus. Graphite fibers have a negative thermal expansion coefficient in the axial direction, which permits the construction of a multi-ply laminate with zero in-plane expansion. This is achieved, however, at the expense of large deformations in thickness.

Graphite fibers are usually subjected to a surface treatment which results in an increase in hydroxyl and amine groups on the fiber surface, thereby improving the adhesion between fiber and matrix and the laminate shear performance. To improve handling, a light sizing is often applied, consisting of a resin that is compatible with the designated matrix. Graphite fibers are available in various forms: continuous, chopped, woven fabric, or mat. Hybrid fabrics (made of a combination of graphite and glass or aramid fibers) are also available. The graphite fibers applicable as reinforcement are offered in high-strength, high-modulus, and ultrahigh-modulus forms. Representative properties are listed in Table 2. Other grades with intermediate properties have also been produced, and production processes continue to be improved.

2.1.3. Aramid Fibers (→ Composite Materials; → High-Performance Fibers)

Aramid (aromatic polyamide) fibers have been used in advanced composites since the early 1970s. The most widely known are Kevlar fibers. Among commercially available aramid fibers, the two types of interest for reinforcing plastics are high-modulus (Kevlar 49) and, to a lesser extent, low-modulus (Kevlar 29) materials.

Aramid fibers have stiffness and strength characteristics intermediate between glass and graphite. Their bonding to polymeric matrices is rather poor. Plasma amination of the fiber surface seems to be the most promising method for improving this property. In addition to high tensile strength, aramid fibers offer excellent toughness, outstanding heat resistance, fairly good chemical resistance, exceptional wear resistance, good frictional properties at elevated temperature, excellent tensile – tensile fatigue resistance, and a negative axial coefficient of thermal expansion.

Aramid fibers are strongly anisotropic (Table 2). Like graphite, their transverse tensile and shear moduli are about an order of magnitude lower than the axial tensile modulus. The fibers are also weak in compression, a result of their fibril microstructure, leading to use predominantly in tension-loaded construction elements. Their very low density makes them ideal for applications requiring critical weight saving, such as components for aircraft, space vehicles, missiles, and sporting goods. However, an elevated water-absorption capacity results in low values for dielectric properties. Aramid fibers are available in a variety of forms, such as continuous-filament yarns, rovings, woven fabrics, discontinuous staple, and pulp.

2.1.4. Other Continuous Fibers

Almost any organic fiber can be incorporated into a composite. For years, natural fibers such as cotton, jute, hemp, and silk have been studied as reinforcements. They are inferior to synthetics in tensile strength and modulus, but exhibit significantly higher elongation, which results in better composite damage tolerance. Low cost makes natural fibers attractive for applications with low load requirements (e.g., housing construction). A major problem, however, is their susceptibility to fungal and insect attack and to degradation by moisture.

From the category of synthetic fibers, cellulosics, polyamides, polyvinyls, polyacrylonitrile, polyethylene, and polyesters have found application in the preparation of composites because of their favorable contribution to impact strength, electrical properties, and chemical resistance.

2.1.5. Other Discontinuous Fibers

Some fibers, such as single-crystal whiskers and alumina and silicon carbide fibers (→ Composite Materials, Section 4.1.3.) lack a corresponding continuous form. They offer very high stiffness and strength, but their full potential for composites has not been realized for two reasons: (1) their original length approaches the critical length, and this low aspect ratio is further reduced during compounding and molding; (2) controlling their alignment in the product is extremely difficult.

2.2. Fillers

Fillers are used to alter certain properties of composites — including price, because they are generally much less expensive than the polymer they replace. Many fillers are available, and in a variety of shapes: fibrous, plate-like, spherical, and irregular. Because of their low aspect ratio, fillers do not improve the tensile strength of a plastic; in fact, strength is usually decreased, especially at higher filler content.

With the addition of most fillers, such as calcium carbonate, kaolin, glass (milled glass fibers, hollow and solid glass spheres, glass flakes), mica, talc, silica, wollastonite, and aluminum oxide, the composite modulus increases, cure shrinkage and thermal expansion decrease, and the heat generated during cure is reduced. Aluminum oxide trihydrate and antimony oxide act both as fillers and as fire retardants; mica and silica improve chemical resistance.

Electrically conductive fillers, such as metallic powders and carbon particles (carbon fibers, carbon black), are used to reduce the accumulation of static charge. Carbon black is used to improve the UV-radiation resistance of polyolefins. Some fillers, such as carbon black, chalk, and titanium dioxide, act as pigments. Addition of extremely fine silica particles affects the thixotrophy of liquid resins. Magnetic properties can be obtained by incorporating such magnetic mineral fillers as barium sulfate.

2.3. Polymeric Matrices (→ Composite Materials, Section 4.2.)

Certain physical and chemical properties of the matrix are particularly significant with respect to the properties of a composite. The main advantages conferred by polymer matrices are low cost, easy processibility, good corrosion resistance, good dielectric properties, and low density. Low strength, low stiffness, and (usually) low service temperatures, on the other hand, limit their applications. The primary functions of a matrix are to bind the reinforcing elements together, distribute the loading among them, and protect the reinforcing system; this in turn presupposes good adhesion between matrix and reinforcement. Stress-transfer capability is generally higher in thermosets than thermoplastics. However, thermosets are more difficult to process, and they require longer molding cycles. They are also high molar mass materials, and therefore inherently relatively brittle. Their creep resistance, particularly at elevated temperature, is generally higher than that of thermoplastic matrices.

Traditionally, thermoplastics were reinforced primarily with discontinuous fibers, but high-temperature resins such as polyether ketone, polysulfone, and polyimide have made thermoplastics competitive with thermosets in the area of high-performance continuous-fiber composites as well.

The most important matrices generally are alkyds, allyl esters, epoxies, furans, melamines, phenolics, polybutadienes, polyesters, polyimides, polyurethanes, silicones, ureas, and vinyl esters. The most widely used, however, are polyesters, epoxies, and vinyl esters. For thermosets, the dominant resins are unsaturated polyesters, followed by acrylic resins, phenolic resins, epoxies, vinyl esters, and unsaturated polyester – urethane hybrids. The complete list of applicable thermoplastic matrices is more extensive, however, and includes acrylonitrile – butadiene – styrenes, acetals, acrylics, fluoropolymers, polyamides, polyamide – imide, polyarylsulfone, polycarbonates, thermoplastic polyesters, polyetherether ketone, polyethersulfones, polyethylenes, polyimides, poly(phenylene sulfides), polypropylenes, polystyrenes, polysulfones, poly(vinyl chloride), and styrene – acrylonitriles.

3. Fabrication of Polymer Composites (→ Composite Materials, Chap. 5.; → Plastics, Processing, 2. Processing of Thermosets)

A typical feature of thermoset composites is that material formation and product fabrication occur simultaneously during final molding (e.g., hand lay-up, spray-up, vacuum-bag molding). In some cases, especially with thermoplastic matrix composites, compounding occurs first, and the shape is formed or molded in a second stage. However, in all cases, final composite properties are influenced significantly by processing conditions. Fabrication processes for thermosetting matrices can be classified broadly as wet forming processes and processes involving premixes or prepregs. The wet processes include hand lay-up, filament winding, pultrusion, bag molding, reinforced reaction injection molding (RRIM), and resin transfer molding (RTM). In processes using premixes, compounding is performed separately to yield bulk molding compounds (BMC), sheet molding compounds (SMC), and prepregs. The use of premixes facilitates automation.

Conventional molding and thermoforming methods, (e.g., injection molding, press molding, stamping) are used when processing thermoplastic matrix composites. However, the addition of fibers or filler particles to thermoplastics significantly alters their rheological properties and thermal conductivity, which can greatly affect molding conditions [2].

3.1. Hand Lay-up Technique

Hand lay-up is an open-mold process based on stacking the plies of reinforcing material in a special orientation and sequence, saturating them with resin, and brushing or rolling them to compact the material and remove entrapped air (Fig. 8). The hand lay-up techniqe is the simplest, most commonly used method for the manufacture of both small and large reinforced products. Typical applications include boats, radomes, pools, and tanks. This process is commonly employed with room-temperature-cure polyester resin and glass reinforcement, but it can also be used with vinyl ester and suitable epoxy systems as well as with carbon or aramid fiber reinforcement. The reinforcement is usually in the form of chopped strand mat or woven roving. A gel coat, formed from a thixotropic

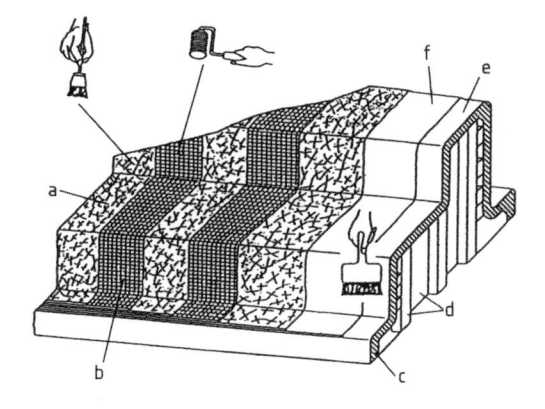

Figure 8. Basic cold-cure contact molding process
a) Strand mat; b) Fabric; c) Mold; d) Mold stiffeners; e) Mold release agent; f) Fine surfacing tissue

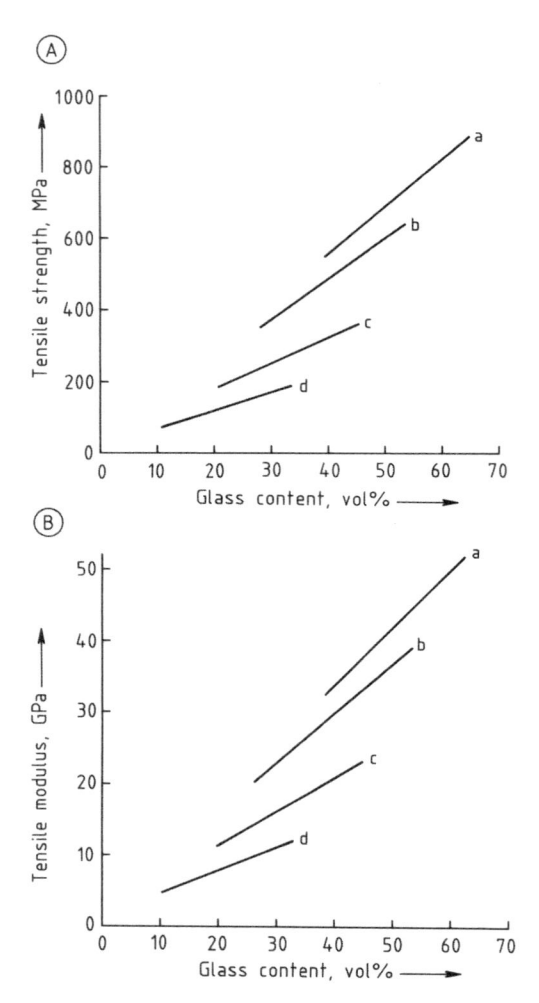

Figure 9. A) Tensile strength and B) stiffness of various glass – polyester laminates as a function of fiber content
a) UD laminate; b) UD fabric; c) Balanced fabric; d) Chopped strand mat

pigmented resin, is widely used to produce a high surface finish.

Performance may be altered greatly by varying the resin-to-reinforcement ratio, the type of resin, the form and direction of reinforcement, and the additive content (Fig. 9).

3.2. Spray-up

Spray-up is a partially automated form of hand lay-up. A spray nozzle is used to propel chopped fiber and liquid resin against the mold surface until the desired thickness is reached (Fig. 10).

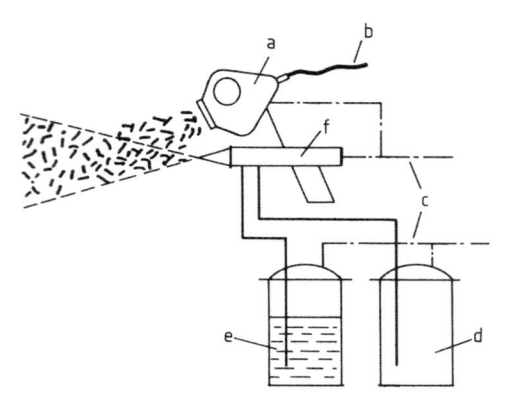

Figure 10. Schematic representation of spray-up process
a) Chopper; b) Fiber strand; c) Pressurized air; d) Resin and accelerator; e) Resin and catalyst; f) Spray nozzle

As in hand lay-up, the quality of the product depends heavily on the skill of the operator, but labor and raw material costs are lower than in hand lay-up operations. Spray-up operations can be automated, and are more practical for fabricating large parts (e.g., boat hulls) or creating protective surface coating.

3.3. Filament Winding

Filament winding is an automated process in which continuous filament (or tape) passes through a resin bath and is then wound on a mandrel. The winding angle and placement of the reinforcement are controlled through specially designed machines in which filament feed is synchronized with mandrel rotation. The device has the capacity to vary the winding angle, filament tension, or resin content in each layer of reinforcement. The winding angle may vary from longitudinal through helical to circumferential or "hoop" (Fig. 11), which permits the design and fabrication of structural parts with a high degree of precision and very

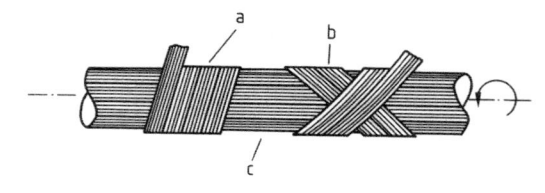

Figure 11. Alternative filament winding patterns
a) Circumferential or hoop; b) Helical; c) Axial

high strength. The two basic techniques for filament winding are

1. Wet winding, in which the reinforcement and matrix resin are applied during the winding stage
2. Dry winding, which uses preimpregnated reinforcement

Resin content in the laminate can be controlled more accurately with prepregs.

Reinforcements are available in a number of forms: rovings, fabrics, tapes, mats, foils. Computer-aided design (CAD) methods are widely used to establish optimal laminate composition and fabrication conditions. Filament-wound applications include components subject to very high loads, such as pressure vessels, rocket engine cases, radomes, storage tanks, pipes, helicopter blades, and other aerospace parts.

3.4. Pultrusion

Pultrusion is a continuous processing method for obtaining composites with a constant cross section. The process consists of pulling continuous reinforcement (usually in the form of a roving or mat) through a resin bath (with removal of superfluous) into a shaping die, where the resin is subsequently cured (Fig. 12). A number of profiles, such as rods, tubes, and various structural shapes, with very high longitudinal strength and stiffness, can be produced via appropriate dies. The process is suitable for thermosetting resins (polyester, epoxy).

3.5. Bag Molding

In this process, a sheetlike reinforcement is laid up in a mold, into which the resin is introduced

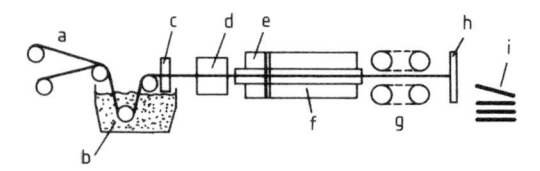

Figure 12. Schematic of the pultrusion process
a) Let-off; b) Impregnation bath; c) Resin control; d) Preformer die; e) Cold die section; f) Heated die section; g) Puller; h) Cut off; i) Product

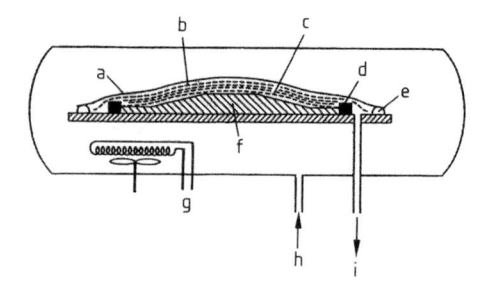

Figure 13. Schematic of an autoclave molding device
a) Bag; b) Bleeder pack; c) Prepreg pack; d) Flexible dam; e) Seal; f) Mold; g) Heater; h) Pressure; i) Vacuum

by spreading or coating. A flexible bag or diaphragm is used to apply pressure uniformly over one surface of the laminate. The general process can be subdivided into three basic molding methods: pressure bag, vacuum bag, and autoclave, the last of which is represented schematically in Figure 13. Heat may be applied in the form of steam in the autoclave itself or through the rigid mold. Large parts can be prepared with high-quality surface finishes. Mats, fabrics, preimpregnated forms of reinforcement, and honeycomb materials are used, and excellent adhesion to composite or sandwich core surfaces can be achieved. These features make bag molding attractive in the aerospace and high-technology areas.

3.6. Sheet Molding Compounds (SMC)

Sheet molding compounds are produced as continuous flat sheets of ready-to-mold composite material containing fibers, usually together with mineral filler, dispersed in a thermosetting resin. Chopped E-glass fibers 20 – 60 mm long are most often used as reinforcement, although carbon or aramid fibers are also applicable, either separately or as a hybrid. Both short (discontinuous) and continuous fibers are used in SMC. Short fibers, randomly oriented in the plane of the sheet, are common in applications requiring isotropic material properties. The resin is in an uncured but highly viscous state, achieved by addition of a thickener (MgO). Curing of the resin takes place during the compression-molding operation. With the application of heat and pressure, SMC first flows into and fills the mold cavity, and then cures into a

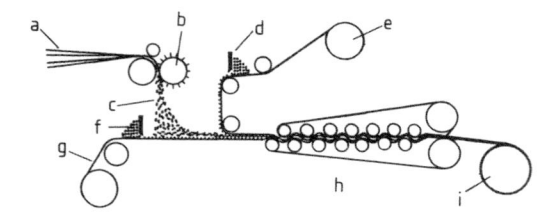

Figure 14. Processing equipment for sheet molding compounds
a) Continuous-strand roving; b) Chopper; c) Chopped roving; d) Doctor blade; e) Carrier film; f) Resin–filler paste; g) Plastic carrier film; h) Compaction section; i) Take-up roll

solid. Volume change on cure can be controlled by the addition of various shrinkage compensators, permitting high dimensional stability to be achieved. A schematic of an SMC device is presented in Figure 14.

Several types of SMC are distinguished as a function of the form of reinforcement (Fig. 15):

SMC-R	randomly oriented fibers
SMC-O	oriented SMC
SMC-CR	combination of continuous, parallel, and random fiber orientation
XMC	criss-crossed continuous fibers in a sheet
HMC	SMC with a high fiber content (no filler)

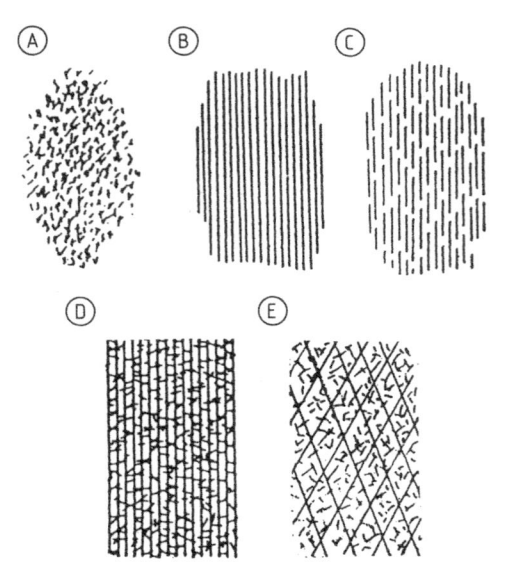

Figure 15. Fiber reinforcement structures of various sheet molding compounds
A) SMC-R; B) SMC-C (HMC); C) SMC-O; D) SMC-CR; E) XMC

A numeral at the end of the letter designation indicates the fiber content in weight percent. For example, SMC-C (HMC) 20 R 15 contains 20 wt % of continuous parallel and 15 wt % of randomly oriented fibers.

The performance of SMC composites depends primarily on a combined effect of the amount and length of the reinforcement as well as the orientation of the reinforcement in the finished part. Many properties can be influenced favorably by the addition of fillers or other additives. SMC composites can be used to produce structural parts with complex shapes in relatively short molding times. For this reason SMC composites have found broad application in mass production, especially in the transportation industry and appliance manufacture (e.g., washing machines, refrigerators, furniture, computer housings).

3.7. Thick Molding Compounds (TMC)

Thick molding compounds differ from sheet molding compounds only in thickness. This may range up to 50 mm with TMC, whereas a thickness of 3 – 6 mm is achieved with conventional SMC. Greater thickness allows chopped fibers to assume a three-dimensional random arrangement. For this reason, strength and stiffness are both lower with TMC than with SMC.

3.8. Bulk (or Dough) Molding Compounds (BMC, DMC)

BMC is prepared by mixing chopped strands with a filled polyester resin paste. For reinforcement, BMC usually contains glass fibers, with lengths ranging from 3 to 12 mm, but cellulose, cotton, and other fibrous materials can also be used. Bulk molding compounds are manufactured by compression molding at 130 – 160 °C, depending on the resin formulation. Temperature control is extremely important in the molding process to reduce internal stresses and ensure reproducible physical properties of the product. Major markets for BMC are the transportation and appliance industries. The products are quickly molded, light, and strong, and may possess complex shapes.

3.9. Prepregs

Prepregs are materials that have been impregnated with partially cured polymer matrix. The matrix differs from that in SMC by the absence of fillers, thickening agents, and pigments. Most prepregs are based on epoxy resins. Reinforcements include glass, carbon, and aramid fibers in the form of roving, woven fabric, or mat. Continuous unidirectional fiber sheets give high-performance multilayer composites, used in high-load-bearing construction parts in the aviation and automobile industries. Prepregs are molded by hand lay-up, bag molding, or autoclave molding.

3.10. Reinforced Reaction Injection Molding (RRIM)

By reaction injection molding (RIM), two chemically reactive liquid streams are injected into a mold where polymerization and formation of the part occur. When short fibers or fillers are added, the process is referred to as reinforced reaction injection molding (RRIM). The fibers and particles must be sufficiently small to fit through the mixing head orifice (< 2 mm). Since viscosities of the essentially monomeric materials used in the RRIM process are low, the injection mixture flows into the mold under relatively low pressure and at low temperature, resulting in savings in energy and equipment.

Polyurethane-based polymers have been widely molded by RRIM. Various mineral fillers (e.g., milled glass fibers, glass fibers, glass flakes and beads, mica, wollastonite, and kaolin) are used to improve the modulus and reduce thermal expansion.

3.11. Resin Transfer Molding (RTM)

In this approach a dry reinforcement material is placed in a mold cavity before the mold is closed and resin is injected. Resin flows through the reinforcement preform, expels air in the cavity, and impregnates the reinforcement. After curing, the molded part is removed and the resin reaction may be allowed to continue to completion through postcuring. RTM resins are low-viscosity polyester epoxy and vinyl ester resins. A broad variety of reinforcements is used, such as continuous strand, cloth, woven roving, long fiber, and chopped fiber mat made of glass, aramid, or carbon.

Resin transfer molding has the potential to become an efficient, low-cost manufacturing process for the fabrication of large, integrated, high-performance products.

4. Physical and Chemical Properties

The properties of a resin – matrix composite depend on many factors, among them the properties of the constituents. The form of the reinforcement (particulate, chopped fibers, woven, etc.); its volume fraction; fiber length, distribution, and orientation; interfacial bond strength; and void content all play important roles. Many of these factors depend strongly on the nature of the fabrication process. The constituents of a composite can be combined in various ways. Therefore, determining the properties of all possible combinations is a practical impossibility. Analytical methods have been developed to permit derivation of a set of predicted properties based on the properties of the composite constituents [9–16].

The fundamental building element upon which fiber-reinforced plastics theory is based is a unidirectionally reinforced layer. Many applications involve components in the form of thin plates or shells [17]. For such structures the components are commonly assumed to be in a state of plane stress; that is, through-thickness deformations and stresses are neglected. As a result, material characteristics in the plane of the structure become of primary interest. In such a case, four elastic constants and two thermal expansion coefficients are necessary to describe the elastic behavior of the UD layer. Like stiffness, strength is a directionally dependent parameter in composites. The strength properties of UD composites are governed by complex internal failure modes and cannot be described in a way similar to that used for elastic characteristics. Although no universally accepted law currently exists for the prediction of composite failure under arbitrary loading conditions, a number of basic strength properties do allow the structural efficiencies of composites to be evaluated. For typical fiber composites, these properties are listed in Table 3. One of the most important factors affecting composite properties

Table 3. Representative mechanical properties of several UD composites (fiber fraction 60 vol %)

Property	E-glass – polyester	E-glass – epoxy	C-HM – epoxy	C-HS – epoxy	Aramid – epoxy
Density, kg/m^3	2000	2000	1650	1600	1400
Young's modulus, GPa					
Longitudinal E_L	43	44	180	140	76
Transverse E_T	12	14	6	8.5	5
Shear G_{LT}	5	6	4	4.5	2.5
Poisson ratio ν_{LT}	0.31	0.30	0.27	0.30	0.36
Tensile strength, MPa					
Longitudinal	800	1000	800	1400	1400
Transverse	25	40	47	50	28
Compression strength, MPa					
Longitudinal	400	450	630	1000	300
Transverse	120	150	190	200	140
Shear strength, MPa					
In-plane	45	80	46	63	60
Interlaminar	40	70	46	70	50
Thermal expansion, 10^{-6} K^{-1}					
Longitudinal	7	7	-0.7	-0.3	-2.7
Transverse	28	26	28	28	54

is the amount of reinforcing agent present (see Fig. 9). Here the important parameter is percentage by volume, not by weight. Therefore, volume fractions of components must be used to compare composites of different composition.

As can be seen from Table 3, maximum strength and stiffness are reached in the fiber direction. To provide improved transverse properties, composites are commonly arranged in laminates such that individual layers have different orientations. In this way, the transverse properties of the laminate are enhanced at the expense of some of the potential inherent in a UD layer. An isotropic arrangement of reinforcement therefore leads to much lower strength and stiffness values (Figs. 9 and 16). Composites reinforced with discontinuous fibers, such as bulk molding compounds, generally have nonuniform properties because local fiber content and orientation are difficult to control. As a result, material properties vary from point to point throughout the material, which contributes to great variation in measured characteristics, especially strength. Relatively high scatter with respect to properties is a typical feature of composites. For generalized comparisons of material weight-savings potential, the strength – density ratio is an adequate guide when loading is uniaxial. This comparison is very favorable for advanced composites (Fig. 16 C). Although reinforcing fibers and fillers are elastic over a wide temperature range, the behavior of a polymer matrix is strongly temperature- and time-dependent. Such behavior can be observed in polymer – matrix composites as well, especially for matrix-governed properties [9], [16–19].

The polymer matrix also plays a dominant role in the chemical resistance of composites. One of the most important tasks of a matrix is to bind the reinforcing elements and protect them against hostile environments. To prevent fiber corrosion, the fiber volume-fraction must be kept low so that absorption and subsequent diffusion of the corrosive medium are minimized. This can be accomplished by protecting the load-carrying layer with a corrosion barrier consisting of thin, resin-rich veil liners. Chemical corrosion of composites can be prevented by careful choice of resin, reinforcement, and additives [20], [21], with the behavior of the resin being most important.

Of the two categories of polymers used as matrices — thermoplastics and thermosets — thermoplastics show excellent resistance to most acids, bases, and solvents. Among thermoset-based composites, polyester-based resins are generally resistant to acids and solvents, whereas epoxies are more appropriate when resistance to bases is necessary. Special resin systems, such as furans, are also available for specific chemical exposures. Phenolics and epoxies provide superior resistance to various forms of radiation [22].

Good resistance of components to environmental effects can be sustained by good manufacturing practice, appropriate material handling, and the application of consistent control standards. Prolonged storage of resins and fibers, or their exposure to high temperature, UV light, or moisture, may result in degradation and subsequent loss of corrosion resistance [23]. In addition to the effects of temperature, time, and environmental exposure, mechanical loading or fatigue also contributes to the acceleration of corrosion. Stress corrosion can be controlled by a reduction of stress levels, choice of the appropriate laminate structure, and the use of resins with high ultimate elongation [21].

Extensive literature on corrosion and environmental resistance is available from resin and fiber manufacturers, research laboratories, and several government agencies. Corrosion data banks are being developed by the manufacturers and users of composites.

5. Quality Specifications

Quality specification is accomplished by tests to determine the suitability of materials, processes, and designs for the intended applications.

The broad range of applications for reinforced plastics and their diverse properties have necessitated the development of a large number of specialized tests for evaluating mechanical, optical, thermal, electrical, physical, chemical, and durability characteristics. Standard test methods and necessary equipment are described in various sources [24], [25], and new testing procedures are constantly being developed [26]. Details about individual methods can be obtained from the literature [27–29]. Specific material features that must be considered when testing polymer composites include:

1. Composition of the constituents. Determination of the fractions of individual components is of prime importance.

2. Anisotropy of properties. The direction of a test (e.g., load application or heat conductivity) should be specified and recorded with the test results.

3. Inhomogeneous structure of the material. With laminates the stacking sequence of layers may play an important role under certain test conditions (e.g., bending, static as well as dynamic fatigue, environmental resistance).

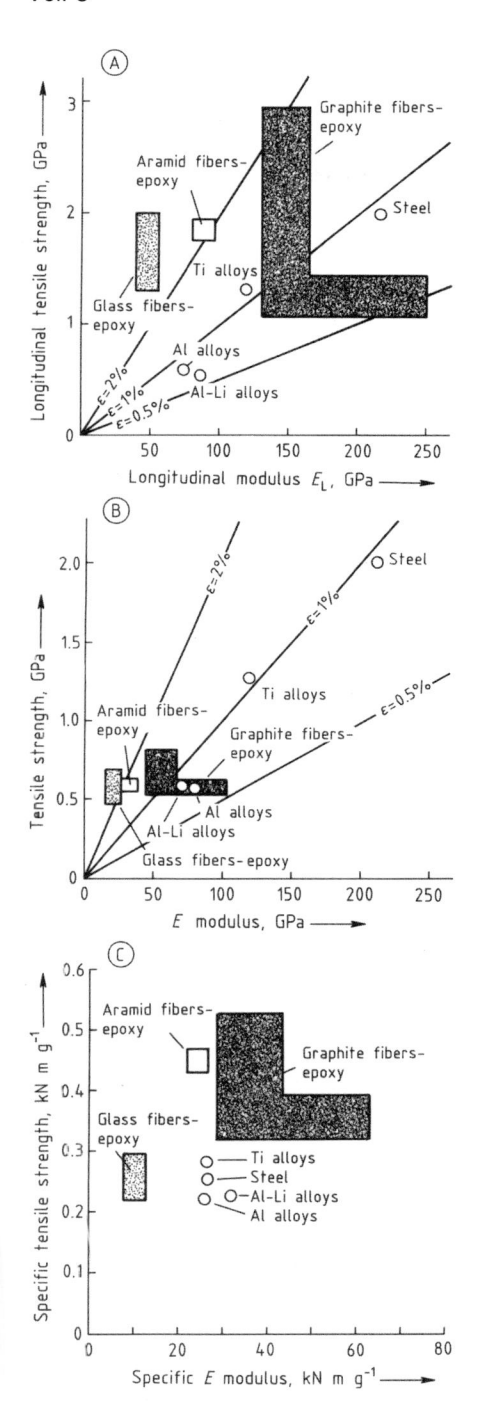

Figure 16. Comparison of strength and stiffness properties of advanced composites with those of high-strength metallic materials
A) UD properties; B, C) Properties of plane-isotropic reinforced composites
ϵ = typical limit elongation

4. Rheological behavior as a function of time. Loading rate and duration, frequency, and loading history must be specified carefully.
5. Relatively great variation in data. A sufficient number of samples must be examined to ensure reasonably reliable test results for the material under evaluation. Statistical analysis of the test data is necessary.
6. Influence of the fabrication process on product properties. Specimens must be prepared and tested under conditions identical to those specified for the product in question.

Tests may be destructive or nondestructive, depending on whether a sample is destroyed or degraded during the test. Nondestructive methods (e.g., ultrasonics, X-ray inspection, acoustic emission, microwave techniques) are becoming increasingly important in ensuring the integrity, reliability, and safety of composite structures [30].

6. Economic Aspects

Based on modern trends in materials technology, the reinforced plastics industry should continue to enjoy rapid growth. The trend can be seen in total shipments of reinforced plastics to the United States, which reached 1.21×10^9 t in 1988 compared to 0.91×10^9 t in 1978, for a mean annual growth rate of nearly 12 %. A similar development can be observed in Europe, where the mean annual growth rate in the 1980s was 9 %. Worldwide growth of composites during 1985 – 1988 averaged 5.5 %, and all predictions for 1990 – 1995 anticipate a growth of ca. 1.5 – 2 times the rate of growth of gross domestic product. Exceptional growth is expected [31] for reinforced plastics with the following characteristics: high performance, high performance – cost ratio, high flame resistance and low smoke index, good adhesion, and corrosion resistance.

Among composites generally, thermosets still dominate thermoplastics as matrix materials. Nevertheless, although thermosets accounted for about 75 % of the demand in the 1980s, this decreased to 70 % in 1990, and the growth in demand for thermoplastics over the first five years of the 1990 s is expected to be double that of composites in general [32].

More than 80 % of the reinforced thermoplastics market belongs to injection-molded resin matrices of nylon, thermoplastic polyesters, olefins, styrene, and polycarbonate. High-performance reinforced thermoplastics in the fastest growing material sector include polyimide, polyetherketone, polybenzimidazole, and polyarylene sulfone. This sector is expected to grow > 10 % annually [33].

The potential exists for significant further advances in composites, in contrast to the relatively minor improvements possible in such wrought sheet-type metals as steel or aluminum. Past and present performance, as well as current areas of research and development, have laid the groundwork for future growth of the composites industry. Effective exploitation of new opportunities is the key to potential large-scale market penetration and consequent profitability for reinforced plastics.

References

General References

1 G. Lubin: *Handbook of Composites*, Van Nostrand Reinhold, New York 1982.
2 P. K. Mallick, S. Newmann: *Composite Materials Technology*, Hanser Verlag, München 1990.
3 W. V. Titow, B. J. Lanham: *Reinforced Thermoplastics*, Applied Science Publ., Barking 1975.
4 C. Zweben, H. T. Hahn, Tsu-Wei Chou: *Mechanical Behavior and Properties of Composite Materials, Delaware Composites Design Encyclopedia (DCDE)*, vol. 1, Technomic Publ., Lancaster 1989.
5 M. G. Bader et al.: *Processing and Fabrication Technology, DCDE*, vol. 3, Technomic Publ., Lancaster 1990.
6 H. S. Katz, J. V. Milewski: *Handbook of Fillers and Reinforcements for Plastics*, Van Nostrand Reinhold, New York 1980.
7 D. W. Rosato, C. S. Grove: *Filament Winding*, J. Wiley & Sons, New York 1964.
8 S. M. Lee (ed): *International Encyclopedia of Composites*, **6 vols.**, VCH Publishers, New York 1990 – 1999.

Specific References

9 B. D. Agarwal, L. J. Broutmann: *Analysis and Performance of Fiber Composites*, J. Wiley & Sons, New York 1990.
10 N. N.: *Engineered Materials Handbook*, vol. 1, "Composites," ASM International, Menlo Park, Ohio, 1987.
11 S. W. Tsai, H. T. Hahn: *Introduction to Composite Materials*, Technomic Publ., Westport 1980.
12 R. Christensen: *Mechanics of Composite Materials*, J. Wiley & Sons, New York 1979.
13 S. C. Halpin: *Primer on Composite Materials: Analysis*, Technomic Publ., Lancaster 1984.

14. R. M. Jones: *Mechanics of Composite Materials*, McGraw-Hill, New York 1967.
15. S. W. Tsai: *Composite Design*, Think Composites Publ., Dayton 1987.
16. S. M. Whitney, R. McCullough: *Micromechanical Materials Modelling, DCDE*, vol. 2, Technomic Publ., Lancaster 1990.
17. J. R. Vinson, R. L. Sierakowski: *The Behavior of Structures Composed of Composite Materials*, Martinus Nijhoff Publ., Dordrecht 1987.
18. J. D. Ferry: *Viscoelastic Properties of Polymers*, 3rd ed., J. Wiley & Sons, New York 1980.
19. A. A. Ogale: "Creep Behavior of Thermoplastic Composites" in L. A. Carlsson, R. B. Pipes (eds.): *Thermoplastic Composite Materials*, Elsevier, Amsterdam 1989.
20. P. Schweitzer (ed.): *Corrosion and Corrosion Protection Handbook*, Marcel Dekker, New York 1983.
21. J. H. Mallinson: *Corrosion-Resistant Plastic Composites in Chemical Plant Design*, Marcel Dekker, New York 1988.
22. B. Dolezel: *Die Beständigkeit von Kunststoffen und Gummi*, Hanser Verlag, München 1978.
23. G. S. Springer (ed.): *Environment Effects on Composite Materials*, vol. **1–3**, Technomic Publ. Co., Lancaster 1981–1988.
24. ASTM, Standards and Literature References for Composite Materials, ASTM, Philadelphia 1987.
25. ASTM, D 3039, Standard Test Methods for Fiber Resin Composites, Philadelphia 1989.
26. "Composite Materials: Testing and Design," *Proceedings of ASTM Conference*, Philadelphia 1969.
27. G. G. Sih, A. M. Skudra: *Fracture Mechanics and Methods of Testing*, North Holland Publ. Co, Amsterdam 1985.
28. L. A. Carlsson, R. B. Pipes: *Experimental Characterization of Advanced Composite Materials*, Prentice Hall, Englewood Cliffs, N. J., 1987.
29. Yu. M. Tarnopolski, T. Kincis: *Static Test Methods for Composites*, Van Nostrand Reinhold, New York 1985.
30. J. Summerscales (ed.): *Nondestructive Testing of Fiber-Reinforced Plastic Composites*, **2 vols.**, Elsevier, Amsterdam 1987–1990.
31. J. S. McDermott: "An Assessment of the Composite Market for the 1990's," *45th Ann. Conf. Comp. Inst.*, SPI, Washington 1990.
32. Frost & Sullivan: Stetiges Wachstum, *Kunststoff-J.* **4**, Europa-Fachpresse-Verlag, München 1992.
33. G. Sterling: "Developments, Trends and Opportunities for the Plastics Business in the 1990's," *45th Ann. Conf. Comp. Inst.*, SPI, Washington 1990.

Further Reading

J. A. Finnegan, B. P. Gersh, J. B. Lang, R. Levy: *Building Materials, Plastic*, Kirk Othmer Encyclopedia of Chemical Technology, 5th edition, John Wiley & Sons, Hoboken, NJ, online DOI: 10.1002/0471238961.1612011906091414.a01.

H. F. Mark (ed.): *Encyclopedia of Polymer Science and Technology*, 3rd ed., Wiley, Hoboken, NJ 2005.

D. V. Rosato, D. V. Rosato: *Reinforced Plastics Handbook*, 3rd ed., Elsevier, Amsterdam 2005.

Specialty Plastics

HANS-GEORG ELIAS, Michigan Molecular Institute, Midland, MI 48640, United States

1.	Introduction	1343	5.1.1.	Introduction	1351
2.	Polyphenylenes	1344	5.1.2.	Polyelectrolytes	1352
2.1.	Chemical Structure	1344	5.1.3.	States of Ions	1352
2.2.	Polybenzenes	1344	**5.2.**	**Acidic Copolymers**	**1353**
2.3.	Polymers from Higher Aromatic Hydrocarbons	1344	5.2.1.	Introduction	1353
			5.2.2.	Structure and Properties	1354
2.4.	Acetylene Group Containing Polymers	1345	5.2.3.	Thermoplastic Acidic Copolymers	1354
			5.2.4.	Elastomeric Acidic Copolymers	1356
3.	Poly(*p*-Xylylenes)	1345	**5.3.**	**Ionomers**	**1356**
3.1.	Introduction	1345	5.3.1.	Introduction	1356
3.2.	Synthesis	1346	5.3.2.	Structure	1357
3.2.1.	Monomer Synthesis	1346	5.3.3.	Properties	1358
3.2.2.	Polymerization	1347	5.3.4.	Ethylene–Methacrylic Acid Ionomers	1360
3.3.	Structure	1348			
3.4.	Manufacture	1349	5.3.5.	Fluorinated Ionomers	1361
3.5.	Properties	1349	5.3.6.	Sulfochlorinated Polyethylene	1362
4.	Polybenzocyclobutenes	1350	5.3.7.	Sulfonated EPDM Polymers	1362
5.	Ionomers and Related Polymers	1351		**References**	**1363**
5.1.	Overview	1351			

1. Introduction

In this article, "specialty plastics" are defined as those plastics which do not fit into other keywords, either because they are prepared from unusual monomers or because they possess unique structures and/or properties. However, the term "specialty plastics" is in general often used as a synonym for "high-performance plastics", which are a special class of engineering plastics (technical plastics, technoplastics; see → Plastics, General Survey). Engineering plastics provide the engineer with a combination of useful properties, whereas highperformance plastics often exhibit only one outstanding property albeit one which surpasses the corresponding property of engineering plastics, e.g., a high continuous service temperature.

"Specialty plastics" are also not necessarily identical with either "functional plastics" or "functional polymers." Functional plastics are those plastics which are useful for one application only; examples are ethylene–vinyl alcohol copolymers with high contents of vinyl alcohol units, which are exclusively used as gas and aroma barriers in the packaging industry. The term "functional polymer" or "functionalized polymer", on the other hand, refers to polymers with functional chemical groups, i.e., specific chemical reactivities.

The polyphenylenes and poly(*p*-xylylenes) described in this article are typical functional plastics since they are used for only one purpose: coatings. The group of polymers known as ionomers, on the other hand, comprises functional plastics, engineering plastics, elastomers, and thermoplastic elastomers.

2. Polyphenylenes

2.1. Chemical Structure

Polyphenylenes are oligomeric to polymeric compounds composed of phenylene units and related units such as naphthalene or anthracene; these units are connected to each other via *ortho*, *meta*, and/or *para* links. Their molecular architecture thus ranges from linear to highly branched compounds; they may or may not become cross-linked on further processing. Polyphenylenes are also known as polyphenyls, oligophenyls, oligophenylenes, or polybenzenes [1–7].

Polyphenylenes exhibit two remarkable properties. As compounds of low hydrogen content and free of readily activated chemical bonds, they show good thermal, oxidative, and hydrolytical stability. As aromatic compounds, they have the potential of becoming electrically conducting materials. Commercial developments are thus along two lines: polyphenylenes are either prepared as thermosetting resins with high service temperatures or as base materials for electrically conducting polymers.

The two classes of polyphenylenes differ in their chemical structures. Polyphenylenes for conductors are linear polymers with phenylene units interconnected in 1,4-position (PPP [*25190-62-9*]), whereas polyphenylenes for thermosets are highly branched oligomers [*9033-83-4*], often with reactive end groups such as acetylene (schematically shown as PXP). This section is exclusively concerned with the latter type of polymers; poly(1,4-phenylenes) are discussed in → Polymers, Electrically Conducting, Section 5.5.

2.2. Polybenzenes

Melting temperatures of simple poly(1,4-phenylenes) increase strongly from biphenyl H $(C_6H_4)_2$H (70 °C) to hepta(1,4-phenylene) H $(p\text{-}C_6H_4)_7$H (ca. 540 °C). Poly(1,4-phenylenes) were thus expected to be very temperature stable compounds. All classic stepwise syntheses of such polyphenylenes are, however, too expensive for the industrial manufacture of plastics but may be utilized for electrically conducting polymers [1], [2], [6], [7].

The first industrial branched polyphenylenes were prepared by oxidative cationic polymerization of benzene with stoichiometric amounts of aluminum trichloride/copper dichloride, iron trichloride, molybdenum pentachloride, etc. [3–5], [8–10]. The arylation reaction proceeds via radical cations [7], [11]. The polymers still contain 1 – 4 % chlorine, < 5 % oxygen, 0.1 – 0.2 % copper, and 0.3 – 0.6 % aluminum after purification [5].

The resulting brown to black masses have densities of ca. 1.25 g/cm^3, tensile strengths of 65 MPa, Young's moduli of 160 MPa, elongations at break of 10 %, and linear expansion coefficients of ca. 3×10^{-5} K^{-1}. They melt above 530 °C and decompose above 700 °C. Their melt viscosities are so high that they cannot be processed by injection molding or similar fabrication techniques. The polymers can, however, be molded at high pressure; the moldings can then be machined. Such materials are produced as Eimac 221 by Eitel–McCollough, United States.

2.3. Polymers from Higher Aromatic Hydrocarbons

The oxidative cationic polymerization of terphenyl isomers, with or without addition of other compounds such as biphenyl, reduces the probability of cross-linking reactions as compared to benzene. The resulting branched prepolymers are soluble; their solutions in chloroform or chlorobenzene are used as impregnating varnish for laminates. Filled prepolymers can be processed in the melt. The prepolymers are cross-linked to give insoluble and unmeltable polymers by the addition of catalysts such as *p*-toluenesulfonic acid, sulfuryl chloride, or boron trifluoride etherate. These techniques are mainly being developed by Hughes Aircraft Co.

2.4. Acetylene Group Containing Polymers

Other types of branched polyphenylenes are the so-called H-resins of Hercules [12–15]. These prepolymers are obtained by cyclotrimerization of *p*-diethynylbenzene (DEB) in aromatic solvents [16] or by copolycyclotrimerization of DEB with phenylacetylene [17]; catalysts are $AlR_3/TiCl_4$, $NiCl_2/P(C_6H_5)_3/NaBH_4$, or $CoBr_2/(C_6H_5O)_3P$.

The prepolymers (B-stage resins) are oligophenylenes with the schematic structure PXP (see Section 2.1). For B-stage resins from the polymerization of DEB by $TiCl_4/(C_2H_5)_2AlCl$ at 86 % monomer conversion, a number-average molar mass of 2300 g/mol and a mass-average molar mass of 8060 g/mol were found. Higher DEB conversions resulted in gelation.

These polymers contain acetylene end groups that form cross-links under the action of titanium(IV) chloride/diethylaluminum chloride, nickel(I) acetyl acetonate, and similar catalysts. The chemical structure of these cross-links is unknown; they are probably not benzene rings.

B-Stage H-resins dissolve in aromatic hydrocarbons, chlorinated hydrocarbons, ketones, and cyclic ethers. They are insoluble in water, aliphatic hydrocarbons, and aliphatic alcohols. The resins can be processed with simultaneous cross-linking by molding, injection molding, extrusion, and powder coating; different grades are provided for these purposes. Molding cycles are 2 – 5 min at 160 – 170 °C [13]. Molding stability starts at 160 °C but improved mechanical stabilities and better solvent resistance require post-curing at 230 – 300 °C.

The cured resins are hard, nonporous materials which are resistant to hot solvents, acids, bases, brine, superheated steam, and salt melts. They become brittle in molten alkali metals. Continuous service temperatures are 200 – 300 °C in air and ca. 400 °C in an oxygen-free atmosphere; the limiting oxygen index is 55 %. Slow carbonization starts at ca. 500 °C.

Molded parts show flexural moduli of several gigapascal and high flexural strengths (Table 1). These properties can be enhanced by fillers such as talc, silica, or asbestos. They are often preserved over a wide temperature range.

Unmodified H-resins exhibit low elongation and low adhesion to polar substrates. The adhesion of H-resins to metal surfaces can be improved by the use of polymer blends and polymeric fillers such as epoxy resins, phenolic resins, polysulfones, and polyimides.

H-Resins serve as corrosion-resistant coatings, for example, for pipes and processing equipment for hot sour petroleum or gas wells, pump linings, high-temperature sodium and lithium batteries, and high-temperature processing equipment. They are also used as nonflammable coatings for electric and electronic components and as self-lubricating coatings for metals.

Table 1. Properties of cured H-resins

Physical property	Temperature, °C	Property value after filling with		
		none	50 wt % talcum	50 wt % silica
Flexural modulus, GPa	23	4.8	9.7	11.4
	250		6.3	5.9
	300	3.5		
	350		6.8	4.6
Flexural strength, MPa	23	<69	<69	<83
	250		<62	62
	350	<55	<55	46
Barco hardness	23	85		
Taber abrasion (17) (1000 g load)	23	0.0015		
Volume resistivity, Ω cm	23	10^{17}		
Dissipation factor (60 Hz)	23	0.002		
	100	0.0009		

3. Poly(*p*-Xylylenes)

3.1. Introduction

Poly(*p*-xylylenes), PPX, are polymers of the unstable *p*-xylylenes (*p*-quinodimethanes) PX; they are produced industrially from the dimeric *p*-xylylenes DPX:

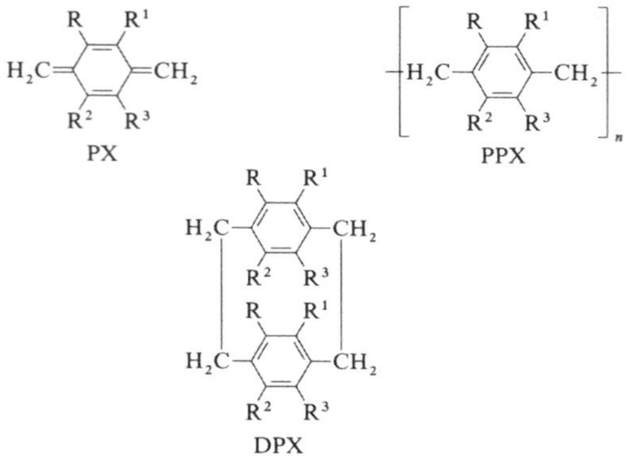

3.2. Synthesis

3.2.1. Monomer Synthesis

Two synthetic routes have been used commercially for the synthesis of dimeric p-xylylenes DPX, also known as [2.2]-paracyclophanes: direct pyrolysis and Hofmann elimination. In direct pyrolysis, p-xylene X is pyrolyzed at ca. 950 °C in the presence of steam. The pyrolysis dehydrogenates p-xylene X to p-xylylene PX-N which is converted into di-p-xylylene DPX-N on quenching with liquid xylene at ca. 20 °C:

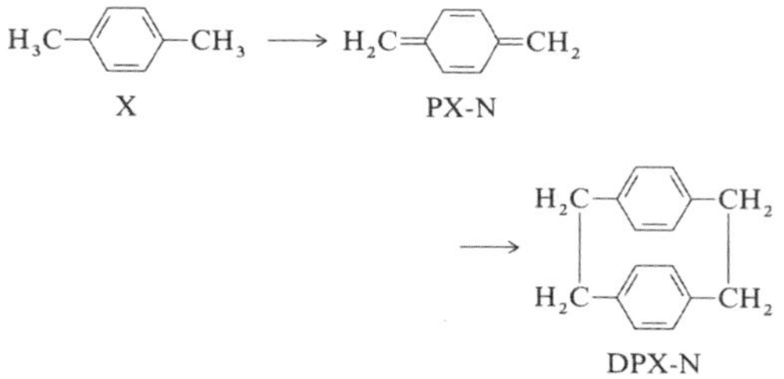

A major byproduct is the polymer PPX-N. Its simultaneous formation can be minimized by using an excess of liquid xylene (Ruggli–Ziegler dilution principle), controlled rates, etc. The polymeric byproduct has no commercial value since it can be neither processed nor recycled.

The Hofmann elimination proceeds under far milder conditions and does not need special equipment, unlike the direct pyrolysis. α-Bromo-p-xylene is first quaternized and the resulting product converted into the quaternary ammonium hydroxide; amine elimination then leads to DPX-N:

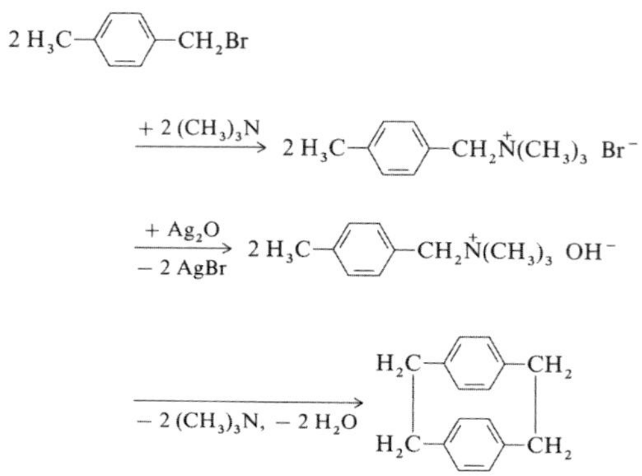

The syntheses, structures, and properties of PPX are described in [18–25]. Although PPXs with many different substituents R, R^1, R^2, and R^3 have been prepared in the laboratory (see the list in [23]), only three of them are industrially important. These polymers are based on DPX as monomer and have no (N) or one chlorine (C) substituent, or are disubstituted (D) with chlorine:

PPX-N ($R = R^1 = R^2 = R^3 = H$) [25722-33-2]
PPX-C ($R = R^2 = R^3 = H$; $R^1 = Cl$) [9052-19-1]
PPX-D ($R = R^3 = H$; $R^1 = R^2 = Cl$) [52261-45-7]

The systematic names are:

PPX-N	poly(1,4-phenylene-1,2-ethanediyl)
PPX-C	poly[(chloro-1,4-phenylene)-1,2-ethanediyl]
PPX-D	poly[(dichloro-1,4-phenylene)-1,2-ethanediyl]

The designations monochloro and dichloro PPX for technical products are misleading since these polymers do not have one or two chlorine substituents on each ring. Monochloro and dichloro rather refer to the average number of chlorine ligands per ring, i.e., there is a distribution of chlorine substituents amongst the rings; each ring in PPX-C or PPX-D may bear between 0 and 4 chlorine substituents.

Poly(p-xylylenes) from the Gorham process (see below) are known as parylenes. Coatings produced by this process are correspondingly named parylene N, parylene C, and parylene D. The sole producer of dimeric p-xylylenes DPX-N, DPX-C, and DPX-D is Union Carbide, which also custom-coats parts with PPX-N, PPX-C, and PPX-D.

The yields of DPX-N vary between 17 and 50 %, depending on the reaction conditions (e.g., use of polymerization inhibitors).

Chlorine-containing dimeric p-xylylenes are produced industrially by chlorination of DPX-N using standard procedures for aromatic ring chlorinations. Chlorination gives mixtures of unreacted DPX-N with various chlorinated compounds (from mono to tetra substitution per ring) and their isomers.

3.2.2. Polymerization

Poly(p-xylylenes) can be synthesized by the Szwarc or the Gorham process. In the Szwarc process [26–28], p-xylene vapor is passed rapidly (residence time < 1 s) in vacuum (< 100 Pa) through quartz tubes heated to 800 – 1150 °C. The pyrolysis gases are subsequently quenched to temperatures below 25 °C. The mechanism was assumed [26–28] and later confirmed [29–32] to consist of the formation of monoradicals (either spontaneously or by radical transfer), followed by disproportionation to give p-xylylene, and finally polycombination of the diradicals to give poly(p-xylylene):

$$H_3C-\langle\bigcirc\rangle-CH_3 \xrightarrow{-H^\cdot} H_3C-\langle\bigcirc\rangle-\dot{C}H_2$$

$$H_3C-\langle\bigcirc\rangle-CH_3 \xrightarrow[-H_2]{+H^\cdot} H_3C-\langle\bigcirc\rangle-\dot{C}H_2$$

$$2\,H_3C-\langle\bigcirc\rangle-\dot{C}H_2$$
$$\longrightarrow H_3C-\langle\bigcirc\rangle-CH_3 + H_2\dot{C}-\langle\bigcirc\rangle-\dot{C}H_2$$

$$H_2\dot{C}-\langle\bigcirc\rangle-\dot{C}H_2 \longrightarrow \sim H_2C-\langle\bigcirc\rangle-CH_2\sim$$

The diradical and not the p-xylylene is the intermediate species since the former is stable in the gas phase between 120 °C and 200 °C and in liquid phases (such as quenched hexane solutions) at low temperatures (half-life 21 h at −78 °C). The resulting poly(p-xylylene) is slightly cross-linked; it always contains 10 – 25 wt % of low molar mass impurities. Furthermore, since the yield never exceeds 25 %, the process was technically and economically useless.

The Gorham process, on the other hand, leads to colorless, linear (uncross-linked) polymers that are free of low molar mass compounds [34], [34]. The process uses di-p-xylylene DPX as monomer and proceeds in three stages: (1) DPX is vaporized at ca. 200 °C and 133 Pa, (2) on heating to 680 °C at ca. 67 Pa, DPX dissociates quantitatively [21] to p-xylene which, after cooling to 25 °C and 13 Pa, (3) polymerizes to poly(p-xylylene) PPX on the surfaces of parts in a deposition chamber at < −70 °C:

$$\begin{array}{c} H_2C-\langle\bigcirc\rangle-CH_2 \\ | \quad\quad\quad\quad\quad | \\ H_2C-\langle\bigcirc\rangle-CH_2 \end{array} \xrightarrow{600\,°C} H_2\dot{C}-\langle\bigcirc\rangle-\dot{C}H_2 \xleftrightarrow{} H_2C=\langle\bigcirc\rangle=CH_2 \xrightarrow{20\,°C} \sim H_2C-\langle\bigcirc\rangle-CH_2\sim$$

p-Xylylene PX can be represented by quinoid (diamagnetic) or benzoid (paramagnetic) structures. The vapor of PX is diamagnetic [24].

The polymerization is thus probably started by paramagnetic diradical structures; such species may be the dimeric diradical from a one-sided ring-opening of DPX-N, the benzoid structure of PX, or the dimeric diradical from the coupling of two p-xylylene molecules. The polymerization seems to be propagated by addition of the diamagnetic quinoid structures to the growing macrodiradicals and not (or only slightly) by the polyrecombination of macrodiradicals. The polymerization proceeds as living polymerization with radical concentrations of $(5 - 10) \times 10^{-4}$ mol of free radicals per mole of xylene units of the final polymers.

Polymer yields are 100 % at −50 °C and drop rapidly at temperatures above ca. 0 °C (Fig. 1). Molar masses are almost independent of polymerization temperature in the range −50 °C to +50 °C [35].

The Gorham process can be applied to substituted [2.2]-paracyclophanes, provided the threshold condensation temperatures for the condensation on surfaces are not exceeded. These threshold temperatures are 30 °C for poly(p-xylylene), 60 °C for poly(2-methyl-p-xylylene), 90 °C for poly(2-chloro-p-xylylene),

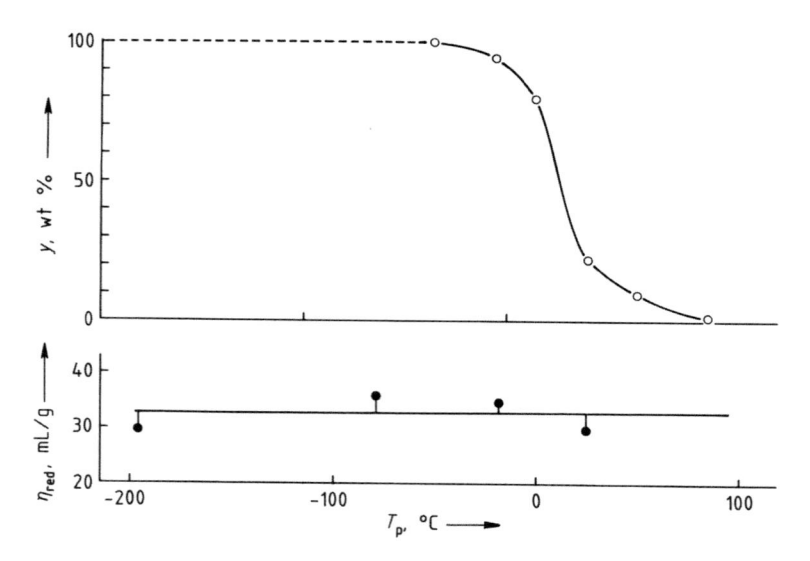

Figure 1. Dependence of polymer yield y and reduced viscosity η_{red} (1×10^{-3} g/mL polymer in α-chloronaphthalene at 238 °C) of poly(p-xylylenes) on the polymerization temperature T_p (data from [35])

and 130 °C for poly(2,5-dichloro-p-xylylene) [34]. The physical meaning (other than the phenomenological) of the threshold temperature is unclear since the most obvious interpretation as thermodynamic ceiling temperatures is made difficult by reports of very high estimated thermodynamic ceiling temperatures of more than 200 °C for a gas pressure of 0.08 torr (11 Pa) [35] and calculated ceiling temperatures of 1149 °C (gas) and 3035 °C (liquid) for PX [25]. However, such calculations suffer from the absence of knowledge about the true nature of the polymerization: true equilibria between all species (monomers, monomeric diradicals, macrodiradicals), equilibria between monomers and macrodiradicals only, or even kinetically-controlled reactions.

Poly(p-xylylenes) can also be obtained by other methods [25], [36], [37], none of which seem to be used industrially since the monomers are too expensive and/or the polymers are obtained in low yield, with low molar masses, or with many low molar mass impurities.

3.3. Structure

Chemical Structure. Poly(p-xylylenes) from the Gorham process are linear polymers, whereas those from the Szwarc process are slightly cross-linked. The number-average molar mass, as calculated from the free-electron concentration of living PPX-N polymers [33], is ca. 200 000 – 400 000 g/mol. Osmometry of soluble fractions in chloroform gave 24 000 g/mol [24].

The commercial chlorinated monomers DPX-C and DPX-D are complex mixtures with 0 – 8 chlorine groups per molecule and 0 – 4 chlorine substituents per ring. On pyrolysis and subsequent vapor deposition polymerization, DPX-C with an average of one chlorine substituent per ring thus gives mainly base units $-CH_2-(p-C_6H_3Cl)-CH_2-$ besides considerable amounts of base units $-CH_2-(p-C_6H_4)-CH_2-$ and $-CH_2-(p-C_6H_2Cl_2)-CH_2-$. Chlorinated units may be arranged head-to-head, head-to-tail, and tail-to-tail so that the resulting constitutional copolymers also show regio-isomerism.

Physical Structure. PPX-N has two crystal modifications [38], [39]. The monoclinic α-modification is formed between -17 °C and $+30$ °C, while the triclinic β-modification exists below -78 °C and above 220 – 260 °C. The unit cell of the α-modification consists of two monomer units. The phenylene units are parallel to each other; the angle between the plane of the benzene rings and the methylene chain is 90°. The $\alpha \to \beta$ transition at 220 °C is accompanied by a marked increase in brittleness. At 270 °C, a reversible smectic transition is observed.

Vapor deposition polymerization is carried out below the melting and near the glass temperatures of the polymers (see Table 2); any crystallization must thus occur simultaneously with the polymerization on surfaces. At polymerization temperatures between 0 °C and 60 °C, the α-modification is formed, whereas mixtures of α and β are found for polymerization temperatures above 80 °C.

The degree of crystallinity of parylene N has been reported to be 57 % [25]. The density of amorphous PPX-N, as calculated from the experimental density of 1.11 g/cm^3 and the theoretical density of the α-modification of 1.185 g/cm^3 (from the unit cell), is thus 1.070 g/cm^3.

Table 2. Properties of poly(p-xylylenes) [25], [34], [40], [41]

Property	Property values of		
	PPX-N	PPX-C	PPX-D
Density at 23 °C, g/cm^3	1.11	1.29	1.42
Refractive index n_D^{23}	1.661	1.639	1.669
Water uptake (24 h, 23 °C), %	0.01	0.06	< 0.1
Melting temperature, °C	420	290	380
Glass temperature, °C	13		
Continuous service temperature (air), °C	100	130	
Continuous service temperature (inert), °C	220		220
Linear expansion coefficient (25 °C), K^{-1}	6.9 × 10^{-5}	3.5 × 10^{-5}	
Specific heat capacity (25 °C), J g^{-1} K^{-1}	1.3	1.0	
Thermal conductivity (25 °C), kW m^{-1} K^{-1}	12	8.2	
Flexural modulus (23 °C), MPa	2450	2800	2800
Modulus of elasticity, MPa			
at 23 °C	2400	3200	2800
at 200 °C	170	170	170
Tensile strength (23 °C), MPa	45	70	75
Yield strength (23 °C), MPa	42	55	60
Elongation at break (23 °C), %	< 30	200	< 10
Rockwell hardness R (23 °C)	85	80	
Relative permittivity (23 °C)			
60 Hz	2.65	3.15	2.84
1 kHz	2.65	3.10	2.82
1 MHz	2.65	2.95	2.80
Dissipation factor (23 °C)			
60 Hz	0.0002	0.020	0.004
1 kHz	0.0002	0.019	0.003
1 MHz	0.0006	0.013	0.002
Surface resistivity, Ω (23 °C)			
50 % RH	1 × 10^{13}	1 × 10^{14}	5 × 10^{16}
90 % RH	9 × 10^{11}	7 × 10^{11}	
Volume resistivity (50 % RH, 23 °C), Ω cm	1.4 × 10^{17}	8.8 × 10^{16}	2 × 10^{16}
Dielectric strength at 23 °C, kV/mm	260 – 280	145 – 220	200 – 215
Permeability (at 25 °C) 10^{14} P. cm^2 s^{-1} Pa^{-1}			
N$_2$	0.35	0.020	0.20
O$_2$	1.76	0.32	1.44
Cl$_2$	3.32	0.016	0.025
CO$_2$	9.64	0.35	0.58
He	18.8	7.4	
H$_2$	24	5.0	
H$_2$S	35.7	0.58	0.065
SO$_2$	85.1	0.49	0.21
H$_2$O	428	150	

3.4. Manufacture

Parylenes are exclusively prepared and applied by vapor deposition polymerization of DPXs on surfaces (Gorham process). Total production is estimated to be several tonnes per year.

Film formation by vapor deposition polymerization differs from other coating techniques in that the coating is formed from the gaseous phase rather than from a liquid one (no surface tension effects on wetting) and that the film grows from the substrate upwards and not downwards as in drying films. The films are self-adherent; coupling agents are usually not necessary.

The technique produces pinhole-free coatings of uniform thickness even on complex, three-dimensional substrates with hidden areas. Films have thicknesses between 0.050 and 100 μm. They are flexible even at low temperatures and possess good dimensional stability.

These features, together with good electrical and mechanical properties and low gas and water vapor permeability, make parylenes especially suitable for surface coating of electrical and electronic components such as printed and integrated circuits, capacitors, transistors, and miniature electrical components. Parylenes are also applied for particle encapsulations, thin membranes, lubricants, and long-term implantation devices.

3.5. Properties

Properties of poly(p-xylylenes) are listed in Table 2.

Solubility. Poly(p-xylylenes) from the Gorham process are sparingly soluble. PPX-N dissolves to 1.5 % and PPX-C to 10 % in methylene chloride at 25 °C [25]. The solubility of

PPX-N in xylene is ca. 10 % at 140 °C. PPX-N and PPX-D can be dissolved in chlorinated biphenyls or in benzyl benzoate at temperatures above 200 °C, and PPX-C in 1-chloronaphthalene at 150 °C [33]. Cooling the solutions to below 150 – 200 °C leads to gel-like precipitation of the polymer. Repeated dissolution and precipitation cause degradation.

Aromatic and halogenated aromatic compounds swell parylenes at room temperature, but stress corrosion has not been observed. Parylenes are resistant to concentrated sulfuric acid and 5N nitric acid.

Thermal Properties. Melting temperatures of symmetrically substituted poly(p-xylylenes) PPX-N and PPX-D are higher than those of unsymmetrically substituted PPX-C. The glass temperature of PPX-N is fairly low (T_g = 13 °C) [25], and lower than the earlier reported 80 °C [21]. Similar low glass temperatures can be expected for PPX-C and PPX-D; earlier reported glass temperatures of 80 – 110 °C are definitely too high since no sudden upswing of relative permittivity (dielectric constant) $_r$ due to the onset of segmental motions is observed in the temperature interval 20 – 190 °C, only a continuous increase of $_r$ from 3.0 to 4.7.

Since the glass temperature is lower than ambient temperature, structural changes occur upon aging and/or annealing of parylenes; freshly deposited polymers have other properties than old ones.

The stability of parylenes is determined by the dibenzyl groups, which are easily oxidized and thermally cleaved. The continuous service temperature of PPX-N is thus only 100 °C in air and 220 °C under an inert atmosphere (see Table 2).

Mechanical Properties. Parylenes have tensile strengths and elastic moduli (Table 2) similar to those of other engineering plastics. The unusually high elongation at break of PPX-C refers to freshly deposited films; such films have high dimensional stability if subjected to high relative humidities. Films remain tough and flexible down to −165 °C. Since further crystallization takes place on annealing, density, hardness, and abrasion resistance also increase.

Permeability. Gas permeabilities (O_2, N_2) of parylenes are similar to those of poly(ethylene terephthalate), and water vapor permeabilities compare favorably with those of high-density polyethylene. PPX-C and PPX-D have also low permeability for SO_2 and H_2S.

Electrical Properties. Parylenes from the Gorham process have low relative permittivity (dielectric constant) and dielectric loss (dissipation factor) due to their chemical structures (only carbon and hydrogen, no hydrophilicity, no dipoles) and the absence of low molar mass impurities and catalyst residues. Dielectric strengths are high compared to other polymers, which is in part due to the thinness of the test films (0.025 mm) since dielectric strengths decrease with increasing thickness and dielectric strengths of other polymers are usually measured at thicknesses of 3.18 mm.

4. Polybenzocyclobutenes

Polybenzocyclobutenes are polymers and copolymers of various benzocyclobutenes [42–46]. On heating, the parent compound, benzocyclobutene (BCB), is converted to the unstable o-quinodimethane, OQDM (o-xylylene, OX), which polymerizes thermally to polybenzocyclobutene (PBCB):

BCB [694-87-1]   OQDM, OX   PBCB (1) [51774-83-5]

The intermediate is not a biradical but has an o-quinone structure (ground-state singlet) according to theoretical calculations. This intermediate is the true monomer; the polymer can thus also be called a poly(o-quinodimethane) or a poly(o-xylylene).

Syntheses of benzocyclobutene derivatives start with either benzocyclobutene or 3-bromobenzocyclobutene. Monobenzocyclobutenes (functionality 2) such as **2** lead to linear polymers, and bis(benzocyclobutenes) (functionality 4) such as to cross-linked polymers:

2
Monomer: [112903-01-2]; Polymer: [112903-02-3]

3
Monomer: [99716-25-3]; Polymer: [99716-26-4]

4
Monomer: [99716-21-9]; Polymer: [99716-22-0]

5
Monomer: [117732-87-3]; Polymer: [124221-30-3]

6
Monomer: [137662-13-6]; Polymer: [137662-14-7]

7
Monomer: [67237-37-0];

The polymerization proceeds without formation of volatile products. It does not require catalysts or initiators and it is first order in benzocyclobutene groups. The polymerization temperature of BCB is 200 – 250 °C; it is lower for substituted benzocyclobutenes, regardless of whether the groups are electron donating or electron withdrawing.

The polymerization mechanism of the bisbenzocyclobutene monomers is not known with certainty: intermolecular reactions **1** may either lead to random cross-linking or to ladder-type structures.

All polymers of have high thermal stability. The good mechanical properties (Table 3) are retained at temperatures up to 200 – 250 °C.

5. Ionomers and Related Polymers

5.1. Overview

5.1.1. Introduction

Terms such as ion-containing polymer, ionic polymer, ionomer, polyelectrolyte, etc., are commonly used in polymer literature. None of these terms are defined by international standardization bodies and their meaning often differs from similar ones in other branches of science.

The term "ion-containing polymer" in its literal sense does not specify whether ions are part of the polymer molecules or whether they are extraneous to the polymer molecules, for example, catalyst residues or other impurities. In the polymer literature, the term "ion-containing polymer" is used as a shorthand notation for

Table 3. Physical properties of polymers from a benzocyclobutene 2 and various bisbenzocyclobutenes 3 – 7

Property	Property values of polymer					
	2	3	4	5	6	7
Young's modulus (25 °C), MPa		2500	2660			
Shear modulus (25 °C), MPa		940	1000			
Flexural modulus (25 °C), MPa	3550*			3310	4250	5150
Flexural strength (25 °C), MPa	200				152	
Elongation at break (25 °C), %	6				4.1	
Glass temperature, °C	310	> 270	> 270	> 350	310	> 350
Expansion coefficient ($\times 10^5$), K^{-1}		4.1	8		3.5	
Water absorption (48 h, 100 °C), %		1.4	0.9	0.25		0.87
Relative permittivity (1 MHz)				2.57		2.7
Dissipation factor (1 MHz)				0.0008		0.0004

*30 °C

"ion-containing polymer molecules"; i.e., it refers exclusively to ionic structures attached to polymer molecules.

Polymer molecules with ionic structures are often called ionic polymers. They are usually subdivided into two groups [47]: water-soluble and water-insoluble. Water solubility is commonly shown by polymer molecules with high contents of ionic groups; these polymers are called polyelectrolytes (Section 5.1.2). Polyelectrolytes may, for example, contain basic groups such as $-NH_2$ or acidic groups such as $-COOH$.

The terminology is less clear for water-insoluble polymers with small contents of ionic groups. If these groups are acid groups such as $-COOH$ or $-SO_3H$, the polymers are termed acidic but never ionic. The term "ionic" is reserved here for polymers with small proportions of salt groups such as $-COONH_4$, $-COONa$, or $(-SO_3)_2Zn$. The designation of these polymers as "ionic polymers" is furthermore not meant to imply that the ionic structures are indeed present as ions, i.e., with charge separation (Section 5.1.3).

The incorporation of small amounts of acidic and/or ionic groups into apolar backbone chains causes these groups to associate ("aggregate") in the solid state. The resulting intermolecular physical cross-links lead to higher moduli and strengths. The cross-links dissociate at higher temperatures and the polymers can then be processed conventionally.

5.1.2. Polyelectrolytes

The ionic groups of polyelectrolytes may be in side chains as in **8** and **9** or in backbones as in:

$$-(CH_2-CH)_n- \quad -(CH_2-CH)_n- \quad -(O-\overset{\overset{O}{\|}}{P})_n-$$
$$\quad\;\;\;|\qquad\qquad\quad\;\;\;| \qquad\qquad\quad\;\;\;|$$
$$\;\;\;COOH \qquad\qquad NH_2 \qquad\qquad\;\; OH$$
$$\quad\;\;\;\mathbf{8} \qquad\qquad\qquad \mathbf{9} \qquad\qquad\qquad \mathbf{10}$$

$$-(CH_2-CH_2-NH)_n- \qquad -(CH_2-CH_2-\underset{nX^-}{\overset{+}{C_5H_4N}})_n-$$
$$\qquad\quad\;\mathbf{11} \qquad\qquad\qquad\qquad \mathbf{12}$$

Polyelectrolytes may be polyacids such as poly(acrylic acid) **8** and poly(phosphoric acid) **10**, or polybases such as poly(vinyl amine) **9** and poly(ethylene imine) **11**. In water and other highly polar solvents, they dissociate into polyions and oppositely charged counterions. Upon dissociation, polyacids lose protons and become polyanions (e.g., $\sim[CH_2-CH(COO^-)]_n\sim$). The salts of these polyanions are thus called polysalts (e.g., $\sim[CH_2-CH(COO^-Na^+)]_n\sim$). Polybases can quaternize by adding protons, methyl groups, etc., and become polycations; the quaternization may be in the substituent as in **9** or in the main chain as in **11**. Polymers which only exist as polycations but not as polybases are called ionenes, an example of which is **12**, the polymer of 4-vinylpyridine by polymerization via the 1,6-positions.

Polyacids, polybases, polyanions, and polycations contain many acidic, basic, anionic, and cationic groups per molecule. They are to be distinguished from macroacids, macrobases, macroanions, and macrocations, which have only one such group (as in, e.g., living anionic polymerizations with monofunctional initiators), and from macrodiacids, etc., with two such functional groups, etc.

Polyampholytes are polymers whose chains carry both acidic and basic (or anionic and cationic) groups; an example is the copolymer of acrylic acid $CH_2=CHCOOH$ and acrylamide $CH_2=CHCONH_2$. Complexes of polyacids and polybases are called polyelectrolyte complexes or symplexes; they may or may not be water soluble, depending on their chemical structure (see also [47]).

5.1.3. States of Ions

In science "ion" denotes an electrically charged atom or molecule (cation or anion) which shows ionic conductivity as a free entity (in solution or as gas) or in its assemblies (in crystals or melts). Each ion possesses an oppositely charged counterion due to the principle of electroneutrality; the electric charges are separated. In polymer science, "ion-containing" does not necessarily refer to charge-separated entities, especially not in the solid state; it is merely necessary that these entities may be able to generate ions under favorable (polar) conditions.

Under less polar conditions, ions can form ion pairs consisting of oppositely charged ions bound together by Coulomb interactions but not by covalent bonds. In contact ion pairs, these ions are in direct contact. In the presence of polar solvents, the ions may be also more loosely held together in solvent-separated ion pairs; such ion pairs may share one solvent molecule or two or more.

$$R-X \rightleftharpoons R^{\delta+}-X^{\delta-} \rightleftharpoons R^+-X^-$$
molecule → polarized molecule → contact ion pair
(polarization) (ionization)

$$\rightleftharpoons R^+/solv/X^- \rightleftharpoons R^+ + X^-$$
solvent-separated ion pair → free ions
(ionization) (dissociation)

Triple ions arising from Coulombic interactions such as $\sim\sim COO^-/M^{2+}/^-OOC\sim\sim$ from two carboxylate groups $\sim\sim COO^-$ and a metal ion M^{2+} are especially stable ionic structures since the sum of the repulsion between two anions and the attraction between the cation and two anions results in a negative potential energy [48]. The same reasoning applies to higher ion aggregates; such structures are responsible for the stability of ion crystals such as NaCl. Triple ions are sometimes called triplets although the latter term should refer exclusively to multiplets of ion pairs.

Ion pairs can associate to form tightly aggregated ion multiplets consisting of associated ion pairs (2 ion pairs), triplets (3 ion pairs), etc.:

$$2 R^+X^- \rightleftharpoons \begin{array}{c} R^+-X^- \\ | \ \ | \\ X^--R^+ \end{array} \xrightarrow{+R^+X^-} \begin{array}{c} X^- \ \ R^+ \ \ X^- \\ | \ \ \ \ \ | \ \ \ \ \ | \\ R^+ \ \ X^- \ \ R^+ \end{array}, \text{etc.}$$
associated ion pair — triplet

The maximum number of ion pairs in such multiplets should be six to eight if the multiplets are spherical. Ion multiplets do not contain polymer chain segments; sizes of spherical multiplets cannot exceed ca. 0.6 nm because of geometrical constraints on the multiplet structures by the chain segments to which the multiplets are attached. Because of their small size, they are not separate phases in solid polymers; they thus directly affect matrix properties such as segment mobilities. No size restriction is imposed on nonspherical geometries.

Multiplets may aggregate into ion clusters [49] which contain significant amounts of polymer segments in addition to the ionic structures [51], [51] (for a review of these and other theories, see [52]). The domains of these clusters are thus considerably larger than those of multiplets; their size is limited to 2 – 10 nm by the elastic forces generated by the backbone chains. These clusters approach the size of true phases. Clusters thus do not markedly affect matrix properties except that they may behave as cross-links with reinforcing action.

Free ions and solvent-separated ion pairs conduct electricity, but contact ion pairs, contact multiplets, and ion clusters do not. Most ion-containing polymers are electrical insulators in the solid state; since obviously no ions are present, they really should not be called ion-containing polymers.

5.2. Acidic Copolymers

5.2.1. Introduction

Acidic copolymers are defined as copolymers of a parent monomer (or several thereof) and a small amount of an acidic comonomer. The polymers may be thermoplastic or elastomeric, depending on the glass transition temperature of the parent monomer segments.

The first acidic elastomers were copolymers of butadiene with small amounts of various unsaturated acids, patented by I. G. Farbenindustrie in 1930 [53]. In 1950, B. F. Goodrich started to introduce a series of acidic elastomers from butadiene, acrylonitrile, and small amounts of acrylic acid [54], [55]; the trade name Hycar now refers to polymers of different compositions and grades with many different CAS numbers. These and other acidic elastomers are also known as carboxylated elastomers or carboxylated rubbers; they are produced by many companies [56].

Hycar rubbers can be neutralized with zinc oxide or zinc salts; these neutralized acidic polymers were the first elastomeric ionomers. However, the ionomeric structures served only as reinforcing entities since the rubbers were still conventionally vulcanized to covalently cross-linked elastomers.

Neutralization of suitable acidic copolymers without chemical cross-linking leads to thermoplastic elastomers. The first thermoplastic elastomers were the cured Hypalon polymers of Du Pont. Hypalons are based on sulfochlorination of polyethylenes [57] and are available in many grades. The curing of these polymers by metal compounds leads to ionomers which behave as thermoplastic elastomers.

In the mid-1960s Du Pont began production of the first thermoplastic acidic copolymers from ethylene (E) with small amounts of methacrylic acid (MAA). These copolymers were not commercially available but were converted in-house to ionomers (Section 5.3.6).

Thermoplastic acidic EAA copolymers of ethylene (E) and up to 15 wt % acrylic acid (AA) were first commercially manufactured in various grades by Dow Chemical and Union Carbide (Section 5.2.3). They were followed by Du Pont's E-MAA copolymers with up to 15 mol % MAA (Nucrel).

Du Pont also manufactures a 1 : 1 copolymer of ethylene and methyl acrylate with small amounts of (meth)acrylic acid (Vamac) (Section 5.2.4).

5.2.2. Structure and Properties

Acidic copolymers can be amorphous or partially crystalline, depending on the crystallizability of the segments of the parent monomer units. The crystalline domains are composed of lamellae formed by folded chains of parent segments and do not contain acidic units, the latter being confined to the amorphous regions where they may be present as dimers. Associated acidic units do not form separate phases. Acidic copolymers are thus one-phase materials if amorphous, and two-phase polymers if crystalline.

Properties of such polymers depend on intramolecular and intermolecular interactions between their segments and groups. COOH groups in polyethylene segments tend to associate; the associated groups act as cross-links. At low COOH content, this association can hardly take place since the COOH groups are not free but bound to polyethylene; the geometrical constraints imposed upon the groups reduce the probability of association. Since the crystallinity of the polymer is not markedly affected, Young's moduli of ethylene–methacrylic acid copolymers remain constant up to 4–5 mol % of methacrylic

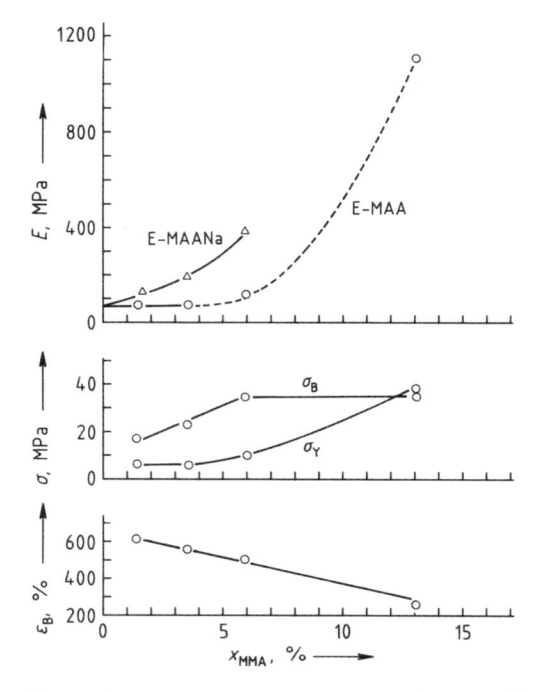

Figure 2. Tensile modulus E, yield strength σ_Y, tensile fracture strength σ_B, and elongation at break ε_B of ethylene – methacrylic acid copolymers (○) and their sodium salts (▲) (100 % neutralization) as function of the mole fraction x_{MAA} of methacrylic acid units (experimental data from [58])

acid groups. At higher COOH group concentrations, intermolecular association is increasingly possible. Intermolecularly associated groups act as cross-links and the moduli and yield strengths increase sharply (Fig. 2).

Polyethylene yields easily and exhibits stress-softening. The increasing association of COOH groups impedes the flow of segments, and the elongation at breaks decreases. Tensile fracture strengths of ethylene–methacrylic acid copolymers increase with increasing COOH content, eventually becoming independent of the proportion of COOH groups.

5.2.3. Thermoplastic Acidic Copolymers

Syntheses. EAA copolymers of ethylene (E) and up to 15 wt % acrylic acid (AA) are produced by Dow Chemical and Union Carbide in various grades. Du Pont manufactures an EMAA copolymer of ethylene and <15 mol % methacrylic acid MAA (Nucrel [59]) and an 1: 1 EMA copolymer of ethylene and

Table 4. Industrial copolymers of ethylene and (meth)-acrylic acid

Parent monomers	Comonomers	Trade name	Manufacturer
Ethylene	acrylic acid	EAA 469	Dow Chemical
Ethylene	acrylic acid		Union Carbide
Ethylene	methacrylic acid	Nucrel	Du Pont
Ethylene + methyl acrylate	(meth)acrylic acid	Vamac	Du Pont

methyl acrylate (MA) with small amounts of (meth)acrylic acid (Table 4).

None of the industrial polymerization processes have been disclosed. They are most likely high-pressure free-radical polymerizations. Since the Q, e values of ethylene ($Q_E = 0.015$; $e_E = -0.200$) and methacrylic acid ($Q_{MAA} = 2.34$; $e_{MAA} = 0.65$) are unfavorable for random arrangements of comonomeric units, copolymerizations are probably limited to ca. 10 % conversion in order to prevent the formation of long blocks of methacrylic acid units. Calculation of copolymerization parameters r from $r_i + (Q_i/Q_j)\exp[-e_i(e_i - e_i)]$ $r_i = (Q_i/Q_j)\exp[-e_i(e_i - e_i)]$ gives $r_{MAA} = 2.07$ and $r_E = 0.234$ and thus number-average sequence lengths $N_{MAA} = 1 + r_{MAA}([MAA]/[E])$ of methacrylic acid units of only $N_{MAA} \approx 1.23$ if the monomer ratio is $[MAA]/[E] = 90/10$. Q,e values of acrylic acid or copolymerization parameters of the ethylene–acrylic acid system have not been published but they should be similarly unfavorable.

Structure and Properties. High-pressure radical polymerization of ethylene leads to polyethylenes with a few short and long branches. The more or less random distribution of fairly bulky acid or salt units along the chains reduces the chain lengths of crystallizable segments of such polyethylenes. Melting temperatures thus decrease with increasing size of the substituents, e.g., for ethylene polymers with ca. 3 mol % comonomer units from 115 °C (LDPE with no comonomer units) to 104 °C (ethylene–acrylic acid), 95 °C (ethylene–sodium acrylate), and 90 °C (ethylene–methyl acrylate). Since the physical cross-linking caused by such comonomer units shortens the segments between cross-links, glass transition temperatures increase from −80 °C (LDPE) to −35 °C (ethylene–methyl acrylate), 0 °C (ethylene–acrylic acid), and 50 – 60 °C (ethylene–sodium acrylate).

The density of EAA copolymers increases linearly with the content of acryl acid units AA, reaching 0.95 g/cm^3 at 15 wt % AA units. Tensile yield strengths first decrease with increasing AA content, pass through a minimum at 12 – 15 wt % AA units, and increase again. The fracture strength exhibits a maximum at approximately the same AA content. Elongations at break are ca. 600 %.

Ethylene–acid copolymers show outstanding toughness since incorporated acrylic acid units reduce the formation of spherulites by polyethylene segments and simultaneously promote intermolecular association. Toughness is maintained at low temperatures; the brittleness temperature is ca. −70 °C for polymers with ca. 3 – 9 wt % acid units.

Ethylene acid copolymers show excellent adhesion to metals, glass, and paper due to the polar COOH groups and the low melt viscosities of the polymers. They also adhere well to polyethylene because of their polyethylene segments. Adhesion to other plastics can be improved by oxidation of EAA and EMAA surfaces.

Processing. EAA and EMAA can be processed similarly to LDPE. Since the acid groups are somewhat corrosive to metals, corrosion-resistant metals or nickel or chromium plating are recommended for processing equipment.

Certain ethylene acid copolymer grades can be dispersed in hot aqueous ammonia solutions to give dispersions stable at room temperature. They lose ammonia upon drying and form surface coatings with the same properties as copolymers processed otherwise.

Applications. Present applications comprise inner laminate layers on aluminum foil for pouches (towelettes, condiments) and tubes (toothpaste), cable sheathing, pipe coatings, and laminates on metals and glass fiber for automotive and building applications, and on polyurethane foams for carpet underlay.

FDA regulations for direct food contact with ethylene acid copolymers exist for EAA resins containing up to 10 % acrylic acid units and EMMA resins with up to 20 % methacrylic acid units.

5.2.4. Elastomeric Acidic Copolymers

Since 1976, Du Pont has offered Vamac, a copolymer with ethylene units $-CH_2-CH_2-$ and methyl acrylate units $-CH_2-CH(COOCH_3)-$ in the molar ratio 1:1 and a small amount of undisclosed cure site units $-R(COOH)-$. The polymer is probably synthesized by free-radical emulsion polymerization. Two carbon-black-filled masterbatches (e.g., type B-124 with 24 parts SRF carbon black and auxiliaries per 100 parts polymer) and two neutral masterbatches (e.g., type N-123 with 23 parts siliceous earth per 100 parts polymer) are available [60].

All Vamac grades can be vulcanized by diamines or hexamethylene diaminocarbamate, and N-123 also by free radicals. Vulcanized Vamacs are EE–EH elastomers according to ASTM D-2000 and SAE J-200 classifications, i.e., elastomers with continuous service temperatures up to 175 °C (E) and swelling between 20 % (H) and 80 % (E) after 70 h in ASTM oil no. 3. Vamac is resistant to hot oils, hydrocarbons, glycols, and engine cooling fluids but not to brake fluids. Its weatherability is excellent.

The heat distortion of Vamac is surpassed only by the more expensive fluoroelastomers and silicones; however, Vamac has better mechanical properties. After 30 min vulcanization at 177 °C, the rebound elasticity of B-124 is 35 %, the tensile strength 14 MPa, the ultimate elongation 440 %, and the 100 % modulus 2.9 MPa. After 7 d at 177 °C, the modulus increases to 5.3 MPa and the tensile strength to 16.5 MPa, whereas the elongation decreases to 240 %, all probably due to an after-vulcanization.

Prices of resins are between those of low-density polyethylenes and Surlyn A ionomers.

Vamac is used for tubes and seals and as a damping material. It can replace sulfochlorinated polyethylene in cables and electrical systems since it burns with little smoke of low toxicity.

5.3. Ionomers

5.3.1. Introduction

The name ionomer was originally introduced by Du Pont for (partial) salts of copolymers of olefins with small amounts of carboxyl group containing monomers, such as ethylene–methacrylic acid copolymers [61]. The term ionomer is now used as a generic term for thermoplastic and elastomeric hydrocarbon polymers with small amounts of partially or completely neutralized pendant acid groups which show typical ionomer properties (Table 5). Total acidic group contents (free or neutralized) are usually under ca. 10 mol %. The syntheses, structures, properties, and applications of ionomers have been described in many books [62–64], reviews [50], [65–71], handbook articles [72–74], and symposia [75–77].

The acidic monomer units of such copolymers may be directly introduced by copolymerization as in ethylene/acrylic acid copolymers, or subsequently by chemical transformation of a primary copolymer as in the sulfonation of copolymers of ethylene, propylene, and a nonconjugated diene (Table 5). In ionomers, acidic monomer units are furthermore partially or completely converted into their salts. Polymer literature restricts the term "ionic" to these salts: polymer molecules with salt groups such as $-COONa$ or $-COONH_4$ are called ionic, whereas those with groups such as

Table 5. Industrial ionomers

Parent monomeric unit	Ionomeric unit	Ions	Trade name	Manufacturer
Thermoplastics				
Ethylene	methacrylic acid	Zn, Na	Surlyn A	Du Pont
Tetrafluoroethylene	perfluoro vinyl ether A *		Nafion	Du Pont
Tetrafluoroethylene	perfluoro vinyl ether B *		Flemion	Asahi Glass
Styrene	styrene *p*-sulfonic acid			Exxon
Elastomers and thermoplastic elastomers				
Ethylene	post-sulfochlorination		Hypalon	Du Pont
Ethylene + propylene+ nonconjugated diene	post-sulfonation		Thionic	Uniroyal
Butadiene + acrylonitrile	acrylic acid	Zn	Hycar	Goodrich

* Perfluoro vinyl ether: $CF_2=CF[O(CF_2CF(CF_3)O)_m R]$ with $R = (CF_2)_2SO_3Na$ (A) or $R = (CF_2)_n COONa$ (B).

—COOH or —SO$_3$H are not considered ionic but are sometimes referred to as acidic copolymers. Ionic and acidic copolymers with small proportions of ionic and/or acidic groups are always insoluble in water.

At present, only three thermoplastic ionomers (Surlyn A, Nafion, Flemion), two thermoplastic elastomers (Hypalon, Thionic), and many carboxylated rubbers (Hycar and others) are manufactured. Many other ionomers have been reported in the literature, e.g., carboxylic, phosphonic, sulfonic, and thioglycolic salts of polypentenamers [78], carboxylated and sulfonated polystyrenes [79], [80], and telechelic poly(isobutylene sulfonate) ionomers [81].

5.3.2. Structure

The first thermoplastic ionomers were copolymers of ethylene and methacrylic acids, partially neutralized with zinc or sodium compounds [82]. These Surlyn A polymers of Du Pont have much better clarity and tensile properties than comparable polyethylenes, which stimulated much research.

The enhanced properties of Surlyn A polymers were first thought to be due to decreased crystallinity and the presence of cross-linking triple ions of one bivalent metal cation (Zn^{2+}) and two carboxyl anions. It was subsequently shown that crystallinities are not markedly lower than those of the parent polymers, and that the good transparency was due to the absence of spherulites. These property changes also occur with monovalent metal cations (such as Na^+) so that other ionic cross-links must be present; these were later found to be ion multiplets and ion clusters.

Ionomers may be amorphous or partially crystalline, depending on the crystallizability of the segments of the parent monomer units. Amorphous ionomers are two-phase polymers. The continuous phase of these ionomers is amorphous and composed of parent segments with interdispersed ion pairs and ion multiplets. The discontinuous phase of amorphous ionomers consists of ion clusters of ion pairs and ion multiplets with interdispersed polymer segments. Crystallizable ionic copolymers possess crystalline regions as a third phase in addition to the amorphous phase and the ion clusters [49] (Fig. 3).

Crystalline domains do not contain acidic or ionic units; the segments in these domains are chain-folded. Acidic units are confined to amorphous regions, whereas ionic units are either in

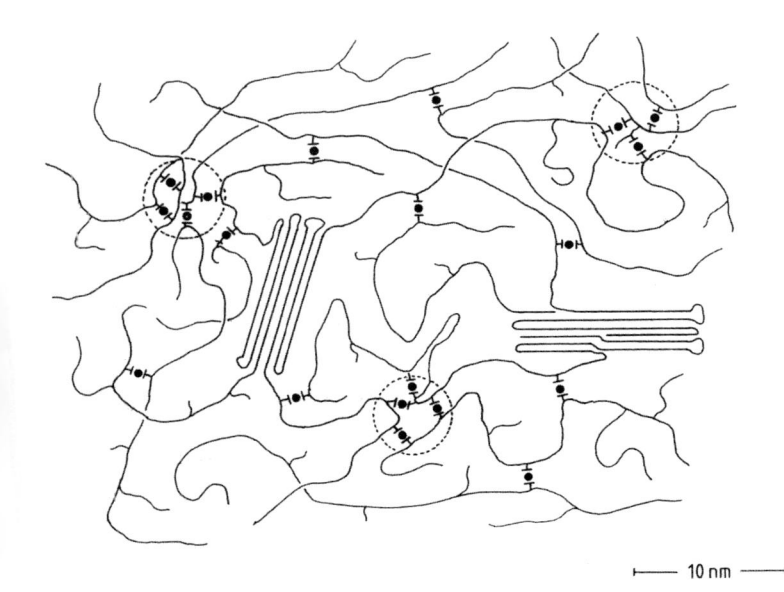

Figure 3. Schematic representation of the structure of crystallizable ionomers with crystalline (lamellar) domains, amorphous regions with ion triplets (and multiplets), and ion clusters with interdispersed chain segments (e.g., zinc-neutralized low-density ethylene – methacrylic acid copolymers)

amorphous regions or in ion clusters. The amorphous regions contain polymer segments with acidic or ionic units, acidic units as dimers, and ionic units as ion pairs and/or ion multiplets in a single phase. Ion clusters are separate phases that contain ion multiplets and polymer segments. The exact structure, size, and size distribution of amorphous and crystalline phases, ion clusters, ion multiplets, ion pairs, and associations of acidic groups are not known.

5.3.3. Properties

Ionic Group Content. Ionomers behave either as physically cross-linked thermoplastics or as thermoplastic elastomers, depending on the segment mobility of the parent monomer units. In amorphous polymers, the mobility is determined by the glass transition temperature, and in crystalline polymers also by the melting temperature, both relative to the service temperatures. Melting temperatures of crystalline polymers are mainly affected by the segments of the parent units and only secondarily by acidic or ionic units, which merely reduce the segment lengths of crystallizable units.

However, the glass transition temperatures T_G reflect the structure and mobility of noncrystalline phases, i.e., amorphous phases and ion clusters. Ionomers thus show two glass transition temperatures [61].

Ion pairs and ion multiplets act as cross-linking agents in amorphous phases. The higher the concentration of these physical cross-links, the shorter are the segments between cross-links. Cooperative movements of the units in segments are affected if the segment lengths are shorter than $N_c = 30 - 50$ chain atoms. At $N_c < 30 - 50$, glass transition temperatures of amorphous phases increase linearly with the concentration of ionic units (Fig. 4).

Glass transition temperatures of ion clusters are strongly influenced by ion pair and multiplet formation. Ions of an ion pair can only move away from each other above T_G; the required work W_{el} must be proportional to kT_G. This work is given by the electrostatic interaction F_{el} and the internuclear distance a. Because $W_{el} \sim kT_G \sim \int F_{el} da$, one obtains $TG \sim q_{cat}q_{an}/a$, where q_{cat} and q_{an} are the cation and anion charges. The anion charge is constant in ionomers and thus $TG \sim q_{cat}/a \sim q/a$ [83]. In

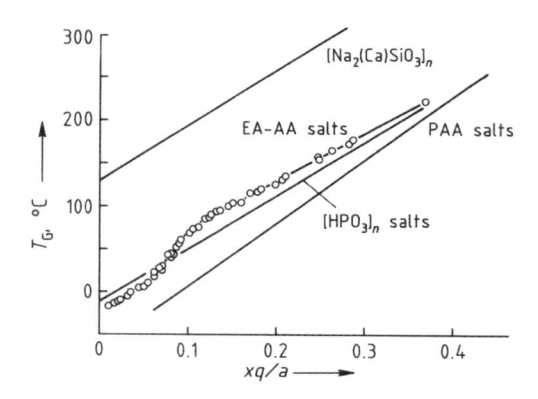

Figure 4. Glass transition temperature T_G as function of the linear charge density xq/a where x is the "metal carboxylate content" (probably in mol % of ionic units), q the cation charge (in "units of electron charges"), and a the distance between centers of charge (probably in 0.1 nm) [85]
○ = T_G of ion clusters of ethyl acrylate – acrylic acid copolymers EA – AA with Li, Na, K, Cs, Ca, and Ba [84]
— = T_G of amorphous domains of $[Na_2(Ca)SiO_3]_n$ glasses, salts of $[HPO_3]_n$, and salts of poly(acrylic acid) PAA

copolymers, only a fraction x of the monomeric units are ionic; the glass transition temperature of ionomers must be therefore proportional xq/a [84] (Fig. 4). The glass transition temperatures of ionic clusters usually show a sigmoidal dependence on either xq/a or x.

Effect of Neutralization. Elastic moduli of ethylene–methacrylic acid polymers with COONa groups are higher than those of the same polymers with COOH groups (Fig. 2). Increasing neutralization of COOH groups thus leads to an increase of moduli up to degrees of neutralization of $f_N \approx 30$ mol % (Fig. 5); moduli become constant at higher degrees of neutralization, which indicates that no new cross-linking sites are formed beyond this concentration.

The situation is quite different for partially neutralized copolymers of ethylene and acrylic acid. Tensile moduli first increase with higher degrees of neutralization f_N, but do not become constant as in ethylene/methacrylic acid copolymers; instead they decrease if $f_N > 0.3$ (Fig. 6). Fracture strengths σ_B increase monotonically with higher degrees of neutralization; they do not become constant. The variation of modulus and fracture strength with degree of neutralization seems to be independent of the type of monovalent cation, whereas the elongation at break depends on the ion type, especially at high

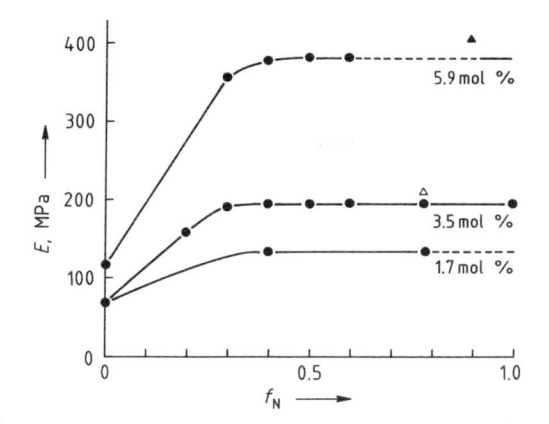

Figure 5. Young's modulus E of ethylene/methacrylic acid copolymers with 1.7, 3.5, and 5.9 mol % methacrylic acid units as function of the degree of neutralization f_N of COOH groups by NaOH (●), Sr(OH)$_2$ (▲), or Cu(OH)$_2$ (△) Reproduced with permission from [58].

degrees of neutralization. These effects are probably due to differences in the structure and/or concentration of ion clusters; experimental data are not available.

Ionic domains are present in neutralized ethylene–methacrylic acid copolymers at temperatures up to 300 °C, far above the crystallite melting temperature, according to small-angle X-ray and neutron-scattering experiments. The number of ionic domains increases with the degree of neutralization f_N. Since they represent physical cross-linking sites, they are responsible for an increase of melt viscosity with increasing

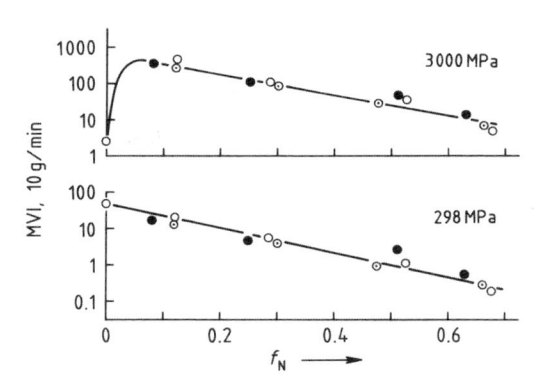

Figure 7. Effect of the degree of neutralization by Li$^+$ (○), Na$^+$ (⊙), and K$^+$ (●) on the melt volume index MVI (at 298 and 3000 MPa) (melt flow index) of copolymers of ethylene with 15 wt % acrylic acid (experimental data from [85])

neutralization, i.e., a decrease in melt volume index (melt flow index) MVI (Fig. 7). The size of these ionic domains is not affected by the type of monovalent ion nor by the applied pressure as the similar slopes of the functions log MVI = f (f_N) show.

Influence of Counterion Type. The association of salt groups depends on the coordination number of the counterions and not on its valency. Monovalent ions such as Li$^+$, Na$^+$, and K$^+$ and bivalent ions such as Mg^{2+}, Ca^{2+}, and Zn^{2+} cause multiplet and cluster formation (Fig. 3). The effects are however quantitatively different: at 25 °C and the same elongations at break, bivalent ions lead to higher tensile fracture strengths than monovalent ions

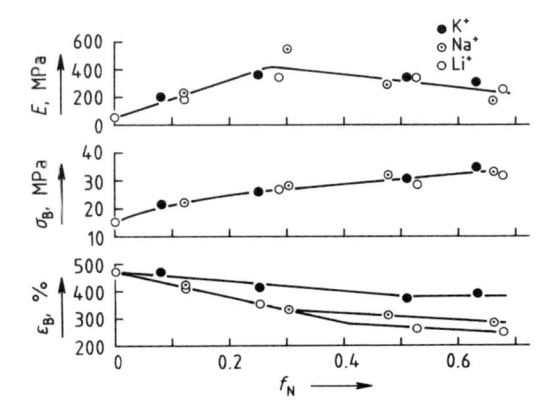

Figure 6. Variation of tensile modulus E, fracture strength σ_B, and elongation at break ε_B with degree of neutralization f_N of ethylene – acrylic acid copolymers (experimental data from [85])

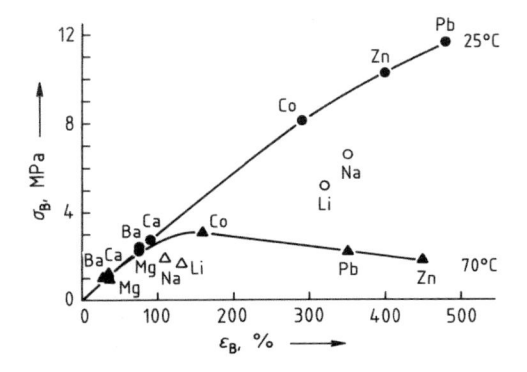

Figure 8. Tensile fracture strength σ_B as a function of elongation at break ε_B of salts of sulfonated EPDM polymers (32 milli-equivalents of sulfone groups per 100 g polymer) at 25 °C (●, ○) and 70 °C (△, ▲) (experimental data from [86]) ●, △ = Bivalent metals; ○, ▲ = Monovalent metals

(Fig. 8). At this temperature, higher strengths are accompanied by higher elongations for the two series of bivalent and monovalent metal cations. In each series, these properties increase with increasing atomic number of the metal.

At the higher test temperature of 70 °C, ionomers with bivalent metals of group 2 of the periodic table (Mg, Ca, Ba) still follow the $\sigma_B = f(_B)$ relationship for 25 °C but the members of higher groups (Co, Zn, Pb) now not only lead to much lower fracture strengths but to a decrease in strength with increasing elongation at break (Fig. 8). These changes indicate that the structures of physical cross-linking sites formed by metals with higher group numbers change at higher temperatures.

5.3.4. Ethylene–Methacrylic Acid Ionomers

Many thermoplastic ionomers have been reported in the literature but only three have been commercialized: Surlyn A (Du Pont) for films, coatings, and molded articles, and two perfluorinated ionomers for membranes (Nafion from Du Pont, Flemion from Asahi Glass).

Synthesis. Surlyn A (Du Pont) is a copolymer of ethylene with ca. 10 wt % methacrylic acid; 50 % of the COOH groups are present as the sodium salt [25608-26-8] or zinc salt [28516-43-0]. CAS registers these copolymers as "2-propanoic acid, 2-methyl-, polymers," i.e., as polymers of methacrylic acid, although methacrylic acid is clearly the minority component.

The commercial synthesis of Surlyn A polymers has not been disclosed. Most likely, it is a high-pressure free-radical copolymerization of ethylene and methacrylic acid. The acid groups are then partially neutralized by rolling the polymers with metal acetates, alkoxides, carbonates, or hydroxides with release of acetic acid, alcohols, carbon dioxide, or water, respectively.

Commercial products contain either sodium or zinc. Polymers with ammonium derivatives or lithium can also be processed but these polymers are not commercial. Magnesium- and barium-containing polymers possess such high viscosities that they can no longer be processed.

Structure. Surlyn A polymers have a multi-phase structure consisting of crystalline regions with chain-folded polyethylene segments, amorphous regions containing ion multiplets, and ion clusters with interdispersed polyethylene segments.

The introduction of acid/salt groups into polyethylene chains prevents the formation of spherulites but does not markedly change the crystallinity. Surlyn A polymers therefore have good optical clarity (haze < 2 %), especially the low-crystallinity grades. Transparency increases with increasing degree of neutralization; it becomes constant at neutralizations >ca. 30 – 50 %.

Surlyn A polymers have chemical structures similar to those of alkali metal salts of long-chain fatty acids. Since the latter are nucleating agents, similar behavior can be expected from Surlyn A polymers. Upon annealing, the crystallinity of these polymers does indeed increase. Spherulites form at temperatures above ca. 90 °C, leading to haziness.

Properties. Some properties of Surlyn A resins are listed in Table 6. Crystalline regions and ion clusters all act as physical cross-linking sites for the rubbery, ion-multiplet-reinforced amorphous domains; Surlyn A polymers thus have the characteristics of high-impact multiphase polymers. Tensile strengths are higher than those of comparable polyethylenes and increase with the proportion of acid units and the degree of neutralization (Fig. 9). Stiffness and hardness are similar to those of LDPE. The various grades retain impact strengths up to their brittleness temperatures, which are between −100 and −130 °C. The continuous service temperatures of ca. 70 °C are fairly low. The elongation at

Table 6. Properties of Surlyn A Resins

Property	Na Resin	Zn Resin
Density, g/cm^3	0.94	0.94
Melting temperature, °C	94	95
Heat deflection temperature (455 kPa), °C	40	41
Thermal expansion coefficient, K^{-1}	14 × 10^{-5}	16 × 10^{-5}
Melt flow index, g/10 min	1.3	5.5
Flexural modulus, MPa	220	130
Tensile strength at yield, MPa	12.4	8.3
Tensile strength at break, MPa	29	21.4
Elongation at break, %	450	500
Tensile impact strength, kJ/m^2	1160	925
Notched impact strength, J/m	610	no break
Hardness (Shore D)	60	54

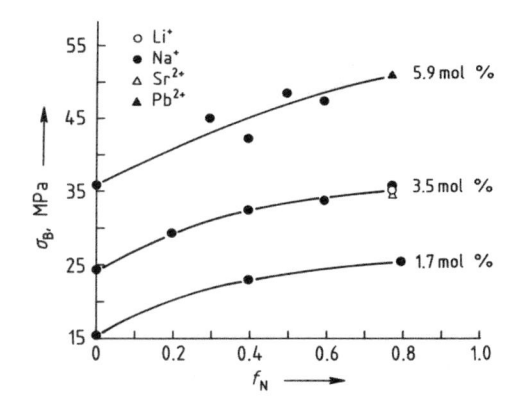

Figure 9. Tensile fracture strength σ_B of copolymers of ethylene with various mole fractions x_{MAA} of methacrylic acid as function of the degree f_N of neutralization
Reproduced with permission from [58].

break is ca. 350 %, and the abrasion resistance is excellent.

The physical cross-linking of chains by ion clusters causes high melt viscosities (low melt volume indices) which are much higher than those of comparable polyethylenes. Melt strengths are high. The cross-linking also reduces the creep under constant load.

At room temperature, Surlyn A ionomers are insoluble in all common organic solvents. They dissolve at temperatures above ca. 70 °C in 60/40 vol/vol mixtures of toluene/isobutanol, toluene/methyl isobutyl ketone, or tetrachloroethylene/butanol, or in a 70/20/10 mixture of decahydronaphthalene/diethylene glycol dimethyl ether/dimethyl sulfoxide. The polymers have good resistance to stress corrosion, except by certain combinations of detergents and alcohols. The permeability for oils and fats is low, while water vapor permeabilities correspond to those of low-density polyethylene. The polymers are readily degraded by light; UV absorbers extend outdoor service lives to ca. 2 years. Surlyn A polymers burn slowly when in contact with flames.

Carboxyl groups provide Surlyn A polymers with good adhesion to metals, glass, and paper; the adhesive properties depend on the carboxyl group content. Adhesion on impact is good, as e.g., required by the painted surfaces of molded golf balls.

Applications. The combination of good impact strength, high melt viscosity, high melt strength, good transparency, and relatively low continuous service temperatures is ideal for the manufacture of extremely thin packaging films. Such films have high tear strengths and are especially well suited for the packaging of sharp objects. Co-extruded ionomer–LDPE films are used for sterilizable packaging of foods and pharmaceuticals. Coatings of zinc-containing Surlyn A grades make glass bottles splinter-resistant. The combination of toughness and good abrasion resistance recommends the polymers for shoe soles and heels. Other applications are sheathing of electric cables and wires, automotive parts, and floor mats. Molded articles can be directly painted.

Closed-cell Surlyn A foams with densities of 50 – 700 kg/m³ can be produced by extrusion or injection molding. Applications include carpet backings, ski boots, hockey sticks, helmets, and tennis court foundations.

5.3.5. Fluorinated Ionomers

Precursors of fluorinated ionomers are manufactured by free-radical copolymerization of tetrafluoroethylene (TFE) with perfluorovinylethers (FVE)

$$F_2C=CF_2 \qquad \qquad F_2C=CF$$
$$\qquad \qquad \qquad \qquad |$$
$$\qquad \qquad \qquad \qquad O-(CF_2CFO)_{\geq 1}(CF_2)_2SO_2F$$
$$\qquad \qquad \qquad \qquad \qquad \qquad |$$
$$\qquad \qquad \qquad \qquad \qquad \qquad CF_3$$
TFE $\qquad \qquad \qquad \qquad \qquad$ VFE-A

$$F_2C=CF$$
$$\quad |$$
$$O-(CF_2CFO)_{0\text{ or }1}(CF_2)_{1-5}COONa$$
$$\qquad \qquad |$$
$$\qquad \qquad CF_3$$
VFE-B

in molar ratios of 5 – 10 mol TFE per mole PVE [88], [88]. VFE-A is synthesized by reaction of hexafluoropropylene oxide (from the oxidation of TFE) with the sultone from the addition of sulfur trioxide to tetrafluoroethylene. Copolymers of TFE and VFE-A are known as XR Resins (Du Pont) whereas copolymers of TFE and VFE-B are called Flemion (Asahi Glass).

XR Resins can be melted and fabricated into membranes and tubes by conventional processing methods for thermoplastics. The $-SO_3F$ end groups are then saponified by hot caustic soda to give $-SO_3Na$ end groups, which in turn are converted into sulfonic acid groups with, for instance, nitric acid.

Du Pont manufactures articles with free sulfonic acid groups as films, tubes, or laminates under the trade name Nafion. Similar carboxylated fluoro ionomers are produced by Asahi Glass under the trade name Flemion. These polymers are permselective [89]: they allow the passage of cations but not of anions and are therefore used as membranes for electrochemical processes, e.g., for the production of sodium hypochlorite. The exchange capacity of such resins is ca. 0.85 mmol/g, the fracture strength of resins with ca. 25 % water content ca. 21 MPa, and the elongation at break ca. 150 %. The strength can be increased by reinforcement with fabrics made from polytetrafluoroethylene fibers [90].

5.3.6. Sulfochlorinated Polyethylene

Simultaneous chlorination by Cl_2 and sulfonation by SO_2 of polyethylenes in homogeneous solution under the action of free-radical initiators or light leads to polymers with $-CH_2-$, $-CHCl-$, and $-CHSO_2Cl-$ groups in the approximate ratio 100 : 30 : 1. The polyethylene segments of these polymers are too short to crystallize, and the polymers thus behave as elastomers because of the low glass transition temperature (actual data unknown). Du Pont manufactures chlorosulfonated polyethylenes (CSM) in various grades (different CAS numbers) under the name Hypalon. Apart from a handbook article [91], little has been published about CSM polymers.

CSM polymers can be cured by way of the SO_2Cl groups [92]. The white polymers need no reinforcing fillers and can be used for many colored products. The presence of polar Cl and SO_2Cl groups provides the elastomers with excellent resistance to oil and ozone compared to standard synthetic rubbers. The maximum service temperature is ca. 120 °C.

5.3.7. Sulfonated EPDM Polymers

Synthesis and Structure. EPDM polymers are produced by Ziegler–Natta copolymerization of ethylene (E), propylene (P), and a small amount of a nonconjugated diene, preferably 5-ethylidene-2-norbornene (EN). The amorphous, rubbery polymers contain the monomeric units E, P and EN; the latter can be sulfonated by $H_2SO_4/(CH_3CO)_2O$ to sulfo units SEN [73]:

$$-CH_2-CH_2- \quad -CH_2-\overset{\overset{\displaystyle CH_3}{|}}{CH}- \quad \underset{\underset{\displaystyle E}{}}{} \quad \underset{\underset{\displaystyle P}{}}{} \quad \underset{\underset{\displaystyle EN}{CH-CH_3}}{} \quad \underset{\underset{\displaystyle SEN}{\overset{\displaystyle CH-CH}{\underset{\displaystyle SO_3H}{|}}}}{}$$

Sulfo-EPDMs can be neutralized with metal alkoxides or other bases, either in solution with subsequent precipitation of the neutralized polymer or by reactive milling of the polymer in an extruder. These thermoplastic elastomers were developed by Exxon and are now being commercialized by Uniroyal under the trade name Thionic.

Properties. Mechanical properties of sulfo-EPDM ionomers depend strongly on the type and concentration of cations (Fig. 8). Tensile strengths of these ionic polymers are typically much more enhanced over those of the acidic base polymers as compared to other ionomers.

The strong associations caused by ion clusters decrease with increasing temperature (Fig. 8). Even then the melt viscosity is very high at the usual processing temperatures. The Mg, Ca, Co, Li, Ba, and Na ionomers of Figure 8 typically have apparent melt viscosities between 55×10^4 Pa · s (Mg) and 51×10^4 Pa · s (Na) at 200 °C and a shear rate of 0.88 s^{-1}; for unknown reasons, melt viscosities of Pb and Zn ionomers are much lower (33×10^4 and 12×10^4 Pa · s, respectively [86]. Lead and zinc ionomers also exhibit melt fracture at higher shear rates (88 and 147 s^{-1}) than the other ionomers (< 0.88 s^{-1}).

The high melt viscosities can be lowered by ionic-domain "plasticizers" such as zinc stearate [86]. Such additives interact preferentially with the ionic clusters, lessen the bonding in the cluster region and thus decrease the melt viscosity. Despite its designation as local or specific "plastification", the phenomenon is not a plastification in the scientific meaning of the word since glass transition temperatures are not affected. "Plasticized" zinc and lead sulfo-

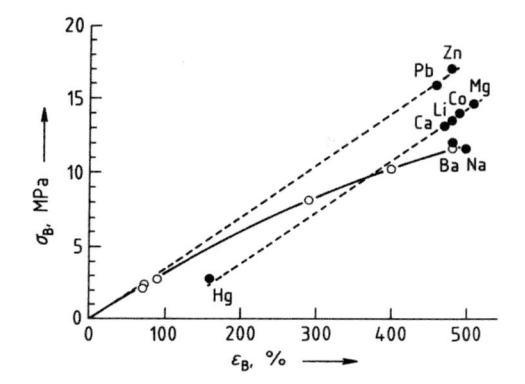

Figure 10. Tensile fracture strength σ_B as function of elongation at break $_B$ of sulfonated EPDM ionomers before (●—●, bivalent ions only) and after (▷ – – – ▷) plasticization with zinc stearate (experimental data from [86])

EPDMs show also higher fracture strengths than the corresponding unplasticized ionomers (Fig. 10). The phenomenon seems to be due to a reduction of the strength of the cross-linking ion clusters upon shearing the melt

References

1 M. S. Shartsberg, I. L. Kotlyarevskii, *Usp. Khim.* **29** (1960) 1439–1473; *Chem. Revs. USSR* **29** (1960) 662.
2 W. Ried, D. Freitag, *Angew. Chem.* **80** (1968) 932–942.
3 A. H. Frazer: *High Temperature Resistant Polymers*, Interscience, New York 1968, pp. 38–49.
4 P. Kovacic, F. W. Koch: "Polyphenylenes," *Encycl. Polym. Sci. Technol.* **11** (1969) 380–389.
5 J. G. Speight, P. Kovacic, F. W. Koch, *J. Macromol. Sci. Rev. Macromol. Chem.* **C 5** (1971) 295–386; *Rev. Macromol. Sci.* **6** (1972) 295–386.
6 K.-U. Bühler: *Spezialplaste*, Akademie Verlag, Berlin 1978, pp. 136–157.
7 H. Naarmann, *Angew. Makromol. Chem.* **109/110** (1982) 295–338.
8 N. Bilow, L. J. Miller, *J. Macromol. Sci. Chem.* **A 3** (1969) 501–525.
9 G. Ensor, *Br. Polym. J.* **2** (1970) 264–269.
10 H.-G. Elias, F. Vohwinkel: *Neue polymere Werkstoffe für die industrielle Anwendung*, 2. Folge, Hanser Verlag, Munich 1983, pp. 78–81; New Commercial Polymers 2, Gordon and Breach, New York 1986, pp. 75 –79.
11 G. G. Engstrom, P. Kovacic, *J. Polym. Sci. Polym. Chem. Ed.* **15** (1977) 2453–2468.
12 L. C. Cessna Jr., H. Jabloner, *J. Elastomers Plast.* **6** (1974) 103–113.
13 T. M. Bednarski, J. H. Del Nero, R. H. Mayer, J. A. Hagan, *SPE Ann. Techn. Papers* **21** (1975) 90–93.
14 J. E. French, *ACS Coat. Plast. Prepr.* **35** (1976) no. 2, 72–76.
15 H.-G. Elias: *New Commercial Polymers 1969–1975*, Gordon and Breach, New York 1977, pp. 26–27.
16 J. Jabloner, L. C. Cessna Jr., *ACS Polymer Prepr.* **17** (1976) no. 1, 169–174.
17 V. A. Sergeev, V. K. Shitikov, Yu. V. Chernomordik, V. V. Korshak, *ACS Polym. Preprints* **16** (1975) no. 1, 328–332.
18 L. A. Errede, M. Szwarc, *Q. Rev., Chem. Soc.* **12** (1958) 301–320.
19 H. L. Lee, D. G. Stoffey, K. Neville: *New Linear Polymers*, McGraw-Hill, New York 1967, pp. 83–100.
20 A. H. Frazer: *High Temperature Resistant Polymers*, Interscience, New York 1968, pp. 53–69.
21 S. N. Nowikow, I. E. Karbasch, A. N. Prawednikow, *Vysokomol. Soedin. Ser. B* **16** (1974) 292–295.
22 M. Szwarc, *Polym. Eng Sci.* **16** (1976) 473–479.
23 L. Baldauf, C. Hamann, L. Libera, *Plaste Kautsch.* **25** (1978) 61–64.
24 *Kirk-Othmer*, 3rd ed., **24**, 744–771.
25 W. F. Beach et al.,"Xylylene Polymers," *Encycl. Polym. Sci. Eng.* **17** (1989) 990–1025.
26 M. Szwarc, *Nature (London)* **160** (1947) 403.
27 M. Szwarc, *Discuss. Faraday Soc.* **2** (1947) 46–49.
28 M. Szwarc, *J. Chem. Phys.* **16** (1948) 128–136; *J. Polym. Sci.* **6** (1951) 319–323.
29 M. H. Kaufman, H. F. Mark, R. B. Mesrobian, *J. Polym. Sci.* **13** (1954) 3–20.
30 J. R. Schaefgen, *J. Polym. Sci.* **15** (1955) 203–219.
31 L. A. Errede, B. F. Landrum, *J. Am. Chem. Soc.* **79** (1957) 4952–4955.
32 L. A. Errede, F. DeMaria, *J. Phys. Chem.* **66** (1962) 2664–2672.
33 W. F. Gorham, *ACS Polymer Preprints* **6** (1965) no. 1, 73–83.
34 W. F. Gorham, *J. Polym. Sci. [A-1]* **4** (1966) 3027–3039.
35 S. Kubo, B. Wunderlich, *J. Polym. Sci. Polym. Phys. Ed.* **10** (1972) 1949–1966.
36 S. D. Ross, D. J. Kelley, *J. Appl. Polym. Sci.* **11** (1967) 1209–1215.
37 K.-U. Bühler: *Spezialplaste*, Akademie Verlag, Berlin 1978, pp. 157–176.
38 C. J. Brown, A. C. Farthing, *J. Chem. Soc.* 1953, 3270–3278.
39 M. Tsuji: "Electron Microscopy,"in C. Booth, C. Price (eds.): *Polymer Characterization*, vol. 1, in G. Allen, J. C. Bevington (eds.): *Comprehensive Polymer Science*, Pergamon Press, Oxford 1989, pp. 813–821.
40 Company Literature, Union Carbide, NewYork 1965 ff.
41 M. Spivack, *Rev. Sci. Instrum.* **41** (1970) 1614–1616.
42 Dow Chemical, US 4 540 763, 1985 (R. A. Kirchhoff).
43 F. E. Arnold, L. S. Tan, 31st International SAMPE Symposium (April 1986), pp. 968–976.
44 Dow Chemical, US 4 711 964, 1987 (L. S. Tan, F. E. Arnold).
45 R. A. Kirchhoff et al., *J. Macromol. Sci. Chem.* **A 28** (1991) 1079.
46 R. A. Kirchhoff, K. Bruza, C. Carriere, N. Rondan, *Makromol. Chem., Macromol. Symp.* **44/45** (1992) 531.
47 H.-G. Elias: Makromoleküle, 5th ed., vol. 1, Hüthig and Wepf, Basle 1990.
48 N. Ise, *Angew. Chem.* **98** (1986) 323–333; *Angew. Chem. Int. Ed. Engl.* **25** (1986) 323–333.
49 R. Longworth, D. J. Vaughan, *Nature (London)* **218** (1968) 85–87.
50 A. Eisenberg, *Macromolecules* **3** (1970) 147–154.
51 B. Dreyfus, *Macromolecules* **18** (1985) 284–292.
52 K. A. Mauritz, *J. Macromol. Sci. Rev. Macromol. Chem. Phys.* **C 28** (1988) 65–98.
53 I.G. Farbenindustrie, DE 667 163, 1930 = FR 710 903, 1930, GB 360 822, 1930 (E. Konrad, W. Bock).
54 B. F. Goodrich, US 2 724 707, 1950 = GB 707 425, 1950; US 2 662 874, 1950 = DE 943 259, 1950 (H. P. Brown).

55. H. P. Brown, *Rubber Chem. Technol.* **30** (1957) 1347–1386; **36** (1963) 931–962.
56. For a list of trade names of carboxylated elastomers up to 1974, see D. K. Jenkins, E. W. Duck, in [62]
57. R. R. Warner, *Rubber Age (N.Y.)* **71** (1952) no. 2, 205–221.
58. R. W. Rees, D. J. Vaughan, *ACS Polymer Preprints* **6** (1965) no. 1, 296–303.
59. Anon., *Modern Plastics International* **13** (April 1983) 54.
60. J. F. Hagman, J. W. Crary, Ethylene–Acrylic Elastomers, *Encycl. Polym. Sci. Eng.*, 2nd ed., **1** (1985) 325–334.
61. R. W. Rees, *Mod. Plast.* **42** (1964) no. 1, 209–210; R. W. Rees, D. J. Vaughan, *ACS Polym. Preprints* **6/1** (1965) 287–295.
62. L. Holliday (ed.): *Ionic Polymers*, Halstead Press, New York 1975.
63. A. Eisenberg, M. King (eds.): *Ion Containing Polymers*, Academic Press, New York 1977.
64. A. Eisenberg, M. Pineri (eds.): *Structure and Properties of Ionomers*, D. Reidel Publ., Den Haag 1987.
65. E. P. Otocka, *J. Macromol. Sci. Rev. Macromol. Chem.* **C 5** (1971) no. 2, 275–294; *Rev. Macromol. Chem.* **6** (1971) 275–294.
66. C. R. Lantman, W. J. MacKnight, R. D. Lundberg, *Ann. Rev. Mater. Sci.* **19** (1989) 295–317.
67. W. J. MacKnight, T. R. Earnest Jr., *J. Polym. Sci. Macromol. Revs.* **16** (1981) 41–122.
68. R. Longworth in A. D. Wilson, H. J. Prosser (eds.): *Dev. Ionic Polym.* **1** (1983) 53–172.
69. W. J. MacKnight, R. D. Lundberg, *Rubber Chem. Technol.* **57** (1984) 652–663.
70. M. R. Tant, G. L. Wilkens, *J. Macromol. Sci. Rev. Macromol. Chem. Phys.* **C 28** (1988) 1–64.
71. J. J. Fitzgerald, R. A. Weiss, *J. Macromol. Sci. Rev. Macromol. Chem. Phys.* **C 28** (1988) 99–185.
72. N. L. Zutty, J. A. Faucher, S. Bonotto: "Ionomers," *Encycl. Polym. Sci. Technol.* **6** (1967) 420–431.
73. R. D. Lundberg, Ionic Polymers, *Encycl. Polym. Sci. Eng.*, 2nd ed., **8** (1987) 393–423.
74. G. W. Lantman, W. J. MacKnight, R. D. Lundberg: "Ionomers," in C. Booth, C. Price (eds.): *Polymer Properties*, vol. 2, G. Allen, J. C. Bevington (eds.): *Comprehensive Polymer Science*, Pergamon Press, Oxford 1989.
75. A. Eisenberg (ed.): *Ions in Polymers* (ACS Adv. Chem. Ser. 187) ACS, Washington, D.C. 1980.
76. A. Eisenberg, F. E. Bailey (eds.): *Coulombic Interactions in Macromolecular Systems (ACS Symp. Ser. 302)* ACS, Washington, D.C. 1986.
77. L. A. Utracki, R. A. Weiss (eds.): *Multiphase Polymers*: Blends and Ionomers (ACS Symp. Ser. 395) ACS, Washington, D.C. 1989.
78. K. Sanui, W. J. MacKnight, R. W. Lenz, *Macromolecules* **7** (1974) 101–105.
79. A. Eisenberg, M. Navratil, *Macromolecules* **6** (1973) 604–612.
80. R. D. Lundberg, H. S. Makowski, *ACS Polymer Preprints* **19** (1978) no. 2, 287–291.
81. Y. Mohajer et al., *Polym. Bull.* **8** (1982) 47–54.
82. Du Pont, *US 3 264 272*, 1966 (R. W. Rees); Du Pont, *US 3 404 134*, 1968 (R. W. Rees).
83. A. Eisenberg, *Macromolecules* **4** (1971) 125–128.
84. H. Matsuura, A. Eisenberg, *J. Polym. Sci. Polym. Phys. Ed.* **14** (1976) 1201–1209.
85. H. Longworth in L. Holliday (ed.): *Ionic Polymers*, Wiley-Interscience, New York 1975, pp. 69–172.
86. H. S. Makowski, R. D. Lundberg, L. Westerman, J. Bock in [75] pp. 3–20; R. D. Lundberg, *Encycl. Polym. Sci. Eng.*, 2nd ed., **8** (1987) 393–423.
87. D. J. Vaughan, *Innovation [Du Pont]* **4** (1973) no. 3, 10.
88. M. F. Hoover, G. B. Butler, *J. Polym. Sci. Polym. Symp.* **45** (1974) 1–37.
89. A. Eisenberg, H. J. Yeager (eds.): *Perfluorinated Ionomer Membranes* (ACS Symp. Ser. 180) ACS, Washington, D.C. 1982.
90. W. Grot, *Chem. Ing. Tech.* **44** (1972) 167–169.
91. G. D. Andrews, R. L. Dawson, *Encycl. Polym. Sci. Eng.*, 2nd ed., **6** (1986) 513–522.
92. J. T. Maynard, P. R. Johnson, *Rubber Chem. Technol.* **36** (1963) 963–974.

Further Reading

M. Chanda, S. K. Roy: *Industrial polymers, specialty polymers, and their applications*, CRC Press, Boca Raton 2009.

F. Mohammad: *Specialty Polymers*, I. K. International Publishing House, New Delhi, 2007

Thermoplastic Elastomers

TRAZ OUHADI, Advanced Elastomer Systems, Brussels, Belgium

SABET ABDOU-SABET, Advanced Elastomer Systems, Akron, United States

HANS-GEORG WUSSOW, Bayer AG, Dormagen, Germany

LARRY M. RYAN, E. I. Du Pont de Nemours and Co., Inc., Engineering Polymers, Wilmington, United States

LAWRENCE PLUMMER, E. I. Du Pont de Nemours and Co., Inc., Engineering Polymers, Wilmington, United States

FRANZ ERICH BAUMANN, Degussa AG - High Performance Polymers, Marl, Germany

JÖRG LOHMAR, Degussa AG - High Performance Polymers, Marl, Germany

HANS F. VERMEIRE, Shell Research SA, Louvain-La-Neuve, Belgium

FRÉDÉRIC L.G. MALET, Arkema, CERDATO, Serquigny, France

1.	Thermoplastic Polyolefin Elastomers	1366
1.1.	Introduction and Definition	1366
1.2.	Thermoplastic Polyolefin Blends	1366
1.2.1.	Morphology	1367
1.2.2.	Elastomer Component, Soft Domain	1368
1.2.3.	Plastic Component, Hard Domain	1368
1.2.4.	Compounds, Trade Names	1368
1.2.5.	Processing	1369
1.3.	Thermoplastic Vulcanizates	1369
1.3.1.	Dynamic Vulcanization	1369
1.3.2.	Production and Morphology	1371
1.3.3.	Types	1373
1.3.4.	Processing	1374
1.3.5.	Performance and Product Positioning	1375
1.3.6.	Uses	1375
1.3.7.	Commercial Products and Trade Names	1375
2.	Thermoplastic Polyurethane Elastomers	1376
2.1.	Introduction	1376
2.2.	Raw Materials	1376
2.3.	Production	1377
2.4.	Properties	1378
2.5.	Processing and Use	1380
2.6.	Economic Aspects	1380
2.7.	Toxicology and Occupational Health	1380
3.	Thermoplastic Copolyester Elastomers	1381
3.1.	Introduction	1381
3.2.	Raw Materials	1381
3.3.	Production	1381
3.4.	Microstructure and Composition of Segments	1382
3.4.1.	Short-Chain Ester Units (Hard Segments)	1382
3.4.2.	Long-Chain Polyether Soft Segments	1383
3.4.3.	Hydrocarbon Soft Segments	1383
3.4.4.	Polyester Soft Segments	1383
3.5.	Properties	1384
3.6.	Blends	1385
3.7.	Uses	1385
3.8.	Producers, Trade Names	1386
4.	Thermoplastic Polyamide Elastomers	1386
4.1.	Introduction	1386
4.2.	Raw Materials	1386
4.2.1.	Oligoamide Unit	1387
4.2.2.	Polyether Units	1387
4.3.	Manufacture of Block Polyetheramides	1387
4.3.1.	Polyetherester Amides	1387
4.3.2.	Synthesis of Polyetheramides	1388
4.4.	Physical and Chemical Properties	1389
4.4.1.	Block Variations	1389
4.4.2.	Morphology	1390
4.4.3.	Outstanding Properties	1391
4.4.4.	Chemical Resistance	1392
4.4.5.	Processing	1392
4.5.	Uses	1393
4.6.	Trade Names	1394
5.	Styrenic Block Copolymers	1394
5.1.	Introduction	1394

Ullmann's Polymers and Plastics: Products and Processes
© 2016 Wiley-VCH Verlag GmbH & Co. KGaA, Weinheim
ISBN: 978-3-527-33823-8 / DOI: DOI: 10.1002/14356007.a26_633.pub4

5.2.	Synthesis	1394	5.4.3. Compounded Products	1399
5.3.	Properties	1395	5.4.4. Modification of Plastics	1400
5.4.	Uses	1397	References	**1401**
5.4.1.	Adhesives, Sealants, and Coatings	1398		
5.4.2.	Modification of Bitumen	1399		

1. Thermoplastic Polyolefin Elastomers

1.1. Introduction and Definition

The significant growth of thermoplastic elastomers in the marketplace since the 1970s can be attributed to the family of thermoplastic polyolefin elastomers because of the wide range of products available, their favorable price–performance relationship, and their acceptance by the automotive industry. Historically, Uniroyal Chemical Company was the first company in 1972 to introduce a thermoplastic polyolefin elastomer to the market.

In 1971, W. K. FISHER [1, 2] filed patent applications on thermoplastic elastomer (TPE) blends based on rubbers that are copolymers of monolefinic monomers—such as ethylene–propene copolymer (EPM) or ethylene–propene–diene terpolymer (EPDM)—with crystalline polyolefin resin, e.g., polypropylene (PP). FISHER employed partial vulcanization of the rubber component to maintain the thermoplastic processibility of the blend by limiting the amount of peroxide. Cross-linking was performed while mixing under dynamic shear, a process known as dynamic vulcanization. However, the dynamic vulcanization process and the first EPM–PP blend were discovered independently by GESSLER et al. [3] (Exxon) and HOLZER et al. [4] (Hercules) in 1958 and 1961, respectively. Two types of thermoplastic rubber (TPR) were produced. One type consisted of simple blends of EPDM and PP; the other contained *partially vulcanized rubber* in the blend. In 1972–1976, most major EPDM producers commercialized products based on simple blends. In general, the TPR products were elastomeric according to DIN 7724 and ISO 1382–1982 definitions, yet these products were not seriously considered for true rubber replacement applications.

Significant improvement in the properties of these blends was achieved in 1975 by CORAN et al. [5] by *fully vulcanizing the rubber phase* under dynamic shear without affecting the thermoplasticity of the blend. This discovery was advanced by ABDOU-SABET et al. [6] in 1977 by the use of phenolic curatives to achieve further improvement in rubber-like properties and processing characteristics.

Therefore, the field of thermoplastic elastomers based on polyolefin rubber–plastic blends grew along two distinctly different product lines or classes: one class consists of a simple blend of α-olefin rubber in a crystalline polyolefin resin and classically meets the definition of thermoplastic polyolefin (TPO). In the production of the other class, the rubber component in the TPO is subjected to dynamic vulcanization to generate a TPE with true rubber-like properties. This class is called thermoplastic vulcanizates (TPVs) or dynamically vulcanized alloys (DVAs).

These two classes of materials share a common basic composition. The plastic phase is a semicrystalline polyolefin such as isotactic and syndiotactic polypropylene, propene copolymer, high-density polyethylene, low-density polyethylene, linear low-density polyethylene, ethylene–vinyl acetate copolymer, ethylene–methacrylate copolymer, or ethylene–ethyl acrylate copolymer. The rubber phase may consist of EPM, EPDM, isobutene–isoprene copolymer (butyl rubber), halobutyl rubber, isobutene–p-methylstyrene copolymers, halogenated isobutene–p-methylstyrene copolymers, natural rubber, styrene–butadiene copolymers, nitrile–butadiene copolymers, styrene–isoprene–styrene block copolymers and their hydrogenated version, etc. (The component polymers are described elsewhere in this encyclopedia).

1.2. Thermoplastic Polyolefin Blends

Thermoplastic polyolefin compositions based on the blends of a polyolefin rubber and a semicrystalline polyolefin have increased dramatically in popularity in recent years. The first thermoplastic polyolefins blends commonly referred to as TPOs were *mechanical blends* of polyolefin polymers. The most common types of TPO are composed of polypropylene and ethylene–propene rubber (EPR). In principle, there is a wide variety of starting materials, an almost unlimited number of formulations for this type of TPOs in major part custom-tailored to meet the varied requirements of the most demanding low temperature high impact applications.

In addition to these mechanical blends, advances in Ziegler–Natta catalysis and reactor design have allowed the preparation of such blends in the polymerization reactor from ethylene and propene monomers. These are commonly called *reactor thermoplastic polyolefins* (RTPO). The amount of EPR content in RTPO can be varied from less than 15 wt%, which is most common, to around 70 wt%. Commercial demand for TPO and RTPO has been particularly strong in the automobile bumper application, especially in Europe and Japan. Such dominance has overshadowed the opportunities for TPO in other market sectors. Advanced compounding technology is being used to generate TPO compounds specific to the application; such technology can also be applied to RTPO.

Mechanically prepared blends are produced by using either batch or continuous mixing techniques. The most common batch mixers are internal intensive mixers, such as the Banbury. Continuous mixing techniques can be performed in either single- or twin-screw extruders. Mixing procedures can be extremely varied because of the wide range of TPO recipes in use, as well as the mechanical characteristics of individual process lines.

Reactor thermoplastic polyolefins are the newest variety of TPO. The amount of rubber (generally EPR) that may be included in a reactor-generated TPO is strongly dependent on the catalyst and process used. The development of a new generation of Ziegler–Natta catalysts (4th generation) with long-lived activity and the capacity of providing PP particles with very high porosity permit the polymerization of a high-rubber component (up to 70 wt%, soft component) inside a shell or skin of crystalline PP. This has been accomplished by the *Cataloy process* [7], which is based on a three-stage reactor. The first stage consists of a tubular loop reactor that produces semicrystalline polypropylene. In the second and third reactors, ethylene and other olefinic monomers are introduced, and ethylene copolymerizes with propene to form the ethylene–propene copolymer rubber inside the semicrystalline homopolymer PP produced in the first reactor; this minimizes reactor fouling. The formation of a continually growing polymer skin and porous particle provides a reaction bed within which the copolymer is produced. This advanced technology in Ziegler–Natta catalysis has been referred to as *Reactor Granules Technology* [8] because the growing polymer particle itself has become the polymerization reactor.

1.2.1. Morphology

Thermoplastic polyolefins possess primarily a two-phase morphology. At least one phase— the thermoplastic component of the blend—is continuous. The second, or rubber, phase can be either co-continuous or dispersed, depending on the composition of the blend. The mixing conditions, surface energy, viscosity, and molecular masses of the two polymers determine at what polymer ratio the two phases are co-continuous [9, 10], since at extreme concentrations (i.e., a 10 : 90 or 90 : 10 ratio) the minor component is expected to be the dispersed phase.

ONOGI and coworkers [11, 12] reported phase inversion in the blends of crystalline polypropylene with EPR or EPDM elastomers when the polypropylene content reaches 50 to 60%.

LOHSE [13] describes the morphology of the simple blend as unstable because the polymers are not miscible, especially at a 50 : 50 polymer ratio. The rubber phase, particularly when present as the dispersed phase, tends to agglomerate or coalesce. Such phase growth is undesirable and would lead to changes in properties and performance as a function of mixing and processing. To stabilize against phase growth, additives or *compatibilizers* are usually added. The

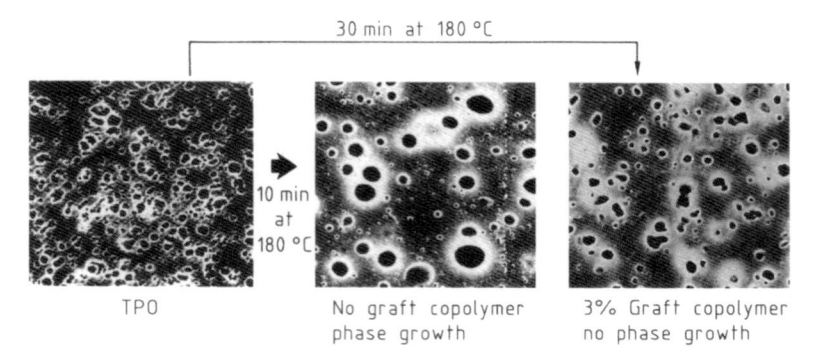

Figure 1. Morphology of EPDM–PP thermoplastic polyolefins containing 0 and 3 mol% compatibilizer

preferred compatibilizer is a graft or block copolymer of isotactic PP and EPR. One approach has been described in [14]. The compatibilizer is generated from EPDM containing vinylnorbornene as the diene; the pendant double bond is then used to graft to the PP. The addition of 3% of a compatibilizer to the blend improves adhesion between the phases and leads to finer dispersion of the phases. This is demonstrated by the smaller size of the rubber phase and the considerably decreased phase growth (Fig. 1).

1.2.2. Elastomer Component, Soft Domain

Either ethylene–propene or ethylene–propene–diene rubber is suitable for these blends. The double bonds in the EPDM are not used as chemically active sites for curing. However, they provide considerably wider choices in raw material selection and undoubtedly affect the green strength properties of the elastomer by changing the number of branch points in the rubber molecule. The diene most commonly used in the EPDM is ethylidene norbornene (ENB), but 1-4 hexadiene (1,4 HD), dicyclopentadiene (DCPD), or vinylnorbornene (VNB) may be used.

EPR rubbers contain 45–85 mol% ethylene, 55–15 mol% propene, and 0–10 mol% diene. EPDM with an ethylene content > 75 mol% exhibits crystallinity that can be examined by differential scanning calorimetry or ^{13}C NMR by measuring the ethylene sequence index [15]. The net effect of ethylene crystallinity is that it greatly strengthens the rubber and, consequently, the blend.

1.2.3. Plastic Component, Hard Domain

In most commercial TPO, the hard domain is isotactic polypropylene homopolymer or a copolymer (random or block) with a small amount of ethylene as comonomer for random PP and until 25 mol% ethylene for block copolymer PP. These resins are semicrystalline. The homopolymer has an mp of 155–165 °C, while the random copolymer has an mp of 140–155 °C. Ethylene decreases the rigidity of the polymer and improves the low-temperature properties at a sacrifice of the upper performance temperature.

1.2.4. Compounds, Trade Names

The two main components of a TPO blend are ethylene–propene rubber and polypropylene. Additional materials, such as other elastomers, processing oils, antioxidants, reinforcing and nonreinforcing fillers, colorants, and other additives to enhance the adhesion or the paintability of the product, are often used. Typical TPO compounds and their properties are listed in Table 1, and the suppliers and trade names of such materials are listed in Table 2.

Automotive and electrical applications are the major markets for TPO. Automotive constitutes by far the largest market because of the excellent weatherability and low cost of TPOs. Some examples are bumper covers, interior trim, fender extensions, etc.

TPOs have good electrical properties in combination with good ozone resistance, which makes these materials a good choice for low voltage wire and cable applications where hot

Table 1. Typical TPO compositions and properties

Components, properties	TPO type		
	1	2	3
EPDM, parts by weight	20	57	85
PP, parts by weight	80	43	15
CaCO$_3$, parts by weight	60	8	8
Irganox 1010, parts by weight	0.2	0.2	0.2
Tinuvin 327, parts by weight		0.15	0.15
N-110 black, parts by weight	3.0		
Calcium stearate, parts by weight	0.10		0.10
Process oil, parts by weight		7.0	64.00
T.S.,* MPa	16.55	13.50	5.50
U.E.,** %	200	740	750
Flexural modulus B, MPa	1240	415	
Shore hardness	62 D	54 D	65 A
Gardner impact (at −29 °C), J	20	36+	

* T.S. = Tensile strength.
** U.E. = Ultimate elongation.

tension set at 200 °C is not a requirement. Such applications can be found in automotive booster cable, flexible cord, and welding cable insulation.

Table 2. Suppliers of TPO

Trade Name	Producer	Location
Vistaflex	AES	Europe
Dexflex, Sequel, Depro, Ontex, Fortilene	Solvay	North America, Europe
Hifax, Adflex	Basell	North America, Europe
Flexathene	Equistar	North America
Kelburon	DSM	Europe
Keltan TP	DSM	Europe
Polytrope	A. Schulman	North America
Telcar	Teknor Apex	North America
Tefabloc	Cousin Tessier	Europe
Multiflex	Multibase	Europe
Vestolen EM	Pawtucket, RI, United States	North America
Vestoprene	Chem. Werke Hüls	Europe
Exxtral	ExxonMobil	North America
Plastomer	Yukong	Asia
Garaflex O	Alpha Gary	North America
Dong II	Deayoung	Asia
Zeonex, Zeonor TPO	Zeon	Asia
Silac	Lucent	North America
WPP TPO	Washington Penn	North America
Vyflex	Lavergne	North America
Novalast	Nova Polymers Inc.	North America
Elastamax	PolyOne	North America
Salflex	ABC Group	North America
J-Prene	J-Von	North America
Tamfer	Mitsui	Asia
Thermoran	Mitsubishi/JSR	Asia
P.E.R.	Tokoyama Soda	Asia

1.2.5. Processing

TPOs can be processed on most common thermoplastic equipment. Products can be formed by injection molding, extrusion, vacuum forming, injection blow molding, extrusion blow molding and so on. The versatility of the compounding TPO products permits designing of compounds to fit the available process and equipment.

1.3. Thermoplastic Vulcanizates

Thermoplastic vulcanizates made via a dynamic vulcanization (TPV) are rubbery engineered polymer blends with fabrication characteristics of a conventional thermoplastic and perform the properties of a conventional thermoset rubber.

TPVs offer a variety of practical advantages over conventional thermoset rubbers. These advantages are:

1. Little or no compounding or blending. Most TPVs are fully formulated and ready for use.
2. Simpler processing with fewer steps. TPVs have the processing simplicity of thermoplastic.
3. Shorter fabrication cycle time and consequently higher productivity for fabrication parts.
4. Recycling of scrap material (regrind). After each step, the scrap from thermoset rubber should be discarded while with TPVs all the eventual scrap is reground and used as virgin material.
5. Lower energy consumption
6. Better quality control and closer tolerances on finished articles
7. In most cases a density lower than a comparable thermoset rubber.

1.3.1. Dynamic Vulcanization

Dynamic vulcanization was discovered by GESSLER et al. [3]. In an attempt to improve the impact properties of PP they partially vulcanized halobutyl rubber in a polypropylene matrix under continuous mixing. Dynamic curing generates the same necessary cross-links or three-dimensional polymer structures as static curing.

In dynamic curing, however, these structures are generated in small rubber particles dispersed in the uncross-linked thermoplastic polymer matrix as a microgel. The formation of such dispersed-phase morphology depends on establishing a high degree of cure (CORAN et al. [5]). The rubber phase is fully vulcanized in a plastic matrix (thermoplastic vulcanizate). The rubber particles are in the order of 1–2 µm in size. Fully curing the rubber phase has led to significant improvement in the material properties including

1. Higher ultimate tensile strength
2. Lower tension and compression set
3. Better property retention at elevated temperature
4. Greater resistance to attack and swelling by fluids
5. Greater flexural fatigue resistance
6. More consistent processibility

Partially cross-linked blends give modest improvement [1, 2].

The effect of the size of the rubber particles and the degree of cure on material properties can be seen in Figures 2 and 3. By proper choice of the curatives, S. ABDOU-SABET et al. [6] demonstrated that rubber-like properties can be improved beyond those achieved by CORAN. Improvements in compression set, oil resistance, and the processing characteristics of the material were achieved by applying phenolic

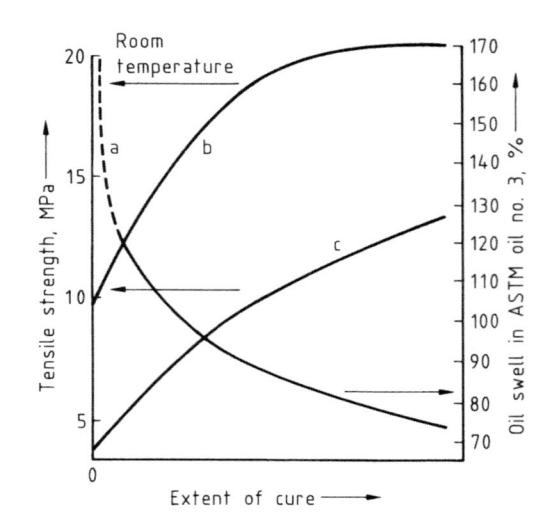

Figure 3. Effect of the degree of cure on mechanical properties
a) Percentage of swell in ASTM oil no. 3; b) Tensile strength at 100 °C; c) Tensile strength at room temperature

curatives (e.g., dimethyloloctylphenol). These improvements are beyond those expected from the state of cure. They probably are due to in situ formation of a graft copolymer between EPDM and PP that optimizes the interfacial adhesion between the rubber particle and the polypropylene matrix, thus allowing for better elastomeric properties. The graft copolymer could have been generated through the functionalization of PP with dimethyloloctylphenol, which then reacts with EPDM [16].

More importantly, significant improvement in processing was achieved over that obtained with the sulfur cure, as shown in Table 3. The use of dimethyloloctylphenol as curative allowed the generation of very soft TPV (as soft as 35 Shore A, with compression set approaching that of a thermoset rubber) while maintaining excellent thermoplastic processing. The poor processing characteristics of sulfur-cured TPV are due to phase growth of the dispersed rubber particles. The polysulfidic bonds generated during vulcanization undergo sulfur exchange reactions leading to significant coalescing of the rubber particles.

Other vulcanization systems are used to cure the rubber phase of TPV. Peroxide with an adequate coagent is the most common curing system after phenolic curatives. The role of

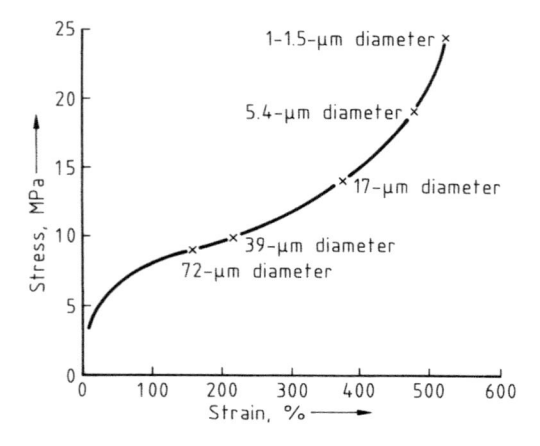

Figure 2. Effect of particle size on stress–strain relationship

Table 3. Effect of cross-linking agent on properties of EPDM rubber–PP TPV

Property	Cross-linking agent			
	None	Sulfur	Dimethylolalkylphenol	Peroxide
Hardness Shore D	36	43	44	39
T.S.,[a] MPa	4.96	24.3	25.6	15.9
M_{100},[b] MPa	4.83	8.00	9.72	8.07
U.E.,[c] %	190	530	350	450
ASTM oil no. 3 swell, %	disintegrated	194	109	225
Comp. set,[d] % (22 h at 100 °C)	91	43	24	32
Processing characteristics[e]	+	−	+++	+

[a]T.S. = tensile strength.
[b]M_{100} = modulus at 100% elongation.
[c]U.E. = ultimate elongation.
[d]Comp. set = compression set.
[e]+ = good; − = poor; +++ = excellent.

coagent is to enhance the cross-linking efficiency for the rubber phase. Apart from cross-linking efficiency, the use of a coagent can also affect the molecular structure of the cross-links. The most common type of coagents are polyfunctional organic molecules.

Another effective vulcanization system reported in several patents [17, 18] uses a hydrosilylation agent with platinum-containing catalysts. This cross-linking system works with rubber containing diene groups, preferably with sterically unhindered carbon−carbon double bonds.

A new cross-linking system to make TPV based on hydrolyzable organosilane-grafted rubber molecules is reported in [19–21]. The organosilane function is hydrolyzed during the dynamic mixing of rubber with the thermoplastic phase (PP) followed by condensation reaction with help of an organotin derivative catalyst such as dibutyltin dilaurate.

1.3.2. Production and Morphology

Melt mixing of the two polymers prior to dynamic vulcanization is the first requirement for a good TPV. The phase morphology of the uncross-linked blend will be influenced by the same variables discussed for TPO. S. ABDOU-SABET et al. [10] demonstrated for an 80 : 20 EPDM–PP blend that PP is the dispersed phase in an EPDM matrix. During dynamic vulcanization of such a blend, EPDM and PP have to undergo a phase inversion to maintain the thermoplasticity of the blend. In the initial stages of dynamic curing, two co-continuous phases are generated, and as the degree of cross-linking increases during mixing, the continuous rubber phase becomes increasingly elongated and then breaks up into polymer droplets. As these droplets form, PP becomes the continuous phase. This can be seen in the SEM electron micrograph (Fig. 4), in which the EPDM phase appears white and the PP phase is black. Photograph 4 B shows that both phases are co-continuous. Figure 5 shows a TEM electron micrograph of a very soft TPV where EPDM phase appears in black and PP in white color.

Reactor TPO can also be used as a raw material to make a TPV [22]. Reactor TPO, however, is not yet commercially produced with a diene function or other functionality in the rubber molecules. Therefore, cross-linking of the rubber phase can be performed only with peroxide. Peroxide curing is carried out while the necessary precautions are taken to protect polypropylene from chain scission and unzipping by the peroxy radical. Protection can be achieved via the incorporation of compounds in the blend that preferentially undergo reaction with the peroxides without affecting PP [23]. Such compounds include polyisobutylene, polybutadiene (PB), and polybutene. Even then, achieving full cure is quite difficult, and only partial curing of the rubber phase is usually accomplished.

To achieve the best improvement in properties of TPV, certain conditions must be satisfied:

1. The higher the crystallinity of the plastic component, the better are the mechanical and elastomeric properties

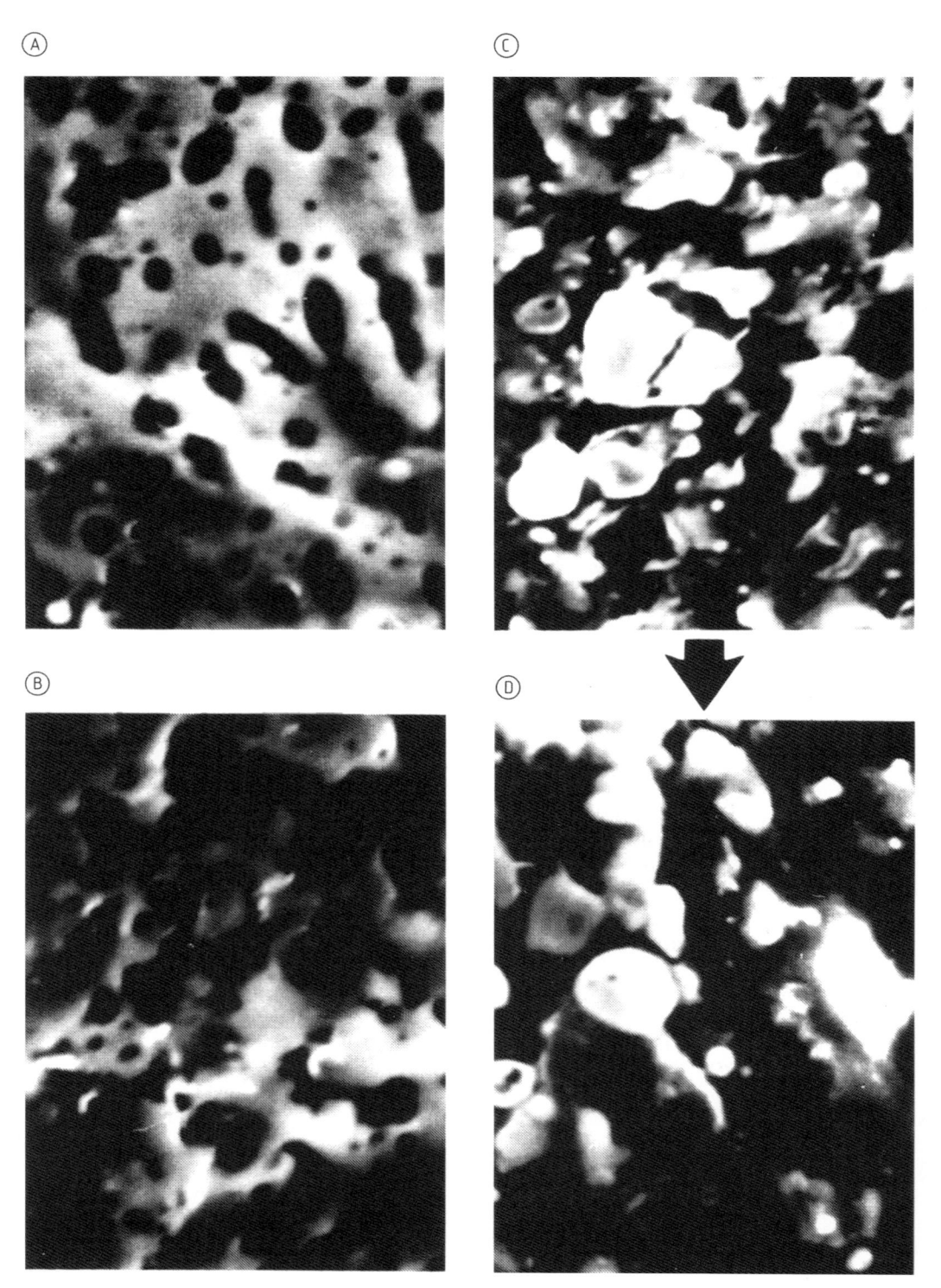

Figure 4. SEM-Electron micrograph of an 80 : 20 EPDM–PP thermoplastic vulcanizate
A) Unvulcanized blend; B) Partially vulcanized; C) Fully vulcanized; D) Fully vulcanized and diluted to 20 : 80 EPDM–PP

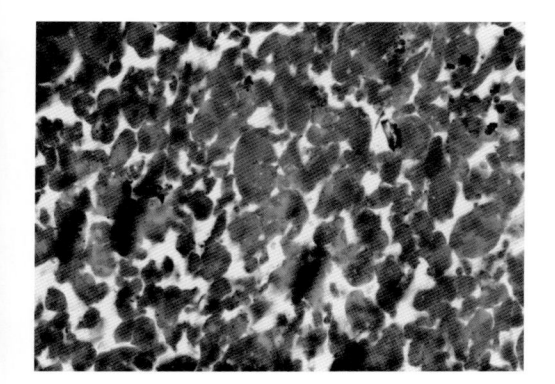

Figure 5. TEM-Electron micrograph of a very soft TPV based on EPDM/PP

2. The smaller the difference in critical surface tension between the rubber and the plastic, the smaller is the size of the dispersed rubber phase
3. The lower the entanglement spacing of the rubber chain, the better are the properties of TPV

1.3.3. Types

Theoretically it should be possible to make a large number of different classes of TPV by combination of at least one type of rubber with one type of thermoplastic. But practically, there is a limitation in the number of commercial TPV families due to limited number of compatible rubber–thermoplastic pairs. Obviously by use of adequate compatibilizing agent between a given rubber and a given thermoplastic, it is possible to expand the number of TPV families.

The discussion in this section is limited to those containing at least one polyolefin polymer.

EP/EPDM rubber–polypropylene based TPVs are the most common types of TPV [24]. The EP/EPDM–PP pair is the mostly used rubber (except for tire application) and thermoplastic material due to the excellent rubbery properties, heat and weathering resistance of EP/EPDM rubber and the excellent versatility, chemical resistance and inertness, and mechanical properties of PP. A broad range of vulcanization systems can be used as mentioned earlier to cross-link EPDM while only peroxide system can be used as cross-linking agent for EP rubber.

Natural rubber (NR)–polyolefin based TPVs [25] have been thoroughly investigated in an attempt to develop commercial products that take advantage of the attributes of natural rubber. Most of the research work was focused on *polypropylene-based material*. These NR–PP TPV thermoplastic natural rubbers (TPNRs) exhibited significant disadvantages compared to their EPDM counterparts and were not acceptable. Drawbacks stemmed from the poor thermal stability and the generation of excessive odor from the natural rubber component on processing at higher temperature. Again, partially cross-linked products exhibited co-continuous morphology, while those cross-linked to 95% or higher of the natural rubber phase showed dispersed-phase morphology. Properties of this type of TPV are listed in Table 4.

Natural rubber–polyethylene-based compositions [26] have also been studied extensively. These TPVs should possess good low-temperature flexibility and elastic recovery for low-temperature applications.

Butyl rubber–polypropylene based TPVs are another family of TPVs where a butyl or halobutyl rubber is used as elastomer component. In 1988, R. C. PUYDAK et al. [27] introduced *butyl rubber-based TPV* for applications requiring impermeability to oxygen and water vapor transmission. Such products are useful in sports bladders and blood vial stoppers.

Polymers with significantly different critical surface tension [28] can still be mixed together with the help of a compatibilizer to improve wetting characteristics or interfacial adhesion [29] between the two phases, thus allowing for the generation of preferred morphology prior to dynamic vulcanization. The application of compatibilization technology allowed the mixing of polar rubbers (e.g., *nitrile rubber* [30, 31] and *acrylate rubber* [32]) with polypropylene, followed by dynamic vulcanization to improve the properties as much as possible. Figure 6 illustrates the effect of the compatibilizer (maleated polypropylene) on polypropylene–ethylene–acrylate rubber TPV.

Table 4. Typical properties of thermoplastic natural rubbers

Property	Rubber type				
	1	2	3	4	5
Shore A hardness	50	60	70	80	90
M_{100},[a] MPa	3.1	4.1	5.3	6.3	8.2
T.S.,[b] MPa	6.5	8.8	11.2	13.2	15.3
U.E. at break,[c] %	285	315	340	370	405
Tear strength, die C (kN/m)	22	31	38	43	51
Tension set, %	9	10	14	18	23
Comp. set,[d] %					
72 h, 23 °C	26	28	31	34	39
24 h, 70 °C	30	35	36	41	55
22 h, 100 °C	38	42	43	47	57
Volume swell after 7 d in					
ASTM oil no. 1, 23 °C	14	9	9	7	
ASTM oil no. 2, 23 °C	19	13	12	9	
ASTM oil no. 3, 23 °C	71	53	47	35	
ASTM oil no. 1, 100 °C	101	80	67	61	
ASTM oil no. 2, 100 °C	151	123	108	82	
ASTM oil no. 3, 100 °C	190	164	139	116	

[a] M_{100} = modulus at 100% elongation.
[b] T.S. = tensile strength.
[c] U.E. = ultimate elongation.
[d] Comp. set = compression set.

Only a small amount of compatibilizers is needed for compatibilization. The function of the compatibilizer is analogous to that of a surfactant that emulsifies and stabilizes a water–oil mixture. Effective compatibilizers are generated by forming a graft or block copolymer; the structures of the blocks must be similar to the phases to be compatibilized. In that case, one block of the copolymer is soluble in the polyolefin resin, while the other is soluble in the polar rubber phase. Graft formation must be controlled to achieve the desired interfacial adhesion while preserving a dispersed phase morphology. Generation of primary chemical bonds between the two phases could lead to an interpenetrating network morphology. Such networks could result in thermoset characteristics and the loss of the thermoplastic processing advantage.

1.3.4. Processing

Thermoplastic vulcanizates or block copolymers that are made of soft and hard blocks are processed or fabricated by the same techniques normally used for rigid plastics. This enables the fabrication of elastomeric articles with improved efficiency and economy over

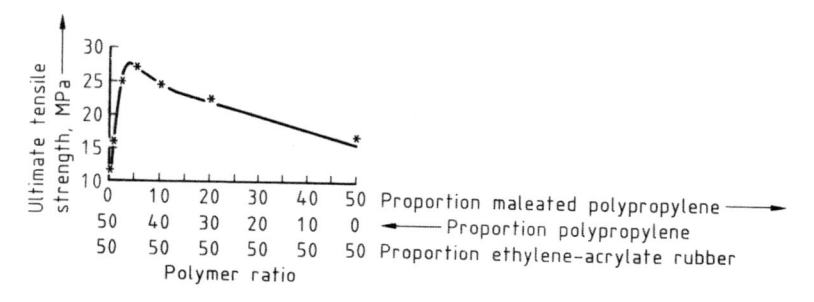

Figure 6. Effect of compatibilizer (maleated polypropylene) on ultimate tensile strength of polypropylene–ethylene–acrylate rubber TPV

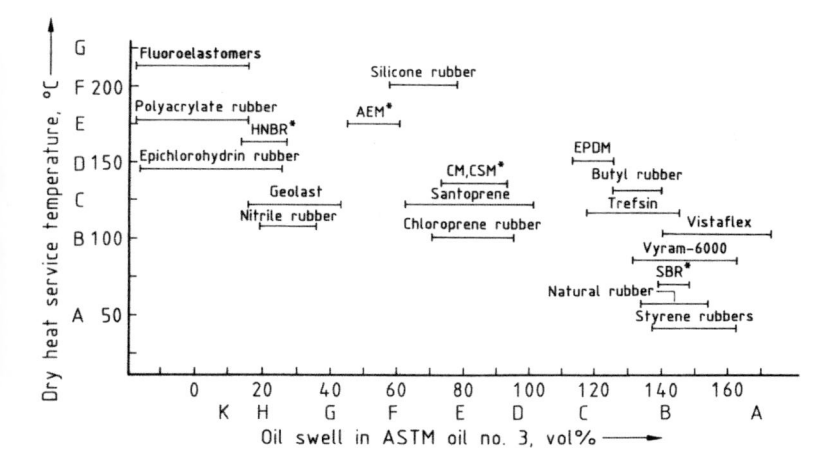

Figure 7. Classification of elastomers according to heat and oil performance
*AEM = ethyl acrylate–ethylene copolymer; CM = chlorinated polyethylene; CSM = chlorosulfonated polyethylene; HNBR = hydrogenated acrylonitrile–butadiene rubber; SBR = styrene–butadiene rubber; Geolast, Santoprene, Trefsin, Vistaflex, and Viram-6000 are registered trade names of commercial TPVs (see Table 5)

those of a conventional thermoset rubber. Scrap from TPV can be recycled readily, whereas thermoset rubber scrap is generally discarded, resulting in a sizeable cost penalty. The thermoplastic nature of TPV enables fabrication by methods unavailable to thermoset rubbers. These include blow molding, thermoforming, and heat welding. Thermoplastic vulcanizates have extremely non-Newtonian flow properties, and their viscosity is highly sensitive to shear [33]. At low shear rates, they flow very poorly; at high shear rates, however, they are quite fluid, enabling rapid injection molding cycles.

1.3.5. Performance and Product Positioning

ASTM D-2000 attempts to classify rubber products according to their performances in dry heat (dry heat service temperature) and ASTM oil no. 3 (see also → Rubber, 3. Synthetic Rubbers, Introduction and Overview). Conventional thermoset rubber occupies certain regions on a performance chart. Such classification is, however, insufficient because other characteristics such as dynamic properties, stress relaxation, barrier properties, UV resistance, adhesion, and colorability, also play an important role in material selection. Nevertheless, according to the ASTM D-2000 classification, EPDM–PP TPVs have performance characteristics superior to chloroprene (Neoprene) rubber. The properties of TPO, however, are considerably less elastomeric, and TPOs occupy a lower-performance category (Fig. 7).

1.3.6. Uses

Since their introduction in 1981, TPVs have found thousands of commercial uses worldwide. Generally speaking, these materials have been used in most rubber applications with the exception of tires. Thermoplastic vulcanizates are employed, e.g., in automotive and construction applications, in foods, hardware, appliances, business machines, mechanical rubber goods, and sporting goods, as well as electrical and medical applications. The uses can vary from the syringe stoppers in medical syringes to the rack-and-pinion boots on the steering mechanism of a front-wheel-drive automobile.

1.3.7. Commercial Products and Trade Names

Advanced Elastomer Systems who commercialized the first TPV (Santoprene) in 1981, is leading the market for this family of products. Global market demand for TPVS is forecast to

Table 5. Suppliers of thermoplastic vulcanizates

Trade name	Supplier	Polymer type	Degree of cure
Santoprene [a]	AES	EPDM–PP	full
Trefsin [b]	AES	butyl–PP	full
Dytron [c]	AES	EPDM–PE	full
Sarlink 3000	DSM	EPDM–PP	partial
Sarlink 2000	DSM	butyl–PP	partial
Hifax-XL	Himont	EPR–PP	partial
Vyram [d] 9000	AES	EPDM–PP	partial
Vyram [d] 6000	AES	natural rubber–PP	full
Milastomer	Mitsui	EPDM–PP	partial
Sumitomo TPE	Sumitomo	EPDM–PP	partial

[a] Santoprene is a registered trademark of Monsanto Company, licensed to Advanced Elastomer Systems, L.P.
[b] Trefsin is a registered trademark of Advanced Elastomer Systems, L.P.
[c] Dytron is a registered trademark of Advanced Elastomer Systems, L.P.
[d] Vyram is a registered trademark of Advanced Elastomer Systems, L.P.

increase 8 to 10% per year. The strong gain anticipated for TPVs results from an intense product development activity led by Advanced Elastomer Systems. Currently, a number of products are on the market (see Table 5). Only fully vulcanized materials achieve the full benefit of dynamic vulcanization.

2. Thermoplastic Polyurethane Elastomers (Polyurethane Rubbers, see → Rubber, 7. Synthesis by Polyaddition, Polycondensation, and Other Mechanisms, Chap. 2; see also → Polyurethanes)

2.1. Introduction

Thermoplastic polyurethanes (TPUs) are a group of elastic materials whose common structural feature is the urethane group –O–CO–NH– arising from the reaction between diisocyanates and long- and short-chain diols. The long-chain polyester or polyether diols form flexible soft segments in the temperature range of use, while the short-chain diols and diisocyanates form rigid or hard segments. Since soft and hard segments are arranged alternately along the molecular chain, TPUs belong to the [AB]$_n$-type block copolymers. Intermolecular interactions between the hard segments lead to the formation of glassy or paracrystalline hard domains, which, as physical cross-linking points, can be reversibly melted during thermoplastic processing. Variation in the ratios of soft and hard segments and monomers permits the production of materials in the modulus range of ca. 5–700 MPa, corresponding to ca. 60 Shore A to 70 Shore D. Because of their very good mechanical properties, especially extraordinary resistance to abrasion, the great choice of monomers available for the production of "tailor-made materials", and efficient thermoplastic processibility, TPUs have found a wide range of applications since their introduction to the market at the beginning of the 1960s. Further information on the production and properties of TPUs can be found in [34–37].

2.2. Raw Materials (→ Isocyanates, Organic; → Polyesters; → Polyoxyalkylenes)

The raw materials used in TPUs are diisocyanates, short-chain diols with molecular masses of 61–ca. 600, and long-chain polyester and polyether diols with average molecular masses between 600 and 4000, so-called polyols.

Diisocyanates. 4,4′-Diisocyanatodiphenylmethane (4,4′-methylenediphenyl diisocyanate, MDI) [101-68-8] is industrially by far the most important diisocyanate. It can be used to produce materials that offer the most favorable compromise between price, thermoplastic processibility, and mechanical properties. Other *aromatic diisocyanates* used are 3,3′-

dimethyl-4,4′-biphenyl diisocyanate (3,3′-tolidene-4,4′-diisocyanate, TODI) [38] and 1,4-phenylene-diisocyanate (PPDI) [104-49-4] [39]. Among the *aliphatic diisocyanates* the following are important: 4,4′-dicyclohexylmethane diisocyanate (H_{12}-MDI) [5124-30-1] (as the isomeric mixture) [40] and 1,6-diisocyanatohexane (hexamethylene diisocyanate, HDI) [822-06-0]. The aliphatic diisocyanates are much less reactive toward alcohols than MDI [41].

Short-Chain Diols. The most important short-chain diol used as a chain extender is 1,4-butanediol. For TPUs with improved dimensional stability at elevated temperature, 1,4-bis(2-hydroxyethoxy)benzene [hydroquinone bis(2-hydroxyethyl) ether] [104-38-1] is also used. The use of diol mixtures (e.g., with ethanediol or 1,6-hexanediol) gives products with a broader processing range [42].

Polyols. *Polyester Polyols.* The most important polyols are polyadipates of short-chain diols with two to ten carbon atoms. Butanediol, hexanediol, ethanediol, and mixtures of these diols are most commonly used. The resistance to hydrolysis of the polyol and TPU increases as the molecular mass of the diol increases. Polycaprolactones and polycarbonate diols with their good to excellent resistance to hydrolysis are also employed. The most widely used products have average molecular masses of 1000–2000. As polycondensation products, polyadipates have a broad molecular mass distribution [43]. They frequently contain unreactive ring esters formed as byproducts. For TPU production the acid number of the ester polyols should be as low as possible, preferably < 1 mg of KOH/g, because carboxyl groups react with isocyanates and impair the resistance to hydrolysis.

Polyether Polyols (→ Poly(propylene oxide) (PPO)raw material for thermoplastic elastomersPolyoxyalkylenes). Poly(tetramethylene oxides) are the most important polyether polyols for the production of TPUs. Poly(propylene oxides) [poly(propylene glycols)] and copolymers with ethylene oxide give products with poorer properties; normally, they are used together with polyester polyols. The molecular mass distribution of the polyether polyols follows the Poisson distribution and is thus considerably narrower than that of the corresponding polyesters. Poly(propylene oxides) have predominantly secondary hydroxyl groups as terminal units and are therefore considerably less reactive than other polyols.

Additives. Internal release agents, such as esters of montanic acid (octacosanoic acid), silicones, or amide waxes in quantities up to 2%, are used as *processing aids*. TPUs based on polyesters can be protected effectively against *hydrolytic degradation* by the addition of carbodiimides [44]. Protection against *microbial degradation*, for example on prolonged contact with soil, is also possible. Sterically hindered phenols and amines are used as *antioxidants*, mainly to protect against discoloration and degradation, particularly in the case of polyether TPUs at the high production and processing temperatures often required. Yellowing because of the effect of light can be slowed considerably through the use of *light stabilizers* such as benzophenones, sterically hindered amines (HALS), or benzotriazoles [45].

Plasticizers are used only to achieve hardness levels below ca. 70 Shore A [46–48]. Short glass fibers are used mainly to *reinforce* TPUs. Many additives, such as colorants, antiblocking agents, antistatic agents, and flame retardants, are supplied as concentrates.

2.3. Production

General. The industrial production of TPUs is carried out virtually exclusively in continuous, solvent-free processes [49]. According to Flory statistics [43] the required high molecular masses are achieved only at high conversions and an essentially stoichiometric ratio of NCO and OH groups. At an NCO : OH ratio < ca. 0.97, products with the required properties are generally not obtained [50]. Too high an excess of NCO can lead to cross-linking through allophanate and biuret formation even after long periods [51]. A prerequisite for maintaining exact stoichiometric ratios is the use of raw materials that are as pure—especially anhydrous—as possible and of exact metering devices. Prepolymer and one-shot processes can be used to carry out the reaction (→ Polyurethanes, → Polyurethanes, Section 5.4). In the one-shot process, OH- and NCO-functional reactants are mixed with each other in one step. In the prepolymer process, a prepolymer with terminal NCO groups is first formed from a long-chain polyol and diisocyanate. This prepolymer is then mixed with the chain extender. Reactivity differences (e.g., between polyether polyols and butanediol) can be overcome with this process. The process used affects the segment length distribution and thus the morphology of the final product [52, 53].

Belt Process [54]. Liquefied raw materials are mixed continuously in a mixing head and poured onto a continuous belt, which passes through an oven heated to ca. 70–130 °C, where the reaction proceeds further. The congealed product is ground to irregular pellets. It can be melted in extruders, possibly along with additives, and formed into regular pellets. According to a modification of this process [55] the reaction is carried out to a desired degree of conversion, generally > 80%, on the belt, and the congealed mixture is then fed immediately into an extruder where the reaction is continued and the mixture is homogenized. Finally, the polymer is pelletized. In the belt process the temperature and residence times in the heated zone can be adapted to the requirements of the particular product.

Reaction Extruder Process. A twin-screw extruder serves as the reactor [56–59]. The screws are conrotating and can be set to the required shearing ratios by a combination of kneading and transporting parts. Temperature is controlled by using barrel sections that can be heated separately. The reactants and additives can be charged at different locations. In this way, a continuous prepolymer or one-shot process is possible without additional equipment. The bulk temperatures are generally > 200 °C. The product is extruded continuously and pelletized immediately after the head or after passage through a cooling area. As in the belt process, subsequent sieving and drying processes occur. In the reaction extruder process, finished pellets can be obtained from starting materials with the addition of all additives (including glass fiber or other thermoplastics) in one process step [60].

Comparison of Products. Because of different shearing and temperature conditions, the type of process used significantly affects the morphology and thus the processing properties of the products. Starting from identical raw materials more opaque products are formed in the belt process with larger hard segment domains that melt at higher temperature, whereas the reaction extruder process gives more transparent, lower-melting products.

2.4. Properties

Morphology [37, 61, 62]. The physical properties of TPUs are determined by their polyphase structure consisting of soft segments with very low glass transition temperatures, hard segments with high melting ranges, and mixed phases. This structure arises from the incompatibility of the hard and soft segments, which, on cooling from the melt during production and processing, leads to partial or complete separation of the phases that are bonded covalently with each other. The extent of separation and the phase structure are determined mainly by the stoichiometric composition of the polymer; the statistical distribution of the hard and soft segments [52]; the structure of the polyol [63], the diisocyanate and the chain extender, and the thermal history. The latter results from the kinetics of the separation and crystallization processes and from the cleavage and transurethanization during production, processing, and possible tempering of the product [64–67]. In polyethers, phase segregation is more pronounced than in polyesters and increases as the molecular mass of the polyol increases. Many investigations of the structure [65, 68, 69], phase interactions [69–71] and structure–property relationships of TPUs have been carried out, mainly on model substances [72–74]. Macroscopically, the soft segments determine the elastic ductility, the low-temperature properties, and the basic chemical properties, while the hard segments are important for dimensional stability at high temperature and the processing properties.

Table 6. Typical mechanical properties of TPUs with different hardness based on polyester polyols (example Desmopan, Bayer)

Property	Test standard (ISO)	Desmopan			
		385	345	359	372
Density, mg/m^3	1183	1.2	1.21	1.23	1.24
Hardness, Shore A/D	868	86 A	47 D	59 D	73 D
Tensile stress at break, MPa	527-1, -3	52	52	60	70
Elongation at break, %	527-1, -3	550	450	400	330
Tear resistance, kN/m	34-1	70	100	160	220
Flexural modulus, MPa	178		70	180	650
Abrasion loss, mm^3	4649	30	30	35	30
Compression set, %					
70 h/22 °C	815	30	25	30	
24 h/70 °C	815	55	42	60	

Technical Properties. By varying the proportion of the hard segment the modulus of elasticity can be varied over a range of ca. 5–700 MPa without the use of plasticizers or reinforcing agents. Typical contents of the long chain diol range from 30–75 wt%. Products containing no [41] or only very small proportions of soft segments can no longer be regarded as thermoplastic elastomers [75]. Table 6 gives an overview of some properties of different types of products.

Figure 8 shows the dependence of tensile modulus on temperature for adipate-based TPU with different hardness; typical stress-strain diagrams are shown in Figure 9.

High extension with very low residual extension, high tensile and tear strength, high moduli of elasticity at a given Shore hardness, and excellent abrasion resistance are common to all TPUs. Polyester TPUs are superior to polyether products with regard to strength and resistance to oxidation or swelling by oils, fats, and water, while they are inferior with regard to resistance to hydrolysis or microbial attack and impact resistance at low temperature. Hybrid types from polyester and polyether polyols combine the advantages of both types to a certain extent [76, 77].

Unstabilized TPUs based on MDI turn yellow through the action of light, initially without the impairment of mechanical properties. TPUs based on aliphatic diisocyanates do not show this discoloration. The lower temperatures of use are determined by the glass transition temperature of the soft phase (−20 to −60 °C), and the upper ones by the onset of softening of the

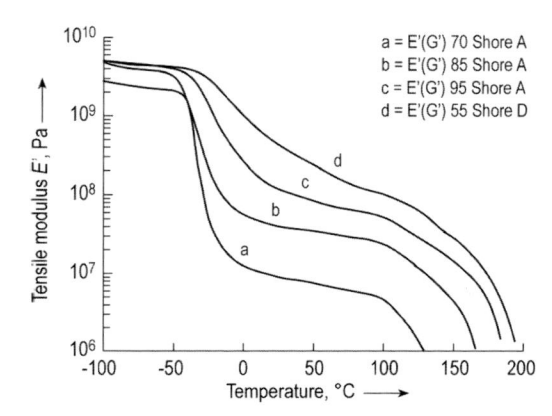

Figure 8. Tensile modulus of elasticity vs. temperature for polyadipate based TPU (example Desmopan, Bayer)

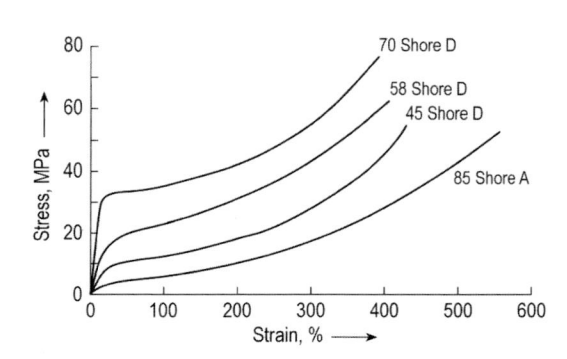

Figure 9. Stress-strain curves of polyadipate-based TPU with different hardness (example Desmopan, Bayer)

hard domains. Depending on the hardness, softening starts at ca. 80–100 °C (see → Polyurethanes). Hard products can be kept at up to ca. 120 °C for short periods. TPUs are soluble in polar, aprotic solvents such as N,N-dimethylformamide, methyl ethyl ketone, or N-methylpyrrolidone.

2.5. Processing and Use [34, 35, 37, 78, 79]

TPUs can be processed by all methods available for thermoplastics, such as injection molding, extrusion, blow molding, sintering, and calendering, although not all types are suited to every processing method. Melt temperatures range between 180 and 250°C, depending on hardness and other parameters. Typically, the melt viscosity of TPU is strongly temperature-sensitive. Generally only products with hardnesses up to ca. 50 Shore D can be extruded. Generally, extrusion becomes more and more difficult with increasing hardness above ca. 50 Shore D, due to formation of gel particles resulting from crystallization effects or side reactions.

TPUs are used in all those areas where their high strength and excellent resistance to abrasion, oil, and many other solvents are important. The large number of uses includes bases for shoe heels, soles and studs, or sports shoes, screens for the classification of minerals, rollers, tool parts, toothed belts, gaskets, electrical plugs, coupling elements, ear tags for animals, parts for car bodies, instrument panels, bearing shells, ski boots, seals, cable sheathings, tubing of all types, blown films, profiles, coatings, and medical equipment such as tubing and catheters. TPUs are also used for melt spinning of elastic fibers and modification of the impact resistance of polyacetals and as blend components for polycarbonates and acrylonitrile–butadiene–styrene (ABS) polymers [80, 81].

2.6. Economic Aspects

TPUs are produced in Europe, the United States, and Japan by the three large producers BASF, Bayer, and Noveon (quantities of > 20 000 t/a) and a number of medium and small producers (Table 7).

Table 7. Important producers of TPU and trade names

Region	Producer	Trade Name
Europe	Bayer	Desmopan
	Elastogran (BASF)	Elastollan
	Noveon	Estane
	Huntsman	Avalon, Irogran
	COIM	Laripur
	Mercquinsa	Pearlthane
USA	Bayer	Texin, Desmopan
	Noveon	Estane
	Huntsman	Krystalflex
	Dow	Pellethane
	BASF	Elastollan
Asia	Dainichi Seika	Resamine
	Nippon Miractran	Miractran
	DIC Bayer Polymers	Pandex, Desmopan
	BASF	Elastollan
	Hosung	Neothane
	SK Chemicals	Skythane

Almost all producers offer materials of differing hardness based on different polyols within several product ranges. In 2000, the estimated market volume was ca. 150 000 t/a. The regional distribution has a peak in Europe (40%), followed by North America (30%) and Eastern Asia (27%). The chances of growth are estimated to be positive in view of the possibilities for replacing rubber and plasticized thermoplastics by TPUs and the versatility of this class of materials.

2.7. Toxicology and Occupational Health

Since the production process involves polyaddition in a melt, very small quantities of byproducts are formed. Correctly dimensioned and efficiently functioning air extraction allows the permitted workplace concentrations for diisocyanates to be adhered to with certainty by producers and processors. Generally neither aromatic amines nor diisocyanates can be extracted from properly produced TPU articles. If only the allowable raw materials are used, TPUs comply with health authority guidelines for contact with foods (FDA: CFR 175.300, 175.105, 177.1680; European Commission Directive 2002/72/EC). Because of their good compatibility with blood and tissue, some products also fulfill the requirements of the ISO standard 10993/1 [82, 83].

3. Thermoplastic Copolyester Elastomers

3.1. Introduction

Thermoplastic copolyester elastomers are segmented block copolymers, consisting of repeating, high-melting, rigid "hard" blocks and amorphous, flexible "soft" blocks, which have a very low glass transition temperature. These polymers exhibit a remarkable combination of properties as a result of microphase separation of the hard block (in the form of crystalline domains) from the matrix of the amorphous soft blocks. The crystalline domains impart excellent mechanical properties (tensile strength, toughness, resistance to creep, good retention of properties at elevated temperature, etc.) and outstanding chemical resistance to the polymer. Concurrently, the amorphous domains provide exceptional flexibility and fatigue resistance over a broad temperature range. By varying the ratio of hard to soft blocks, products ranging from relatively rigid plastics to very flexible elastomers can be produced.

Copolyester elastomers are typically produced by condensation of an aromatic dicarboxylic acid or ester with a low molecular mass aliphatic diol and a polyalkylene ether glycol (molecular mass generally between 1000 and 4000) [84–86]. Reaction of the aromatic dicarboxylic acid with the diol leads to the crystalline hard segment phase, while the soft segment is the product of the diacid (or diester) and the long-chain glycol. Other combinations are possible, however (see Sections 3.4.2, 3.4.3).

3.2. Raw Materials

Polyester elastomers are produced from aromatic dicarboxylic acids or diesters, polyoxyalkylenes, and an excess of short-chain diols [84]. The most common raw materials follow:

Dicarboxylic Acids. *Terephthalic Acid* [100-21-0]. For production and properties, see → Terephthalic Acid, Dimethyl Terephthalate, and Isophthalic Acid.

2,6-Naphthalenedicarboxylic Acid [1141-38-4]. For production, see → Carboxylic Acids, Aromatic.

Polyoxyalkylenes. *Poly(tetramethylene oxide)*; *poly(propylene glycol)* [25322-69-4]; *poly-(ethylene glycol)* [25322-68-3]; for production and properties, see → Polyoxyalkylenes.

Short-Chain Diols. *1,4-Butanediol* [513-85-9]. For production and properties, see → Butanediols, Butenediol, and Butynediol, Section 1.3.

Ethylene glycol [107-21-1]. For production and properties, see → Ethylene Glycol.

3.3. Production

(see also → Polyesters, Section 3.2.1, → Polyesters, Section 3.2.2, → Polyesters, Section 3.2.3, → Polyesters, Section 3.2.4).

Polyester elastomers are manufactured by a *multistage melt polymerization process*. Soluble titanium alkoxides are generally used as catalyst. In the first stage of the process, all reactants are mixed and heated at atmospheric pressure to 190–220 °C to form a low molecular mass prepolymer. In the latter stages the prepolymer is converted to high molecular mass polymer by increasing the temperature to 250–260 °C while lowering the pressure to < 133 Pa to distill off the excess short-chain diol. The degree of polymerization is monitored by the viscosity change of the molten reaction mass by measuring the torque on the reactor agitator. To protect the polyether ester from oxidative degradation during synthesis and in end-use applications, polymerization is usually carried out in the presence of an antioxidant. Hindered phenols such as Irganox 1098 (Ciba–Geigy) and Ethanox 330 (Ethyl Corp) or an aromatic secondary amine [i.e., Naugard 445 (Uniroyal)] are examples of antioxidants that are added during the preparation stage of the prepolymer. At the end of polymerization, the polymer is extruded, quenched, and cut into pellets. Predominantly hydroxyl and carboxylic acid end groups are produced by this process. The acid end groups are formed by the thermal cleavage of polymer chains or by degradation of a butanol end group with simultaneous elimination of tetrahydrofuran.

Both batch and continuous processes produce materials of sufficient viscosity to be fabricated by either extrusion or injection molding. However, a polymer with ultrahigh viscosity is required for blow molding operations and in extrusion processes where very precise dimensions are required. Generally, the polymer viscosities required for these methods of processing cannot be obtained by a melt condensation process. One technique used to manufacture ultrahigh-viscosity copolyester elastomers is *solid-phase polymerization* [87–90]. In this process the solid polymer is heated to just below its melting point under reduced pressure. This leads to the reaction of acid and hydroxyl end groups or of two hydroxyl end groups with elimination of water or short-chain glycols. They diffuse out of the polymer, and the molecular mass increases. A second method of increasing viscosity is by polymer chain extension in a postcompounding step. In numerous examples [91–93], diisocyanates have been used as chain extenders to increase the viscosity of low and moderate molecular mass polyesters through the reaction of the polyester hydroxyl end group to form a urethane.

Thermoplastic copolyester elastomers are stable for an indefinite period of time. However, they will absorb moisture, reaching equilibrium with existing atmospheric conditions very rapidly. Since the ester group hydrolyzes at processing temperature, these products must be dried thoroughly prior to fabrication. The moisture content should not exceed 0.1 wt% when molding or extruding [94].

3.4. Microstructure and Composition of Segments

3.4.1. Short-Chain Ester Units (Hard Segments)

The hard segment employed in the polyester elastomer strongly influences the load-bearing capabilities and processing characteristics of the polymer. Because of its high melting point, rapid crystallization, and low cost, the predominant hard segment in commercial use is tetramethylene terephthalate (4GT), an ester of 1,4-butanediol (4G) and terephthalic acid (T). Hard segments based on ethylene glycol (2G) and terephthalic acid (2GT) are high melting and yield polymers with excellent physical properties. However, crystallization of this hard segment is very slow, requiring 36 h to completely develop the physical properties of the polymer. This limitation greatly reduces the usefulness of 2GT thermoplastic elastomers in injection molding and extrusion applications [95]. WOLFE [96] has prepared a series of copolyester elastomers with a wide range of short-chain diols and compared their properties (Table 8). The properties of the elastomers resulting from 2G, 3G, (1,3-propanediol), and 4G are very similar except for tensile and tear strength. Hard segments based on diols with five, six, and ten carbon atoms do not form high-melting crystals, and as a result, copolyesters made with these diols generally have inferior properties.

Substituting a cycloaliphatic glycol, 1,4-cyclohexanedimethanol, for 1,4-butanediol produces a high-melting polyether ester elastomer that crystallizes somewhat slower than one containing tetramethylene terephthalate but rapidly enough to be blow- or injection molded. Because of their slower crystallization characteristics, finished parts made of these elastomers are clear [97, 98].

Naphthalenedicarboxylic acid, in place of terephthalic acid, in combination with 1,4-butanediol provides products with improved thermal properties compared to their 4GT counterparts [99]. Some mechanical properties such as tear strength are also improved. A major limitation in using naphthalenedicarboxylic acid in

Table 8. Properties of 50% alkylene terephthalate–PTMO terephthalate copolymer as a function of diol structure [96]

Properties	Diol					
	2G	3G	4G	5G	6G	10G
Melting point, °C	224	198	189	106	122	106
Shore D hardness	46	48	48	32	33	35
Stress at 100% elongation, MPa	11.4	11.7	11.7	4.5	5.2	6.1
Tensile strength, MPa	45.5	22.8	48.4	15.4	13.5	15.5
Elongation, %	675	660	755	880	750	640
Tear strength, kN/m	42	15	48	25	18	7.8

commercial thermoplastic copolyester elastomers, however, is its high price.

3.4.2. Long-Chain Polyether Soft Segments

The long-chain polyether soft segments are formed by reaction of poly(ether glycols) and diacids. The predominant poly(ether glycols) used in commercial thermoplastic copolyester elastomers are poly(tetramethylene oxide) (PTMO) and poly(propylene glycol) end-capped with ethylene oxide (EOPPG). In general, the chemical resistance and certain physical properties of thermoplastic elastomers, such as tear strength, are influenced by the poly(ether glycol) segment. For example, products based on PTMO have much higher tensile and tear strengths and significantly better resistance toward oxidative degradation than polymers made with EOPPG. Melting point, modulus, and hardness are generally independent of the type of polyether employed.

Because of their polarity, polymers made using poly(ethylene glycol) (PEO) show the greatest water swell, yet the least swell in nonpolar oils [100]. These products have high moisture vapor transmission rates.

A comparison of typical properties of thermoplastic polyester elastomers containing soft segments based on PTMO, EOPPG, or PEO is shown in Table 9 [101].

Through a novel set of reactions, polyethers end-capped with amino groups [102] also have been utilized to manufacture copolyester elastomers. These polyoxyalkylenamines are first converted to acid-terminated polyethers by reaction with benzene-1,2,4-tricarboxylic acid, followed by copolymerization with terephthalic acid and a short-chain diol. Since the backbone of the polyoxyalkylenamines is based on ethylene and propylene oxides, the resulting thermoplastic elastomers have properties similar to EOPPG-based products [103].

3.4.3. Hydrocarbon Soft Segments

Unsaturated straight-chain aliphatic carboxylic acids can be dimerized to form high molecular mass dicarboxylic acids, generally referred to as dimer acids [104]. When purified, these dibasic hydrocarbons have been used in place of poly(ether glycols) as soft segments in the preparation of thermoplastic copolyester elastomers. HOESCHELE [105] prepared several of these polymers and demonstrated the striking similarities with their polyether counterparts.

One major difference is that products based on dimer acid exhibit poor low-temperature properties because of their high glass transition temperatures. However, due to the hydrocarbon nature of the backbone, these polymers have much better heat aging characteristics and resistance to UV degradation. MCCREADY [106] has noted that the heat aging of thermoplastic polyether esters can be improved by using a mixed soft segment of dimer acid and poly(ether glycol).

3.4.4. Polyester Soft Segments

The use of low molecular mass polyester glycols as soft segments gives thermoplastic copolyester elastomers with improved heat aging characteristics and stability toward UV degradation compared to their polyether counterparts. However, the preparation of these products presents special problems because of the randomization that can occur between the hard and soft ester segments during synthesis. This negatively affects formation of blocks, and the resulting polymers are low melting with poor mechanical properties [107].

One approach to solving this problem is to react preformed hard and soft polyester segments in the presence of a titanate catalyst until the reaction mass becomes clear. At that point the catalyst is deactivated by addition of a phosphorus compound [108]. On further

Table 9. Properties of 57% 4 GT polyether elastomers as functions of polyoxyalkylene structure [101]

Properties	Polyoxyalkylene		
	PTMO	EOPPG	PEO
Nominal molecular mass	1000	1000	1000
Stress at 100% elongation, MPa	13.4	11.7	13.2
Tensile strength, MPa	47.2	27.2	42.6
Elongation, %	660	675	705
Tear strength, kN/m	62	23	31
Volume swell, % [97]			
Water (3 d/24 °C)	0.5	1.0	17.0
ASTM oil no. 3 (7 d/100 °C)	13.0	12.0	4.0

processing into end use articles, very little ester exchange occurs between the hard and soft blocks, and as a result, parts with good mechanical properties are obtained. Generally, better results are obtained by chemical coupling of the blocks with diisocyanates [109] or acylating agents such as diacylbis(N-lactams). A detailed description of diisocyanate coupling between poly(tetramethylene terephthalate) and various aliphatic polyester diols has been published by VAN BERKEL [107]. After deactivating the catalyst used for formation of 4GT, coupling is carried out by using either toluene diisocyanate or methylene diisocyanate. A slight excess of isocyanate is used to obtain optimum mechanical properties. These thermoplastic polyester ester products show improved heat aging and UV resistance, but also have poor hydrolytic stability and stiffen much faster than polyether ester elastomers at low temperature.

3.5. Properties

Physical and Chemical Properties. Polymers prepared by the method described in Section 3.3 have the molecular mass distribution expected from a condensation polymerization. A number-average molecular mass in the range of 20 000–25 000 can be obtained. The two-phase hard–soft segment structure of these segmented polyesters has been demonstrated clearly by differential scanning calorimetry [110]. By varying the ratio of these two segments, thermoplastic polyesters ranging from soft, transparent rubber-like elastomers to semirigid plastics can be prepared. Melting points ranging from 140 to 220 °C have been reported for polymers containing 20–80 wt% 4GT [111]. As expected, the glass transition temperatures of this same series of polymers increase from −70 to 25 °C.

Thermoplastic copolyester elastomers have good to excellent resistance to hydrocarbons. Products containing > 50% 4GT are particularly well-suited for contact with oil, grease, or hydraulic fluids. These polymers offer better resistance to hot hydrocarbons than nitrile rubber.

Resistant to

Nonpolar fluids
    Oil
    Fuels
    Hydraulic fluids
Polar fluids at room temperature
    Glycols
    High molecular mass alcohols
    Weak acids and bases

Not resistant to

Hot glycols
Hot alcohols
Strong acids and bases
Amines

Resistance to polar fluids, including water, acids, bases, amines, and glycols, is a function of both the composition of the polymer and the temperature of use. At > 70 °C, polar fluids attack the polymer.

Mechanical Properties. The predominant hard segment in thermoplastic copolyester elastomers is poly(tetramethylene terephthalate). WOLFE [96] has shown that a hard segment content of 30 to 80 wt% produces a wide range of useful products (Table 10). A complete

Table 10. Properties of tetramethylene terephthalate–PTMO terephthalate copolymer as a function of tetramethylene terephthalate content [96]

Properties	Percentage 4GT				
	30	40	50	57.4	80
Melting point, °C	152	172	189	197	212
Shore D hardness	34	44	48	54	70
Stress at 100% elongation, MPa	5.9	8.3	11.7	13.4	24.4
Tensile strength, MPa	24.6	31.0	48.4	46.7	54.5
Elongation, %	900	830	755	660	535
Tear strength, kN/m	13	26	48	63	108

summary of mechanical properties for these elastomers has been published [112].

3.6. Blends

The low melt viscosity and good processing characteristics of thermoplastic elastomers facilitate their use in polymer blends to give products that are otherwise not easily obtainable. In many instances the properties of these blends are an average of the properties of the individual components, for example, the combination of poly(vinyl chloride) (PVC) and a polyester based on 4GT and PTMO [113, 114]. The addition of 10–30% of thermoplastic polyester elastomer provides PVC with improved heat resistance, low-temperature flexibility, and resistance to oils and fuels. Another important advantage is that the polyester elastomer cannot be extracted from the blend, which extends the service life of PVC molded parts.

In other instances, blends offer improved physical properties that are not obtainable by direct synthesis. Blending poly(butylene terephthalate) with a thermoplastic copolyester elastomer containing 58 wt% hard segment provides a polymer with improved low-temperature impact resistance compared to a directly synthesized polymer of identical composition [113]. The improvement in low-temperature properties is said to result from the incompatibility of the components of the resultant blend.

High-melting, crystalline thermoplastic copolyester elastomers can be softened (thus improving their low-temperature properties) by blending with an amorphous or very low melting polyester made from PTMO, 4G, and a diacid [115]. Since the melting point of the resultant blend is similar to that of the unblended crystalline thermoplastic, evidence that the two components of the blend are not miscible is good. Soft thermoplastic copolyester elastomers can also be prepared by blending the polyester with ethylene copolymers. Shih [116] reports blends with EPDM that retain the melting point of the polyester yet exhibit reduced stress–strain properties. In an extension of this work, blends of polyesters have been prepared with ethylene–carboxylic acid copolymers that had been partially neutralized [117]. This novel combination of polymers provides blends with enhanced melt viscosity, which greatly improves the processibility by blow molding and film extrusion of the thermoplastic copolyester elastomers.

Very efficient mixing is required to achieve good dispersion of polyester homopolymers and thermoplastic copolyester elastomers because of their incompatibility [113]. If the materials being blended differ greatly in polarity, homogeneous blends can sometimes not be achieved unless *compatibilizers* are employed. These compatibilizers generally contain reactive functional groups to assist in dispersing the components of the blend. In two separate examples, polycarbonate resins having improved flexibility and impact strength were obtained by blending with copolyester elastomers. To obtain optimum properties, polystyrene grafted with organosiloxanes or styrene–butadiene copolymerized with acrylic acid derivatives has been used to compatibilize the blend components [118, 119].

3.7. Uses

Thermoplastic copolyester elastomers are available in grades ranging from soft, rubber-like to semirigid plastics. This provides customers with products covering a wide spectrum of load-bearing characteristics, flexibility, and low-temperature properties. These materials are simple to process and can be made into high-performance products by a variety of polymer-forming processes, including injection molding, extrusion, blow molding, rotational molding, and melt casting. All thermoplastic copolyester elastomers have sharp, well-defined melting points, crystallize rapidly, and have good melt stability.

Grades having a Shore D hardness of 50–80 have excellent chemical resistance, which makes them suitable replacements for specialty rubber such as nitrile and epichlorohydrin rubbers in hose and tubing applications. They can be extruded into long, continuous lengths with relatively low wall thickness, thereby providing lightweight constructions. Compared to competing flexible polyamide compounds, no plasticizer is required, so the material remains flexible over long periods of time. Overall, hose construction costs using thermoplastic copolyester elastomers are comparable to those using conventional rubber. This is due to the less expensive plastics processing techniques involved and the fact that thermoplastic copolyester elastomers require no postcuring [112].

The good impact strength and toughness, dimensional stability, and good paintability of thermoplastic copolyester elastomers have been utilized in automotive applications to replace metal in exterior automotive bumpers and front and rear end spoilers. The use of thermoplastic copolyester elastomers is not limited, however, to the exterior of automobiles. Overmolded interior parts such as door handles having good "feel" properties also have been commercialized [120].

In a very demanding application requiring excellent flexural fatigue resistance, thermoplastic polyesters have replaced polychloroprene rubber in constant-velocity joint boot bellows for front-wheel-drive automobiles. In addition to the necessary flexibility, this application also requires a product with good thermal properties, high tear resistance, very good low-temperature properties, and resistance to lubricating greases [121].

Thermoplastic copolyester elastomers containing either EOPPG or PEO soft segments absorb significant quantities of water. Because of this characteristic, thin films produced from these materials have high moisture vapor transmission rates. In addition to their use in laminated films for moisture-proof sports and rain wear [122, 123], these materials have also found application in the medical field [124]. By varying the EOPPG or PEO content, a wide range of moisture vapor transmission rates can be obtained. These products are suitable for medical wound care treatment, transdermal patches, and surgical drapes [124].

3.8. Producers, Trade Names

Thermoplastic copolyester elastomers were originally introduced by DuPont in 1972 under the trade name Hytrel and are manufactured by DuPont in the United States and Luxembourg, and by DuPont–Toray in Japan. Other U.S. producers include General Electric (trade name Lomod), Hoechst–Celanese (trade name Riteflex), and Eastman Chemical Products (trade name Ecdel). Both copolyether esters and copolyester esters are manufactured by DSM in the Netherlands, under the trade name Arnitel, and by Toboyo in Japan, under the trade name Pelprene. Pibiflex is manufactured in Europe by Montedison. More recent manufacturers entering this field include two South Korean producers, Kolon Industries (Kopel) and Cheil Synthetics (Esrel).

4. Thermoplastic Polyamide Elastomers

4.1. Introduction

Polyamides have been highly regarded thermoplastics for more than 50 years because of their high heat deflection temperatures, rigidity, and excellent resistance against solvents and aggressive chemicals. Their low melt viscosity permits easy processing, e.g., in injection molding. These and other positive properties have been transferred successfully to a greater or lesser extent to intrinsically softened polyamide elastomers.

In the early 1980s, Chemische Werke Hüls (today Evonik) and Ato-Chimie (today Arkema) independently developed processes for the synthesis of multiblock copolymers containing sequences of different polyamides and various kinds of polyethers linked by either ester or amide bonds. These polyamide elastomers (PAE, TPE-A or TPA)–more often referred to as Poly Ether Block Amides (PEBA)–behave like thermoplastic elastomers, i.e., they combine elastomeric properties over a wide range of temperature with an easy thermoplastic processibility. Because their properties are closely related to the basic polyamides, PEBA can be assigned to the high-performance class of thermoplastic elastomers.

4.2. Raw Materials

Polyamide elastomers consist formally of difunctional crystallizable oligoamides (hard segments) and difunctional soft segments having a low glass transition temperature. In many cases, this sequential co-condensation is also a viable preparative procedure. Various embodiments of this "internal plastification" concept have been described such as the use of (co-) polyolefin derivatives [125–131], polysiloxane soft segments [132], polyester soft segments [133, 134] and, in 2002, polycarbonate diols as soft segments [135]. However only polyether derivatives have become technologically significant on a commercial scale so far [136–140].

Polyether diols and diamines undergo condensation with dicarboxylic acid terminated oligoamides to form sequential multiblock copolymers via polyesterification or polyamidation, respectively [139, 140]. The typical structural patterns of these copolymers can be simply visualized as shown below.

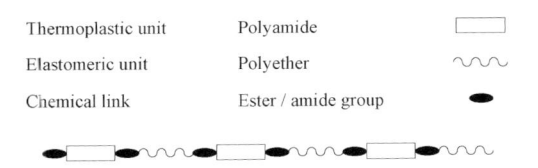

Thermoplastic unit	Polyamide	▭
Elastomeric unit	Polyether	∿
Chemical link	Ester / amide group	●

●▭∿▭∿▭●∿

Typical PEBA building blocks are polyamides with carboxylic acid terminal groups, e.g., poly(laurolactam) (PA12), and polyethers with hydroxyl terminal groups or polyethers with amino terminal groups, e.g., poly(tetramethylene glycol) (PTMG).

4.2.1. Oligoamide Unit

Oligoamides having carboxylic terminal groups are obtained by reacting a dicarboxylic acid with a lactam, e.g., Lactam 12 (laurolactam) or Lactam 6 (caprolactam), or an aminocarboxylic acid, e.g., Amino 11. In the case where a nylon salt is used, an excess of the dicarboxylic acid is employed [140]. The ratio between monomers and dicarboxylic acid excess controls the number average molar mass, M_n, of the oligoamide, usually around 600 to 4000 g/mol.

4.2.2. Polyether Units

Polyether diols applied in commercial PEBA grades are based on poly(ethylene glycol) (PEG), polypropylene glycol (PPG), and most frequently on poly(tetramethylene glycol) (PTMG or PTMEG), also named polytetrahydrofuran (PTHF) or poly(tetramethylene oxide) (PTMO). Poly(tetramethylene glycol) diols are commercialized worldwide, e.g., by Invista (Terathane) or BASF (PTHF).

Polyether diamines have been commercialized by Huntsman–formerly by Texaco–as Jeffamine D-grades (polypropylene glycol diamine) and as Jeffamine ED-grades (polyethylene glycol diamines), but are now commercialized under the tradename Elastamine R or H depending on the nature of the chain ends. PTMG-based diamines are also available.

4.3. Manufacture of Block Polyetheramides

The most important group of PEBA is derived from polyamide 12. The monomer basis is laurolactam which is converted to polyether block amides (PEBA) usually in a two-step process [136, 140]. The intermediate is an oligoamide dicarboxylic acid, which is then condensed to a PEBA by forming ester or amide links (Fig. 10).

4.3.1. Polyetherester Amides

The first step—ring opening polymerization of laurolactam—requires either acidolytic conditions, i.e., sufficiently high amounts of dicarboxylic acids [141] or the presence of water at high temperatures [142]. It may be advantageous to enhance the miscibility of oligoamide and oligoether by pressurizing the reaction mixture with water vapor [143].

The second polyesterification stage usually requires addition of a catalyst. Recommended classes of catalysts are dialkyltin(IV) compounds [140], phosphoric acid [144], tetraalkyl titanates [144, 145], tetraalkyl zirconates [146], tin(II) salts [147], or antimony oxide [148], the latter also being applied as a mixture with tin and phosphorus compounds [149].

A *single-step batch synthesis* has been described [144], in which both lactam polymerization and polyesterification are performed in the same reactor: Usual polyamidation equipment can be employed, such as stirred autoclaves, provided that sufficient water vapor pressure up to 25 bar (25×10^5 Pa)

Dicarboxylic acid
(e.g. dodecanedioic acid)

$HOOC-(CH_2)_{10}-COOH$

(Temp. 250 °C - 290 °C)

\+

Polyamide monomer
(e.g. laurolactam)

$y \; \overline{HN-(CH_2)_{11}-CO}$

(Pressure 1 - 20 bar)

↓

$HO-\{OC[(CH_2)_{11}NH]_{y1}-CO-(CH_2)_{10}-CO-[NH(CH_2)_{11}-CO]_{y2}\}-OH$

\+ Polyether diol
(e.g. PTHF), $HO-[O(CH_2)_4]_{\overline{x}}OH$

(Vacuum, catalyst)
(Temp. 200 °C - 270 °C)

↓

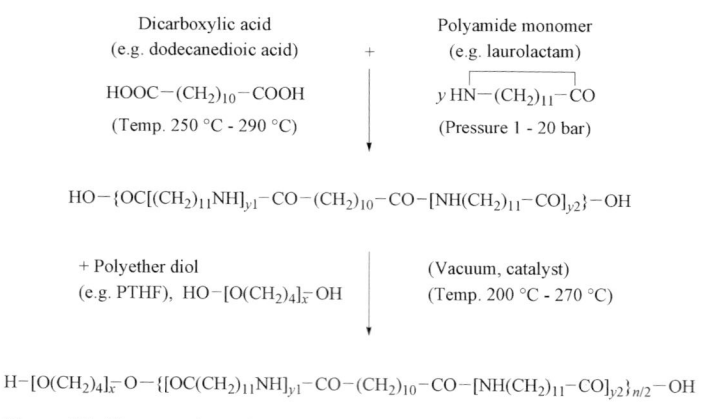

$H-[O(CH_2)_4]_{\overline{x}}-O-\{[OC(CH_2)_{11}NH]_{y1}-CO-(CH_2)_{10}-CO-[NH(CH_2)_{11}-CO]_{y2}\}_{n/2}-OH$

Figure 10. Two step formation of polyetheresteramide type PEBA

for the lactam cleavage can be applied and the reaction vessel can be evacuated for the second stage.

The one-pot synthesis, i.e., charging the reaction vessel with both polyamide forming monomers, including excess dicarboxylic acid and polyether diols or diamines is always the method of choice when polyamide monomers more reactive than laurolactam are used: Drastic reaction conditions necessary for the ring opening of laurolactam—high temperatures over an extended period of time may cause thermal degradation of polyethers—can be avoided in these cases, in particular for caprolactam [150], aminocarboxylic acids [151], nylon salts, or equimolar mixtures of diamines and dicarboxylic acids [152] as polyamide-forming monomers.

A continuous process applying *thin layer evaporation* has been described by Atofina (now Arkema) [153–155], it requires, however, a separate synthesis of oligoamide dicarboxylic acid. This dicarboxylic acid-terminated oligoamide is fed with a stoichiometrically equivalent amount of polyether diol via a static mixer into one or several thin-layer reactors equipped with rotating film-forming wipers. These elements are inclined to promote a reverse flow in the initial phase of the polyesterification, whereas in the final stages they are positioned in a way to promote downstream flow of the higher viscous melt [153]. The published procedure makes use of three sequential thin-layer reactors [153].

4.3.2. Synthesis of Polyetheramides

Condensation of oligoamide dicarboxylic acids with polyetherdiamines can be performed without the necessity of applying vacuum in the final phase of polycondensation [156, 159].

Polyamide elastomers based on this structural pattern have been developed by Ube, Ems (Grilamid ELY), and former Rhône Poulenc (Dynyl) [156, 160]. An additional benefit of the amide link instead of the ester group is the increased hydrolytic stability. The synthesis is usually performed batchwise in pressurized polyamide reactors. Both a single-step process [161] and a two-step process comprising a separate preparation of oligoamide dicarboxylic acids [162] have been described by Ube. Cocondensation of polyetherdiamines with low molar mass aliphatic diamines has also been described [163]. In the Ube synthesis a dimeric fatty acid is preferred both as a nylon chain terminator and as a link between polyetherdiamine segments.

Hybrid elastomers based on polyetheramides terminated with both amino and hydroxyl groups have been described [164, 165]: Reaction with isocyanates converts these polyetheramides into polyurethanes or polyureas, respectively.

Polyetheramides with very high melting points derived from nylon 46 segments have been prepared [166, 167] employing polyethers terminated by two primary amine moieties.

Table 11. Variation of block types and correlations with PEBA properties

Variable	Influenced property
Type of polyamide block:	melting range (upper service temperature)
	water absorption
PA 12, PA 11, PA 6, PA 66, PA 612	chemical stability
	density
PA 11	enhanced elastomeric properties
Length of polyamide block	melting range (upper service temperature)
	crystallinity
	transparency
Type of polyether block:	glass transition temperature (lower service temperature)
	water absorption, water permeability
PTMG (or PTHF), PEG, PPG	electrostatic properties
	chemical and thermal stability
Block composition Polyamide/polyether	rigidity, hardness, flexibility

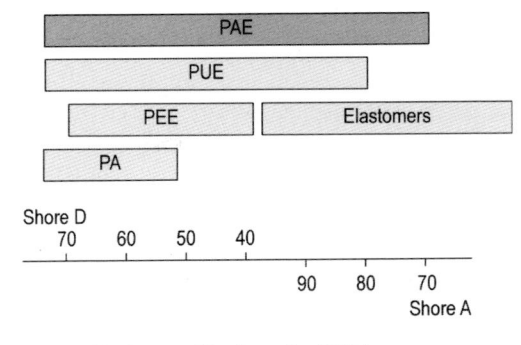

Figure 11. Range of hardness for PEBA
PEBA = Polyamide elastomer; TPU = Polyurethane elastomer; PEE = Polyester elastomer; PA = Polyamide

4.4. Physical and Chemical Properties

4.4.1. Block Variations

The properties of PEBA can be varied and determined by the selection of both hard and soft block types (Table 11). The choice of hard blocks defines the upper service temperature, corresponding with the melting temperature of the base polyamide. Also water absorption, chemical stability, and density are determined by the chosen polyamide. The length of the hard blocks is controlled by the amount of chain-limiting dicarboxylic acid and affects the melting range, the crystallinity, and the transparency of the PEBA.

The choice of the soft segment type determines the glass transition temperature and, hence, the lower service temperature. Using hydrophilic soft segments, such as polyethylene glycol, the moisture absorption and permeability is enhanced and permanent antielectrostatic as well as breathable properties induced. Using polyethers with amino end groups, the resulting polyetheretheramides have higher hydrolytic stability.

Finally, the block composition of polyamide and polyether blocks can be varied in a wide range and the resulting products encompass an outstanding range of rigidity. This is demonstrated in Figure 11 in terms of Shore hardness. PEBA with high polyamide block contents have a Shore hardness D up to 72, i.e., close to polyamides themselves. With lowest polyamide block contents PEBA can be very soft with Shore hardness A down to 60, hence, in the domain of soft rubbers. Such soft grades cannot be achieved using other TPEs.

Figure 12 shows a series of tensile stress–strain curves of PA 12–PEBA with varied PA 12 block content. PEBAs rich in PA 12 exhibit the

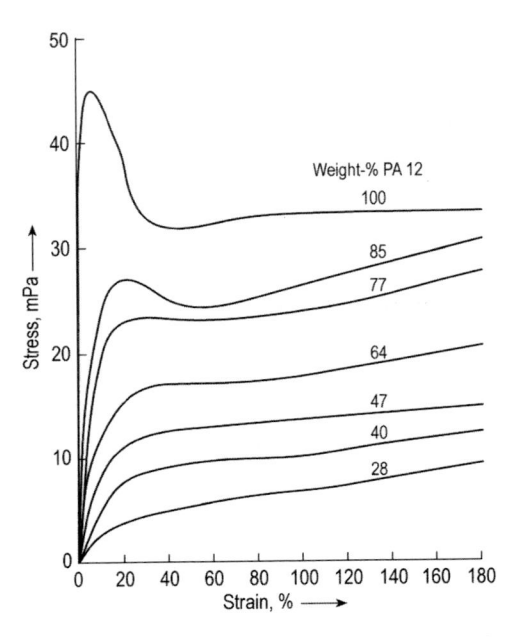

Figure 12. Stress–strain diagrams of PA 12–PEBA from tensile tests according to ISO 527

typical behavior of thermoplastics with yielding. With increasing polyether content there is a transition to more and more elastomeric characteristics. Better elastomeric properties for rigid grades can also be obtained when using PA 11 blocks (PEBAX, R = new range from Arkema) [168]. Indeed, PA 11–PEBA usually crystallizes in two arrangements: a triclinic one as well as an hexagonal one. Since the triclinic phase can change into a pseudo-hexagonal one, under thermal or mechanical stress, PA11–PEBA grades are more elastic, have better impact resistance and even the more rigid grade does not show any yield on their tensile stress-strain curve.

4.4.2. Morphology

The specific mechanical properties of PEBA are strongly determined by the phase morphology of the principally incompatible polyamide and polyether blocks. Four different phases have been identified by thermo-analytical methods and electron microscopy [169–171]. Differential scanning calorimetry revealed two melting signals, the dominant one corresponding with the melting of PA 12 blocks, the other one with a little content of crystallized PTMG blocks (Fig. 13). With decreasing level of PA 12 hard blocks, the melting temperature drops indicating that the size of crystallites is becoming smaller. Amorphous phases have been detected by torsional vibration analysis. Figure 14 shows the loss modulus plots of the investigated PA 12–PEBA series. The maximum values of the plots correspond with the glass transition temperatures of amorphous phases. In a range of higher PA 12 block content, glass transition temperatures of two mixed amorphous phases with different block compositions are observed. At higher PTMG block content, the loss modulus plots are dominated by the signal of a PTMG-rich mixed phase.

With the aid of highly sophisticated contrast-enhancing techniques, transmission electron microscopy (TEM) images were obtained, demonstrating the unique morphology in PEBA. Crystallized lamellae of PA 12 blocks form the same spherolitic superstructures as known from polyamides (Fig. 15). These crystalline superstructures form an interpenetrating reinforcing "physical network" for the amorphous mixed phase of PA 12 and PTMG blocks between the lamellae. At lower PA 12 content, only fragments of spherolites or single lamellae can be formed with a weak reinforcing effect on the soft amorphous PTMG-rich mixed phase (Fig. 16). For completeness, it should be mentioned that PTMG-rich PEBA molecules build

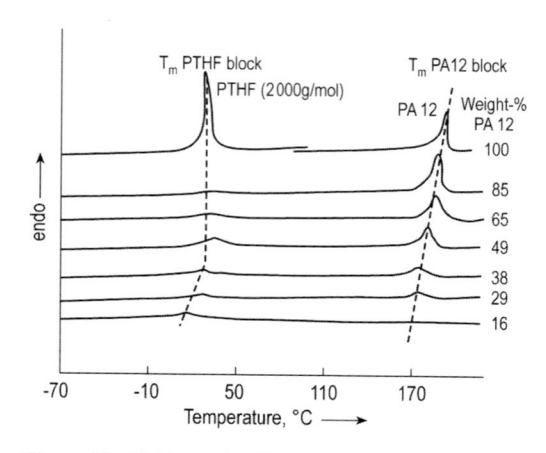

Figure 13. Melting point diagrams (DSC) indicating two crystalline phases in PA 12–PTMG–PEBA (T_m = melting temperature)

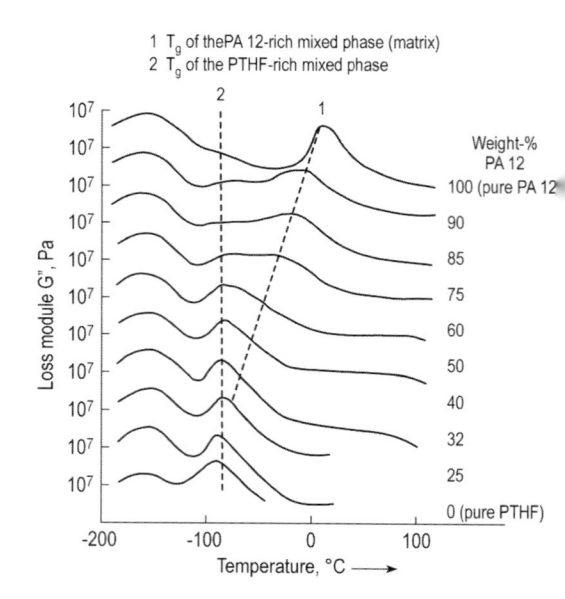

Figure 14. Loss modulus curves from torsional vibration analyses indicating two amorphous mixed phases in PA 12–PTMG–PEBA (T_g = glass transition temperature)

up a separate disperse phase due to the strong incompatibility between both blocks. This is the second amorphous mixed phase with the low glass transition temperature observed in torsional vibration analysis.

A number of the unique mechanical properties which distinguish PEBA from all other TPE can be deduced from the interpenetrating network of crystallized polyamide blocks. This is valid, for example, for the lower temperature dependence of stiffness and elasticity compared to other TPEs.

4.4.3. Outstanding Properties

The specific advantages of PEBA (mainly PA 12–PTMG, unless otherwise mentioned) compared with other TPE can be summarized as follows:

- Low density: 1.01–1.03 g/cm^3 (TPU, copolyester elastomer: 1.14–1.25 g/cm^3),
- High stability against chemicals and solvents,
- Easy processing, coloring, and overmolding,
- Excellent ability to be decorated by thermodiffusion printing,
- Excellent cold impact strength,
- Wide range of hardness and flexibility,
- High elasticity and resilience,
- Low temperature dependence of mechanical properties,
- Excellent abrasion resistance (only surpassed by TPU),
- Highest service temperatures (especially PEBA with PA 6 blocks).

Some of the outstanding properties of PEBA are listed in Table 12 for a series of commercial products (PEBAX® from ARKEMA) based on PA 12 and PTMG: The density (close to 1 g/cm^3) is up to 20% lower than that of competing TPE. The flexural modulus of PEBA comprises a full decade of magnitude, from about 10 to 500 MPa. The cold impact strength of PEBA is excellent, as demonstrated by Charpy notched impact data at −30 °C in Table 12. Very soft and medium soft PEBA do not break at all, stiffer

Figure 15. Transmission electron micrographs showing spherolitic superstructures of crystallized lamella in PA 12–PTMG–PEBA (PA 12-rich PEBA).

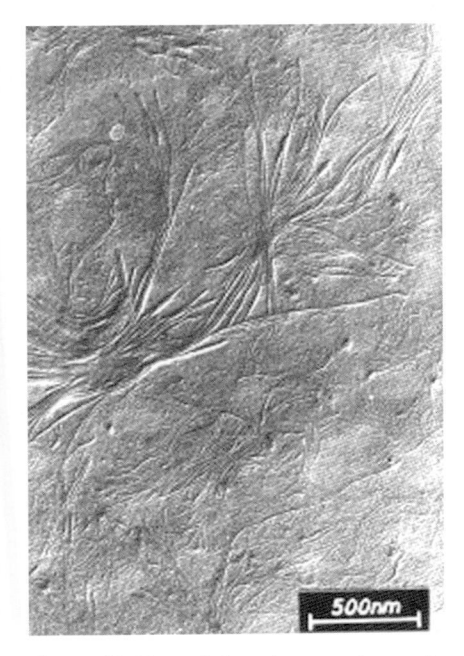

Figure 16. Transmission electron micrographs showing spherolitic superstructures of crystallized lamella in PA 12–PTMG–PEBA (PTMG-rich PEBA).

Table 12. Major properties of PA 12-PEBA compared with PA 12

Properties	Test-method	Unit	PA 12-PEBA (PEBAX) with Shore D hardness (ISO 868)						
			27	33	42	54	64	69	69
Density at 23 °C	ISO 1183	g/cm^3	1.00	1.00	1.00	1.01	1.01	1.01	1.01
Flexural modulus	ISO 178 -1/-2	MPa	12	21	77	170	285	390	513
Charpy notched impact strength at −30 °C	ISO 179	kJ/mm^2	No Break	No Break	No Break	No Break	20	20	10
Vicat softening temperature Method A/10 N	ISO 306	°C	58	77	131	142	157	164	164

grades with Shore hardness D > 60 and tensile moduli above 250 MPa have still excellent toughness data.

The uptake of water is a crucial property for many PEBA applications. In this regard PEBA based on PA 12 and PTMG blocks do not differ significantly from PA 12 which has the lowest water absorption of all polyamides. With increasing polarity of polyamide and polyether building blocks the water uptake of PEBA increases dramatically (Table 13). This creates a strong influence on the electrical behavior of the PEBA. PEBA with PA12 and PTMG blocks are good insulators, while, e.g., PEBA based on PA 6 and poly(ethylene glycol) blocks are permanent antielectrostatic due to their fast and high moisture uptake from the ambient air. High water uptake of PEBA is not tolerable for applications where dimensional stability of parts is crucial. On the other hand PEBA with higher absorption and permeability of moisture have found applications as, for instance, antistatic modifier and for moisture-breathing film.

4.4.4. Chemical Resistance

Due to their close relationship to polyamides, PEBA exhibit high stability against many chemicals and solvents compared with other TPE. The stability of PEBA against diluted hydrochloric and sulfuric acid as well as bases should be especially noted. Rigid PEBA grades show only low swelling in ASTM oil, even at 100 °C. This is also valid for hydraulic liquids, aliphatic solvents, and alcohols. With softer PEBA grades, the swelling in aromatic solvents and fuels is rather high; however, the mechanical properties are not affected dramatically.

4.4.5. Processing

PEBA can be processed on standard equipment and with processing conditions usual for polyamides. For injection molding of PA 12-PEBA melt temperatures of 190–230 °C are recommended for rigid grades and 170–210 °C for softer grades. Mold temperatures of 20–40 °C are suitable for all grades. Because of the very low melt viscosity and rapid crystallization, short injection cycle times are possible.

At extrusion temperatures of 210–230 °C, thin sheets, hoses, tubes, profiles, plates, and monofilaments can be processed. Other processing techniques applied for PEBA are blow molding, rotational molding, thermoforming, and powder coating.

Table 13. Density and water absorption of various PEBA in comparison

TPE-type	Density	Water absorption (saturated)	
	ISO 1183 g/cm^3	20 °C/65% R.H., wt%	20 °C/24 h water storage, wt%
PA 12-PTMG	1.01–1.03	0.5	1.2
PA 6-PPG	1.10	2.4–2.8	6.1–6.4
PA 6-PEG	1.14	4.5	120
TPU[*]	1.14–1.25	0.1–5	0.5–15
Copolyester TPE[*]	1.17–1.25	0.2–0.4	0.6–0.8
Rubber[*]	>1.20	0.6–0.8	0.1–15

[*] Strongly varying for different types.

4.5. Uses

Since the polyamide block building monomers are rather expensive because of their multistep synthesis, PEBA are ranking at the high end of the scale of prices of TPE together with TPU and copolyester TPE. As a consequence, they are only used where one or more of their outstanding product and processing properties are required.

Polyamide elastomers are especially acknowledged for the construction of lightweight soles for specialists sport shoes. PEBA are used primarily because of their low density, but also because of their good resilience, cold impact strength, and low temperature dependence of flexibility. The excellent processing and overmolding properties allow high sophisticated sole constructions. This has been applied for designs with sole parts of different stiffness and for decorative effects in overmolding of parts with different colors (Fig. 17). Even PA 12-based composites and PA 12-coated metal sheets have been incorporated in sole constructions. Recently, the combination of overmolding and decorative printing techniques have been applied for new sports shoe soles constructions.

Another main application field for PEBA are winter sports goods such as lightweight ski boots or PEBA films for decorative ski and snowboard surfaces. By means of thermodiffusion printing processes brilliant decorations are applied often on the bottom side of a PEBA film. The colorants penetrate only about a 100 µm

Figure 18. Design of skies with decorative PA12–PEBA films, sublimation printed from the bottom side

into the film of typical 500 µm total thickness. The film itself then acts as a protective layer for the decoration and the print motifs are almost indestructible (Fig. 18).

Soft PEBA grades, with Shore hardness D below 40, have been used in such applications as straps of watches. Beside flexural properties, the dermatological compatibility, UV stability, and translucence of PEBA is striking. For frames of sunglasses, dermatological compatibility, easy processing, and coating stability of these elastomers is important.

Medical uses are, e.g., catheters, or controlled drug release carriers. Hydrophilic grades are applied for moisture breathable film in surgical drapes and garments.

The excellent processing and demolding properties and the dimensional stability of PEBA make them well suited for manufacturing precise moldings. PEBA gear wheels (Fig. 19), for example, contribute to the noiseless operation of tape recorders or CD players.

For permanent antistatic housings, hydrophobic PEBA can be compounded with special conductive carbon blacks which have still good impact resistance in spite of the necessary high loadings. Hydrophilic PEBA grades can also be used as antistatic additives to reduce the surface resistivity of various thermoplastic resins (PE, PP, ABS, PS, PMMA, PA etc). Applications are permanently antielectrostatic parts, for example, castor wheels (Fig. 20), conveyor belts, rollers, and paint spray hoses, clean room equipment, surface and flooring, containers for flammable or

Figure 17. Sport shoe soles with overmolded PEBA parts and PA 12 composites

Figure 19. Precisely molded PEBA wheels for noiseless gears

explosive goods, carried/reel tapes, medical (inhalers, bags).

4.6. Trade Names

PEBA play only a niche role in the TPE market due to their high-end positioning. The main and historical suppliers of PEBA are Evonik AG/Germany (Vestamid E-grades, Shore hardness D 30–70, PA 12 base) and Arkema/France (Pebax, Shore hardness D 25–72, PA 12, PA 11, and PA 6 base, including hydrophilic and transparent grades). Other suppliers are EMS Chemie/Switzerland (Grilamid ELY, Grilon ELX, Shore hardness D 50–63, PA 12 and PA 6 base), Ube/Japan (Ubesta, selected grades), and Sanyo/Japan (Pelestat, hydrophilic grades).

Figure 20. Tough antielectrostatic PEBA parts highly loaded with conductive carbon blacks

5. Styrenic Block Copolymers

5.1. Introduction

Thermoplastic elastomers known as styrenic block copolymers (SBC) have emerged from developments and research in the organometallic-initiated polymerization of diene monomers in hydrocarbon solvents (→ Rubber, 5. Solution Rubbers). Commercial block copolymers are essentially based on styrene, butadiene, and isoprene monomers. The first commercial quantities were produced by Shell in 1964 in its production plant in Torrance, Ca., and marketed under the trade name Kraton [172]. Worldwide capacity for styrenic block copolymers in 1993 was estimated at 500 000 t/a.

Styrenic block copolymers combine rubber performance with thermoplastic processibility on the basis of their specific structure. Anionic polymerization allows the selective polymerization of butadiene or isoprene and styrene in such a way that the resulting polymer is essentially a polybutadiene or polyisoprene molecule tipped at both ends with a polystyrene molecule [173]. Such a structure is commonly identified as ABA, where A represents the plastic and B the elastomer. More specific notations used are SBS or SIS polymers. Styrenic block copolymers with saturated rubber blocks are produced from SIS or SBS precursors by selective hydrogenation. This process converts the polyisoprene into poly(ethylene–propene) and the polybutadiene into poly(ethylene–butene). To provide rubber performance, the volume fraction of plastic (polystyrene) in these structures must be < 50%, and commercial hydrogenated grades usually contain around 30%. Newer variants of the styrene–ethylene/butene–styrene S–E/B–S or styrene–ethylene/propene–styrene S–E/P–S polymers are functionalized with maleic anhydride or other reactive comonomers. Functionalization levels are between 0.2 and 5% and the reactive group is grafted onto the rubber mid block [174].

5.2. Synthesis

The first step in the production of SBCs is activation of purified styrene in a hydrocarbon solvent with butyllithium and subsequent

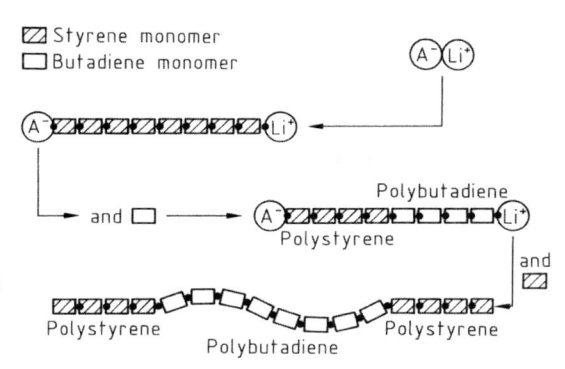

Figure 21. Formation of block copolymers by anionic polymerization (A$^-$ = butyl anion)

anionic polymerization. The molecular mass of the first polystyrene block is determined by the butyllithium : styrene ratio (see Fig. 21). The formed living polymer, PS–Li, is then reacted with purified butadiene to form the second block. When all the diene has been converted, two routes can be followed to complete the formation of SBC:

1. Addition of a coupling agent to produce linear or branched copolymers
2. Addition of a second amount of styrene monomer to form the third block

Technically feasible coupling agents include di-, tri-, tetra-, or multifunctional compounds that allow formation of a multitude of different polymer structures [175].

The procedure for production of the precursor for *saturated block polymers* is the same. To ensure a rubbery mid block after hydrogenation in the S–E/B–S polymer, a modifier is added during polymerization of the polybutadiene mid block to increase its vinyl content (1,2-content) (→ Rubber, 5. Solution Rubbers) to approximately 40% which will ensure optimum elastic properties after hydrogenation. This allows optimization of the ratio of ethylene and butene units in the hydrogenated polymer. After preparation of the precursor, polymer solution is transferred to the hydrogenation section where the double bonds of the polydiene are reacted with hydrogen in the presence of a catalyst.

Functionalized saturated block copolymers are made by extruder grafting of an S–E/B–S or S–E/P–S molecule. The polymer is contacted in an extruder with a peroxide to create free radical sites along the rubber mid block. These react in the extrusion step with a compound containing the functional group and a group able to react with the free radical sites (e.g., maleic anhydride). The process requires careful optimization of the amount of peroxide and reaction conditions to minimize polymer degradation.

The result of the polymerization process is a solution of polymer in hydrocarbon solvent. At this stage, additives such as stabilizers, antioxidants, or plasticizers can be added to the solution which is subsequently converted into solid polymer during finishing. The following systems are used industrially:

Direct Desolventizing. The polymer solution (cement) is introduced directly into a special vacuum extruder; the solvent is evaporated into a recovery system and the extruded polymer melt is face-cut and cooled at the die head of the extruder to form compact transparent pellets.

Coagulation with Steam. Cement is pumped to a vessel where steam is passed into the solution. The resulting coagulate separates from the solvent, and the polymer–water mixture is dried to remove water. The usual form resulting from the coagulation–drying route is referred to as crumb.

5.3. Properties

Thermoplastic rubbers based on styrenic block copolymer can be considered polybutadiene [or polyisoprene, poly(ethylene–butene), or poly (ethylene–propene)] tipped with polystyrene end blocks. Polybutadiene (butadiene rubber, BR) and polystyrene of sufficiently high molecular mass are incompatible, which forms the basis for the performance of these thermoplastic rubbers. At temperatures above the softening point of polystyrene, the SBC molecule is quite mobile and can be melt-processed by, e.g., extrusion or injection molding. On cooling, the PS end blocks separate from the BR mid blocks, and two distinct phases are formed. The separate, discrete phase of the PS domains acts both as a cross-linking material and as a reinforcing filler analogous to the sulfur and carbon black in vulcanized rubber. Figure 22

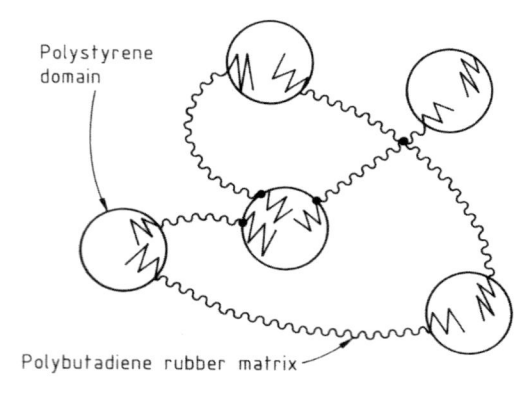

Figure 22. Schematic phase structure of an SBS block copolymer

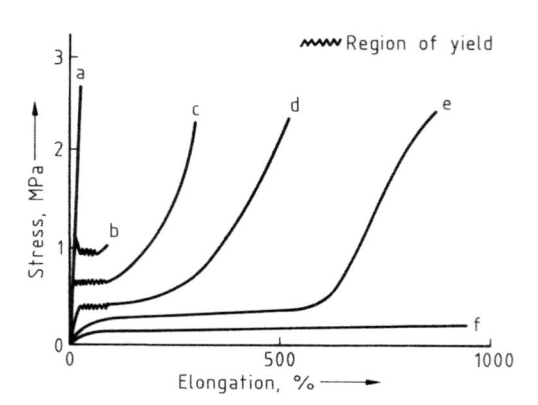

Figure 23. Stress–strain curves for various SBS copolymers
a) 80% styrene; b) 65% styrene; c) 50% styrene; d) 40% styrene; e) 30% styrene; f) 10% styrene

shows an SBC where the PS domains hold together a rubbery BR network. The tensile strength (Table 14) and resilience data (Table 15) illustrate the truly rubbery behavior of a linear SBS polymer with a polystyrene content of 30%.

The *molecular structure* of an SBC can be varied by changing the proportion of PS in the base polymer. Furthermore the total molecular mass can be varied, but practical limits exist to the processibility in the melt. Figure 23 shows the effect of altering polystyrene content in a series of linear polymers. Polymers can be made that range from very low strength at low (10%) PS content, through rubbery materials of about 30% PS, to those with 80% PS content that behave much like toughened polystyrene. SBC polymers can be linear or branched; branched types are usually higher in molecular mass than linear ones and have a higher melt viscosity at equal strength levels. Furthermore, a minimum molecular mass of the end block exists below which no phase separation occurs, no domain structure forms, and hence unsatisfactory properties are obtained [176]. A comparison of properties of polymers differing in the molecular mass of the PS blocks and in total molecular mass is given in Table 16.

The *processing properties* of SBCs can be derived from their basic chemical structure. The molecular mass of commercially available grades varies from about 75 000 to 400 000. Their behavior in the molten state is similar to that of polyethylene and polystyrene, for example. However, even in the molten state the tendency to form PS domains persists, especially for those types with a saturated rubber mid block. *Melt processing SBCs* at higher shear rates leads to lower apparent viscosity; this effect is relatively independent of temperature (see Fig. 24). Reprocessing of SBCs at < 200 °C has little effect on their properties; saturated

Table 14. Comparison of tensile strengths of BR and SBS

Polymer	Tensile strength, MPa
Unvulcanized BR	0.2–0.5
Vulcanized BR	4.0–5.0
Vulcanized + carbon black reinforced BR	5.0–20.0
Unvulcanized pure SBS	30.0–35.0

Table 15. Resilience (Lüpke-rebound) of different polymers

Polymer	Resilience, %
Unvulcanized SBS	65
Vulcanized BR	65
Unvulcanized SIS	60
Vulcanized NR/IR	60
Vulcanized SBR	50
Vulcanized EPDM	45
Unvulcanized EVA (28% VA)	40

Table 16. Properties of styrenic block copolymer structures with different PS block and total molecular mass

Polymer structure	Molecular mass		Viscosity	Strength
	PS block	Total		
Linear	12 000	80 000	low	high
Branched	12 000	160 000	medium	high
Branched	6000	80 000	very low	low

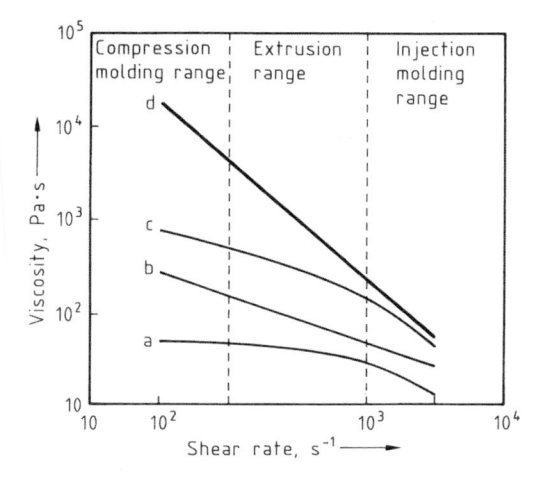

Figure 24. Viscosity–shear rate relationship at 200 °C
a) Clear polystyrene; b) SBS; c) polypropylene melt flow rate (MFR) = 5; d) S–E/B–S

grades can be processed and reprocessed at up to 280 °C.

Solution processing of SBCs is possible, provided that solvents of sufficiently high solubility are used to dissolve the PS end blocks. Solvents with solubility parameters between 8 and 10 are usually employed. Because of the easy solubility of SBCs, adhesive, sealant, and coating systems can be designed that have low viscosity and high solids content. A consequence of their good solubility in organic solvents is that the oil and solvent resistance of parts based on SBC is limited. However, their resistance to dilute aqueous solutions of the most commonly encountered chemicals is generally satisfactory.

The *service performance* of SBCs and the products derived from them is also related to their basic chemical structure. Because of their moderate molecular mass and their physical and reversible cross-linking, they flow under pressure and exhibit creep. Block copolymers with *BR or IR mid blocks* are unsaturated and sensitive to aging. Exposure to elevated temperature can result in polymer degradation with simultaneous decrease in viscosity and tensile strength. With the exception of black compounds, exposure to UV light for any length of time will result in discoloration and surface cracking, which becomes more pronounced on prolonged exposure. These block copolymers are also susceptible to ozone attack, and in shaped products at points with local stresses, microcracks develop on exposure to atmosphere containing ozone. The use of antioxidants, UV stabilizers, and antiozonants reduces these degradation effects to workable limits in end uses.

The *hydrogenated S–E/B–S or S–E/P–S types*, in contrast, have a saturated rubber mid block and are therefore highly resistant to degradation by oxidation, ozone, and UV light. They can be processed over a much wider processing temperature range.

5.4. Uses

Because of their rubbery characteristics (especially at low temperature), high strength, compound ability, and thermoplastic nature, SBCs are used in a variety of applications as highlighted in Figure 25.

On the left-hand side of Figure 25 the list includes outlets that use SBC as an additive or modifier for other materials. The polymers are delivered to the end user who will incorporate them in his existing process. The right-hand column lists applications that utilize a compounded product with performance tailored to the specific end use and plastic conversion technology. A comprehensive review of the applications of thermoplastic rubber is given in [177, 178].

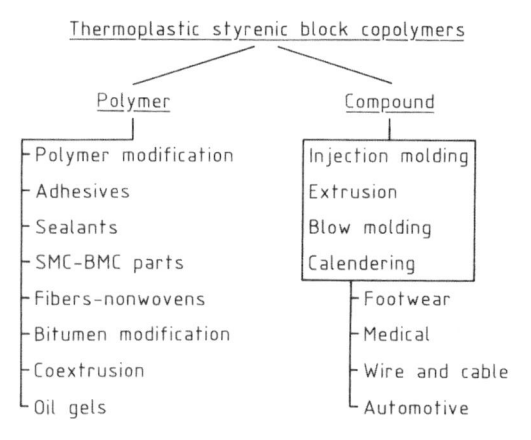

Figure 25. Applications of styrenic block copolymers
SMC = sheet molding compounds; BMC = bulk molding compounds

5.4.1. Adhesives, Sealants, and Coatings

Adhesive properties are determined primarily by the styrenic block copolymer type chosen for a typical application. SBS polymers with high cohesive strength (high polystyrene content) are commonly used in contact adhesives, whereas SIS polymers with low polystyrene content that can be tackified easily are preferred for pressure-sensitive adhesives. For applications in hot melt form, low molecular mass polymers with a high melt flow rate are selected. When high UV resistance is required, hydrogenated block copolymers should be considered as base polymer. Furthermore, adhesive properties are also determined by the nature and the concentration of the various resins and plasticizers used and their degree of compatibility with the respective mid or end block phase of the block copolymer [179].

Pressure-sensitive adhesives are materials that remain permanently tacky and adhere instantaneously to solid surfaces on the application of light pressure. Major applications are tapes and labels. Tapes require high shear strength and holding power, while labels require less shear strength but good convertibility. For general-purpose solvent-based pressure-sensitive adhesives, SIS block copolymers are combined with resins that are compatible with the mid block, such as wood rosins, polyterpenes, or synthetic hydrocarbon resins. A typical composition is shown in Table 17, formulation no. 1. The formulation for an adhesive used for permanent labels is given in Table 17, formulation no. 2. Pressure-sensitive adhesives for peelable labels have a relatively high cohesive strength and a low peel strength (Table 17, no. 3). Labels that are applied under cold conditions must have a low T_g, and formulations usually contain low amounts of high-melting resin but large proportions of plasticizer (Table 17, no. 4).

In outdoor applications where the adhesive must withstand UV light, saturated block copolymers are employed. Pressure-sensitive adhesives with very high temperature resistance can be obtained by UV or electron beam curing (→ Kraton D-1320X Radiation Chemistry). A special polymer developed by Shell for this application is Kraton D-1320X. The shear adhesion failure temperature (SAFT) of a pressure-sensitive adhesive based on this polymer is increased from 80 °C to ca. 140 °C by either UV or electron beam curing [180].

Contact adhesives are applied as liquid onto one or two substrates; after a certain time to allow partial evaporation of the solvent (open time) the coated substrates are pressed together to form a strong bond. SBCs are ideal for these types of adhesives because they offer high cohesive strength without vulcanization and are readily soluble in many solvents. For some applications, the contact adhesive is allowed to dry completely to a tack-free non-blocking adhesive film. Exposure to heat reactivates the adhesive, and a firm bond forms under pressure. Large surfaces are commonly coated by spraying with a uniform adhesive layer (e.g., on flexible foams used in the furniture and automotive industries). Because of their molecular structures, SBC-based adhesives combine high strength with low viscosity and ease of spraying. A variety of melt adhesives are used in the construction of hygienic products such as disposable diapers, sanitary napkins, etc. These adhesives can be applied by spray or slot-die coating techniques. Selection of the styrenic block copolymer depends on the properties required and the application technique used.

Table 17. Pressure-sensitive adhesive formulations and properties

Composition, property	Formulation number			
	1	2	3	4
Parts by weight				
SIS polymer*	100	100	100	100
Piccotac 95	90		25	
Escorez 1310		120		90
Hercures C 10	10	50	125	
Wingtack 10				40
Irganox 1010	1			3
ZDBC		3	3	
Properties				
Melt viscosity (175 °C), Pa·s		20		60
Rolling ball tack, cm	2	3	3	1
Probe tack, N	10	>10		10
Loop tack, N/25 mm	25	30	8	27
180° peel adhesion, N/25 mm	17	15	5	16
SAFT, °C	115	100	80	100
T_g (Fox), °C	−20	−19	−33	−29

* Kraton D-1107 from Shell Chemicals.

Sealants. Styrenic block copolymers with *saturated rubber mid block* are ideally suited as base polymers for both hot-melt and solvent-based sealants that require resistance to long-term degradation and weathering [181, 182].

Coatings. In several applications, SBCs are used to formulate *protective coatings* based on solvent systems, but the trend is toward solvent-free systems, for example, in chemical milling.

5.4.2. Modification of Bitumen

The addition of SBCs to bitumen increases its upper and decreases its lower operating service temperature, which results in substantial performance improvement in applications such as roofing felts, waterproofing membranes, and asphalt pavement. Selective development of SBS polymers has enabled bitumen to maintain its strength and integrity at elevated temperature without flowing and to remain flexible and ductile at temperatures well below the freezing point [183]. Moreover, SBCs greatly enhance the lifetime of bitumen-based products (e.g., roofing felts). This is achieved by development of a three-dimensional network. To optimize the effect of SBS addition, bitumen with the right degree of compatibility must be selected to ensure effective formation of the elastomer network. For roofing felts, SBS addition levels are commonly between 8 and 12%. Blending into the bitumen is carried out at 170–190 °C. For fast dissolution, especially when low-shear paddle mixers are employed, SBS polymer is ground to a powder with particle size of ca. 1000 μm. For most applications, high-shear mixers are used with regular polymer crumb [184, 185].

The use of polymer-modified bitumen for asphalt pavements is growing rapidly and offers large potentials for SBC in the future. The SBS addition levels for road applications are lower than for roofing: typically 3–7%. In addition to improved deformation and flexibility, laboratory data clearly demonstrate that the durability of asphalt roads is increased by addition of SBS polymer [186]. Many test roads have been installed that are being monitored to gain experience and evidence for performance under practical conditions. For documentation of the most important practical experiences so far, see [187–193].

5.4.3. Compounded Products

As mentioned earlier, end use parts that are made via established plastic processing techniques do not employ pure SBC polymers but rather—like traditional vulcanized rubber—compounded and end use-tailored compositions. Depending on their requirements, such compounds have hardnesses in the range from 35 Shore A to as high as 55 Shore D. The actual SBC content can be as low as 15 wt% but is usually between 25 and 40 wt%. SIS block copolymers are rarely used as base polymer for compounds. The main classes of compounding ingredients include the following:

Mineral oil is used as plasticizer to increase flow and reduce cost. Paraffinic or naphthenic oils are used that compatibilize with the rubber mid block. Aromatic oils or polar plasticizers such as dioctyl phthalate (DOP) are avoided because they strongly plasticize the end blocks and reduce strength.

Polystyrenes are important compounding ingredients to increase hardness, strength, and wear resistance, particularly in SBS polymers. Both clear and high-impact polystyrenes are used.

Polyolefins, particularly polypropylene, are important ingredients in S–E/B–S and S–E/P–S block copolymers.

Fillers, mainly cheapeners and nonreinforcing types, are used to decrease cost and increase hardness. They also increase density, which is exploited in formulations for sound-deadening applications.

Additives such as antioxidants, UV stabilizers, and antiozonants, as well as additives that induce special effects, can be used. Examples are pigments, surface-active slip agents, blowing agents, and antistatics.

Compounding is carried out at elevated temperature and at shear conditions that result in a fine, stable dispersion of the various ingredients. Compounding equipment originates from both the rubber industry with two-roll mills, Banbury-type internal mixers, and Farrel continuous mixers, and the plastic processing industry with mainly single- and double-screw compounding extruders. Modern compounding plants are fully instrumented robots in which a twin-screw conrotating compounder with a length : diameter ratio ($L : D$) up to 42 is the key unit. Screw types with feeding, kneading, homogenization, and venting sections are designed for the specific compounding operation. Ingredients are added either as preblended mixture or via separate weigh feeders and oil injection pumps. The granulating section at the end of the extruder is important because it is responsible for formation of the homogeneous, free-flowing pellets required in subsequent plastics conversion equipment. Depending on line capacity and product characteristics (i.e., soft or hard), die plates with knife cutters, water-ring pelletizers, underwater pelletizers, or a strand die with water cooling bath and rotating knife granulator can be used.

Detailed compositions and selection of base polymers have been published by SBC suppliers in the form of technical bulletins or formulating guides. Compounding know-how and formulations based on SEBS or SEPS polymers is rather specific [194]. Table 18 lists properties of typical SBC-based compounds tailored for selected end uses. The largest volume of SBCs in compounded form is consumed by the footwear industry. Here, SBS-based compounds are established mainly as high-quality soling materials. An important aspect is the multitude of colors that can be combined in shoe design. Compounding aims to optimize the end products with respect to melt flow rate (for ease of processing), high abrasion resistance (for durability), and good resistance to repeated flexing. Hardness for most typical soling materials falls within the Shore A 55 to 75 range.

Compounds designed for use in medical devices and articles intended for food contact are also important outlets and take advantage of the high purity of polymers made by anionic polymerization. SBS- and SEBS-based compounds which meet the requirements set in the United States by the FDA and in Germany by the BfArM ■correct?■, are used for syringes, tubing, bottle nipples, stoppers, i.v. bag systems, etc. Contact of SBS and SEBS with fatty food is restricted.

Thermoplastic rubber products can be recycled, and scrap, trimmings, and off-specification parts can to a certain extent be reintroduced in the manufacturing stream.

5.4.4. Modification of Plastics

Most plastics are brittle and lack impact resistance, particularly below 0 °C. Rubber is added to achieve

1. Increased impact strength (Izod, Charpy, FWIS, Dart, etc.) especially at low temperature
2. Increased environmental stress corrosion resistance

The addition of a rubber to a thermoplastic also changes properties such as stiffness, tensile strength, yield modulus, surface friction, softening point, clarity, electrical properties, vapor transmission, and gloss.

With many plastics, an impact resistance improving rubber fraction can be incorporated in the polymerization reaction. Examples are propene–ethylene copolymers, high-impact polystyrene (HIPS), and ABS. Nevertheless, styrenic block copolymers are used as melt-blended component in plastics that do not allow reactor modification or to obtain impact performance in excess of that achievable in the reactor. Polystyrenes are often compounded with flame-retarding additives and then require the addition of SBS polymer to compensate for the loss in impact strength due to addition of the flame retardant [195]. Engineering plastics require processing temperatures beyond the operating range of SBS. Here, saturated SEBS and SEPS block polymers are used, especially

Table 18. Properties of compounded SBCs

Property	Reference no[*], base SBC polymer				
	1 SEBS	2 SBS	3 SEBS	4 SEBS	5 SEBS
MFR (E or G), dg/min	50 G	35 E		3 G	17 G
Density, g/mL	0.91	1.0	1.94	0.91	1.01
Hardness, Shore A/D	45 A	65 A	36 D	60 A	82 A
Tensile strength, MPa	5.0	7.6	4.8	7.5	17.3
Elongation, %	610	650	250	480	670
DIN abrasion, mm^3		180			
Haze, %	7				
UL 62[**], Class 34 rating, °C					105
Compression set 22 h at 100 °C, %				42	

[*] 1 = Transparent medical molding; 2 = soling material; 3 = automotive sound deadening; 4 = high-temperature performance; 5 = electrical wire insulation.
[**] UL = Underwriters Laboratories code.

as modifiers in polycarbonate (PC) and poly(phenylene ether) (PPE). The more polar engineering plastics such as polyamides and polyesters can be adequately impact modified by SEBS or SEPS functionalized with maleic anhydride [196].

Styrenic block copolymers are also capable of compatibilizing different and dissimilar plastics into useful and workable blends. Addition of SBS or SEBS to a blend of polystyrene and polypropylene or polyethylene, for example, creates a ductile material that can be extruded or molded without the delamination or phase separation encountered in the absence of block polymer addition [197]. Similar results are obtained when SEBS is used in blends with polycarbonate and polypropylene, as well as polyethylene–poly(ethylene terephthalate) (PETP) blends. Functionalized block polymers are good compatibilizers for, e.g., polyamides and polypropylenes. SBCs can also be used in thermoset resin systems, particularly as additives for sheet molding or bulk molding processes with unsaturated polyesters. Moldings modified with specifically developed SBS or SEBS polymers to control shrinkage show high-quality surface appearance [198, 199].

The use of SBCs as compatibilizers for mixed plastic streams is becoming more and more important with respect to waste management and upgrading and compatibilization of used plastics [200].

Abbreviations for polymers and monomers:

2G	ethylene glycol
2GT	dimethylene terephthalate, ethylene terephthalate
3G	1,3-propanediol
4G	1,4-butanediol
4GT	tetramethylene terephthalate, butylene terephthalate
5G	1,5-pentanediol
6G	1,6-hexanediol
10G	1,10-decanediol
ABS	acrylonitrile–butadiene–styrene copolymer
BR	butadiene rubber
DOP	dioctyl phthalate
DVA	dynamically vulcanized alloys
EOPPG	poly(propylene glycol) capped with ethylene oxide
EPDM	ethylene–propene–diene terpolymer
EPM	ethylene–propene copolymer
EPR	ethylene–propene rubber
EVA	ethylene–vinyl acetate copolymer
H_{12}-MDI	4,4'-dicyclohexylmethane diisocyanate
HDI	hexamethylene diisocyanate
HIPS	high-impact polystyrene
IR	isoprene rubber
MDI	4,4'-methylene diphenyl diisocyanate
NDI	1,5-naphthalene diisocyanate
NR	natural rubber
PEBA	thermoplastic polyamide elastomer
PB	polybutadiene
PC	polycarbonate
PEO	poly(ethylene glycol); poly(ethylene oxide)
PETP	poly(ethylene terephthalate)
PP	polypropylene
PPE	poly(phenylene ether)
PS	polystyrene
PTHF	poly(tetrahydrofuran)
PTMG	poly(tetramethylene glycol)
PTMO	poly(tetramethylene oxide)
PVC	poly(vinyl chloride)
RTPO	reactor thermoplastic polyolefins
S–E/B–S, SEBS	styrene–ethylene/butene–styrene copolymer
S–E/P–S, SEPS	styrene–ethylene/propene–styrene copolymer
SBC	styrenic block copolymers

SBS	styrene–butadiene–styrene block copolymer
SIS	styrene–isoprene–styrene block copolymer
T	terephthalic acid
TODI	3,3′-tolidine-4,4′-diisocyanate
TPE	thermoplastic elastomer
TPNR	thermoplastic natural rubber
TPO	thermoplastic polyolefin elastomer
TPR	thermoplastic rubber
TPU	thermoplastic polyurethane elastomer
TPV	thermoplastic vulcanizates
VA	vinyl acetate

Other abbreviations used

BMC	bulk molding compound
MFR	melt flow rate
SAFT	shear adhesion failure temperature
SMC	sheet molding compound
T. S.	tensile strength
U. E.	ultimate elongation
UL	Underwriters Laboratories code

References

1. Uniroyal, US 3758643, 1979 (W.K. Fisher).
2. Uniroyal, US 3806558, 1974 (W.K. Fisher).
3. Exxon, US 3037954, 1962 (A.M. Gessler, W.H. Haslett).
4. Hercules, US 3262992, 1966 (R. Holzer, O. Taunus, K. Mehnert).
5. Monsanto, US 4130535, 1978 (A.Y. Coran, B. Das, R.P. Patel).
6. Monsanto, US 4311628, 1982 (S. Abdou-Sabet, M.A. Fath).
7. SRI Int. Report no. 104 A, Thermoplastic Elastomers, June 1993.
8. P. Galli, J.C. Haylock: "*Polyolefins VII International Conference*," Society of Plastics Engineers, Houston, Texas, Feb. 24–27, 1991.
9. J. Yin et al., *Sci. Sin. Ser. B (Engl. Ed.)* **29** (1986) no. 12, 1233.
10. S. Abdou-Sabet, R.P. Patel, *Rubber Chem. Technol.* **64** (1991) 769.
11. S. Onogi, T. Asada, A. Tanaka, *J. Pol. Sci. Part A-2* **7** (1969) 171.
12. S. Onogi, T. Asada, *Prog. Polym. Sci. Jpn.*, **2** (1971) 261.
13. D.M. Lohse, *Annu. Tech. Conf. Soc. Plast. Eng.* **43rd** (1985) 301.
14. D.M. Lohse, S. Datta, E.N. Kresge, *Macromolecules* **24** (1991) 561.
15. B.F. Goodrich, US 4036912, 1977 (J. Sticharczuk).
16. A.Y. Coran, R.P. Patel, D. Williams-Headd, *Rubber Chem. Technol.* **58** (1985) 1014.
17. Union Carbide, US 4803244, 1989 (J. Umpleby).
18. Advanced Elastomer Systems, US 5672660, 1997 (R.E. Medsker, R. Patel).
19. Continental, EP 0510559 B1, 1992 (S. Lüpfert, R. Anderlik, H.-G. Fritz, F. Röthemeyer).
20. Bernhard Rustige; EP 0913427 A1, 1997 (C.-P. Kirchner, J. Pac, H.-G. Fritz).
21. Chemplast Marketing Services, WO 00/68287, 1999 (C.-P. Kirchner, Q. Cai, H.-G. Fritz).
22. Himont, US 5196462, 1993 (D.A. Berta).
23. Mitsui, US 4212787, 1980 (A. Matsuda, S. Shimizu, S. Abe).
24. C.P. Rader, S. Abdou-Sabet in S.K. De, A.K. Bhowmick (eds.): *Thermoplastic Elastomers from Rubber-Plastic Blends*, Ellis Horwood, Chichester 1990, p. 159.
25. D.J. Elliott in S.K. De, A.K. Bhowmick (eds.): *Thermoplastic Elastomers from Rubber-Plastic Blends*, Ellis Horwood, Chichester 1990, p. 102.
26. N.R. Choudhury, P.P. De, A.K. Bhowmick in S.K. De, A.K. Bhowmick (eds.): *Thermoplastic Elastomers from Rubber-Plastic Blends*, Ellis Horwood, Chichester 1990, p. 71.
27. R.C. Puydak, D.R. Hazelton, *Plast. Eng.* **44** (1988) no. 9, 38.
28. W.A. Zigman, *Adv. Chem. Ser.* **43** (1964) 1.
29. S. Wu: *Polymer Interface Adhesion*, Marcel Dekker, New York 1982.
30. A.Y. Coran, R.P. Patel, *Rubber Chem. Technol.* **56** (1983) 1045.
31. S. Abdou-Sabet, Y.L. Wang, E.F. Chu, *Rubber Plast. News* 1985, Nov. 4, 20.
32. Monsanto, US 4654402, 1987 (R.P. Patel).
33. L.A. Goettler, J.R. Richwine, F.J. Wille, *Rubber Chem. Technol.* **55** (1982) 1448.
34. G. Oertel (ed.): *Polyurethane Handbook*, 2nd ed., Hanser Publ., München–Wien–New York 1993.
35. C.S. Schollenberger in A.K. Bhowmick, H.L. Stephens (eds.): *Handbook of Elastomers*, Marcel Dekker, New York–Basel 1988.
36. E.C. Ma in B.M. Walker, C.P. Rader (eds.): *Handbook of Thermoplastic Elastomers*, 2nd ed., Van Nostrand Reinhold, New York 1988.
37. W. Meckel, W. Goyert, W. Wieder in G. Holden, N.R. Legge, R.P. Quirk, H.E. Schroeder (eds.): *Thermoplastic Elastomers*, 2nd ed., Hanser Verlag, München 1996.
38. Upjohn, US 3899467, 1974 (H.W. Bonk, T.M. Shah).
39. A. Singh, *Advances in Urethane Science and Technology* **13** (1995) 112–139.
40. H. Hespe et al., *J. Appl. Polym. Sci.* **44** (1992) 2029–2035.
41. J.H. Saunders, K.C. Frisch: "Chemistry," *Polyurethanes*, Part I, Interscience, New York 1962.
42. Bayer, EP 4393, 1978 (B. Quiring, H.G. Niederdellmann, W. Goyert, H. Wagner).
43. P.J. Flory: *Principles of Polymer Chemistry*, Cornell University Press, Ithaca, N.Y., 1953.
44. W. Neumann, P. Fischer, *Angew. Chem.* **24** (1962) 806.
45. F. Gugumus in H. Zwiefel (ed.): *Plastics Additives Handbook*, 5th ed., Hanser Verlag, München 2001.
46. BASF, EP 0134455, 1984 (G. Zeitler, G. Bittner, K. Faehndrich, H.M. Rombrecht).
47. BASF Schwarzheide, DD 300900, 1987–1992 (R. Krech et al.).
48. Elastogran, DE 4112805, 1991 (A. Chlosta, F. Lehrich, J. Sadlowski, L. Thil).

49. U. Barth (eds.): *Polymerreaktionen und reaktives Aufbereiten in kontinuierlichen Maschinen*, VDI-Verlag, Düsseldorf 1988.
50. C.S. Schollenberger, K. Dinbergs, *J. Elastoplast.* **5** (1973) 222; **7** (1975) 65.
51. H.J. Maaß et al., *Plaste Kautsch.* **34** (1987) 251–254.
52. J.A. Miller et al., *Macromolecules* **18** (1985) 32–44.
53. S. Abouzaar, G.L. Wilkes, *J. Appl. Polym. Sci.* **29** (1984) 2695–2711.
54. ICI, GB 1057018, 1964 (J.P. Brown, G. Trappe).
55. BASF, EP 922552, 1998 (P. Bartholomaus, H. Dauns, T. Friedl, G. Scholz, H. Loock, W. Lukat).
56. Mobay, US 3233025, 1966 (B.F. Frye, K.A. Piggot, J.H. Saunders).
57. A. Bouilloux, C.W. Macosko, T. Kotnour, *Ind. Eng. Chem. Prod. Res. Dev.* **30** (1991) 2431–2436.
58. Bayer, DE 4217367, 1993 (F. Mueller, W. Braeuer, K. Ott, H. G. Hoppe).
59. Bayer, DE 4406948, 1995 (W. Braeuer, F. Mueller, H. Heidingsfeld, B. Schulte, J. Winkler).
60. Bayer, DE 2854406, 1978 (W. Goyert et al.).
61. C. Li, J.G. Homan, R.A. Phillips, S.L. Cooper in K.C. Frisch, Jr. (ed.): *Recent Developments in Polyurethanes and Interpenetrating Polymer Networks*, Technomic, Lancaster–Basel 1988.
62. Z. Petrovic, J. Ferguson, *Prog. Polym. Sci.* **16** (1991) 695–836.
63. C.G. Seefried, Jr., J.V. Koleske, F.E. Critchfield, *J. Appl. Polym. Sci.* **19** (1975) 2493–2502, 2503–2513, 3185–3191.
64. W.P. Yang, C.W. Macosko, S.T. Wellinghoff, *Polymer* **27** (1986) 1235–1240.
65. L. Born, H. Hespe, J. Crone, K.H. Wolf, *Colloid. Polym. Sci.* **260** (1982) 819–828.
66. G. Pohl, D. Joel, H. Goering, H.E. Carius, *Plaste Kautsch.* **40** (1993) 357–362.
67. *Houben-Weyl Methoden der organischen Chemie*, ■edition■E 20, Georg Thieme Verlag, Stuttgart, p. 534. ■please check and insert Chapter title, and year)
68. J. Foks, I. Naumann, G.H. Michler, *Angew. Makromol. Chem.* **189** (1991) 63–76.
69. E. Tocha, H. Janik, M. Debowski, G.J. Vancso, *J. Macromol. Sci. B* **41** (2002) 1291–1304.
70. S. Hwang, D.J. Hemker, S.L. Cooper, *Macromolecules* **17** (1984) 307–315.
71. L.M. Leung, J.T. Koberstein, *Macromolecules* **19** (1986) 706–713.
72. J.W.C. Van Bogart, P.E. Gibson, S.L. Cooper, *J. Polym. Sci. Polym. Phys. Ed.* **21** (1983) 65–95.
73. C.P. Christenson et al., *J. Polym. Sci. Polym. Phys. Ed.* **24** (1986) 1401–1439.
74. C.D. Eisenbach, T. Heinemann, A. Ribbe, E. Stadler, *Angew. Makromol. Chem.* **202/203** (1992) 221–241.
75. Dow Chemical, EP 80031, 1983 (D.J. Goldwasser, K. Onder).
76. B.F. Goodrich,DE 1720843, 1967 (E.G. Kolycheck).
77. Bayer, DE 1940181, 1969 (E. Meisert, A. Awater, C. Muehlhausen, U.J. Doebereiner).
78. W. Goyert, H. Hespe, *Kunststoffe* **68** (1978) 2–8.
79. H.G. Hoppe, *Plastverarbeiter* **44** (1993) no. 9, 75–80; no. 10, 40–44.
80. E.C. Ma, *Rubber World* **199** (1989) 30–35.
81. A.T. Chen et al., *Elastomerics* **122** (1990) no. 9, 19–22.
82. S. Gogolewski, *Colloid Polym. Sci.* **267** (1989) 757–785.
83. R. Zdrahala, I. Zdrahala, *J. Biomat. Appl.* **14** (1999) no. 1, 67–90.
84. DuPont, US 3651014, 1972 (W.K. Witsiepe).
85. DuPont, US 3763109, 1973 (W.K. Witsiepe).
86. DuPont, US 3766146, 1973 (W.K. Witsiepe).
87. L.H. Buxbaum, *J. Appl. Polym. Sci.* **35** (1979) 59–66.
88. DuPont, US 3801547, 1974 (G.K. Hoeschele).
89. BASF, US 4056514, 1977 (H. Strehler).
90. General Electric, US 4732948, 1988 (R.J. McCready, J.A. Tyrell).
91. Kuraray, JP 6181419, 1985 (M. Ishiguro, H. Hira).
92. DuPont, US 5004748, 1991 (G. Huynh-Ba).
93. DuPont, US 3726 569, 1973 (G.K. Hoeschele).
94. DuPont: "Hytrel," Product Bulletin H-33430–1, 1994, p. 6.
95. G.K. Hoeschele, W.K. Witsiepe, *Angew. Makromol. Chem.* **29/30** (1973) 267.
96. J.R. Wolfe, Jr., *ACS Adv. Chem. Ser.* **176** (1979) 129.
97. Eastman Kodak, US 4256860, 1981 (B. Davis, R.B. Barbee, H. R. Musser).
98. General Electric, US 4711947, 1987 (R.J. McCready, J.A. Tyrell).
99. DuPont, US 3775374, 1973 (J.R. Wolfe Jr.).
100. DuPont, US 3784540, 1974 (G.K. Hoeschele).
101. J.R. Wolfe, Jr., *Rubber Chem. Technol.* **50** (1977) 688.
102. Texaco Chemical Co.: Polyoxyalkyleneamines, product bulletin, 1991.
103. General Electric, US 4544734, 1985; US 4556705, 1985 (R.J. McCready).
104. E.C. Leonard, *Dimer Acids*, 1975, 1.
105. DuPont, US 3954689, 1976 (G.K. Hoeschele).
106. General Electric, US 4594377, 1986 (R.J. McCready).
107. R.W.M. van Berkel, S.A.G. de Graaf, F.J. Huntjens, C.M.F. Vrouenraets:Development Series, Developments in Block Copolymers-1, p. 261.
108. Teijin, BE 834004, 1975.
109. Goodyear, US 3446778, 1969 (R.C. Waller, M.H. Keck).
110. R.J. Cella, *J. Polym. Sci. Symp.* **42** (1973) no. 2, 727.
111. R.J. Cella, *Polymer Symposia*, No. 42, Helsinki1972.
112. S.C. Wells in B.M. Walker, C.P. Rader (eds.): *Handbook of Thermoplastic Elastomers*, 2nd ed., Van Nostrand Reinhold, New York 1988, Chap. 4
113. M. Brown, *Rubber Ind. (London)* **9** (1975) 102.
114. DuPont, US 3718715, 1973 (R.W. Crawford, W.K. Witsiepe).
115. DuPont, US 3917743, 1975 (H.E. Schroeder, J.R. Wolf, Jr.).
116. DuPont, US 3963802, 1976 (C. Shih).
117. DuPont, US 4010222, 1977 (C. Shih).
118. General Electric, US 4939205, 1989 (Nan-I Liu).
119. General Electric, US 4814380, 1990 (Nan-I Liu, R.J. McCready).
120. M. Veldstra: *Thermoplastic Elastomers-II*, Rapra Technology Ltd., London 1989.
121. H. Reinhardt, M. Negri, *Thermoplastic Elastomers-II*, Rapra Technology Ltd., London 1989.
122. Teijin, JP 50-35623, 1976.
123. AKZO, US 4493870, 1985 (C.M.F. Vrouenraets).
124. M.H. Horn, K.P. Schodt, G.J. Ostapchenko, *Annu. Tech. Conf. Soc. Plast. Eng.* 1992.
125. M. Tessier, E. Marechal, *J. Polym. Sci. A1*, **27** (1989) 539.
126. M. Tessier, E. Marechal, *Eur. Polym. J.* **20** (1984) 281.
127. M. Tessier, E. Marechal, *Eur. Polym. J.* **22** (1986) 877.
128. Bayer, EP 83434, 1983 (U. Grigo, K.-H. Köhler, R. Binsack, L. Morbitzer, J. Merten, L. Trabert, W. Heitz).
129. Daicel-Degussa, EP-PS 73838, 1983 (K. Arita, H. Suzuki).
130. Bayer, EP-A 244601, 1987 (K.-H. Köhler, K. Reinking, G. Weber, R. Prinz).
131. Bayer, DE-OS 3105365, 1982 (U. Grigo, K.-H. Köhler, R. Binsack, L. Morbitzer, J. Merten, L. Bottenbruch, W. Heitz).
132. A. Koichi, *J. Polym. Sci. A* **25** (1987) 1591.
133. J.M. Huet, E. Marechal, *Eur Polym J.* **10** (1974) 771.

134 Daicel Chemical Ind, JP 61171731, 1985 (K. Okitsu, K. Murabayashi).
135 Ube, EP-A 1235165, 2002 (H. Okushita, T. Muramatsu).
136 G. Deleens, P. Foy, E. Marechal, *Eur. Polym. J.* **13** (1977) 337.
137 G. Deleens, P. Foy, E. Marechal, *Eur. Polym. J.* **13** (1977) 343.
138 G. Deleens, P. Foy, E. Marechal, *Eur. Polym. J.* **13** (1977) 353.
139 G. Deleens, Ph.D. Thesis, University Rouen 1975.
140 S. Mumcu, K. Burzin, R. Feldmann, R. Feinauer, *Angew. Makromol. Chem.* **74** (1978) 49.
141 Degussa, DE-OS 2936976, 1981 (S. Mumcu).
142 Degussa, DE-PS 2936977, 1981 (S. Mumcu).
143 Degussa, DE-OS 2932234, 1979 (H.-J. Panoch, S. Mumcu).
144 Degussa, DE-PS 2712987, 1978 (K. Burzin, S. Mumcu, R. Feldmann, H. Jadamus, R. Feinauer).
145 Atofina, DE-PS 2523991, 1975 (P. Foy, C. Jungblut, G. Deleens).
146 Atofina, DE-PS 2837687, 1979 (G. Deleens, J. Ferlampin, M. Gonnet).
147 Ems-Chemie, DE-PS 3428404, 1985 (H.J. Liedlof).
148 Atofina, DE-OS 2632120, 1977 (P. Foy, R. Kern).
149 Toray, EP-AS 163902, 1985 (C. Tanaka, M. Kondou, Y. Yamamoto).
150 Asahi, EP-A 221188, 1987 (Y. Suzuki, M. Nakamura, A. Aochiama).
151 Toray, DE-OS 3145998, 1982 (C. Tanaka, A. Chita, S. Nakashima, N. Shimobu).
152 Toray, EP 95893, 1983 (C. Tanaka, S. Nakashima, M. Kon Dou).
153 Atofina, DE-OS 2856787, 1979 (G. Deleens, J. Ferlampin, C. Prulain).
154 F. Widmer, *Adv. Chem. Ser.* **128** (1973) 51–67.
155 B. Loretan, E. Heimgartner, *Chem. Ing. Tech.* (1979) *MS* **467/77**.
156 Ems-Chemie, DE-PS 3006961, 1980 (W. Isler, E. Schmid).
157 Rhone-Poulenc, EP-PS 130927, 1986 (I. Coquard; J. Goletto).
158 Rhone-Poulenc, EP-PS 187607, 1986 (I. Coquard; J. Goletto).
159 Rhone-Poulenc, EP-PS 201434, 1986 (I. Coquard; J. Goletto).
160 H.G. Elias, R. Vohwinkel: *Neue Polymere Werkstoffe für die industrielle Anwendung II*, C. Hanser Verlag, München 1983, p. 125–134.
161 Ube, JP 59133224, (1983) (H. Okamoto, Y. Okushita).
162 Ube, JP 59131628, (1983) (H. Okamoto, Y. Okushita).
163 Ube, JP 59193923, (1983) (H. Okamoto, Y. Okushita).
164 Huntsman, EP-PS 432943, 1991 (G.P. Speranza, W.-Y. Su).
165 Huntsman, EP-A 449419, 1991 (G.P. Speranza, W.-Y. Su).
166 R.J. Gaymans, P. Schwering, J.L. Haan, *Polymer* **30** (1989) 974.
167 DSM, EP-A 360311, 1990 (R.J. Gaymans, E. Roerdink, P.J. Schwering, E. Walch).
168 R.P. Eustache, G. Lé, D. Silagy, F.L.G. Malet: "PEBA made from Renewable Resources or how to offer simultaneously sustainability and high performance", *Rubber World*, April2009, Akron, OH.
169 E. Bornschlegl, G. Goldbach, K. Meyer, *Progr. Colloid Polym. Sci.* **71** (1985) 119–124.
170 J. Lohmar, K. Meyer, G. Goldbach, *Makromol. Chem.* **189** (1988) 2053–2065.
171 G. Goldbach, M. Kita, K. Meyer, K.P. Richter, *Progr. Colloid Polym. Sci.* **72** (1986) 83–96.
172 N.R. Legge: "Thermoplastic Elastomers Based on Three-Block Polymers," *Chemtech.* **13** (1983) 630.
173 Shell Oil Co., US 3149182, 1964 (L.M. Porter).
174 R.J. Ceresa: *Block and Graft Copolymers*, Butterworth, Washington, D.C., 1962.
175 P. Dreyfuss, L.J. Fetters, D.R. Hansen, *Rubber Chem. Technol.* **53** (1980) 728.
176 E.V. Gouinlock, R.S. Porter, *Polym. Eng. Sci.* **17** (1977) 535.
177 B.M. Walker, C.P. Rader (eds.): *Handbook of Thermoplastic Elastomers*, 2nd ed., Van Nostrand Reinhold, New York 1988.
178 A.D. Thorn: *Thermoplastic Rubbers–A Review of Current Information*, RAPRA, Shawsbury, UK, 1980.
179 D.J. St. Clair, *Rubber Chem. Technol.* **55** (1982) 208.
180 J.R. Erickson, *Rubber Plast. News*1985, Sept. 9, 16.
181 G. Holden, S.S. Chin,*Paper to the Adhesives and Sealants Conference*, Washington, D.C., March1986.
182 G. Holden, Paper to the Adhesives and Sealants Council Seminar, Chicago, Il., Oct. 1984.
183 E.J. Van Beem, P. Brasser, *J. Inst. Pet.* **59** (1973) 91.
184 Shell Int. Chem. Co., Bulletin TR-18, London1984.
185 *The Shell Bitumen Handbook*, Shell Bitumen UK Ltd., London 1990.
186 J.H. Collins, W.J. Mikols, *60th Meeting of Asphalt Paving Technologists*,San Antonio, Tx., Feb. 1985.
187 D.M. Colwill, M.E. Daines:Progress in the Trials of Pervious Macadam, *Highways*, Jan. 1989, p. 15.
188 A. Dinnen: "Bitumen Thermoplastic Rubber Blends in Road Applications, "*3rd Eurobitumen Symposium*,The Hague, Sept. 1985.
189 M.J.W. Downes, R.C. Koole, E.A. Mulder, W.E. Graham: "Some Proven New Binders and Their Cost Effectiveness," *7th AAPA Int. Asphalt Conf.*,Brisbane, Aug. 1988.
190 K.H. Kolb, T. Pallay, J.P. Serfass, O. Ruud: "Asphalt for Better Roads," Symposium, E.A.P.A., Copenhagen1986.
191 L. Laitinen: "Rubberised Bitumen in Road Surfacing Applications," *Asfaltti* **43** (1988) 18.
192 J.P. Marchand: "15 Years of Experience in Thick Bitumen Pavements," *Petrol. Inf.* **1987**, Nov., 62.
193 S.H. Carpenter, T. Van Dam: "Modified and Unmodified Asphalt Concrete Mixtures," *Transp. Res. Rec.* **1115** (1987) p. 62–74.
194 H.F. Vermeire, *Kautsch. Gummi Kunstst.* **43** (1990) 11.
195 A.L. Bull, G. Holden, *J. Elastomers Plast.* **9** (1981) 281.
196 Shell Int. Res. Co., EP 0215501, 1982 (W.P. Gergen, R.G. Lutz, M.K. Martin).
197 D.W. Bartlett, D.R. Paul, J.W. Barlow, *Mod. Plast.* **58** (1981) no. 12, 60.
198 C.L. Willis, W.M. Halper, D.L. Handlin, Jr., *Polym. Plast. Technol. Eng.* **23** (1984) no. 2, 207.
199 Shell Int. Chem. Co., Technical Bulletin TR.3.6, London1991.
200 J. Schneider,Paper presented at Recycle '92, Davos, Apr. 1992.

Further Reading

M. Biron: *Thermoplastics and Thermoplastic Composites*, Elsevier, Amsterdam 2006.

J.G. Drobny: *Handbook of Thermoplastic Elastomers*, William Andrew Publ., Norwich, NY, 2007.

J.K. Fink: *Handbook of Engineering and Specialty Thermoplastics*, Wiley, Hoboken 2010.

G. Holden: "Thermoplastic Elastomers", *Kirk Othmer Encyclopedia of Chemical Technology*, 5th ed., John Wiley & Sons, Hoboken, NJ.

G. Holden, H.R. Kricheldorf, R.P. Quirk (eds.): *Thermoplastic Elastomers*, 3rd ed., Hanser, München 2004.

C.C. Ibeh: *Thermoplastic Materials: Properties, Manufacturing Methods, and Applications*, CRC Press, Baton Rouge 2010.